LINEAR ALGEBRA

WITH APPLICATIONS

Seventh Edition

The Jones and Bartlett Publishers Series in Mathematics

Geometry

Geometry with an Introduction to Cosmic Topology
Hitchman (978-0-7637-5457-0) © 2009

Euclidean and Transformational Geometry: A Deductive Inquiry
Libeskind (978-0-7637-4366-6) © 2008

A Gateway to Modern Geometry: The Poincaré Half-Plane, Second Edition
Stahl (978-0-7637-5381-8) © 2008

Understanding Modern Mathematics
Stahl (978-0-7637-3401-5) © 2007

Lebesgue Integration on Euclidean Space, Revised Edition
Jones (978-0-7637-1708-7) © 2001

Precalculus

Precalculus: A Functional Approach to Graphing and Problem Solving, Sixth Edition
Smith (978-0-7637-5177-7) © 2012

Precalculus with Calculus Previews (Expanded Volume), Fourth Edition
Zill/Dewar (978-0-7637-6631-3) © 2010

Precalculus with Calculus Previews (Essentials Version), Fourth Edition
Zill/Dewar (978-0-7637-3779-5) © 2007

Calculus

Single Variable Calculus: Early Transcendentals, Fourth Edition
Zill/Wright (978-0-7637-4965-1) © 2011

Multivariable Calculus, Fourth Edition
Zill/Wright (978-0-7637-4966-8) © 2011

Calculus: Early Transcendentals, Fourth Edition
Zill/Wright (978-0-7637-5995-7) © 2011

Multivariable Calculus
Damiano/Freije (978-0-7637-8247-4) © 2011

Calculus: The Language of Change
Cohen/Henle (978-0-7637-2947-9) © 2005

Applied Calculus for Scientists and Engineers
Blume (978-0-7637-2877-9) © 2005

Calculus: Labs for Mathematica
O'Connor (978-0-7637-3425-1) © 2005

Calculus: Labs for MATLAB
O'Connor (978-0-7637-3426-8) © 2005

Linear Algebra

Linear Algebra with Applications, Seventh Edition
Williams (978-0-7637-8248-1) © 2011

Linear Algebra with Applications, Alternate Seventh Edition
Williams (978-0-7637-8249-8) © 2011

Linear Algebra: Theory and Applications
Cheney/Kincaid (978-0-7637-5020-6) © 2009

Advanced Engineering Mathematics

A Journey into Partial Differential Equations
Bray (978-0-7637-7256-7) © 2011

Advanced Engineering Mathematics, Fourth Edition
Zill/Wright (978-0-7637-7966-5) © 2011

An Elementary Course in Partial Differential Equations, Second Edition
Amaranath (978-0-7637-6244-5) © 2009

Complex Analysis

A First Course in Complex Analysis with Applications, Second Edition
Zill/Shanahan (978-0-7637-5772-4) © 2009

Complex Analysis for Mathematics and Engineering, Fifth Edition
Mathews/Howell (978-0-7637-3748-1) © 2006

Classical Complex Analysis
Hahn (978-0-8672-0494-0) © 1996

Real Analysis

Real Analysis
Denlinger (978-0-7637-7947-4) © 2011

An Introduction to Analysis, Second Edition
Bilodeau/Thie/Keough (978-0-7637-7492-9) © 2010

Basic Real Analysis
Howland (978-0-7637-7318-2) © 2010

Closer and Closer: Introducing Real Analysis
Schumacher (978-0-7637-3593-7) © 2008

The Way of Analysis, Revised Edition
Strichartz (978-0-7637-1497-0) © 2000

Topology

Foundations of Topology, Second Edition
Patty (978-0-7637-4234-8) © 2009

Discrete Math and Logic

Discrete Structures, Logic, and Computability, Third Edition
Hein (978-0-7637-7206-2) © 2010

Essentials of Discrete Mathematics
Hunter (978-0-7637-4892-0) © 2009

Logic, Sets, and Recursion, Second Edition
Causey (978-0-7637-3784-9) © 2006

Numerical Methods

Numerical Mathematics
Grasselli/Pelinovsky (978-0-7637-3767-2) © 2008

Exploring Numerical Methods: An Introduction to Scientific Computing Using MATLAB
Linz (978-0-7637-1499-4) © 2003

Advanced Mathematics

Mathematical Modeling with Excel®
Albright (978-0-7637-6566-8) © 2010

Clinical Statistics: Introducing Clinical Trials, Survival Analysis, and Longitudinal Data Analysis
Korosteleva (978-0-7637-5850-9) © 2009

Harmonic Analysis: A Gentle Introduction
DeVito (978-0-7637-3893-8) © 2007

Beginning Number Theory, Second Edition
Robbins (978-0-7637-3768-9) © 2006

A Gateway to Higher Mathematics
Goodfriend (978-0-7637-2733-8) © 2006

For more information on this series and its titles, please visit us online at http://www.jbpub.com/math. Qualified instructors, contact your Publisher's Representative at 1-800-832-0034 or info@jbpub.com to request review copies for course consideration.

The Jones and Bartlett Publishers International Series in Mathematics

For more information on this series and its titles, please visit us online at http://www.jbpub.com/math. Qualified instructors, contact your Publisher's Representative at 1-800-832-0034 or info@jbpub.com to request review copies for course consideration.

LINEAR ALGEBRA

WITH APPLICATIONS

Seventh Edition

GARETH WILLIAMS

Stetson University

JONES AND BARTLETT PUBLISHERS

Sudbury, Massachusetts

BOSTON   TORONTO   LONDON   SINGAPORE

World Headquarters
Jones and Bartlett Publishers
40 Tall Pine Drive
Sudbury, MA 01776
978-443-5000
info@jbpub.com
www.jbpub.com

Jones and Bartlett Publishers
Canada
6339 Ormindale Way
Mississauga, Ontario L5V 1J2
Canada

Jones and Bartlett Publishers
International
Barb House, Barb Mews
London W6 7PA
United Kingdom

Jones and Bartlett's books and products are available through most bookstores and online booksellers. To contact Jones and Bartlett Publishers directly, call 800-832-0034, fax 978-443-8000, or visit our website, www.jbpub.com.

Photo Credits
Chapter Opener 1, page 2 © Aditya Kok/Dreamstime.com; **Chapter Opener 2, page 34** © Carlos Neto/ShutterStock, Inc.; **Chapter Opener 3, page 107** © Timage/ShutterStock, Inc.; **Chapter Opener 4, page 134** © Best View Stock/age fotostock; **Chapter Opener 5, page 209** © Cahir Davitt/age fotostock; **Chapter Opener 6, page 244** © Darleen Stry/Dreamstime.com; **Chapter Opener 7, page 312** © Nikolay Okhitin/Dreamstime.com; **Chapter Opener 8, page 346** © Styve Reineck/ShutterStock, Inc.; **Chapter Opener 9, page 385** © laurent dambies/ShutterStock, Inc.; **Appendix Opener, page 406** © Pres Panayotov/ShutterStock, Inc.

Unless otherwise indicated, all photographs are under copyright of Jones and Bartlett Publishers, LLC.

Production Credits
Chief Executive Officer: Clayton Jones
Chief Operating Officer: Don W. Jones, Jr.
President, Higher Education and Professional Publishing:
    Robert W. Holland, Jr.
V.P., Sales: William J. Kane
V.P., Design and Production: Anne Spencer
V.P., Manufacturing and Inventory Control: Therese Connell
Publisher: David Pallai
Acquisitions Editor: Timothy Anderson
Editorial Assistant: Melissa Potter
Production Director: Amy Rose

Production Assistant: Ashlee Hazeltine
Senior Marketing Manager: Andrea DeFronzo
Composition: Northeast Compositors
Cover and Title Page Design: Kristin E. Parker
Photo Research and Permissions Manager: Kimberly Potvin
Cover and Title Page Image: © 2009 Santiago Calatrava/
    Artists Rights Society (ARS), New York/VEGAP, Madrid.
    Photo provided © Antony Mcaulay/Dreamstime.com
Printing and Binding: Courier Kendallville
Cover Printing: Courier Kendallville

Library of Congress Cataloging-in-Publication Data
Williams, Gareth, 1937-
    Linear algebra with applications / Gareth Williams. — 7th ed., alternate ed.
      p. cm.
    Includes index.
    ISBN-13: 978-0-7637-8249-8 (casebd.)
    ISBN-10: 0-7637-8249-1 (casebd.)
    1. Algebras, Linear—Textbooks. I. Title.
    QA184.2.W55 2011b
    512'.5—dc22
                            2009041375

6048
Printed in the United States of America
13 12 11 10 09   10 9 8 7 6 5 4 3 2 1

Dedication

I dedicate this book to Ryan

Contents

*Sections marked with an asterisk are optional. The instructor can use these sections to build around the core material to give the course the desired flavor.

Preface

*L*inear Algebra with Applications, Alternate Seventh Edition, is an introduction to Linear Algebra suitable for a course usually offered at the sophomore level. Linear algebra is fundamental to a large part of modern mathematics. While abstraction should remain a key part of the introductory linear algebra course, an important goal should also be to include numerical methods and applications. This opens up the course to the large group of students who need to use linear algebra. This book is a flexible blend of theory, important numerical techniques, and interesting applications. The book is arranged around 27 core sections. These sections include topics that I think are essential to an introductory linear algebra course. There is then ample time for the instructor to select further topics that give the course the desired flavor.

Alternate 7th Edition The sophomore-level linear algebra course can be taught in many ways—the order in which topics are presented can vary. There are merits to various approaches, often depending on the needs of the students. There are two versions of this book now available. The main difference between the two versions of the 7th Edition is in when the vector space $\mathbf{R}^n$ is introduced and used—early or later. This, the *Alternate 7th Edition*, is built upon the sequence of topics of the 5th Edition favored by certain users. The first three chapters cover systems of linear equations, matrices, and determinants—the more abstract material starts later. The vector space $\mathbf{R}^n$ is introduced in Chapter 4, leading directly into general vector spaces. Such topics as eigenvectors and linear transformations then follow. This alternate version is especially appropriate for students who are taking the course with the intent of using linear algebra in those fields. They develop a solid knowledge of the basics of systems of equations and matrices that will serve them well in their own fields, before meeting the more abstract vector space structures that give added insight and knowledge.

The regular version of the 7th Edition, on the other hand, introduces the vector space $\mathbf{R}^n$, subspaces, bases, and dimension early on, in Chapter 1. These concepts are then used in the discussions of systems of linear equations, matrices, linear transformations, and eigenspaces leading up to the chapter on general vector spaces—a more theoretical approach.

The Goals of This Text

- To provide a solid foundation in the mathematics of linear algebra.
- To introduce some of the important numerical aspects of the field.
- To discuss interesting applications so that students may know when and how to apply linear algebra. Applications are taken from such areas as archaeology, demography, coding theory, fractal geometry, and relativity.

The Mathematics Linear algebra is a central subject in undergraduate mathematics. Many important topics must be included in this course. For example, linear dependence, basis, eigenvalues and eigenvectors, and linear transformations should be covered carefully. Not only are such topics important in linear algebra, they are usually a prerequisite for other courses such as differential equations. A great deal of attention has been given in this book to presenting the "standard" linear algebra topics.

This course is often the student's first course in abstract mathematics. The student should not be overwhelmed with proofs, but should nevertheless be taught how to prove theorems. When considered instructive, proofs of theorems are provided or given as exercises. Other proofs are given in outline form and some have been omitted. Students should be introduced slowly and carefully to the art of developing and writing proofs. This is at the heart of mathematics. The student should be trained to think "mathematically." For example, the idea of "if and only if" is extremely important in mathematics. It arises very naturally in linear algebra.

Computation While linear algebra has its abstract side, it also has its numerical side. Students should feel comfortable with the term "algorithm" by the end of the course. The student participates in the process of determining exactly where certain algorithms are more efficient than others.

For those who wish to integrate the computer into the course, a MATLAB manual has been included in Appendix D. MATLAB is the most widely used software for working with matrices. The manual consists of 27 sections that tie into the regular course material. A brief summary of the relevant mathematics is given at the beginning of each section. The built-in functions of MATLAB—such as inv(A) for finding the inverse of a matrix A—are introduced, and programs written in the MATLAB language also are available and can be downloaded from www.stetson.edu/~gwilliam/mfiles.htm. The programs include not only computational programs such as Gauss-Jordan elimination with an all-steps option, but also applications such as digraphs, Markov chains, and a simulated space–time voyage. While this manual is presented in terms of MATLAB, the ideas should be of general interest. The exercises can be implemented on any matrix algebra software package.

A graphing calculator also can be used in linear algebra. Calculators are available for performing matrix computation and for computing reduced echelon forms. A calculator manual for the course has been included in Appendix C.

Applications Linear algebra is a subject of great breadth. Its spectrum ranges from the abstract through numerical techniques to applications. In this book I have attempted to give the reader a glimpse of many interesting applications. These applications range from theoretical applications, such as the use of linear algebra in differential equations, difference equations, and least squares analyses, to many practical applications in fields such as archaeology, demography, electrical engineering, traffic analysis, fractal geometry, relativity, and history. All such discussions are self-contained. There should be something here to interest most people! I have tried to involve the reader in the applications by using exercises that extend the discussions given. Students have to be trained in the art of applying mathematics. Where better than in the linear algebra course with its wealth of applications?

Time is always a challenge when teaching. It becomes important to tap that out-of-class time as much as possible. A good way to do this is with group application projects. The instructor can select those applications that are of most interest to the class.

The Flow of Material

This book contains mathematics with interesting applications integrated into the main body of the text. The student meets the application in its "natural setting." The approach that I have used is to develop the mathematics first and then provide applications. I believe that this makes for the clearest text presentation. However, some instructors may prefer to look ahead with the class to an application and use it to motivate the mathematics. Historically, mathematics has developed through interplay with applications. For example, the analysis of the long-term behavior of a Markov chain model for analyzing population movement between U.S. cities and suburbs can be used to motivate eigenvalues and eigenvectors. This type of approach can be very instructive but should not be overdone.

Chapter 1 Linear Equations The reader is led from solving systems of two linear equations to solving general systems. The method of Gauss-Jordan elimination is introduced as the standard method for the course, because it is a "clean, uncomplicated" algorithm for the small systems encountered. It is important that the student master the method at this time, as it will be used frequently throughout the course. The chapter closes with three optional applications. Fitting a polynomial of degree $n - 1$ to n data points leads to a system of linear equations that has a unique solution. The analyses of electrical networks and traffic flow give rise to systems that have unique solutions and many solutions. The model for traffic flow is similar to that of electrical networks, but has fewer restrictions, leading to more freedom and thus many solutions in place of a unique solution.

Chapter 2 Matrices In the first chapter, matrices were used to handle systems of equations. This application motivates the algebraic development of the theory of matrices in this chapter. In addition to the standard development of the properties of matrix addition, scalar multiplication, matrix multiplication, partition of matrices, and matrix inverse, interesting discussions of computation are included. A brief but beautiful application of matrices in archaeology that illustrates the importance of matrix multiplication, transpose, and symmetric matrices, is included. The reader can anticipate, for physical reasons, why the product of a matrix and its transpose has to be symmetric, and then arrive at the result mathematically. This is mathematics at its best! The chapter closes with three optional sections on applications that should have broad appeal. The Leontief Input-Output Model in Economics is used to analyze the interdependence of industries. Wassily Leontief received a Nobel prize in 1973 for his work in this area. A Markov chain model is used in demography and genetics, and digraphs are used in communication and sociology. Instructors who cannot fit these sections into their formal class schedule should encourage readers to browse through them. All discussions are self contained. These sections can be given as out-of-class projects or as reading assignments.

Chapter 3 Determinants Determinants and their properties are introduced as quickly and painlessly as possible. Some proofs are included for the sake of completeness, but can be skipped if the instructor so desires.

Chapter 4 General Vector Spaces This chapter starts with an introduction to $\mathbf{R}^n$. Vector addition and scalar multiplication are introduced, laying the foundation for the later introduction of the abstract vector space. The dot product leads to the geometry of $\mathbf{R}^n$; the ideas of angles, magnitudes, and distances are developed from the dot product. The groundwork has now been laid for a more theoretical turn. The structure of the abstract vector space is developed. The concepts of subspace, linear dependence, basis, and dimension are introduced. These ideas are given geometrical interpretation whenever possible. The concept of rank brings together many of the earlier concepts. The reader will see that matrix inverse, determinant, rank, and uniqueness of solutions are all related. The chapter closes with a discussion of orthonormal sets, orthogonal matrices, and projections.

Chapter 5 Eigenvalues and Eigenvectors Eigenvalues, eigenvectors, and applications are introduced. Eigenspaces and the diagonalization of matrices are covered. Students see the role of eigenvalues and eigenvectors in such applications as analyzing long-term trends of population movements, in weather prediction, in determining the normal modes of oscillating systems, and in the implementation of Google.

Chapter 6 Linear Transformations A discussion of matrix transformations leads to linear transformations. Rotations, dilations, orthogonal, and affine transformations are introduced. Readers explore projection, scaling, and shear in the exercises. Matrices are used to draw a fractal fern. Core topics such as kernel, range, and the rank/nullity theorem are presented. Linear transformations, kernel, and range are used to give the reader a geometrical picture of the sets of solutions to systems of linear equations, both homogeneous and non-homogeneous.

Chapter 7 Inner Product Spaces This chapter is a natural continuation of the thought process developed in the previous chapters. The axioms of inner product spaces are given. The theory is applied to the problem of approximating functions by polynomials, and the importance of such approximations to computer software is discussed. I could not resist including a discussion of the use of vector space theory to detect errors in codes. The Hamming code, whose elements are vectors over a finite field, is introduced. The reader is also introduced to non-Euclidean geometry, leading to a self-contained discussion of the special relativity model of space–time. Having developed the general inner product space, the reader finds that the framework is not appropriate for the mathematical description of space–time. The positive definite axiom is discarded, opening up the door first for the pseudo inner product that is used in special relativity, and later for one that describes gravity in general relativity. It is appropriate at this time to discuss the importance of first mastering standard mathematical structures, such as inner product spaces, and then to indicate that mathematical research often involves changing the axioms of such standard structures. The chapter closes with a discussion of the use of a pseudoinverse to determine least squares curves for given data.

Chapter 8 Numerical Methods This chapter on numerical methods is important to the practitioner of linear algebra in today's computing environment. I have included Gaussian Elimination, *LU* decomposition, and the Jacobi and Gauss-Seidel iterative methods. The merits of the various methods for solving linear systems are discussed. In addition to discussing the standard topics of round-off error, pivoting, and scaling, I felt it important and well within the scope of the course to introduce the concept of ill-conditioning. It is very interesting to return to some of the systems of equations that have arisen earlier in the course and find out how dependable the solutions are! The matrix of coefficients of a least squares problem, for example, is very often a Vandermonde matrix, leading to an ill-conditioned system. The chapter concludes with the iterative method for finding dominant eigenvalues and eigenvectors. This discussion leads very naturally into a discussion of techniques used by geographers to measure the relative accessibility of nodes in a network.

Chapter 9 Linear Programming This final chapter gives the student a brief introduction to the ideas of linear programming. The field, developed by George Dantzig and his associates at the U.S. Department of the Air Force in 1947, is now widely used in industry and has its foundation in linear algebra. Problems are described by systems of linear inequalities. The reader sees how small systems can be solved in a geometrical manner, but that large systems are solved using row operations on matrices using the simplex algorithm.

Chapter Features

- Each section begins with a motivating introduction, which ties the material to previously learned topics.

- The pace of the book gradually increases. As the student matures mathematically, the explanations gradually become more sophisticated.
- Carefully developed notation. It is important that notation at this level be standard; but there is some flexibility. Good notation helps understanding; poor notation clouds the picture.
- Much attention has been given to the layout of the text. Readability is vital.
- Many carefully explained examples illustrate the concepts.
- There is an abundance of exercises. Initial exercises are usually of a computational nature, then become more theoretical in flavor.
- Many, but not all, exercises are based on examples given in the text. It is important that students have the maximum opportunity to develop their creative abilities.
- Review exercises at the end of each chapter have been carefully selected to give the student an overview of material covered in that chapter.

Supplements

- *Complete Solutions Manual,* with detailed solutions to all exercises.
- *Student Solutions Manual,* with complete answers to selected exercises.
- MATLAB programs for those who wish to integrate MATLAB into the course are available from www.stetson.edu/~gwilliam/mfiles.htm.
- WebAssign online homework and assessment.

Designated instructor's materials are for qualified instructors only. Jones and Bartlett reserves the right to evaluate all requests.

Acknowledgments

It is a pleasure to acknowledge the help that made this book possible. My deepest thanks go to my friend Dennis Kletzing for sharing his many insights into the teaching of linear algebra. A special thanks to my colleague Lisa Coulter of Stetson University for her conversations on linear algebra and her collaboration on software development. A number of Lisa's M-files appear in the MATLAB Appendix. Thanks to Janet Beery of the University of Redlands for constructive comments on my books over a period of many years. Thanks to Gloria Child of Rollins College for valuable feedback on the book. I am grateful to Michael Branton, Erich Friedman, Margie Hale, Will Miles, and Hari Pulapaka of Stetson University for the discussions and suggestions that made this a better book.

My deep thanks goes to Amy Rose, Production Director, of Jones and Bartlett Publishers who oversaw the production of this book in such an efficient, patient, and understanding manner. I am especially grateful to the Mathematics Editor Tim Anderson for his continued enthusiastic backing and encouragement. Thanks, also, to Melissa Potter, Editorial Assistant; Ashlee Hazeltine, Production Assistant; and Andrea DeFronzo, Marketing Manager, for their support and hard work.

I thank the National Science Foundation for grants under its Curriculum Development Program to develop much of the MATLAB software. I am grateful to The MathWorks for their continued support of this project, and to my contact person in the company, Courtney Esposito.

I am, as usual, grateful to my wife Donna for all her mathematical and computing input, and for her continued support of my writing. This book would not have been possible without her involvement and encouragement.

City Hall in London, England, stands on the left side of the
River Thames, in the More London development by Tower
Bridge. The building, designed by British architect Norman
Foster, was opened in 2002. Its bulbous shape reduces
surface area and thus improves its energy efficiency.

Linear Equations

Mathematics is, of course, a discipline in its own right. It is, however, more than that—it is a tool used in many other fields. Linear algebra is a branch of mathematics that plays a central role in modern mathematics, and also is of importance to engineers and physical, social, and behavioral scientists. In this course the reader will learn mathematics, will learn to think mathematically, and will be instructed in the art of applying mathematics. The course is a blend of theory, numerical techniques, and interesting applications.

When mathematics is used to solve a problem it often becomes necessary to find a solution to a so-called system of linear equations. Historically, linear algebra developed from studying methods for solving such equations. This chapter introduces methods for solving systems of linear equations. We shall discuss two applications of systems of linear equations. We shall determine currents through electrical networks and analyze traffic flows through road networks.

1.1 Matrices and Systems of Linear Equations

An equation in the variables x and y that can be written in the form $ax + by = c$, where a, b, and c are real constants (a and b not both zero), is called a *linear equation*. The graph of such an equation is a straight line in the xy plane. Consider the system of two linear equations,

$$x + y = 5$$
$$2x - y = 4$$

A pair of values of x and y that satisfies both equations is called a **solution**. It can be seen by substitution that $x = 3$, $y = 2$ is a solution to this system. A solution to such a system will be a point at which the graphs of the two equations intersect. The following examples, Figures 1.1, 1.2, and 1.3, illustrate that three possibilities can arise for such systems of equations. There can be a unique solution, no solution, or many solutions. We use the point/slope form $y = mx + b$, where m is the slope and b is the y-intercept, to graph these lines.

Unique solution	No solution	Many solutions

$$x + y = 5$$
$$2x - y = 4$$

Write as $y = -x + 5$ and $y = 2x - 4$. The lines have slopes -1 and 2, and y-intercepts 5 and -4. They intersect at a point, the solution. There is a unique solution, $x = 3$, $y = 2$.

$$-2x + y = 3$$
$$-4x + 2y = 2$$

Write as $y = 2x + 3$ and $y = 2x + 1$. The lines have slope 2, and y-intercepts 3 and 1. They are parallel. There is no point of intersection. No solution.

$$4x - 2y = 6$$
$$6x - 3y = 9$$

Each equation can be written as $y = 2x - 3$. The graph of each equation is a line with slope 2 and y-intercept -3. Any point on the line is a solution. Many solutions

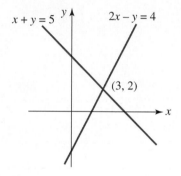

Figure 1.1

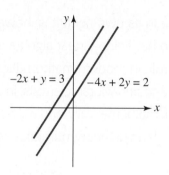

Figure 1.2

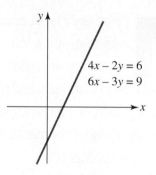

Figure 1.3

Our aim in this chapter is to analyze larger systems of linear equations. A **linear equation** *in n variables* x_1, x_2, x_3, ..., x_n is one that can be written in the form

$$a_1x_1 + a_2x_2 + a_3x_3 + \ldots + a_nx_n = b$$

where the coefficients a_1, a_2, ..., a_n and b are constants. The following is an example of a system of three linear equations.

$$x_1 + x_2 + x_3 = 2$$
$$2x_1 + 3x_2 + x_3 = 3$$
$$x_1 - x_2 - 2x_3 = -6$$

It can be seen on substitution that $x_1 = -1$, $x_2 = 1$, $x_3 = 2$ is a solution to this system. (We arrive at this solution in Example 1 of this section.)

A linear equation in three variables corresponds to a plane in three-dimensional space. Solutions to a system of three such equations will be points that lie on all three planes. As for systems of two equations there can be a unique solution, no solution, or many solutions. We illustrate some of the various possibilities in Figure 1.4.

As the number of variables increases, a geometrical interpretation of such a system of equations becomes increasingly complex. Each equation will represent a space embedded in a larger space. Solutions will be points that lie on all the embedded spaces. While a general geometrical way of thinking about a problem is often useful, we rely on algebraic methods for arriving at and interpreting the solution. We introduce a method for solving systems

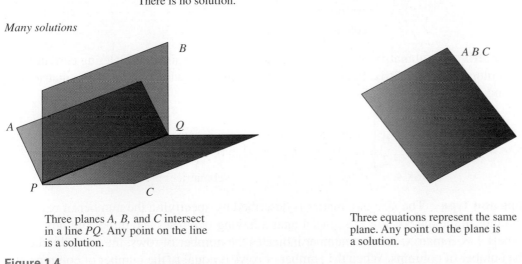

Unique solution

Three planes *A*, *B*, and *C* intersect at a single point *P*.
P corresponds to a unique solution.

No solution

Planes *A*, *B*, and *C* have no points in common.
There is no solution.

Many solutions

Three planes *A*, *B*, and *C* intersect
in a line *PQ*. Any point on the line
is a solution.

Three equations represent the same
plane. Any point on the plane is
a solution.

Figure 1.4

of linear equations called **Gauss-Jordan elimination**.[1] This method involves systematically eliminating variables from equations. In this section we shall see how this method applies to systems of equations that have a unique solution. In the following section we shall extend the method to more general systems of linear equations.

We shall use rectangular arrays of numbers called matrices to describe systems of linear equations. At this time we introduce the necessary terminology.

---

[1]Carl Friedrich Gauss (1777–1855) was one of the greatest mathematical scientists ever. Among his discoveries was a way to calculate the orbits of asteroids. He taught for forty-seven years at the University of Göttingen, Germany. He made contributions to many areas of mathematics, including number theory, probability, and statistics. Gauss has been described as "not really a physicist in the sense of searching for new phenomena, but rather a mathematician who attempted to formulate in exact mathematical terms the experimental results of others." Gauss had a turbulent personal life, suffering financial and political problems because of revolutions in Germany.

Wilhelm Jordan (1842–1899) taught geodesy at the Technical College of Karlsruhe, Germany. His most important work was a handbook on geodesy that contained his research on systems of equations. Jordan was recognized as being a master teacher and an excellent writer.

> **DEFINITION**   A **matrix** is a rectangular array of numbers. The numbers in the array are called the *elements* of the matrix.

Matrices are usually denoted by capital letters. Examples of matrices in standard notation are

$$A = \begin{bmatrix} 2 & 3 & -4 \\ 7 & 5 & -1 \end{bmatrix}, \qquad B = \begin{bmatrix} 7 & 1 \\ 0 & 5 \\ -8 & 3 \end{bmatrix}, \qquad C = \begin{bmatrix} 3 & 5 & 6 \\ 0 & -2 & 5 \\ 8 & 9 & 12 \end{bmatrix}$$

Rows and Columns   Matrices consist of rows and columns. Rows are labeled from the top of the matrix, columns from the left. The following matrix has two rows and three columns.

$$\begin{bmatrix} 2 & 3 & -4 \\ 7 & 5 & -1 \end{bmatrix}$$

The rows are:

$$\begin{bmatrix} 2 & 3 & -4 \end{bmatrix}, \qquad \begin{bmatrix} 7 & 5 & -1 \end{bmatrix}$$
$$\text{row 1} \qquad\qquad \text{row 2}$$

The columns are:

$$\begin{bmatrix} 2 \\ 7 \end{bmatrix}, \qquad \begin{bmatrix} 3 \\ 5 \end{bmatrix}, \qquad \begin{bmatrix} -4 \\ -1 \end{bmatrix}$$
$$\text{column 1} \qquad \text{column 2} \qquad \text{column 3}$$

Submatrix   A submatrix of a given matrix is an array obtained by deleting certain rows and columns of the matrix. For example, consider the following matrix A. The matrices P, Q, and R are submatrices of A.

$$A = \begin{bmatrix} 1 & 7 & 4 \\ 2 & 3 & 0 \\ 5 & 1 & -2 \end{bmatrix} \qquad P = \begin{bmatrix} 1 & 7 \\ 2 & 3 \\ 5 & 1 \end{bmatrix} \qquad Q = \begin{bmatrix} 7 \\ 3 \\ 1 \end{bmatrix} \qquad R = \begin{bmatrix} 1 & 4 \\ 5 & -2 \end{bmatrix}$$
$$\text{matrix } A \qquad\qquad\qquad\qquad \text{submatrices of } A$$

Size and Type   The size of a matrix is described by specifying the number of rows and columns in the matrix. For example, a matrix having two rows and three columns is said to be a 2×3 matrix; the first number indicates the number of rows, the second indicates the number of columns. When the number of rows is equal to the number of columns, the matrix is said to be a **square matrix**. A matrix consisting of one row is called a **row matrix**. A matrix consisting of one column is a **column matrix**. The following matrices are of the stated sizes and types.

$$\begin{bmatrix} 1 & 0 & 3 \\ -2 & 4 & 5 \end{bmatrix} \qquad \begin{bmatrix} 2 & 5 & 7 \\ -9 & 0 & 1 \\ -3 & 5 & 8 \end{bmatrix} \qquad \begin{bmatrix} 4 & -3 & 8 & 5 \end{bmatrix} \qquad \begin{bmatrix} 8 \\ 3 \\ 2 \end{bmatrix}$$
$$2 \times 3 \text{ matrix} \qquad\quad 3 \times 3 \text{ matrix} \qquad\quad 1 \times 4 \text{ matrix} \qquad\quad 3 \times 1 \text{ matrix}$$
$$\qquad\qquad\qquad\quad \text{a square matrix} \qquad\quad \text{a row matrix} \qquad\quad \text{a column matrix}$$

Location   The location of an element in a matrix is described by giving the row and column in which the element lies. For example, consider the following matrix.

$$\begin{bmatrix} 2 & 3 & -4 \\ 7 & 5 & -1 \end{bmatrix}$$

The element 7 is in row 2, column 1. We say that it is in location (2, 1).

The element in location $(1, 3)$ is -4. Note that the convention is to give the row in which the element lies, followed by the column.

Identity Matrices   An identity matrix is a square matrix with 1s in the *diagonal* locations $(1, 1)$, $(2, 2)$, $(3, 3)$, etc., and zeros elsewhere. We write I_n for the $n \times n$ identity matrix. The following matrices are identity matrices.

$$I_2 = \begin{bmatrix} 1 & 0 \\ 0 & 1 \end{bmatrix}, \qquad I_3 = \begin{bmatrix} 1 & 0 & 0 \\ 0 & 1 & 0 \\ 0 & 0 & 1 \end{bmatrix}$$

We are now ready to continue the discussion of systems of linear equations. We use matrices to describe systems of linear equations. There are two important matrices associated with every system of linear equations. The coefficients of the variables form a matrix called the **matrix of coefficients** of the system. The coefficients, together with the constant terms, form a matrix called the **augmented matrix** of the system. For example, the matrix of coefficients and the augmented matrix of the following system of linear equations are as shown:

$$\begin{array}{rcl} x_1 + x_2 + x_3 &=& 2 \\ 2x_1 + 3x_2 + x_3 &=& 3 \\ x_1 - x_2 - 2x_3 &=& -6 \end{array} \qquad \begin{bmatrix} 1 & 1 & 1 \\ 2 & 3 & 1 \\ 1 & -1 & -2 \end{bmatrix} \qquad \begin{bmatrix} 1 & 1 & 1 & 2 \\ 2 & 3 & 1 & 3 \\ 1 & -1 & -2 & -6 \end{bmatrix}$$

$$\text{matrix of coefficients} \qquad\qquad \text{augmented matrix}$$

Observe that the matrix of coefficients is a submatrix of the augmented matrix. The augmented matrix completely describes the system.

Transformations called **elementary transformations** can be used to change a system of linear equations into another system of linear equations that has the same solution. These transformations are used to solve systems of linear equations by eliminating variables. In practice it is simpler to work in terms of matrices using analogous transformations called **elementary row operations**. It is not necessary to write down the variables x_1, x_2, x_3, at each stage. Systems of linear equations are in fact described and manipulated on computers in terms of such matrices. These transformations are as follows.

Elementary Transformations	**Elementary Row Operations**
1. Interchange two equations.	1. Interchange two rows of a matrix.
2. Multiply both sides of an equation by a nonzero constant.	2. Multiply the elements of a row by a nonzero constant.
3. Add a multiple of one equation to another equation.	3. Add a multiple of the elements of one row to the corresponding elements of another row.

Systems of equations that are related through elementary transformations are called **equivalent systems**. Matrices that are related through elementary row operations are called **row equivalent matrices**. The symbol $\approx$ is used to indicate equivalence in both cases.

Elementary transformations preserve solutions since the order of the equations does not affect the solution, multiplying an equation throughout by a nonzero constant does not change the truth of the equality, and adding equal quantities to both sides of an equality results in an equality.

The method of Gauss-Jordan elimination uses elementary transformations to eliminate variables in a systematic manner, until we arrive at a system that gives the solution. We illustrate Gauss-Jordan elimination using equations and the analogous matrix implementation of the method side by side in the following example. The reader should note the way in which the variables are eliminated in the equations in the left column. At the same time observe how this is accomplished in terms of matrices in the right column by creating zeros in certain locations. We shall henceforth be using the matrix approach.

EXAMPLE 1 Solve the system of linear equations

$$\begin{aligned} x_1 + x_2 + x_3 &= 2 \\ 2x_1 + 3x_2 + x_3 &= 3 \\ x_1 - x_2 - 2x_3 &= -6 \end{aligned}$$

SOLUTION

Equation Method
Initial System

$$\begin{aligned} x_1 + x_2 + x_3 &= 2 \\ 2x_1 + 3x_2 + x_3 &= 3 \\ x_1 - x_2 - 2x_3 &= -6 \end{aligned}$$

Eliminate x_1 from 2nd and 3rd equations.

$$\approx \qquad \begin{aligned} x_1 + x_2 + x_3 &= 2 \\ x_2 - x_3 &= -1 \\ -2x_2 - 3x_3 &= -8 \end{aligned}$$

Eq2 + (−2)Eq1
Eq3 + (−1)Eq1

Eliminate x_2 from 1st and 3rd equations.

$$\approx \qquad \begin{aligned} x_1 + 2x_3 &= 3 \\ x_2 - x_3 &= -1 \\ -5x_3 &= -10 \end{aligned}$$

Eq1 + (−1)Eq2
Eq3 + (2)Eq2

Make coefficient of x_3 in 3rd equation 1 (i.e., solve for x_3).

$$\approx \qquad \begin{aligned} x_1 + 2x_3 &= 3 \\ x_2 - x_3 &= -1 \\ x_3 &= 2 \end{aligned}$$

(−1/5)Eq3

Eliminate x_3 from 1st and 2nd equations.

$$\approx \qquad \begin{aligned} x_1 &= -1 \\ x_2 &= 1 \\ x_3 &= 2 \end{aligned}$$

Eq1 + (−2)Eq3
Eq2 + Eq3

The solution is $x_1 = -1$, $x_2 = 1$, $x_3 = 2$.

Analogous Matrix Method
Augmented Matrix

$$\begin{bmatrix} 1 & 1 & 1 & 2 \\ 2 & 3 & 1 & 3 \\ 1 & -1 & -2 & -6 \end{bmatrix}$$

Create zeros in column 1.

$$\approx \qquad \begin{bmatrix} 1 & 1 & 1 & 2 \\ 0 & 1 & -1 & -1 \\ 0 & -2 & -3 & -8 \end{bmatrix}$$

R2 + (−2)R1
R3 + (−1)R1

Create appropriate zeros in column 2.

$$\approx \qquad \begin{bmatrix} 1 & 0 & 2 & 3 \\ 0 & 1 & -1 & -1 \\ 0 & 0 & -5 & -10 \end{bmatrix}$$

R1 + (−1)R2
R3 + (2)R2

Make the (3, 3) element 1 (called "normalizing" the element).

$$\approx \qquad \begin{bmatrix} 1 & 0 & 2 & 3 \\ 0 & 1 & -1 & -1 \\ 0 & 0 & 1 & 2 \end{bmatrix}$$

(−1/5)R3

Create zeros in column 3.

$$\approx \qquad \begin{bmatrix} 1 & 0 & 0 & -1 \\ 0 & 1 & 0 & 1 \\ 0 & 0 & 1 & 2 \end{bmatrix}$$

R1 + (−2)R3
R2 + R3

Matrix corresponds to the system.

$$\begin{aligned} x_1 &= -1 \\ x_2 &= 1 \\ x_3 &= 2 \end{aligned}$$

Solution is $x_1 = -1$, $x_2 = 1$, $x_3 = 2$.

Geometrically, each of the three original equations in this example represents a plane in three-dimensional space. The fact that there is a unique solution means that these three planes intersect at a single point. The solution $(-1, 1, 2)$ gives the coordinates of this point where the three planes intersect. We now give another example to reinforce the method.

EXAMPLE 2 Solve the following system of linear equations.

$$x_1 - 2x_2 + 4x_3 = 12$$
$$2x_1 - x_2 + 5x_3 = 18$$
$$-x_1 + 3x_2 - 3x_3 = -8$$

SOLUTION

Start with the augmented matrix and use the first row to create zeros in the first column. (This corresponds to using the first equation to eliminate x_1 from the second and third equations.)

$$\begin{bmatrix} 1 & -2 & 4 & 12 \\ 2 & -1 & 5 & 18 \\ -1 & 3 & -3 & -8 \end{bmatrix} \quad \begin{matrix} \approx \\ R2 + (-2)R1 \\ R3 + R1 \end{matrix} \quad \begin{bmatrix} 1 & -2 & 4 & 12 \\ 0 & 3 & -3 & -6 \\ 0 & 1 & 1 & 4 \end{bmatrix}$$

Next multiply row 2 by $\frac{1}{3}$ to make the (2, 2) element 1. (This corresponds to making the coefficient of x_2 in the second equation 1.)

$$\begin{matrix} \approx \\ (\frac{1}{3})R2 \end{matrix} \quad \begin{bmatrix} 1 & -2 & 4 & 12 \\ 0 & 1 & -1 & -2 \\ 0 & 1 & 1 & 4 \end{bmatrix}$$

Create zeros in the second column as follows. (This corresponds to using the second equation to eliminate x_2 from the first and third equations.)

$$\begin{matrix} \approx \\ R1 + (2)R2 \\ R3 + (-1)R2 \end{matrix} \quad \begin{bmatrix} 1 & 0 & 2 & 8 \\ 0 & 1 & -1 & -2 \\ 0 & 0 & 2 & 6 \end{bmatrix}$$

Multiply row 3 by $\frac{1}{2}$. (This corresponds to making the coefficient of x_3 in the third equation 1.)

$$\begin{matrix} \approx \\ (\frac{1}{2})R3 \end{matrix} \quad \begin{bmatrix} 1 & 0 & 2 & 8 \\ 0 & 1 & -1 & -2 \\ 0 & 0 & 1 & 3 \end{bmatrix}$$

Finally, create zeros in the third column. (This corresponds to using the third equation to eliminate x_3 from the first and second equations.)

$$\begin{matrix} \approx \\ R1 + (-2)R3 \\ R2 + R3 \end{matrix} \quad \begin{bmatrix} 1 & 0 & 0 & 2 \\ 0 & 1 & 0 & 1 \\ 0 & 0 & 1 & 3 \end{bmatrix}$$

This matrix corresponds to the system

$$
\begin{aligned}
x_1 \quad\quad\quad &= 2 \\
x_2 \quad\quad &= 1 \\
x_3 &= 3
\end{aligned}
$$

The solution is $x_1 = 2$, $x_2 = 1$, $x_3 = 3$.

This Gauss-Jordan method of solving a system of linear equations using matrices involves creating 1s and 0s in certain locations of matrices. These numbers are created in a systematic manner, column by column. The following example illustrates that it may be necessary to interchange two rows at some stage in order to proceed in the preceding manner.

EXAMPLE 3 Solve the system

$$
\begin{aligned}
4x_1 + 8x_2 - 12x_3 &= 44 \\
3x_1 + 6x_2 - 8x_3 &= 32 \\
-2x_1 - x_2 \quad\quad &= -7
\end{aligned}
$$

SOLUTION

We start with the augmented matrix and proceed as follows. (Note the use of zero in the augmented matrix as the coefficient of the missing variable x_3 in the third equation.)

$$
\begin{bmatrix} 4 & 8 & -12 & 44 \\ 3 & 6 & -8 & 32 \\ -2 & -1 & 0 & -7 \end{bmatrix}
\underset{(\frac{1}{4})\text{R1}}{\approx}
\begin{bmatrix} 1 & 2 & -3 & 11 \\ 3 & 6 & -8 & 32 \\ -2 & -1 & 0 & -7 \end{bmatrix}
$$

$$
\underset{\substack{\text{R2} + (-3)\text{R1} \\ \text{R3} + (2)\text{R1}}}{\approx}
\begin{bmatrix} 1 & 2 & -3 & 11 \\ 0 & 0 & 1 & -1 \\ 0 & 3 & -6 & 15 \end{bmatrix}
$$

At this stage we need a nonzero element in the location (2, 2) in order to continue. To achieve this we interchange the second row with the third row (a *later* row) and then proceed.

$$
\underset{\text{R2} \leftrightarrow \text{R3}}{\approx}
\begin{bmatrix} 1 & 2 & -3 & 11 \\ 0 & 3 & -6 & 15 \\ 0 & 0 & 1 & -1 \end{bmatrix}
\underset{(\frac{1}{3})\text{R2}}{\approx}
\begin{bmatrix} 1 & 2 & -3 & 11 \\ 0 & 1 & -2 & 5 \\ 0 & 0 & 1 & -1 \end{bmatrix}
$$

$$
\underset{\text{R1} + (-2)\text{R2}}{\approx}
\begin{bmatrix} 1 & 0 & 1 & 1 \\ 0 & 1 & -2 & 5 \\ 0 & 0 & 1 & -1 \end{bmatrix}
\underset{\substack{\text{R1} + (-1)\text{R3} \\ \text{R2} + (2)\text{R3}}}{\approx}
\begin{bmatrix} 1 & 0 & 0 & 2 \\ 0 & 1 & 0 & 3 \\ 0 & 0 & 1 & -1 \end{bmatrix}
$$

The solution is $x_1 = 2$, $x_2 = 3$, $x_3 = -1$.

Summary

We now summarize the method of Gauss-Jordan elimination for solving a system of n linear equations in n variables that has a unique solution. The augmented matrix is made up of a matrix of coefficients A and a column matrix of constant terms B. Let us write $[A : B]$ for this matrix. Use row operations to gradually transform this matrix, column by column, into a matrix $[I_n : X]$, where I_n is the identity $n \times n$ matrix.

$$[A : B] \approx \cdots \approx [I_n : X]$$

This final matrix $[I_n : X]$ is called the **reduced echelon form** of the original augmented matrix. The matrix of coefficients of the final system of equations is I_n and X is the column matrix of constant terms. This implies that the elements of X are the unique solution. Observe that as $[A : B]$ is being transformed to $[I_n : X]$, A is being changed to I_n. Thus:

> If A is the matrix of coefficients of a system of n equations in n variables that has a unique solution, then it is row equivalent to I_n.

If $[A : B]$ cannot be transformed in this manner into a matrix of the form $[I_n : X]$, the system of equations does not have a unique solution. More will be said about such systems in the next section.

Many Systems

Certain applications involve solving a number of systems of linear equations, all having the same square matrix of coefficients A. Let the systems be

$$[A : B_1], [A : B_2], \ldots, [A : B_k]$$

The constant terms $B_1, B_2, \ldots, B_k$, might for example be test data, and one wants to know the solutions that would lead to these results. The situation often dictates that the solutions be unique. One could of course go through the method of Gauss-Jordan elimination for each system, solving each system independently. This procedure would lead to the reduced echelon forms

$$[I_n : X_1], [I_n : X_2], \ldots, [I_n : X_k]$$

and the solutions would be $X_1, X_2, \ldots, X_k$. However, the same reduction of A to I_n would be repeated for each system; this involves a great deal of unnecessary duplication. The systems can be represented by one large augmented matrix $[A : B_1 \ B_2 \cdots B_k]$, and the Gauss-Jordan method can be applied to this one matrix. We would get

$$[A : B_1 \ B_2 \cdots B_k] \approx \cdots \approx [I_n : X_1 \ X_2 \cdots X_k]$$

leading to the solutions $X_1, X_2, \ldots, X_k$.

EXAMPLE 4 Solve the following three systems of linear equations, all of which have the same matrix of coefficients.

$$\begin{array}{rcl} x_1 - x_2 + 3x_3 &=& b_1 \\ 2x_1 - x_2 + 4x_3 &=& b_2 \\ -x_1 + 2x_2 - 4x_3 &=& b_3 \end{array} \quad \text{for} \quad \begin{bmatrix} b_1 \\ b_2 \\ b_3 \end{bmatrix} = \begin{bmatrix} 8 \\ 11 \\ -11 \end{bmatrix}, \begin{bmatrix} 0 \\ 1 \\ 2 \end{bmatrix}, \begin{bmatrix} 3 \\ 3 \\ -4 \end{bmatrix} \quad \text{in turn.}$$

SOLUTION

Construct the large augmented matrix that describes all three systems and determine the reduced echelon form as follows.

$$
\begin{bmatrix}
1 & -1 & 3 & 8 & 0 & 3 \\
2 & -1 & 4 & 11 & 1 & 3 \\
-1 & 2 & -4 & -11 & 2 & -4
\end{bmatrix}
\quad \approx \atop {R2 + (-2)R1 \atop R3 + R1}
\begin{bmatrix}
1 & -1 & 3 & 8 & 0 & 3 \\
0 & 1 & -2 & -5 & 1 & -3 \\
0 & 1 & -1 & -3 & 2 & -1
\end{bmatrix}
$$

$$
\approx \atop {R1 + R2 \atop R3 + (-1)R2}
\begin{bmatrix}
1 & 0 & 1 & 3 & 1 & 0 \\
0 & 1 & -2 & -5 & 1 & -3 \\
0 & 0 & 1 & 2 & 1 & 2
\end{bmatrix}
$$

$$
\approx \atop {R1 + (-1)R3 \atop R2 + 2R3}
\begin{bmatrix}
1 & 0 & 0 & 1 & 0 & -2 \\
0 & 1 & 0 & -1 & 3 & 1 \\
0 & 0 & 1 & 2 & 1 & 2
\end{bmatrix}
$$

The solutions to the three systems of equations are given by the last three columns of the reduced echelon form. They are

$$x_1 = 1, x_2 = -1, x_3 = 2$$
$$x_1 = 0, x_2 = 3, x_3 = 1$$
$$x_1 = -2, x_2 = 1, x_3 = 2$$

In this section we have limited our discussion to systems of n linear equations in n variables that have a unique solution. In the following section we shall extend the method of Gauss-Jordan elimination to accommodate other systems that have a unique solution, and also to include systems that have many solutions or no solutions.

EXERCISE SET 1.1

Matrices

1. Give the sizes of the following matrices.

(a) $\begin{bmatrix} 1 & 2 & 3 \\ 0 & 1 & 2 \\ 4 & 5 & 3 \end{bmatrix}$ (b) $\begin{bmatrix} 0 & 9 \\ -6 & 4 \\ -3 & 2 \end{bmatrix}$

(c) $\begin{bmatrix} 1 & 2 & 3 & 0 \\ 1 & 2 & 4 & 5 \end{bmatrix}$ (d) $\begin{bmatrix} -7 \\ 4 \\ 3 \end{bmatrix}$

(e) $\begin{bmatrix} 1 & 2 & 9 & -8 & 7 \\ 4 & 2 & 5 & 7 & 2 \\ 4 & -6 & 4 & 0 & 0 \end{bmatrix}$

(f) $\begin{bmatrix} 2 & -3 & 4 & 7 \end{bmatrix}$

2. Give the (1, 1), (2, 2), (3, 3), (1, 5), (2, 4), (3, 2) elements of the following matrix.

$$\begin{bmatrix} 1 & 2 & 3 & 0 & -1 \\ -2 & 4 & -5 & 3 & 6 \\ 5 & 8 & 9 & 2 & 3 \end{bmatrix}$$

3. Give the (2, 3), (3, 2), (4, 1), (1, 3), (4, 4), (3, 1) elements of the following matrix.

$$\begin{bmatrix} 1 & 2 & 7 & 0 \\ -1 & 2 & 4 & 5 \\ 3 & 5 & 0 & -1 \\ 6 & 9 & 0 & 2 \end{bmatrix}$$

4. Write down the identity matrix I_4.

---

*Answers to exercises marked in red are provided in the back of the book.

Matrices and Systems of Equations

5. Determine the matrix of coefficients and augmented matrix of each of the following systems of equations.

(a) $\quad x_1 + 3x_2 = \quad 7$

$\quad 2x_1 - 5x_2 = -3$

(b) $\quad 5x_1 + 2x_2 - 4x_3 = 8$

$\quad x_1 + 3x_2 + 6x_3 = 4$

$\quad 4x_1 + 6x_2 - 9x_3 = 7$

(c) $\quad -x_1 + 3x_2 - 5x_3 = -3$

$\quad 2x_1 - 2x_2 + 4x_3 = \quad 8$

$\quad x_1 + 3x_2 \quad\quad = \quad 6$

(d) $\quad 5x_1 + 4x_2 = \quad 9$

$\quad 2x_1 - 8x_2 = -4$

$\quad x_1 + 2x_2 = \quad 3$

(e) $\quad 5x_1 + 2x_2 - 4x_3 = 8$

$\quad\quad\quad 4x_2 + 3x_3 = 0$

$\quad x_1 \quad\quad - x_3 = 7$

(f) $\quad -x_1 + 3x_2 - 9x_3 = -4$

$\quad x_1 \quad\quad - 4x_3 = \quad 11$

$\quad x_1 + 8x_2 \quad\quad = \quad 1$

(g) $\quad x_1 \quad\quad\quad = -3$

$\quad\quad x_2 \quad\quad = 12$

$\quad\quad\quad x_3 = \quad 8$

(h) $\quad -4x_1 + 2x_2 - 9x_3 + \quad x_4 = -1$

$\quad x_1 + 6x_2 - 8x_3 - 7x_4 = \quad 15$

$\quad\quad -x_2 + 3x_3 - 5x_4 = \quad 0$

6. Interpret the following matrices as augmented matrices of systems of equations. Write down each system of equations.

(a) $\begin{bmatrix} 1 & 2 & 3 \\ 4 & 5 & 6 \end{bmatrix}$ **(b)** $\begin{bmatrix} 7 & 9 & 8 \\ 6 & 4 & -3 \end{bmatrix}$

(c) $\begin{bmatrix} 1 & 9 & -3 \\ 5 & 0 & 2 \end{bmatrix}$ **(d)** $\begin{bmatrix} 8 & 7 & 5 & -1 \\ 4 & 6 & 2 & 4 \\ 9 & 3 & 7 & 6 \end{bmatrix}$

(e) $\begin{bmatrix} 2 & -3 & 6 & 4 \\ 7 & -5 & -2 & 3 \\ 0 & 2 & 4 & 0 \end{bmatrix}$ **(f)** $\begin{bmatrix} 0 & -2 & 4 \\ 5 & 7 & -3 \\ 6 & 0 & 8 \end{bmatrix}$

(g) $\begin{bmatrix} 1 & 0 & 0 & 3 \\ 0 & 1 & 0 & 8 \\ 0 & 0 & 1 & 4 \end{bmatrix}$ **(h)** $\begin{bmatrix} 1 & 2 & -1 & 6 \\ 0 & 1 & 4 & 5 \\ 0 & 0 & 1 & -2 \end{bmatrix}$

Elementary Row Operations

7. In the following exercises you are given a matrix followed by an elementary row operation. Determine each resulting matrix.

(a) $\begin{bmatrix} 2 & 6 & -4 & 0 \\ 1 & 2 & -3 & 6 \\ 8 & 3 & 2 & 5 \end{bmatrix}$ $\begin{array}{c} \approx \\ (\tfrac{1}{2})R1 \end{array}$

(b) $\begin{bmatrix} 0 & -8 & 4 & 3 \\ 2 & 7 & 5 & 1 \\ 3 & -5 & 8 & 9 \end{bmatrix}$ $\begin{array}{c} \approx \\ R1 \leftrightarrow R2 \end{array}$

(c) $\begin{bmatrix} 1 & 2 & 3 & -1 \\ -1 & 1 & 7 & 1 \\ 2 & -4 & 5 & -3 \end{bmatrix}$ $\begin{array}{c} \approx \\ R2 + R1 \\ R3 + (-2)R1 \end{array}$

(d) $\begin{bmatrix} 1 & 2 & 3 & -4 \\ 0 & 1 & 2 & 1 \\ 0 & -4 & 3 & -5 \end{bmatrix}$ $\begin{array}{c} \approx \\ R1 + (-2)R2 \\ R3 + (4)R2 \end{array}$

(e) $\begin{bmatrix} 1 & 0 & 4 & -3 \\ 0 & 1 & -3 & 2 \\ 0 & 0 & 1 & 5 \end{bmatrix}$ $\begin{array}{c} \approx \\ R1 + (-4)R3 \\ R2 + (3)R3 \end{array}$

(f) $\begin{bmatrix} 1 & 0 & 2 & 7 \\ 0 & 1 & 5 & -3 \\ 0 & 0 & -2 & 8 \end{bmatrix}$ $\begin{array}{c} \approx \\ (-\tfrac{1}{2})R3 \end{array}$

8. Interpret each of the following row operations as a stage in arriving at the reduced echelon form of a matrix. Why have the indicated operations been selected? What particular aims do they accomplish in terms of the systems of linear equations that are described by the matrices?

(a)

$\begin{bmatrix} 1 & -4 & 3 & 5 \\ -2 & 1 & 7 & 5 \\ 4 & 0 & -3 & 6 \end{bmatrix} \begin{array}{c} \approx \\ R2 + (2)R1 \\ R3 + (-4)R1 \end{array} \begin{bmatrix} 1 & -4 & 3 & 5 \\ 0 & -7 & 13 & 15 \\ 0 & 16 & -15 & -14 \end{bmatrix}$

(b) $\begin{bmatrix} 1 & 2 & -4 & 7 \\ 0 & 3 & 9 & -6 \\ 0 & 4 & 7 & -8 \end{bmatrix} \begin{array}{c} \approx \\ (\tfrac{1}{3})R2 \end{array} \begin{bmatrix} 1 & 2 & -4 & 7 \\ 0 & 1 & 3 & -2 \\ 0 & 4 & 7 & -8 \end{bmatrix}$

(c) $\begin{bmatrix} 1 & 3 & -4 & 5 \\ 0 & 0 & -2 & 6 \\ 0 & 1 & 3 & -8 \end{bmatrix} \begin{array}{c} \approx \\ R2 \leftrightarrow R3 \end{array} \begin{bmatrix} 1 & 3 & -4 & 5 \\ 0 & 1 & 3 & -8 \\ 0 & 0 & -2 & 6 \end{bmatrix}$

(d)

$\begin{bmatrix} 1 & 2 & 5 & 0 \\ 0 & 1 & 2 & -3 \\ 0 & -3 & 1 & -2 \end{bmatrix} \begin{array}{c} \approx \\ R1 + (-2)R2 \\ R3 + (3)R2 \end{array} \begin{bmatrix} 1 & 0 & 1 & 6 \\ 0 & 1 & 2 & -3 \\ 0 & 0 & 7 & -11 \end{bmatrix}$

9. Interpret each of the following row operations as a stage in arriving at the reduced echelon form of a matrix. Why have these operations been selected?

(a) $\begin{bmatrix} 1 & 0 & 2 & 6 \\ 0 & 1 & -1 & 3 \\ 0 & 0 & 1 & 2 \end{bmatrix} \begin{array}{c} \\ \text{R1} + (-2)\text{R3} \\ \text{R2} + \text{R3} \end{array} \approx \begin{bmatrix} 1 & 0 & 0 & 2 \\ 0 & 1 & 0 & 5 \\ 0 & 0 & 1 & 2 \end{bmatrix}$

(b) $\begin{bmatrix} 0 & 2 & 4 & -1 \\ 4 & 3 & 2 & -8 \\ 5 & -7 & 1 & 2 \end{bmatrix} \begin{array}{c} \\ \text{R1} \leftrightarrow \text{R2} \end{array} \approx \begin{bmatrix} 4 & 3 & 2 & -8 \\ 0 & 2 & 4 & -1 \\ 5 & -7 & 1 & 2 \end{bmatrix}$

(c) $\begin{bmatrix} 1 & 0 & 3 & 7 \\ 0 & 1 & 4 & 2 \\ 0 & 0 & -2 & 6 \end{bmatrix} \begin{array}{c} \\ (-\frac{1}{2})\text{R3} \end{array} \approx \begin{bmatrix} 1 & 0 & 3 & 7 \\ 0 & 1 & 4 & 2 \\ 0 & 0 & 1 & -3 \end{bmatrix}$

(d)

$\begin{bmatrix} 1 & 0 & -2 & 4 \\ 0 & 1 & 3 & -4 \\ 0 & 0 & 1 & -3 \end{bmatrix} \begin{array}{c} \\ \text{R1} + (2)\text{R3} \\ \text{R2} + (-3)\text{R3} \end{array} \approx \begin{bmatrix} 1 & 0 & 0 & -2 \\ 0 & 1 & 0 & 5 \\ 0 & 0 & 1 & -3 \end{bmatrix}$

Solving Systems of Linear Equations

10. The following systems of equations all have unique solutions. Solve these systems using the method of Gauss-Jordan elimination with matrices.

(a) $\begin{aligned} x_1 - 2x_2 &= -8 \\ 2x_1 - 3x_2 &= -11 \end{aligned}$

(b) $\begin{aligned} 2x_1 + 2x_2 &= 4 \\ 3x_1 + 2x_2 &= 3 \end{aligned}$

(c) $\begin{aligned} x_1 \qquad + x_3 &= 3 \\ 2x_2 - 2x_3 &= -4 \\ x_2 - 2x_3 &= 5 \end{aligned}$

(d) $\begin{aligned} x_1 + x_2 + 3x_3 &= 6 \\ x_1 + 2x_2 + 4x_3 &= 9 \\ 2x_1 + x_2 + 6x_3 &= 11 \end{aligned}$

(e) $\begin{aligned} x_1 - x_2 + 3x_3 &= 3 \\ 2x_1 - x_2 + 2x_3 &= 2 \\ 3x_1 + x_2 - 2x_3 &= 3 \end{aligned}$

(f) $\begin{aligned} -x_1 + x_2 - x_3 &= -2 \\ 3x_1 + x_2 + x_3 &= 10 \\ 4x_1 + 2x_2 + 3x_3 &= 14 \end{aligned}$

11. The following systems of equations all have unique solutions. Solve these systems using the method of Gauss-Jordan elimination with matrices.

(a) $\begin{aligned} x_1 + 2x_2 + 3x_3 &= 14 \\ 2x_1 + 5x_2 + 8x_3 &= 36 \\ x_1 - x_2 \qquad &= -4 \end{aligned}$

(b) $\begin{aligned} x_1 - x_2 - x_3 &= -1 \\ -2x_1 + 6x_2 + 10x_3 &= 14 \\ 2x_1 + x_2 + 6x_3 &= 9 \end{aligned}$

(c) $\begin{aligned} 2x_1 + 2x_2 - 4x_3 &= 14 \\ 3x_1 + x_2 + x_3 &= 8 \\ 2x_1 - x_2 + 2x_3 &= -1 \end{aligned}$

(d) $\begin{aligned} 2x_2 + 4x_3 &= 8 \\ 2x_1 + 2x_2 \qquad &= 6 \\ x_1 + x_2 + x_3 &= 5 \end{aligned}$

(e) $\begin{aligned} x_1 \qquad - x_3 &= 3 \\ -x_1 \qquad + 2x_3 &= -8 \\ 3x_1 + x_2 - x_3 &= 0 \end{aligned}$

12. The following systems of equations all have unique solutions. Solve these systems using the method of Gauss-Jordan elimination with matrices.

(a) $\begin{aligned} \tfrac{3}{2}x_1 \qquad + 3x_3 &= 15 \\ -x_1 + 7x_2 - 9x_3 &= -45 \\ 2x_1 \qquad + 5x_3 &= 22 \end{aligned}$

(b) $\begin{aligned} -3x_1 - 6x_2 - 15x_3 &= -3 \\ 2x_1 + 3x_2 + 9x_3 &= 1 \\ -4x_1 - 7x_2 - 17x_3 &= -4 \end{aligned}$

(c) $\begin{aligned} 3x_1 + 6x_2 \qquad - 3x_4 &= 3 \\ x_1 + 3x_2 - x_3 - 4x_4 &= -12 \\ x_1 - x_2 + x_3 + 2x_4 &= 8 \\ 2x_1 + 3x_2 \qquad &= 8 \end{aligned}$

(d) $\begin{aligned} x_1 + 2x_2 + 2x_3 + 5x_4 &= 11 \\ 2x_1 + 4x_2 + 2x_3 + 8x_4 &= 14 \\ x_1 + 3x_2 + 4x_3 + 8x_4 &= 19 \\ x_1 - x_2 + x_3 \qquad &= 2 \end{aligned}$

(e) $\begin{aligned} x_1 + x_2 + 2x_3 + 6x_4 &= 11 \\ 2x_1 + 3x_2 + 6x_3 + 19x_4 &= 36 \\ 3x_2 + 4x_3 + 15x_4 &= 28 \\ x_1 - x_2 - x_3 - 6x_4 &= -12 \end{aligned}$

13. The following exercises involve many systems of linear equations with unique solutions that have the same matrix of coefficients. Solve the systems by applying the method of Gauss-Jordan elimination to a large augmented matrix that describes many systems.

(a) $x_1 + 2x_2 = b_1$
$3x_1 + 5x_2 = b_2$

for $\begin{bmatrix} b_1 \\ b_2 \end{bmatrix} = \begin{bmatrix} 3 \\ 8 \end{bmatrix}, \begin{bmatrix} 4 \\ 9 \end{bmatrix}, \begin{bmatrix} 3 \\ 7 \end{bmatrix}$ in turn.

(b) $x_1 + x_2 = b_1$
$2x_1 + 3x_2 = b_2$

for $\begin{bmatrix} b_1 \\ b_2 \end{bmatrix} = \begin{bmatrix} 0 \\ 1 \end{bmatrix}, \begin{bmatrix} 5 \\ 13 \end{bmatrix}, \begin{bmatrix} 1 \\ 2 \end{bmatrix}$ in turn.

(c) $x_1 - 2x_2 + 3x_3 = b_1$
$x_1 - x_2 + 2x_3 = b_2$
$2x_1 - 3x_2 + 6x_3 = b_3$

for $\begin{bmatrix} b_1 \\ b_2 \\ b_3 \end{bmatrix} = \begin{bmatrix} 6 \\ 5 \\ 14 \end{bmatrix}, \begin{bmatrix} -5 \\ -3 \\ -8 \end{bmatrix}, \begin{bmatrix} 4 \\ 3 \\ 9 \end{bmatrix}$ in turn.

(d) $x_1 + 2x_2 - x_3 = b_1$
$-x_1 - x_2 + x_3 = b_2$
$3x_1 + 7x_2 - x_3 = b_3$

for $\begin{bmatrix} b_1 \\ b_2 \\ b_3 \end{bmatrix} = \begin{bmatrix} -1 \\ 1 \\ -1 \end{bmatrix}, \begin{bmatrix} 6 \\ -4 \\ 18 \end{bmatrix}, \begin{bmatrix} 0 \\ -2 \\ -4 \end{bmatrix}$ in turn.

1.2 Gauss-Jordan Elimination

In the previous section we used the method of Gauss-Jordan elimination to solve systems of n equations in n variables that had a unique solution. We shall now discuss the method in its more general setting, where the number of equations can differ from the number of variables and where there can be a unique solution, many solutions, or no solutions. Our approach again will be to start from the augmented matrix of the given system and to perform a sequence of elementary row operations that will result in a simpler matrix (the reduced echelon form), which leads directly to the solution.

We now give the general definition of reduced echelon form. The reader will observe that the reduced echelon forms discussed in the previous section all conform to this definition.

DEFINITION   A matrix is in **reduced echelon form** if:
1. Any rows consisting entirely of zeros are grouped at the bottom of the matrix.
2. The first nonzero element of each other row is 1. This element is called a **leading 1**.
3. The leading 1 of each row after the first is positioned to the right of the leading 1 of the previous row.
4. All other elements in a column that contains a leading 1 are zero.

The following matrices are all in reduced echelon form.

$$\begin{bmatrix} 1 & 0 & 8 \\ 0 & 1 & 2 \\ 0 & 0 & 0 \end{bmatrix} \quad \begin{bmatrix} 1 & 0 & 0 & 7 \\ 0 & 1 & 0 & 3 \\ 0 & 0 & 1 & 9 \end{bmatrix} \quad \begin{bmatrix} 1 & 4 & 0 & 0 \\ 0 & 0 & 1 & 0 \\ 0 & 0 & 0 & 1 \end{bmatrix} \quad \begin{bmatrix} 1 & 2 & 3 & 0 \\ 0 & 0 & 0 & 1 \\ 0 & 0 & 0 & 0 \end{bmatrix}$$

$$\begin{bmatrix} 1 & 0 & 5 & 0 & 0 & 8 \\ 0 & 1 & 7 & 0 & 0 & 9 \\ 0 & 0 & 0 & 1 & 0 & 5 \\ 0 & 0 & 0 & 0 & 1 & 4 \end{bmatrix} \quad \begin{bmatrix} 1 & 2 & 0 & 3 & 0 & 4 \\ 0 & 0 & 1 & 2 & 0 & 7 \\ 0 & 0 & 0 & 0 & 1 & 6 \\ 0 & 0 & 0 & 0 & 0 & 0 \end{bmatrix}$$

The following matrices are not in reduced echelon form for the reasons stated.

$$\begin{bmatrix} 1 & 2 & 0 & 4 \\ 0 & 0 & 0 & 0 \\ 0 & 0 & 1 & 3 \end{bmatrix} \quad \begin{bmatrix} 1 & 2 & 0 & 3 & 0 \\ 0 & 0 & 3 & 4 & 0 \\ 0 & 0 & 0 & 0 & 1 \end{bmatrix} \quad \begin{bmatrix} 1 & 0 & 0 & 2 \\ 0 & 0 & 1 & 4 \\ 0 & 1 & 0 & 3 \end{bmatrix} \quad \begin{bmatrix} 1 & 7 & 0 & 8 \\ 0 & 1 & 0 & 3 \\ 0 & 0 & 1 & 2 \\ 0 & 0 & 0 & 0 \end{bmatrix}$$

| Row of zeros not at bottom of matrix | First nonzero element in row 2 is not 1 | Leading 1 in row 3 not to the right of leading 1 in row 2 | Nonzero element above leading 1 in row 2 |

There are usually many sequences of row operations that can be used to transform a given matrix to reduced echelon form—they all, however, lead to the same reduced echelon form. We say that *the reduced echelon form of a matrix is unique*. The method of Gauss-Jordan elimination is an important systematic way (called an algorithm) for arriving at the reduced echelon form. It can be programmed on a computer. We now summarize the method, then give examples of its implementation.

Gauss-Jordan Elimination
1. Write down the augmented matrix of the system of linear equations.
2. Derive the reduced echelon form of the augmented matrix using elementary row operations. This is done by creating leading 1s, then zeros above and below each leading 1, column by column, starting with the first column.
3. Write down the system of equations corresponding to the reduced echelon form. This system gives the solution.

We stress the importance of mastering this algorithm. Not only is getting the correct solution important, the method of arriving at the solution is important. We shall, for example, be interested in the efficiency of this algorithm (the number of additions and multiplications used) and comparing it with other algorithms that can be used to solve systems of linear equations.

EXAMPLE 1 Use the method of Gauss-Jordan elimination to find the reduced echelon form of the following matrix.

$$\begin{bmatrix} 0 & 0 & 2 & -2 & 2 \\ 3 & 3 & -3 & 9 & 12 \\ 4 & 4 & -2 & 11 & 12 \end{bmatrix}$$

SOLUTION

Step 1 Interchange rows, if necessary, to bring a nonzero element to the top of the first nonzero column. This nonzero element is called a **pivot**.

$$\underset{R1 \leftrightarrow R2}{\approx} \begin{bmatrix} ③ & 3 & -3 & 9 & 12 \\ 0 & 0 & 2 & -2 & 2 \\ 4 & 4 & -2 & 11 & 12 \end{bmatrix} \quad \text{pivot}$$

Step 2 Create a 1 in the pivot location by multiplying the pivot row by $\frac{1}{\text{pivot}}$.

$$\underset{(\frac{1}{3})R1}{\approx} \begin{bmatrix} 1 & 1 & -1 & 3 & 4 \\ 0 & 0 & 2 & -2 & 2 \\ 4 & 4 & -2 & 11 & 12 \end{bmatrix}$$

Step 3 Create zeros elsewhere in the pivot column by adding suitable multiples of the pivot row to all other rows of the matrix.

$$\underset{R3 + (-4)R1}{\approx} \begin{bmatrix} 1 & 1 & -1 & 3 & 4 \\ 0 & 0 & 2 & -2 & 2 \\ 0 & 0 & 2 & -1 & -4 \end{bmatrix}$$

Step 4 Cover the pivot row and all rows above it. Repeat Steps 1 and 2 for the remaining submatrix. Repeat step 3 for the whole matrix. Continue thus until the reduced echelon form is reached.

$$\begin{bmatrix} 1 & 1 & -1 & 3 & 4 \\ 0 & 0 & 2 & -2 & 2 \\ 0 & 0 & 2 & -1 & -4 \end{bmatrix} = \begin{bmatrix} 1 & 1 & -1 & 3 & 4 \\ 0 & 0 & ② & -2 & 2 \\ 0 & 0 & 2 & -1 & -4 \end{bmatrix}$$

first nonzero column of the submatrix.   pivot

$$\underset{(\frac{1}{2})R2}{\approx} \begin{bmatrix} 1 & 1 & -1 & 3 & 4 \\ 0 & 0 & 1 & -1 & 1 \\ 0 & 0 & 2 & 1 & 4 \end{bmatrix} \qquad \underset{\substack{R1 + R2 \\ R3 + (-2)R2}}{\approx} \begin{bmatrix} 1 & 1 & 0 & 2 & 5 \\ 0 & 0 & 1 & -1 & 1 \\ 0 & 0 & 0 & ① & -6 \end{bmatrix}$$

pivot

$$\underset{\substack{R1 + (-2)R3 \\ R2 + R3}}{\approx} \begin{bmatrix} 1 & 1 & 0 & 0 & 17 \\ 0 & 0 & 1 & 0 & -5 \\ 0 & 0 & 0 & 1 & -6 \end{bmatrix}$$

This matrix is the reduced echelon form of the given matrix. ∎

We now illustrate how this method is used to solve various systems of equations. The following example illustrates how to solve a system of linear equations that has many solutions. The reduced echelon form is derived. It then becomes necessary to interpret the reduced echelon form, expressing the many solutions in a clear manner.

EXAMPLE 2 Solve, if possible, the system of equations

$$3x_1 - 3x_2 + 3x_3 = 9$$
$$2x_1 - x_2 + 4x_3 = 7$$
$$3x_1 - 5x_2 - x_3 = 7$$

SOLUTION

Start with the augmented matrix and follow the Gauss-Jordan algorithm. Pivots and leading ones are circled.

$$\begin{bmatrix} ③ & -3 & 3 & 9 \\ 2 & -1 & 4 & 7 \\ 3 & -5 & -1 & 7 \end{bmatrix} \underset{(\frac{1}{3})R1}{\approx} \begin{bmatrix} ① & -1 & 1 & 3 \\ 2 & -1 & 4 & 7 \\ 3 & -5 & -1 & 7 \end{bmatrix}$$

$$\begin{array}{c} \approx \\ \text{R2} + (-2)\text{R1} \\ \text{R3} + (-3)\text{R1} \end{array} \begin{bmatrix} 1 & -1 & 1 & 3 \\ 0 & ① & 2 & 1 \\ 0 & -2 & -4 & -2 \end{bmatrix} \qquad \begin{array}{c} \approx \\ \text{R1} + \text{R2} \\ \text{R3} + (2)\text{R2} \end{array} \begin{bmatrix} 1 & 0 & 3 & 4 \\ 0 & 1 & 2 & 1 \\ 0 & 0 & 0 & 0 \end{bmatrix}$$

We have arrived at the reduced echelon form. The corresponding system of equations is

$$x_1 + 3x_3 = 4$$
$$x_2 + 2x_3 = 1$$

There are many values of x_1, x_2, and x_3 that satisfy these equations. This is a system of equations that has many solutions. x_1 is called the **leading variable** of the first equation and x_2 is the leading variable of the second equation. To express these many solutions, we write the leading variables in each equation in terms of the remaining variables. We get

$$x_1 = -3x_3 + 4$$
$$x_2 = -2x_3 + 1$$

Let us assign the arbitrary value r to x_3. The **general solution** to the system is

$$x_1 = -3r + 4, \, x_2 = -2r + 1, \, x_3 = r$$

As r ranges over the set of real numbers we get many solutions. r is called a **parameter**. We can get specific solutions by giving r different values. For example,

$$r = 1 \qquad \text{gives} \qquad x_1 = 1, \, x_2 = -1, \, x_3 = 1$$
$$r = -2 \qquad \text{gives} \qquad x_1 = 10, \, x_2 = 5, \, x_3 = -2$$

■

EXAMPLE 3 This example illustrates that the general solution can involve a number of parameters. Solve the system of equations

$$x_1 + 2x_2 - x_3 + 3x_4 = 4$$
$$2x_1 + 4x_2 - 2x_3 + 7x_4 = 10$$
$$-x_1 - 2x_2 + x_3 - 4x_4 = -6$$

SOLUTION

On applying the Gauss-Jordan algorithm we get

$$\begin{bmatrix} 1 & 2 & -1 & 3 & 4 \\ 2 & 4 & -2 & 7 & 10 \\ -1 & -2 & 1 & -4 & -6 \end{bmatrix} \begin{array}{c} \approx \\ \text{R2} + (-2)\text{R1} \\ \text{R3} + \text{R1} \end{array} \begin{bmatrix} 1 & 2 & -1 & 3 & 4 \\ 0 & 0 & 0 & 1 & 2 \\ 0 & 0 & 0 & -1 & -2 \end{bmatrix}$$

$$\begin{array}{c} \approx \\ \text{R1} + (-3)\text{R2} \\ \text{R3} + \text{R2} \end{array} \begin{bmatrix} 1 & 2 & -1 & 0 & -2 \\ 0 & 0 & 0 & 1 & 2 \\ 0 & 0 & 0 & 0 & 0 \end{bmatrix}$$

We have arrived at the reduced echelon form. The corresponding system of equations is

$$x_1 + 2x_2 - x_3 \quad\quad = -2$$
$$x_4 = \quad 2$$

Expressing the leading variables in terms of the remaining variables we get

$$x_1 = -2x_2 + x_3 - 2, \ x_4 = 2$$

Let us assign the arbitrary values r to x_2 and s to x_3. The general solution is

$$x_1 = -2r + s - 2, \ x_2 = r, \ x_3 = s, \ x_4 = 2$$

Specific solutions can be obtained by giving r and s various values.

■

EXAMPLE **4** This example illustrates a system that has no solution. Let us try to solve the system

$$x_1 + \quad x_2 + 5x_3 = \quad 3$$
$$x_2 + 3x_3 = -1$$
$$x_1 + 2x_2 + 8x_3 = \quad 3$$

SOLUTION

Starting with the augmented matrix we get

$$\begin{bmatrix} 1 & 1 & 5 & 3 \\ 0 & 1 & 3 & -1 \\ 1 & 2 & 8 & 3 \end{bmatrix} \underset{\text{R3}+(-1)\text{R1}}{\approx} \begin{bmatrix} 1 & 1 & 5 & 3 \\ 0 & 1 & 3 & -1 \\ 0 & 1 & 3 & 0 \end{bmatrix}$$

$$\underset{\substack{\text{R1}+(-1)\text{R2} \\ \text{R3}+(-1)\text{R2}}}{\approx} \begin{bmatrix} 1 & 0 & 2 & 4 \\ 0 & 1 & 3 & -1 \\ 0 & 0 & 0 & 1 \end{bmatrix} \underset{\substack{\text{R1}+(-4)\text{R3} \\ \text{R2}+\text{R3}}}{\approx} \begin{bmatrix} 1 & 0 & 2 & 0 \\ 0 & 1 & 3 & 0 \\ 0 & 0 & 0 & 1 \end{bmatrix}$$

The last row of this reduced echelon form gives the equation

$$0x_1 + 0x_2 + 0x_3 = 1$$

This equation cannot be satisfied for any values of x_1, x_2, and x_3. Thus the system has no solution. (This information was in fact available from the next-to-last matrix.)

■

Homogeneous Systems of Linear Equations

A system of linear equations is said to be **homogeneous** if all the constant terms are zeros. As we proceed in the course we shall find that homogeneous systems of linear equations have many interesting properties and play a key role in our discussions.

The following system is a homogeneous system of linear equations.

$$x_1 + 2x_2 - 5x_3 = 0$$
$$-2x_1 - 3x_2 + 6x_3 = 0$$

Observe that $x_1 = 0$, $x_2 = 0$, $x_3 = 0$, is a solution to this system. It is apparent that this result can be extended as follows to any homogeneous system of equations.

A homogeneous system of linear equations in n variables always has the solution $x_1 = 0, x_2 = 0, \ldots, x_n = 0$. This solution is called the **trivial solution**.

Let us see if the preceding homogeneous system has any other solutions. We solve the system using Gauss-Jordan elimination.

$$\begin{bmatrix} 1 & 2 & -5 & 0 \\ -2 & -3 & 6 & 0 \end{bmatrix} \quad \approx \quad \begin{bmatrix} 1 & 2 & -5 & 0 \\ 0 & 1 & -4 & 0 \end{bmatrix}$$
$$\text{R2} + (2)\text{R1}$$
$$\approx \quad \begin{bmatrix} 1 & 0 & 3 & 0 \\ 0 & 1 & -4 & 0 \end{bmatrix}$$
$$\text{R1} + (-2)\text{R2}$$

This reduced echelon form gives the system

$$x_1 \qquad + 3x_3 = 0$$
$$x_2 - 4x_3 = 0$$

Expressing the leading variables in terms of the remaining variable we get

$$x_1 = -3x_3$$
$$x_2 = \quad 4x_3$$

Letting $x_3 = r$ we see that the system has many solutions,

$$x_1 = -3r, \, x_2 = 4r, \, x_3 = r$$

Observe that the solution $x_1 = 0, x_2 = 0, x_3 = 0$ is obtained by letting $r = 0$.

In a similar manner, the augmented matrix of any homogeneous system of linear equations that has more variables than equations will have a reduced echelon form that has more nonzero columns than rows with the last column being zero. The corresponding system of equations, and thus the original system, will have many solutions, one of which is the trivial solution as in the last example. We summarize this important observation in the following theorem.

A homogeneous system of linear equations that has more variables than equations has many solutions. One of these solutions is the trivial solution.

In the first two sections of this chapter we introduced the method of Gauss-Jordan elimination for solving systems of linear equations. As we proceed in the course we shall introduce other methods and compare the merits of the methods. There is another popular elimination method for solving systems of linear equations, for example, called **Gaussian elimination**. We introduce that method in Section 8.1.[2] The following discussion reveals some of the numerical concerns when solving systems of equations.

Numerical Considerations

In practice, systems of linear equations are solved on computers. Numbers are represented on computers in the form $\pm 0. \, a_1 \ldots a_n \times 10^r$, where $a_1, \ldots, a_n$ are integers between 0

---

[2]Gaussian elimination can in fact be used in place of Gauss-Jordan elimination as the standard method for this course if so desired.

and 9 and r is an integer (positive or negative). Such a number is called a **floating-point number**. The quantity $a_1, \ldots, a_n$ is called the **mantissa**, and r is the **exponent**. For example, the number 125.6 is written in floating-point form as 0.1256×10^3. An arithmetic operation of multiplication, division, addition, or subtraction on floating-point numbers is called a **floating point operation**, or **flop**.

Computers can handle only a limited number of integers in the mantissa of a number. The mantissa is rounded to a certain number of places during each operation and consequently errors called **round-off errors** occur in methods such as Gauss-Jordan elimination. The fewer flops that are performed during computation the faster and more accurate the result will be. (Ways of minimizing these errors are discussed in Chapter 8.) To compute the reduced echelon form of a system of n equations in n variables, the method of Gauss-Jordan elimination requires $\frac{1}{2}n^3 + \frac{1}{2}n^2$ multiplications and $\frac{1}{2}n^3 - \frac{1}{2}n$ additions (Section 8.1). The number of multiplications required to solve a system of, say, ten equations in ten variables ($n = 10$) is 550, and the number of additions is 495. The total number of flops is the sum of these, namely 1045. Algorithms are usually measured and compared using such data.

EXERCISE SET 1.2

Reduced Echelon Form of a Matrix

1. Determine whether the following matrices are in reduced echelon form. If a matrix is not in reduced echelon form give a reason.

(a) $\begin{bmatrix} 1 & 0 & 2 \\ 0 & 1 & 3 \end{bmatrix}$ (b) $\begin{bmatrix} 1 & 2 & 0 & 4 \\ 0 & 0 & 1 & 7 \end{bmatrix}$

(c) $\begin{bmatrix} 1 & 2 & 5 & 6 \\ 0 & 1 & 3 & -7 \end{bmatrix}$ (d) $\begin{bmatrix} 1 & 4 & 0 & 5 \\ 0 & 0 & 2 & 9 \end{bmatrix}$

(e) $\begin{bmatrix} 1 & 0 & 0 \\ 0 & 1 & 0 \\ 0 & 0 & 1 \end{bmatrix}$ (f) $\begin{bmatrix} 1 & 5 & 0 \\ 0 & 0 & 1 \\ 0 & 0 & 0 \end{bmatrix}$

(g) $\begin{bmatrix} 1 & 0 & 0 & 4 \\ 0 & 1 & 0 & 5 \\ 0 & 0 & 1 & 9 \end{bmatrix}$ (h) $\begin{bmatrix} 1 & 0 & 0 & 3 & 2 \\ 0 & 2 & 0 & 6 & 1 \\ 0 & 0 & 1 & 2 & 3 \end{bmatrix}$

(i) $\begin{bmatrix} 1 & 0 & 3 & 0 \\ 0 & 1 & 6 & 0 \\ 0 & 0 & 0 & 1 \end{bmatrix}$

2. Determine whether the following matrices are in reduced echelon form. If a matrix is not in reduced echelon form give a reason.

(a) $\begin{bmatrix} 1 & 0 & 3 & -2 \\ 0 & 0 & 1 & 8 \\ 0 & 1 & 4 & 9 \end{bmatrix}$ (b) $\begin{bmatrix} 1 & 2 & 0 & 0 & 4 \\ 0 & 0 & 1 & 0 & 6 \\ 0 & 0 & 0 & 1 & 5 \end{bmatrix}$

(c) $\begin{bmatrix} 1 & 5 & 0 & 2 & 0 \\ 0 & 0 & 1 & 9 & 0 \\ 0 & 0 & 0 & 0 & 1 \\ 0 & 0 & 0 & 0 & 0 \end{bmatrix}$ (d) $\begin{bmatrix} 1 & 0 & 4 & 2 & 6 \\ 0 & 1 & 2 & 3 & 4 \\ 0 & 0 & 0 & 1 & 2 \\ 0 & 0 & 0 & 0 & 1 \end{bmatrix}$

(e) $\begin{bmatrix} 1 & 0 & 2 & 0 & 3 \\ 0 & 0 & 0 & 0 & 0 \\ 0 & 1 & 2 & 0 & 7 \\ 0 & 0 & 0 & 1 & 3 \end{bmatrix}$ (f) $\begin{bmatrix} 1 & 0 & 4 & 0 & 0 \\ 0 & 1 & 2 & 0 & 0 \\ 0 & 0 & 0 & 1 & 0 \\ 0 & 0 & 0 & 0 & 1 \end{bmatrix}$

(g) $\begin{bmatrix} 1 & 0 & 0 & 5 & 3 \\ 0 & 0 & 1 & 0 & 3 \\ 0 & 1 & 2 & 3 & 7 \end{bmatrix}$ (h) $\begin{bmatrix} 0 & 0 & 1 & 0 & 4 \\ 0 & 0 & 0 & 1 & 5 \\ 0 & 1 & 0 & 0 & 3 \end{bmatrix}$

(i) $\begin{bmatrix} 1 & 5 & -3 & 0 & 7 \\ 0 & 0 & 0 & 1 & 4 \\ 0 & 0 & 0 & 0 & 0 \end{bmatrix}$

3. Each of the following matrices is the reduced echelon form of the augmented matrix of a system of linear equations. Give the solution (if it exists) to each system of equations.

(a) $\begin{bmatrix} 1 & 0 & 0 & 2 \\ 0 & 1 & 0 & 4 \\ 0 & 0 & 1 & -3 \end{bmatrix}$ (b) $\begin{bmatrix} 1 & 0 & -3 & 4 \\ 0 & 1 & 2 & 8 \\ 0 & 0 & 0 & 0 \end{bmatrix}$

(c) $\begin{bmatrix} 1 & 3 & 0 & 6 \\ 0 & 0 & 1 & -2 \\ 0 & 0 & 0 & 0 \end{bmatrix}$ (d) $\begin{bmatrix} 1 & 0 & 5 & 0 \\ 0 & 1 & -7 & 0 \\ 0 & 0 & 0 & 1 \end{bmatrix}$

(e) $\begin{bmatrix} 1 & 0 & 0 & 5 & 3 \\ 0 & 1 & 0 & 6 & -2 \\ 0 & 0 & 1 & 2 & -4 \end{bmatrix}$

(f) $\begin{bmatrix} 1 & 3 & 0 & 0 & 2 \\ 0 & 0 & 1 & 0 & 4 \\ 0 & 0 & 0 & 1 & 5 \end{bmatrix}$

4. Each of the following matrices is the reduced echelon form of the augmented matrix of a system of linear equations. Give the solution (if it exists) to each system of equations.

(a) $\begin{bmatrix} 1 & 0 & 2 & 4 & 1 \\ 0 & 1 & -3 & 5 & -6 \\ 0 & 0 & 0 & 0 & 0 \end{bmatrix}$

(b) $\begin{bmatrix} 1 & -3 & 2 & 0 & 4 \\ 0 & 0 & 0 & 1 & -7 \\ 0 & 0 & 0 & 0 & 0 \end{bmatrix}$

(c) $\begin{bmatrix} 1 & -2 & 0 & 3 & 0 & 4 \\ 0 & 0 & 1 & 2 & 0 & 9 \\ 0 & 0 & 0 & 0 & 1 & 8 \end{bmatrix}$

(d) $\begin{bmatrix} 1 & 0 & 2 & 0 & 3 & 6 \\ 0 & 1 & 5 & 0 & 4 & 7 \\ 0 & 0 & 0 & 1 & 9 & -3 \end{bmatrix}$

Solving Systems of Linear Equations

5. Solve (if possible) each of the following systems of three equations in three variables using the method of Gauss-Jordan elimination.

(a) $\begin{aligned} x_1 + 4x_2 + 3x_3 &= 1 \\ 2x_1 + 8x_2 + 11x_3 &= 7 \\ x_1 + 6x_2 + 7x_3 &= 3 \end{aligned}$

(b) $\begin{aligned} x_1 + 2x_2 + 4x_3 &= 15 \\ 2x_1 + 4x_2 + 9x_3 &= 33 \\ x_1 + 3x_2 + 5x_3 &= 20 \end{aligned}$

(c) $\begin{aligned} x_1 + x_2 + x_3 &= 7 \\ 2x_1 + 3x_2 + x_3 &= 18 \\ -x_1 + x_2 - 3x_3 &= 1 \end{aligned}$

(d) $\begin{aligned} x_1 + 4x_2 + x_3 &= 2 \\ x_1 + 2x_2 - x_3 &= 0 \\ 2x_1 + 6x_2 &= 3 \end{aligned}$

(e) $\begin{aligned} x_1 - x_2 + x_3 &= 3 \\ 2x_1 - x_2 + 4x_3 &= 7 \\ 3x_1 - 5x_2 - x_3 &= 7 \end{aligned}$

(f) $\begin{aligned} 3x_1 - 3x_2 + 9x_3 &= 24 \\ 2x_1 - 2x_2 + 7x_3 &= 17 \\ -x_1 + 2x_2 - 4x_3 &= -11 \end{aligned}$

6. Solve (if possible) each of the following systems of three equations in three variables using the method of Gauss-Jordan elimination.

(a) $\begin{aligned} 3x_1 + 6x_2 - 3x_3 &= 6 \\ -2x_1 - 4x_2 - 3x_3 &= -1 \\ 3x_1 + 6x_2 - 2x_3 &= 10 \end{aligned}$

(b) $\begin{aligned} x_1 + 2x_2 + x_3 &= 7 \\ x_1 + 2x_2 + 2x_3 &= 11 \\ 2x_1 + 4x_2 + 3x_3 &= 18 \end{aligned}$

(c) $\begin{aligned} x_1 + 2x_2 - x_3 &= 3 \\ 2x_1 + 4x_2 - 2x_3 &= 6 \\ 3x_1 + 6x_2 + 2x_3 &= -1 \end{aligned}$

(d) $\begin{aligned} x_1 + 2x_2 + 3x_3 &= 8 \\ 3x_1 + 7x_2 + 9x_3 &= 26 \\ 2x_1 \qquad\ + 6x_3 &= 11 \end{aligned}$

(e) $\begin{aligned} x_2 + 2x_3 &= 5 \\ x_1 + 2x_2 + 5x_3 &= 13 \\ x_1 \qquad\ + 2x_3 &= 4 \end{aligned}$

(f) $\begin{aligned} x_1 + 2x_2 + 8x_3 &= 7 \\ 2x_1 + 4x_2 + 16x_3 &= 14 \\ x_2 + 3x_3 &= 4 \end{aligned}$

7. Solve (if possible) each of the following systems of equations using the method of Gauss-Jordan elimination.

(a) $\begin{aligned} x_1 + x_2 - 3x_3 &= 10 \\ -3x_1 - 2x_2 + 4x_3 &= -24 \end{aligned}$

(b) $\begin{aligned} 2x_1 - 6x_2 - 14x_3 &= 38 \\ -3x_1 + 7x_2 + 15x_3 &= -37 \end{aligned}$

(c) $\begin{aligned} x_1 + 2x_2 - x_3 - x_4 &= 0 \\ x_1 + 2x_2 \qquad\ + x_4 &= 4 \\ -x_1 - 2x_2 + 2x_3 + 4x_4 &= 5 \end{aligned}$

(d) $\begin{aligned} x_1 + 2x_2 \qquad\ + 4x_4 &= 0 \\ -2x_1 - 4x_2 + 3x_3 - 2x_4 &= 0 \end{aligned}$
(A homogeneous system)

(e) $\begin{aligned} x_2 - 3x_3 + x_4 &= 0 \\ x_1 + x_2 - x_3 + 4x_4 &= 0 \\ -2x_1 - 2x_2 + 2x_3 - 8x_4 &= 0 \end{aligned}$
(A homogeneous system)

8. Solve (if possible) each of the following systems of equations using the method of Gauss-Jordan elimination.

(a)
$$x_1 + x_2 + x_3 - x_4 = -3$$
$$2x_1 + 3x_2 + x_3 - 5x_4 = -9$$
$$x_1 + 3x_2 - x_3 - 6x_4 = -7$$
$$-x_1 - x_2 - x_3 = 1$$

(b)
$$x_2 + 2x_3 = 7$$
$$x_1 - 2x_2 - 6x_3 = -18$$
$$-x_1 - x_2 - 2x_3 = -5$$
$$2x_1 - 5x_2 - 15x_3 = -46$$

(c)
$$2x_1 - 4x_2 + 16x_3 - 14x_4 = 10$$
$$-x_1 + 5x_2 - 17x_3 + 19x_4 = -2$$
$$x_1 - 3x_2 + 11x_3 - 11x_4 = 4$$
$$3x_1 - 4x_2 + 18x_3 - 13x_4 = 17$$

(d)
$$x_1 - x_2 + 2x_3 = 7$$
$$2x_1 - 2x_2 + 2x_3 - 4x_4 = 12$$
$$-x_1 + x_2 - x_3 + 2x_4 = -4$$
$$-3x_1 + x_2 - 8x_3 - 10x_4 = -29$$

(e)
$$x_1 + 6x_2 - x_3 - 4x_4 = 0$$
$$-2x_1 - 12x_2 + 5x_3 + 17x_4 = 0$$
$$3x_1 + 18x_2 - x_3 - 6x_4 = 0$$
(A homogeneous system)

(f)
$$4x_1 + 8x_2 - 12x_3 = 28$$
$$-x_1 - 2x_2 + 3x_3 = -7$$
$$2x_1 + 4x_2 - 8x_3 = 16$$
$$-3x_1 - 6x_2 + 9x_3 = -21$$

(g)
$$x_1 + x_2 = 2$$
$$2x_1 + 3x_2 = 3$$
$$x_1 + 3x_2 = 0$$
$$x_1 + 2x_2 = 1$$

Understanding Systems of Linear Equations

9. Construct examples of the following:

(a) A system of linear equations with more variables than equations, having no solution.

(b) A system of linear equations with more equations than variables, having a unique solution.

10. The reduced echelon forms of the matrices of systems of two equations in two variables, and the types of solutions they represent can be classified as follows. (• corresponds to possible nonzero elements.)

$$\begin{bmatrix} 1 & 0 & \bullet \\ 0 & 1 & \bullet \end{bmatrix} \quad \begin{bmatrix} 1 & \bullet & 0 \\ 0 & 0 & 1 \end{bmatrix} \quad \begin{bmatrix} 1 & \bullet & \bullet \\ 0 & 0 & 0 \end{bmatrix}$$

unique solution   no solutions   many solutions

Classify in a similar manner the reduced echelon forms of the matrices, and the types of solutions they represent, of

(a) systems of three equations in two variables,

(b) systems of three equations in three variables.

11. Consider the homogeneous system of linear equations

$$ax + by = 0$$
$$cx + dy = 0$$

(a) Show that if $x = x_0$, $y = y_0$ is a solution, then $x = kx_0$, $y = ky_0$, is also a solution, for any value of the constant k.

(b) Show that if $x = x_0$, $y = y_0$, and $x = x_1$, $y = y_1$, are any two solutions, then $x = x_0 + x_1$, $y = y_0 + y_1$, is also a solution.

12. Show that $x = 0$, $y = 0$ is a solution to the homogeneous system of linear equations

$$ax + by = 0$$
$$cx + dy = 0$$

Prove that this is the only solution if and only if $ad - bc \neq 0$.

13. Consider two systems of linear equations having augmented matrices $[A : B_1]$ and $[A : B_2]$, where the matrix of coefficients of both systems is the same 3×3 matrix A.

(a) Is it possible for $[A : B_1]$ to have a unique solution and $[A : B_2]$ to have many solutions?

(b) Is it possible for $[A : B_1]$ to have a unique solution and $[A : B_2]$ to have no solutions?

(c) Is it possible for $[A : B_1]$ to have many solutions and $[A : B_2]$ to have no solutions?

14. Solve the following systems of linear equations by applying the method of Gauss-Jordan elimination to a large augmented matrix that represents two systems with the same matrix of coefficients.

(a)
$$x_1 + x_2 + 5x_3 = b_1$$
$$x_1 + 2x_2 + 8x_3 = b_2$$
$$2x_1 + 4x_2 + 16x_3 = b_3$$

$$\text{for } \begin{bmatrix} b_1 \\ b_2 \\ b_3 \end{bmatrix} = \begin{bmatrix} 2 \\ 5 \\ 10 \end{bmatrix}, \begin{bmatrix} 3 \\ 2 \\ 4 \end{bmatrix}, \text{ in turn.}$$

(b) $x_1 + 2x_2 + 4x_3 = b_1$

$\quad\quad x_1 + x_2 + 2x_3 = b_2$

$\quad 2x_1 + 3x_2 + 6x_3 = b_3$

$$\text{for}\ \begin{bmatrix} b_1 \\ b_2 \\ b_3 \end{bmatrix} = \begin{bmatrix} 8 \\ 5 \\ 13 \end{bmatrix},\ \begin{bmatrix} 5 \\ 3 \\ 11 \end{bmatrix},\ \text{in turn.}$$

15. Write down a 3 × 3 matrix at random. Find its reduced echelon form. The reduced echelon form is probably the identity matrix I_3! Explain this. [*Hint*: Think about the geometry.]

16. If a 3 × 4 matrix is written down at random, what type of reduced echelon form is it likely to have and why?

17. Computers can only carry a finite number of digits. This causes errors called round-off errors to occur when numbers are truncated. Because of this phenomenon computers can give incorrect results. Much research goes into developing algorithms that minimize such round-off errors. (Readers who are interested in these algorithms should read Section 8.3.) A computer is used to determine the reduced echelon form of an augmented matrix of a system of linear equations. Which of the following is most likely to happen, and why?

    (a) The computer gives a solution to the system, when in fact a solution does not exist.

    (b) The computer gives that a solution does not exist, when in fact a solution does exist.

1.3 Curve Fitting, Electrical Networks, and Traffic Flow

Systems of linear equations are used in such diverse fields as electrical engineering, economics, and traffic analysis. We now discuss applications in some of these fields.

Curve Fitting

The following problem occurs in many different branches of science. A set of data points

$$(x_1, y_1), (x_2, y_2), \dots, (x_n, y_n)$$

is given and it is necessary to find a polynomial whose graph passes through the points. The points are often measurements in an experiment. The x-coordinates are called **base points**. It can be shown that if the base points are all distinct, then a unique polynomial of degree $n - 1$ (or less)

$$y = a_0 + a_1 x + \cdots + a_{n-2}x^{n-2} + a_{n-1}x^{n-1}$$

can be **fitted** to the points. See Figure 1.5.

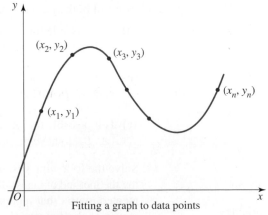

Fitting a graph to data points

Figure 1.5

The coefficients $a_0, a_1, \dots, a_{n-2}, a_{n-1}$ of the appropriate polynomial can be found by substituting the points into the polynomial equation and then solving a system of linear equations. (It is usual to write the polynomial in terms of ascending powers of x for the pur-

---

*Sections and chapters marked with an asterisk are optional. The instructor can use these sections to build around the core material to give the course the desired flavor.

pose of finding these coefficients. The columns of the matrix of coefficients of the system of equations then often follow a pattern. More will be said about this later.)

We now illustrate the procedure by fitting a polynomial of degree two, a parabola, to a set of three such data points.

EXAMPLE 1 Determine the equation of the polynomial of degree two whose graph passes through the points (1, 6), (2, 3), (3, 2).

SOLUTION

Observe that in this example we are given *three* points and we want to find a polynomial of degree *two* (one less than the number of data points). Let the polynomial be

$$y = a_0 + a_1 x + a_2 x^2$$

We are given three points and shall use these three sets of information to determine the three unknowns a_0, a_1, and a_2. Substituting

$$x = 1, y = 6; x = 2, y = 3; x = 3, y = 2$$

in turn into the polynomial leads to the following system of three linear equations in a_0, a_1, and a_2.

$$a_0 + a_1 + a_2 = 6$$
$$a_0 + 2a_1 + 4a_2 = 3$$
$$a_0 + 3a_1 + 9a_2 = 2$$

Solve this system for a_2, a_1, and a_0 using Gauss-Jordan elimination.

$$\begin{bmatrix} 1 & 1 & 1 & 6 \\ 1 & 2 & 4 & 3 \\ 1 & 3 & 9 & 2 \end{bmatrix} \begin{matrix} \\ R2 + (-1)R1 \\ R3 + (-1)R1 \end{matrix} \approx \begin{bmatrix} 1 & 1 & 1 & 6 \\ 0 & 1 & 3 & -3 \\ 0 & 2 & 8 & -4 \end{bmatrix}$$

$$\begin{matrix} \approx \\ R1 + (-1)R2 \\ R3 + (-2)R2 \end{matrix} \begin{bmatrix} 1 & 0 & -2 & 9 \\ 0 & 1 & 3 & -3 \\ 0 & 0 & 2 & 2 \end{bmatrix} \begin{matrix} \approx \\ (\frac{1}{2})R3 \end{matrix} \begin{bmatrix} 1 & 0 & -2 & 9 \\ 0 & 1 & 3 & -3 \\ 0 & 0 & 1 & 1 \end{bmatrix}$$

$$\begin{matrix} \approx \\ R1 + (2)R3 \\ R2 + (-3)R3 \end{matrix} \begin{bmatrix} 1 & 0 & 0 & 11 \\ 0 & 1 & 0 & -6 \\ 0 & 0 & 1 & 1 \end{bmatrix}$$

We get $a_0 = 11$, $a_1 = -6$, $a_2 = 1$. The parabola that passes through these points is $y = 11 - 6x + x^2$. See Figure 1.6.

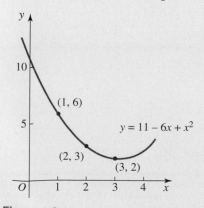

Figure 1.6

Electrical Network Analysis

Systems of linear equations are used to determine the currents through various branches of electrical networks. The following two laws, which are based on experimental verification in the laboratory, lead to the equations.

Kirchhoff's Laws*
1. **Junctions**: All the current flowing into a junction must flow out of it.
2. **Paths**: The sum of the IR terms (I denotes current, R resistance) in any direction around a closed path is equal to the total voltage in the path in that direction.

EXAMPLE 2 Consider the electrical network of Figure 1.7. Let us determine the currents through each branch of this network.

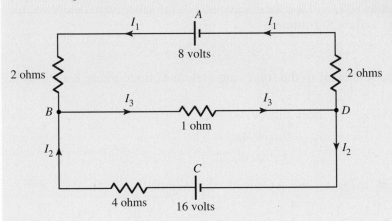

Figure 1.7

SOLUTION

The batteries, (denoted ⊣⊢) are 8 volts and 16 volts. The following convention is used in electrical engineering to indicate the terminal of the battery out of which the current flows: ⟐. The resistances (denoted ∿) are one 1-ohm, one 4-ohm, and two 2-ohm. The current entering each battery will be the same as that leaving it.

Let the currents in the various branches of the above circuit be I_1, I_2, and I_3. Kirchhoff's laws refer to junctions and closed paths. There are two junctions in this circuit, namely the points B and D. There are three closed paths, namely $ABDA$, $CBDC$, and $ABCDA$. Apply the laws to the junctions and paths.

Junctions:

$$\text{Junction } B, I_1 + I_2 = I_3$$
$$\text{Junction } D, I_3 = I_1 + I_2$$

These two equations result in a single linear equation

$$I_1 + I_2 - I_3 = 0$$

Paths:

$$\text{Path } ABDA, \quad 2I_1 + 1I_3 + 2I_1 = 8$$
$$\text{Path } CBDC, \quad 4I_2 + 1I_3 = 16$$

*Gustav Rubert Kirchhoff (1824–1887) was educated at the University of Königsberg and did most of his teaching at the University of Heidelberg. His most important contributions were in the discovery and analysis of the laws of electromagnetic radiation. Kirchhoff was an excellent teacher and writer whose books influenced teaching in German universities. Kirchhoff was described as "not easily drawn out but of a cheerful and obliging disposition."

It is not necessary to look further at path *ABCDA*. We now have a system of three linear equations in three unknowns, I_1, I_2, and I_3. Path *ABCDA* in fact leads to an equation that is a combination of the last two equations; there is no new information.

The problem thus reduces to solving the following system of three linear equations in three variables.

$$I_1 + I_2 - I_3 = 0$$
$$4I_1 \qquad + I_3 = 8$$
$$4I_2 + I_3 = 16$$

Using the method of Gauss-Jordan elimination, we get

$$\begin{bmatrix} 1 & 1 & -1 & 0 \\ 4 & 0 & 1 & 8 \\ 0 & 4 & 1 & 16 \end{bmatrix} \begin{matrix} \approx \\ \text{R2} + (-4)\text{R1} \end{matrix} \begin{bmatrix} 1 & 1 & -1 & 0 \\ 0 & -4 & 5 & 8 \\ 0 & 4 & 1 & 16 \end{bmatrix}$$

$$\begin{matrix} \approx \\ (-\frac{1}{4})\text{R2} \end{matrix} \begin{bmatrix} 1 & 1 & -1 & 0 \\ 0 & 1 & -\frac{5}{4} & -2 \\ 0 & 4 & 1 & 16 \end{bmatrix} \begin{matrix} \approx \\ \text{R1} + (-1)\text{R2} \\ \text{R3} + (-4)\text{R2} \end{matrix} \begin{bmatrix} 1 & 0 & \frac{1}{4} & 2 \\ 0 & 1 & -\frac{5}{4} & -2 \\ 0 & 0 & 6 & 24 \end{bmatrix}$$

$$\begin{matrix} \approx \\ (\frac{1}{6})\text{R3} \end{matrix} \begin{bmatrix} 1 & 0 & \frac{1}{4} & 2 \\ 0 & 1 & -\frac{5}{4} & -2 \\ 0 & 0 & 1 & 4 \end{bmatrix} \begin{matrix} \approx \\ \text{R1} + (-\frac{1}{4})\text{R3} \\ \text{R2} + (\frac{5}{4})\text{R3} \end{matrix} \begin{bmatrix} 1 & 0 & 0 & 1 \\ 0 & 1 & 0 & 3 \\ 0 & 0 & 1 & 4 \end{bmatrix}$$

The currents are $I_1 = 1$, $I_2 = 3$, $I_3 = 4$. The units are amps. The solution is unique, as is to be expected in this physical situation.

EXAMPLE 3 Determine the currents through the various branches of the electrical network in Figure 1.8. This example illustrates how one has to be conscious of direction in applying law 2 for closed paths.

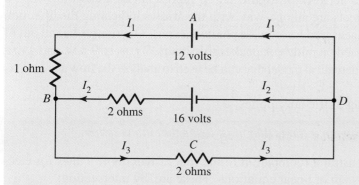

Figure 1.8

SOLUTION

Junctions:

$$\text{Junction } B, I_1 + I_2 = I_3$$
$$\text{Junction } D, I_3 = I_1 + I_2$$

giving $I_1 + I_2 - I_3 = 0$.

Paths:

$$\text{Path } ABCDA, 1I_1 + 2I_3 = 12$$
$$\text{Path } ABDA, 1I_1 + 2(-I_2) = 12 + (-16)$$

Observe that we have selected the direction $ABDA$ around this last path. The current along the branch BD in this direction is $-I_2$, and the voltage is -16. We now have three equations in the three variables I_1, I_2, and I_3.

$$\begin{aligned} I_1 + I_2 - I_3 &= 0 \\ I_1 \qquad\;\; + 2I_3 &= 12 \\ I_1 - 2I_2 \qquad &= -4 \end{aligned}$$

Solving these equations, we get $I_1 = 2$, $I_2 = 3$, $I_3 = 5$ amps.

In practice, electrical networks can involve many resistances and circuits; determining currents through branches involves solving large systems of equations on a computer.

Traffic Flow

Network analysis, as we saw in the previous discussion, plays an important role in electrical engineering. In recent years, the concepts and tools of network analysis have been found to be useful in many other fields, such as information theory and the study of transportation systems. The following analysis of traffic flow that was mentioned in the introduction illustrates how systems of linear equations with many solutions can arise in practice.

Consider the typical road network of Figure 1.9. It represents an area of downtown Jacksonville, Florida. The streets are all one-way with the arrows indicating the direction of traffic flow. The traffic is measured in vehicles per hour (vph). The figures in and out of the network given here are based on midweek peak traffic hours, 7 A.M. to 9 A.M. and 4 P.M. to 6 P.M. Let us construct a mathematical model that can be used to analyze the flow $x_1, \ldots, x_4$ within the network.

Assume that the following traffic law applies.

All traffic entering an intersection must leave that intersection.

This conservation of flow constraint (compare it to the first of Kirchhoff's laws for electrical networks) leads to a system of linear equations. These are, by intersection:

$$A: \text{Traffic in} = x_1 + x_2. \text{ Traffic out} = 400 + 225. \text{ Thus } x_1 + x_2 = 625.$$
$$B: \text{Traffic in} = 350 + 125. \text{ Traffic out} = x_1 + x_4. \text{ Thus } x_1 + x_4 = 475.$$
$$C: \text{Traffic in} = x_3 + x_4. \text{ Traffic out} = 600 + 300. \text{ Thus } x_3 + x_4 = 900.$$
$$D: \text{Traffic in} = 800 + 250. \text{ Traffic out} = x_2 + x_3. \text{ Thus } x_2 + x_3 = 1050.$$

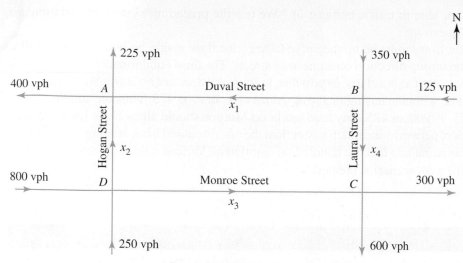

Figure 1.9

The constraints on the traffic are described by the following system of linear equations.

$$
\begin{aligned}
x_1 + x_2 \quad\quad &= 625 \\
x_1 \quad\quad\quad + x_4 &= 475 \\
x_3 + x_4 &= 900 \\
x_2 + x_3 \quad\quad &= 1050
\end{aligned}
$$

The method of Gauss-Jordan elimination is used to solve this system of equations. The augmented matrix and reduced echelon form of the preceding system are as follows:

$$
\begin{bmatrix}
1 & 1 & 0 & 0 & 625 \\
1 & 0 & 0 & 1 & 475 \\
0 & 0 & 1 & 1 & 900 \\
0 & 1 & 1 & 0 & 1050
\end{bmatrix}
\approx \cdots \approx
\begin{bmatrix}
1 & 0 & 0 & 1 & 475 \\
0 & 1 & 0 & -1 & 150 \\
0 & 0 & 1 & 1 & 900 \\
0 & 0 & 0 & 0 & 0
\end{bmatrix}
$$

The system of equations that corresponds to this reduced echelon form is

$$
\begin{aligned}
x_1 \quad\quad\quad + x_4 &= 475 \\
x_2 \quad\quad - x_4 &= 150 \\
x_3 + x_4 &= 900
\end{aligned}
$$

Expressing each leading variable in terms of the remaining variable, we get

$$
\begin{aligned}
x_1 &= -x_4 + 475 \\
x_2 &= x_4 + 150 \\
x_3 &= -x_4 + 900
\end{aligned}
$$

As was perhaps to be expected the system of equations has many solutions—there are many traffic flows possible. One does have a certain amount of choice at intersections. Let us now use this mathematical model to arrive at information. Suppose it becomes necessary to perform road work on the stretch DC of Monroe Street. It is desirable to have as small a flow x_3 as possible along this stretch of road. The flows can be controlled along various branches by means of traffic lights. What is the minimum value of x_3 along DC

that would not lead to traffic congestion? We use the preceding system of equations to answer this question.

All traffic flows must be nonnegative (a negative flow would be interpreted as traffic moving in the wrong direction on a one-way street). The third equation tells us that x_3 will be a minimum when x_4 is as large as possible, as long as it does not go above 900. The largest value x_4 can be without causing negative values of x_1 or x_2 is 475. Thus the smallest value of x_3 is $-475 + 900$, or 425. Any road works on Monroe should allow for at least 425 vph.

In practice, networks are much vaster than the one discussed here, leading to larger systems of linear equations that are handled on computers. Various values of variables can be fed in and different scenarios created.

EXERCISE SET 1.3

Curve Fitting

In Exercises 1–5, determine the equations of the polynomials of degree two whose graphs pass through the given points.

1. $(1, 2), (2, 2), (3, 4)$
2. $(1, 14), (2, 22), (3, 32)$
3. $(1, 5), (2, 7), (3, 9)$
4. $(1, 8), (3, 26), (5, 60)$. What is the value of y when $x = 2$?
5. $(-1, -1), (0, 1), (1, -3)$. What is the value of y when $x = 3$?
6. Find the equation of the polynomial of degree three whose graph passes through the points $(1, -3), (2, -1), (3, 9), (4, 33)$.

Electrical Networks

In Exercises 7–14, determine the currents in the various branches of the electrical networks. The units of current are amps and the units of resistance are ohms. (*Hint*: In Exercise 14 it is difficult to decide the direction of the current along AB. Make a guess. A negative result for the current means that your guess was the wrong one—the current is in the opposite direction. However, the magnitude will be correct. There is no need to rework the problem.)

7.

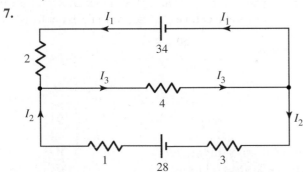

8.

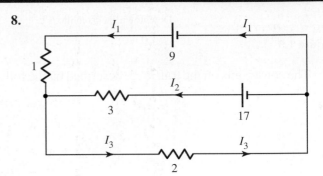

9.

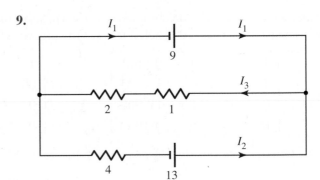

10.

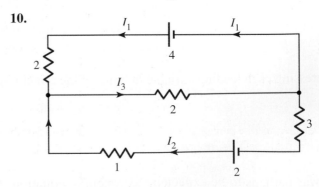

11.

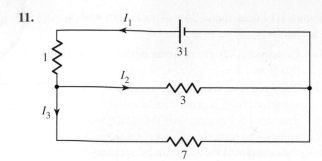

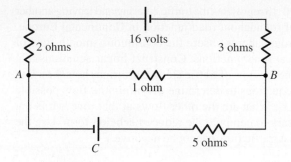

Figure 1.10

12.

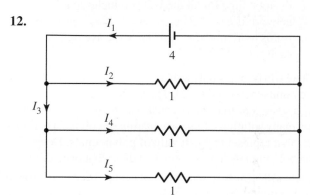

Traffic Flow

16. Construct a system of linear equations that describes the traffic flow in the road network of Figure 1.11. All streets are one-way streets in the directions indicated. The units are vehicles per hour. Give two distinct possible flows of traffic. What is the minimum possible flow that can be expected along branch AB?

13.

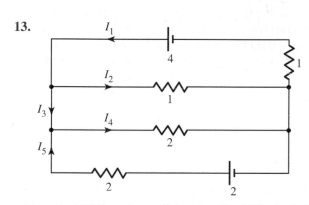

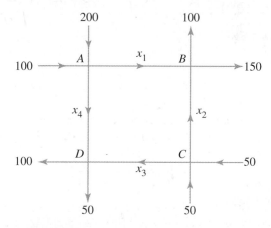

Figure 1.11

17. Figure 1.12 represents the traffic entering and leaving a "roundabout" road junction. Such junctions are very common in Europe. Construct a system of equations that describes the flow of traffic along the various branches. What is the minimum flow possible along the branch BC? What are the other flows at that time? (Units are vehicles per hour.)

14.

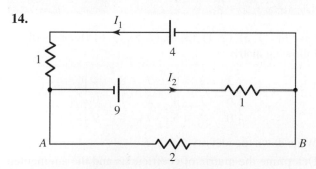

15. Determine the currents through the various branches of the electrical network in Figure 1.10.

  (a) when battery C is 9 volts

  (b) when battery C is 23 volts

Note how the current through the branch AB is reversed in (b). What would the voltage of C have to be for no current to pass through AB?

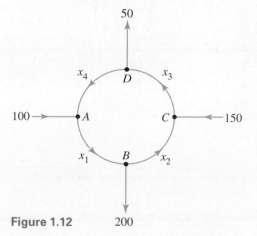

Figure 1.12   200

18. Figure 1.13 represents the traffic entering and leaving another type of roundabout road junction in Continental Europe. Such roundabouts ensure the continuous smooth flow of traffic at road junctions. Construct linear equations that describe the flow of traffic along the various branches. Use these equations to determine the minimum flow possible along x_1. What are the other flows at that time? (It is not necessary to compute the reduced echelon form. Use the fact that traffic flow cannot be negative.)

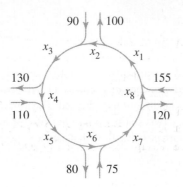

Figure 1.13

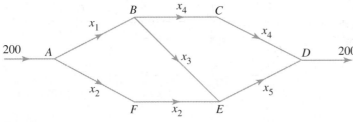

Figure 1.14

19. Figure 1.14 describes a flow of traffic, the units being vehicles per hour.

(a) Construct a system of linear equations that describes this flow.

(b) The total time it takes the vehicles to travel any stretch of road is proportional to the traffic along that stretch. For example, the total time it takes x_1 vehicles to traverse AB is kx_1 minutes. Assuming that the constant is the same for all sections of road, the total time for all these 200 vehicles to be in this network is $kx_1 + 2kx_2 + kx_3 + 2kx_4 + kx_5$. What is this total time if $k = 4$? Give the average time for each car.

20. There will be many polynomials of degree 2 that pass through the points $(1, 2)$ and $(3, 4)$. The situation can be described by a system of two linear equations in three variables that has many solutions. Find an equation (involving a parameter) that represents this family of polynomials. Determine the polynomials that open up and those that open down.

21. There will be many polynomials of degree 3 that pass through the points $(1, 2)$, $(3, 4)$, and $(4, 8)$. The situation can be described by a system of three linear equations in four variables that has many solutions. Find an equation (involving a parameter) that represents this family of polynomials. Determine the unique polynomial of degree 3 passing through these points for which the coefficient of x^3 is 1.

CHAPTER 1 REVIEW EXERCISES*

1. Give the sizes of the following matrices.

(a) $\begin{bmatrix} 4 & 3 & -2 \\ 1 & 5 & 7 \end{bmatrix}$ 　　 (b) $\begin{bmatrix} 0 & 2 \\ 4 & 6 \end{bmatrix}$

(c) $\begin{bmatrix} 4 & 3 & 2 & 7 \end{bmatrix}$ 　　 (d) $\begin{bmatrix} -2 \\ 3 \\ 6 \end{bmatrix}$

(e) $\begin{bmatrix} 8 & 5 & 3 & -7 & 5 & 9 \\ -2 & 3 & 5 & 7 & 0 & 2 \\ 4 & -3 & 5 & 1 & 2 & 3 \\ 0 & -8 & -1 & 5 & 3 & 8 \end{bmatrix}$

2. Give the $(1, 3)$, $(2, 1)$, $(3, 3)$, $(2, 5)$, $(3, 6)$ elements of the following matrix.

$$\begin{bmatrix} 3 & 2 & 0 & 7 & 8 & 4 \\ 6 & 7 & 4 & 2 & 1 & 0 \\ 0 & 2 & 5 & 7 & 8 & 9 \end{bmatrix}$$

3. Write down the identity matrix I_5.

4. Determine the matrix of coefficients and the augmented matrix of each of the following systems of equations.

(a) $\begin{aligned} x_1 + 2x_2 &= 6 \\ 4x_1 - 3x_2 &= -1 \end{aligned}$

(b) $\begin{aligned} 2x_1 + x_2 - 4x_3 &= 1 \\ x_1 - 2x_2 + 8x_3 &= 0 \\ 3x_1 + 5x_2 - 7x_3 &= -3 \end{aligned}$

*Answers to all review exercises are provided in the back of the book.

(c) $-x_1 + 2x_2 - 7x_3 = -2$
$\quad\ \ 3x_1 - \ x_2 + 5x_3 = \quad 3$
$\quad\ \ 4x_1 + 3x_2 \qquad\quad = \quad 5$

(d) $x_1 \qquad\qquad = \quad 1$
$\qquad x_2 \qquad\quad = \quad 5$
$\qquad\qquad x_3 = -3$

(e) $-2x_1 + 3x_2 - 8x_3 + 5x_4 = -2$
$\quad\ \ x_1 + 5x_2 \qquad - 6x_4 = \quad 0$
$\qquad\ -x_2 + 2x_3 + 3x_4 = \quad 5$

5. Interpret the following matrices as augmented matrices of systems of equations. Write down each system of equations.

(a) $\begin{bmatrix} 4 & 2 & 0 \\ -3 & 7 & 8 \end{bmatrix}$ (b) $\begin{bmatrix} 1 & 9 & -3 \\ 0 & 3 & 2 \end{bmatrix}$

(c) $\begin{bmatrix} 1 & 2 & 3 & 4 \\ 5 & 0 & -3 & 6 \end{bmatrix}$ (d) $\begin{bmatrix} 1 & 0 & 0 & 5 \\ 0 & 1 & 0 & -8 \\ 0 & 0 & 1 & 2 \end{bmatrix}$

(e) $\begin{bmatrix} 1 & 4 & -1 & 7 \\ 0 & 1 & 3 & 8 \\ 0 & 0 & 1 & -5 \end{bmatrix}$

6. Determine whether the following matrices are in reduced echelon form. If a matrix is not in reduced echelon form give a reason.

(a) $\begin{bmatrix} 1 & 0 & 4 \\ 0 & 1 & 7 \end{bmatrix}$ (b) $\begin{bmatrix} 1 & 3 & 0 & 5 \\ 0 & 0 & 1 & 9 \\ 0 & 0 & 0 & 0 \end{bmatrix}$

(c) $\begin{bmatrix} 1 & 2 & 0 & 6 \\ 0 & 1 & 0 & -7 \\ 0 & 0 & 1 & 9 \end{bmatrix}$ (d) $\begin{bmatrix} 1 & 3 & 0 & 0 & 2 \\ 0 & 0 & 1 & 0 & 5 \\ 0 & 0 & 0 & 1 & 7 \end{bmatrix}$

(e) $\begin{bmatrix} 1 & 8 & 0 & 0 & 2 \\ 0 & 0 & 0 & 1 & -7 \\ 0 & 0 & 1 & 0 & 3 \end{bmatrix}$

7. The following systems of equations have unique solutions. Solve these systems using the method of Gauss-Jordan elimination with matrices.

(a) $2x_1 + 4x_2 = 2$
$\quad 3x_1 + 7x_2 = 2$

(b) $x_1 - 2x_2 - \ 6x_3 = -17$
$\quad 2x_1 - 6x_2 - 16x_3 = -46$
$\quad x_1 + 2x_2 - \quad x_3 = \ -5$

(c) $\qquad x_2 + 2x_3 + 6x_4 = \ 21$
$\quad x_1 - \ x_2 + \ x_3 + 5x_4 = \ 12$
$\quad x_1 - \ x_2 - \ x_3 - 4x_4 = -9$
$\quad 3x_1 - 2x_2 \qquad\ - 6x_4 = -4$

8. Solve (if possible) the following systems of equations using the method of Gauss-Jordan elimination.

(a) $\quad x_1 - \ x_2 + \quad x_3 = \quad 3$
$\quad -2x_1 + 3x_2 + \quad x_3 = -8$
$\quad\ 4x_1 - 2x_2 + 10x_3 = \ 10$

(b) $\quad x_1 + 3x_2 + \ 6x_3 - 2x_4 = \ -7$
$\quad -2x_1 - 5x_2 - 10x_3 + 3x_4 = \ 10$
$\quad\ x_1 + 2x_2 + \ 4x_3 \qquad\quad = \quad 0$
$\qquad\ x_2 + \ 2x_3 - 3x_4 = -10$

9. Let A be an $n \times n$ matrix in reduced echelon form. Show that if $A \neq I_n$, then A has a row consisting entirely of zeros.

10. Let A and B be row equivalent matrices. Show that A and B have the same reduced echelon form.

11. Determine the equation of the polynomial of degree two whose graph passes through the points $(1, 3), (2, 6), (3, 13)$.

12. Determine the currents through the branches of the network in Figure 1.15.

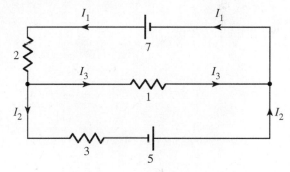

Figure 1.15

13. Consider the traffic entering and leaving a roundabout road junction in Britain, shown in Figure 1.16. (Observe that the traffic goes around this roundabout in the opposite direction to the one on the Continent in Exercise 18, Section 1.3. They drive on the left of the road in Britain.) Construct a system of linear equations that describe the flow of traffic along the various branches. Determine the minimum flow possible along x_8. What are the other flows at that time?

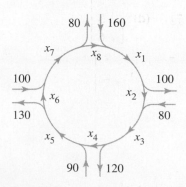

Figure 1.16

Built in the shape of an eye, the L'Hemisfèric houses an Imax Cinema, a Planetarium, and a Laserium. This modern structure is a part of the Valencia City of Arts and Science in Valencia, Spain. L'Hemisferic is considered to be Europe's largest planetarium.

Matrices

In this chapter we introduce operations of addition and multiplication for matrices. We discuss the algebraic properties of these operations. We define powers of matrices and inverses of matrices. These tools lead to further methods for solving linear systems and insights into their behavior. We lay the foundation for using matrices to define functions, called linear transformations. These transformations will include rotations, expansions, and reflections.

The reader will see how matrices are used in a wide range of applications. They are used in archaeology to determine the chronological order of artifacts, in cryptography to ensure security, and in demography to predict population movement. The inverse of a matrix is used in a model for analyzing the interdependence of economies. Wassily Leontief received a Nobel Prize for his work in this field. This model is now a standard tool for investigating economic structures ranging from cities and corporations to states and countries.

Throughout these discussions we shall be conscious of numerical implications. We shall be aware of the need for efficiency and accuracy in implementing matrix models.

2.1 Addition, Scalar Multiplication, and Multiplication of Matrices

A convenient notation has been developed for working with matrices. Matrices consist of rows and columns. Rows are labeled from the top of the matrix, columns from the left. The location of an element in a matrix is described by giving the row and column in which it lies. The element in row i, column j of the matrix A is denoted a_{ij}.

$$a_{ij}$$

1st subscript ——— 2nd subscript
indicates row   indicates column

We refer to a_{ij} as the (i, j)th element of the matrix A. We can visualize an arbitrary $m \times n$ matrix A as in Figure 2.1.

If the number of rows m is equal to the number of columns n, A is said to be a **square matrix**. The elements of a square matrix A where the subscripts are equal, namely $a_{11}, a_{22}, \ldots, a_{nn}$, form the **main diagonal**. See Figure 2.2.

main diagonal

$$A = \begin{bmatrix} a_{11} & a_{12} & \cdots & a_{1n} \\ a_{21} & a_{22} & \cdots & a_{2n} \\ \vdots & \vdots & \cdots & \vdots \\ a_{m1} & a_{m2} & \cdots & a_{mn} \end{bmatrix}$$

$m \times n$ matrix A

$$A = \begin{bmatrix} a_{11} & a_{12} & \cdots & a_{1n} \\ a_{21} & a_{22} & \cdots & a_{2n} \\ \vdots & \vdots & \cdots & \vdots \\ a_{n1} & a_{n2} & \cdots & a_{nn} \end{bmatrix}$$

square $n \times n$ matrix A

Figure 2.1   **Figure 2.2**

For example, consider the matrix

$$A = \begin{bmatrix} 1 & -2 & -1 \\ 3 & -3 & 4 \\ 2 & 7 & 5 \end{bmatrix}$$

a_{12} is the element in row 1, column 2. Thus $a_{12} = -2$. We see that $a_{23} = 4$ and $a_{31} = 2$. A is a square matrix. The main diagonal of A consists of the elements $a_{11} = 1, a_{22} = -3, a_{33} = 5$.

We now begin our development of an algebraic theory of matrices by defining a concept of equality of matrices.

> **DEFINITION**   Two matrices are *equal* if they are of the same size and if their corresponding elements are equal. Thus $A = B$ if they are of the same size, and $a_{ij} = b_{ij}$ for all i and j.

This definition will enable us to introduce equations involving matrices. It immediately allows us to define an operation of addition for matrices.

Addition of Matrices

> **DEFINITION**   Let A and B be matrices of the same size. Their *sum $A + B$* is the matrix obtained by adding together the corresponding elements of A and B. The matrix $A + B$ will be of the same size as A and B. If A and B are not of the same size they cannot be added, and we say that *the sum does not exist*.
> Thus if $C = A + B$ then $c_{ij} = a_{ij} + b_{ij}$.

For example, if $A = \begin{bmatrix} 1 & 4 & 7 \\ 0 & -2 & 3 \end{bmatrix}$, $B = \begin{bmatrix} 2 & 5 & -6 \\ -3 & 1 & 8 \end{bmatrix}$, and $C = \begin{bmatrix} -5 & 4 \\ 2 & 7 \end{bmatrix}$, then

$$A + B = \begin{bmatrix} 1+2 & 4+5 & 7-6 \\ 0-3 & -2+1 & 3+8 \end{bmatrix} = \begin{bmatrix} 3 & 9 & 1 \\ -3 & -1 & 11 \end{bmatrix}$$

A and C are not of the same size. $A + C$ does not exist.

Scalar Multiplication of Matrices

When working with matrices, it is customary to refer to numbers as *scalars*. We shall use uppercase letters to denote matrices and lowercase letters for scalars. The next step in the development of a theory of matrices is to introduce a rule for multiplying matrices by scalars.

> **DEFINITION** Let A be a matrix and c be a scalar. The *scalar multiple* of A by c, denoted cA, is the matrix obtained by multiplying every element of A by c. The matrix cA will be the same size as A.
> Thus if $B = cA$, then $b_{ij} = ca_{ij}$.

For example, if $A = \begin{bmatrix} 1 & -2 & 4 \\ 7 & -3 & 0 \end{bmatrix}$, then $3A = \begin{bmatrix} 3 & -6 & 12 \\ 21 & -9 & 0 \end{bmatrix}$.

Negation and Subtraction

The matrix $(-1)C$ is written $-C$ and is called the *negative of C*. Thus, for example, if

$$C = \begin{bmatrix} 1 & 0 & -7 \\ -3 & 6 & 2 \end{bmatrix}, \text{ then } -C = \begin{bmatrix} -1 & 0 & 7 \\ 3 & -6 & -2 \end{bmatrix}$$

We now define subtraction in terms of addition and scalar multiplication. Let

$$A - B = A + (-1)B$$

This definition implies that *subtraction is performed between matrices of the same size by subtracting corresponding elements*. Thus if $C = A - B$, then $c_{ij} = a_{ij} - b_{ij}$.

Suppose $A = \begin{bmatrix} 5 & 0 & -2 \\ 3 & 6 & -5 \end{bmatrix}$ and $B = \begin{bmatrix} 2 & 8 & -1 \\ 0 & 4 & 6 \end{bmatrix}$. Then

$$A - B = \begin{bmatrix} 5-2 & 0-8 & -2-(-1) \\ 3-0 & 6-4 & -5-6 \end{bmatrix} = \begin{bmatrix} 3 & -8 & -1 \\ 3 & 2 & -11 \end{bmatrix}$$

Matrix Multiplication

The most natural way of multiplying two matrices might seem to be to multiply corresponding elements when the matrices are of the same size, and to say that the product does not exist if they are of different size. However, mathematicians have introduced an alternative rule that is more useful. It involves multiplying the rows of the first matrix times the columns of the second matrix in a systematic manner. You may be interested in glancing ahead to see how such multiplication is used in planning electrical circuits (Section 2.2), predicting population movements (Section 2.6), and providing information about communication networks (Section 2.7).

We first give examples to illustrate the rule, then an application to appreciate its relevance. Interpret the first matrix of a product in terms of its rows and the second in terms of its columns. Multiply rows times columns by multiplying corresponding elements and adding. Consider first the product of a 2×2 matrix and a 2×1 matrix (column matrix).

$$\begin{bmatrix} 1 & 2 \\ 3 & 4 \end{bmatrix}\begin{bmatrix} 2 \\ 5 \end{bmatrix} = \begin{bmatrix} [1 \quad 2]\begin{bmatrix} 2 \\ 5 \end{bmatrix} \\ [3 \quad 4]\begin{bmatrix} 2 \\ 5 \end{bmatrix} \end{bmatrix} = \begin{bmatrix} (1 \times 2) + (2 \times 5) \\ (3 \times 2) + (4 \times 5) \end{bmatrix} = \begin{bmatrix} 12 \\ 26 \end{bmatrix}$$

multiply each row   elementwise
times the column   multiplication

We now illustrate the product of a 2×2 and 2×3 matrix.

$$\begin{bmatrix} 1 & 3 \\ 2 & 0 \end{bmatrix}\begin{bmatrix} 5 & 0 & 1 \\ 3 & -2 & 6 \end{bmatrix} = \begin{bmatrix} [1 & 3]\begin{bmatrix} 5 \\ 3 \end{bmatrix} & [1 & 3]\begin{bmatrix} 0 \\ -2 \end{bmatrix} & [1 & 3]\begin{bmatrix} 1 \\ 6 \end{bmatrix} \\ [2 & 0]\begin{bmatrix} 5 \\ 3 \end{bmatrix} & [2 & 0]\begin{bmatrix} 0 \\ -2 \end{bmatrix} & [2 & 0]\begin{bmatrix} 1 \\ 6 \end{bmatrix} \end{bmatrix} = \begin{bmatrix} 14 & -6 & 19 \\ 10 & 0 & 2 \end{bmatrix}$$

multiply 1st row times each column in turn;
then 2nd row times each column in turn.

Let us now see the importance of this type of matrix multiplication in graphics. Consider the square *PQRO* (Figure 2.3) whose vertices, written as column matrices, are the points

$$P\begin{bmatrix} 0 \\ 1 \end{bmatrix}, Q\begin{bmatrix} 1 \\ 1 \end{bmatrix}, R\begin{bmatrix} 1 \\ 0 \end{bmatrix}, O\begin{bmatrix} 0 \\ 0 \end{bmatrix}$$

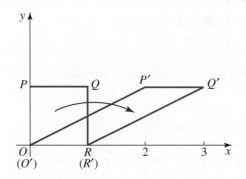

Figure 2.3

Consider a *transformation* of this square defined by the matrix *A*

$$A = \begin{bmatrix} 1 & 2 \\ 0 & 1 \end{bmatrix}$$

This transformation moves (or *maps*) a point $X\begin{bmatrix} x \\ y \end{bmatrix}$ into the point $X'\begin{bmatrix} x' \\ y' \end{bmatrix}$. It is defined by

$$\begin{bmatrix} x' \\ y' \end{bmatrix} = \begin{bmatrix} 1 & 2 \\ 0 & 1 \end{bmatrix}\begin{bmatrix} x \\ y \end{bmatrix}$$

X' is called the *image* of *X*. The images of the points *P*, *Q*, *R*, and *O* under matrix multiplication are found to be

$$P': \begin{bmatrix} 1 & 2 \\ 0 & 1 \end{bmatrix}\begin{bmatrix} 0 \\ 1 \end{bmatrix} = \begin{bmatrix} 2 \\ 1 \end{bmatrix}, \qquad Q': \begin{bmatrix} 1 & 2 \\ 0 & 1 \end{bmatrix}\begin{bmatrix} 1 \\ 1 \end{bmatrix} = \begin{bmatrix} 3 \\ 1 \end{bmatrix},$$

$$R': \begin{bmatrix} 1 & 2 \\ 0 & 1 \end{bmatrix}\begin{bmatrix} 1 \\ 0 \end{bmatrix} = \begin{bmatrix} 1 \\ 0 \end{bmatrix}, \qquad O': \begin{bmatrix} 1 & 2 \\ 0 & 1 \end{bmatrix}\begin{bmatrix} 0 \\ 0 \end{bmatrix} = \begin{bmatrix} 0 \\ 0 \end{bmatrix}$$

We get

$$P'\begin{bmatrix} 2 \\ 1 \end{bmatrix}, Q'\begin{bmatrix} 3 \\ 1 \end{bmatrix}, R'\begin{bmatrix} 1 \\ 0 \end{bmatrix}, O'\begin{bmatrix} 0 \\ 0 \end{bmatrix}$$

(The points *R* and *O* remain fixed for this transformation.) Such a matrix transformation maps line segments into line segments, $OP \rightarrow O'P'$, $PQ \rightarrow P'Q'$, $QR \rightarrow Q'R'$,

$RO \rightarrow R'O'$. It *deforms* the square $PQRO$ into the parallelogram $P'Q'R'O'$. When loads are applied to bodies such deformations occur. Bridges, for example, are modeled and analyzed on computers using these techniques. This particular transformation is called a *sheer*.

Such transformations as rotations, reflections, projections, and magnifications can be similarly described by matrices. We shall discuss these transformations later in the course. An important aspect of this matrix approach is that it lends itself to describing a sequence of such transformations, used, for example, to perform motion in computer graphics. Consider a rotation $X' = AX$ that maps the point X into X', followed by a magnification $X'' = BX'$ that maps X' into X''. These can be combined to get the composite transformation $X'' = B(AX)$. We see in the next section that matrix multiplication is associative so that $B(AX) = (BA)X$. Thus $X'' = (BA)X$; the composite transformation is defined by the product matrix BA. Matrix multiplication was in fact first introduced into mathematics because it enables us to define composite transformations by a single product matrix. (This was a long time before computer graphics!)

The following example illustrates that it may not be possible to multiply matrices.

EXAMPLE 1 Consider the following matrices A and B.

$$A = \begin{bmatrix} 3 & 1 & 2 \\ 4 & 1 & 5 \end{bmatrix}, B = \begin{bmatrix} 7 & 2 \\ 6 & 3 \end{bmatrix}$$

Let us attempt to compute AB using the above matrix multiplication rule. We get

$$AB = \begin{bmatrix} 3 & 1 & 2 \\ 4 & 1 & 5 \end{bmatrix}\begin{bmatrix} 7 & 2 \\ 6 & 3 \end{bmatrix} = \begin{bmatrix} [3 \ 1 \ 2]\begin{bmatrix} 7 \\ 6 \end{bmatrix} & [3 \ 1 \ 2]\begin{bmatrix} 2 \\ 3 \end{bmatrix} \\ [4 \ 1 \ 5]\begin{bmatrix} 7 \\ 6 \end{bmatrix} & [4 \ 1 \ 5]\begin{bmatrix} 2 \\ 3 \end{bmatrix} \end{bmatrix}$$

If we try to compute $\begin{bmatrix} 3 & 1 & 2 \end{bmatrix}\begin{bmatrix} 7 \\ 6 \end{bmatrix}$ the elements do not match, and the product does not exist. The same shortcoming applies to all the other elements of AB. We say that *the product AB does not exist*. We have arrived at the following observation. If the number of columns in A is equal to the number of rows in B (elements match), the product AB exists. If the number of columns in A is not equal to the number of rows in B (the elements do not match), the product does not exist.

We state this rule of matrix multiplication by giving a method of arriving at an arbitrary element of the product matrix AB.

DEFINITION Let the number of columns in a matrix A be the same as the number of rows in a matrix B. The product AB then exists. The element in row i and column j of AB is obtained by multiplying the corresponding elements of row i of A and column j of B and adding the products.

If the number of columns in A does not equal the number of rows in B, *the product does not exist*.

Let A have n columns and B have n rows. The elements of the ith row of A are $a_{i1}, \ldots, a_{in}$, and the elements of the jth column of B are $b_{1j}, \ldots, b_{nj}$. Thus if $C = AB$, then

$$c_{ij} = a_{i1}b_{1j} + \cdots + a_{in}b_{nj}$$

EXAMPLE 2 Let $A = \begin{bmatrix} 1 & 3 \\ 2 & 0 \end{bmatrix}$, $B = \begin{bmatrix} 5 & 0 & 1 \\ 3 & -2 & 6 \end{bmatrix}$. Determine AB and BA, if the products exist.

SOLUTION

A has two columns and B has two rows; thus AB exists. Interpret A in terms of its rows and B in terms of its columns and multiply the rows by the columns. We find that

$$AB = \begin{bmatrix} 1 & 3 \\ 2 & 0 \end{bmatrix}\begin{bmatrix} 5 & 0 & 1 \\ 3 & -2 & 6 \end{bmatrix} = \begin{bmatrix} 14 & -6 & 19 \\ 10 & 0 & 2 \end{bmatrix}$$

BA does not exist because B has three columns and A has two rows.

We see that the order in which two matrices are multiplied is important. Unlike multiplication of real numbers, matrix multiplication is not commutative. In general, for two matrices A and B, $AB \neq BA$.

Matrix multiplication is not commutative.

The following example illustrates that we can use the definition of matrix multiplication to compute any desired element in a product matrix without computing the whole product.

EXAMPLE 3 Let $C = AB$ for the following matrices A and B. Determine the element c_{23} of C.

$$A = \begin{bmatrix} 2 & 1 \\ -3 & 4 \end{bmatrix} \text{ and } B = \begin{bmatrix} -7 & 3 & 2 \\ 5 & 0 & 1 \end{bmatrix}$$

SOLUTION

c_{23} is the element in row 2, column 3 of C. It will be the product of row 2 of A and column 3 of B. We get

$$c_{23} = \begin{bmatrix} -3 & 4 \end{bmatrix}\begin{bmatrix} 2 \\ 1 \end{bmatrix} = (-3 \times 2) + (4 \times 1) = -2$$

Size of a Product Matrix

Let us now discuss the size of a product matrix. Let A be an $m \times r$ matrix and B be an $r \times n$ matrix. A has r columns and B has r rows. AB thus exists. The first row of AB is obtained by multiplying the first row of A by each column of B in turn. Thus the number of columns in AB is equal to the number of columns in B. The first column of AB results from multiplying each row of A in turn with the first column of B. Thus the number of rows in AB is equal to the number of rows in A. AB will be an $m \times n$ matrix.

If A is an $m \times r$ matrix and B is an $r \times n$ matrix, then AB will be an $m \times n$ matrix.

We can picture this result as follows:

$$
\begin{array}{ccc}
A & B & = \quad AB \\
m \times r & r \times n & m \times n
\end{array}
$$

insides match
outsides give size of AB

For example, suppose A is a 5×6 matrix and B is a 6×7 matrix. Matrix A has six columns, whereas B has six rows. Thus AB exists. AB will be a 5×7 matrix.

Special Matrices

We now define three classes of matrices that play an important role in matrix theory.

> **DEFINITION**   A *zero matrix* is a matrix all of whose elements are zeros. A *diagonal matrix* is a square matrix in which all the elements not on the main diagonal are zeros. An *identity matrix* is a diagonal matrix in which every diagonal element is 1. See the following figure.

$$
O_{mn} = \begin{bmatrix} 0 & 0 & \cdots & 0 \\ 0 & 0 & \cdots & 0 \\ \vdots & \vdots & \cdots & \vdots \\ 0 & 0 & \cdots & 0 \end{bmatrix}
\qquad
A = \begin{bmatrix} a_{11} & 0 & \cdots & 0 \\ 0 & a_{22} & \cdots & 0 \\ \vdots & \vdots & \cdots & \vdots \\ 0 & 0 & \cdots & a_{nn} \end{bmatrix}
\qquad
I_n = \begin{bmatrix} 1 & 0 & \cdots & 0 \\ 0 & 1 & \cdots & 0 \\ \vdots & \vdots & \cdots & \vdots \\ 0 & 0 & \cdots & 1 \end{bmatrix}
$$

Zero matrix O_{mn} Diagonal matrix A Identity matrix I_n

Zero matrices play a role in matrix theory similar to the role of the number 0 for real numbers, and identity matrices play a role similar to the number 1. These roles are described in the following theorem, which we illustrate by means of an example.

THEOREM 2.1

Let A be an $m \times n$ matrix and O_{mn} be the zero $m \times n$ matrix. Let B be an $n \times n$ square matrix, O_n and I_n be the zero and identity $n \times n$ matrices. Then

$$A + O_{mn} = O_{mn} + A = A$$
$$BO_n = O_nB = O_n$$
$$BI_n = I_nB = B$$

EXAMPLE 4 Let $A = \begin{bmatrix} 2 & 1 & -3 \\ 4 & 5 & 8 \end{bmatrix}$ and $B = \begin{bmatrix} 2 & 1 \\ -3 & 4 \end{bmatrix}$.

We see that

$$A + O_{23} = \begin{bmatrix} 2 & 1 & -3 \\ 4 & 5 & 8 \end{bmatrix} + \begin{bmatrix} 0 & 0 & 0 \\ 0 & 0 & 0 \end{bmatrix} = \begin{bmatrix} 2 & 1 & -3 \\ 4 & 5 & 8 \end{bmatrix} = A$$

$$BO_2 = \begin{bmatrix} 2 & 1 \\ -3 & 4 \end{bmatrix}\begin{bmatrix} 0 & 0 \\ 0 & 0 \end{bmatrix} = \begin{bmatrix} 0 & 0 \\ 0 & 0 \end{bmatrix} = O_2$$

$$BI_2 = \begin{bmatrix} 2 & 1 \\ -3 & 4 \end{bmatrix}\begin{bmatrix} 1 & 0 \\ 0 & 1 \end{bmatrix} = \begin{bmatrix} 2 & 1 \\ -3 & 4 \end{bmatrix} = B$$

Similarly, $O_{23} + A = A$, $O_2B = O_2$, $I_2B = B$.

We have introduced matrix multiplication by giving the rule for determining individual elements of the product. This element approach is the most useful for computing products of small matrices by hand. There are however numerous ways of looking at a product. We now introduce column approaches that are useful in theory and in applications.

Matrix Multiplication in Terms of Columns

(a) Consider the product AB where A is an $m \times n$ matrix and B is an $n \times r$ matrix (so that AB exists). Let the columns of B be the matrices $B_1, B_2, \ldots, B_r$. Write B as $[B_1 \quad B_2 \ldots B_r]$. Thus

$$AB = A[B_1 \quad B_2 \ldots B_r]$$

Matrix multiplication implies that the columns of the product are $AB_1, AB_2, \ldots, AB_r$. We can write

$$AB = [AB_1 \quad AB_2 \ldots AB_r]$$

For example, suppose $A = \begin{bmatrix} 2 & 0 \\ 1 & 5 \end{bmatrix}$ and $B = \begin{bmatrix} 4 & 1 & 3 \\ 0 & 2 & -1 \end{bmatrix}$. Then

$$AB = \begin{bmatrix} \begin{bmatrix} 2 & 0 \\ 1 & 5 \end{bmatrix}\begin{bmatrix} 4 \\ 0 \end{bmatrix} & \begin{bmatrix} 2 & 0 \\ 1 & 5 \end{bmatrix}\begin{bmatrix} 1 \\ 2 \end{bmatrix} & \begin{bmatrix} 2 & 0 \\ 1 & 5 \end{bmatrix}\begin{bmatrix} 3 \\ -1 \end{bmatrix} \end{bmatrix}$$

$$= \begin{bmatrix} 8 & 2 & 6 \\ 4 & 11 & -2 \end{bmatrix}$$

(b) The matrix product AB, where B is a column matrix, often occurs in practice. Consider the general case where A is an $m \times n$ matrix and B is an $n \times 1$ matrix. Write A in terms of its columns $[A_1 \quad A_2 \ldots A_n]$. Then

$$AB = [A_1 \quad A_2 \ldots A_n] \begin{bmatrix} b_1 \\ \vdots \\ b_n \end{bmatrix}$$

Matrix multiplication gives

$$AB = b_1 A_1 + b_2 A_2 + \cdots + b_n A_n$$

The expression $b_1 A_1 + b_2 A_2 + \cdots + b_n A_n$ is called a *linear combination* of $A_1, A_2, \ldots, A_n$. It is computed by performing the scalar multiples and then adding corresponding elements of the resulting matrices.

For example, suppose $A = \begin{bmatrix} 2 & 3 & 1 \\ -4 & 8 & 5 \end{bmatrix}$ and $B = \begin{bmatrix} 3 \\ -2 \\ 5 \end{bmatrix}$. Then

$$AB = 3\begin{bmatrix} 2 \\ -4 \end{bmatrix} - 2\begin{bmatrix} 3 \\ 8 \end{bmatrix} + 5\begin{bmatrix} 1 \\ 5 \end{bmatrix} = \begin{bmatrix} 6 \\ -12 \end{bmatrix} - \begin{bmatrix} 6 \\ 16 \end{bmatrix} + \begin{bmatrix} 5 \\ 25 \end{bmatrix}$$

$$= \begin{bmatrix} 5 \\ -3 \end{bmatrix}$$

This type of fluency in expressing the product in various ways is valuable when working with matrices. We shall, for example, use the first of these approaches in arriving at a way for finding the inverse of a matrix later in this chapter.

Partitioning of Matrices

In the previous discussion we subdivided matrices into column matrices. We now extend these ideas. A matrix can be subdivided into a number of *submatrices* (or *blocks*). For example the following matrix A can be subdivided into the submatrices P, Q, R, and S.

$$A = \left[\begin{array}{c|cc} 0 & -1 & 2 \\ 3 & 1 & 4 \\ \hline -2 & 5 & -3 \end{array}\right] = \begin{bmatrix} P & Q \\ R & S \end{bmatrix}$$

where $P = \begin{bmatrix} 0 \\ 3 \end{bmatrix}$, $Q = \begin{bmatrix} -1 & 2 \\ 1 & 4 \end{bmatrix}$, $R = [-2]$, and $S = \begin{bmatrix} 5 & 3 \end{bmatrix}$

Provided appropriate rules are followed, matrix addition and multiplication can be applied to submatrices as if they were elements of an ordinary matrix. Partitioning is used to reduce memory requirements and also to speed up computation on computers, especially when matrices have large blocks of zeros.[1]

Let us look at addition. Let A and B be matrices of the same kind. If A and B are partitioned in the same way, into $P, \ldots, U$ and $H, \ldots, M$, as follows for example, their sum is the sum of the corresponding submatrices.

$$A + B = \begin{bmatrix} P & Q & R \\ S & T & U \end{bmatrix} + \begin{bmatrix} H & I & J \\ K & L & M \end{bmatrix} = \begin{bmatrix} P+H & Q+I & R+J \\ S+K & T+L & U+M \end{bmatrix}$$

In multiplication, any partition of the first matrix in a product determines the row partition of the second matrix. For example, let us consider the product of the following matrices AB,

$$A = \begin{bmatrix} 1 & 2 & -1 \\ 3 & 0 & -2 \\ 4 & -3 & 2 \end{bmatrix} \text{ and } B = \begin{bmatrix} -1 & 0 \\ 2 & 1 \\ 5 & 4 \end{bmatrix}$$

Let A be subdivided

$$A = \left[\begin{array}{cc|c} 1 & 2 & -1 \\ 3 & 0 & -2 \\ 4 & -3 & 2 \end{array}\right] = \begin{bmatrix} P & Q \\ R & S \end{bmatrix}$$

A is interpreted as having two columns in this form. B must be subdivided into a suitable form having two rows for matrix multiplication to be possible. The following would be a suitable partition for B.

$$B = \left[\begin{array}{cc} -1 & 0 \\ 2 & 1 \\ \hline 5 & 4 \end{array}\right] = \begin{bmatrix} M \\ N \end{bmatrix}$$

Then we get,

$$AB = \begin{bmatrix} P & Q \\ R & S \end{bmatrix}\begin{bmatrix} M \\ N \end{bmatrix} = \begin{bmatrix} PM + QN \\ RM + SN \end{bmatrix}$$

---

[1] We illustrate the ideas for small matrices. Results can be quickly checked using the standard element ways of adding and multiplying matrices.

In general, if A is subdivided into a form having j columns then B must be divided into a suitable form having j rows to obtain a product AB. There is freedom in the column subdivision of B. If the partition of A is an $i \times j$ form, and B is a $j \times k$ form, then AB will be partitioned into i rows and k columns.

EXAMPLE 5 Let $A = \begin{bmatrix} 1 & -1 \\ 3 & 0 \\ 2 & 4 \end{bmatrix}$ and $B = \begin{bmatrix} 1 & 2 & -1 \\ 1 & 3 & 1 \end{bmatrix}$. Consider the following partition of A.

$$A = \left[\begin{array}{c|c} 1 & -1 \\ 3 & 0 \\ \hline 2 & 4 \end{array}\right] = \begin{bmatrix} P & Q \\ R & S \end{bmatrix}$$

Under this partition A is interpreted as a 2×2 matrix. For the product AB to exist, B must be partitioned into a matrix having two rows. Let

$$B = \begin{bmatrix} H \\ J \end{bmatrix}$$

The number of columns of P and Q determine the number of rows of H and J, since PH and QJ must exist:

$$AB = \begin{bmatrix} P & Q \\ R & S \end{bmatrix}\begin{bmatrix} H \\ J \end{bmatrix} = \begin{bmatrix} PH + QJ \\ RH + SJ \end{bmatrix}$$

One appropriate partition of B is

$$B = \left[\begin{array}{ccc} 1 & 2 & -1 \\ 1 & 3 & 1 \end{array}\right]$$

B is interpreted as a 2×1 matrix.

Let us check these partitions to see that they do indeed work.

$$AB = \left[\begin{array}{c|c} 1 & -1 \\ 3 & 0 \\ \hline 2 & 4 \end{array}\right]\left[\begin{array}{ccc} 1 & 2 & -1 \\ 1 & 3 & 1 \end{array}\right]$$

Multiply the submatrices,

$$\begin{bmatrix} 1 \\ 3 \end{bmatrix}\begin{bmatrix} 1 & 2 & -1 \end{bmatrix} + \begin{bmatrix} -1 \\ 0 \end{bmatrix}\begin{bmatrix} 1 & 3 & 1 \end{bmatrix} = \begin{bmatrix} 1 & 2 & -1 \\ 3 & 6 & -3 \end{bmatrix} + \begin{bmatrix} -1 & -3 & -1 \\ 0 & 0 & 0 \end{bmatrix}$$

$$= \begin{bmatrix} 0 & -1 & -2 \\ 3 & 6 & -3 \end{bmatrix}$$

and

$$\begin{bmatrix} 2 \end{bmatrix}\begin{bmatrix} 1 & 2 & -1 \end{bmatrix} + \begin{bmatrix} 4 \end{bmatrix}\begin{bmatrix} 1 & 3 & 1 \end{bmatrix} = \begin{bmatrix} 2 & 4 & -2 \end{bmatrix} + \begin{bmatrix} 4 & 12 & 4 \end{bmatrix} = \begin{bmatrix} 6 & 16 & 2 \end{bmatrix}$$

Thus the product is

$$AB = \begin{bmatrix} 0 & -1 & -2 \\ 3 & 6 & -3 \\ 6 & 16 & 2 \end{bmatrix}$$

It can be verified using elementwise multiplication that this is indeed the product of A and B. There are three other possible partitions of B that can be used to compute AB for this partition of A, namely

$$\left[\begin{array}{cc|c} 1 & 2 & -1 \\ \hline 1 & 3 & 1 \end{array}\right], \left[\begin{array}{cc|c} 1 & 2 & -1 \\ \hline 1 & 3 & 1 \end{array}\right], \text{ and } \left[\begin{array}{cc|c} 1 & 2 & -1 \\ 1 & 3 & 1 \end{array}\right]$$

■

EXERCISE SET 2.1

Matrix Operations

1. Let $A = \begin{bmatrix} 5 & 4 \\ -1 & 7 \\ 9 & -3 \end{bmatrix}$, $B = \begin{bmatrix} -3 & 0 \\ 4 & 2 \\ 5 & -7 \end{bmatrix}$,

$C = \begin{bmatrix} 1 & 2 \\ 3 & 4 \end{bmatrix}$, and $D = \begin{bmatrix} 9 & -5 \\ 3 & 0 \end{bmatrix}$

Compute the following (if they exist).

(a) $A + B$ (b) $2B$ (c) $-D$

(d) $C + D$ (e) $A + D$ (f) $2A + B$

(g) $A - B$

2. Let $A = \begin{bmatrix} 9 \\ 2 \\ -1 \end{bmatrix}$, $B = \begin{bmatrix} 0 & -1 & 4 \\ 6 & -8 & 2 \\ -4 & 5 & 9 \end{bmatrix}$,

$C = \begin{bmatrix} 1 & 2 & -5 \\ -7 & 9 & 3 \\ 5 & -4 & 0 \end{bmatrix}$, and $D = \begin{bmatrix} -3 \\ 0 \\ 2 \end{bmatrix}$

Compute the following (if they exist).

(a) $A + B$ (b) $4B$ (c) $-3D$

(d) $B - 3C$ (e) $-A$ (f) $3A + 2D$

(g) $A + D$

3. Let $A = \begin{bmatrix} 1 & 0 \\ 0 & 1 \end{bmatrix}$, $B = \begin{bmatrix} 0 & 1 \\ -2 & 5 \end{bmatrix}$,

$C = \begin{bmatrix} 2 \\ 3 \end{bmatrix}$, and $D = \begin{bmatrix} -1 & 0 & 3 \\ 5 & 7 & 2 \end{bmatrix}$

Compute the following (if they exist).

(a) AB (b) BA (c) AC

(d) CA (e) AD (f) DC

(g) BD (h) A^2 [where $A^2 = AA$]

4. Let $A = \begin{bmatrix} -1 \\ 2 \\ 5 \end{bmatrix}$, $B = \begin{bmatrix} 0 & 1 & 5 \\ 3 & -7 & 8 \\ 2 & 3 & 1 \end{bmatrix}$,

$C = \begin{bmatrix} -2 & 0 & 5 \end{bmatrix}$, and $D = \begin{bmatrix} 9 & -5 \\ 3 & 0 \\ -4 & 2 \end{bmatrix}$

Compute the following (if they exist).

(a) BA (b) AB (c) CB

(d) CA (e) DA (f) DB

(g) AC (h) B^2

5. Let $A = \begin{bmatrix} 0 & 1 \\ 0 & 3 \\ 5 & 6 \end{bmatrix}$, $B = \begin{bmatrix} -1 & 0 \\ 3 & 5 \\ 2 & 6 \end{bmatrix}$,

$C = \begin{bmatrix} -4 & 0 \\ 3 & 2 \end{bmatrix}$, and $D = \begin{bmatrix} 5 & 0 \\ -2 & 1 \end{bmatrix}$

Compute the following (if they exist).

(a) $2A - 3(BC)$ (b) AB (c) $AC - BD$

(d) $CD - 2D$ (e) BA (f) $AD + 2(DC)$

(g) $C^3 + 2(D^2)$ (where $C^3 = CCC$).

6. Let $A = \begin{bmatrix} 1 & -8 & 4 \\ 5 & -6 & 3 \\ 2 & 0 & -1 \end{bmatrix}$, and $B = \begin{bmatrix} 0 & 2 & -3 \\ 5 & 6 & 7 \\ -1 & 0 & 4 \end{bmatrix}$.

Let O_3 and I_3 be the 3×3 zero and identity matrices. Show that

$$A + O_3 = O_3 + A = A,$$
$$BO_3 = O_3B = O_3,$$

and

$$BI_3 = I_3B = B.$$

7. (a) Let A be a $n \times n$ matrix and X be an $n \times 1$ column matrix of 1s. What can you say about the rows of A if $AX = X$? (We call such a matrix X an *eigenvector* of A. We shall study eigenvectors in Chapter 5.)

(b) Let A be an $n \times n$ matrix and X be a $1 \times n$ row matrix of 1s. What can you say about the columns of A if $XA = X$?

Sizes of Matrices

8. Let A be a 3×5 matrix, B a 5×2 matrix, C a 3×4 matrix, D a 4×2 matrix, and E a 4×5 matrix. Determine which of the following matrix expressions exist and give the size of the resulting matrices when they do exist.

(a) AB **(b)** EB

(c) AC **(d)** $AB + CD$

(e) $3(EB) + 4D$ **(f)** $CD - 2(CE)B$

(g) $2(EB) + DA$

9. Let A be a 2×2 matrix, B a 2×2 matrix, C a 2×3 matrix, D a 3×2 matrix, and E a 3×1 matrix. Determine which of the following matrix expressions exist and give the size of the resulting matrices when they do exist.

(a) AB **(b)** $(A^2)C$

(c) $B^3 + 3(CD)$ **(d)** $DC + BA$

(e) $DA - 2(DB)$ **(f)** $C + 3D$

(g) $3(BA)(CD) + (4A)(BC)$

Computing Certain Elements of Matrices

10. Let $C = AB$ and $D = BA$ for the following matrices A and B.

$$A = \begin{bmatrix} 0 & 3 & -5 \\ 2 & 6 & 3 \\ 1 & 0 & -2 \end{bmatrix} \text{ and } B = \begin{bmatrix} -1 & 2 & -3 \\ 5 & 7 & 2 \\ 0 & 1 & 6 \end{bmatrix}$$

Determine the following elements of C and D, without computing the complete matrices.

(a) c_{31} **(b)** c_{23} **(c)** d_{12} **(d)** d_{22}

11. Let $R = PQ$ and $S = QP$, where

$$P = \begin{bmatrix} 1 & -2 \\ 4 & 6 \\ -1 & 3 \end{bmatrix} \text{ and } Q = \begin{bmatrix} 0 & 1 & 3 \\ 0 & -1 & 4 \end{bmatrix}.$$

Determine the following elements (if they exist) of R and S, without computing the complete matrices.

(a) r_{21} **(b)** r_{33} **(c)** s_{11} **(d)** s_{23}

12. If $A = \begin{bmatrix} 1 & -3 \\ 0 & 4 \end{bmatrix}$, $B = \begin{bmatrix} 1 & 2 & -3 \\ 5 & 0 & -1 \end{bmatrix}$, and

$C = \begin{bmatrix} 2 & -4 & 5 \\ 7 & 1 & 0 \end{bmatrix}$ determine the following elements of

$D = AB + 2C$, without computing the complete matrix.

(a) d_{12} **(b)** d_{23}

13. If $A = \begin{bmatrix} 1 & -3 & 0 \\ 4 & 5 & 1 \\ 3 & 8 & 0 \end{bmatrix}$, $B = \begin{bmatrix} 1 & 1 & -2 \\ 3 & 0 & 4 \\ -1 & 3 & 2 \end{bmatrix}$, and

$C = \begin{bmatrix} 2 & 0 & -2 \\ 4 & 7 & -5 \\ 1 & 0 & -1 \end{bmatrix}$, determine the following elements of

$D = 2(AB) + C^2$, without computing the complete matrix.

(a) d_{11} **(b)** d_{21} **(c)** d_{32}

Using Columns and Rows of Matrices

14. Let $A = \begin{bmatrix} 1 & 2 \\ 3 & 0 \end{bmatrix}$, $B = \begin{bmatrix} -2 & 3 \\ 4 & 1 \end{bmatrix}$, $C = \begin{bmatrix} 1 & -2 & 3 \\ 4 & 0 & 5 \end{bmatrix}$

Compute the following products using the columns of B and C.

(a) AB **(b)** AC **(c)** BC

15. Let $A = \begin{bmatrix} 3 & -2 & 0 \\ 4 & 2 & 7 \\ 8 & -5 & 6 \end{bmatrix}$, $B = \begin{bmatrix} 4 \\ 3 \\ -5 \end{bmatrix}$;

$P = \begin{bmatrix} 3 & 0 & 2 & 1 \\ 5 & 6 & 7 & 3 \end{bmatrix}$, $Q = \begin{bmatrix} -3 \\ 2 \\ 1 \\ 5 \end{bmatrix}$

(a) Express the product AB as a linear combination of the columns of A.

(b) Express PQ as a linear combination of the columns of P.

16. Let A and B be the following matrices. Compute row 2 of the matrix AB without computing the whole product.

$$A = \begin{bmatrix} 2 & -3 & 1 \\ 4 & 0 & 3 \\ 5 & 1 & 0 \end{bmatrix}, B = \begin{bmatrix} 8 & 1 & 3 \\ 2 & 1 & 0 \\ 4 & 6 & 3 \end{bmatrix}$$

17. Let A be a matrix whose third row is all zeros. Let B be any matrix such that the product AB exists. Prove that the third row of AB is all zeros.

18. Let D be a matrix whose second column is all zeros. Let C be any matrix such that CD exists. Prove that the second column of CD is all zeros.

19. Let A be an $m \times r$ matrix, B an $r \times n$ matrix, and $C = AB$. Let the column submatrices of B be $B_1, B_2, \ldots, B_n$ and of C be $C_1, C_2, \ldots, C_n$. We can write B in the form $\begin{bmatrix} B_1 & B_2 \ldots B_n \end{bmatrix}$ and C as $\begin{bmatrix} C_1 & C_2 \ldots C_n \end{bmatrix}$. Prove that $C_j = AB_j$.

20. Let A and B be the following matrices. Use the result of Exercise 19 to compute column 3 of the matrix AB without computing the whole product.

$$A = \begin{bmatrix} 1 & 2 & 3 \\ 0 & 4 & 1 \\ 2 & 5 & 0 \end{bmatrix}, B = \begin{bmatrix} 2 & 1 & 4 \\ 6 & 0 & 1 \\ 2 & 3 & 5 \end{bmatrix}$$

Partitioning of Matrices

21. Use the given partitions of A and B below to compute AB.

(a) $A = \left[\begin{array}{c|c} 2 & 1 \\ \hline -1 & 0 \\ \hline 3 & 1 \end{array}\right], B = \left[\begin{array}{c|c} 3 & 0 \\ 2 & 1 \end{array}\right]$

(b) $A = \left[\begin{array}{cc|c} 1 & 2 & -1 \\ 3 & 0 & 1 \end{array}\right], B = \left[\begin{array}{cc} 2 & 4 \\ 0 & -1 \\ \hline 1 & 3 \end{array}\right]$

(c) $A = \left[\begin{array}{cc|c} 1 & 2 & 0 \\ 3 & -1 & 1 \\ \hline 4 & -2 & 0 \end{array}\right], B = \left[\begin{array}{cc} 3 & -1 \\ 2 & 5 \\ \hline 0 & 1 \end{array}\right]$

22. Let $A = \begin{bmatrix} 1 & 2 & 3 \\ -1 & 1 & 4 \\ 0 & 1 & 2 \end{bmatrix}$ and $B = \begin{bmatrix} -1 & -2 \\ 0 & 3 \\ 4 & 1 \end{bmatrix}$.

For each partition of A given below find all the partitions of B that can be used to calculate AB.

(a) $A = \left[\begin{array}{c|cc} 1 & 2 & 3 \\ \hline -1 & 1 & 4 \\ 0 & 1 & 2 \end{array}\right]$

(b) $A = \left[\begin{array}{c|cc} 1 & 2 & 3 \\ \hline -1 & 1 & 4 \\ \hline 0 & 1 & 2 \end{array}\right]$

(c) $A = \left[\begin{array}{cc|c} 1 & 2 & 3 \\ -1 & 1 & 4 \\ \hline 0 & 1 & 2 \end{array}\right]$

23. Let $A = \begin{bmatrix} 2 & 0 & 3 & -1 \\ 6 & 2 & -5 & 9 \\ 1 & 1 & 1 & 1 \end{bmatrix}$ and $B = \begin{bmatrix} -1 & -2 & 0 \\ 3 & 4 & 1 \\ 5 & 7 & 2 \\ 4 & 5 & -8 \end{bmatrix}$.

For each partition of B given below find all the partitions of A that can be used to calculate AB.

(a) $B = \left[\begin{array}{c|cc} -1 & -2 & 0 \\ \hline 3 & 4 & 1 \\ 5 & 7 & 2 \\ 4 & 5 & -8 \end{array}\right]$

(b) $B = \left[\begin{array}{cc|c} -1 & -2 & 0 \\ \hline 3 & 4 & 1 \\ 5 & 7 & 2 \\ 4 & 5 & -8 \end{array}\right]$

(c) $B = \left[\begin{array}{cc|c} -1 & -2 & 0 \\ \hline 3 & 4 & 1 \\ \hline 5 & 7 & 2 \\ 4 & 5 & -8 \end{array}\right]$

24. Suggest suitable partitions involving zero and identity submatrices for computing the following products. Compute the products using these partitions (show your work). Check your answers using standard elementwise multiplication.

(a) $\begin{bmatrix} 2 & 3 & 2 & 1 \\ 4 & 0 & 0 & 0 \\ 1 & 0 & 0 & 0 \\ 5 & 0 & 0 & 0 \end{bmatrix} \begin{bmatrix} 1 & 2 \\ -1 & 3 \\ 4 & 0 \\ 2 & 5 \end{bmatrix}$

(b) $\begin{bmatrix} 1 & 0 & 0 & 2 \\ 0 & 1 & 0 & 3 \\ 0 & 0 & 1 & 4 \\ 1 & 1 & 1 & -2 \end{bmatrix} \begin{bmatrix} 1 & 2 \\ 3 & 4 \\ 5 & 6 \\ -1 & 3 \end{bmatrix}$

(c) $\begin{bmatrix} 1 & 0 & 0 & 1 \\ 0 & 1 & 0 & 2 \\ 0 & 0 & 1 & 3 \\ 0 & 0 & 0 & 4 \\ 0 & 0 & 0 & 5 \end{bmatrix} \begin{bmatrix} -1 \\ 2 \\ 6 \\ 3 \end{bmatrix}$

Miscellaneous Results

25. State (with a brief explanation) whether the following statements are true or false for matrices A, B, and C.

(a) If the sums $A + B$ and $B + C$ exist then $A + C$ exists.

(b) If the products AB and BC exist then AC exists.

(c) AB is never equal to BA.

(d) Let A be a column matrix and B a row matrix, both with the same number of elements. Then AB is a square matrix.

(e) If the element a_{ij} of a square matrix A lies below the main diagonal, then $i > j$.

2.2 Properties of Matrix Operations

We have defined operations of addition, scalar multiplication, and multiplication of matrices. We now list the most important properties of these operations.

THEOREM 2.2

Let A, B, and C be matrices and r and s be scalars. Assume that the sizes of the matrices are such that the operations can be performed.

Properties of Matrix Addition and Scalar Multiplication

1. $A + B = B + A$ — *Commutative Property of Addition*
2. $A + (B + C) = (A + B) + C$ — *Associative Property of Addition*
3. $A + O = O + A = A$ — *(where O is the appropriate zero matrix)*
4. $r(A + B) = rA + rB$
5. $(r + s)C = rC + sC$ — *Distributive Properties*
6. $r(sC) = (rs)C$

Properties of Matrix Multiplication

1. $A(BC) = (AB)C$ — *Associative Property of Multiplication*
2. $A(B + C) = AB + AC$
3. $(A + B)C = AC + BC$ — *Distributive Properties of Multiplication*
4. $AI = IA = A$ — *(where I is the appropriate identity matrix)*
5. $r(AB) = (rA)B = A(rB)$ — *Distributive Property*

Note: $AB \neq BA$ *in general.*   *Multiplication of matrices is not commutative.*

Each one of these results asserts an equality between matrices. We know that two matrices are equal if they are of the same size and their corresponding elements are equal. Each result is verified by showing this to be the case. We illustrate the method for the commutative property of addition. The reader is asked to use the same approach to prove some of the other results in the exercises that follow.

$A + B = B + A$ By the rule of matrix addition we know that $A + B$ and $B + A$ are both matrices of the same size. It remains to show that their corresponding elements are equal. Consider the (i, j)th element of each matrix.

$$(i, j)\text{th element of } A + B = a_{ij} + b_{ij}$$
$$(i, j)\text{th element of } B + A = b_{ij} + a_{ij}$$
$$= a_{ij} + b_{ij} \quad \text{(addition of real numbers is commutative)}$$

The corresponding elements of $A + B$ and $B + A$ are equal. Thus $A + B = B + A$.

These properties enable us to extend addition, scalar multiplication, and multiplication to more than two matrices. We can write sums and products such as $aA + bB + cC$ and ABC without parentheses. The following examples illustrate these concepts. We remind you that we call an expression such as $aA + bB + cC$ a *linear combination* of the matrices A, B, and C.

EXAMPLE **1** Compute the linear combination $2A + 3B - 5C$ for the following three matrices.

$$A = \begin{bmatrix} 1 & 3 \\ -4 & 5 \end{bmatrix}, \quad B = \begin{bmatrix} 3 & -7 \\ 2 & 1 \end{bmatrix}, \quad \text{and} \quad C = \begin{bmatrix} 0 & 2 \\ 3 & -1 \end{bmatrix}$$

SOLUTION

We compute the scalar multiples first, then add or subtract corresponding elements in a natural way.

$$2\begin{bmatrix} 1 & 3 \\ -4 & 5 \end{bmatrix} + 3\begin{bmatrix} 3 & -7 \\ 2 & 1 \end{bmatrix} - 5\begin{bmatrix} 0 & 2 \\ 3 & -1 \end{bmatrix} = \begin{bmatrix} 2 & 6 \\ -8 & 10 \end{bmatrix} + \begin{bmatrix} 9 & -21 \\ 6 & 3 \end{bmatrix} - \begin{bmatrix} 0 & 10 \\ 15 & -5 \end{bmatrix}$$

$$= \begin{bmatrix} 11 & -25 \\ -17 & 18 \end{bmatrix}$$

Certain products of matrices such as $ABCD$ will of course exist while other products will not. We can determine whether a product exists by comparing the numbers of rows and columns in adjacent matrices of the product, to see if they match.

If the product of a chain of matrices exists, the product matrix will have the same number of rows as the first matrix in the chain and the same number of columns as the last matrix.

EXAMPLE **2** Compute the product ABC of the following three matrices.

$$A = \begin{bmatrix} 1 & 2 \\ 3 & -1 \end{bmatrix}, \quad B = \begin{bmatrix} 0 & 1 & 3 \\ -1 & 0 & -2 \end{bmatrix}, \quad C = \begin{bmatrix} 4 \\ -1 \\ 0 \end{bmatrix}$$

SOLUTION

Let us check to see if the product ABC exists before we start spending time multiplying matrices. We get

$$
\begin{array}{cccc}
A & B & C & = & ABC \\
2 \times 2 & 2 \times 3 & 3 \times 1 & & 2 \times 1
\end{array}
$$

match — match

size of product is 2×1

The product exists and will be a 2×1 matrix. Since matrix multiplication is associative, the matrices in the product ABC can be grouped together in any manner for multiplying, as long as the order is maintained. Let us use the grouping $(AB)C$. This is probably the most natural. We get

$$AB = \begin{bmatrix} 1 & 2 \\ 3 & -1 \end{bmatrix} \begin{bmatrix} 0 & 1 & 3 \\ -1 & 0 & -2 \end{bmatrix} = \begin{bmatrix} -2 & 1 & -1 \\ 1 & 3 & 11 \end{bmatrix},$$

and

$$(AB)C = \begin{bmatrix} -2 & 1 & -1 \\ 1 & 3 & 11 \end{bmatrix} \begin{bmatrix} 4 \\ -1 \\ 0 \end{bmatrix} = \begin{bmatrix} -9 \\ 1 \end{bmatrix}$$

Caution

In algebra we know that the following cancellation laws apply.

- If $ab = ac$ and $a \neq 0$ then $b = c$.
- If $pq = 0$ then $p = 0$ or $q = 0$.

However the corresponding results are not true for matrices.

- $AB = AC$ does not imply that $B = C$.
- $PQ = O$ does not imply that $P = O$ or $Q = O$.

We demonstrate these possibilities by means of examples.

Consider the matrices $A = \begin{bmatrix} 1 & 2 \\ 2 & 4 \end{bmatrix}$, $B = \begin{bmatrix} -1 & 2 \\ 2 & 1 \end{bmatrix}$, $C = \begin{bmatrix} -3 & 8 \\ 3 & -2 \end{bmatrix}$.

Observe that $AB = AC = \begin{bmatrix} 3 & 4 \\ 6 & 8 \end{bmatrix}$, but $B \neq C$.

Consider the matrices $P = \begin{bmatrix} 1 & -2 \\ -2 & 4 \end{bmatrix}$, $Q = \begin{bmatrix} 2 & -6 \\ 1 & -3 \end{bmatrix}$. Observe that $PQ = O$, but $P \neq O$ and $Q \neq O$.

Powers of Matrices

A similar notation is used for the powers of matrices as for powers of real numbers. If A is a square matrix then A multiplied by itself k times is written A^k.

$$A^k = \underbrace{AA \ldots A}_{k \text{ times}}$$

Familiar rules of exponents of real numbers hold for matrices.

THEOREM 2.3

If A is an $n \times n$ square matrix and r and s are nonnegative integers, then
1. $A^r A^s = A^{r+s}$
2. $(A^r)^s = A^{rs}$
3. $A^0 = I_n$ (By definition.)

We verify the first rule. The proof of the second rule is similar.

$$A^r A^s = \underbrace{A \ldots A}_{r \text{ times}} \underbrace{A \ldots A}_{s \text{ times}} = \underbrace{A \ldots A}_{r + s \text{ times}} = A^{r+s}$$

EXAMPLE **3** If $A = \begin{bmatrix} 1 & -2 \\ -1 & 0 \end{bmatrix}$, compute A^4.

SOLUTION

This example illustrates how the preceding rules can be used to reduce the amount of computation involved in multiplying matrices. We know that $A^4 = AAAA$. We could perform three matrix multiplications to arrive at A^4. However we can apply rule 2 to write $A^4 = (A^2)^2$ and thus arrive at the result using two products. We get

$$A^2 = \begin{bmatrix} 1 & -2 \\ -1 & 0 \end{bmatrix} \begin{bmatrix} 1 & -2 \\ -1 & 0 \end{bmatrix} = \begin{bmatrix} 3 & -2 \\ -1 & 2 \end{bmatrix}.$$

$$A^4 = \begin{bmatrix} 3 & -2 \\ -1 & 2 \end{bmatrix} \begin{bmatrix} 3 & -2 \\ -1 & 2 \end{bmatrix} = \begin{bmatrix} 11 & -10 \\ -5 & 6 \end{bmatrix}.$$

The following example illustrates that the properties of matrix operations can be used to simplify matrix expressions in a similar way to the simplification of ordinary algebraic expressions.

EXAMPLE **4** Simplify the following matrix expression.
$$A(A + 2B) + 3B(2A - B) - A^2 + 7B^2 - 5AB$$

SOLUTION

Using the properties of matrix operations we get
$$A(A + 2B) + 3B(2A - B) - A^2 + 7B^2 - 5AB = A^2 + 2AB + 6BA - 3B^2 - A^2$$
$$+ 7B^2 - 5AB$$
$$= -3AB + 6BA + 4B^2$$

Resist the temptation to simplify $-3AB + 6BA$; matrix multiplication is not commutative!

We now introduce a valuable way of writing a system of linear equations as a single matrix equation. We will see how this technique is used in designing equipment for controlling currents and voltages in electrical circuits. Furthermore it opens up the possibility of using matrix algebra to further discuss the *behavior* of systems of linear equations.

Systems of Linear Equations

We can write a general system of m linear equations in n variables using matrix notation as follows:

$$\begin{matrix} a_{11}x_1 + \cdots + a_{1n}x_n = b_1 \\ \vdots \quad \cdots \quad \vdots \quad \vdots \\ a_{m1}x_1 + \cdots + a_{mn}x_n = b_m \end{matrix}$$

Represent each side of this equation as a column matrix,

$$\begin{bmatrix} a_{11}x_1 + \cdots + a_{1n}x_n \\ \vdots \quad\cdots\quad \vdots \\ a_{m1}x_1 + \cdots + a_{mn}x_n \end{bmatrix} = \begin{bmatrix} b_1 \\ \vdots \\ b_m \end{bmatrix}$$

The left matrix can be written as a product of the matrix of coefficients A and a column matrix of variables X. Let the column matrix of constants be B.

$$\overset{A}{\begin{bmatrix} a_{11} & \cdots & a_{1n} \\ \vdots & \cdots & \vdots \\ a_{m1} & \cdots & a_{mn} \end{bmatrix}} \overset{X}{\begin{bmatrix} x_1 \\ \vdots \\ x_n \end{bmatrix}} = \overset{B}{\begin{bmatrix} b_1 \\ \vdots \\ b_m \end{bmatrix}}$$

Thus we can write the system of equations in matrix form

$$\boxed{AX = B}$$

For example

$$\begin{array}{rcr} 3x_1 + 2x_2 - 5x_3 &=& 7 \\ x_1 - 8x_2 + 4x_3 &=& 9 \\ 2x_1 + 6x_2 - 7x_3 &=& -2 \end{array} \quad \text{can be written} \quad \begin{bmatrix} 3 & 2 & -5 \\ 1 & -8 & 4 \\ 2 & 6 & -7 \end{bmatrix} \begin{bmatrix} x_1 \\ x_2 \\ x_3 \end{bmatrix} = \begin{bmatrix} 7 \\ 9 \\ -2 \end{bmatrix}$$

We now use this notation and the properties of matrices to examine sums and scalar multiples of solutions to systems of linear equations.

Solutions to Systems of Linear Equations

Consider a homogeneous system of linear equations $AX = 0$. Let X_1 and X_2 be solutions. Then

$$AX_1 = 0 \quad \text{and} \quad AX_2 = 0$$

Adding these equations we get

$$AX_1 + AX_2 = 0, \text{giving } A(X_1 + X_2) = 0$$

Thus $X_1 + X_2$ satisfies the equation $AX = 0$. This means that $X_1 + X_2$, the sum of two solutions, is a solution. We say that the *set of solutions is closed under addition.*

Furthermore, if c is a scalar, multiplying $AX_1 = 0$ by c,

$$cAX_1 = 0, \text{giving } A(cX_1) = 0$$

Thus cX_1, the scalar multiple of a solution, is a solution. *The set of solutions is closed under scalar multiplication.*

> The set of solutions to a homogeneous system of linear equations is closed under addition and under scalar multiplication.

You are asked to show that the set of solutions to a nonhomogeneous system is not closed under addition or scalar multiplication (Exercise 37). The concept of closure under addition and scalar multiplication is extremely important in linear algebra. You will meet it many times in different contexts in the course. We reinforce this discussion with the following example.

EXAMPLE 5 Consider the following homogeneous system of linear equations:

$$x_1 + x_2 - x_3 + 4x_4 = 0$$
$$x_2 - 3x_3 + x_4 = 0$$
$$x_1 - x_2 + 5x_3 + 2x_4 = 0$$

It can be shown that there are many solutions, $x_1 = -2r - 3s$, $x_2 = 3r - s$, $x_3 = r$, $x_4 = s$. Consider two solutions, say $x_1 = -5$, $x_2 = 2$, $x_3 = 1$, $x_4 = 1$, and $x_1 = 5$, $x_2 = 9$, $x_3 = 2$, $x_4 = -3$, corresponding to $r = 1$, $s = 1$, and $r = 2$, $s = -3$.

Write these solutions as matrices.

$$X_1 = \begin{bmatrix} -5 \\ 2 \\ 1 \\ 1 \end{bmatrix} \quad \text{and} \quad X_2 = \begin{bmatrix} 5 \\ 9 \\ 2 \\ -3 \end{bmatrix}$$

It can be seen that their sum $X_1 + X_2 = \begin{bmatrix} 0 \\ 11 \\ 3 \\ -2 \end{bmatrix}$ is a solution corresponding to $r = 3$,

$s = -2$. Furthermore, if we consider a scalar multiple of one of these solutions, say $3X_1$, we find that

$$3X_1 = \begin{bmatrix} -15 \\ 6 \\ 3 \\ 3 \end{bmatrix} \text{ is a solution corresponding to } r = 3, s = 3.$$

■

EXAMPLE 6 In this example we analyze a *two-port* in an electrical circuit.

Many networks are designed to accept signals at certain points and to deliver a modified version of the signals. The usual arrangement is illustrated in Figure 2.4. A current I_1 at voltage V_1 is delivered into a two-port and it in some way determines the output current I_2 at voltage V_2. In practice the relationship between the input and output currents and voltages is usually linear—they are related by a matrix equation:

$$\begin{bmatrix} V_2 \\ I_2 \end{bmatrix} = \begin{bmatrix} a_{11} & a_{12} \\ a_{21} & a_{22} \end{bmatrix} \begin{bmatrix} V_1 \\ I_1 \end{bmatrix}$$

The matrix $\begin{bmatrix} a_{11} & a_{12} \\ a_{21} & a_{22} \end{bmatrix}$ is called the *transmission matrix* of the port. This matrix defines the two-port.

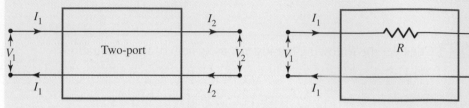

Figure 2.4   **Figure 2.5**

Figure 2.5 is an example of a two-port. The interior consists of a resistance R connected as shown. Let us show that the currents and voltages do indeed behave in a linear manner and determine the transmission matrix. Our approach will be to construct two equations, one expressing V_2 in terms of V_1 and I_1, and the other expressing I_2 in terms of V_1 and I_1 and then combine these two equations into a single matrix equation. We use the following law.

Ohm's Law
The voltage drop across a resistance is equal to the current times the resistance.

The voltage drop across the resistance R is $V_1 - V_2$. The current through the resistance is I_1. Thus Ohm's Law implies that $V_1 - V_2 = I_1 R$. The current I_1 passes through the resistance R unchanged and exits unchanged as I_1. Thus $I_2 = I_1$. Write these two equations in the following standard form.

$$V_2 = V_1 - RI_1$$
$$I_2 = 0V_1 + I_1$$

Combine the two equations into a single matrix equation,

$$\begin{bmatrix} V_2 \\ I_2 \end{bmatrix} = \begin{bmatrix} 1 & -R \\ 0 & 1 \end{bmatrix} \begin{bmatrix} V_1 \\ I_1 \end{bmatrix}$$

The transmission matrix is $\begin{bmatrix} 1 & -R \\ 0 & 1 \end{bmatrix}$. Thus for example, if R is 2 ohms and the input voltage and current are $V_1 = 5$ volts, $I_1 = 1$ amp, we get

$$\begin{bmatrix} V_2 \\ I_2 \end{bmatrix} = \begin{bmatrix} 1 & -2 \\ 0 & 1 \end{bmatrix} \begin{bmatrix} 5 \\ 1 \end{bmatrix} = \begin{bmatrix} 3 \\ 1 \end{bmatrix}$$

The output voltage and current are 3 volts and 1 amp.

In practice a number of standard two-ports such as the one above are placed in series to provide a desired voltage and current change. Consider the three two-ports of Figure 2.6, with transmission matrices A, B, and C.

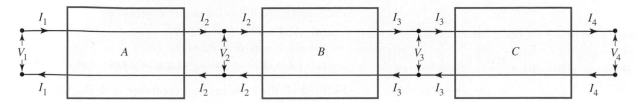

Figure 2.6

Considering each port separately, we have

$$\begin{bmatrix} V_2 \\ I_2 \end{bmatrix} = A \begin{bmatrix} V_1 \\ I_1 \end{bmatrix}, \quad \begin{bmatrix} V_3 \\ I_3 \end{bmatrix} = B \begin{bmatrix} V_2 \\ I_2 \end{bmatrix}, \quad \begin{bmatrix} V_4 \\ I_4 \end{bmatrix} = C \begin{bmatrix} V_3 \\ I_3 \end{bmatrix}$$

Substituting for $\begin{bmatrix} V_2 \\ I_2 \end{bmatrix}$ from the first equation into the second gives $\begin{bmatrix} V_3 \\ I_3 \end{bmatrix} = BA \begin{bmatrix} V_1 \\ I_1 \end{bmatrix}$.

Substituting this $\begin{bmatrix} V_3 \\ I_3 \end{bmatrix}$ into the third equation gives

$$\begin{bmatrix} V_4 \\ I_4 \end{bmatrix} = CBA \begin{bmatrix} V_1 \\ I_1 \end{bmatrix}$$

The three ports are thus equivalent to a single two-port. The transmission matrix of this two-port is the product CBA of the individual ports. Note that the placement of each port in the sequence is important since matrices are not commutative under multiplication.

EXERCISE SET 2.2

Computation

1. Let $A = \begin{bmatrix} 1 & 2 \\ -1 & 0 \end{bmatrix}$, $B = \begin{bmatrix} 0 & 5 & 4 \\ 2 & 1 & 3 \end{bmatrix}$,

$C = \begin{bmatrix} 2 & 3 \\ 6 & 1 \end{bmatrix}$, $D = \begin{bmatrix} 2 & -2 \\ 1 & 3 \end{bmatrix}$. Calculate, if possible,

(a) AB and BA (b) AC and CA

(c) AD and DA

Observe that $AB \neq BA$ since BA does not exist, $AC \neq CA$ and $AD = DA$, illustrating the different possibilities when order is reversed in matrix multiplication.

2. Compute $A(BC)$ and $(AB)C$ for the matrices

$$A = \begin{bmatrix} 1 & 2 \\ -1 & 0 \\ 1 & 1 \end{bmatrix}, \quad B = \begin{bmatrix} 2 & 4 \\ -2 & 3 \end{bmatrix}, \quad C = \begin{bmatrix} 1 \\ 2 \end{bmatrix}$$

Observe that these products are equal, illustrating the associative property of matrix multiplication.

3. Compute the product ABC for the following three matrices in two distinct ways.

$$A = \begin{bmatrix} 1 & 2 \\ -1 & 3 \end{bmatrix}, \quad B = \begin{bmatrix} 3 & 1 & 2 \\ 4 & 3 & 1 \end{bmatrix}, \quad C = \begin{bmatrix} 1 & 2 \\ 3 & 4 \\ 1 & 0 \end{bmatrix}$$

4. Compute each of the following linear combinations for

$$A = \begin{bmatrix} 1 & 2 \\ 3 & 4 \end{bmatrix}, \quad B = \begin{bmatrix} 2 & -3 \\ 0 & 1 \end{bmatrix}, \quad \text{and} \quad C = \begin{bmatrix} -2 & 0 \\ 3 & 4 \end{bmatrix}.$$

   (a) $2A + 3B$ (b) $A + 2B + 4C$
   (c) $3A + B - 2C$

5. Compute each of the following expressions for

$$A = \begin{bmatrix} 2 & 0 \\ -1 & 5 \end{bmatrix}, \quad B = \begin{bmatrix} -1 & 1 \\ 2 & 4 \end{bmatrix}, \quad \text{and} \quad C = \begin{bmatrix} 3 & 4 \\ 0 & 2 \end{bmatrix}.$$

   (a) $(AB)^2$ (b) $A - 3B^2$
   (c) $A^2B + 2C^3$ (d) $2A^2 - 2A + 3I_2$

Sizes of Matrix Products

6. Given that A is a 4×2 matrix, B is 2×6, C is 3×4, and D is 6×3, determine the sizes of the following products, if they exist.

   (a) ABC (b) ABD (c) CAB
   (d) $DCAB$ (e) A^2BDC

7. If P is 3×2, Q is 2×1, R is 1×3, S is 3×1, and T is 3×3, determine the sizes of the following matrix expressions, if they exist.

   (a) PQR (b) $PQ + TPQ$
   (c) $5QR - 2TPR$ (d) $4SPQ + 3PQ$
   (e) $QRSR + QR$

Matrix Operations

8. Let A be an $m \times n$ matrix. Prove that AB and BA both exist only if B is an $n \times m$ matrix.

9. Verify the following properties of matrix operations given in this section:

   (a) the associative property of matrix addition
       $A + (B + C) = (A + B) + C$
   (b) the distributive property $c(A + B) = cA + cB$
   (c) $AI_n = I_nA = A$ if A is an $n \times n$ matrix

10. Let A be an $m \times n$ matrix. Show that $AI_n = A$.

11. Let A be any $m \times n$ matrix, O_{mn} be the $m \times n$ zero matrix, and c be a scalar. Show that if $cA = O_{mn}$ then either $c = 0$ or $A = O_{mn}$.

12. Simplify the following matrix expressions.
    (a) $A(A - 4B) + 2B(A + B) - A^2 + 7B^2 + 3AB$
    (b) $B(2I_n - BA) + B(4I_n + 5A)B - 3BAB + 7B^2A$
    (c) $(A - B)(A + B) - (A + B)^2$

13. Simplify the following matrix expressions.
    (a) $A(A + B) - B(A + B)$
    (b) $A(A - B)B + B2AB - 3A^2$
    (c) $(A + B)^3 - 2A^3 - 3ABA - A3B^2 - B^3$

14. Find all the matrices that commute with the following matrices.

    (a) $\begin{bmatrix} 1 & 0 \\ -1 & 0 \end{bmatrix}$ (b) $\begin{bmatrix} 1 & 0 \\ 0 & 2 \end{bmatrix}$ (c) $\begin{bmatrix} 0 & 1 & 0 \\ 0 & 0 & 1 \\ 0 & 0 & 0 \end{bmatrix}$

15. What is incorrect about the following proof? Let $AX = B$ be a system of linear equations with solutions X_1 and X_2. Thus

$$AX_1 = B \quad \text{and} \quad AX_2 = B$$
$$AX_1 = AX_2$$
$$X_1 = X_2$$

Thus every system of linear equations has at most one solution.

Powers of Matrices

16. (a) Let A be an $n \times n$ matrix. Prove that A^2 is an $n \times n$ matrix.
    (b) Let A be an $m \times n$ matrix, with $m \neq n$. Prove that A^2 does not exist.

Thus one can only talk about powers of square matrices.

17. If A and B are square matrices of the same size, prove that in general

$$(A + B)^2 \neq A^2 + 2AB + B^2$$

Under what condition does equality hold?

18. If A and B are square matrices of the same size such that $AB = BA$, prove that $(AB)^2 = A^2B^2$. By constructing an example, show that this result does not hold for all square matrices of the same size.

19. If n is a nonnegative integer and A and B are square matrices of the same size such that $AB = BA$, prove that $(AB)^n = A^nB^n$. By constructing an example, show that this identity does not hold in general for all square matrices of the same size.

20. Show that nonnegative integer powers of the same matrix commute. $(A^rA^s = A^sA^r)$

Diagonal Matrices

21. Let A and B be diagonal matrices of the same size and c a scalar. Prove that **(a)** $A + B$ is diagonal **(b)** cA is diagonal **(c)** AB is diagonal.

22. If A and B are diagonal matrices of the same size, prove that $AB = BA$.

23. Prove that if a matrix A commutes with a diagonal matrix which has no two diagonal elements the same, then A is a diagonal matrix.

Idempotent and Nilpotent Matrices

A square matrix A is said to be **idempotent** if $A^2 = A$. A square matrix A is said to **nilpotent** if there is a positive integer p such that $A^P = 0$. The least integer such that $A^P = 0$ is called the **degree of nilpotency** of the matrix.

24. Determine whether the following matrices are idempotent.

(a) $\begin{bmatrix} 1 & 0 \\ 0 & 1 \end{bmatrix}$ **(b)** $\begin{bmatrix} 1 & 0 \\ 0 & 0 \end{bmatrix}$

(c) $\begin{bmatrix} 0 & 1 \\ 1 & 0 \end{bmatrix}$ **(d)** $\begin{bmatrix} 3 & -6 \\ 1 & -2 \end{bmatrix}$

(e) $\begin{bmatrix} 1 & 2 & 2 \\ 0 & 0 & -1 \\ 0 & 0 & 1 \end{bmatrix}$ **(f)** $\begin{bmatrix} 1 & 3 & 0 \\ 0 & 0 & 1 \\ 0 & 0 & 0 \end{bmatrix}$

25. Determine b, c, and d such that $\begin{bmatrix} 1 & b \\ c & d \end{bmatrix}$ is idempotent.

26. Determine a, c, and d such that $\begin{bmatrix} a & 0 \\ c & d \end{bmatrix}$ is idempotent.

27. Prove that if A and B are idempotent and $AB = BA$, then AB is idempotent.

28. Show that if A is idempotent, and if n is a positive integer, then $A^n = A$.

29. Show that the following matrices are nilpotent with degree of nilpotency 2.

(a) $\begin{bmatrix} 1 & 1 \\ -1 & -1 \end{bmatrix}$ **(b)** $\begin{bmatrix} -4 & 8 \\ -2 & 4 \end{bmatrix}$ **(c)** $\begin{bmatrix} 3 & -9 \\ 1 & -3 \end{bmatrix}$

30. Show that the following matrix is nilpotent with degree of nilpotency 3.

$$\begin{bmatrix} 0 & 1 & 0 \\ 0 & 0 & 1 \\ 0 & 0 & 0 \end{bmatrix}$$

Systems of Linear Equations

31. Write each of the following systems of linear equations as a single matrix equation $AX = B$.

(a) $\begin{aligned} 2x_1 + 3x_2 &= 4 \\ 3x_1 - 8x_2 &= -1 \end{aligned}$ **(b)** $\begin{aligned} 4x_1 + 7x_2 &= -2 \\ -2x_1 + 3x_2 &= -4 \end{aligned}$

(c) $\begin{aligned} -9x_1 - 3x_2 &= -4 \\ 6x_1 - 2x_2 &= 7 \end{aligned}$

32. Write each of the following systems of linear equations as a single matrix equation $AX = B$.

(a) $\begin{aligned} x_1 + 8x_2 - 2x_3 &= 3 \\ 4x_1 - 7x_2 + x_3 &= -3 \\ -2x_1 - 5x_2 - 2x_3 &= 1 \end{aligned}$

(b) $\begin{aligned} 5x_1 + 2x_2 &= 6 \\ 4x_1 - 3x_2 &= -2 \\ 3x_1 + x_2 &= 9 \end{aligned}$

(c) $\begin{aligned} x_1 - 3x_2 + 6x_3 &= 2 \\ 7x_1 + 5x_2 + x_3 &= -9 \end{aligned}$

(d) $\begin{aligned} 2x_1 + 5x_2 - 3x_3 + 4x_4 &= 4 \\ x_1 \qquad\quad + 9x_3 + 5x_4 &= 12 \\ 3x_1 - 3x_2 - 8x_3 + 5x_4 &= -2 \end{aligned}$

33. Prove that if X_1 and X_2 are solutions to the homogeneous system of linear equations $AX = 0$, then the linear combination $aX_1 + bX_2$ is a solution for all scalars a and b.

34. Consider the following system of equations. You are given two solutions, X_1 and X_2. Generate four other solutions using the operations of matrix addition and scalar multiplication. Use the result of Exercise 33 to find a solution for which $x_1 = -3$ and $x_2 = 2$.

$$\begin{aligned} x_1 + 6x_2 - x_3 - 4x_4 &= 0 \\ -2x_1 - 12x_2 + 5x_3 + 17x_4 &= 0 \\ 3x_1 + 18x_2 - x_3 - 6x_4 &= 0 \end{aligned}$$

$$X_1 = \begin{bmatrix} -4 \\ 1 \\ -6 \\ 2 \end{bmatrix}, \quad X_2 = \begin{bmatrix} -17 \\ 3 \\ -3 \\ 1 \end{bmatrix}$$

35. Consider the following system of equations. You are given two solutions, X_1 and X_2. Generate four other solutions using the operations of addition and scalar multiplication. Use the result of Exercise 33 to find a solution for which $x_1 = 1$ and $x_2 = 0$.

$$\begin{aligned} x_1 + 2x_2 - x_3 - 2x_4 &= 0 \\ 2x_1 + 5x_2 \qquad\quad - 2x_4 &= 0 \\ 4x_1 + 9x_2 - 2x_3 - 6x_4 &= 0 \\ x_1 + 3x_2 + x_3 \qquad\quad &= 0 \end{aligned}$$

$$X_1 = \begin{bmatrix} 5 \\ -2 \\ 1 \\ 0 \end{bmatrix}, \quad X_2 = \begin{bmatrix} 11 \\ -4 \\ 1 \\ 1 \end{bmatrix}$$

36. Consider the following system of four equations. You are given two solutions X_1 and X_2. Generate four other solutions using the operations of addition and scalar multiplication. Use the result of Exercise 33 to find a solution for which $x_1 = 6$ and $x_2 = 9$.

$$x_1 - x_2 + x_3 + 2x_4 = 0$$
$$3x_1 - 2x_2 + x_3 + 3x_4 = 0$$
$$5x_1 - 4x_2 + 3x_3 + 7x_4 = 0$$
$$2x_1 - x_2 \qquad + x_4 = 0$$

$$X_1 = \begin{bmatrix} 0 \\ -1 \\ 1 \\ -1 \end{bmatrix}, \quad X_2 = \begin{bmatrix} 3 \\ 7 \\ 2 \\ 1 \end{bmatrix}$$

37. Show that a set of solutions to a system of nonhomogeneous linear equations is not closed under addition or under scalar multiplication.

38. Prove that a system of linear equations $AX = B$ has a solution if B is a linear combination of the columns of the matrix A.

Miscellaneous Results

39. State (with a brief explanation) whether the following statements are true or false for matrices A, B, C, and D.

(a) $A^2 - B^2 = (A + B)(A - B)$

(b) $A(B + C + D) = AB + AC + AD$

(c) If A is an $m \times n$ matrix, B is $n \times r$, and C is $r \times q$ then ABC has $m + q$ elements.

(d) If $A^2 = A$ then $A^r = A$ for all positive integer values of r.

(e) If X_1 is a solution of $AX = B_1$ and X_2 is a solution of $AX = B_2$ then $X_1 + X_2$ is a solution of $AX = B_1 + B_2$.

Two-Ports

40. Determine the transmission matrices of the two-ports in Figure 2.7.

Hints:

(a) $V_1 = V_2$ since terminals are connected directly. Current through resistance R is $I_1 - I_2$. Drop in voltage across R is V_1.

(b) Current through R_1 is $I_1 - I_2$. Drop in voltage across R_1 is V_1. Current through R_2 is I_2. Drop in voltage across R_2 is $V_1 - V_2$.

(c) Current through R_1 is I_1. Drop in voltage across R_1 is $V_1 - V_2$. Current through R_2 is $I_1 - I_2$. Drop in voltage across R_2 is V_2.

41. The two-port in Figure 2.8 consists of three two-ports placed in series. The transmission matrices are indicated. **(a)** What is the transmission matrix of the composite two-port? **(b)** If the input voltage is three volts and the current is two amps, determine the output voltage and current.

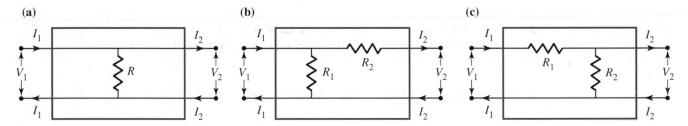

Figure 2.7

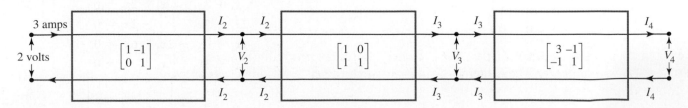

Figure 2.8

2.3 Symmetric Matrices and Seriation in Archaeology

In this section we continue the algebraic development of matrices and we see how the theory developed is used by archaeologists to determine the chronological order of graves and artifacts.

> **DEFINITION** The **transpose** of a matrix A, denoted A^t, is the matrix whose columns are the rows of the given matrix A.

The first row of A becomes the first column of A^t, the second row of A becomes the second column of A^t, and so on. The (i, j)th element of A becomes the (j, i)th element of A^t. If A is an $m \times n$ matrix then A^t is an $n \times m$ matrix.

EXAMPLE 1 Determine the transpose of each of the following matrices.

$$A = \begin{bmatrix} 2 & 7 \\ -8 & 0 \end{bmatrix}, \quad B = \begin{bmatrix} 1 & 2 & -7 \\ 4 & 5 & 6 \end{bmatrix}, \quad C = \begin{bmatrix} -1 & 3 & 4 \end{bmatrix}$$

SOLUTION

Writing rows as columns we get

$$A^t = \begin{bmatrix} 2 & -8 \\ 7 & 0 \end{bmatrix}, \quad B^t = \begin{bmatrix} 1 & 4 \\ 2 & 5 \\ -7 & 6 \end{bmatrix}, \quad C^t = \begin{bmatrix} -1 \\ 3 \\ 4 \end{bmatrix}$$

Observe that the $(1, 3)$ element of B namely -7, becomes the $(3, 1)$ element of B^t. Note also that B is a 2×3 matrix while B^t is 3×2, C is a 1×3 matrix while C^t is 3×1.

There are three operations that we have defined for matrices, namely addition, scalar multiplication, and multiplication. The following theorem tells us how transpose works in conjunction with these operations.

THEOREM 2.4

Properties of Transpose

Let A and B be matrices and c be a scalar. Assume that the sizes of the matrices are such that the operations can be performed.

1. $(A + B)^t = A^t + B^t$ *Transpose of a sum*
2. $(cA)^t = cA^t$ *Transpose of a scalar multiple*
3. $(AB)^t = B^t A^t$ *Transpose of a product*
4. $(A^t)^t = A$

We demonstrate the techniques that are used in verifying results involving transposes by verifying the third property. The reader is asked to derive the other results in the exercises that follow.

$(AB)^t = B^tA^t$ The expressions $(AB)^t$ and B^tA^t will be matrices upon carrying out all the products and transposes. We prove that these matrices are equal by showing that corresponding elements are equal.

$$(i, j)\text{th element of } (AB)^t = (j, i)\text{th element of } AB$$
$$(i, j)\text{th element of } B^tA^t = (\text{row } i \text{ of } B^t) \times (\text{column } j \text{ of } A^t)$$
$$= (\text{column } i \text{ of } B) \times (\text{row } j \text{ of } A)$$
$$= (j, i)\text{th element of } AB$$

The corresponding elements of $(AB)^t$ and B^tA^t are equal, proving the result.

Remark: The results for the transpose of a sum and a product can be extended to any number of matrices. For example, for three matrices A, B, and C,

$$(A + B + C)^t = A^t + B^t + C^t \quad \text{and} \quad (ABC)^t = C^tB^tA^t$$

Note the reversal of the order of the matrices in the transpose of a product. (See the following exercises for the proofs.)

We now introduce *symmetric matrices*. They are probably the single most important class of matrices. They are used in areas of mathematics such as geometry, and in fields such as theoretical physics, mechanical and electrical engineering, and sociology.

DEFINITION   A **symmetric matrix** is a square matrix that is equal to its transpose.

The following are examples of symmetric matrices. Note the symmetry of these matrices about the main diagonal. All nondiagonal elements occur in pairs symmetrically located about the main diagonal.

$$\begin{bmatrix} 2 & 5 \\ 5 & -4 \end{bmatrix} \qquad \begin{bmatrix} 0 & 1 & -4 \\ 1 & 7 & 8 \\ -4 & 8 & 3 \end{bmatrix} \qquad \overset{\text{match}}{\begin{bmatrix} 1 & 0 & -2 & 4 \\ 0 & 7 & 3 & 9 \\ -2 & 3 & 2 & -3 \\ 4 & 9 & -3 & 6 \end{bmatrix}}$$

match

> **EXAMPLE 2** Consider the following matrix, which represents distances between various U.S. cities.
>
> | | Chicago | LA | Miami | NY |
> |---|---|---|---|---|
> | Chicago | 0 | 2092 | 1374 | 841 |
> | Los Angeles | 2092 | 0 | 2733 | 2797 |
> | Miami | 1374 | 2733 | 0 | 1336 |
> | New York | 841 | 2797 | 1336 | 0 |
>
> Observe that the matrix is symmetric. All elements occur in pairs, symmetrically located about the main diagonal. There is a reason for this, namely that the distance from city X to city Y is the same as the distance from city Y to city X. For example, the distance from Chicago to Miami, which is 1374 miles (row 1, column 3) will be the same as the distance from Miami to Chicago (row 3, column 1). All such mileage matrices will be symmetric.

EXAMPLE 3 Let A and B be symmetric matrices of the same size. Let C be a linear combination of A and B. Prove that C is symmetric.

Proof Let $C = aA + bB$, where a and b are scalars. We now use the results of Theorem 2.4 to prove that C is symmetric.

$$
\begin{aligned}
C^t &= (aA + bB)^t \\
&= (aA)^t + (bB)^t \quad \text{\textit{Transpose of sum}} \\
&= aA^t + bB^t \quad \text{\textit{Transpose of scalar multiple}} \\
&= aA + bB \quad \text{\textit{A and B are symmetric}} \\
&= C
\end{aligned}
$$

Thus C is symmetric.

The Expression "If and Only If"

The expressions *"if and only if"* and *"necessary and sufficient"* (they mean the same thing) are frequently used in mathematics, and we shall use them periodically in this course. Let p and q be statements. Suppose that p implies q, written $p \Rightarrow q$, and that also $q \Rightarrow p$. The second implication is called the **converse** of the first. We say that "p if and only if q" or "p is necessary and sufficient for q." The next example states a result using this language.

EXAMPLE 4 Let A and B be symmetric matrices of the same size. Prove that the product AB is symmetric if and only if $AB = BA$.

Proof Every "if and only if" situation such as this consists of two parts. We have to show that (**a**) if AB is symmetric then $AB = BA$, and then conversely, (**b**) if $AB = BA$ then AB is symmetric.

(**a**) Let AB be symmetric. Then

$$
\begin{aligned}
AB &= (AB)^t \quad \text{\textit{Definition of symmetric matrix}} \\
&= B^t A^t \quad \text{\textit{Transpose of a product}} \\
&= BA \quad \text{\textit{A and B are symmetric matrices}}
\end{aligned}
$$

(**b**) Let $AB = BA$. Then

$$
\begin{aligned}
(AB)^t &= (BA)^t \\
&= A^t B^t \quad \text{\textit{Transpose of a product}} \\
&= AB \quad \text{\textit{A and B are symmetric matrices}}
\end{aligned}
$$

Therefore AB is symmetric.

Thus, given two symmetric matrices of the same size, A and B, the product AB is symmetric if and only if $AB = BA$.

This result could also have been stated as follows: "Given two symmetric matrices of the same size, A and B, then a necessary and sufficient condition for the product AB to be symmetric is that $AB = BA$."

We now introduce a number that is associated with every square matrix, called the *trace of the matrix*.

> **DEFINITION** Let A be a square matrix. The **trace** of A, denoted $tr(A)$ is the sum of the diagonal elements of A. Thus if A is an $n \times n$ matrix,
>
> $$tr(A) = a_{11} + a_{22} + \cdots + a_{nn}$$

EXAMPLE 5 Determine the trace of the matrix $A = \begin{bmatrix} 4 & 1 & -2 \\ 2 & -5 & 6 \\ 7 & 3 & 0 \end{bmatrix}$.

SOLUTION

We get,

$$tr(A) = 4 + (-5) + 0 = -1$$

The trace of a matrix plays an important role in matrix theory and matrix applications because of its properties and the ease with which it can be evaluated. It is important in fields such as statistical mechanics, general relativity, and quantum mechanics, where it has physical significance.

The following theorem tells us how the operation of trace interacts with the operations of matrix addition, scalar multiplication, multiplication, and transpose.

THEOREM 2.5

Properties of Trace

Let A and B be matrices and c be a scalar. Assume that the sizes of the matrices are such that the operations can be performed.

1. $tr(A + B) = tr(A) + tr(B)$
2. $tr(AB) = tr(BA)$
3. $tr(cA) = c\,tr(A)$
4. $tr(A^t) = tr(A)$

Proof We prove the first property, leaving the proofs of the other properties for the reader to complete in the exercises. Since the diagonal elements of $A + B$ are $(a_{11} + b_{11}), (a_{22} + b_{22}), \ldots, (a_{nn} + b_{nn})$ we get

$$tr(A + B) = (a_{11} + b_{11}) + (a_{22} + b_{22}) + \cdots + (a_{nn} + b_{nn})$$
$$= (a_{11} + a_{22} + \cdots + a_{nn}) + (b_{11} + b_{22} + \cdots + b_{nn})$$
$$= tr(A) + tr(B)$$

Matrices with Complex Elements (Optional)

The elements of a matrix may be complex numbers. A **complex number** is of the form

$$z = a + bi$$

where a and b are real numbers and $i = \sqrt{-1}$. a is called the **real part** and b the **imaginary part** of z.

The rules of arithmetic for complex numbers are as follows.

Let $z_1 = a + bi$, $z_2 = c + di$ be complex numbers.

Equality: $\qquad\qquad z_1 = z_2$ if $a = c$ and $b = d$

Addition: $\qquad z_1 + z_2 = (a + c) + (b + d)i$

Subtraction: $\qquad z_1 - z_2 = (a - c) + (b - d)i$

Multiplication: $\qquad z_1 z_2 = (a + bi)(c + di) = a(c + di) + bi(c + di)$

$$= ac + adi + bci + bdi^2$$

$$= ac + bdi^2 + (ad + bc)i = (ac - bd) + (ad + bc)i$$

The **conjugate** of a complex number $z = a + bi$ is defined and written $\bar{z} = a - bi$.

EXAMPLE **6** Consider the complex numbers $z_1 = 2 + 3i$ and $z_2 = 1 - 2i$. Compute $z_1 + z_2$, $z_1 z_2$, and $\bar{z}_1$.

SOLUTION

Using the above definitions we get

$$z_1 + z_2 = (2 + 3i) + (1 - 2i) = (2 + 1) + (3 - 2)i = 3 + i$$

$$z_1 z_2 = (2 + 3i)(1 - 2i) = 2(1 - 2i) + 3i(1 - 2i) = 2 - 4i + 3i - 6i^2 = 8 - i$$

$$\bar{z}_1 = 2 - 3i$$

Matrices having complex elements are added, subtracted, and multiplied using the same rules as matrices having real elements.

EXAMPLE **7** Let $A = \begin{bmatrix} 2 + i & 3 - 2i \\ 4 & 5i \end{bmatrix}$ and $B = \begin{bmatrix} 3 & 2i \\ 1 + i & 2 + 3i \end{bmatrix}$. Compute $A + B$, $2A$, and AB.

SOLUTION

We get

$$A + B = \begin{bmatrix} 2 + i & 3 - 2i \\ 4 & 5i \end{bmatrix} + \begin{bmatrix} 3 & 2i \\ 1 + i & 2 + 3i \end{bmatrix}$$

$$= \begin{bmatrix} 2 + i + 3 & 3 - 2i + 2i \\ 4 + 1 + i & 5i + 2 + 3i \end{bmatrix} = \begin{bmatrix} 5 + i & 3 \\ 5 + i & 2 + 8i \end{bmatrix}$$

$$2A = 2 \begin{bmatrix} 2 + i & 3 - 2i \\ 4 & 5i \end{bmatrix} = \begin{bmatrix} 4 + 2i & 6 - 4i \\ 8 & 10i \end{bmatrix}$$

$$AB = \begin{bmatrix} 2 + i & 3 - 2i \\ 4 & 5i \end{bmatrix} \begin{bmatrix} 3 & 2i \\ 1 + i & 2 + 3i \end{bmatrix}$$

$$= \begin{bmatrix} (2 + i)3 + (3 - 2i)(1 + i) & (2 + i)(2i) + (3 - 2i)(2 + 3i) \\ (4)(3) + (5i)(1 + i) & 4(2i) + (5i)(2 + 3i) \end{bmatrix}$$

$$= \begin{bmatrix} 11 + 4i & 10 + 9i \\ 7 + 5i & -15 + 8i \end{bmatrix}$$

The **conjugate** of a matrix A is denoted $\overline{A}$ and is obtained by taking the conjugate of each element of the matrix. The **conjugate transpose** of A is written and defined by $A^* = \overline{A}\,^t$.

For example if $A = \begin{bmatrix} 2 + 3i & 1 - 4i \\ 6 & 7i \end{bmatrix}$ then

$$\overline{A} = \begin{bmatrix} 2 - 3i & 1 + 4i \\ 6 & -7i \end{bmatrix} \quad \text{and} \quad A^* = \overline{A}\,^t = \begin{bmatrix} 2 - 3i & 6 \\ 1 + 4i & -7i \end{bmatrix}$$

A square matrix C is said to be **hermitian** if $C = C^*$.

Let us show that the matrix $C = \begin{bmatrix} 2 & 3 - 4i \\ 3 + 4i & 6 \end{bmatrix}$ is hermitian. We get

$$\overline{C} = \begin{bmatrix} 2 & 3 + 4i \\ 3 - 4i & 6 \end{bmatrix}; C^* = \overline{C}\,^t = \begin{bmatrix} 2 & 3 - 4i \\ 3 + 4i & 6 \end{bmatrix} = C$$

Hermitian matrices are more important than symmetric matrices for matrices having complex elements.

The properties of conjugate transpose are similar to those of transpose. We list these properties in the following theorem, leaving the proofs for the reader to do in the exercises that follow.

THEOREM 2.6

Properties of Conjugate Transpose
Let A and B be matrices with complex elements and let z be a complex number.

1. $(A + B)^* = A^* + B^*$ *Conjugate transpose of a sum*
2. $(zA)^* = \overline{z}A^*$ *Conjugate transpose of a scalar multiple*
3. $(AB)^* = B^*A^*$ *Conjugate transpose of a product*
4. $(A^*)^* = A$

The time is now ripe for a good application of matrix algebra!

Seriation in Archaeology

A problem confronting archaeologists is that of placing sites and artifacts in proper chronological order. This branch of archaeology, called **sequence dating** or **seriation**, began with the work of Sir Flinders Petrie in the late nineteenth century. Petrie studied graves in the cemeteries of Nagada, Ballas, and Hu, all located in what was prehistoric Egypt. (Recent carbon dating shows that all the graves ranged from 6000 B.C. to 2500 B.C.) Petrie used the data from approximately 900 graves to order them and assign a time period to each type of pottery found.

Let us look at this general problem of seriation in terms of graves and varieties of pottery found in graves. An assumption usually made in archaeology is that two graves that have similar contents are more likely to lie close together in time than are two graves that have little in common. The mathematical model that we now construct leads to information concerning the common contents of graves and thus to the chronological order of the graves.

We construct a matrix A, all of whose elements are either 1 or 0, that describes the pottery content of the graves. Label the graves $1, 2, \ldots$, and the types of pottery $1, 2, \ldots$. Let the matrix A be defined by

$$a_{ij} = \begin{cases} 1 & \text{if grave } i \text{ contains pottery type } j \\ 0 & \text{if grave } i \text{ does not contain pottery type } j \end{cases}$$

The matrix A contains all the information about the pottery content of the various graves. The following result now tells us how information is extracted from A.

> *The element g_{ij} of the matrix $G = AA^t$ is equal to the number of types of pottery common to both grave i and grave j.*

Thus the larger g_{ij} the closer grave i and grave j are in time. By examining the elements of G the archaeologist can arrive at the chronological order of the graves. Let us verify this result.

$$
\begin{aligned}
g_{ij} &= \text{element in row } i, \text{ column } j \text{ of } G \\
&= (\text{row } i \text{ of } A) \times (\text{column } j \text{ of } A^t) \\
&= [a_{i1}\ a_{i2} \ldots a_{in}] \begin{bmatrix} a_{j1} \\ a_{j2} \\ \vdots \\ a_{jn} \end{bmatrix} \\
&= a_{i1}a_{j1} + a_{i2}a_{j2} + \cdots + a_{in}a_{jn}
\end{aligned}
$$

Each term in this sum will be either 1 or 0. For example the term $a_{i2}a_{j2}$ will be 1 if a_{i2} and a_{j2} are both 1; that is if pottery type 2 is common to graves i and j. It will be 0 if pottery type 2 is not common to graves i and j. Thus the number of 1's in this expression for g_{ij}, (the actual value of g_{ij}), is the number of types of pottery common to graves i and j.

The matrix $P = A^tA$ leads in an analogous manner to information about the sequence dating of the pottery. The assumption is made that the larger the number of graves in which two types of pottery appear, the closer they are chronologically. The element p_{ij} of the matrix $P = A^tA$ gives the number of graves in which the ith and jth types of pottery both appear. Thus the larger p_{ij} the closer pottery types i and j are in time. By examining the elements of P we can arrive at the chronological order of the pottery (see Exercise 30).

It can be shown mathematically that the matrices $G(= AA^t)$ and $P(= A^tA)$ are symmetric matrices. Furthermore it can be argued from the physical interpretation that G and P should be symmetric matrices (see Exercise 28). This illustrates the compatibility of the mathematics and the interpretation. The implication of this symmetry is that all the information is contained in the elements above the main diagonals of these matrices. The information is just duplicated in the elements below the main diagonals.

We now illustrate this method by means of an example. Let the following matrix A represent the three pottery contents of four graves.

$$A = \begin{bmatrix} 1 & 0 & 1 \\ 1 & 0 & 0 \\ 0 & 1 & 1 \\ 0 & 1 & 0 \end{bmatrix}$$

Thus, for example, $a_{13} = 1$ implies that grave 1 contains pottery type 3; $a_{23} = 0$ implies that grave 2 does not contain pottery type 3. G is calculated:

$$G = AA^t = \begin{bmatrix} 1 & 0 & 1 \\ 1 & 0 & 0 \\ 0 & 1 & 1 \\ 0 & 1 & 0 \end{bmatrix} \begin{bmatrix} 1 & 1 & 0 & 0 \\ 0 & 0 & 1 & 1 \\ 1 & 0 & 1 & 0 \end{bmatrix} = \begin{bmatrix} 2 & 1 & 1 & 0 \\ 1 & 1 & 0 & 0 \\ 1 & 0 & 2 & 1 \\ 0 & 0 & 1 & 1 \end{bmatrix}$$

Observe that G is indeed symmetric. The information contained in the elements above the main diagonal is duplicated in the elements below it. We systematically look at the elements above the main diagonal.

$g_{12} = 1$—graves 1 and 2 have one type of pottery in common.

$g_{13} = 1$—graves 1 and 3 have one type of pottery in common.

$g_{14} = 0$—graves 1 and 4 have no pottery in common.

$g_{23} = 0$—graves 2 and 3 have no pottery in common.

$g_{24} = 0$—graves 2 and 4 have no pottery in common.

$g_{34} = 1$—graves 3 and 4 have one type of pottery in common.

Graves 1 and 2 have pottery in common; they are close together in time. Let us start with graves 1 and 2 and construct a diagram.

$$1 - 2$$

Next add grave 3 to this diagram. $g_{13} = 1$ while $g_{23} = 0$. Thus grave 3 is close to grave 1 but not close to grave 2. We get

$$3 - 1 - 2$$

Finally add grave 4 to the diagram. $g_{34} = 1$ while $g_{14} = 0$ and $g_{24} = 0$. Grave 4 is close to grave 3 but not close to grave 1 or to grave 2. We get

$$4 - 3 - 1 - 2$$

The mathematics does not tell us which way time flows in this diagram. There are two possibilities:

$$4 \rightarrow 3 \rightarrow 1 \rightarrow 2 \quad \text{and} \quad 4 \leftarrow 3 \leftarrow 1 \leftarrow 2$$

The archaeologist usually knows from other sources which of the two extreme graves (4 and 2 in our case) came first. Thus the chronological order of the graves is known.

The matrices G and P contain information about the chronological order of the graves and pottery, through the relative magnitudes of their elements. These matrices are, in practice, large and the information cannot be sorted out as easily or give results that are as unambiguous as in the above illustration. For example, Petrie examined 900 graves; his matrix G would be a 900×900 matrix. Special mathematical techniques have been developed for extracting information from these matrices, these methods now being executed on computers. Readers who are interested in pursuing this topic further should consult "Some Problems and Methods in Statistical Archaeology" by David G. Kendall, *World Archaeology*, 1, 61–76, 1969.

In these sections we have developed the algebraic theory of matrices. This theory is extremely important in applications of mathematics. We shall for example use matrices in mathematical models of communication and of population movements. Albert Einstein used a matrix equation to describe the relationship between geometry and matter in his general theory of relativity:

$$T = R - \tfrac{1}{2}rG$$

T is a matrix that represents matter, R is a matrix that represents geometry, r is a scalar, and G is a matrix that describes gravity. All the matrices are symmetric 4×4 matrices. The theory involves solving this matrix equation to determine a gravitational field. The German physicist Werner Heisenberg made use of matrices in his development of quantum mechanics, for which he received a Nobel Prize. The theory of matrices is a cornerstone of two of the foremost physical theories of the twentieth century.

EXERCISE SET 2.3

Computation

1. Determine the transpose of each of the following matrices. Indicate whether or not the matrix is symmetric.

(a) $A = \begin{bmatrix} -1 & 2 \\ 2 & -3 \end{bmatrix}$

(b) $B = \begin{bmatrix} 1 & 2 \\ 0 & 3 \end{bmatrix}$

(c) $C = \begin{bmatrix} 3 & -1 \\ 2 & 4 \end{bmatrix}$

(d) $D = \begin{bmatrix} 4 & 5 \\ -2 & 3 \\ 7 & 0 \end{bmatrix}$

(e) $E = \begin{bmatrix} 4 & 5 & 6 \\ -1 & 2 & 3 \\ 0 & 1 & 2 \end{bmatrix}$

(f) $F = \begin{bmatrix} 1 & -1 & 3 \\ -1 & 2 & 0 \\ 3 & 0 & 4 \end{bmatrix}$

(g) $G = \begin{bmatrix} -2 & 4 & 5 & 7 \\ 1 & 0 & 3 & -7 \end{bmatrix}$

(h) $H = \begin{bmatrix} 1 & -2 & 3 \\ 4 & 5 & 6 \\ -2 & 6 & 7 \end{bmatrix}$

(i) $K = \begin{bmatrix} 7 & 0 & 0 \\ 0 & -3 & 0 \\ 0 & 0 & 9 \end{bmatrix}$

2. Each of the following matrices is to be symmetric. Determine the elements indicated with a •.

(a) $\begin{bmatrix} 1 & 2 & 4 \\ • & 6 & • \\ 4 & 5 & 2 \end{bmatrix}$

(b) $\begin{bmatrix} 3 & 5 & • \\ • & 8 & 4 \\ -3 & • & 3 \end{bmatrix}$

(c) $\begin{bmatrix} -3 & • & 8 & 9 \\ -4 & 7 & • & 7 \\ • & 2 & 6 & 4 \\ • & 7 & • & 9 \end{bmatrix}$

3. If A is 4×1, B is 2×3, C is 2×4, and D is 1×3, determine the sizes of the following matrices, if they exist.

(a) ADB^t
(b) $C^tB - 5AD$
(c) $4CA - (CA)^2$
(d) $(ADB^tC)^2 + I_4$
(e) $(B^t C)^t - AD$

Transpose

4. Prove the following properties of transpose given in Theorem 2.4.

(a) $(A + B)^t = A^t + B^t$
(b) $(cA)^t = cA^t$
(c) $(A^t)^t = A$

5. Prove the following properties of transpose using the results of Theorem 2.4.

(a) $(A + B + C)^t = A^t + B^t + C^t$
(b) $(ABC)^t = C^tB^tA^t$

6. Let A be a diagonal matrix. Prove that $A = A^t$.

7. Let A be a square matrix. Prove that $(A^n)^t = (A^t)^n$.

Symmetric Matrices

8. Prove that a square matrix A is symmetric if and only if $a_{ij} = a_{ji}$ for all elements of the matrix.

9. Let A be a symmetric matrix. Prove that A^t is symmetric.

10. (a) Prove that the sum of two symmetric matrices of the same size is symmetric.

(b) Prove that the scalar multiple of a symmetric matrix is symmetric. Thus, *the set of symmetric matrices of the same size is closed under addition and under scalar multiplication.*

Antisymmetric Matrices

11. A square matrix A is said to be **antisymmetric** if $A = -A^t$.

(a) Give an example of an antisymmetric matrix.
(b) Prove that the diagonal elements of an antisymmetric matrix are zero.
(c) Prove that the sum of two antisymmetric matrices of the same size is an antisymmetric matrix.
(d) Prove that the scalar multiple of an antisymmetric matrix is antisymmetric.

Thus, *the set of antisymmetric matrices of the same size is closed under addition and under scalar multiplication.*

12. If A is a square matrix prove that

(a) $A + A^t$ is symmetric. **(b)** $A - A^t$ is antisymmetric.

13. Prove that any square matrix A can be decomposed into the sum of a symmetric matrix B and an antisymmetric matrix C: $A = B + C$.

14. (a) Prove that if A is idempotent, then A^t is also idempotent.
(b) Prove that if A^t is idempotent, then A is idempotent.

Trace of a Matrix

15. Determine the trace of each of the following matrices.

(a) $\begin{bmatrix} 2 & 3 \\ -1 & -4 \end{bmatrix}$ (b) $\begin{bmatrix} 5 & 1 & 2 \\ 4 & -3 & 5 \\ -7 & 2 & 8 \end{bmatrix}$

(c) $\begin{bmatrix} 0 & -1 & 2 & 3 \\ -4 & 5 & 3 & 2 \\ 1 & 6 & -7 & 2 \\ 3 & 9 & 2 & 1 \end{bmatrix}$

16. Prove the following properties of trace given in Theorem 2.5.

(a) $tr(cA) = ctr(A)$ (b) $tr(AB) = tr(BA)$
(c) $tr(A^t) = tr(A)$

17. Prove the following property of trace using the results of Theorem 2.5. $tr(A + B + C) = tr(A) + tr(B) + tr(C)$

If and Only If Condition

18. Consider two matrices A and B of the same size. Prove that $A = B$ if and only if $A^t = B^t$.

19. Prove that the matrix product AB exists if and only if the number of columns in A is equal to the number of rows in B.

20. Prove that $AB = O_n$ for all $n \times n$ matrices B if and only if $A = O_n$.

Complex Matrices

21. Compute $A + B, AB$, and BA for the matrices

$$A = \begin{bmatrix} 5 & 3 - i \\ 2 + 3i & -5i \end{bmatrix}, B = \begin{bmatrix} -2 + i & 5 + 2i \\ 3 - i & 4 + 3i \end{bmatrix}$$

22. Compute $A + B, AB$, and BA for the matrices

$$A = \begin{bmatrix} 4 + i & 2 - 3i \\ 6 + 2i & 1 - i \end{bmatrix}, B = \begin{bmatrix} 2 + i & -3 \\ 2 & 4 - 5i \end{bmatrix}$$

23. Find the conjugate and conjugate transpose of each of the following matrices. Determine which matrices are hermitian.

$$A = \begin{bmatrix} 2 - 3i & 5i \\ 2 & 5 - 4i \end{bmatrix}, B = \begin{bmatrix} 4 & 5 - i \\ 5 + i & 6 \end{bmatrix},$$

$$C = \begin{bmatrix} 7i & 4 - 3i \\ 6 + 8i & -9 \end{bmatrix}, D = \begin{bmatrix} -2 & 3 - 5i \\ 3 + 5i & 9 \end{bmatrix}$$

24. Find the conjugate and conjugate transpose of each of the following matrices. Determine which matrices are hermitian.

$$A = \begin{bmatrix} 3 & 7 + 2i \\ 7 - 2i & 5 \end{bmatrix}, B = \begin{bmatrix} 3 + 5i & 1 - 2i \\ 1 + 2i & 5 + 6i \end{bmatrix},$$

$$C = \begin{bmatrix} 1 & 2 \\ 2 & 4 \end{bmatrix}, D = \begin{bmatrix} 9 & -3i \\ 3i & 8 \end{bmatrix}$$

25. Prove the following four properties of conjugate transpose.
(a) $(A + B)^* = A^* + B^*$ (b) $(zA)^* = \bar{z}A^*$
(c) $(AB)^* = B^*A^*$ (d) $(A^*)^* = A$

26. Prove that the diagonal elements of a hermitian matrix are real numbers.

Applications

27. The following matrices describe the pottery contents of various graves. For each situation determine possible chronological orderings of the graves and then the pottery types.

(a) $\begin{bmatrix} 1 & 0 \\ 0 & 1 \\ 1 & 1 \end{bmatrix}$ (b) $\begin{bmatrix} 0 & 0 & 1 \\ 1 & 1 & 0 \\ 1 & 0 & 1 \\ 0 & 1 & 0 \end{bmatrix}$

(c) $\begin{bmatrix} 1 & 0 & 1 & 0 \\ 0 & 1 & 1 & 1 \\ 1 & 1 & 1 & 1 \end{bmatrix}$ (d) $\begin{bmatrix} 0 & 0 & 0 & 1 \\ 1 & 1 & 0 & 0 \\ 0 & 0 & 1 & 1 \\ 1 & 0 & 1 & 0 \end{bmatrix}$

(e) $\begin{bmatrix} 1 & 0 & 1 & 0 \\ 1 & 0 & 0 & 0 \\ 0 & 1 & 0 & 1 \\ 0 & 1 & 1 & 0 \end{bmatrix}$ (f) $\begin{bmatrix} 0 & 1 & 0 & 0 \\ 1 & 0 & 1 & 1 \\ 0 & 0 & 1 & 0 \\ 0 & 1 & 0 & 0 \\ 1 & 1 & 0 & 1 \end{bmatrix}$

28. Let $G = AA^t$ and $P = A^tA$, for an arbitrary matrix A.
(a) Prove that G and P are both symmetric matrices.
(b) G and P both have physical interpretation in the archaeological model. Use this physical interpretation to reason that G and P should be symmetric. The mathematical result and the physical interpretation are compatible.

29. Let A be an arbitrary matrix. What information does the ith diagonal element of the matrix AA^t normally give? Discuss.

30. Derive the result for analyzing the pottery in graves. Let A describe the pottery contents of various graves. Prove that:

The element p_{ij} of the matrix $P = A^tA$ gives the number of graves in which the ith and jth types of pottery both appear.

Thus the larger p_{ij}, the closer pottery types i and j are in time. By examining the elements of P the archaeologist can arrive at the chronological order of the pottery.

31. The model introduced here in archaeology is used in sociology to analyze relationships within a group of people. For example, consider the relationship of "friendship" within a group. Assume that all friendships are mutual. Label the people $1, \ldots, n$ and define a square matrix A as follows: $a_{ii} = 0$ for all i (diagonal elements of A are zero)

$$a_{ij} = \begin{cases} 1 & \text{if } i \text{ and } j \text{ are friends} \\ 0 & \text{if } i \text{ and } j \text{ are not friends} \end{cases}$$

(a) Prove that if $F = AA^t$, then f_{ij} is the number of friends that i and j have in common.
(b) Suppose that all friendships are not mutual. How does this affect the model?

2.4 The Inverse of a Matrix and Cryptography

In this section we introduce the concept of matrix inverse. We shall see how an inverse can be used to solve certain systems of linear equations and we shall see the role it plays in implementing color on computer monitors, and in cryptography, the study of codes.

We motivate the idea of the inverse of a matrix by looking at the multiplicative inverse of a real number. If number b is the inverse of a then

$$ab = 1 \quad \text{and} \quad ba = 1.$$

For example $\frac{1}{4}$ is the inverse of 4 and we have

$$4\left(\tfrac{1}{4}\right) = \left(\tfrac{1}{4}\right)4 = 1.$$

These are the ideas that we extend to matrices.

DEFINITION   Let A be an $n \times n$ matrix. If a matrix B can be found such that $AB = BA = I_n$, then A is said to be **invertible** and B is called an **inverse** of A. If such a matrix B does not exist, then A has no inverse.

EXAMPLE 1   Prove that the matrix $A = \begin{bmatrix} 1 & 2 \\ 3 & 4 \end{bmatrix}$ has an inverse $B = \begin{bmatrix} -2 & 1 \\ \frac{3}{2} & -\frac{1}{2} \end{bmatrix}$.

SOLUTION

We have that

$$AB = \begin{bmatrix} 1 & 2 \\ 3 & 4 \end{bmatrix}\begin{bmatrix} -2 & 1 \\ \frac{3}{2} & -\frac{1}{2} \end{bmatrix} = \begin{bmatrix} 1 & 0 \\ 0 & 1 \end{bmatrix} = I_2$$

and

$$BA = \begin{bmatrix} -2 & 1 \\ \frac{3}{2} & -\frac{1}{2} \end{bmatrix}\begin{bmatrix} 1 & 2 \\ 3 & 4 \end{bmatrix} = \begin{bmatrix} 1 & 0 \\ 0 & 1 \end{bmatrix} = I_2$$

Thus $AB = BA = I_2$, proving that the matrix A has an inverse B.

We know that a real number can have at most one inverse. We now see that this is also the case for a matrix.

THEOREM 2.7

If a matrix has an inverse, that inverse is unique.

Proof   Let B and C be inverses of A. Thus $AB = BA = I_n$ and $AC = CA = I_n$. Multiply both sides of the equation $AB = I_n$ by C and use the algebraic properties of matrices.

$$C(AB) = CI_n$$
$$(CA)B = C$$
$$I_n B = C$$
$$B = C$$

Thus an invertible matrix has only one inverse.

Notation

The notation for the inverse of a matrix is similar to that used for the inverse of a real number. Let A be an invertible matrix. We denote its inverse A^{-1}. Thus

$$AA^{-1} = A^{-1}A = I_n$$

Let k be a positive integer. We define A^{-k} to be $(A^{-1})^k$. Therefore

$$A^{-k} = \underbrace{A^{-1}A^{-1}\ldots A^{-1}}_{k \text{ times}}$$

Determining the Inverse of a Matrix

We now derive a method for finding the inverse of a matrix. The method is based on the Gauss-Jordan algorithm. Let A be an invertible matrix. Then $AA^{-1} = I_n$. Let the columns of A^{-1} be $X_1, X_2, \ldots, X_n$, and the columns of I_n be $C_1, C_2, \ldots, C_n$. Express A^{-1} and I_n in terms of their columns,

$$A^{-1} = [X_1 \ X_2 \ldots X_n] \quad \text{and} \quad I_n = [C_1 \ C_2 \ldots C_n]$$

We shall find A^{-1} by finding $X_1, X_2, \ldots, X_n$. Write the equation $AA^{-1} = I_n$ in the form

$$A[X_1 \ X_2 \ldots X_n] = [C_1 \ C_2 \ldots C_n]$$

Using the column form of matrix multiplication,

$$[AX_1 \ AX_2 \ldots AX_n] = [C_1 \ C_2 \ldots C_n]$$

Thus

$$AX_1 = C_1,\ AX_2 = C_2, \ldots,\ AX_n = C_n$$

Therefore $X_1, X_2, \ldots, X_n$ are solutions to the systems $AX = C_1, AX = C_2, \ldots, AX = C_n$, all of which have the same matrix of coefficients A. Solve these systems by using Gauss-Jordan elimination on the large augmented matrix $[A : C_1 C_2 \ldots C_n]$. Since the solutions $X_1, X_2, \ldots, X_n$ are unique (they are the columns of A^{-1}),

$$[A : C_1 \ C_2 \ldots C_n] \approx \cdots \approx [I_n : X_1 \ X_2 \ldots X_n]$$

Thus, when A^{-1} exists,

$$[A : I_n] \approx \cdots \approx [I_n : B] \text{ where } B = A^{-1}.$$

On the other hand, if the reduced echelon form of $[A : I_n]$ is computed and the first part is not of the form I_n then A has no inverse.

We now summarize the results of this discussion.

Finding the Inverse of a Matrix Using Elimination

Let A be an $n \times n$ matrix.

1. Adjoin the identity $n \times n$ matrix I_n to A to form the matrix $[A : I_n]$.
2. Compute the reduced echelon form of $[A : I_n]$. If the reduced echelon form is of the type $[I_n : B]$, then B is the inverse of A. If the reduced echelon form is not of the type $[I_n : B]$, in that the first $n \times n$ submatrix is not I_n, then A has no inverse.

This discussion also leads to a result about the reduced echelon form of an invertible matrix. Suppose A is invertible. Then as $[A : I_n]$ is transformed to $[I_n : B]$ A is transformed to I_n. A is row equivalent to I_n. Conversely, if A is not row equivalent to I_n then $[A : I_n]$ is not row equivalent to a matrix of the form $[I_n : B]$ and is not invertible. Thus

An $n \times n$ matrix A is invertible if and only if it is row equivalent to I_n.

The following example illustrates this method for finding the inverse of a matrix.

EXAMPLE 2 Determine the inverse of the matrix

$$A = \begin{bmatrix} 1 & -1 & -2 \\ 2 & -3 & -5 \\ -1 & 3 & 5 \end{bmatrix}$$

SOLUTION

Applying the method of Gauss-Jordan elimination we get

$$[A : I_n] = \begin{bmatrix} 1 & -1 & -2 & 1 & 0 & 0 \\ 2 & -3 & -5 & 0 & 1 & 0 \\ -1 & 3 & 5 & 0 & 0 & 1 \end{bmatrix} \underset{R3 + R1}{\overset{R2 + (-2)R1}{\approx}} \begin{bmatrix} 1 & -1 & -2 & 1 & 0 & 0 \\ 0 & -1 & -1 & -2 & 1 & 0 \\ 0 & 2 & 3 & 1 & 0 & 1 \end{bmatrix}$$

$$\underset{(-1)R2}{\approx} \begin{bmatrix} 1 & -1 & -2 & 1 & 0 & 0 \\ 0 & 1 & 1 & 2 & -1 & 0 \\ 0 & 2 & 3 & 1 & 0 & 1 \end{bmatrix} \underset{R3 + (-2)R2}{\overset{R1 + R2}{\approx}} \begin{bmatrix} 1 & 0 & -1 & 3 & -1 & 0 \\ 0 & 1 & 1 & 2 & -1 & 0 \\ 0 & 0 & 1 & -3 & 2 & 1 \end{bmatrix}$$

$$\underset{R2 + (-1)R3}{\overset{R1 + R3}{\approx}} \begin{bmatrix} 1 & 0 & 0 & 0 & 1 & 1 \\ 0 & 1 & 0 & 5 & -3 & -1 \\ 0 & 0 & 1 & -3 & 2 & 1 \end{bmatrix}$$

Thus

$$A^{-1} = \begin{bmatrix} 0 & 1 & 1 \\ 5 & -3 & -1 \\ -3 & 2 & 1 \end{bmatrix}$$

The following example illustrates what happens when the method is used for a matrix that does not have an inverse.

EXAMPLE 3 Determine the inverse of the following matrix, if it exists.

$$A = \begin{bmatrix} 1 & 1 & 5 \\ 1 & 2 & 7 \\ 2 & -1 & 4 \end{bmatrix}$$

SOLUTION

Applying the method of Gauss-Jordan elimination we get

$$[A: I_3] = \begin{bmatrix} 1 & 1 & 5 & 1 & 0 & 0 \\ 1 & 2 & 7 & 0 & 1 & 0 \\ 2 & -1 & 4 & 0 & 0 & 1 \end{bmatrix} \begin{matrix} \\ R2 + (-1)R1 \\ R3 + (-2)R1 \end{matrix} \approx \begin{bmatrix} 1 & 1 & 5 & 1 & 0 & 0 \\ 0 & 1 & 2 & -1 & 1 & 0 \\ 0 & -3 & -6 & -2 & 0 & 1 \end{bmatrix}$$

$$\begin{matrix} \\ R1 + (-1)R2 \\ R3 + (3)R2 \end{matrix} \approx \begin{bmatrix} 1 & 0 & 3 & 2 & -1 & 0 \\ 0 & 1 & 2 & -1 & 1 & 0 \\ 0 & 0 & 0 & -5 & 3 & 1 \end{bmatrix}$$

There is no need to proceed further. The reduced echelon form cannot have a one in the $(3, 3)$ location. The reduced echelon form cannot be of the form $[I_n: B]$. Thus A^{-1} does not exist.

■

We now summarize some of the algebraic properties of matrix inverse.

Properties of Matrix Inverse

Let A and B be invertible matrices and c a nonzero scalar. Then

1. $(A^{-1})^{-1} = A$
2. $(cA)^{-1} = \frac{1}{c}A^{-1}$
3. $(AB)^{-1} = B^{-1}A^{-1}$
4. $(A^n)^{-1} = (A^{-1})^n$
5. $(A^t)^{-1} = (A^{-1})^t$

We verify results 1 and 3 to illustrate the techniques involved, leaving the remaining results for the reader to verify in the exercises that follow.

$(A^{-1})^{-1} = A$ This result follows directly from the definition of inverse of a matrix. Since A^{-1} is the inverse of A we have

$$AA^{-1} = A^{-1}A = I_n$$

This statement also tells us that A is the inverse of A^{-1}. Thus $(A^{-1})^{-1} = A$.

$(AB)^{-1} = B^{-1}A^{-1}$ We want to show that the matrix $B^{-1}A^{-1}$ is the inverse of the matrix AB. We get, using the properties of matrices,

$$(AB)(B^{-1}A^{-1}) = A(BB^{-1})A^{-1}$$
$$= AI_nA^{-1}$$
$$= AA^{-1}$$
$$= I_n$$

Similarly, it can be shown that $(B^{-1}A^{-1})(AB) = I_n$. Thus $B^{-1}A^{-1}$ is the inverse of the matrix AB.

EXAMPLE 4 If $A = \begin{bmatrix} 4 & 1 \\ 3 & 1 \end{bmatrix}$, then it can be shown that $A^{-1} = \begin{bmatrix} 1 & -1 \\ -3 & 4 \end{bmatrix}$. Use this information to compute $(A^t)^{-1}$.

SOLUTION

Result 5 above tells us that if we know the inverse of a matrix we also know the inverse of its transpose. We get

$$(A^t)^{-1} = (A^{-1})^t = \begin{bmatrix} 1 & -1 \\ -3 & 4 \end{bmatrix}^t = \begin{bmatrix} 1 & -3 \\ -1 & 4 \end{bmatrix}$$

Systems of Linear Equations

We now see that matrix inverse enables us to conveniently express the solutions to certain systems of linear equations.

THEOREM 2.8

Let $AX = Y$ be a system of n linear equations in n variables. If A^{-1} exists, the solution is unique and is given by $X - A^{-1}Y$.

Proof We first prove that $X = A^{-1}Y$ is a solution.
Substitute $X = A^{-1}Y$ into the matrix equation. Using the properties of matrices we get

$$AX = A(A^{-1}Y) = (AA^{-1})Y = I_nY = Y$$

$X = A^{-1}Y$ satisfies the equation; thus it is a solution.

We now prove the uniqueness of the solution. Let X_1 be any solution. Thus $AX_1 = Y$. Multiplying both sides of this equation by A^{-1} gives

$$A^{-1}AX_1 = A^{-1}Y$$
$$I_nX_1 = A^{-1}Y$$
$$X_1 = A^{-1}Y$$

Thus there is a unique solution $X_1 = A^{-1}Y$.

EXAMPLE 5 Solve the following system of equations using the inverse of the matrix of coefficients.

$$\begin{aligned} x_1 - x_2 - 2x_3 &= 1 \\ 2x_1 - 3x_2 - 5x_3 &= 3 \\ -x_1 + 3x_2 + 5x_3 &= -2 \end{aligned}$$

SOLUTION

This system can be written in the following matrix form,

$$\begin{bmatrix} 1 & -1 & -2 \\ 2 & -3 & -5 \\ -1 & 3 & 5 \end{bmatrix} \begin{bmatrix} x_1 \\ x_2 \\ x_3 \end{bmatrix} = \begin{bmatrix} 1 \\ 3 \\ -2 \end{bmatrix}$$

If the matrix of coefficients is invertible, the unique solution is

$$\begin{bmatrix} x_1 \\ x_2 \\ x_3 \end{bmatrix} = \begin{bmatrix} 1 & -1 & -2 \\ 2 & -3 & -5 \\ -1 & 3 & 5 \end{bmatrix}^{-1} \begin{bmatrix} 1 \\ 3 \\ -2 \end{bmatrix}$$

This inverse has already been found in Example 2. Using that result we get

$$\begin{bmatrix} x_1 \\ x_2 \\ x_3 \end{bmatrix} = \begin{bmatrix} 0 & 1 & 1 \\ 5 & -3 & -1 \\ -3 & 2 & 1 \end{bmatrix} \begin{bmatrix} 1 \\ 3 \\ -2 \end{bmatrix} = \begin{bmatrix} 1 \\ -2 \\ 1 \end{bmatrix}$$

The unique solution is $x_1 = 1$, $x_2 = -2$, $x_3 = 1$.

Numerical Considerations

The knowledge that if A is invertible the solution to the system of equations $AX = Y$ is $X = A^{-1}Y$, is primarily of theoretical importance. It gives an algebraic expression for the solution. It is not usually used in practice to solve a specific system of equations. An elimination method such as Gauss-Jordan elimination or Gaussian elimination of Section 8.1 are more efficient. Most systems of equations are solved on a computer. As mentioned earlier, two factors that are important when using a computer are efficiency and accuracy. To solve a system of n equations, the matrix inverse method requires $n^3 + n^2$ multiplications and $n^3 - n^2$ additions, while Gauss-Jordan elimination requires $\frac{1}{2}n^3 + \frac{1}{2}n^2$ multiplications (half as many) and $\frac{1}{2}n^3 - \frac{1}{2}n$ additions. Thus, for example, for a system of ten equations in ten variables, the matrix inverse method would involve 1,100 multiplications and 900 additions, while Gauss-Jordan elimination would involve 550 multiplications and 495 additions. Furthermore, the more operations that are performed, the larger the possible round-off error. Thus Gauss-Jordan elimination is in general also more accurate than the matrix inverse method.

One may be tempted to assume that given a number of linear systems $AX = Y_1$, $AX = Y_2, \ldots, AX = Y_k$, all having the same invertible matrix of coefficients A, that it would be efficient to calculate A^{-1} and then compute the solutions using $X_1 = A^{-1}Y_1$, $X_2 = A^{-1}Y_2, \ldots, X_n = A^{-1}Y_k$. This approach would involve only one computation of A^{-1} and then a number of matrix multiplications. In general however, it is more efficient and accurate to solve such systems using a large augmented matrix that represents all systems, and an elimination method such as Gauss-Jordan elimination, as discussed earlier.

In certain instances the matrix inverse method is used to arrive at specific solutions. In Section 2.5 we illustrate such a situation in a model for analyzing the interdependence of industries. The elements of the matrix of coefficients in that example lend themselves to an efficient algorithm for computing the inverse.

Elementary Matrices

We now introduce a very useful class of matrices called *elementary matrices*. Row operations and their inverses can be performed using these matrices. This way of implementing row operations is particularly appropriate for computers.

An elementary matrix is one that can be obtained from the identity matrix I_n through a single elementary row operation.

Illustration   Consider the following three row operations T_1, T_2, and T_3 on I_3 (one representing each kind of row operation). They lead to the three elementary matrices E_1, E_2, and E_3.

$$I_3 = \begin{bmatrix} 1 & 0 & 0 \\ 0 & 1 & 0 \\ 0 & 0 & 1 \end{bmatrix}$$

Elementary Row Operation	Corresponding Elementary Matrix
T_1: interchange rows 2 and 3 of I_3.	$E_1 = \begin{bmatrix} 1 & 0 & 0 \\ 0 & 0 & 1 \\ 0 & 1 & 0 \end{bmatrix}$
T_2: multiply row 2 of I_3 by 5.	$E_2 = \begin{bmatrix} 1 & 0 & 0 \\ 0 & 5 & 0 \\ 0 & 0 & 1 \end{bmatrix}$
T_3: add 2 times row1 of I_3 to row 2.	$E_3 = \begin{bmatrix} 1 & 0 & 0 \\ 2 & 1 & 0 \\ 0 & 0 & 1 \end{bmatrix}$

Suppose we want to perform a row operation T on an m × n matrix A. Let E be the elementary matrix obtained from I_n through the operation T. This row operation can be performed by multiplying A by E.

Illustration   Let $A = \begin{bmatrix} a & b & c \\ d & e & f \\ g & h & i \end{bmatrix}$ be an arbitrary 3 × 3 matrix. Consider the three row operations above. Let us show that the corresponding elementary matrices can indeed be used to peform these operations.

Interchange rows 2 and 3 of A: $\begin{bmatrix} 1 & 0 & 0 \\ 0 & 0 & 1 \\ 0 & 1 & 0 \end{bmatrix}\begin{bmatrix} a & b & c \\ d & e & f \\ g & h & i \end{bmatrix} = \begin{bmatrix} a & b & c \\ g & h & i \\ d & e & f \end{bmatrix}$

Multiply row 2 by 5: $\begin{bmatrix} 1 & 0 & 0 \\ 0 & 5 & 0 \\ 0 & 0 & 1 \end{bmatrix}\begin{bmatrix} a & b & c \\ d & e & f \\ g & h & i \end{bmatrix} = \begin{bmatrix} a & b & c \\ 5d & 5e & 5f \\ g & h & i \end{bmatrix}$

Add 2row1 to row 2: $\begin{bmatrix} 1 & 0 & 0 \\ 2 & 1 & 0 \\ 0 & 0 & 1 \end{bmatrix}\begin{bmatrix} a & b & c \\ d & e & f \\ g & h & i \end{bmatrix} = \begin{bmatrix} a & b & c \\ d + 2a & e + 2b & f + 2c \\ g & h & i \end{bmatrix}$

Each row operation has an inverse, namely the row operation that returns the original matrix. The elementary matrices of a row operation and its inverse operation are inverse matrices.

Each elementary matrix is square and invertible.

We now illustrate a way elementary matrices are used to arrive at theoretical results. Matrices that can be obtained from one another by a finite sequence of elementary row operations are said to be *row equivalent*.

If A and B are row equivalent matrices and A is invertible then B is invertible.

Let us prove this result. Since A and B are row equivalent there exists a sequence of row operations $T_1, \ldots, T_n$ such that $B = T_n \circ \cdots \circ T_1(A)$. Let the elementary matrices of these operations be $E_1, \ldots, E_n$. Thus

$$B = E_n \ldots E_1 A$$

The matrices $A, E_1, \ldots, E_n$ are all invertible. Repeatedly applying the property of matrix inverse of a product to the following expression we get

$$A^{-1}E_1^{-1}E_2^{-1}E_3^{-1} \ldots E_n^{-1} = (E_1 A)^{-1}E_2^{-1}E_3^{-1} \ldots E_n^{-1}$$
$$= (E_2 E_1 A)^{-1}E_3^{-1} \ldots E_n^{-1} = \cdots$$
$$= (E_n \ldots E_1 A)^{-1} = B^{-1}$$

Thus B is invertible and the inverse is given by

$$B^{-1} = A^{-1}E_1^{-1}E_2^{-1} \ldots E_n^{-1}$$

Elementary matrices are used in arriving at the so-called *LU* decomposition of certain square matrices. These are decompositions into products of lower (L) and upper (U) triangular matrices; matrices that have zeros above or below the main diagonal. The importance of this decomposition lies in the fact that once it is accomplished for a matrix A, the *LU* form provides a powerful starting point for performing many matrix tasks such as solving equations, computing matrix inverses, and finding determinants of matrices. (See Section 8.2.) *LU* decomposition is for example used extensively in MATLAB (discussed in Appendix D). MATLAB is probably the most widely used matrix software package.

Color Models

A color model in the context of graphics is a method of implementing colors. There are numerous models that are used in practice, such as the RGB model (Red, Green, Blue) used in computer monitors, and the YIQ model used in television screens. An RGB computer signal can be converted to a YIQ television signal using what is known as an NTSC encoder. (NTSC stands for National Television System Committee.) The conversion is accomplished by using the following matrix transformation*

$$\begin{bmatrix} Y \\ I \\ Q \end{bmatrix} = \begin{bmatrix} .299 & .587 & .114 \\ .596 & -.275 & -.321 \\ .212 & -.523 & .311 \end{bmatrix} \begin{bmatrix} R \\ G \\ B \end{bmatrix}$$

Let us look at the RGB model for Microsoft Word. The default text color is black. Let us find the RGB values for black, change the text color to a purple, and find the RGB values for this color. On the right of the tool bar of Microsoft Word observe A▾. The bar under the A is black, indicating the current text color. Point the cursor at this bar. It shows "Font Color (RGB(0, 0, 0))". The RGB setting for black is (0, 0, 0). To change the color select

---

*Numbers in this field are usually written to three decimal places.

the sequence "▼→ More Colors → Custom". A spectrum of colors is displayed. Select a purple hue. The corresponding RGB values are seen to be R = 213, G = 77, B = 187. The bar under the A has now changed to purple and any text entered at the keyboard is in purple. The range of values for each of R, G, and B is 0 to 255, the set of numbers that can be represented by a byte on a computer (note that $2^8 = 256$). You are asked to use the matrix transformation to find the range of Y, I, and Q values in the exercises that follow.

If we enter the RGB values for black, namely R = 0, G = 0, B = 0, into the preceding transformation we find that Y = 0, I = 0, Q = 0. Black has the same RGB and YIQ values. The RGB values R = 213, G = 77, B = 187 for purple become Y = 130.204, I = 45.746, Q = 63.042. These are the YIQ values that would be used to duplicate this purple color on a television screen.

A signal is converted from a television screen to a computer monitor using the inverse of the above matrix,

$$\begin{bmatrix} R \\ G \\ B \end{bmatrix} = \begin{bmatrix} .299 & .587 & .114 \\ .596 & -.275 & -.321 \\ .212 & -.523 & .311 \end{bmatrix}^{-1} \begin{bmatrix} Y \\ I \\ Q \end{bmatrix}$$

That is,

$$\begin{bmatrix} R \\ G \\ B \end{bmatrix} = \begin{bmatrix} 1 & .956 & .620 \\ 1 & -.272 & -.647 \\ 1 & -1.108 & 1.705 \end{bmatrix} \begin{bmatrix} Y \\ I \\ Q \end{bmatrix}$$

Cryptography

In the previous application we talked about two different ways colors are coded. We now turn our attention to coding messages. Cryptography is the process of coding and decoding messages. The word comes from the Greek "kryptos" meaning "hidden." The technique can be traced back to the ancient Greeks. Today governments use sophisticated methods of coding and decoding messages. One type of code that is extremely difficult to break makes use of a large invertible matrix to encode a message. The receiver of the message decodes it using the inverse of the matrix. This first matrix is called the *encoding matrix* and its inverse is called the *decoding matrix*. We illustrate the method for a 3 × 3 matrix.

Let the message be

BUY IBM STOCK

and the encoding matrix be

$$\begin{bmatrix} -3 & -3 & -4 \\ 0 & 1 & 1 \\ 4 & 3 & 4 \end{bmatrix}$$

We assign a number to each letter of the alphabet. For convenience, let us associate each letter with its position in the alphabet. A is 1, B is 2, and so on. Let a space between words be denoted by the number 27. The digital form of the message is

B	U	Y	–	I	B	M	–	S	T	O	C	K
2	21	25	27	9	2	13	27	19	20	15	3	11

Since we are going to use a 3×3 matrix to encode the message we break the digital message up into a sequence of 3×1 column matrices as follows.

$$\begin{bmatrix} 2 \\ 21 \\ 25 \end{bmatrix}, \begin{bmatrix} 27 \\ 9 \\ 2 \end{bmatrix}, \begin{bmatrix} 13 \\ 27 \\ 19 \end{bmatrix}, \begin{bmatrix} 20 \\ 15 \\ 3 \end{bmatrix}, \begin{bmatrix} 11 \\ 27 \\ 27 \end{bmatrix}$$

Observe that it was necessary to add two spaces at the end of the message in order to complete the last matrix. We now put the message into code by multiplying each of the above column matrices by the encoding matrix. This can be conveniently done by writing the given column matrices as columns of a matrix and premultiplying that matrix by the encoding matrix. We get

$$\begin{bmatrix} -3 & -3 & -4 \\ 0 & 1 & 1 \\ 4 & 3 & 4 \end{bmatrix} \begin{bmatrix} 2 & 27 & 13 & 20 & 11 \\ 21 & 9 & 27 & 15 & 27 \\ 25 & 2 & 19 & 3 & 27 \end{bmatrix} = \begin{bmatrix} -169 & -116 & -196 & -117 & -222 \\ 46 & 11 & 46 & 18 & 54 \\ 171 & 143 & 209 & 137 & 233 \end{bmatrix}$$

The columns of this matrix give the encoded message. The message is transmitted in the following linear form.

$$-169, 46, 171, -116, 11, 143, -196, 46, 209, -117, 18, 137, -222, 54, 233$$

To decode the message, the receiver writes this string as a sequence of 3×1 column matrices and repeats the technique using the inverse of the encoding matrix. The inverse of this encoding matrix, the decoding matrix is

$$\begin{bmatrix} 1 & 0 & 1 \\ 4 & 4 & 3 \\ -4 & -3 & -3 \end{bmatrix}$$

Thus, to decode the message

$$\begin{bmatrix} 1 & 0 & 1 \\ 4 & 4 & 3 \\ -4 & -3 & -3 \end{bmatrix} \begin{bmatrix} -169 & -116 & -196 & -117 & -222 \\ 46 & 11 & 46 & 18 & 54 \\ 171 & 143 & 209 & 137 & 233 \end{bmatrix} = \begin{bmatrix} 2 & 27 & 13 & 20 & 11 \\ 21 & 9 & 27 & 15 & 27 \\ 25 & 2 & 19 & 3 & 27 \end{bmatrix}$$

The columns of this matrix, written in linear form, give the original message.

2	21	25	27	9	2	13	27	19	20	15	3	11
B	U	Y	–	I	B	M	–	S	T	O	C	K

Readers who are interested in an introduction to cryptography are referred to *Coding Theory and Cryptography* edited by David Joyner, Springer-Verlag, 2000. This is an excellent collection of articles that contain historical, elementary, and advanced discussions.

EXERCISE SET 2.4

Checking for Matrix Inverse

1. Use the definition $AB = BA = I_2$ of inverse to check whether B is the inverse of A, for each of the following 2×2 matrices A and B.

(a) $A = \begin{bmatrix} 7 & -3 \\ 5 & -2 \end{bmatrix}$, $\quad B = \begin{bmatrix} -2 & 3 \\ -5 & 7 \end{bmatrix}$

(b) $A = \begin{bmatrix} 3 & 4 \\ 5 & 7 \end{bmatrix}$, $\quad B = \begin{bmatrix} 7 & -4 \\ -5 & 3 \end{bmatrix}$

(c) $A = \begin{bmatrix} 2 & -4 \\ -5 & 3 \end{bmatrix}$, $\quad B = \begin{bmatrix} 3 & 1 \\ 5 & 2 \end{bmatrix}$

(d) $A = \begin{bmatrix} 7 & 6 \\ 8 & 7 \end{bmatrix}$, $\quad B = \begin{bmatrix} 7 & -6 \\ -8 & 7 \end{bmatrix}$

2. Use the definition $AB = BA = I_3$ of inverse to check whether B is the inverse of A, for each of the following 3×3 matrices A and B.

(a) $A = \begin{bmatrix} 5 & 0 & 0 \\ 0 & \frac{1}{3} & 0 \\ 0 & 0 & -2 \end{bmatrix}$, $\quad B = \begin{bmatrix} \frac{1}{5} & 0 & 0 \\ 0 & 3 & 0 \\ 0 & 0 & -\frac{1}{2} \end{bmatrix}$

(b) $A = \begin{bmatrix} 1 & 1 & -1 \\ -3 & 2 & -1 \\ 3 & -3 & 2 \end{bmatrix}$, $\quad B = \begin{bmatrix} 1 & 1 & 1 \\ 3 & 5 & 4 \\ 3 & 6 & 5 \end{bmatrix}$

(c) $A = \begin{bmatrix} 0 & 1 & -1 \\ 2 & -2 & -1 \\ -1 & 1 & 1 \end{bmatrix}$, $\quad B = \begin{bmatrix} 1 & 2 & 3 \\ 1 & 1 & 2 \\ 0 & 1 & 1 \end{bmatrix}$

Finding the Inverse of a Matrix

3. Determine the inverse of each of the following 2×2 matrices, if it exists, using the method of Gauss-Jordan elimination.

(a) $\begin{bmatrix} 1 & 0 \\ 2 & 1 \end{bmatrix}$ (b) $\begin{bmatrix} 1 & 2 \\ 9 & 4 \end{bmatrix}$

(c) $\begin{bmatrix} 2 & 1 \\ 4 & 3 \end{bmatrix}$ (d) $\begin{bmatrix} 0 & 1 \\ 1 & 3 \end{bmatrix}$

(e) $\begin{bmatrix} 1 & 2 \\ 3 & 6 \end{bmatrix}$ (f) $\begin{bmatrix} 2 & -3 \\ 6 & -7 \end{bmatrix}$

4. Determine the inverse of each of the following 3×3 matrices, if it exists, using the method of Gauss-Jordan elimination.

(a) $\begin{bmatrix} 1 & 2 & 3 \\ 0 & 1 & 2 \\ 4 & 5 & 3 \end{bmatrix}$ (b) $\begin{bmatrix} 2 & 0 & 4 \\ -1 & 3 & 1 \\ 0 & 1 & 2 \end{bmatrix}$

(c) $\begin{bmatrix} 1 & 2 & -3 \\ 1 & -2 & 1 \\ 5 & -2 & -3 \end{bmatrix}$ (d) $\begin{bmatrix} 1 & 2 & -1 \\ 3 & -1 & 0 \\ 2 & -3 & 1 \end{bmatrix}$

5. Determine the inverse of each of the following 3×3 matrices, if it exists, using the method of Gauss-Jordan elimination.

(a) $\begin{bmatrix} 1 & 2 & 3 \\ 2 & -1 & 4 \\ 0 & -1 & 1 \end{bmatrix}$ (b) $\begin{bmatrix} 1 & 2 & -1 \\ 2 & 4 & -3 \\ 1 & -2 & 0 \end{bmatrix}$

(c) $\begin{bmatrix} 1 & -2 & -1 \\ -2 & 4 & 6 \\ 0 & 0 & 5 \end{bmatrix}$ (d) $\begin{bmatrix} 7 & 0 & 0 \\ 0 & -3 & 0 \\ 0 & 0 & \frac{1}{5} \end{bmatrix}$

6. Determine the inverse of each of the following 4×4 matrices, if it exists, using the method of Gauss-Jordan elimination.

(a) $\begin{bmatrix} -3 & -1 & 1 & -2 \\ -1 & 3 & 2 & 1 \\ 1 & 2 & 3 & -1 \\ -2 & 1 & -1 & -3 \end{bmatrix}$ (b) $\begin{bmatrix} 1 & 1 & 0 & 0 \\ 0 & 1 & 1 & 0 \\ 1 & 0 & 0 & 1 \\ 0 & 0 & 1 & 1 \end{bmatrix}$

(c) $\begin{bmatrix} -1 & 0 & -1 & -1 \\ -3 & -1 & 0 & -1 \\ 5 & 0 & 4 & 3 \\ 3 & 0 & 3 & 2 \end{bmatrix}$

7. If $A = \begin{bmatrix} a & b \\ c & d \end{bmatrix}$, show that $A^{-1} = \dfrac{1}{(ad - bc)}\begin{bmatrix} d & -b \\ -c & a \end{bmatrix}$.

This formula can be quicker than Gauss-Jordan elimination to compute the inverse of a 2×2 matrix. Compute the inverses of the following 2×2 matrices using both methods to see which you prefer.

(a) $\begin{bmatrix} 3 & 8 \\ 1 & 3 \end{bmatrix}$ (b) $\begin{bmatrix} 3 & 5 \\ 2 & 4 \end{bmatrix}$

(c) $\begin{bmatrix} 5 & 6 \\ 3 & 4 \end{bmatrix}$ (d) $\begin{bmatrix} 4 & -6 \\ 2 & -2 \end{bmatrix}$

Systems of Linear Equations

8. Solve the following systems of two equations in two variables by determining the inverse of the matrix of coefficients and then using matrix multiplication.

(a) $\begin{aligned} x_1 + 2x_2 &= 2 \\ 3x_1 + 5x_2 &= 4 \end{aligned}$ (b) $\begin{aligned} x_1 + 5x_2 &= -1 \\ 2x_1 + 9x_2 &= 3 \end{aligned}$

(c) $\begin{aligned} x_1 + 3x_2 &= 5 \\ 2x_1 + x_2 &= 10 \end{aligned}$ (d) $\begin{aligned} 2x_1 + x_2 &= 4 \\ 4x_1 + 3x_2 &= 6 \end{aligned}$

(e) $\begin{aligned} 2x_1 + 4x_2 &= 6 \\ 3x_1 + 8x_2 &= 1 \end{aligned}$ (f) $\begin{aligned} 3x_1 + 9x_2 &= 9 \\ 2x_1 + 7x_2 &= 4 \end{aligned}$

9. Solve the following systems of three equations in three variables by determining the inverse of the matrix of coefficients and then using matrix multiplication.

   (a) $x_1 + 2x_2 - x_3 = 2$
   $x_1 + x_2 + 2x_3 = 0$
   $x_1 - x_2 - x_3 = 1$

   (b) $x_1 - x_2 = 1$
   $x_1 + x_2 + 2x_3 = 2$
   $x_1 + 2x_2 + x_3 = 0$

   (c) $x_1 + 2x_2 + 3x_3 = 1$
   $2x_1 + 5x_2 + 3x_3 = 3$
   $x_1 + 8x_3 = 15$

   (d) $x_1 - 2x_2 + 2x_3 = 3$
   $-x_1 + x_2 + 3x_3 = 2$
   $x_1 - x_2 - 4x_3 = -1$

   (e) $-x_1 + x_2 = 5$
   $-x_1 + x_3 = -2$
   $6x_1 - 2x_2 - 3x_3 = 1$

10. Solve the following system of four equations in four variables by determining the inverse of the matrix of coefficients and then using matrix multiplication.

$$x_1 + x_2 + 2x_3 + x_4 = 5$$
$$2x_1 + 2x_3 + x_4 = 6$$
$$x_2 + 3x_3 - x_4 = 1$$
$$3x_1 + 2x_2 + 2x_4 = 7$$

11. Solve the following systems of equations, all having the same matrix of coefficients, using the matrix inverse method,

$$x_1 + 2x_2 - x_3 = b_1$$
$$x_1 + x_2 + 2x_3 = b_2 \quad \text{for}$$
$$x_1 - x_2 - x_3 = b_3$$

$$\begin{bmatrix} b_1 \\ b_2 \\ b_3 \end{bmatrix} = \begin{bmatrix} 1 \\ 2 \\ 3 \end{bmatrix}, \begin{bmatrix} 0 \\ 1 \\ 4 \end{bmatrix}, \begin{bmatrix} 5 \\ 2 \\ -3 \end{bmatrix} \text{ in turn.}$$

Miscellaneous Results

12. Prove the following properties of matrix inverse that were listed in this section.

   (a) $(cA)^{-1} = \frac{1}{c}A^{-1}$
   (b) $(A^n)^{-1} = (A^{-1})^n$
   (c) $(A^t)^{-1} = (A^{-1})^t$

13. If $A^{-1} = \begin{bmatrix} 2 & 1 \\ 4 & 3 \end{bmatrix}$, find A.

14. If $A^{-1} = \frac{1}{2}\begin{bmatrix} -3 & 2 \\ -10 & 6 \end{bmatrix}$, find A.

15. Consider the matrix $A = \begin{bmatrix} 3 & 1 \\ 5 & 2 \end{bmatrix}$, having inverse $\begin{bmatrix} 2 & -1 \\ -5 & 3 \end{bmatrix}$. Determine

   (a) $(3A)^{-1}$ (b) $(A^2)^{-1}$ (c) A^{-2} (d) $(A^t)^{-1}$

   [*Hint*: Use the algebraic properties of matrix inverse.]

16. If $A = \begin{bmatrix} 5 & 1 \\ 9 & 2 \end{bmatrix}$ then $A^{-1} = \begin{bmatrix} 2 & -1 \\ -9 & 5 \end{bmatrix}$. Use this information to determine

   (a) $(2A^t)^{-1}$ (b) A^{-3} (c) $(AA^t)^{-1}$

17. Find x such that $\begin{bmatrix} 2x & 7 \\ 1 & 2 \end{bmatrix}^{-1} = \begin{bmatrix} 2 & -7 \\ -1 & 4 \end{bmatrix}$.

18. Find x such that $2\begin{bmatrix} 2x & x \\ 5 & 3 \end{bmatrix}^{-1} = \begin{bmatrix} 3 & -2 \\ -5 & 4 \end{bmatrix}$.

19. Find A such that $(4A^t)^{-1} = \begin{bmatrix} 2 & 3 \\ -4 & -4 \end{bmatrix}$.

20. Prove that $(ABC)^{-1} = C^{-1}B^{-1}A^{-1}$.

21. Prove that $(A^tB^t)^{-1} = (A^{-1}B^{-1})^t$.

22. Prove that, in general, $(A + B)^{-1} \neq A^{-1} + B^{-1}$.

23. Prove that if A has no inverse then A^t also has no inverse.

24. Prove that if A is an invertible matrix such that
   (a) $AB = AC$, then $B = C$ (b) $AB = 0$, then $B = 0$

25. Prove that a matrix has no inverse if
   (a) two rows are equal.
   (b) two columns are equal. [*Hint*: Use the transpose.]
   (c) it has a column of zeros.

26. Prove that a diagonal matrix is invertible if and only if all its diagonal elements are nonzero. Can you find a quick way for determining the inverse of an invertible diagonal matrix?

27. Consider the set of invertible 2×2 matrices. Prove that, in general, the sum of two of the matrices is not invertible, and that the scalar multiple is not invertible. (*Hint*: Give examples.) Thus the set of 2×2 invertible matrices is not closed under either addition or scalar multiplication.

28. State (with a brief explanation) whether the following statements are true or false for a square matrix A.
   (a) If A is invertible A^{-1} is invertible.
   (b) If A is invertible A^2 is invertible.
   (c) If A has a zero on the main diagonal it is not invertible.
   (d) If A is not invertible then AB is not invertible.
   (e) A^{-1} is row equivalent to I_n.

Numerical Considerations

29. Let $AX = Y$ be a system of 25 linear equations in 25 variables, where A is invertible. Find the number of multiplications and additions needed to solve this system using (a) Gauss-Jordan elimination, (b) the matrix inverse method.

30. Let A be an invertible 2×2 matrix. Show that it takes the same amount of computation to find the solution to the system of equations $AX = Y$ using Gauss-Jordan elimination as it does to find the first column of the inverse of A. This exercise emphasizes the fact that it is more efficient to solve a system of equations by using Gauss-Jordan elimination than by using the inverse of the matrix of coefficients.

Elementary Matrices

31. Let A be a 3×3 matrix and T_1, T_2, T_3, be the following row operations. T_1: interchange rows 1 and 2; T_2: multiply row 3 by -2; T_3: add 4row 1 to row 3. Show that T_1, T_2, T_3 can be performed using the following elementary matrices E_1, E_2, E_3.

$$E_1 = \begin{bmatrix} 0 & 1 & 0 \\ 1 & 0 & 0 \\ 0 & 0 & 1 \end{bmatrix}, \quad E_2 = \begin{bmatrix} 1 & 0 & 0 \\ 0 & 1 & 0 \\ 0 & 0 & -2 \end{bmatrix},$$

$$E_3 = \begin{bmatrix} 1 & 0 & 0 \\ 0 & 1 & 0 \\ 4 & 0 & 1 \end{bmatrix}$$

32. Let T_1, T_2 be the following row operations. T_1: multiply row 1 by -3. T_2: add 3 times row 2 to row 1. Find the elementary 3×3 matrices of T_1, T_2.

33. Determine the row operation defined by each of the following elementary matrices. Find the inverse of that row operation and use it to find the inverse of the elementary matrix. Arrive at rules that will enable you to quickly write down the inverse of any given elementary matrix.

(a) $\begin{bmatrix} 0 & 0 & 1 \\ 0 & 1 & 0 \\ 1 & 0 & 0 \end{bmatrix}$,   **(b)** $\begin{bmatrix} 0 & 1 & 0 \\ 1 & 0 & 0 \\ 0 & 0 & 1 \end{bmatrix}$,

(c) $\begin{bmatrix} 1 & 0 & 0 \\ 0 & 1 & 0 \\ 0 & 0 & 4 \end{bmatrix}$,   **(d)** $\begin{bmatrix} \frac{1}{3} & 0 & 0 \\ 0 & 1 & 0 \\ 0 & 0 & 1 \end{bmatrix}$,

(e) $\begin{bmatrix} 1 & 0 & 0 \\ 0 & 1 & 0 \\ -4 & 0 & 1 \end{bmatrix}$,   **(f)** $\begin{bmatrix} 1 & 0 & 0 \\ 0 & 1 & 0 \\ 0 & 3 & 1 \end{bmatrix}$

Color Model

34. In the color model discussion we indicated that the range of each of the RGB values is 0 to 255. The interval for each of Y, I, and Q, however, is different. Use the encoding matrix equation to find the range of each of Y, I, and Q.

35. Consider the following YIQ values: $(176, -111, -33)$, $(184, 62, -18)$, $(171, 5, -19)$, and $(165, -103, -23)$.

(a) Use the inverse matrix transformation to find the corresponding RGB values. **(b)** Use Microsoft Word to find the colors corresponding to these YIQ values; describe them in your own terms.

36. Black-and-white television monitors use only the Y signal. **(a)** Show that every YIQ signal of the form $(s, 0, 0)$ transforms into an RGB signal of the form (s, s, s). What is there about the matrix of the transformation from YIQ to RGB that makes this happen? **(b)** Find the RGB signals correponding to the YIQ signals $(255, 0, 0)$, $(200, 0, 0)$, $(150, 0, 0)$, $(100, 0, 0)$, $(0, 0, 0)$ of a black-and-white set. Use Microsoft Word to investigate these signals. Describe the effect of decreasing a in the television signal $(a, 0, 0)$ from 255 to zero.

Cryptography

In Exercises 37–42 associate each letter with its position in the alphabet.

37. Encode the message RETREAT using the matrix $\begin{bmatrix} 4 & -3 \\ 3 & -2 \end{bmatrix}$.

38. Encode the message THE BRITISH ARE COMING using the matrix

$$\begin{bmatrix} 1 & 2 & 1 \\ 2 & 3 & 1 \\ -2 & 0 & 1 \end{bmatrix}.$$

39. Decode the message 49, 38, -5, -3, -61, -39, which was encoded using the matrix of Exercise 37.

40. Decode the message 71, 100, -1, 28, 43, -5, 84, 122, -11, 63, 98, -27, 69, 102, -12, 88, 126, -3, which was encoded using the matrix of Exercise 38.

41. Intelligence sources give the information that the message BOSTON CAFE AT TWO was sent as 32, 47, 59, 79, 43, 57, 33, 36, 13, 19, 59, 86, 41, 61, 67, 87, 53, 68. **(a)** Find the encoding matrix. **(b)** Decode the message 43, 64, 49, 70, 59, 79, 39, 45, 45, 63, 59, 79.

42. Base station sends messages to an agent using the encoding matrix $A = \begin{bmatrix} 4 & -3 \\ 3 & -2 \end{bmatrix}$. The agent sends messages to an informer using the encoding matrix $B = \begin{bmatrix} 3 & 8 \\ 4 & 11 \end{bmatrix}$. Find the encoding matrix that is consistent with this communication circle, that enables base to send messages directly to the informer.

*2.5 The Leontief Input-Output Model in Economics

In this section we introduce the Leontief model that is used to analyze the interdependence of economies. The importance of this model to current economic planning was mentioned in the introduction to this chapter.

Consider an economic situation that involves n interdependent industries. The output of any one industry is needed as input by other industries, and even possibly by the industry itself. We shall see how a mathematical model involving a system of linear equations can be constructed to analyze such a situation. Let us assume, for the sake of simplicity, that each industry produces one commodity. Let a_{ij} denote the amount of input of a certain commodity i to produce unit output of commodity j. In our model let the amounts of input and output be measured in dollars. Thus, for example, $a_{34} = 0.45$ means that 45 cents' worth of commodity 3 is required to produce one dollar's worth of commodity 4.

$$a_{ij} = \text{amount of commodity } i \text{ in one dollar of commodity } j$$

The elements a_{ij}, called **input coefficients**, define a matrix A called the **input-output matrix**, which describes the interdependence of the industries.

EXAMPLE 1 National input-output matrices are used to describe interindustry relations that constitute the economic fabric of countries. We now display part of the matrix that describes the interdependency of the U.S. economy for 1972. The economic structure is actually described in terms of the flow among 79 producing sectors; the matrix is thus a 79×79 matrix. We cannot of course display the whole matrix. We list 10 sectors to give the reader a feel for the catagories involved.

1. Livestock and livestock products
2. Agricultural crops
3. Forestry and fishery products
4. Agricultural, forestry, and fishery services
5. Iron and ferroalloy ores mining
6. Nonferrous metal ores mining
7. Coal mining
8. Crude petroleum and natural gas
9. Stone and clay mining and quarrying
10. Chemical and fertilizer mineral mining

The matrix A based on these sectors is

	1	2	3	4 ...
1	0.26110	0.02481	0	0.05278...
2	0.23277	0.03218	0	0.01444
3	0	0	0.00467	0.00294
4	0.02821	0.03673	0.02502	0.02959
5	0	0	0	0
6	0	0	0	0
7	0	0.00002	0	0
8	0	0	0	0
9	0.00001	0.00251	0	0.00034
10	0	0.00130	0	0
⋮	⋮			

Thus, for example, $a_{72} = 0.00002$ implies that \$0.00002 from the coal mining sector (sector 7) goes into producing each \$1 from the the agricultural crops sector (sector 2).

■

We now extend the model to include an open sector. The products of industries may go not only into other producing industries, but also into other nonproducing sectors of the economy such as consumers and governments. All such nonproducing sectors are grouped into what is called the **open sector**. The open sector in the above model of the 1972 U.S. economy included for example, federal, state, and local government purchases. Let

d_i = demand of the open sector from industry i.

x_i = total output of industry i necessary to meet demands of all n industries and the open sector.

a_{ij} is the amount required from industry i to produce unit output in industry j. Thus $a_{ij}x_i$ will be the amount required to produce x_i units of output in industry j. We get

$$x_i \quad = \quad a_{i1}x_1 \quad + \quad a_{i2}x_2 \quad + \cdots + \quad a_{in}x_n \quad + \quad d_i$$

total output of industry i	demand of industry 1	demand of industry 2	demand of industry n	demand of open sector

The output levels required of the entire set of n industries in order to meet these demands are given by the system of n linear equations

$$x_1 = a_{11}x_1 + a_{12}x_2 + \cdots + a_{1n}x_n + d_1$$
$$x_2 = a_{21}x_1 + a_{22}x_2 + \cdots + a_{2n}x_n + d_2$$
$$\vdots$$
$$n_n = a_{n1}x_1 + a_{n2}x_2 + \cdots + a_{nn}x_n + d_n$$

This system of equations can be written in matrix form

$$\begin{bmatrix} x_1 \\ x_2 \\ \vdots \\ x_n \end{bmatrix} = \begin{bmatrix} a_{11} & a_{12} & \cdots & a_{1n} \\ a_{21} & a_{22} & \cdots & a_{2n} \\ \vdots & & & \\ a_{n1} & a_{n2} & \cdots & a_{nn} \end{bmatrix} \begin{bmatrix} x_1 \\ x_2 \\ \vdots \\ x_n \end{bmatrix} + \begin{bmatrix} d_1 \\ d_2 \\ \vdots \\ d_n \end{bmatrix}$$

Let us introduce the following notation.

$$X = \begin{bmatrix} x_1 \\ x_2 \\ \vdots \\ x_n \end{bmatrix} \quad \text{and} \quad D = \begin{bmatrix} d_1 \\ d_2 \\ \vdots \\ d_n \end{bmatrix}$$

the output matrix   the demand matrix

The system of equations can now be written as a single matrix equation with the terms having the following significance:

$$X \quad = \quad AX \quad + \quad D$$

total output	interindustry portion of output	open sector portion of output

When the model is applied to the economy of a country, X represents the total output of each of the producing sectors of the economy and AX describes the contributions made by the various sectors to fullfilling the intersectional input requirements of the economy. D is equal to $(X - AX)$, the difference between total output X and industry transaction AX.

D is thus the GNP of the economy

In practice the equation $X = AX + D$ is applied in a variety of ways, depending on which variables are considered known and which are not known. For example, an analyst seeking to determine the implications of a change in government purchases or consumer demands on the economy described by A might assign values to D and solve the equation for X. The equation could be used in this manner to predict the amount of outputs from each sector needed to attain various GNPs. (Example 2 illustrates this application of the model.) On the other hand, an economist knowing the limited production capacity of an economic system described by A would consider X as known and solve the equation for D, to predict the maximum GNP the system can achieve. (Exercises 7, 8, and 9 following illustrate this application of the model.)

EXAMPLE **2** Consider an economy consisting of three industries having the following input-output matrix A. Determine the output levels required of the industries to meet the demands of the other industries and of the open sector in each case.

$$A = \begin{bmatrix} 0.2 & 0.2 & 0.4 \\ 0.6 & 0.6 & 0 \\ 0 & 0 & 0.2 \end{bmatrix}, \quad D = \begin{bmatrix} 9 \\ 12 \\ 16 \end{bmatrix}, \quad \begin{bmatrix} 6 \\ 9 \\ 8 \end{bmatrix}, \text{ and } \begin{bmatrix} 12 \\ 18 \\ 32 \end{bmatrix} \text{ in turn}$$

The units of D are millions of dollars.

SOLUTION

We wish to compute the output levels X that correspond to the various open sector demands D. X is given by the equation $X = AX + D$. Rewrite as follows.

$$X - AX = D$$
$$(I - A)X = D$$

To solve this equation for X we can use either Gauss-Jordan elimination or the matrix inverse method. In practice the matrix inverse method is used—a discussion of the merits of this approach is given below. We get

$$X = (I - A)^{-1}D$$

This is the equation that is used to determine X when A and D are known. For our matrix A we get

$$I - A = \begin{bmatrix} 1 & 0 & 0 \\ 0 & 1 & 0 \\ 0 & 0 & 1 \end{bmatrix} - \begin{bmatrix} 0.2 & 0.2 & 0.4 \\ 0.6 & 0.6 & 0 \\ 0 & 0 & 0.2 \end{bmatrix} = \begin{bmatrix} 0.8 & -0.2 & -0.4 \\ -0.6 & 0.4 & 0 \\ 0 & 0 & 0.8 \end{bmatrix}$$

$(I - A)^{-1}$ is computed using Gauss-Jordan elimination.

$$(I - A)^{-1} = \begin{bmatrix} 2 & 1 & 1 \\ 3 & 4 & 1.5 \\ 0 & 0 & 1.25 \end{bmatrix}$$

We can efficiently compute $X = (I - A)^{-1}D$ for each of the three values of D by forming a matrix having the various values of D as columns:

$$X = \begin{bmatrix} 2 & 1 & 1 \\ 3 & 4 & 1.5 \\ 0 & 0 & 1.25 \end{bmatrix} \begin{bmatrix} 9 & 6 & 12 \\ 12 & 9 & 18 \\ 16 & 8 & 32 \end{bmatrix} = \begin{bmatrix} 46 & 29 & 74 \\ 99 & 66 & 156 \\ 20 & 10 & 40 \end{bmatrix}$$

$$\underset{(I-A)^{-1}}{\uparrow} \qquad \underset{\substack{\text{various values} \\ \text{of } D}}{\uparrow \quad \uparrow \quad \uparrow} \qquad \underset{\substack{\text{corresponding} \\ \text{outputs}}}{\uparrow \quad \uparrow \quad \uparrow}$$

The output levels necessary to meet the demands

$$\begin{bmatrix} 9 \\ 12 \\ 16 \end{bmatrix}, \quad \begin{bmatrix} 6 \\ 9 \\ 8 \end{bmatrix}, \quad \text{and} \quad \begin{bmatrix} 12 \\ 18 \\ 32 \end{bmatrix} \quad \text{are} \quad \begin{bmatrix} 46 \\ 99 \\ 20 \end{bmatrix}, \quad \begin{bmatrix} 29 \\ 66 \\ 10 \end{bmatrix}, \quad \text{and} \quad \begin{bmatrix} 74 \\ 156 \\ 40 \end{bmatrix}$$

respectively. The units are millions of dollars.

$\blacksquare$

Numerical Considerations   In practice, analyses of this type usually involve many sectors (as we saw in the example of the U.S. economy), implying large input-output matrices. There is usually a great deal of computation involved in implementing the model and an efficient algorithm is needed. The elements of an input-output matrix A are usually zero or very small. This characteristic of A has led to an appropriate numerical method for computing $(I - A)^{-1}$ that makes the matrix inverse method more efficient for solving the system of equations $(I - A)X = D$ than an elimination method. We now describe this method for computing $(I - A)^{-1}$. Consider the following matrix multiplication for any positive integer m.

$$\begin{aligned} (I - A)(I &+ A + A^2 + \cdots + A^m) \\ &= I(I + A + A^2 + \cdots + A^m) - A(I + A + A^2 + \cdots + A^m) \\ &= (I + A + A^2 + \cdots + A^m) - (A + A^2 + A^3 + \cdots + A^{m+1}) \\ &= I - A^{m+1} \end{aligned}$$

The elements of successive powers of A become small rapidly and A^{m+1} approaches the zero matrix. Thus, for an appropriately large m,

$$(I - A)(I + A + A^2 + \cdots + A^m) = I$$

This implies that

$$(I - A)^{-1} = I + A + A^2 + \cdots + A^m$$

This expression is used on a computer to compute $(I - A)^{-1}$ in this model.

Readers who are interested in finding out more about applications of this model should read "The World Economy of the Year 2000" by Wassily W. Leontief, page 166, *Scientific American*, September 1980. The article describes the application of this model to a world economy. The model was commissioned by the United Nations with special financial support from the Netherlands. In the model the world is divided into 15 distinct geographic regions, each one described by an individual input-output matrix. The regions are then linked by a larger matrix that is used in an input-output model. Overall more than 200 economic sectors are included in the model. By feeding in various values economists use the model to create scenarios of future world economic conditions.

EXERCISE SET 2.5

1. Consider the following input-output matrix that defines the interdependency of five industries.

	1	2	3	4	5
1. Auto	0.15	0.10	0.05	0.05	0.10
2. Steel	0.40	0.20	0.10	0.10	0.10
3. Electricity	0.10	0.25	0.20	0.10	0.20
4. Coal	0.10	0.20	0.30	0.15	0.10
5. Chemical	0.05	0.10	0.05	0.02	0.05

Determine

(a) the amount of electricity consumed in producing $1 worth of steel.

(b) the amount of steel consumed in producing $1 worth in the auto industry.

(c) the largest consumer of coal.

(d) the largest consumer of electricity.

(e) On which industry is the auto industry most dependent?

In Exercises 2–6 consider the economies consisting of either two or three industries. Determine the output levels required of each industry in each situation to meet the demands of the other industries and of the open sector.

2. $A = \begin{bmatrix} 0.20 & 0.60 \\ 0.40 & 0.10 \end{bmatrix}$,

$D = \begin{bmatrix} 24 \\ 12 \end{bmatrix}$, $\begin{bmatrix} 8 \\ 6 \end{bmatrix}$, and $\begin{bmatrix} 0 \\ 12 \end{bmatrix}$ in turn

3. $A = \begin{bmatrix} 0.10 & 0.40 \\ 0.30 & 0.20 \end{bmatrix}$,

$D = \begin{bmatrix} 6 \\ 12 \end{bmatrix}$, $\begin{bmatrix} 18 \\ 6 \end{bmatrix}$, and $\begin{bmatrix} 24 \\ 12 \end{bmatrix}$ in turn

4. $A = \begin{bmatrix} 0.30 & 0.60 \\ 0.35 & 0.10 \end{bmatrix}$,

$D = \begin{bmatrix} 42 \\ 84 \end{bmatrix}$, $\begin{bmatrix} 0 \\ 10 \end{bmatrix}$, $\begin{bmatrix} 14 \\ 7 \end{bmatrix}$, and $\begin{bmatrix} 42 \\ 42 \end{bmatrix}$ in turn

5. $A = \begin{bmatrix} 0.20 & 0.20 & 0.10 \\ 0 & 0.40 & 0.20 \\ 0 & 0.20 & 0.60 \end{bmatrix}$,

$D = \begin{bmatrix} 4 \\ 8 \\ 8 \end{bmatrix}$, $\begin{bmatrix} 0 \\ 8 \\ 16 \end{bmatrix}$, and $\begin{bmatrix} 8 \\ 24 \\ 8 \end{bmatrix}$ in turn

6. $A = \begin{bmatrix} 0.20 & 0.20 & 0 \\ 0.40 & 0.40 & 0.60 \\ 0.40 & 0.10 & 0.40 \end{bmatrix}$,

$D = \begin{bmatrix} 36 \\ 72 \\ 36 \end{bmatrix}$, $\begin{bmatrix} 36 \\ 0 \\ 18 \end{bmatrix}$, $\begin{bmatrix} 36 \\ 0 \\ 0 \end{bmatrix}$, and $\begin{bmatrix} 0 \\ 18 \\ 18 \end{bmatrix}$ in turn

In Exercises 7–9 consider the economies consisting of either two or three industries. The output levels of the industries are given. Determine the amounts available for the open sector from each industry.

7. $A = \begin{bmatrix} 0.20 & 0.40 \\ 0.50 & 0.10 \end{bmatrix}$, $X = \begin{bmatrix} 8 \\ 10 \end{bmatrix}$

8. $A = \begin{bmatrix} 0.10 & 0.20 & 0.30 \\ 0 & 0.10 & 0.40 \\ 0.50 & 0.40 & 0.20 \end{bmatrix}$, $X = \begin{bmatrix} 10 \\ 10 \\ 20 \end{bmatrix}$

9. $A = \begin{bmatrix} 0.10 & 0.10 & 0.20 \\ 0.20 & 0.10 & 0.30 \\ 0.40 & 0.30 & 0.15 \end{bmatrix}$, $X = \begin{bmatrix} 6 \\ 4 \\ 5 \end{bmatrix}$

10. Let a_{ij} be an arbitrary element of an input-output matrix. Why would you expect a_{ij} to satisfy the condition $0 \le a_{ij} \le 1$?

11. In an economically feasible situation the sum of the elements of each column of the input-output matrix is less than or equal to unity. Explain why this should be so.

12. Consider a two-industry economy described by an input-output matrix A whose columns add up to one. We can express such a matrix in the form

$$A = \begin{bmatrix} a & 1-b \\ 1-a & b \end{bmatrix}.$$

(a) Show that the matrix $I - A$ has no inverse.

(b) Illustrate this result for the matrix $A = \begin{bmatrix} 0.2 & 0.7 \\ 0.8 & 0.3 \end{bmatrix}$.

(c) What is the implication for an economy described by such a matrix A? (*Hint*: Consider the equation $X = (I - A)^{-1}D$.)

*2.6 Markov Chains, Population Movements, and Genetics

Certain matrices, called **stochastic matrices**, are important in the study of random phenomena where the exact outcome is not known but probabilities can be determined. In this section, we introduce stochastic matrices, derive some of their properties, and give examples of their application. One example is an analysis of population movement between cities and suburbs in the United States. The second example illustrates the use of stochastic matrices in genetics.

At this time we remind the reader of some basic ideas of probability. If the outcome of an event is *sure to occur*, we say that the probability of that outcome is 1. On the other hand, if it *will not occur* we say that the probability is 0. Other probabilities are represented by fractions between 0 and 1; *the larger the fraction, the greater the probability p of that outcome occurring*. Thus we have the restriction $0 \le p \le 1$ on a probability p.

If any one of n completely independent outcomes is equally likely to happen, and if m of these outcomes are of interest to us, then the probability p that one of these outcomes will occur is defined to be the fraction m/n.

As an example, consider the event of drawing a single card from a deck of 52 playing cards. What is the probability that the outcome will be an ace or a king? First of all we see that there are 52 possible outcomes. There are 4 aces and 4 kings in the deck; there are 8 outcomes of interest. Thus the probability of drawing an ace or a king is $\frac{8}{52}$, or $\frac{2}{13}$.

We now introduce matrices whose elements are probabilities.

DEFINITION   A **stochastic matrix** is a square matrix whose elements are probabilities and whose columns add up to 1.

The following matrices are stochastic matrices.

$$\begin{bmatrix} \frac{1}{2} & \frac{1}{3} \\ \frac{1}{2} & \frac{2}{3} \end{bmatrix} \quad \begin{bmatrix} 0 & \frac{3}{4} \\ 1 & \frac{1}{4} \end{bmatrix} \quad \begin{bmatrix} 1 & 0 & \frac{3}{4} \\ 0 & \frac{1}{2} & \frac{1}{8} \\ 0 & \frac{1}{2} & \frac{1}{8} \end{bmatrix}$$

The following matrices are not stochastic.

$$\begin{bmatrix} \frac{1}{2} & 0 \\ \frac{3}{4} & 1 \end{bmatrix} \qquad\qquad \begin{bmatrix} 0 & 2 \\ 1 & \frac{3}{4} \end{bmatrix}$$

the sum of the elements in the   the 2 in the 1st row is not a proba-
first column is not 1   bility since it is greater than 1

A general 2×2 stochastic matrix can be written

$$\begin{bmatrix} x & y \\ 1-x & 1-y \end{bmatrix}$$

where $0 \le x \le 1$ and $0 \le y \le 1$.

Stochastic matrices have the following useful property. (The reader is asked to prove this result for 2×2 stochastic matrices in the exercises that follow.)

THEOREM 2.9

If A and B are stochastic matrices of the same size, then AB is a stochastic matrix.

■

Thus if A is stochastic, then A^2, A^3, A^4, ... are all stochastic.

EXAMPLE 1 Stochastic matrices are used by city planners to analyze trends in land use. Such a matrix has been used by the city of Toronto, for example. The researchers collect data and write them in the form of a stochastic matrix P. The rows and columns of P represent land uses. We illustrate typical categories for a five-year period in the matrix that follows. The element P_{ij} is the probability that land that was in use j in 2000 was in use i in 2005.

Use in 2000

$$\begin{array}{ccccc} 1 & 2 & 3 & 4 & 5 \end{array} \qquad \text{Use in 2005}$$

$$\begin{bmatrix} .4 & .15 & .1 & .05 & .05 \\ .1 & .35 & .3 & .15 & .35 \\ .15 & .15 & .5 & .35 & .20 \\ .1 & .30 & .1 & .4 & .25 \\ .25 & .05 & 0 & .05 & .15 \end{bmatrix} \quad \begin{array}{l} \text{1. Residential} \\ \text{2. Office} \\ \text{3. Commercial} \\ \text{4. Parking} \\ \text{5. Vacant} \end{array}$$

Let us interpret some of the information contained in this matrix. For example, $p_{42} = 0.30$. This tells us that land that was office space in 2000 had a probability of 0.30 of becoming a parking area by 2005. The fourth row of P gives the probabilities that various areas of the city have become parking areas by 2005. These relatively large figures reveal the increasingly dominant role of parking in land use.

The diagonal elements give the probabilities that land use remained in the same category. For example, $p_{22} = 0.35$ is the probability that office land remained office land. The relatively high figures of these diagonal elements reflect the tendency for land to remain in the same broad category of usage.

Perhaps the most interesting statistic in the preceding matrix is that office land within the city in 2000 has such a high probability of becoming residential in 2005; $p_{12} = 0.15$.

EXAMPLE 2 In this example we develop a model of population movement between cities and surrounding suburbs in the United States. The numbers given are based on statistics in *Statistical Abstract of the United States*.

It is estimated that the number of people living in cities in the U.S. during 2007 was 82 million. The number of people living in the surrounding suburbs was 163 million. Let us represent this information by the matrix $X_0 = \begin{bmatrix} 82 \\ 163 \end{bmatrix}$.

Consider the population flow from cities to suburbs. During 2007 the probability of a person staying in the city was 0.96. Thus the probability of moving to the suburbs was 0.04 (assuming that all those who moved went to the suburbs). Consider now the reverse population flow, from suburbia to the city. The probability of a person moving to the city was 0.01; the probability of remaining in suburbia was 0.99. These probabilities can be written as the elements of a stochastic matrix P:

$$\begin{array}{cc} \text{(from)} & \text{(to)} \\ \text{city} \quad \text{suburb} & \end{array}$$

$$P = \begin{bmatrix} 0.96 & 0.01 \\ 0.04 & 0.99 \end{bmatrix} \begin{array}{l} \text{city} \\ \text{suburb} \end{array}$$

The probability of moving from location A to location B is given by the element in column A and row B. In this context, the stochastic matrix is called a **matrix of transition probabilities**.

Now consider the population distribution in 2008, one year later:

city population in 2008 = people who remained + people who moved
                          from 2007   in from the suburbs

$$= (0.96 \times 82) + (0.01 \times 163)$$
$$= 80.35 \text{ million.}$$

suburban population in 2008 = people who moved in + people who stayed
                              from the city   from 2007

$$= (0.04 \times 82) + (0.99 \times 163)$$
$$= 164.65 \text{ million.}$$

Note that we can arrive at these numbers using matrix multiplication

$$\begin{bmatrix} 0.96 & 0.01 \\ 0.04 & 0.99 \end{bmatrix} \begin{bmatrix} 82 \\ 163 \end{bmatrix} = \begin{bmatrix} 80.35 \\ 164.65 \end{bmatrix}$$

Using 2007 as the base year, let X_1 be the population in 2008, one year later. We can write

$$X_1 = PX_0$$

Assume that the population flow represented by the matrix P is unchanged over the years. The population distribution X_2 after 2 years is given by

$$X_2 = PX_1$$

After 3 years the population distribution is given by

$$X_3 = PX_2$$

After n years we get

$$X_n = PX_{n-1}$$

The predictions of this model (to four decimal places) are

$$X_0 = \begin{bmatrix} 82 \\ 163 \end{bmatrix} \begin{matrix} \text{city} \\ \text{suburb} \end{matrix}, \qquad X_1 = \begin{bmatrix} 80.35 \\ 164.65 \end{bmatrix}, \qquad X_2 = \begin{bmatrix} 78.7825 \\ 166.2175 \end{bmatrix},$$

$$X_3 = \begin{bmatrix} 77.2934 \\ 167.7066 \end{bmatrix}, \qquad X_4 = \begin{bmatrix} 75.8787 \\ 169.1213 \end{bmatrix},$$

and so on.

Observe how the city population is decreasing annually, while that of the suburbs is increasing. We return to this model in Section 5.2. There we find that the sequence $X_0, X_1,$ $X_2, \ldots$ approaches $\begin{bmatrix} 49 \\ 196 \end{bmatrix}$. If conditions do not change, city population will gradually approach 49 million, while the population of suburbia will approach 196 million.

Further, note that the sequence $X_1, X_2, X_3, \ldots X_n$ can be directly computed from X_0, as follows:

$$X_1 = PX_0, \qquad X_2 = P^2X_0, \qquad X_3 = P^3X_0, \ldots, X_n = P^nX_0$$

The matrix P^n is a stochastic matrix that takes X_0 into X_n, in n steps. This result can be generalized. That is, P^n can be used in this manner to predict the distribution n stages later, from any given distribution.

$$X_{i+n} = P^nX_i$$

P^n is called the **n-step transition matrix**. The (i, j)th element of P^n gives the probability of going from state j to state i in n steps. For example, it can be shown that (writing to 2 decimal places)

$$P^4 = \begin{array}{c} \text{(from)} \\ \begin{array}{cc} \text{city} & \text{suburb} \end{array} \\ \begin{bmatrix} 0.85 & 0.04 \\ 0.15 & 0.96 \end{bmatrix} \end{array} \begin{array}{l} \text{(to)} \\ \\ \text{city} \\ \text{suburb} \end{array}$$

Thus, for instance, the probability of living in the city in 2008 and being in the suburbs 4 years later is 0.15.

The probabilities in this model depend only on the current state of a person—whether the person is living in the city or in suburbia. This type of model, where the probability of going from one state to another depends only on the current state rather than on a more complete historical description, is called a *Markov Chain.**

A modification that allows for possible annual population growth or decrease would give improved estimates of future population distributions. The reader is asked to build such a factor into the model in the exercises that follow.

These concepts can be extended to Markov processes involving more than two states. The following example illustrates a Markov chain involving three states.

EXAMPLE 3 Markov chains are useful tools for scientists in many fields. We now discuss the role of Markov chains in *genetics*.

Genetics is the branch of biology that deals with heredity. It is the study of units called **genes**, which determine the characteristics living things inherit from their parents. The inheritance of such traits as sex, height, eye color, and hair color of human beings, and such traits as petal color and leaf shape of plants, are governed by genes. Because many diseases are inherited, genetics is important in medicine. In agriculture, breeding methods based on genetic principles led to important advances in both plant and animal breeding. High-yield hybrid corn ranks as one of the most important contributions of genetics to increasing food production. We shall discuss a mathematical model developed for analyzing the behavior of traits involving a pair of genes. We illustrate the concepts involved in terms of crossing a pair of guinea pigs.

The traits that we shall study in guinea pigs are the traits of long hair and short hair. The length of hair is governed by a pair of genes that we shall denote A and a. A guinea pig may have any one of the combinations AA, Aa, or aa. (aA is genetically the same as Aa.) Each of these classes is called a **genotype**. The AA type of guinea pig is indistinguishable in appearance from the Aa type—both have long hair—while the aa type has short hair. The A gene is said to **dominate** the a gene. An animal is called **dominant** if it has AA genes, **hybrid** with Aa genes, and **recessive** with aa genes.

---

*Andrei Andreevich Markov (1856–1922) was educated and taught at the University of St. Petersburg, Russia. He made contributions to the mathematical fields of number theory, probability, and function theory. It was said that "he gave distinguished lectures with irreproachable strictness of argument, and developed in his students that mathematical cast of mind that takes nothing for granted." Markov was personally interested in his students, being faculty advisor to a math circle. He developed chains to analyze literary texts, where the states were vowels and consonants. Markov was a man of strong opinions who was involved in politics. When the establishment celebrated the 300th anniversary of the House of Romanov, Markov organized his own celebration of the 200th anniversary of the law of large numbers!

When two guinea pigs are crossed the offspring inherits one gene from each parent in a random manner. Given the genotypes of the parents, we can determine the probabilities of the genotype of the offspring. Consider a given population of guinea pigs. Let us perform a series of experiments in which we *keep crossing offspring with dominant animals only.* Thus we keep crossing AA, Aa, and aa with AA. What are the probabilities of the offspring being AA, Aa, or aa in each of these cases?

Consider the crossing of AA with AA. The offspring will have one gene from each parent, so it will be of type AA. Thus the probabilities of AA, Aa, and aa resulting are 1, 0, and 0 respectively. All offspring have long hair.

Next consider the crossing of Aa with AA. Taking one gene from each parent, we have the possibilities of AA, AA (taking A from the first parent and each A in turn from the second parent), aA, and aA (taking the a from the first parent and each A in turn from the second parent). Thus the probabilities of AA, Aa, and aa, respectively, are $\frac{1}{2}$, $\frac{1}{2}$, and 0, respectively. All offspring again have long hair.

Finally, on crossing aa with AA there is only one possibility, namely aA. Thus the probabilities of AA, Aa, and aa are 0, 1, and 0, respectively.

All offspring resulting from these experiments have long hair. This series of experiments is a Markov chain having transition matrix

$$P = \begin{array}{c} \\ \\ \\ \\ \end{array} \begin{array}{ccc} AA & Aa & aa \\ \begin{bmatrix} 1 & \frac{1}{2} & 0 \\ 0 & \frac{1}{2} & 1 \\ 0 & 0 & 0 \end{bmatrix} & \begin{array}{c} AA \\ Aa \\ aa \end{array} \end{array}$$

Consider an initial population of guinea pigs made up of an equal number of each genotype. Let the initial distribution be $X_0 = \begin{bmatrix} \frac{1}{3} \\ \frac{1}{3} \\ \frac{1}{3} \end{bmatrix}$, representing the fraction of guinea pigs of each type initially. The components of X_1, X_2, X_3, . . . will give the fractions of following generations that are of types AA, Aa, and aa, respectively. We get

$$X_0 = \begin{bmatrix} \frac{1}{3} \\ \frac{1}{3} \\ \frac{1}{3} \end{bmatrix} \begin{array}{c} AA \\ Aa, \\ aa \end{array} \quad X_1 = \begin{bmatrix} \frac{1}{2} \\ \frac{1}{2} \\ 0 \end{bmatrix}, \quad X_2 = \begin{bmatrix} \frac{3}{4} \\ \frac{1}{4} \\ 0 \end{bmatrix}, \quad X_3 = \begin{bmatrix} \frac{7}{8} \\ \frac{1}{8} \\ 0 \end{bmatrix}, \quad X_4 = \begin{bmatrix} \frac{15}{16} \\ \frac{1}{16} \\ 0 \end{bmatrix},$$

and so on.

Observe that the aa type disappears after the initial generation and that the Aa type becomes a smaller and smaller fraction of each successive generation. The sequence in fact approaches the matrix

$$X = \begin{bmatrix} 1 \\ 0 \\ 0 \end{bmatrix} \begin{array}{c} AA \\ Aa \\ aa \end{array}$$

The genotype AA in this model is called an **absorbing state**.

Here we have considered the case of crossing offspring with a dominant animal. The reader is asked to construct a similar model that describes the crossing of offspring with a hybrid in the exercises that follow. Some of the offspring will have long hair and some short hair in that series of experiments.

EXERCISE SET 2.6

Stochastic Matrices

1. State which of the following matrices are stochastic and which are not. Explain why a matrix is not stochastic.

(a) $\begin{bmatrix} \frac{1}{4} & 0 \\ \frac{3}{4} & 1 \end{bmatrix}$ (b) $\begin{bmatrix} \frac{1}{2} & -1 \\ \frac{1}{2} & 2 \end{bmatrix}$

(c) $\begin{bmatrix} \frac{1}{3} & \frac{1}{7} \\ \frac{2}{3} & \frac{5}{7} \end{bmatrix}$ (d) $\begin{bmatrix} 1 & 0 & 0 \\ 0 & 1 & 0 \\ 0 & 0 & 1 \end{bmatrix}$

(e) $\begin{bmatrix} 0 & \frac{3}{8} & 0 \\ \frac{1}{2} & \frac{1}{8} & 1 \\ \frac{1}{2} & \frac{1}{2} & 0 \end{bmatrix}$ (f) $\begin{bmatrix} 0 & \frac{1}{5} & \frac{3}{4} \\ \frac{5}{6} & \frac{2}{5} & \frac{3}{4} \\ \frac{1}{6} & \frac{2}{5} & -\frac{1}{2} \end{bmatrix}$

2. Prove that the product of two 2×2 stochastic matrices is a stochastic matrix.

3. A stochastic matrix, the sum of whose rows is 1, is called a **doubly stochastic matrix**. Give examples of 2×2 and 3×3 doubly stochastic matrices. Is the product of two doubly stochastic matrices doubly stochastic?

Population Movement Models

4. Use the stochastic matrix of Example 1 of this section to answer the following questions:

   (a) What is the probability that land used for residential in 2000 was used for offices in 2005?

   (b) What is the probability that land used for parking in 2000 was in a residential area in 2005?

   (c) Vacant land in 2000 had the highest probability of becoming what kind of land in 2005?

   (d) Which was the most stable usage of land over the period 2000–2005?

5. In the model of Example 2 determine

   (a) the probability of moving from a city to a suburb in two years.

   (b) the probability of moving from a suburb to a city in three years.

6. Construct a model of population flow between metropolitan and nonmetropolitan areas of the United States, given that their respective populations in 2007 were 245 million and 52 million. The probabilities are given by the matrix

$$\begin{matrix} & \text{(from)} & \text{(to)} \\ & \text{city} \quad \text{nonmetro} & \\ \begin{bmatrix} 0.99 & 0.02 \\ 0.01 & 0.98 \end{bmatrix} & \begin{matrix} \text{metro} \\ \text{nonmetro} \end{matrix} \end{matrix}$$

Predict the population distributions of metropolitan and nonmetropolitan areas for the years 2008 through 2010 (in millions, to four decimal places). If a person was living in a metropolitan area in 2007, what is the probability that the person will still be living in a metropolitan area in 2010?

7. Construct a model of population flows between cities, suburbs, and nonmetropolitan areas of the United States. Their respective populations in 2007 were 82 million, 163 million, and 52 million. The stochastic matrix giving the probabilities of the moves is

$$\begin{matrix} & \text{(from)} & & \text{(to)} \\ \text{city} & \text{suburb} & \text{nonmetro} & \\ \begin{bmatrix} 0.96 & 0.01 & 0.015 \\ 0.03 & 0.98 & 0.005 \\ 0.01 & 0.01 & 0.98 \end{bmatrix} & & & \begin{matrix} \text{city} \\ \text{suburb} \\ \text{nonmetro} \end{matrix} \end{matrix}$$

This model is a refinement on the model of the previous exercise in that the metropolitan population is broken down into city and suburb. It is also a more complete model than that of Example 2 of this section, which did not allow for any population outside cities and suburbs.

Predict the populations of city, suburban, and nonmetropolitan areas for 2008, 2009, and 2010. If a person was living in the city in 2007 what is the probability that the person will be living in a nonmetropolitan area in 2009?

8. In the period 2000 to 2007 the total population of the United States increased by 1% per annum. Assume that the population increases annually by 1% during the years immediately following. Build this factor into the model of Example 2 and predict the populations of city and suburbia in 2010.

9. Assume that the populations of U.S. cities due to births, deaths, and immigration increased by 1.2% during the period 2007 to 2010, and that the populations of the suburbs increased by .8% due to these factors. Allow for these increases in the model of Example 2 and predict the population for the year 2010.

10. Consider the population movement model for flow between cities and suburbs of Example 2 of this section. Determine the population distributions for 2002 to 2006— prior to 2007. Is the chain going from 2007 into the past a Markov Chain? What are the characteristics of the matrix that takes one from distribution to distribution into the past.

Miscellaneous Stochastic Models

11. The following stochastic matrix gives occupational transition probabilities.

$$\begin{matrix} \text{(initial generation)} \\ \text{white-collar} \quad \text{manual} \\ \begin{bmatrix} 1 & 0.2 \\ 0 & 0.8 \end{bmatrix} \begin{matrix} \text{white-collar} \\ \text{manual} \end{matrix} \quad \text{(next generation)} \end{matrix}$$

(a) If the father is a manual worker, what is the probability that the son will be a white-collar worker?

(b) If there are 10,000 in the white-collar category and 20,000 in the manual category, what will the distribution be one generation later?

12. The following matrix gives occupational transition probabilities.

$$
\begin{array}{cc}
\text{(initial generation)} \\
\text{nonfarming} \quad \text{farming}
\end{array}
$$

$$
\begin{bmatrix} 1 & 0.4 \\ 0 & 0.6 \end{bmatrix} \begin{array}{l} \text{nonfarming} \\ \text{farming} \end{array} \quad \text{(next generation)}
$$

(a) If the father is a farmer, what is the probability that the son will be a farmer?

(b) If there are 10,000 in the nonfarming category and 1,000 in the farming category at a certain time, what will the distribution be one generation later? Four generations later?

(c) If the father is a farmer, what is the probability that the grandson will be a farmer?

13. A market analysis of car purchasing trends in a certain region has concluded that a family purchases a new car once every 3 years on an average. The buying patterns are described by the matrix

$$
\begin{array}{cc}
\text{small} & \text{large}
\end{array}
$$
$$
P = \begin{bmatrix} 80\% & 40\% \\ 20\% & 60\% \end{bmatrix} \begin{array}{l} \text{small} \\ \text{large} \end{array}
$$

The elements of P are to be interpreted as follows: The first column indicates that of the current small cars, 80% will be replaced with small cars, 20% with a large car. The second column implies that 40% of the current large cars will be replaced with small cars while 60% will be replaced with large cars. Write the elements of P as follows to get a stochastic matrix that defines a Markov chain.

$$
P = \begin{bmatrix} \frac{80}{100} & \frac{40}{100} \\ \frac{20}{100} & \frac{60}{100} \end{bmatrix} = \begin{bmatrix} 0.8 & 0.4 \\ 0.2 & 0.6 \end{bmatrix}
$$

If there are currently 40,000 small cars and 50,000 large cars in the region, what is your prediction of the distribution in 12 years' time?

14. The conclusion of an analysis of voting trends in a certain state is that the voting patterns of successive generations are described by the following matrix P.

$$
\begin{array}{ccc}
\text{Dem.} & \text{Rep.} & \text{Ind.}
\end{array}
$$
$$
P = \begin{bmatrix} 80\% & 20\% & 60\% \\ 15\% & 70\% & 30\% \\ 5\% & 10\% & 10\% \end{bmatrix} \begin{array}{l} \text{Democrat} \\ \text{Republican} \\ \text{Independent} \end{array}
$$

Among the Democrats of one generation, 80% of the next generation are Democrats, 15% are Republican, and 5% are Independents. Express P as a stochastic matrix that defines a Markov chain model of the voting patterns. If there are 2.5 million registered Democrats, 1.5 million registered Republicans, and 0.25 million registered Independents at a certain time, what is the distribution likely to be in the next generation?

15. Determine the transition matrix for a Markov chain that describes the crossing of offspring of guinea pigs with hybrids only. There is no absorbing state in this model. Let the initial matrix $X_0 = \begin{bmatrix} \frac{1}{3} \\ \frac{1}{3} \\ \frac{1}{3} \end{bmatrix}$ be the fraction of guinea pigs of each type initially. Determine the distributions for the next three generations.

16. Let P be a 2×2 symmetric stochastic matrix. Prove that P must be of the form $\begin{bmatrix} x & 1-x \\ 1-x & x \end{bmatrix}$, where $0 \le x \le 1$. If P describes population movement between two states what can you say about the movement?

*2.7 A Communication Model and Group Relationships in Sociology

Many branches of the physical sciences, social sciences, and business use models from *graph theory* to analyze relationships. We introduce the reader to this important area of mathematics, which uses linear algebra, with an example from the field of communication.

Consider a communication network involving five stations, labeled $P_1, \ldots, P_5$. The communication links could be roads, phone lines, or Internet links for example. Certain stations are linked by two-way communication, others by one-way links. Still others may have only indirect communication by way of intermediate stations. Suppose the network of interest is described in Figure 2.9. Lines joining stations represent direct communication links; the arrows give the directions of those links. For example, stations P_1 and

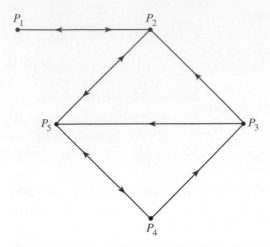

Figure 2.9

P_2 have two-way direct communication. Station P_4 can send a message to P_1 by way of stations P_3 and P_2 or by way of P_5 and P_2. This communication network is an example of a digraph.

DEFINITION   A **digraph** is a finite collection of **vertices** $P_1, P_2, \ldots, P_n$, together with **directed arcs** joining certain pairs of vertices. A **path** between vertices is a sequence of arcs that allows one to proceed in a continuous manner from one vertex to another. The **length** of a path is its number of arcs. A path of length n is called an ***n*-path**.

In the above communication network there are five vertices, namely $P_1, \ldots, P_5$. Suppose that we are interested in sending a message from P_3 to P_1. From the figure we see that there are various paths that can be taken. The path $P_3 \rightarrow P_2 \rightarrow P_1$ is of length 2 (a 2-path), while the path $P_3 \rightarrow P_5 \rightarrow P_2 \rightarrow P_1$ is of length 3 (a 3-path). The path $P_3 \rightarrow P_5 \rightarrow P_4 \rightarrow P_3 \rightarrow P_2 \rightarrow P_1$, a 5-path, takes in the vertex P_3 twice. Such paths could be of interest if, for example, P_3 wanted to consult with P_5 and P_4 before sending a message to P_1. In this section however we shall be interested in finding the shortest route to send a message from one station to another.

Communication networks can be vast, involving many stations. It is impractical to get information about large networks from diagrams, as we have done here. The mathematical theory that we now present, from graph theory, can be used to give information about large networks. The mathematics can be implemented on the computer.

A digraph can be described by a matrix A, called its adjacency matrix. This matrix consists of zeros and ones and is defined as follows.

DEFINITION   Consider a digraph with vertices $P_1, \ldots, P_n$. The **adjacency matrix** A of the digraph is such that

$$a_{ij} = \begin{cases} 1 & \text{if there is an arc from vertex } P_i \text{ to } P_j \\ 0 & \text{if there is no arc from vertex } P_i \text{ to } P_j \\ 0 & \text{if } i = j \end{cases}$$

The adjacency matrix of the communication network is

$$A = \begin{bmatrix} 0 & 1 & 0 & 0 & 0 \\ 1 & 0 & 0 & 0 & 1 \\ 0 & 1 & 0 & 0 & 1 \\ 0 & 0 & 1 & 0 & 1 \\ 0 & 1 & 0 & 1 & 0 \end{bmatrix}$$

For example, $a_{12} = 1$ since there is an arc from P_1 to P_2; $a_{13} = 0$ since there is no arc from P_1 to P_3.

The network is completely described by the adjacency matrix. This matrix is a mathematical "picture" of the network that can be given to a computer. We could look at the sketch of the network given in Figure 2.9 and decide how best to send a message from P_3 to P_1. How can we extract such information from the adjacency matrix? The following theorem from graph theory gives information about paths within digraphs.

THEOREM 2.10

If A is the adjacency matrix of a digraph let $a_{ij}^{(m)}$ be the element in row i and column j of A^m.

$$\text{The number of } m\text{-paths from } P_i \text{ to } P_j = a_{ij}^{(m)}$$

Proof Consider a digraph with n vertices. a_{i1} is the number of arcs from P_i to P_1, and a_{1j} is the number of arcs from P_1 to P_j. Thus $a_{i1} a_{1j}$ is the number of 2-paths from P_i to P_j, passing through P_1. Summing up over all such possible intermediate stations we see that the total number of 2-paths from P_i to P_j is

$$a_{i1}a_{1j} + a_{i2}a_{2j} + \cdots + a_{in}a_{nj}$$

This is the element in row i, column j of A^2. Thus $a_{ij}^{(2)}$ is the number of 2-paths from P_i to P_j.

Let us now look at 3-paths. Interpret a 3-path as a 2-path followed by an arc. The number of 2-paths from P_i to P_1 followed by an arc from P_1 to P_j is $a_{i1}^{(2)} a_{1j}$. Summing up over all such possible intermediate stations we see that the total number of 3-paths from P_i to P_j is

$$a_{i1}^{(2)}a_{1j} + a_{i2}^{(2)}a_{2j} + \cdots + a_{in}^{(2)}a_{nj}$$

This is the element in row i, column j, of the matrix product A^2A; that is of A^3. Thus $a_{ij}^{(3)}$ is the number of 3-paths from P_i to P_j.

Continuing thus, we can interpret a 4-path as a 3-path followed by an arc, and so on, arriving at the result that $a_{ij}^{(m)}$ is the number of m-paths from P_i to P_j.

We now illustrate the application of this theorem to the communication network of Figure 2.9. Successive powers of the adjacency matrix are determined.

$$A = \begin{bmatrix} 0 & 1 & 0 & 0 & 0 \\ 1 & 0 & 0 & 0 & 1 \\ 0 & 1 & 0 & 0 & 1 \\ 0 & 0 & 1 & 0 & 1 \\ 0 & 1 & 0 & 1 & 0 \end{bmatrix} \qquad A^2 = \begin{bmatrix} 1 & 0 & 0 & 0 & 1 \\ 0 & 2 & 0 & 1 & 0 \\ 1 & 1 & 0 & 1 & 1 \\ 0 & 2 & 0 & 1 & 1 \\ 1 & 0 & 1 & 0 & 2 \end{bmatrix}$$

$$A^3 = \begin{bmatrix} 0 & 2 & 0 & 1 & 0 \\ 2 & 0 & 1 & 0 & 3 \\ 1 & 2 & 1 & 1 & 2 \\ 2 & 1 & 1 & 1 & 3 \\ 0 & 4 & 0 & 2 & 1 \end{bmatrix} \quad A^4 = \begin{bmatrix} 2 & 0 & 1 & 0 & 3 \\ 0 & 6 & 0 & 3 & 1 \\ 2 & 4 & 1 & 2 & 4 \\ 1 & 6 & 1 & 3 & 3 \\ 4 & 1 & 2 & 1 & 6 \end{bmatrix} \quad A^5 = \cdots$$

Let us use these matrices to discuss paths from P_4 to P_1.

A gives $a_{41} = 0$. There is no direct communication.

A^2 gives $a_{41}^{(2)} = 0$. There are no 2-paths from P_4 to P_1.

A^3 gives $a_{41}^{(3)} = 2$. There are two distinct 3-paths from P_4 to P_1.

These are the shortest paths from P_4 to P_1. If we check with Figure 2.9, we see that this is the case. The two 3-paths are

$$P_4 \to P_3 \to P_2 \to P_1 \quad \text{and} \quad P_4 \to P_5 \to P_2 \to P_1$$

As a second example, let us determine the length of the shortest path from P_1 to P_3.

A gives $a_{13} = 0$. There is no direct communication.

A^2 gives $a_{13}^{(2)} = 0$. There are no 2-paths from P_1 to P_3.

A^3 gives $a_{13}^{(3)} = 0$. There are no 3-paths from P_1 to P_3.

A^4 gives $a_{13}^{(4)} = 1$. There is a single 4-path from P_1 to P_3.

This result is confirmed when we examine the digraph. The quickest way to send a message from P_1 to P_3 is the 4-path

$$P_1 \to P_2 \to P_5 \to P_4 \to P_3$$

This model that we have discussed gives the lengths of the shortest paths of a digraph; it does not give the intermediate stations that make up that path. Mathematicians have not, as yet, been able to derive this information from the adjacency matrix. An algorithm for finding the shortest paths for a specific digraph, using a search procedure, has been developed by a Dutch computer scientist named Edsger Dijkstra. See, for example, *Algorithms, Practice and Theory* by Gilles Brassard and Paul Bratley, 87, Prentice Hall 1988 for a discussion of this algorithm. The following discussion leads to an application where the lengths of the shortest paths, not the actual paths, are important.

Distance in a Digraph

The **distance** from one vertex of a digraph to another is the length of the shortest path from that vertex to the other. If there is no path from the one vertex to the other we say that the distance is **undefined**. The distances between the various vertices of a digraph form the elements of a matrix. The **distance matrix** D is defined as follows:

$$d_{ij} = \begin{cases} \text{number of arcs in shortest path from vertex } P_i \text{ to vertex } P_j \\ 0 \quad \text{if } i = j \\ x \quad \text{if there is no path from } P_i \text{ to } P_j \end{cases}$$

If the digraph is small the distance matrix can be constructed by observation. Powers of the adjacency matrix are used to construct the distance matrix of a large digraph. The distance matrix of the previous communication network in Figure 2.9 is

$$D = \begin{bmatrix} 0 & 1 & 4 & 3 & 2 \\ 1 & 0 & 3 & 2 & 1 \\ 2 & 1 & 0 & 2 & 1 \\ 3 & 2 & 1 & 0 & 1 \\ 2 & 1 & 2 & 1 & 0 \end{bmatrix}$$

Note that the distance from P_i to P_j is not necessarily equal to the distance from P_j to P_i in a digraph, implying that a distance matrix in graph theory is not necessarily symmetric.

We now illustrate how a digraph and a distance matrix can be used to analyze group relations in sociology.

Group Relationships in Sociology

Consider a group of five people. A sociologist is interested in finding out which one of the five has the most influence over, or dominates, the other members. The group is asked to fill out the following questionnaire:

- Your name   _____

- Person whose opinion you value most   _____

These answers are then tabulated. Let us for convenience label the group members $M_1, M_2, \ldots, M_5$. Suppose the results are as given in Table 2.1.

Table 2.1

Group member	Person whose opinion valued
M_1	M_4
M_2	M_1
M_3	M_2
M_4	M_2
M_5	M_4

The sociologist makes the assumption that the person whose opinion a member values most is the person who influences that member most. Thus influence goes from the right column to the left column in the above table. We can represent these results by a digraph. The group members are represented by vertices, direct influence by an arc, the direction of influence being the direction of the arc. See Figure 2.10. Construct the distance matrix of this digraph, and add up all the elements in each row.

$$D = \begin{bmatrix} 0 & 1 & 2 & 2 & 3 \\ 2 & 0 & 1 & 1 & 2 \\ x & x & 0 & x & x \\ 1 & 2 & 3 & 0 & 1 \\ x & x & x & x & 0 \end{bmatrix} \quad \begin{matrix} \text{row sums} \\ 8 \\ 6 \\ 4x \\ 7 \\ 4x \end{matrix}$$

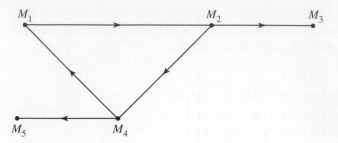

Figure 2.10

In this graph arcs correspond to direct influence, 2-paths, 3-paths, etc. correspond to indirect influence. Thus, presumably, the smaller the distance from M_i to M_j the greater the influence M_i has on M_j. The sum of the elements in row i gives the total distance of M_i to the other vertices. This leads to the following interpretation of row sums.

The smaller row sum i, the greater the influence of person M_i on the group.

We see that the smallest row sum is 6; for row 2. Thus M_2 is the most influential person in the group, followed by M_4 and then M_1. M_3 and M_5 have no influence on other members.

Readers who are interested in learning more about graph theory are referred to the following two books: *Introduction to Graph Theory* by Robin J. Wilson, John Wiley and Sons, 1987. *Discrete Mathematical Structures* by Fred S. Roberts, Prentice Hall, 1976. The former book is a beautiful introduction to the mathematics of graph theory; the latter has a splendid collection of applications. "Predicting Chemistry by Topology" by Dennis H. Rouvray, *Scientific American*, 40, September 1986, contains a fascinating account of how graph theoretical methods are being used to predict chemical properties of molecules that have not yet been synthesized.

We complete this section with a discussion of research that is currently being done in the fields of transportation and sociology.

The Worldwide Air Transportation Network

The global structure of the worldwide air transportation network has been analyzed by a team of researchers led by Luis Amaral of Northwestern University and published in *The Proceedings of the National Academy of Sciences*, Volume 102, pages 7794–7799, 2005. The network is interpreted as a digraph with airports as nodes and nonstop passenger flights between airports as arcs. The study involved 3,883 airports with 531,574 flights. (When there is more than one major airport for a region they are all grouped together. For example Newark, JFK, and LaGuardia airports are all assigned to New York.)

The lengths of the shortest paths (distances) for pairs of airports were computed. The average minimum number of flights that one needs to take to get from any airport to any other airport in the world was found to be 4.4. The farthest cities in the network are Mount Pleasant in the Falkland Islands and Wasu, Papua, New Guinea. A journey in either direction involves 15 different flights. (The *diameter* of the digraph is 15.)

The report discusses two other indexes for this network. The *degree* of an airport is the number of nonstop flights leaving that airport (the number of arcs leaving the node). This is taken to be a measure of the connectedness of the airport. The *betweenness* index of an airport is the number of shortest paths connecting two other airports that involve a transfer at the given airport. This is a measure of the centrality of the airport. It is found that the most connected airports are not necessarily the most central. Although most connected airports are located in western Europe and North America, the most central airports are distributed uniformly across all of the continents. Significantly, each continent has at least one central airport, e.g., Johannesburg in Africa, and Buenos Aires and Sao Paulo in South America. Interestingly Anchorage (Alaska) and Port Moresby, (Papua, New Guinea) have small degrees, as might be expected, but are among the most central airports in the world—Anchorage has few flights out, but many shortest paths between other airports would be disrupted if Anchorage closed down. A list of the most central airports and their betweenness index and degree follows.

Rank	Airport	Betweenness	Degree
1	Paris	58.8	250
2	Anchorage	55.2	39
3	London	54.7	242
4	Singapore	47.5	92
5	New York	47.2	179
6	Los Angeles	44.8	133
7	Port Moresby	43.4	38
8	Frankfurt	41.5	237
9	Tokyo	39.1	111
10	Moscow	34.5	186

Chicago is ranked 13th and Miami 25th, with betweenness indexes of 28.8 and 20.1, respectively, and degrees of 184 and 110. Surprisingly, Atlanta is not listed in the top 25 airports. Readers who are interested in reading more about connectivity of networks should read Section 8.5.

Distance in Social Networks[*]

A team of scientists at Columbia University—Duncan J. Watts, Peter Dodds, and Ruby Muhamad—are studying communication on the Web. The *Small World* project is an online experiment to test the idea that any two people in the world can be connected via "six degrees of separation." Their initial results indicate that most anyone can reach a distant stranger in an average of six relays. Participants were given basic facts (name, location, profession, some educational background) about 18 target individuals in 13 countries and asked to send a message to those targets. The participants sent the message to someone they thought was "closer" (in a message sense) to the target. These recipients were urged to do the same. Calculations based on the results of the experiment, which allowed for attrition, predicted that on average it takes 5 message relays to reach a target in the same country, while it takes 7 to find a target in another country and 6 worldwide. Thinking in terms of a graph, where the vertices are people and directed arcs are transmitted messages, the average distance between two people in the same country is 5, the average distance between two people in different countries is 7, while the worldwide average is 6. The experiment is now being refined and will be repeated. Readers can find the report of this experiment in the journal *Science*, Volume 301, pages 827–829, 2003. The web address for this project is www.smallworld.columbia.edu.

---

[*]I am grateful to Duncan J. Watts for helpful comments on this project.

EXERCISE SET 2.7

Digraphs and Adjacency Matrices

1. Determine the adjacency matrix and the distance matrix of each of the digraphs in Figure 2.11.

2. The **diameter** of a digraph is the largest of the distances between the vertices. Determine the diameters of the digraphs in Figure 2.11.

(a)
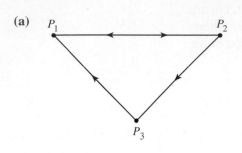

(b)
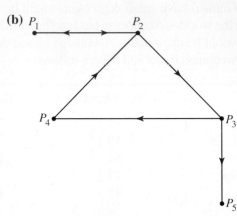

(c)
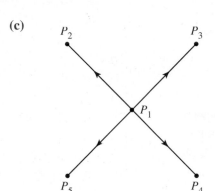

(d)
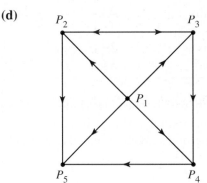

Figure 2.11

3. Sketch the digraphs that have the following adjacency matrices.

(a) $\begin{bmatrix} 0 & 1 & 1 & 1 \\ 1 & 0 & 0 & 0 \\ 1 & 0 & 0 & 0 \\ 1 & 0 & 0 & 0 \end{bmatrix}$
(b) $\begin{bmatrix} 0 & 1 & 1 & 0 \\ 0 & 0 & 1 & 1 \\ 0 & 0 & 0 & 1 \\ 1 & 0 & 0 & 0 \end{bmatrix}$

(c) $\begin{bmatrix} 0 & 1 & 1 & 0 \\ 1 & 0 & 1 & 0 \\ 0 & 0 & 0 & 1 \\ 1 & 1 & 0 & 0 \end{bmatrix}$
(d) $\begin{bmatrix} 0 & 1 & 1 & 0 & 0 \\ 0 & 0 & 1 & 0 & 1 \\ 0 & 0 & 0 & 1 & 0 \\ 0 & 0 & 0 & 0 & 1 \\ 1 & 1 & 0 & 0 & 0 \end{bmatrix}$

(e) $\begin{bmatrix} 0 & 1 & 0 & 1 & 0 \\ 1 & 0 & 1 & 0 & 0 \\ 0 & 1 & 0 & 1 & 0 \\ 1 & 0 & 0 & 0 & 0 \\ 1 & 1 & 1 & 1 & 0 \end{bmatrix}$

Applications of Digraphs

4. The network given in Figure 2.12 describes a system of streets in a city downtown area. Many of the streets are one-way. Interpret the network as a digraph. Give its adjacency matrix and distance matrix.

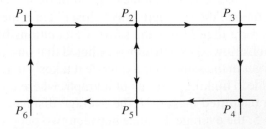

Figure 2.12

5. Graph theory is being used in mathematical models to better understand the delicate balance of nature. Figure 2.13 gives the digraph that describes the food web of an ecological community in the Ocala National Forest, in central Florida. Determine the adjacency matrix.

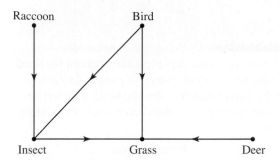

Raccoon   Bird

Insect   Grass   Deer

Figure 2.13

6. When all arcs in a network are two-way, it is customary not to include arrows, since they are not necessary. The term **graph** is then used. Scientists are using graphs to predict the chemical properties of molecules. The graphs in Figure 2.14 describe the molecular structures of butane and isobutane. Both have the same chemical formula C_4H_{10}. Determine the adjacency matrices of these graphs. Note that the matrices are symmetric. Would you expect the adjacency matrix of any graph to be symmetric?

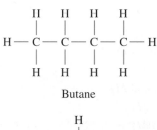

Butane

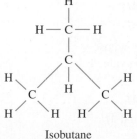

Isobutane

Figure 2.14

7. The following matrix defines a communication network. Sketch the network. Determine the shortest path for sending a message from

(a) P_2 to P_5 **(b)** P_3 to P_2

Find the distance matrix of the digraph.

$$\begin{bmatrix} 0 & 1 & 1 & 0 & 0 \\ 1 & 0 & 1 & 0 & 0 \\ 0 & 0 & 0 & 1 & 0 \\ 0 & 0 & 0 & 0 & 1 \\ 1 & 0 & 0 & 0 & 0 \end{bmatrix}$$

8. In each of the following exercises the matrix A is the adjacency matrix for a communication network. Sketch the networks. Powers of the adjacency matrices are given. Interpret the elements that have been circled.

(a) $A = \begin{bmatrix} 0 & 1 & 0 & 0 \\ 1 & 0 & 0 & 0 \\ 0 & 1 & 0 & 0 \\ 0 & 0 & 1 & 0 \end{bmatrix}$ **(b)** $A = \begin{bmatrix} 0 & 0 & 1 & 0 \\ 0 & 0 & 1 & 0 \\ 0 & 1 & 0 & 0 \\ 0 & 0 & 1 & 0 \end{bmatrix}$

$A^2 = \begin{bmatrix} 1 & 0 & 0 & 0 \\ 0 & ① & 0 & ⓪ \\ ① & 0 & 0 & 0 \\ 0 & ① & 0 & 0 \end{bmatrix}$ $A^2 = \begin{bmatrix} 0 & 1 & 0 & 0 \\ 0 & 1 & 0 & ⓪ \\ 0 & 0 & ① & 0 \\ 0 & ① & 0 & 0 \end{bmatrix}$

$A^3 = \begin{bmatrix} 0 & ① & 0 & 0 \\ 1 & 0 & 0 & ⓪ \\ 0 & ① & 0 & 0 \\ ① & 0 & 0 & 0 \end{bmatrix}$ $A^3 = \begin{bmatrix} 0 & 0 & ① & 0 \\ 0 & 0 & 1 & 0 \\ 0 & ① & 0 & ⓪ \\ 0 & 0 & 1 & 0 \end{bmatrix}$

(c) $A = \begin{bmatrix} 0 & 1 & 0 & 1 \\ 0 & 0 & 1 & 0 \\ 1 & 0 & 0 & 0 \\ 0 & 0 & 1 & 0 \end{bmatrix}$ **(d)** $A = \begin{bmatrix} 0 & 1 & 0 & 0 & 1 \\ 0 & 0 & 1 & 1 & 0 \\ 0 & 0 & 0 & 0 & 0 \\ 0 & 0 & 1 & 0 & 0 \\ 0 & 0 & 0 & 1 & 0 \end{bmatrix}$

$A^2 = \begin{bmatrix} 0 & 0 & ② & 0 \\ ① & 0 & 0 & 0 \\ 0 & 1 & 0 & ① \\ 1 & 0 & 0 & 0 \end{bmatrix}$ $A^2 = \begin{bmatrix} 0 & 0 & 1 & ② & 0 \\ 0 & 0 & ① & 0 & 0 \\ 0 & 0 & 0 & 0 & 0 \\ 0 & 0 & 0 & 0 & 0 \\ 0 & 0 & ① & 0 & 0 \end{bmatrix}$

$A^3 = \begin{bmatrix} ② & 0 & 0 & 0 \\ 0 & 1 & 0 & ① \\ 0 & 0 & ② & 0 \\ 0 & ① & 0 & 1 \end{bmatrix}$ $A^3 = \begin{bmatrix} 0 & 0 & ② & 0 & 0 \\ 0 & 0 & 0 & 0 & 0 \\ ⓪ & 0 & 0 & 0 & 0 \\ 0 & 0 & 0 & 0 & 0 \\ 0 & 0 & 0 & 0 & 0 \end{bmatrix}$

(e) $A = \begin{bmatrix} 0 & 0 & 0 & 0 & 0 \\ 1 & 0 & 0 & 1 & 0 \\ 0 & 1 & 0 & 1 & 0 \\ 0 & 0 & 1 & 0 & 1 \\ 1 & 1 & 0 & 0 & 0 \end{bmatrix}$ $A^2 = \begin{bmatrix} 0 & 0 & 0 & 0 & 0 \\ 0 & 0 & 1 & 0 & ① \\ ① & 0 & 1 & 1 & 1 \\ 1 & ② & 0 & 1 & 0 \\ ① & 0 & ⓪ & ① & 0 \end{bmatrix}$

$A^3 = \begin{bmatrix} 0 & 0 & 0 & 0 & 0 \\ ① & ② & 0 & 1 & 0 \\ 1 & ② & 1 & ① & ① \\ ② & 0 & 1 & 2 & 1 \\ 0 & 0 & ① & 0 & 1 \end{bmatrix}$

Information about Digraphs

9. Let A be the adjacency matrix of a digraph. What do you know about the digraph in each of the following cases?

   (a) The third row of A is all zcros.

   (b) The fourth column of A is all zeros.

   (c) The sum of the elements in the fifth row of A is 3.

   (d) The sum of the elements in the second column of A is 2.

   (e) The second row of A^3 is all zeros.

   (f) The third column of A^4 is all zeros.

10. Consider digraphs with adjacency matrices having the following characteristics. What can you tell about each digraph?

    (a) The second row is all zeros.

    (b) The third column is all zeros.

    (c) The fourth row has all ones except for zero in the diagonal location.

    (d) The fifth column has all ones except for zero in the diagonal location.

    (e) The sum of the elements in the third row is 5.

    (f) The sum of the elements in the second column is 4.

    (g) The number of ones in the matrix is 7.

    (h) The sum of the elements in row 2 of the fourth power is 3.

    (i) The sum of the elements in column 3 of thc fifth powcr is 4.

    (j) The fourth row of the square of the matrix is all zeros.

    (k) The third column of the fourth power is all zeros.

11. Let A be the adjacency matrix of a digraph. Sketch the digraph if A^2 is as follows. Use the digraph to find A^3.

$$A^2 = \begin{bmatrix} 0 & 1 & 0 & 0 \\ 0 & 0 & 1 & 0 \\ 0 & 0 & 0 & 1 \\ 0 & 1 & 0 & 0 \end{bmatrix}$$

[*Hint*: You are given all the 2-paths in the digraph.]

12. Let A be the adjacency matrix of a digraph. Sketch all the possible digraphs described by A if

$$A^2 = \begin{bmatrix} 0 & 0 & 1 & 1 \\ 0 & 0 & 0 & 0 \\ 0 & 0 & 0 & 0 \\ 0 & 0 & 0 & 0 \end{bmatrix}$$

Group Relationships

13. The following tables represent information obtained from questionaires given to groups of people. In each case construct the digraph that describes the leadership structure within the group. Rank the members according to their influence on the group.

(a)

Group member	Person whose opinion valued
M_1	M_4
M_2	M_1
M_3	M_2
M_4	M_2

(b)

Group member	Person whose opinion valued
M_1	M_5
M_2	M_1
M_3	M_2
M_4	M_3
M_5	M_4

(c)

Group member	Person whose opinion valued
M_1	M_4
M_2	M_1 and M_5
M_3	M_2
M_4	M_3
M_5	M_1

(d)

Group member	Person whose opinion valued
M_1	M_5
M_2	M_1
M_3	M_1 and M_4
M_4	M_5
M_5	M_3

14. The following matrices describe the relationship "friendship" between groups of people. $a_{ij} = 1$ if M_i is a friend of M_j; $a_{ij} = 0$ otherwise. Draw the digraphs that describe these relationships. Note that all the matrices are symmetric. What is the significance of this symmetry? Can such a relationship be described by a matrix that is not symmetric?

(a) $A = \begin{bmatrix} 0 & 1 & 0 & 0 & 0 \\ 1 & 0 & 0 & 0 & 1 \\ 0 & 0 & 0 & 1 & 0 \\ 0 & 0 & 1 & 0 & 0 \\ 0 & 1 & 0 & 0 & 0 \end{bmatrix}$

(b) $A = \begin{bmatrix} 0 & 0 & 0 & 0 & 0 & 1 \\ 0 & 0 & 1 & 0 & 0 & 0 \\ 0 & 1 & 0 & 0 & 1 & 0 \\ 0 & 0 & 0 & 0 & 0 & 1 \\ 0 & 0 & 1 & 0 & 0 & 0 \\ 1 & 0 & 0 & 1 & 0 & 0 \end{bmatrix}$

15. A structure in a digraph that is of interest to social scientists is a clique. A **clique** is the largest subset of a digraph consisting of three or more vertices, each pair of which is mutually related. The application of this concept to the relationship "friendship" is immediate: three or more people form a clique if they are all friends and if they do not all have any mutual friendships with any single person outside that set. Give an example of a digraph that contains a clique.

Miscellaneous Results

16. Prove that the adjacency matrix of a digraph is necessarily square.

17. Let A be the adjacency matrix of a digraph. The matrix A^t is a square matrix consisting of zeros and ones. It is also the adjacency matrix of a digraph. How are the digraphs of A and A^t related?

18. If the adjacency matrix of a digraph is symmetric what does this tell you about the digraph?

19. Prove that the shortest path from one vertex of a digraph to another vertex cannot contain any repeated vertices.

20. In a graph with n vertices what is the greatest possible distance between two vertices.

21. Let A be the adjacency matrix of a communication digraph. Let $C = AA^t$. Show that $c_{ij} =$ number of stations that can receive a message directly from both stations i and j.

22. The **reachability matrix** R of a digraph is defined as follows:

$$r_{ij} = \begin{cases} 1 & \text{if there is a path from vertex } P_i \text{ to } P_j \\ 1 & \text{if } i = j \\ 0 & \text{if there is no path from } P_i \text{ to } P_j \end{cases}$$

Determine the reachability matrices of the digraphs of Exercise 8.

23. The reachability matrix of a digraph can be constructed using information from the adjacency matrix and its various powers. How many powers of the adjacency matrix of a digraph having n vertices would have to be calculated to get all the information needed?

24. (a) If the adjacency matrix of a digraph is symmetric does this mean that the reachability matrix is symmetric?

(b) If the reachability matrix of a digraph is symmetric does this mean that the adjacency matrix is symmetric?

25. Let R be the reachability matrix of a communication digraph. Let $r(i)$ be the sum of the elements of row i of R and $c(j)$ be the sum of the elements of column j of R. What information about the digraph do $r(i)$ and $c(j)$ give?

26. Let R be the reachability matrix of a digraph. What information about the digraph does the element in row i, column j of R^2 give?

27. Let R be the reachability matrix of a digraph. What information about the digraph does R^t give?

28. The adjacency matrix A and reachability matrix R of a digraph are both made up of elements that are either zero or one. Can

(a) A and R have the same number of ones?

(b) A have more ones than R?

(c) A have fewer ones than R?

CHAPTER 2 REVIEW EXERCISES

1. Let $A = \begin{bmatrix} 2 & 0 \\ 7 & -1 \end{bmatrix}$, $B = \begin{bmatrix} 7 & 0 \\ -1 & 3 \end{bmatrix}$, $C = \begin{bmatrix} 6 & -1 & 3 \\ 5 & 0 & -2 \end{bmatrix}$, and $D = \begin{bmatrix} 6 \\ 4 \end{bmatrix}$.

Compute the following (if they exist).

(a) $2AB$ **(b)** $AB + C$ **(c)** $BA + AB$
(d) $AD - 3D$ **(e)** $AC + BC$ **(f)** $2DA + B$

2. Let A be a 2×2 matrix, B a 2×2 matrix, C a 2×3 matrix, D a 3×2 matrix, and E a 3×1 matrix. Determine which of the following matrix expressions exist and give the sizes of the resulting matrices when they do exist.

(a) AB **(b)** $(A^2)C$
(c) $B^3 + 3(CD)$ **(d)** $DC + BA$
(e) $DA - 2(DB)$ **(f)** $C - 3D$
(g) $3(BA)(CD) + (4A)(BC)D$

3. If $A = \begin{bmatrix} 1 & -3 \\ 0 & 4 \end{bmatrix}$, $B = \begin{bmatrix} 1 & 2 & -3 \\ 5 & 0 & -1 \end{bmatrix}$, and

$C = \begin{bmatrix} 2 & -4 & 5 \\ 7 & 1 & 0 \end{bmatrix}$, determine the following elements of

$D = 2AB - 3C$, without computing the complete matrix.

(a) d_{12} (b) d_{23}

4. **(a)** Let $A = \begin{bmatrix} 3 & 1 \\ 7 & 2 \end{bmatrix}$ and $B = \begin{bmatrix} 1 & 3 & 6 \\ 2 & 0 & -1 \end{bmatrix}$.

Compute the product AB using the columns of B.

(b) Let $P = \begin{bmatrix} 1 & 2 & 3 \\ 5 & -1 & 4 \end{bmatrix}$, $Q = \begin{bmatrix} 2 \\ -3 \\ 5 \end{bmatrix}$.

Express the product PQ as a linear combination of the columns of P.

(c) Let $A = \begin{bmatrix} 3 & -1 & 2 \\ 0 & 4 & 1 \\ 2 & 5 & 0 \end{bmatrix}$ and $B = \begin{bmatrix} 2 & 4 & 2 \\ -1 & -3 & 7 \\ 0 & 1 & -2 \end{bmatrix}$.

Find all the partitions of B that can be used to calculate AB for the partition of A given here:

$$A = \begin{bmatrix} 3 & -1 & 2 \\ 0 & 4 & 1 \\ 2 & 5 & 0 \end{bmatrix}$$

5. If $A = \begin{bmatrix} 3 & 1 \\ 0 & 2 \end{bmatrix}$, $B = \begin{bmatrix} -2 & 1 \\ 3 & 1 \end{bmatrix}$, and $C = \begin{bmatrix} 0 & 1 \\ 3 & 2 \end{bmatrix}$,

compute each of the following.

(a) $(A^t)^2$ (b) $A^t - B^2$

(c) $AB^3 - 2C^2$ (d) $A^2 - 3A + 4I_2$

6. Consider the following system of equations. You are given two solutions, X_1 and X_2. Generate four other solutions using the operations of addition and scalar multiplication. Find a solution for which $x = 1$, $y = 9$.

$$\begin{aligned} x + 2y + 5z - 3w &= 0 \\ x - 2y - 3z - w &= 0 \\ 5x - 4y - 3z - 8w &= 0 \\ -4x + 8y + 12z + 4w &= 0 \end{aligned}$$

$$X_1 = \begin{bmatrix} 5 \\ 3 \\ -1 \\ 2 \end{bmatrix}, \quad X_2 = \begin{bmatrix} 3 \\ -1 \\ 1 \\ 2 \end{bmatrix}$$

7. Determine the inverse of each of the following matrices, if it exists, using the method of Gauss-Jordan elimination.

(a) $\begin{bmatrix} 1 & 4 \\ 2 & -1 \end{bmatrix}$ (b) $\begin{bmatrix} 0 & 3 & 3 \\ 1 & 2 & 3 \\ 1 & 4 & 6 \end{bmatrix}$ (c) $\begin{bmatrix} 1 & 2 & 3 \\ 2 & 5 & 3 \\ 1 & 0 & 8 \end{bmatrix}$

8. Use the matrix inverse method to solve the following system of equations.

$$\begin{aligned} x_1 + 3x_2 - 2x_3 &= 1 \\ 2x_1 + 5x_2 - 3x_3 &= 5 \\ 23x_1 + 2x_2 - 4x_3 &= 7 \end{aligned}$$

9. Find A such that $3A^{-1} = \begin{bmatrix} 5 & -6 \\ -2 & 3 \end{bmatrix}$.

10. Verify the associative property of multiplication $A(BC) = (AB)C$.

11. **(a)** Let T_1 and T_2 be the following row operations. T_1: interchange rows 1 and 3 of I_3. T_2: add -4 times row 2 of I_3 to row 1. Give the elementary matrices corresponding to T_1 and T_2.

(b) Determine the row operations defined by the elementary matrices

$$\begin{bmatrix} 1 & 0 & 0 \\ 0 & 1 & 0 \\ 2 & 0 & 1 \end{bmatrix} \quad \text{and} \quad \begin{bmatrix} 1 & 0 & 0 \\ 0 & 0 & 1 \\ 0 & 1 & 0 \end{bmatrix}.$$

12. If n is a nonnegative integer and c is a scalar, prove that $(cA)^n = c^n A^n$.

13. Let A be a matrix such that $AA^t = O$. Show that $A = O$.

14. A matrix is said to be *normal* if $AA^t = A^t A$. Prove that all symmetric matrices are normal.

15. A matrix A is *idempotent* if $A^2 = A$. Prove that if A is idempotent then A^t is also idempotent.

16. A matrix A is *nilpotent* if $A^p = 0$ for some positive integer p. The least such integer p is called the *degree of nilpotency*. Prove that if A is nilpotent, then A^t is also nilpotent with the same degree of nilpotency.

17. Prove that if A is symmetric and invertible, then A^{-1} is also symmetric.

18. Prove that a matrix with a row of zeros or a column of zeros has no inverse.

19. Compute $A + B$ and AB for the following matrices, and show that A is hermitian.

$$A = \begin{bmatrix} 2 & 4 - 3i \\ 4 + 3i & -1 \end{bmatrix}, B = \begin{bmatrix} 3 + i & 1 - 2i \\ 2 + 7i & -2 + i \end{bmatrix}$$

20. Prove that every real symmetric matrix is hermitian.

21. The following matrix A describes the pottery contents of various graves. Determine possible chronological orderings of the graves and then the pottery types.

$$A = \begin{bmatrix} 1 & 0 & 0 & 0 \\ 1 & 1 & 0 & 0 \\ 0 & 0 & 1 & 1 \\ 0 & 0 & 0 & 1 \\ 0 & 1 & 1 & 0 \end{bmatrix}$$

22. The following stochastic matrix P gives the probabilities for a certain region of college and noncollege educated households having at least one college educated child. By a college educated household we understand that at least one parent is college educated, while by noncollege educated we mean that neither parent is college educated.

$$P = \begin{array}{cc} & \begin{array}{cc} \text{household} \\ \text{coll ed} \quad \text{noncoll ed} \end{array} \\ \begin{bmatrix} 0.9 & 0.25 \\ 0.1 & 0.75 \end{bmatrix} & \begin{array}{l} \text{college educated} \\ \text{noncollege educated} \end{array} \end{array} \text{child}$$

If there are currently 300,000 college educated households and 750,000 noncollege educated households what is the predicted distribution for two generations hence? What is the probability that a couple that has no college education will have at least one grandchild with a college education?

23. Let A be the adjacency matrix of a digraph. What do you know about the digraph in each of the following cases?

(a) All the elements in the fourth column of A are zero.

(b) The sum of the elements in the third row of A is 2.

(c) The sum of the elements in the second row of A^3 is 4.

(d) The third column of A^2 is all zeros.

(e) The element in the $(4, 4)$ location of A^3 is 2.

(f) The number of nonzero elements in A^4 is 3.

Beijing National Stadium, the world's largest steel structure, was built for the track and field events for the 2008 Summer Olympics and Paralympics held in Beijing, China. More commonly known as the Bird's Nest, the stadium also hosted the Opening and Closing Ceremonies.

CHAPTER

Determinants

3

Associated with every square matrix is a number called its *determinant*. The determinant of a matrix is a tool used in many branches of mathematics, science, and engineering. While the method of Gauss-Jordan elimination enables us to find the inverse of a specific matrix, for example, it does not give us an algebraic formula for the inverse of an arbitrary matrix—a formula that can be used in theoretical work. Determinants provide us with such formulas. Furthermore, criteria for when certain systems of linear equations have unique, none, or many solutions can be stated in terms of determinants.

In the first section we define determinants and use the definition to compute their values. In following sections we find that elementary row operations are useful for working with determinants, and that these operations can in fact be extended to columns. These operations are used to find the properties of determinants.

3.1 Introduction to Determinants

We commence our discussion of determinants by defining the determinant of a 2 × 2 matrix.

DEFINITION   The **determinant** of a 2 × 2 matrix A is denoted $|A|$ and is given by

$$\begin{vmatrix} a_{11} & a_{12} \\ a_{21} & a_{22} \end{vmatrix} = a_{11}a_{22} - a_{12}a_{21}$$

Observe that the determinant of a 2 × 2 matrix is given by *the difference of the products of the two diagonals* of the matrix.

The notation $\det(A)$ is also used for the determinant of A.

EXAMPLE 1 Find the determinant of the matrix

$$A = \begin{bmatrix} 2 & 4 \\ -3 & 1 \end{bmatrix}$$

SOLUTION

Applying the above definition we get

$$\begin{vmatrix} 2 & 4 \\ -3 & 1 \end{vmatrix} = (2 \times 1) - (4 \times (-3)) = 2 + 12 = 14$$

The determinant of a 3×3 matrix is defined in terms of determinants of 2×2 matrices. The determinant of a 4×4 matrix is defined in terms of determinants of 3×3 matrices, and so on. For these definitions we need the following concepts of *minor* and *cofactor*.

DEFINITION   Let A be a square matrix.

The **minor** of the element a_{ij} is denoted M_{ij} and is the determinant of the matrix that remains after deleting row i and column j of A.

The **cofactor** of a_{ij} is denoted C_{ij} and is given by

$$C_{ij} = (-1)^{i+j} M_{ij}$$

Note that the minor and cofactor differ in at most sign.

EXAMPLE 2 Determine the minors and cofactors of the elements a_{11} and a_{32} of the following matrix A.

$$A = \begin{bmatrix} 1 & 0 & 3 \\ 4 & -1 & 2 \\ 0 & -2 & 1 \end{bmatrix}$$

SOLUTION

Applying the above definitions we get the following.

Minor of a_{11}: $M_{11} = \begin{vmatrix} 1 & 0 & 3 \\ 4 & -1 & 2 \\ 0 & -2 & 1 \end{vmatrix} = \begin{vmatrix} -1 & 2 \\ -2 & 1 \end{vmatrix} = (-1 \times 1) - (2 \times (-2)) = 3$

Cofactor of a_{11}: $C_{11} = (-1)^{1+1} M_{11} = (-1)^2(3) = 3$

Minor of a_{32}: $M_{32} = \begin{vmatrix} 1 & 0 & 3 \\ 4 & -1 & 2 \\ 0 & -2 & 1 \end{vmatrix} = \begin{vmatrix} 1 & 3 \\ 4 & 2 \end{vmatrix} = (1 \times 2) - (3 \times 4) = -10$

Cofactor of a_{32}: $C_{32} = (-1)^{3+2} M_{32} = (-1)^5(-10) = 10$

We now define determinants of larger matrices.

> **DEFINITION** The **determinant of a square matrix** is the sum of the products of the elements of the first row and their cofactors.
>
> $$\text{If } A \text{ is } 3 \times 3, \; |A| = a_{11}C_{11} + a_{12}C_{12} + a_{13}C_{13}$$
> $$\text{If } A \text{ is } 4 \times 4, \; |A| = a_{11}C_{11} + a_{12}C_{12} + a_{13}C_{13} + a_{14}C_{14}$$
> $$\vdots$$
> $$\text{If } A \text{ is } n \times n, \; |A| = a_{11}C_{11} + a_{12}C_{12} + a_{13}C_{13} + \cdots + a_{1n}C_{1n}$$
>
> These equations are called **cofactor expansions** of |A|.

EXAMPLE 3 Evaluate the determinant of the following matrix A.

$$A = \begin{bmatrix} 1 & 2 & -1 \\ 3 & 0 & 1 \\ 4 & 2 & 1 \end{bmatrix}$$

SOLUTION

Using the elements of the first row and their corresponding cofactors we get

$$|A| = a_{11}C_{11} + a_{12}C_{12} + a_{13}C_{13}$$

$$= 1(-1)^2 \begin{vmatrix} 0 & 1 \\ 2 & 1 \end{vmatrix} + 2(-1)^3 \begin{vmatrix} 3 & 1 \\ 4 & 1 \end{vmatrix} + (-1)(-1)^4 \begin{vmatrix} 3 & 0 \\ 4 & 2 \end{vmatrix}$$

$$= [(0 \times 1) - (1 \times 2)] - 2[(3 \times 1) - (1 \times 4)] - [(3 \times 2) - (0 \times 4)]$$

$$= -2 + 2 - 6 = -6$$

We have defined the determinant of a matrix in terms of its first row. It can be shown that the determinant can be found according to the following rules, using any row or column.

THEOREM 3.1

The determinant of a square matrix is the sum of the products of the elements of any row or column and their cofactors.

$$\text{ith row expansion:} \qquad |A| = a_{i1}C_{i1} + a_{i2}C_{i2} + \cdots + a_{in}C_{in}$$
$$\text{jth column expansion:} \qquad |A| = a_{1j}C_{1j} + a_{2j}C_{2j} + \cdots + a_{nj}C_{nj}$$

There is a useful rule that can be used to give the sign part, $(-1)^{i+j}$, of the cofactors in these expansions. The rule is summarized in the following array.

$$\begin{bmatrix} + & - & + & - & \cdots \\ - & + & - & + & \cdots \\ + & - & + & - & \cdots \\ \vdots & & & & \end{bmatrix}$$

If, for example, one expands in terms of the second row the signs will be $-\;+\;-$, etc. The signs alternate as one goes along any row or column.

EXAMPLE 4 Find the determinant of the following matrix using the second row.

$$A = \begin{bmatrix} 1 & 2 & -1 \\ 3 & 0 & 1 \\ 4 & 2 & 1 \end{bmatrix}$$

SOLUTION

Expanding the determinant in terms of the second row we get

$$|A| = a_{21}C_{21} + a_{22}C_{22} + a_{23}C_{23}$$

$$= -3\begin{vmatrix} 2 & -1 \\ 2 & 1 \end{vmatrix} + 0\begin{vmatrix} 1 & -1 \\ 4 & 1 \end{vmatrix} - 1\begin{vmatrix} 1 & 2 \\ 4 & 2 \end{vmatrix}$$

$$= -3[(2 \times 1) - (-1 \times 2)] + 0[(1 \times 1) - (-1 \times 4)]$$

$$- 1[(1 \times 2) - (2 \times 4)]$$

$$= -12 + 0 + 6 = -6$$

Note that we have already evaluated this determinant in terms of the first row in Example 3. As is to be expected, we obtained the same value.

The computation in evaluating a determinant can be minimized by expanding in terms of the row or column that contains the most 0's. The following example illustrates this.

EXAMPLE 5 Evaluate the determinant of the following 4×4 matrix.

$$A = \begin{bmatrix} 2 & 1 & 0 & 4 \\ 0 & -1 & 0 & 2 \\ 7 & -2 & 3 & 5 \\ 0 & 1 & 0 & -3 \end{bmatrix}$$

SOLUTION

The third column of this matrix contains the most zeros. Expand the determinant in terms of the third column, we get

$$|A| = a_{13}C_{13} + a_{23}C_{23} + a_{33}C_{33} + a_{43}C_{43}$$

$$= 0(C_{13}) + 0(C_{23}) + 3(C_{33}) + 0(C_{43})$$

$$= +3\begin{vmatrix} 2 & 1 & 4 \\ 0 & -1 & 2 \\ 0 & 1 & -3 \end{vmatrix}$$

The first column of this determinant contains the most zeros. Expand in terms of the first column.

$$|A| = 3(2)\begin{vmatrix} -1 & 2 \\ 1 & -3 \end{vmatrix} = 6(3 - 2) = 6$$

The elements of determinants can be variables. Equations can involve determinants. The following example illustrates such a **determinantal equation**.

EXAMPLE 6 Solve the following equation for the variable x.

$$\begin{vmatrix} x & x+1 \\ -1 & x-2 \end{vmatrix} = 7$$

SOLUTION

Expand the determinant to get the equation

$$x(x-2) - (x+1)(-1) = 7$$

Proceed to simplify this equation and solve for x.

$$x^2 - 2x + x + 1 = 7$$
$$x^2 - x - 6 = 0$$
$$(x+2)(x-3) = 0$$
$$x = -2 \text{ or } 3.$$

There are two solutions to this equation, namely $x = -2$ and $x = 3$.

Computing Determinants of 2×2 and 3×3 Matrices

The determinants of 2×2 and 3×3 matrices can be found quickly using diagonals. For a 2×2 matrix the actual diagonals are used while in the case of a 3×3 matrix the diagonals of an array consisting of the matrix with the two first columns added to the right are used. A determinant is equal to the sum of the diagonal products that go from left to right minus the sum of the diagonal products that go from right to left, as follows.

2 × 2 matrix A 3 × 3 matrix A

 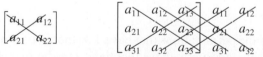

2×2 Matrix:

$$|A| = a_{11}a_{22} - a_{12}a_{21}$$

3×3 Matrix:

$$|A| = a_{11}a_{22}a_{33} + a_{12}a_{23}a_{31} + a_{13}a_{21}a_{32} - a_{13}a_{22}a_{31} - a_{11}a_{23}a_{32} - a_{12}a_{21}a_{33}$$
$$\text{(diagonal products from left to right)} \qquad \text{(diagonal products from right to left)}$$

(We leave it to the reader to show that the expansion of a 3×3 arbitrary determinant can be written thus.)

For example,

A B

$$|A| = (2-12) = -10 \qquad |B| = 0 + 10 + 24 - 0 - 2 - 48 = -16$$

There are no such short cuts for computing determinants of larger matrices.

*Alternative Definition of Determinant

We now introduce an alternative, equivalent definition of determinant that involves permutations. This definition is used mainly for theoretical purposes. We first give the necessary background about permutations.

Let $S = \{1, 2, \ldots, n\}$ be the set of integers from 1 to n. A rearrangement $i_1 i_2 \ldots i_n$ of the elements of S is called a **permutation** of S. Thus, for example, 2314 is a permutation of the set $\{1, 2, 3, 4\}$. $i_1 i_2 \ldots i_n$ is called an **even permutation** if it can be rearranged in the form $123 \ldots n$ through an even number of interchanges. $i_1 i_2 \ldots i_n$ is an **odd permutation** if it can be rearranged in the form $123 \ldots n$ through an odd number of interchanges. (Zero is considered an even number.)

EXAMPLE 7 Show that **(a)** 13425 is an even permutation and **(b)** 13452 is an odd permutation.

SOLUTION

(a) 13425 can be rearranged as follows.

$$13425 \rightarrow 13245 \rightarrow 12345$$

There are two interchanges; an even number. Thus 13425 is an even permutation.

(b) 13452 can be rearranged as follows.

$$13452 \rightarrow 13425 \rightarrow 13245 \rightarrow 12345$$

There are three interchanges; an odd number. 13452 is an odd permutation.

EXAMPLE 8 Give all the permutations of $\{1, 2, 3\}$ and indicate whether the permutation is even or odd.

SOLUTION

Systematically consider the permutations that have 1 as the first element, then those that have 2 as the first element, and finally those that have 3 as the first element. We get

$$123 \text{ (even)}$$
$$132 \rightarrow 123 \text{ (odd)}$$
$$213 \rightarrow 123 \text{ (odd)}$$
$$231 \rightarrow 213 \rightarrow 123 \text{ (even)}$$
$$312 \rightarrow 132 \rightarrow 123 \text{ (even)}$$
$$321 \rightarrow 231 \rightarrow 213 \rightarrow 123 \text{ (odd)}$$

These are all the permutations of the set $\{1, 2, 3\}$.

If the determinant of a matrix is expanded repeatedly, using cofactors, one eventually arrives at a sum of products of elements. The following definition is based on that expression.

DEFINITION   Let A be an $n \times n$ matrix. Then

$$|A| = \sum \pm a_{1j_1} a_{2j_2} \ldots a_{nj_n}$$

Σ means that the terms are to be summed over all permutations $j_1 j_2 \ldots j_n$ of the set $\{1, 2, \ldots, n\}$. The sign $+$ or $-$ is selected for each term according to whether the permutation is even or odd.

EXAMPLE 9 We have given two definitions of determinant, one in terms of cofactors, the other in terms of permutations. Show that these definitions are equivalent when applied to a 2×2 matrix.

SOLUTION

Let A be a 2×2 matrix.
Cofactor definition: This definition gives

$$|A| = \begin{vmatrix} a_{11} & a_{12} \\ a_{21} & a_{22} \end{vmatrix} = a_{11}a_{22} - a_{12}a_{21}$$

Permutation definition: The permutations of $\{1, 2\}$ are 12 (even) and 21 (odd). Thus this definition gives

$$|A| = \sum \pm a_{1j_1}a_{2j_2} = a_{11}a_{22} - a_{12}a_{21}$$

The two definitions agree for 2×2 matrices. You are asked to show that the two definitions agree for 3×3 matrices in Exercise 21.

■

EXERCISE SET 3.1

Determinants of 2 × 2 Matrices

1. Evaluate the determinants of the following 2×2 matrices.

   (a) $\begin{bmatrix} 2 & 1 \\ 3 & 5 \end{bmatrix}$ (b) $\begin{bmatrix} 3 & -2 \\ 1 & 2 \end{bmatrix}$

   (c) $\begin{bmatrix} 4 & 1 \\ -2 & 3 \end{bmatrix}$ (d) $\begin{bmatrix} 5 & -2 \\ -3 & -4 \end{bmatrix}$

2. Evaluate the determinants of the following 2×2 matrices.

   (a) $\begin{bmatrix} 1 & -5 \\ 0 & 3 \end{bmatrix}$ (b) $\begin{bmatrix} 1 & 2 \\ 3 & 4 \end{bmatrix}$

   (c) $\begin{bmatrix} -3 & 1 \\ 2 & -5 \end{bmatrix}$ (d) $\begin{bmatrix} 3 & -2 \\ -1 & 0 \end{bmatrix}$

Minors and Cofactors

3. Let

   $$A = \begin{bmatrix} 1 & 2 & -3 \\ 5 & 0 & 6 \\ 7 & 1 & -4 \end{bmatrix}$$

   Find the following minors and cofactors of A.

   (a) M_{11} and C_{11} (b) M_{21} and C_{21}
   (c) M_{23} and C_{23} (d) M_{33} and C_{33}

4. Let

   $$A = \begin{bmatrix} 5 & 0 & 1 \\ -2 & 3 & 7 \\ 0 & -6 & 2 \end{bmatrix}$$

Find the following minors and cofactors of A.

   (a) M_{13} and C_{13} (b) M_{22} and C_{22}
   (c) M_{31} and C_{31} (d) M_{33} and C_{33}

5. Let

   $$A = \begin{bmatrix} 2 & 0 & 1 & -5 \\ 8 & -1 & 2 & 1 \\ 4 & -3 & -5 & 0 \\ 1 & 4 & 8 & 2 \end{bmatrix}$$

   Find the following minors and cofactors of A.

   (a) M_{12} and C_{12} (b) M_{24} and C_{24}
   (c) M_{33} and C_{33} (d) M_{43} and C_{43}

Determinants of 3 × 3 Matrices

6. Evaluate the determinants of the following matrices (i) using the first row (ii) using the "diagonals" method.

   (a) $\begin{bmatrix} 1 & 2 & 4 \\ 4 & -1 & 5 \\ -2 & 2 & 1 \end{bmatrix}$ (b) $\begin{bmatrix} 3 & 1 & 4 \\ -7 & -2 & 1 \\ 9 & 1 & -1 \end{bmatrix}$

   (c) $\begin{bmatrix} 4 & 1 & -2 \\ 5 & 3 & -1 \\ 2 & 4 & 1 \end{bmatrix}$

7. Evaluate the determinants of the following matrices (i) using the first row (ii) using the "diagonals" method.

(a) $\begin{bmatrix} 2 & 0 & 7 \\ 8 & -1 & -2 \\ 5 & 6 & 1 \end{bmatrix}$ (b) $\begin{bmatrix} 2 & -1 & 3 \\ 4 & 0 & 2 \\ 1 & 1 & 1 \end{bmatrix}$

(c) $\begin{bmatrix} 0 & 0 & 5 \\ 1 & 1 & 1 \\ 2 & 2 & 2 \end{bmatrix}$

8. Evaluate the determinants of each of the following matrices in two ways, using the indicated rows and columns. Observe that you get the same answers both ways.

(a) $\begin{bmatrix} 0 & 3 & 2 \\ 1 & 5 & 7 \\ -2 & -6 & -1 \end{bmatrix}$ (b) $\begin{bmatrix} 4 & 2 & 1 \\ -6 & 3 & -2 \\ 7 & 1 & -1 \end{bmatrix}$

  2nd row and 1st column   3rd row and 2nd column

(c) $\begin{bmatrix} 5 & -1 & 2 \\ 3 & 0 & 6 \\ -4 & 3 & 1 \end{bmatrix}$ (d) $\begin{bmatrix} 6 & 3 & 0 \\ -2 & -1 & 5 \\ 4 & 6 & -2 \end{bmatrix}$

  1st row and 3rd row   2nd column and 3rd column

9. Evaluate the determinants of each of the following matrices in two ways, using the indicated rows and columns. Observe that you get the same answers both ways.

(a) $\begin{bmatrix} 1 & 3 & -1 \\ 2 & 0 & 5 \\ 1 & 4 & 3 \end{bmatrix}$ (b) $\begin{bmatrix} -2 & -1 & 1 \\ 9 & 3 & 2 \\ 4 & 0 & 0 \end{bmatrix}$

  2nd row and 1st column   1st row and 3rd column

(c) $\begin{bmatrix} 1 & 0 & 2 \\ 3 & -2 & 1 \\ 4 & 0 & 2 \end{bmatrix}$ (d) $\begin{bmatrix} 1 & 2 & 3 \\ -1 & -4 & 0 \\ 0 & 0 & 4 \end{bmatrix}$

  1st column and 2nd column   3rd row and 1st column

10. Evaluate the determinants of the following matrices using the row or column that involves least computation.

(a) $\begin{bmatrix} 1 & -2 & 3 \\ 1 & 4 & 0 \\ 2 & -1 & 0 \end{bmatrix}$ (b) $\begin{bmatrix} 3 & -1 & 2 \\ 0 & 4 & 0 \\ -5 & 1 & 9 \end{bmatrix}$

(c) $\begin{bmatrix} 9 & 2 & 1 \\ -3 & 2 & 6 \\ 0 & 0 & -3 \end{bmatrix}$

11. Evaluate the determinants of the following matrices using as little computation as possible.

(a) $\begin{bmatrix} 1 & -2 & 3 & 0 \\ 4 & 0 & 5 & 0 \\ 7 & -3 & 8 & 4 \\ -3 & 0 & 4 & 0 \end{bmatrix}$

(b) $\begin{bmatrix} 1 & 4 & 5 & 9 \\ 2 & 3 & -7 & 1 \\ 0 & 0 & 0 & -3 \\ 0 & 1 & 0 & 8 \end{bmatrix}$

(c) $\begin{bmatrix} 9 & 3 & 7 & -8 \\ 1 & 0 & 4 & 2 \\ 1 & 0 & 0 & -1 \\ -2 & 0 & -1 & 3 \end{bmatrix}$

Equations Involving Determinants

12. Solve the following equation for x.

$$\begin{vmatrix} x+1 & x \\ 3 & x-2 \end{vmatrix} = 3$$

13. Solve the following equation for x.

$$\begin{vmatrix} 2x & -3 \\ x-1 & x+2 \end{vmatrix} = 1$$

14. Find all the values of x that make the following determinant zero.

$$\begin{vmatrix} x-1 & -2 \\ x-2 & x-1 \end{vmatrix}$$

15. Find all the values of x that make the following determinant zero.

$$\begin{vmatrix} x & 0 & 2 \\ 2x & x-1 & 4 \\ -x & x-1 & x+1 \end{vmatrix}$$

Miscellaneous Results

16. Why would you expect the following two determinants to have the same value?

$$\begin{vmatrix} 4 & -1 & 0 \\ 5 & 6 & 3 \\ 2 & 1 & 0 \end{vmatrix} \text{ and } \begin{vmatrix} 4 & -1 & 0 \\ 9 & 2 & 3 \\ 2 & 1 & 0 \end{vmatrix}$$

17. Why would you expect the following determinant to have the same value whatever values are given to a and b?

$$\begin{vmatrix} 3 & 5 & a \\ 9 & 1 & b \\ 0 & 0 & 2 \end{vmatrix}$$

18. Determine whether the following permutations are even or odd.

(a) 4213   (b) 3142   (c) 3214
(d) 2413   (e) 4321

19. Determine whether the following permutations are even or odd.

    (a) 35241   **(b)** 43152   **(c)** 54312

    (d) 25143   **(e)** 32514

20. Give all the permutations of $\{1, 2, 3, 4\}$ and indicate whether they are even or odd.

21. We have given two definitions of determinant in this section, namely in terms of minors and in terms of permutations. Show that these definitions are equivalent when applied to a 3×3 matrix.

22. Evaluate the determinants of the matrices in Exercise 6 using the permutation definition of determinant.

3.2 Properties of Determinants

In this section we discuss the algebraic properties of determinants. Many properties of determinants hold for both rows and columns. When this is the case we write "row (column)" to indicate that *column* may be substituted for *row* in the statement. Proofs are either given or suggested as exercises only when we feel that they illustrate useful techniques.

The following theorem tells us how elementary row operations affect determinants. It also tells us that these operations can be extended to columns.

THEOREM 3.2

Let A be an $n \times n$ matrix and c be a nonzero scalar.
(a) If a matrix B is obtained from A by multiplying the elements of a row (column) by c then $|B| = c|A|$.
(b) If a matrix B is obtained from A by interchanging two rows (columns) then $|B| = -|A|$.
(c) If a matrix B is obtained from A by adding a multiple of one row (column) to another row (column), then $|B| = |A|$.

Proof
(a) Let the matrix B be obtained by multiplying the kth row of A by c. The kth row of B is thus

$$ca_{k1} \ ca_{k2} \ ca_{k3} \ldots ca_{kn}$$

Expand $|B|$ in terms of this row.

$$
\begin{aligned}
|B| &= ca_{k1}C_{k1} + ca_{k2}C_{k2} + \cdots + ca_{kn}C_{kn} \\
&= c(a_{k1}C_{k1} + a_{k2}C_{k2} + \cdots + a_{kn}C_{kn}) \\
&= c|A|
\end{aligned}
$$

The corresponding result for columns is obtained by expanding B in terms of the kth column.

The following example illustrates how row and column operations can be used to create zeros in a determinant, making it easier to compute the determinant.

EXAMPLE 1 Evaluate the determinant

$$
\begin{vmatrix}
3 & 4 & -2 \\
-1 & -6 & 3 \\
2 & 9 & -3
\end{vmatrix}
$$

SOLUTION

We examine the rows and columns of the determinant to see if we can create zeros in a row or column using the above operations. Note that we can create zeros in the second column by adding twice the third column to it:

$$\begin{vmatrix} 3 & 4 & -2 \\ -1 & -6 & 3 \\ 2 & 9 & -3 \end{vmatrix} \underset{C2 + 2C3}{=} \begin{vmatrix} 3 & 0 & -2 \\ -1 & 0 & 3 \\ 2 & 3 & -3 \end{vmatrix}$$

Expand this determinant in terms of the second column to take advantage of the zeros.

$$= (-3)\begin{vmatrix} 3 & 2 \\ -1 & 3 \end{vmatrix} = (-3)(9 - 2) = -21$$

■

EXAMPLE 2 If $A = \begin{bmatrix} 1 & 4 & 3 \\ 0 & 2 & 5 \\ -2 & -4 & 10 \end{bmatrix}$, then it can be shown, by expanding in terms of cofactors, that $|A| = 12$. Use this information, together with the above row and column properties of determinants to evaluate the determinants of the following matrices.

(a) $B_1 = \begin{bmatrix} 1 & 12 & 3 \\ 0 & 6 & 5 \\ -2 & -12 & 10 \end{bmatrix}$ (b) $B_2 = \begin{bmatrix} 1 & 4 & 3 \\ -2 & -4 & 10 \\ 0 & 2 & 5 \end{bmatrix}$

(c) $B_3 = \begin{bmatrix} 1 & 4 & 3 \\ 0 & 2 & 5 \\ 0 & 4 & 16 \end{bmatrix}$

SOLUTION

(a) Observe that B_1 can be obtained from A by multiplying column 2 by 3. Thus $|B_1| = 3|A| = 36$.

(b) B_2 can be obtained from A by interchanging row 2 and row 3. Thus $|B_2| = -|A| = -12$.

(c) B_3 can be obtained from A by adding 2 times row 1 to row 3. Thus $|B_3| = |A| = 12$.

■

We shall find that matrices that have zero determinant play a significant role in the theory of matrices.

DEFINITION   A square matrix A is said to be **singular** if $|A| = 0$. A is **nonsingular** if $|A| \neq 0$.

The following theorem gives information about some of the circumstances under which we can expect a matrix to be singular.

THEOREM 3.3

Let A be a square matrix. A is singular if
(a) all the elements of a row (column) are zero.
(b) two rows (columns) are equal.
(c) two rows (columns) are proportional.
[Note that (b) is a special case of (c), but we list it to give it special emphasis.]

Proof

(a) Let all the elements of the kth row of A be zero. Expand $|A|$ in terms of the kth row.

$$|A| = a_{k1}C_{k1} + a_{k2}C_{k2} + \cdots + a_{kn}C_{kn}$$
$$= 0C_{k1} + 0C_{k2} + \cdots + 0C_{kn}$$
$$= 0$$

The corresponding result can be seen to hold for columns by expanding the determinant in terms of the column of zeros.

(b) Interchange the equal rows of A to get a matrix B that is equal to A. Thus $|B| = |A|$. We know that interchanging two rows of a matrix negates the determinant. Thus $|B| = -|A|$.

The two results $|B| = |A|$ and $|B| = -|A|$ combine to give $|A| = -|A|$. Thus $2|A| = 0$, implying that $|A| = 0$.
The proof for columns is similar.

■

EXAMPLE 3 Show that the following matrices are singular.

(a) $A = \begin{bmatrix} 2 & 0 & -7 \\ 3 & 0 & 1 \\ -4 & 0 & 9 \end{bmatrix}$ (b) $B = \begin{bmatrix} 2 & -1 & 3 \\ 1 & 2 & 4 \\ 2 & 4 & 8 \end{bmatrix}$

SOLUTION

(a) All the elements in column 2 of A are zero. Thus $|A| = 0$.
(b) Observe that every element in row 3 of B is twice the corresponding element in row 2. We write

$$(\text{row } 3) = 2(\text{row } 2)$$

Row 2 and row 3 are proportional. Thus $|B| = 0$.

■

The following theorem tells us how determinants interact with various matrix operations.

THEOREM 3.4

Let A and B be $n \times n$ matrices and c be a nonzero scalar.
(a) Determinant of a scalar multiple: $|cA| = c^n|A|$
(b) Determinant of a product: $|AB| = |A||B|$
(c) Determinant of a transpose: $|A^t| = |A|$

(d) Determinant of an inverse: $|A^{-1}| = \dfrac{1}{|A|}$ (Assuming A^{-1} exists)

Proof

(a) Each row of cA is a row of A multiplied by c. Apply Theorem 3.2(a) to each of the n rows of cA to get $|cA| = c^n |A|$.

(d) $AA^{-1} = I_n$. Thus $|AA^{-1}| = |I_n|$. $|A||A^{-1}| = 1$, by Part (b). $|A^{-1}| = \dfrac{1}{|A|}$.

EXAMPLE 4 If A is a 2×2 matrix with $|A| = 4$, use the properties of determinants to compute the following determinants.

(a) $|3A|$

(b) $|A^2|$

(c) $|5A^t A^{-1}|$, assuming A^{-1} exists

SOLUTION

(a) $|3A| = (3^2)|A|$, by Theorem 3.4(a)

$\qquad = 9 \times 4 = 36$

(b) $|A^2| = |AA|$

$\qquad = |A||A|$, by Theorem 3.4(b)

$\qquad = 4 \times 4 = 16$

(c) $|5A^t A^{-1}| = (5^2)|A^t A^{-1}|$, by Theorem 3.4(a), since $A^t A^{-1}$ is a 2×2 matrix

$\qquad = 25|A^t||A^{-1}|$, by Theorem 3.4(b)

$\qquad = 25|A|\dfrac{1}{|A|}$, by Theorem 3.4(c) and (d)

$\qquad = 25$

EXAMPLE 5 Prove that $|A^{-1}A^t A| = |A|$

SOLUTION

By the properties of matrices, determinants, and real numbers we get

$$|A^{-1}A^t A| = |(A^{-1}A^t)A| = |A^{-1}A^t||A| = |A^{-1}||A^t||A| = \frac{1}{|A|}|A||A| = |A|$$

We ask the reader to justify each step in this proof.

EXAMPLE 6 Prove that if A and B are square matrices of the same size, with A being singular, then AB is also singular. Is the converse true?

SOLUTION

The matrix A is singular. Thus $|A| = 0$. Applying the properties of determinants we get

$$|AB| = |A||B| = 0$$

Thus the matrix AB is singular.

We now investigate the converse: Does AB being singular mean that A is singular? We get

$$|AB| = 0 \Rightarrow |A||B| = 0 \Rightarrow |A| = 0 \text{ or } |B| = 0$$

Thus AB being singular implies that either A or B is singular. (We do not exclude the possibility of both being singular.) The converse is not true.

■

EXAMPLE 7 Let S be the set of 2×2 singular matrices. Prove that the sum of two such matrices is not always singular, but that the scalar multiple is always singular. S is not closed under addition but is closed under scalar multiplication.

SOLUTION

To prove that S is not closed under addition, one example will suffice. Let $A = \begin{bmatrix} 1 & 1 \\ 1 & 1 \end{bmatrix}$ and $B = \begin{bmatrix} 1 & 2 \\ 2 & 4 \end{bmatrix}$. Then $|A| = 0$ and $|B| = 0$; A and B are both singular. Their sum is $A + B = \begin{bmatrix} 2 & 3 \\ 3 & 5 \end{bmatrix}$. Observe that $|A + B| = 1 \neq 0$; $A + B$ is nonsingular. Thus S is not closed under addition.

To prove that S is closed under scalar multiplication we have to consider the general case. Let k be a scalar and C be a singular matrix. Thus $|C| = 0$. Then $|kC| = k^2|C| = 0$. The matrix kC is also singular. Therefore S is closed under scalar multiplication.

This example illustrates that closure under addition and under scalar multiplication are independent conditions. It is possible to have the one condition hold without the other holding.

■

Row operations can be used in a systematic manner similar to Gauss-Jordan elimination to compute determinants. We lead up to this method with a discussion of the determinants of upper triangular matrices.

Determinant of an Upper Triangular Matrix

Consider the following matrix A. This matrix is called an *upper triangular matrix* because all nonzero elements lie on or above the main diagonal. Let us evaluate its determinant.

$$A = \begin{bmatrix} 2 & -1 & 9 & 4 \\ 0 & 3 & 0 & 6 \\ 0 & 0 & -5 & 3 \\ 0 & 0 & 0 & 1 \end{bmatrix}$$

Expand $|A|$ and each succeeding determinant in terms of its first column.

$$\begin{vmatrix} 2 & -1 & 9 & 4 \\ 0 & 3 & 0 & 6 \\ 0 & 0 & -5 & 3 \\ 0 & 0 & 0 & 1 \end{vmatrix} = 2\begin{vmatrix} 3 & 0 & 6 \\ 0 & -5 & 3 \\ 0 & 0 & 1 \end{vmatrix} - 0\begin{vmatrix} -1 & 9 & 4 \\ 0 & -5 & 3 \\ 0 & 0 & 1 \end{vmatrix} - 0\begin{vmatrix} -1 & 9 & 4 \\ 3 & 0 & 6 \\ 0 & 0 & 1 \end{vmatrix} - 0\begin{vmatrix} -1 & 9 & 4 \\ 3 & 0 & 6 \\ 0 & -5 & 3 \end{vmatrix}$$

$$= 2\left(3\begin{vmatrix} -5 & 3 \\ 0 & 1 \end{vmatrix} - 0\begin{vmatrix} 0 & 6 \\ 0 & 1 \end{vmatrix} + 0\begin{vmatrix} 0 & 6 \\ -5 & 3 \end{vmatrix}\right) = 2 \times 3 \times -5 \times 1 = -30$$

Observe that the determinant is the product of the diagonal elements of A. By expanding determinants in terms of first columns, it can similarly be seen that the determinant of any upper triangular matrix is the product of its diagonal elements.

A *lower triangular matrix* is one with all nonzero elements on or below the main diagonal. Its determinant is also the product of the diagonal elements.

> *The determinant of a triangular matrix is the product of its diagonal elements.*

This discussion leads to the following method for evaluating a determinant.

Elimination Method for Evaluating a Determinant

Transform the given determinant into upper triangular form using two types of elementary row operations:

1. Add a multiple of one row to another row. This transformation leaves the determinant unchanged.
2. Interchange two rows of the determinant. This transformation multiplies the determinant by -1.

The zeros *below* the main diagonal are created systematically in the columns, from left to right, according to the Gauss-Jordan pattern.

The final determinant, in upper triangular form, is the product of the diagonal elements. The given determinant will be either this number or its negative, depending on the number of row interchanges performed.

EXAMPLE 8 Evaluate the determinant

$$\begin{vmatrix} 2 & 4 & 1 \\ -2 & -5 & 4 \\ 4 & 9 & 10 \end{vmatrix}$$

SOLUTION

We create zeros below the main diagonal, column by column.

$$\begin{vmatrix} 2 & 4 & 1 \\ -2 & -5 & 4 \\ 4 & 9 & 10 \end{vmatrix} \begin{array}{c} = \\ \text{R2 + R1} \\ \text{R3 + (−2)R1} \end{array} \begin{vmatrix} 2 & 4 & 1 \\ 0 & -1 & 5 \\ 0 & 1 & 8 \end{vmatrix} \begin{array}{c} = \\ \text{R3 + R2} \end{array} \begin{vmatrix} 2 & 4 & 1 \\ 0 & -1 & 5 \\ 0 & 0 & 13 \end{vmatrix}$$

$$= 2 \times (-1) \times 13 = -26$$

It sometimes becomes necessary to interchange rows to obtain the upper triangular matrix, as the following example illustrates. Whenever rows are interchanged, remember to negate the determinant.

EXAMPLE 9 Evaluate the determinant

$$\begin{vmatrix} 1 & -2 & 4 \\ -1 & 2 & -5 \\ 2 & -2 & 11 \end{vmatrix}$$

SOLUTION

We get

$$\begin{vmatrix} 1 & -2 & 4 \\ -1 & 2 & -5 \\ 2 & -2 & 11 \end{vmatrix} \quad \underset{\substack{R2 + R1 \\ R3 + (-2)R1}}{=} \quad \begin{vmatrix} 1 & -2 & 4 \\ 0 & 0 & -1 \\ 0 & 2 & 3 \end{vmatrix}$$

At this time we need a nonzero element in the $(2, 2)$ location, if possible. Interchange row 2 and row 3 and then proceed.

$$\underset{R2 \leftrightarrow R3}{=} \quad (-1) \begin{vmatrix} 1 & -2 & 4 \\ 0 & 2 & 3 \\ 0 & 0 & -1 \end{vmatrix}$$

$$= -1 \times 1 \times 2 \times (-1) = 2.$$

If at any stage the diagonal element is zero, and all elements below it in that column are also zero, it is not possible to interchange rows to obtain a nonzero diagonal element at that stage. The final upper triangular matrix will have a zero diagonal element. It is not necessary to continue beyond this stage, the determinant is zero. The following example illustrates this situation.

EXAMPLE **10** Evaluate the determinant

$$\begin{vmatrix} 1 & -1 & 0 & 2 \\ -1 & 1 & 2 & 3 \\ 2 & -2 & 3 & 4 \\ 6 & -6 & 5 & 1 \end{vmatrix}$$

SOLUTION

Creating zero in the first column we get

$$\begin{vmatrix} 1 & -1 & 0 & 2 \\ -1 & 1 & 2 & 3 \\ 2 & -2 & 3 & 4 \\ 6 & -6 & 5 & 1 \end{vmatrix} \quad \underset{\substack{R2 + R1 \\ R3 + (-2)R1 \\ R4 + (-6)R1}}{=} \quad \begin{vmatrix} 1 & -1 & 0 & 2 \\ 0 & 0 & 2 & 5 \\ 0 & 0 & 3 & 0 \\ 0 & 0 & 5 & -11 \end{vmatrix} = 0.$$

diagonal element is zero,
and all elements below this
diagonal element are zero

EXERCISE SET 3.2

Row Operations

1. Simplify the determinants of the following matrices by creating zeros in a single row or column, then evaluate the determinant by expanding in terms of that row or column.

(a) $\begin{bmatrix} 1 & 2 & 3 \\ 2 & 4 & 1 \\ 1 & 1 & 1 \end{bmatrix}$ (b) $\begin{bmatrix} 0 & 1 & 5 \\ 1 & 1 & 6 \\ 2 & 2 & 7 \end{bmatrix}$

(c) $\begin{bmatrix} 2 & 1 & -1 \\ 3 & -1 & 1 \\ 1 & 4 & -4 \end{bmatrix}$ (d) $\begin{bmatrix} 3 & -1 & 0 \\ 4 & 2 & 1 \\ 1 & 1 & 2 \end{bmatrix}$

2. Simplify the determinants of the following matrices by creating zeros in a single row or column, then evaluate the determinant by expanding in terms of that row or column.

(a) $\begin{bmatrix} 2 & -1 & 2 \\ 1 & 2 & -4 \\ 3 & 1 & 2 \end{bmatrix}$ (b) $\begin{bmatrix} 5 & 1 & 3 \\ 1 & 2 & 4 \\ -1 & 1 & -4 \end{bmatrix}$

(c) $\begin{bmatrix} 1 & -2 & 3 \\ -3 & 6 & -9 \\ 4 & 5 & 7 \end{bmatrix}$ (d) $\begin{bmatrix} -1 & 3 & 2 \\ 2 & 5 & -4 \\ 4 & 1 & -8 \end{bmatrix}$

3. If $A = \begin{bmatrix} 1 & -1 & -3 \\ 2 & 0 & -4 \\ -1 & 1 & 2 \end{bmatrix}$ then $|A|$ can be expanded using cofactors to get $|A| = -2$. Use this information, together with the properties of determinants, to compute the determinants of the following matrices.

(a) $\begin{bmatrix} 1 & -1 & -3 \\ 2 & 0 & -4 \\ -2 & 2 & 4 \end{bmatrix}$ (b) $\begin{bmatrix} 2 & 0 & -4 \\ 1 & -1 & -3 \\ -1 & 1 & 2 \end{bmatrix}$

(c) $\begin{bmatrix} 1 & -1 & -3 \\ 4 & -2 & -10 \\ -1 & 1 & 2 \end{bmatrix}$

4. If $A = \begin{bmatrix} 8 & 5 & 2 \\ 1 & 3 & 2 \\ -1 & -2 & -1 \end{bmatrix}$ then $|A| = 5$. Use this information, together with the row and column properties of determinants, to compute determinants of the following matrices.

(a) $\begin{bmatrix} 8 & 2 & 5 \\ 1 & 2 & 3 \\ -1 & -1 & -2 \end{bmatrix}$ (b) $\begin{bmatrix} 8 & 1 & 2 \\ 1 & -1 & 2 \\ -1 & 0 & -1 \end{bmatrix}$

(c) $\begin{bmatrix} 8 & 1 & -1 \\ 5 & 3 & -2 \\ 2 & 2 & -1 \end{bmatrix}$

5. In the following we have used elementary row operations to compute a determinant in two different ways and obtained two different answers. One approach is correct, while the other is incorrect. Which is the correct answer? [The mistake illustrated by this exercise is a very common one. Beware!]

$$\begin{vmatrix} 1 & 2 & 3 \\ -1 & 2 & 4 \\ 2 & 4 & 7 \end{vmatrix} \underset{2R1 - R3}{=} \begin{vmatrix} 0 & 0 & -1 \\ -1 & 2 & 4 \\ 2 & 4 & 7 \end{vmatrix}$$
$$= (-1)\begin{vmatrix} -1 & 2 \\ 2 & 4 \end{vmatrix}$$
$$= (-1)(-4-4) = 8$$

$$\begin{vmatrix} 1 & 2 & 3 \\ -1 & 2 & 4 \\ 2 & 4 & 7 \end{vmatrix} \underset{R3 - 2R1}{=} \begin{vmatrix} 1 & 2 & 3 \\ -1 & 2 & 4 \\ 0 & 0 & 1 \end{vmatrix}$$
$$= (1)\begin{vmatrix} 1 & 2 \\ -1 & 2 \end{vmatrix}$$
$$= (1)(2+2) = 4$$

Singular Matrices

6. The following matrices are singular because of certain column or row properties. Give the reason.

(a) $\begin{bmatrix} 1 & -2 & 3 \\ 7 & 5 & 4 \\ 0 & 0 & 0 \end{bmatrix}$ (b) $\begin{bmatrix} 1 & 5 & 5 \\ 0 & -2 & -2 \\ 3 & 1 & 1 \end{bmatrix}$

(c) $\begin{bmatrix} 1 & -2 & 4 \\ 0 & -1 & 3 \\ -3 & 6 & -12 \end{bmatrix}$ (d) $\begin{bmatrix} 1 & 2 & 0 \\ 0 & 0 & 3 \\ 0 & 0 & 0 \end{bmatrix}$

7. The following matrices are singular because of certain column or row properties. Give the reason.

(a) $\begin{bmatrix} 7 & 9 & 0 \\ -2 & 3 & 0 \\ 4 & 5 & 0 \end{bmatrix}$ (b) $\begin{bmatrix} 1 & 0 & 4 \\ 0 & 1 & 9 \\ 0 & 0 & 0 \end{bmatrix}$

(c) $\begin{bmatrix} 1 & -2 & 3 \\ 0 & 4 & 1 \\ 3 & -6 & 9 \end{bmatrix}$ (d) $\begin{bmatrix} 2 & 3 & -3 \\ -4 & 1 & 6 \\ 6 & 2 & -9 \end{bmatrix}$

Using Properties of Determinants

8. If A is a 2×2 matrix with $|A| = 3$, use the properties of determinants to compute the following determinants.

(a) $|2A|$ (b) $|3A^t|$ (c) $|A^2|$
(d) $|(A^t)^2|$ (e) $|(A^2)^t|$ (f) $|4A^{-1}|$

9. If A and B are 3×3 matrices and $|A| = -3$, $|B| = 2$, compute the following determinants.

(a) $|AB|$ (b) $|AA^t|$ (c) $|A^tB|$
(d) $|3A^2B|$ (e) $|2AB^{-1}|$ (f) $|(A^2B^{-1})^t|$

10. If $A = \begin{bmatrix} a & b & c \\ d & e & f \\ g & h & i \end{bmatrix}$ and $|A| = 3$, compute the determinants of the following matrices.

(a) $\begin{bmatrix} d & e & f \\ g & h & i \\ a & b & c \end{bmatrix}$ (b) $\begin{bmatrix} d & a & g \\ e & b & h \\ f & c & i \end{bmatrix}$

(c) $\begin{bmatrix} d & f & e \\ a & c & b \\ g & i & h \end{bmatrix}$ (d) $\begin{bmatrix} d & e & f \\ a & b & c \\ 2g & 2h & 2i \end{bmatrix}$

11. Prove the following identity without evaluating the determinants.

$$\begin{vmatrix} a+b & c+d & e+f \\ p & q & r \\ u & v & w \end{vmatrix} = \begin{vmatrix} a & c & e \\ p & q & r \\ u & v & w \end{vmatrix} + \begin{vmatrix} b & d & f \\ p & q & r \\ u & v & w \end{vmatrix}$$

Triangular Matrices

12. Find the determinants of the following triangular matrices.

(a) $\begin{bmatrix} 3 & 9 & 6 \\ 0 & -1 & 2 \\ 0 & 0 & 4 \end{bmatrix}$ (b) $\begin{bmatrix} 2 & 5 & 1 & 0 \\ 0 & 3 & 2 & -7 \\ 0 & 0 & 5 & 1 \\ 0 & 0 & 0 & -2 \end{bmatrix}$

(c) $\begin{bmatrix} 3 & 0 & 0 & 0 \\ 1 & -2 & 0 & 0 \\ -7 & 9 & 5 & 0 \\ 6 & 4 & 2 & 1 \end{bmatrix}$

Elimination Method

13. Evaluate the following 3×3 determinants using the elimination method.

(a) $\begin{vmatrix} 1 & 0 & -1 \\ 2 & 1 & 2 \\ -1 & 1 & 1 \end{vmatrix}$ (b) $\begin{vmatrix} 1 & -2 & 3 \\ -1 & 2 & 1 \\ 2 & 1 & 3 \end{vmatrix}$

(c) $\begin{vmatrix} 2 & 3 & 8 \\ -2 & -3 & 4 \\ 4 & 6 & -2 \end{vmatrix}$ (d) $\begin{vmatrix} 2 & -1 & 4 \\ -2 & 1 & -3 \\ 0 & 3 & -2 \end{vmatrix}$

14. Evaluate the following 4×4 determinants using the elimination method.

(a) $\begin{vmatrix} 2 & 1 & 3 & 1 \\ -2 & 3 & -1 & 2 \\ 2 & 1 & 2 & 3 \\ -4 & -2 & 0 & -1 \end{vmatrix}$

(b) $\begin{vmatrix} 1 & 2 & -1 & 0 \\ 1 & 4 & 2 & 1 \\ -1 & 2 & 6 & 6 \\ 2 & 2 & -4 & -2 \end{vmatrix}$

(c) $\begin{vmatrix} 1 & -1 & 0 & 2 \\ -1 & 1 & 0 & 0 \\ 2 & -2 & 0 & 1 \\ 3 & 1 & 5 & -1 \end{vmatrix}$

Miscellaneous Results

15. Prove that the determinant of a diagonal matrix is the product of the diagonal elements.

16. Prove that if a matrix B is obtained from a square matrix A by adding a multiple of one row (column) to another row (column), then $|B| = |A|$.

17. Prove that if two rows (columns) of a square matrix A are proportional then $|A| = 0$.

18. Prove that if the sum of the elements of each row (column) of a square matrix A is zero, then $|A| = 0$.

19. Let A be a square matrix. Prove that if there exists a positive integer n such that $A^n = 0$ then $|A| = 0$.

20. Prove that if A and B are square matrices of the same size then $|AB| = |BA|$.

21. Prove that if A, B, and C are square matrices of the same size, then $|ABC| = |A||B||C|$.

22. Let A and B be square matrices of the same size. Is the result $|A + B| = |A| + |B|$ true in general? [*Hint:* Construct an example.]

23. Let A be a square matrix. Let B be a matrix obtained from A using an elementary row operation. Prove that $|B| \neq 0$ if and only if $|A| \neq 0$.

24. (a) Let A be a 2×2 matrix. Show that the only reduced echelon form that has no zero rows is I_2.

(b) Let A be a 3×3 matrix. Show that the only reduced echelon form that has no zero rows is I_3.

(c) Let A be an $n \times n$ matrix. Show that the only reduced echelon form that has no zero rows is I_n.

(d) Let A be an $n \times n$ matrix with reduced echelon form E. Show that $|A| \neq 0$ if and only if $E = I_n$.

25. Let S be the set of 2×2 nonsingular matrices. Prove that S is not closed under addition or under scalar multiplication. (*Hint:* Construct examples.)

3.3 Determinants, Matrix Inverses, and Systems of Linear Equations

We have introduced the concept of the determinant, discussed various ways of computing determinants, and have looked at the algebraic properties of determinants. We shall now see how a determinant can give information about the inverse of a matrix and about solutions to systems of equations.

We first introduce the tools necessary for developing a formula for the inverse of a nonsingular matrix.

DEFINITION   Let A be an $n \times n$ matrix and C_{ij} be the cofactor of a_{ij}. The matrix whose (i, j)th element is C_{ij} is called the **matrix of cofactors** of A. The transpose of this matrix is called the **adjoint** of A and is denoted adj(A).

$$
\begin{bmatrix}
C_{11} & C_{12} & \cdots & C_{1n} \\
C_{21} & C_{22} & \cdots & C_{2n} \\
\vdots & \vdots & & \vdots \\
C_{n1} & C_{n2} & \cdots & C_{nn}
\end{bmatrix}
\qquad
\begin{bmatrix}
C_{11} & C_{12} & \cdots & C_{1n} \\
C_{21} & C_{22} & \cdots & C_{2n} \\
\vdots & \vdots & & \vdots \\
C_{n1} & C_{n2} & \cdots & C_{nn}
\end{bmatrix}^{t}
$$

$$\text{matrix of cofactors} \qquad\qquad \text{adjoint matrix}$$

EXAMPLE 1 Give the matrix of cofactors and the adjoint matrix of the following matrix A.

$$
A = \begin{bmatrix}
2 & 0 & 3 \\
-1 & 4 & -2 \\
1 & -3 & 5
\end{bmatrix}
$$

SOLUTION

The cofactors of A are as follows.

$$
C_{11} = \begin{vmatrix} 4 & -2 \\ -3 & 5 \end{vmatrix} = 14 \qquad
C_{12} = -\begin{vmatrix} -1 & -2 \\ 1 & 5 \end{vmatrix} = 3 \qquad
C_{13} = \begin{vmatrix} -1 & 4 \\ 1 & -3 \end{vmatrix} = -1
$$

$$
C_{21} = -\begin{vmatrix} 0 & 3 \\ -3 & 5 \end{vmatrix} = -9 \qquad
C_{22} = \begin{vmatrix} 2 & 3 \\ 1 & 5 \end{vmatrix} = 7 \qquad
C_{23} = -\begin{vmatrix} 2 & 0 \\ 1 & -3 \end{vmatrix} = 6
$$

$$
C_{31} = \begin{vmatrix} 0 & 3 \\ 4 & -2 \end{vmatrix} = -12 \qquad
C_{32} = -\begin{vmatrix} 2 & 3 \\ -1 & -2 \end{vmatrix} = 1 \qquad
C_{33} = \begin{vmatrix} 2 & 0 \\ -1 & 4 \end{vmatrix} = 8
$$

The matrix of cofactors of A is

$$
\begin{bmatrix}
14 & 3 & -1 \\
-9 & 7 & 6 \\
-12 & 1 & 8
\end{bmatrix}
$$

The adjoint of A is the transpose of this matrix.

$$
\text{adj}(A) = \begin{bmatrix}
14 & -9 & -12 \\
3 & 7 & 1 \\
-1 & 6 & 8
\end{bmatrix}
$$

Determinants and Matrix Inverses

The following theorem gives a formula for the inverse of a nonsingular matrix.

THEOREM 3.5

Let A be a square matrix with $|A| \neq 0$. A is invertible with

$$A^{-1} = \frac{1}{|A|} \, \text{adj}(A)$$

Proof Consider the matrix product $A \, \text{adj}(A)$. The (i, j)th element of this product is

$$(i, j)\text{th element} = (\text{row } i \text{ of } A) \times (\text{column } j \text{ of adj}(A))$$

$$= [a_{i1}, a_{i2} \ldots a_{in}] \begin{bmatrix} C_{j1} \\ C_{j2} \\ \vdots \\ C_{jn} \end{bmatrix}$$

$$= a_{i1}C_{j1} + a_{i2}C_{j2} + \cdots + a_{in}C_{jn}$$

If $i = j$ this is the expansion of $|A|$ in terms of the ith row. If $i \neq j$ then it is the expansion of the determinant of a matrix in which the jth row of A has been replaced by the ith row of A, a matrix having two identical rows. Therefore

$$(i, j)\text{th element} = \begin{cases} |A| & \text{if } i = j \\ 0 & \text{if } i \neq j \end{cases}$$

The product $A \, \text{adj}(A)$ is thus a diagonal matrix with the diagonal elements all being $|A|$. Factor out all the diagonal elements to get $|A| \, I_n$. Thus

$$A \, \text{adj}(A) = |A| \, I_n$$

Since $|A| \neq 0$, we can rewrite this equation

$$A\left(\frac{1}{|A|} \, \text{adj}(A) \right) = I_n$$

Similarly, it can be shown that

$$\left(\frac{1}{|A|} \, \text{adj}(A) \right)A = I_n$$

Thus

$$A^{-1} = \frac{1}{|A|} \, \text{adj}(A)$$

proving the theorem.

The importance of this result lies in that it give us a *formula* for the inverse of an arbitrary nonsingular matrix that can be used in theoretical work. The Gauss-Jordan algorithm presented earlier is much more efficient than this formula for computing the inverse of a specific matrix. However the Gauss-Jordan method cannot be used to describe the inverse of an arbitrary matrix.

The following theorem complements the previous one. It tells us that the nonsingular matrices are in fact the only matrices that have inverses.

THEOREM 3.6

A square matrix A is invertible if and only if $|A| \neq 0$.

Proof Assume that A is invertible. Thus $AA^{-1} = I_n$. This implies that $|AA^{-1}| = |I_n|$. Properties of determinants give $|A||A^{-1}| = 1$. Thus $|A| \neq 0$.

Conversly, Theorem 3.5 tells us that if $|A| \neq 0$ then A is invertible.

$$A^{-1} \text{ exists if and only if } |A| \neq 0$$

EXAMPLE 2 Use a determinant to find out which of the following matrices are invertible.

$$A = \begin{bmatrix} 1 & -1 \\ 3 & 2 \end{bmatrix}, \quad B = \begin{bmatrix} 4 & 2 \\ 2 & 1 \end{bmatrix}, \quad C = \begin{bmatrix} 2 & 4 & -3 \\ 4 & 12 & -7 \\ -1 & 0 & 1 \end{bmatrix}, \quad D = \begin{bmatrix} 1 & 2 & -1 \\ -1 & 1 & 2 \\ 2 & 8 & 0 \end{bmatrix}$$

SOLUTION

Compute the determinant of each matrix and apply the previous theorem. We get

$|A| = 5 \neq 0$ A is invertible.

$|B| = 0$ B is singular. The inverse does not exist.

$|C| = 0$ C is singular. The inverse does not exist.

$|D| = 2 \neq 0$ D is invertible.

EXAMPLE 3 Use the formula for the inverse of a matrix to compute the inverse of the matrix

$$A = \begin{bmatrix} 2 & 0 & 3 \\ -1 & 4 & -2 \\ 1 & -3 & 5 \end{bmatrix}$$

SOLUTION

$|A|$ is computed and found to be 25. Thus the inverse of A exists. This matrix was discussed in Example 1. There we found that

$$\text{adj}(A) = \begin{bmatrix} 14 & -9 & -12 \\ 3 & 7 & 1 \\ -1 & 6 & 8 \end{bmatrix}$$

The formula for the inverse of a matrix gives

$$A^{-1} = \frac{1}{25} \text{adj}(A) = \begin{bmatrix} \frac{14}{25} & -\frac{9}{25} & -\frac{12}{25} \\ \frac{3}{25} & \frac{7}{25} & \frac{1}{25} \\ -\frac{1}{25} & \frac{6}{25} & \frac{8}{25} \end{bmatrix}$$

Determinants and Systems of Linear Equations

We now discuss the relationship between the existence and uniqueness of the solutions to a system of n linear equations in n variables and the determinant of the matrix of coefficients of the system.

THEOREM 3.7

Let $AX = B$ be a system of n linear equations in n variables. If $|A| \neq 0$ there is a unique solution. If $|A| = 0$ there may be many or no solutions.

Proof If $|A| \neq 0$ we know that A^{-1} exists (Theorem 3.6) and that there is then a unique solution given by $X = A^{-1}B$ (Theorem 2.8).

If $|A| = 0$ the determinant of every matrix leading up to, and including the reduced echelon form of A is zero (by Theorem 3.2). Thus the reduced echelon form of A is not I_n. The solution to the system $AX = B$ is not unique. The following two systems of linear equations, each of which has a singular matrix of coefficients, illustrate that there may be many or no solutions.

$$x_1 - 2x_2 + 3x_3 = 1 \qquad\qquad x_1 + 2x_2 + 3x_3 = 3$$
$$3x_1 - 4x_2 + 5x_3 = 3 \qquad\qquad 2x_1 + x_2 + 3x_3 = 3$$
$$2x_1 - 3x_2 + 4x_3 = 2 \qquad\qquad x_1 + x_2 + 2x_3 = 0$$

many solutions: $\qquad\qquad\qquad\qquad$ no solutions

$$x_1 = r + 1, x_2 = 2r, x_3 = r$$

EXAMPLE 4 Determine whether or not the following system of equations has a unique solution.

$$3x_1 + 3x_2 - 2x_3 = 2$$
$$4x_1 + x_2 + 3x_3 = -5$$
$$7x_1 + 4x_2 + x_3 = 9$$

SOLUTION

We compute the determinant of the matrix of coefficients. We get

$$\begin{vmatrix} 3 & 3 & -2 \\ 4 & 1 & 3 \\ 7 & 4 & 1 \end{vmatrix} = 0$$

Thus the system does not have a unique solution.

We now introduce a result called *Cramer's Rule* for solving a system of n linear equations in n variables that has a unique solution. This rule is of theoretical importance in that it gives us a formula for the solution of a system of equations.

THEOREM 3.8

Cramer's Rule* Let $AX = B$ be a system of n linear equations in n variables such that $|A| \neq 0$. The system has a unique solution given by

$$x_1 = \frac{|A_1|}{|A|}, \quad x_2 = \frac{|A_2|}{|A|}, \dots, \quad x_n = \frac{|A_n|}{|A|},$$

where A_i is the matrix obtained by replacing column i of A with B.

Proof Since $|A| \neq 0$ the solution to the system $AX = B$ is unique and is given by

$$X = A^{-1}B$$

$$= \frac{1}{|A|} \text{adj}(A)B$$

x_i, the ith element of X, is given by

$$x_i = \frac{1}{|A|} \left[\text{row } i \text{ of adj}(A) \right] \times B$$

$$= \frac{1}{|A|} \left[C_{1i} \; C_{2i} \; \dots \; C_{ni} \right] \begin{bmatrix} b_1 \\ b_2 \\ \vdots \\ b_n \end{bmatrix}$$

$$= \frac{1}{|A|} (b_1 C_{1i} + b_2 C_{2i} + \dots + b_n C_{ni})$$

The expression in parentheses is the cofactor expansion of $|A_i|$ in terms of the ith column. Thus

$$x_i = \frac{|A_i|}{|A|}$$

proving the theorem.

EXAMPLE 5 Solve the following system of equations using Cramer's Rule.

$$\begin{aligned} x_1 + 3x_2 + x_3 &= -2 \\ 2x_1 + 5x_2 + x_3 &= -5 \\ x_1 + 2x_2 + 3x_3 &= 6 \end{aligned}$$

---

*Gabriel Cramer (1704–1752) was educated and lived in Geneva, Switzerland, and traveled widely in Europe. He made contributions to the fields of geometry and probability and was described as "a great poser of stimulating problems." It has long been stated that Cramer's Rule was not in fact his work but that he explained and encouraged its use. New research ("Cramer's Rule is due to Cramer," A. A. Kosinski, *Mathematical Magazine*, vol 74, No 4, 310, October 2001), however, seems to indicate that the rule is in fact due to Cramer. Nevertheless, perhaps Cramer's greatest contribution to the mathematical community was editing and distributing the works of others. Cramer's interests were broad, being involved in civic life, the restoration of cathedrals, and in excavations. He was said to be "friendly, good-humored, pleasant in voice and appearance, possessing a good memory, judgment, and health."

SOLUTION

The matrix of coefficients A and column matrix of constants B are

$$A = \begin{bmatrix} 1 & 3 & 1 \\ 2 & 5 & 1 \\ 1 & 2 & 3 \end{bmatrix} \quad \text{and} \quad B = \begin{bmatrix} -2 \\ -5 \\ 6 \end{bmatrix}$$

It is found that $|A| = -3 \neq 0$. Thus Cramer's Rule can be applied. We get

$$A_1 = \begin{bmatrix} -2 & 3 & 1 \\ -5 & 5 & 1 \\ 6 & 2 & 3 \end{bmatrix}, \quad A_2 = \begin{bmatrix} 1 & -2 & 1 \\ 2 & -5 & 1 \\ 1 & 6 & 3 \end{bmatrix}, \quad A_3 = \begin{bmatrix} 1 & 3 & -2 \\ 2 & 5 & -5 \\ 1 & 2 & 6 \end{bmatrix}$$

giving $|A_1| = -3, |A_2| = 6, |A_3| = -9$.

Cramer's Rule now gives

$$x_1 = \frac{|A_1|}{|A|} = \frac{-3}{-3} = 1, \qquad x_2 = \frac{|A_2|}{|A|} = \frac{6}{-3} = -2, \qquad x_3 = \frac{|A_3|}{|A|} = \frac{-9}{-3} = 3$$

The unique solution is $x_1 = 1, x_2 = -2, x_3 = 3$.

Numerical Considerations

It is important that the reader get to understand Cramer's Rule by solving systems of equations like this. However we stress again that the importance of Cramer's Rule lies in that it gives a *formula* for the solution of a system of equations that has a unique solution. It is not used in practice to solve specific systems; the methods of Gauss-Jordan elimination (or Gaussian elimination, Section 8.1) is more efficient for doing that. Gauss-Jordan elimination requires $\frac{1}{2}n^3 + \frac{1}{2}n^2$ multiplications and $\frac{1}{2}n^3 - \frac{1}{2}n$ additions, while it can be shown that Cramer's Rule uses $\frac{1}{3}n^4 + \frac{1}{3}n^3 + \frac{2}{3}n^2 - \frac{1}{3}n - 1$ multiplications and $\frac{1}{3}n^4 - \frac{1}{6}n^3 - \frac{1}{3}n^2 + \frac{1}{6}n$ additions. This means, for example, that a system of ten equations in ten variables can be solved using 550 multiplications and 495 additions using Gauss-Jordan elimination in contrast to 3,729 multiplications and 3,135 additions using Cramer's Rule.

Homogeneous Systems of Linear Equations

Theorem 3.7 implies that *the solution to a linear system of n equations in n variables $AX = B$ will be unique if and only if $|A| \neq 0$.* Consider a homogeneous system $AX = 0$. $X = 0$ is a solution, called the **trivial solution**. If we are going to have nontrivial solutions to such a system, as is often the case in applications, we must have $|A| = 0$. The following example illustrates this situation.

EXAMPLE 6 Determine values of λ for which the following system of equations has nontrivial solutions. Find the solutions for each value of λ.

$$(\lambda + 2)x_1 + (\lambda + 4)x_2 = 0$$
$$2x_1 + (\lambda + 1)x_2 = 0$$

SOLUTION

This system is a homogeneous system of linear equations. It thus has the trivial solution $x_1 = 0$, $x_2 = 0$. Theorem 3.7 tells us that there is the possibility of other solutions only if the determinant of the matrix of coefficients is zero. Equating this determinant to zero we get

$$\begin{vmatrix} \lambda + 2 & \lambda + 4 \\ 2 & \lambda + 1 \end{vmatrix} = 0$$

$$(\lambda + 2)(\lambda + 1) - 2(\lambda + 4) = 0$$

$$\lambda^2 + \lambda - 6 = 0$$

$$(\lambda - 2)(\lambda + 3) = 0$$

The determinant is zero if $\lambda = -3$ or $\lambda = 2$.
   $\lambda = -3$ results in the system

$$\begin{aligned} -x_1 + x_2 &= 0 \\ 2x_1 - 2x_2 &= 0 \end{aligned}$$

This system has many solutions, $x_1 = r$, $x_2 = r$.
   $\lambda = 2$ results in the system

$$\begin{aligned} 4x_1 + 6x_2 &= 0 \\ 2x_1 + 3x_2 &= 0 \end{aligned}$$

This system has many solutions, $x_1 = -3r/2$, $x_2 = r$.

These values of λ often arise in applications as scalars called *eigenvalues*. We shall study eigenvalues later in the course and see their importance for the search engine Google, in demography, and in weather prediction.

EXERCISE SET 3.3

Determining If an Inverse Exists

1. Use determinants to find out whether the following matrices have inverses. You need not compute inverses.

(a) $\begin{bmatrix} 4 & 7 \\ 1 & 3 \end{bmatrix}$ (b) $\begin{bmatrix} -3 & 1 \\ 6 & -2 \end{bmatrix}$

(c) $\begin{bmatrix} 6 & 4 \\ 3 & 2 \end{bmatrix}$ (d) $\begin{bmatrix} 7 & 2 \\ 3 & 1 \end{bmatrix}$

2. Determine whether the following matrices have inverses. You need not compute inverses.

(a) $\begin{bmatrix} 0 & 2 \\ 3 & 4 \end{bmatrix}$ (b) $\begin{bmatrix} 10 & 5 \\ 4 & 2 \end{bmatrix}$

(c) $\begin{bmatrix} 5 & 3 \\ 1 & 2 \end{bmatrix}$ (d) $\begin{bmatrix} 4 & -1 \\ -8 & 2 \end{bmatrix}$

3. Determine whether the following matrices have inverses. You need not compute inverses.

(a) $\begin{bmatrix} 2 & 4 & -7 \\ 0 & 1 & 3 \\ 0 & 0 & 9 \end{bmatrix}$ (b) $\begin{bmatrix} 4 & -2 & 9 \\ 0 & 0 & 3 \\ 0 & 0 & 6 \end{bmatrix}$

(c) $\begin{bmatrix} -3 & 0 & 0 \\ 1 & 7 & 0 \\ -2 & 8 & 5 \end{bmatrix}$ (d) $\begin{bmatrix} 1 & 2 & 3 \\ 2 & 4 & 6 \\ 7 & 3 & -1 \end{bmatrix}$

4. Determine whether the following matrices have inverses. You need not compute inverses.

(a) $\begin{bmatrix} 1 & -2 & 3 \\ 2 & 7 & 6 \\ -3 & 5 & -9 \end{bmatrix}$ (b) $\begin{bmatrix} 4 & 0 & 5 \\ 1 & 3 & 7 \\ 2 & 0 & 6 \end{bmatrix}$

(c) $\begin{bmatrix} 7 & 1 & 3 \\ -1 & 2 & 0 \\ 5 & 4 & 1 \end{bmatrix}$ (d) $\begin{bmatrix} 1 & -2 & 3 \\ 4 & -3 & 2 \\ 1 & -1 & 1 \end{bmatrix}$

Finding the Inverse of a Matrix

5. Determine whether the following matrices have inverses. If a matrix has an inverse, find the inverse using the formula for the inverse of a matrix.

(a) $\begin{bmatrix} 1 & 4 \\ 3 & 2 \end{bmatrix}$ (b) $\begin{bmatrix} -2 & -1 \\ 7 & 3 \end{bmatrix}$

(c) $\begin{bmatrix} 1 & 2 \\ 2 & 4 \end{bmatrix}$ (d) $\begin{bmatrix} 2 & 1 \\ 4 & 3 \end{bmatrix}$

6. Determine whether the following matrices have inverses. If a matrix has an inverse, find the inverse using the formula for the inverse of a matrix.

(a) $\begin{bmatrix} 1 & 2 & 3 \\ 0 & 1 & 2 \\ 4 & 5 & 3 \end{bmatrix}$ (b) $\begin{bmatrix} 0 & 3 & 3 \\ 1 & 2 & 3 \\ 1 & 4 & 6 \end{bmatrix}$

(c) $\begin{bmatrix} 1 & 2 & -1 \\ 2 & 4 & -3 \\ 1 & -2 & 0 \end{bmatrix}$ (d) $\begin{bmatrix} 5 & 1 & 3 \\ 6 & 1 & 4 \\ 1 & 0 & 1 \end{bmatrix}$

7. Determine whether the following matrices have inverses. If a matrix has an inverse, find the inverse using the formula for the inverse of a matrix.

(a) $\begin{bmatrix} 5 & 2 & 4 \\ 2 & 1 & 2 \\ 4 & 2 & 3 \end{bmatrix}$ (b) $\begin{bmatrix} -3 & -2 & -5 \\ 3 & 4 & 3 \\ 1 & 1 & 1 \end{bmatrix}$

(c) $\begin{bmatrix} 7 & 8 & 3 \\ 3 & 6 & 3 \\ 1 & 2 & 1 \end{bmatrix}$ (d) $\begin{bmatrix} 2 & 2 & 1 \\ 4 & -1 & 4 \\ 7 & 4 & 5 \end{bmatrix}$

Cramer's Rule

8. Solve the following systems of equations using Cramer's Rule.

(a) $x_1 + 2x_2 = 8$
$2x_1 + 5x_2 = 19$

(b) $x_1 + 2x_2 = 3$
$3x_1 + x_2 = -1$

(c) $x_1 + 3x_2 = 11$
$-2x_1 + x_2 = -1$

9. Solve the following systems of equations using Cramer's Rule.

(a) $3x_1 + x_2 = -1$
$x_1 + x_2 = 3$

(b) $3x_1 + 2x_2 = 11$
$2x_1 + 3x_2 = 14$

(c) $2x_1 + x_2 = -1$
$-2x_1 + 2x_2 = 10$

10. Solve the following systems of equations using Cramer's Rule.

(a) $x_1 + 3x_2 + 4x_3 = 3$
$2x_1 + 6x_2 + 9x_3 = 5$
$3x_1 + x_2 - 2x_3 = 7$

(b) $x_1 + 2x_2 + x_3 = 9$
$x_1 + 3x_2 - x_3 = 4$
$x_1 + 4x_2 - x_3 = 7$

(c) $2x_1 + x_2 + 3x_3 = 2$
$3x_1 - 2x_2 + 4x_3 = 2$
$x_1 + 4x_2 - 2x_3 = 1$

11. Solve the following systems of equations using Cramer's Rule.

(a) $x_1 + 4x_2 + 2x_3 = 5$
$x_1 + 4x_2 - x_3 = 2$
$2x_1 + 6x_2 + x_3 = 7$

(b) $2x_1 - x_2 + 3x_3 = 7$
$x_1 + 4x_2 + 2x_3 = 10$
$3x_1 + 2x_2 + x_3 = 0$

(c) $8x_1 - 2x_2 + x_3 = 1$
$2x_1 - x_2 + 6x_3 = 3$
$6x_1 + x_2 + 4x_3 = 3$

12. Solve the following systems of equations using Cramer's Rule (if possible).

(a) $2x_1 + 7x_2 + 3x_3 = 7$
$x_1 + 2x_2 + x_3 = 2$
$x_1 + 5x_2 + 2x_3 = 5$

(b) $3x_1 + x_2 - x_3 = 7$
$x_1 + 2x_2 + x_3 = 3$
$2x_1 + 6x_2 = -4$

(c) $3x_1 + 6x_2 - x_3 = 3$
$x_1 - 2x_2 + 3x_3 = 2$
$4x_1 - 2x_2 + 5x_3 = 5$

Checking for a Unique Solution

13. By examining the determinant of the matrix of coefficients, decide whether or not the following systems of equations have a unique solution.

(a) $2x_1 - 3x_2 = -1$
$-4x_1 + 6x_2 = 3$

(b) $3x_1 + x_2 = 11$
$4x_1 - 2x_2 = 14$

(c) $6x_1 + 9x_2 = -1$
$-2x_1 - 3x_2 = 10$

14. By examining the determinant of the matrix of coefficients, decide whether or not the following systems of equations have a unique solution.

(a) $3x_1 + 2x_2 + x_3 = 4$
$x_1 - 3x_2 + 2x_3 = -11$
$2x_1 + x_2 - 3x_3 = -5$

(b) $2x_1 + 7x_2 + x_3 = 8$
$-x_1 - 5x_2 = 3$
$x_1 + 2x_2 + x_3 = -4$

(c) $5x_1 + 6x_2 + 4x_3 = 3$
$7x_1 + 8x_2 + 6x_3 = 1$
$6x_1 + 7x_2 + 5x_3 = 0$

Constructing Nontrivial Solutions

15. Determine the values of λ for which the following system of equations has nontrivial solutions. Find the solutions for each value of λ.

$$(1 - \lambda)x_1 + \quad\quad 6x_2 = 0$$
$$5x_1 + (2 - \lambda)x_2 = 0$$

16. Determine the values of λ for which the following system of equations has nontrivial solutions. Find the solutions for each value of λ.

$$(\lambda + 4)x_1 + (\lambda - 2)x_2 = 0$$
$$4x_1 + (\lambda - 3)x_2 = 0$$

17. Determine the values of λ for which the following system of equations has nontrivial solutions. Find the solutions for each value of λ.

$$(5 - \lambda)x_1 + \quad\quad 4x_2 + \quad\quad 2x_3 = 0$$
$$4x_1 + (5 - \lambda)x_2 + \quad\quad 2x_3 = 0$$
$$2x_1 + \quad\quad 2x_2 + (2 - \lambda)x_3 = 0$$

18. Let $AX = \lambda X$ be a system of n linear equations in n variables. Prove that this system has a nontrivial solution if and only if $|A - \lambda I_n| = 0$.

Miscellaneous Results

19. Prove that if A is a symmetric matrix, then $\text{adj}(A)$ is also symmetric.

20. Prove that if the square matrix A is not invertible, then $\text{adj}(A)$ is not invertible.

21. Prove that if a matrix A is invertible then $\text{adj}(A)$ is also invertible, with inverse given by

$$[\text{adj}(A)]^{-1} = \frac{1}{|A|}A$$

22. Prove that $[\text{adj}(A)]^{-1} = \text{adj}(A^{-1})$.

23. Prove that if A and B are square matrices of the same size, with AB being invertible, then A and B are invertible. Is the converse true?

24. Prove that if A is an invertible upper triangular matrix then A^{-1} is also an upper triangular matrix.

25. Let A be a matrix all of whose elements are integers. Show that if $|A| = \pm 1$, then all the elements of A^{-1} are integers.

26. Let $AX = 0$ be a homogeneous system of n linear equations in n variables that has only the trivial solution. If k is a positive integer, show that the system $A^k X = 0$ also has only the trivial solution.

27. Let $AX = B_1$ and $AX = B_2$ be two systems of n linear equations in n variables, having the same matrix of coefficients A. Prove that $AX = B_2$ has a unique solution if and only if $AX = B_1$ has a unique solution.

28. State (with a brief explanation) whether the following statements are true or false for square matrices A, B, and C of the same size.

(a) $|A^2| = (|A|)^2$.

(b) If a diagonal matrix A is singular then at least one diagonal element is zero.

(c) If $|A| = 1$ then $A^{-1} = \text{adj}(A)$.

(d) If $|A - B| = 0$ then $|A| = |B|$.

(e) If A is nonsingular it is row equivalent to the identity matrix.

CHAPTER 3 REVIEW EXERCISES

1. Evaluate the determinants of the following 2×2 matrices.

(a) $\begin{bmatrix} 3 & 2 \\ 5 & 1 \end{bmatrix}$ (b) $\begin{bmatrix} -3 & 0 \\ 1 & 6 \end{bmatrix}$ (c) $\begin{bmatrix} 9 & 7 \\ 1 & 4 \end{bmatrix}$

2. Let $A = \begin{bmatrix} 2 & 1 & 0 \\ -3 & 4 & 1 \\ 7 & 9 & 2 \end{bmatrix}$. Find the following minors and cofactors.

(a) M_{12} and C_{12} (b) M_{31} and C_{31} (c) M_{22} and C_{22}

3. Evaluate the determinants of each of the following matrices in two ways, using the indicated rows and columns. Observe that you get the same answers both ways.

(a) $\begin{bmatrix} 1 & 2 & -3 \\ 0 & 2 & 5 \\ 4 & 1 & 2 \end{bmatrix}$ (b) $\begin{bmatrix} 0 & 5 & 3 \\ 2 & -3 & 1 \\ 2 & 7 & 3 \end{bmatrix}$

1st row and   3rd row and
1st column   2nd column

4. Solve the following equation for x.

$$\begin{vmatrix} x & x \\ 2 & x-3 \end{vmatrix} = -6$$

5. Simplify the determinants of the following matrices by creating zeros in a single row or column, then evaluate the determinant by expanding in terms of that row or column.

(a) $\begin{bmatrix} 1 & 2 & -1 \\ 3 & 1 & 1 \\ 2 & 4 & 1 \end{bmatrix}$ (b) $\begin{bmatrix} 5 & 3 & 4 \\ 4 & 6 & 1 \\ 2 & -3 & 7 \end{bmatrix}$

(c) $\begin{bmatrix} 1 & 4 & -2 \\ 2 & 3 & 1 \\ -1 & 5 & 6 \end{bmatrix}$

6. If $A = \begin{bmatrix} 2 & -1 & 3 \\ 4 & -1 & 6 \\ 2 & -3 & 4 \end{bmatrix}$ then $|A| = 2$. Use this information, together with the row and column properties of determinants, to compute determinants of the following matrices.

(a) $\begin{bmatrix} 2 & -1 & 3 \\ 12 & -3 & 18 \\ 2 & -3 & 4 \end{bmatrix}$ (b) $\begin{bmatrix} 2 & -1 & 3 \\ 0 & 1 & 0 \\ 2 & -3 & 4 \end{bmatrix}$

(c) $\begin{bmatrix} 4 & -2 & 6 \\ -4 & 1 & -6 \\ 6 & -9 & 12 \end{bmatrix}$

7. Evaluate the following 3×3 determinants using elementary row operations.

(a) $\begin{vmatrix} 1 & 2 & 4 \\ -1 & 4 & 3 \\ 2 & 0 & 5 \end{vmatrix}$ (b) $\begin{vmatrix} -1 & 3 & 2 \\ 0 & 5 & 2 \\ 1 & 7 & 6 \end{vmatrix}$

(c) $\begin{vmatrix} 2 & -3 & 5 \\ 4 & 0 & 6 \\ 1 & 2 & 7 \end{vmatrix}$

8. If A is a 3×3 matrix with $|A| = -2$, compute the following determinants.

(a) $|3A|$ (b) $|2AA^t|$ (c) $|A^3|$

(d) $|(A^tA)^2|$ (e) $|(A^t)^3|$ (f) $|2A^t(A^{-1})^2|$

9. Prove that if A and B are square matrices of the same size, with $|B| \neq 0$, then there exists a matrix C such $CB = A$.

10. Let A and C be matrices of the same size with C being invertible. Prove that $|A| = |C^{-1}AC|$.

11. If A is a triangular matrix, prove that A^2 is also triangular.

12. Show that if A is an invertible matrix such that $A^2 = A$, then $|A| = 1$.

13. Determine whether the following matrices have inverses. If a matrix has an inverse, find the inverse using the formula for the inverse of a matrix.

(a) $\begin{bmatrix} 3 & 5 \\ 1 & 2 \end{bmatrix}$ (b) $\begin{bmatrix} 3 & 2 \\ -1 & 5 \end{bmatrix}$

(c) $\begin{bmatrix} 1 & 4 & -1 \\ 0 & 2 & 0 \\ 1 & 6 & -1 \end{bmatrix}$ (d) $\begin{bmatrix} 2 & 1 & 3 \\ 0 & 2 & 9 \\ 4 & 2 & 11 \end{bmatrix}$

14. Solve the following systems of equations using Cramer's Rule.

(a) $2x_1 + x_2 = -1$
     $3x_1 - 5x_2 = 18$

(b) $x_1 + x_2 + x_3 = 1$
     $2x_1 - x_2 + 3x_3 = 5$
     $4x_1 + 5x_2 + x_3 = 3$

15. Prove that if the square matrix A is not invertible then $A[\text{adj}(A)]$ is the zero matrix.

16. Prove that if A is an invertible triangular matrix then all the diagonal elements must be nonzero.

17. Let $AX = B$ be a system of n linear equations in n variables with the elements of both A and B being integers. Prove that if $|A| = \pm 1$, the solution X has all integer components.

Beijing Capital International Airport is the largest airport in China. Construction of Terminal 3, seen in this photo, started on March 28, 2004, and is larger than London's Heathrow Airport's five terminals combined. Currently the second largest airport passenger terminal building in the world, Terminal 3 has dozens of windows with adjustable angles to ensure adequate lighting.

Vector Spaces

At this time the course becomes geometrical in nature. We discuss two- dimensional and three-dimensional spaces, leading up to *n*-dimensional spaces. The results that we derive are powerful and the way in which they are developed is beautiful. The thinking that goes into this process illustrates an important aspect of mathematics, that of setting up and building upon axioms. We shall find that points, directed line segments, matrices, and functions all have certain mathematical structures in common and can be interpreted as elements of spaces called vector spaces.

4.1 The Vector Space R''

The locations of points in a plane are usually discussed in terms of a coordinate system. For example, in Figure 4.1, the location of each point in the plane can be described using a **rectangular coordinate system**. The point A is the point $(5, 3)$.

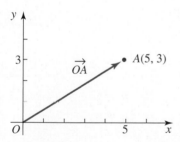

Figure 4.1

Furthermore, point A is a certain distance in a certain direction from the **origin** $(0, 0)$. The distance and direction are characterized by the length and direction of the line segment from the origin, O, to A. We call such a directed line segment a **position vector** and denote it $\overrightarrow{OA}$. O is called the **initial point** of $\overrightarrow{OA}$ and A is called the **terminal point**. There are thus two ways of interpreting $(5, 3)$; it defines the location of a point in a plane, and it also defines the position vector $\overrightarrow{OA}$.

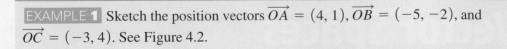

Sketch the position vectors $\overrightarrow{OA} = (4, 1)$, $\overrightarrow{OB} = (-5, -2)$, and $\overrightarrow{OC} = (-3, 4)$. See Figure 4.2.

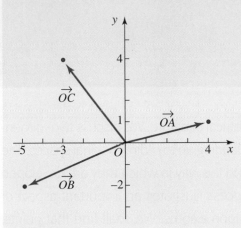

Figure 4.2

Denote the set of all ordered pairs of real numbers by $\mathbf{R}^2$. ($\mathbf{R}$ stands for real number and 2 stands for the number of entries; it is pronounced "r-two.") Note the significance of "ordered" here; for example, the point $(5, 3)$ is not the same point as $(3, 5)$. The order is important.

These concepts can be extended to the set of ordered triples, denoted by $\mathbf{R}^3$. Elements of this set such as $(2, 4, 3)$ can be interpreted in two ways: as the location of a point in three-space relative to an xyz coordinate system, or as a position vector. These interpretations are shown in Figure 4.3.

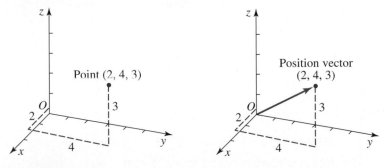

Figure 4.3

We now generalize these concepts. Let $(u_1, u_2, \ldots, u_n)$ be a sequence of n real numbers. The set of all such sequences is called n-space and is denoted $\mathbf{R}^n$. u_1 is the **first component** of $(u_1, u_2, \ldots, u_n)$, u_2 is the **second component**, and so on.

For example, $\mathbf{R}^4$ is the set of sequences of four real numbers; $(1, 2, 3, 4)$ and $(-1, 3, 5.2, 0)$ are in $\mathbf{R}^4$. $\mathbf{R}^5$ is the set of sequences of five real numbers; $(-1, 2, 0, 3, 9)$ is in this set.

Many of the results and techniques that we develop for $\mathbf{R}^n$ with $n > 3$ will be useful mathematical tools, without direct geometrical significance. The elements of $\mathbf{R}^n$ can, however, be interpreted as points in n-space or as position vectors in n-space. It is difficult to visualize an n-space for $n > 3$, but the reader is encouraged to try to form an intuitive picture. A geometrical "feel" for what is taking place often makes an algebraic discussion easier to follow. The mathematics that we shall develop on **R**^{n} will be motivated by the geometry that we are familiar with on **R**^{2} and **R**^{3}.

In keeping with this notation let **R** denote the set of real numbers.

Addition and Scalar Multiplication

We begin the development of an algebraic theory of vectors by introducing equality of vectors.

Let $\mathbf{u} = (u_1, \ldots, u_n)$ and $\mathbf{v} = (v_1, \ldots, v_n)$ be two elements of **R**^{n}. We say that **u** and **v** are **equal** if $u_1 = v_1, \ldots, u_n = v_n$. Thus two elements of **R**^{n} are equal if their **corresponding components** are equal.

When working with elements of **R**^{n} it is customary to refer to numbers as **scalars**. We now define addition and scalar multiplication. They are similar to those for matrices.

DEFINITION   Let $\mathbf{u} = (u_1, \ldots, u_n)$ and $\mathbf{v} = (v_1, \ldots, v_n)$ be elements of **R**^{n} and let c be a scalar. Addition and scalar multiplication are performed as follows.

Addition: $\qquad\qquad\qquad \mathbf{u} + \mathbf{v} = (u_1 + v_1, \ldots, u_n + v_n)$
Scalar multiplication: $\qquad\qquad c\mathbf{u} = (cu_1, \ldots, cu_n)$

To add two elements of **R**^{n} we add corresponding components. To multiply an element of **R**^{n} by a scalar we multiply every component by that scalar. Observe that the resulting elements are in **R**^{n}. We say that **R**^{n} is **closed under addition and under scalar multiplication**.

R^n with operations of componentwise addition and scalar multiplication is an example of a **vector space**, and its elements are called **vectors**. (We will use boldface for vectors and plain text for scalars.)

*We shall henceforth in this course interpret **R**^{n} to be a vector space.*

EXAMPLE 2 Let $\mathbf{u} = (-1, 4, 3, 7)$ and $\mathbf{v} = (-2, -3, 1, 0)$ be vectors in **R**^{4}. Find $\mathbf{u} + \mathbf{v}$ and $3\mathbf{u}$.

SOLUTION

We get

$$\mathbf{u} + \mathbf{v} = (-1, 4, 3, 7) + (-2, -3, 1, 0) = (-3, 1, 4, 7)$$
$$3\mathbf{u} = 3(-1, 4, 3, 7) = (-3, 12, 9, 21)$$

Note that the resulting vector under each operation is in the original vector space **R**^{4}.

We now give examples to illustrate geometrical interpretations of these vectors and their operations.

EXAMPLE 3 This example gives us a geometrical interpretation of vector addition. Consider the sum of the vectors $(4, 1)$ and $(2, 3)$. We get

$$(4, 1) + (2, 3) = (6, 4)$$

In Figure 4.4 we interpret these vectors as position vectors. Construct the parallelogram having the vectors $(4, 1)$ and $(2, 3)$ as adjacent sides. The vector $(6, 4)$, the sum, will be the diagonal of the parallelogram.

Such vectors are used in the physical sciences to describe forces. In this example $(4, 1)$ and $(2, 3)$ might be forces, acting on a body at the origin, O. The vectors would give the directions of the forces and their lengths (using the Pythagorean Theorem) would be the magnitudes of the forces, $\sqrt{4^2 + 1^2} = 4.12$ and $\sqrt{2^2 + 3^2} = 3.61$. The vector sum $(6, 4)$ is the resultant force, a single force that would be equivalent to the two forces. The magnitude of the resultant force would be $\sqrt{6^2 + 4^2} = 7.21$.

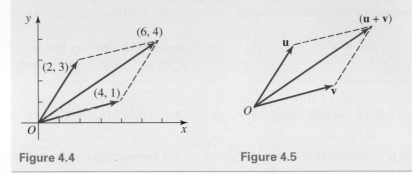

Figure 4.4   **Figure 4.5**

In general, if $\mathbf{u}$ and $\mathbf{v}$ are vectors in the same vector space, then $\mathbf{u} + \mathbf{v}$ is the diagonal of the parallelogram defined by $\mathbf{u}$ and $\mathbf{v}$. See Figure 4.5. This way of visualizing vector addition is useful in all vector spaces.

EXAMPLE **4** This example gives us a geometrical interpretation of scalar multiplication. Consider the scalar multiple of the vector $(3, 2)$ by 2. We get

$$2(3, 2) = (6, 4)$$

Observe in Figure 4.6 that $(6, 4)$ is a vector in the same direction as $(3, 2)$, and 2 times it in length.

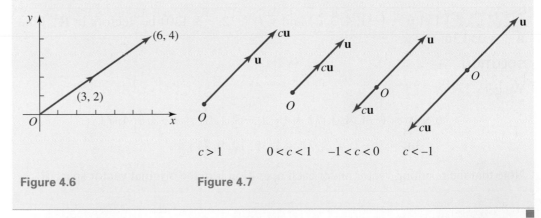

Figure 4.6   **Figure 4.7**

In general, if $\mathbf{u}$ is a vector in any vector space, and c is a nonzero scalar, the direction of $c\mathbf{u}$ will be the same as the direction of $\mathbf{u}$ if $c > 0$, and the opposite direction to $\mathbf{u}$ if $c < 0$. The length of $c\mathbf{u}$ is $|c|$ times the length of $\mathbf{u}$. See Figure 4.7.

Special Vectors

The vector $(0, 0, \ldots, 0)$, having n zero components, is called the **zero vector** of $\mathbf{R}^n$ and is denoted $\mathbf{0}$. For example, $(0, 0, 0)$ is the zero vector of $\mathbf{R}^3$. We shall find that zero vectors play a central role in the development of vector spaces.

The vector $(-1)\mathbf{u}$ is written $-\mathbf{u}$ and is called the **negative of u**. It is a vector having the same magnitude as $\mathbf{u}$, but lies in the opposite direction to $\mathbf{u}$. For example, $(-2, 3, -1)$ is the negative of $(2, -3, 1)$.

Subtraction   Subtraction is performed on elements of $\mathbf{R}^n$ by subtracting corresponding components. For example, in $\mathbf{R}^3$,

$$(5, 3, -6) - (2, 1, 3) = (3, 2, -9)$$

Observe that this is equivalent to

$$(5, 3, -6) + (-1)(2, 1, 3) = (3, 2, -9)$$

Thus subtraction is not a new operation on $\mathbf{R}^n$—it is the sum of the first vector and the negative of the second. *There are only two independent operations on the vector space* $\mathbf{R}^n$, *namely addition and scalar multiplication.*

We now summarize some of the properties of vector addition and scalar multiplication. (Many of the properties of vectors and matrices are the same.)

THEOREM 4.1

Properties of Vector Addition and Scalar Multiplication
Let $\mathbf{u}$, $\mathbf{v}$, and $\mathbf{w}$ be vectors in $\mathbf{R}^n$ and let c and d be scalars.

(a) $\mathbf{u} + \mathbf{v} = \mathbf{v} + \mathbf{u}$ Commutative property
(b) $\mathbf{u} + (\mathbf{v} + \mathbf{w}) = (\mathbf{u} + \mathbf{v}) + \mathbf{w}$ Associative property
(c) $\mathbf{u} + \mathbf{0} = \mathbf{0} + \mathbf{u} = \mathbf{u}$ Property of the zero vector
(d) $\mathbf{u} + (-\mathbf{u}) = \mathbf{0}$ Property of the negative vector
(e) $c(\mathbf{u} + \mathbf{v}) = c\mathbf{u} + c\mathbf{v}$
(f) $(c + d)\mathbf{u} = c\mathbf{u} + d\mathbf{u}$ } Distributive properties
(g) $c(d\mathbf{u}) = (cd)\mathbf{u}$
(h) $1\mathbf{u} = \mathbf{u}$ Scalar multiplication by 1

Proof   These results are proved by writing the vectors in terms of components and using the definitions of vector addition and scalar multiplication, and the properties of real numbers. We give the proofs of (a) and (e).

$\mathbf{u} + \mathbf{v} = \mathbf{v} + \mathbf{u}$:

Let $\mathbf{u} = (u_1, \ldots, u_n)$ and $\mathbf{v} = (v_1, \ldots, v_n)$. Then

$$
\begin{aligned}
\mathbf{u} + \mathbf{v} &= (u_1, \ldots, u_n) + (v_1, \ldots, v_n) \\
&= (u_1 + v_1, \ldots, u_n + v_n) \\
&= (v_1 + u_1, \ldots, v_n + u_n) \\
&= (v_1, \ldots, v_n) + (u_1, \ldots, u_n) \\
&= \mathbf{v} + \mathbf{u}
\end{aligned}
$$

$c(\mathbf{u} + \mathbf{v}) = c\mathbf{u} + c\mathbf{v}$:

$$
\begin{aligned}
c(\mathbf{u} + \mathbf{v}) &= c((u_1, \ldots, u_n) + (v_1, \ldots, v_n)) \\
&= c(u_1 + v_1, \ldots, u_n + v_n) \\
&= (c(u_1 + v_1), \ldots, c(u_n + v_n)) \\
&= (cu_1 + cv_1, \ldots, cu_n + cv_n) \\
&= (cu_1, \ldots, cu_n) + (cv_1, \ldots, cv_n) \\
&= c(u_1, \ldots, u_n) + c(v_1, \ldots, v_n) \\
&= c\mathbf{u} + c\mathbf{v}
\end{aligned}
$$

Some of the preceding properties can be illustrated geometrically. The commutative property of vector addition is illustrated in Figure 4.8. Note that we get the same diagonal to the parallelogram regardless of whether we add the vectors in the order **u** + **v** or in the order **v** + **u**.

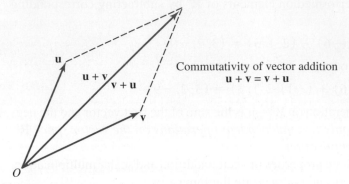

Commutativity of vector addition
u + **v** = **v** + **u**

Figure 4.8

Linear Combinations of Vectors

One implication of the preceding properties is that we can write certain algebraic expressions, such as $a\mathbf{u} + b\mathbf{v} + c\mathbf{w}$, without parentheses. We call $a\mathbf{u} + b\mathbf{v} + c\mathbf{w}$ a **linear combination** of the vectors **u**, **v**, and **w**.

EXAMPLE 5 Let $\mathbf{u} = (2, 5, -3)$, $\mathbf{v} = (-4, 1, 9)$, and $\mathbf{w} = (4, 0, 2)$. Determine the linear combination $2\mathbf{u} - 3\mathbf{v} + \mathbf{w}$.

SOLUTION

We get

$$
\begin{aligned}
2\mathbf{u} - 3\mathbf{v} + \mathbf{w} &= 2(2, 5, -3) - 3(-4, 1, 9) + (4, 0, 2) \\
&= (4, 10, -6) - (-12, 3, 27) + (4, 0, 2) \\
&= (4 + 12 + 4, 10 - 3 + 0, -6 - 27 + 2) \\
&= (20, 7, -31)
\end{aligned}
$$

We say that the vector $(20, 7, -31)$ is a linear combination

$$(20, 7, -31) = 2(2, 5, -3) - 3(-4, 1, 9) + (4, 0, 2)$$

of the three vectors $(2, 5, -3)$, $(-4, 1, 9)$, and $(4, 0, 2)$.

Column Vectors

To this time we have defined only **row vectors**; that is, the components of a vector were written in row form. We shall find that it is more suitable at times to use **column vectors**. Matrix notation is often used for column vectors—many of the properties of vectors and matrices are the same and we shall often find it useful to interpret column vectors as matrices. We define addition and scalar multiplication of column vectors in $\mathbf{R}^n$ in a componentwise manner:

$$
\begin{bmatrix} u_1 \\ \vdots \\ u_n \end{bmatrix} + \begin{bmatrix} v_1 \\ \vdots \\ v_n \end{bmatrix} = \begin{bmatrix} u_1 + v_1 \\ \vdots \\ u_n + v_n \end{bmatrix} \text{ and } c\begin{bmatrix} u_1 \\ \vdots \\ u_n \end{bmatrix} = \begin{bmatrix} cu_1 \\ \vdots \\ cu_n \end{bmatrix}
$$

For example, in $\mathbf{R}^2$,

$$\begin{bmatrix} 1 \\ 2 \end{bmatrix} + \begin{bmatrix} 4 \\ 7 \end{bmatrix} = \begin{bmatrix} 5 \\ 9 \end{bmatrix} \quad \text{and} \quad 4\begin{bmatrix} 2 \\ 1 \end{bmatrix} = \begin{bmatrix} 8 \\ 4 \end{bmatrix}$$

EXERCISE SET 4.1

Position Vectors

1. Sketch the position vectors $(1, 0)$ and $(0, 1)$ in $\mathbf{R}^2$. The notation $\mathbf{i}$ and $\mathbf{j}$ is often used in science for these vectors.

2. Sketch the position vectors $(1, 0, 0), (0, 1, 0), (0, 0, 1)$ in $\mathbf{R}^3$. The notation $\mathbf{i}$, $\mathbf{j}$, and $\mathbf{k}$ is often used in science for these vectors.

3. Sketch the following position vectors in $\mathbf{R}^2$.

   (a) $\overrightarrow{OA} = (5, 6), \overrightarrow{OB} = (-3, 2), \overrightarrow{OC} = (1, -3)$

   (b) $\overrightarrow{OP} = (2, 4), \overrightarrow{OQ} = (-4, 5), \overrightarrow{OR} = (3, -3)$

   (c) $\mathbf{u} = (1, 1), \mathbf{v} = (-1, -4), \mathbf{w} = (5, 3)$

4. Sketch the following position vectors in $\mathbf{R}^3$.

   (a) $\overrightarrow{OA} = (2, 3, 1), \overrightarrow{OB} = (0, 5, -1), \overrightarrow{OC} = (-1, 3, 4)$

   (b) $\mathbf{u} = (1, 1, 1), \mathbf{v} = (-1, -2, -4), \mathbf{w} = (0, 0, -3)$

Addition and Scalar Multiplication

5. Multiply the following vectors by the given scalars.

   (a) $(1, 4)$ by 3   (b) $(-1, 3)$ by -2

   (c) $(2, 6)$ by $\frac{1}{2}$ (d) $(2, 4, 2)$ by $-\frac{1}{2}$

   (e) $(-1, 2, 3)$ by 3   (f) $(-1, 2, 3, -2)$ by 4

   (g) $(1, -4, 3, -2, 5)$ by -5

   (h) $(3, 0, 4, 2, -1)$ by 3

6. Compute the following linear combinations for $\mathbf{u} = (1, 2)$, $\mathbf{v} = (4, -1)$, and $\mathbf{w} = (-3, 5)$.

   (a) $\mathbf{u} + \mathbf{w}$ (b) $\mathbf{u} + 3\mathbf{v}$

   (c) $\mathbf{v} + \mathbf{w}$ (d) $2\mathbf{u} + 3\mathbf{v} - \mathbf{w}$

   (e) $-3\mathbf{u} + 4\mathbf{v} - 2\mathbf{w}$

7. Compute the following linear combinations for $\mathbf{u} = (2, 1, 3), \mathbf{v} = (-1, 3, 2)$, and $\mathbf{w} = (2, 4, -2)$.

   (a) $\mathbf{u} + \mathbf{w}$ (b) $2\mathbf{u} + \mathbf{v}$

   (c) $\mathbf{u} + 3\mathbf{w}$ (d) $5\mathbf{u} - 2\mathbf{v} + 6\mathbf{w}$

   (e) $2\mathbf{u} - 3\mathbf{v} - 4\mathbf{w}$

8. If $\mathbf{u}$, $\mathbf{v}$, and $\mathbf{w}$ are the following column vectors in $\mathbf{R}^2$, determine the linear combinations.

   (a) $\mathbf{u} + \mathbf{v}$ (b) $2\mathbf{v} - 3\mathbf{w}$

   (c) $2\mathbf{u} + 4\mathbf{v} - \mathbf{w}$ (d) $-3\mathbf{u} - 2\mathbf{v} + 4\mathbf{w}$

   $$\mathbf{u} = \begin{bmatrix} 2 \\ 3 \end{bmatrix}, \quad \mathbf{v} = \begin{bmatrix} -1 \\ -4 \end{bmatrix}, \quad \mathbf{w} = \begin{bmatrix} 4 \\ -6 \end{bmatrix}$$

9. If $\mathbf{u}$, $\mathbf{v}$, and $\mathbf{w}$ are the following column vectors in $\mathbf{R}^3$, determine the linear combinations.

   (a) $\mathbf{u} + 2\mathbf{v}$ (b) $-4\mathbf{v} + 3\mathbf{w}$

   (c) $3\mathbf{u} - 2\mathbf{v} + 4\mathbf{w}$ (d) $2\mathbf{u} + 3\mathbf{v} - 8\mathbf{w}$

   $$\mathbf{u} = \begin{bmatrix} 1 \\ 2 \\ -1 \end{bmatrix}, \quad \mathbf{v} = \begin{bmatrix} 3 \\ 0 \\ 1 \end{bmatrix}, \quad \mathbf{w} = \begin{bmatrix} -1 \\ 0 \\ 5 \end{bmatrix}$$

10. Prove the following properties of vector addition and scalar multiplication that were introduced in this section.

    (a) $\mathbf{u} + (\mathbf{v} + \mathbf{w}) = (\mathbf{u} + \mathbf{v}) + \mathbf{w}$

    (b) $\mathbf{u} + (-\mathbf{u}) = \mathbf{0}$

    (c) $(c + d)\mathbf{u} = c\mathbf{u} + d\mathbf{u}$

    (d) $1\mathbf{u} = \mathbf{u}$

4.2 Dot Product, Norm, Angle, and Distance

In this section we develop a geometry for the vector space $\mathbf{R}^n$. As we construct this geometry the reader should pay close attention to the approach we use. Although the results are of course important, the way we arrive at the results is also very important. The magnitude of a vector, the angle between two vectors, and the distance between two points are defined in $\mathbf{R}^n$ by generalizing these concepts from $\mathbf{R}^2$. We extend the Pythagorean Theorem to $\mathbf{R}^n$. This process of gradually extending familiar concepts to more general surroundings is fundamental to mathematics.

We start the discussion with the definition of the dot product of two vectors. The *dot product* is a tool that is used to build the geometry of $\mathbf{R}^n$.

DEFINITION   Let $\mathbf{u} = (u_1, \ldots, u_n)$ and $\mathbf{v} = (v_1, \ldots, v_n)$ be two vectors in $\mathbf{R}^n$. The **dot product** of $\mathbf{u}$ and $\mathbf{v}$ is denoted $\mathbf{u} \cdot \mathbf{v}$ and is defined by

$$\mathbf{u} \cdot \mathbf{v} = u_1 v_1 + \cdots + u_n v_n$$

The dot product assigns a real number to each pair of vectors.

For example, if $\mathbf{u} = (1, -2, 4)$ and $\mathbf{v} = (3, 0, 2)$ their dot product is

$$\mathbf{u} \cdot \mathbf{v} = (1 \times 3) + (-2 \times 0) + (4 \times 2) = 3 + 0 + 8 = 11$$

Properties of the Dot Product

Let $\mathbf{u}$, $\mathbf{v}$, and $\mathbf{w}$ be vectors in $\mathbf{R}^n$ and let c be a scalar. Then

1. $\mathbf{u} \cdot \mathbf{v} = \mathbf{v} \cdot \mathbf{u}$
2. $(\mathbf{u} + \mathbf{v}) \cdot \mathbf{w} = \mathbf{u} \cdot \mathbf{w} + \mathbf{v} \cdot \mathbf{w}$
3. $c\mathbf{u} \cdot \mathbf{v} = c(\mathbf{u} \cdot \mathbf{v}) = \mathbf{u} \cdot c\mathbf{v}$
4. $\mathbf{u} \cdot \mathbf{u} \geq 0$, and $\mathbf{u} \cdot \mathbf{u} = 0$ if and only if $\mathbf{u} = \mathbf{0}$

Proof   We prove Parts 1 and 4, leaving the other parts for the reader to prove in the exercises that follow.

1. Let $\mathbf{u} = (u_1, \ldots, u_n)$ and $\mathbf{v} = (v_1, \ldots, v_n)$. We get

$$
\begin{aligned}
\mathbf{u} \cdot \mathbf{v} &= u_1 v_1 + \cdots + u_n v_n \\
&= v_1 u_1 + \cdots + v_n u_n, \text{ by the commutative property of real numbers} \\
&= \mathbf{v} \cdot \mathbf{u}
\end{aligned}
$$

4. $\mathbf{u} \cdot \mathbf{u} = u_1 u_1 + \cdots + u_n u_n = (u_1)^2 + \cdots + (u_n)^2$.

$(u_1)^2 + \cdots + (u_n)^2 \geq 0$, since it is the sum of squares. Thus $u \cdot u \geq 0$.

Further, $(u_1)^2 + \cdots + (u_n)^2 = 0$ if and only if $u_1 = 0, \ldots, u_n = 0$.

Thus $\mathbf{u} \cdot \mathbf{u} = 0$ if and only if $\mathbf{u} = \mathbf{0}$.

We shall use these properties to simplify expressions involving dot products. In this section, for example, we shall use the properties in arriving at a suitable expression for the angle between two vectors in $\mathbf{R}^n$. In later sections we shall use these properties to generalize the concept of a dot product to spaces of matrices and functions.

Norm of a Vector in $\mathbf{R}^n$

Let $\mathbf{u} = (u_1, u_2)$ be a vector in $\mathbf{R}^2$. We know by the Pythagorean Theorem that the length of this vector is $\sqrt{(u_1)^2 + (u_2)^2}$. See Figure 4.9. We generalize this result to define the length (the technical name is *norm*) of a vector in $\mathbf{R}^n$.

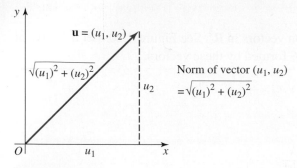

Figure 4.9

> **DEFINITION** The **norm** (**length** or **magnitude**) of a vector $\mathbf{u} = (u_1, \ldots, u_n)$ in $\mathbf{R}^n$ is denoted $\|\mathbf{u}\|$ and defined by
>
> $$\|\mathbf{u}\| = \sqrt{(u_1)^2 + \cdots + (u_n)^2}$$
>
> *Note*: The norm of a vector can also be written in terms of the dot product
>
> $$\|\mathbf{u}\| = \sqrt{\mathbf{u} \cdot \mathbf{u}}$$

For example, if $\mathbf{u} = (1, 3, 5)$ in $\mathbf{R}^3$ and $\mathbf{v} = (3, 0, 1, 4)$ in $\mathbf{R}^4$, then

$$\|\mathbf{u}\| = \sqrt{(1)^2 + (3)^2 + (5)^2} = \sqrt{1 + 9 + 25} = \sqrt{35}, \text{ and}$$

$$\|\mathbf{v}\| = \sqrt{(3)^2 + (0)^2 + (1)^2 + (4)^2} = \sqrt{9 + 0 + 1 + 16} = \sqrt{26}$$

> **DEFINITION** A **unit vector** is a vector whose norm is one. If $\mathbf{v}$ is a nonzero vector then the vector
>
> $$\mathbf{u} = \frac{1}{\|\mathbf{v}\|} \mathbf{v}$$
>
> is a unit vector in the direction of $\mathbf{v}$. (See Exercise 24.)
>   This procedure of constructing a unit vector in the same direction as a given vector is called **normalizing** the vector.

EXAMPLE 1 Find the norm of the vector $(2, -1, 3)$. Normalize this vector.

SOLUTION

$\|(2, -1, 3)\| = \sqrt{2^2 + (-1)^2 + 3^2} = \sqrt{14}$. The norm of $(2, -1, 3)$ is $\sqrt{14}$. The normalized vector is

$$\frac{1}{\sqrt{14}}(2, -1, 3)$$

This vector may also be written $\left(\frac{2}{\sqrt{14}}, \frac{-1}{\sqrt{14}}, \frac{3}{\sqrt{14}}\right)$. This vector is a unit vector in the direction of $(2, -1, 3)$.

Angle Between Vectors

Let $\mathbf{u} = (a, b)$ and $\mathbf{v} = (c, d)$ be position vectors in $\mathbf{R}^2$. See Figure 4.10. Let us arrive at an expression for the cosine of the angle θ formed by these vectors.

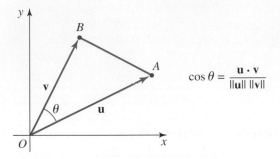

$$\cos\theta = \frac{\mathbf{u} \cdot \mathbf{v}}{\|\mathbf{u}\|\,\|\mathbf{v}\|}$$

Figure 4.10

The **law of cosines** gives

$$AB^2 = OA^2 + OB^2 - 2(OA)(OB)\cos\theta$$

Solve for $\cos\theta$.

$$\cos\theta = \frac{OA^2 + OB^2 - AB^2}{2(OA)(OB)}$$

We get that

$$
\begin{aligned}
OA^2 + OB^2 - AB^2 &= \|\mathbf{u}\|^2 + \|\mathbf{v}\|^2 - \|\mathbf{v} - \mathbf{u}\|^2 \\
&= a^2 + b^2 + c^2 + d^2 - [(c - a)^2 + (d - b)^2] \\
&= 2ac + 2db = 2\mathbf{u} \cdot \mathbf{v}
\end{aligned}
$$

Further,

$$2(OA)(OB) = 2\|\mathbf{u}\|\,\|\mathbf{v}\|$$

Thus, the angle θ between two vectors in $\mathbf{R}^2$ is given by

$$\cos\theta = \frac{\mathbf{u} \cdot \mathbf{v}}{\|\mathbf{u}\|\,\|\mathbf{v}\|}$$

We know that the cosine of any angle such as θ has to satisfy the condition $|\cos\theta| \leq 1$. The Cauchy-Schwarz inequality, which we discuss later in this section, assures us that

$$\left|\frac{\mathbf{u} \cdot \mathbf{v}}{\|\mathbf{u}\|\,\|\mathbf{v}\|}\right| \leq 1$$

We can thus extend this concept of angle between two vectors to $\mathbf{R}^n$ as follows.

DEFINITION   Let $\mathbf{u}$ and $\mathbf{v}$ be two nonzero vectors in $\mathbf{R}^n$. The cosine of the angle θ between these vectors is

$$\cos\theta = \frac{\mathbf{u} \cdot \mathbf{v}}{\|\mathbf{u}\|\,\|\mathbf{v}\|}, \qquad 0 \leq \theta \leq \pi$$

EXAMPLE 2 Determine the angle between the vectors $\mathbf{u} = (1, 0, 0)$ and $\mathbf{v} = (1, 0, 1)$ in $\mathbf{R}^3$.

SOLUTION

We get

$$\mathbf{u} \cdot \mathbf{v} = (1, 0, 0) \cdot (1, 0, 1) = 1, \qquad \|\mathbf{u}\| = \sqrt{1^2 + 0^2 + 0^2} = 1,$$

$$\|\mathbf{v}\| = \sqrt{1^2 + 0^2 + 1^2} = \sqrt{2}.$$

Thus

$$\cos\theta = \frac{\mathbf{u} \cdot \mathbf{v}}{\|\mathbf{u}\| \|\mathbf{v}\|} = \frac{1}{\sqrt{2}}$$

The angle between $\mathbf{u}$ and $\mathbf{v}$ is $\pi/4$ (or $45°$).

We say that two nonzero vectors are **orthogonal** if the angle between them is a right angle. The following theorem, which the reader is asked to prove in Exercise 25, gives us a condition for orthogonality.

THEOREM 4.2

Two nonzero vectors $\mathbf{u}$ and $\mathbf{v}$ are orthogonal if and only if $\mathbf{u} \cdot \mathbf{v} = 0$.

For example, the vectors $(2, -3, 1)$ and $(1, 2, 4)$ are orthogonal since

$$(2, -3, 1) \cdot (1, 2, 4)) = (2 \times 1) + (-3 \times 2) + (1 \times 4) = 2 - 6 + 4 = 0$$

EXAMPLE 3 Determine a vector in $\mathbf{R}^2$ that is orthogonal to $(3, -1)$. Show that there are many such vectors and that they all lie on a line.

SOLUTION

Let the vector (a, b) be orthogonal to $(3, -1)$. We get

$$(a, b) \cdot (3, -1) = 0$$
$$(a \times 3) + (b \times -1) = 0$$
$$3a - b = 0$$
$$b = 3a$$

Thus any vector of the form $(a, 3a)$ is orthogonal to the vector $(3, -1)$.

Any vector of this form can be written

$$a(1, 3)$$

The set of all such vectors lie on the line defined by the vector $(1, 3)$. See Figure 4.11.

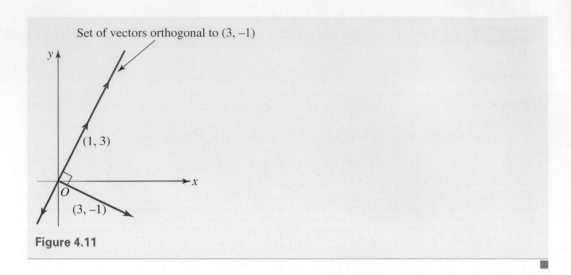

Figure 4.11

The following example illustrates how these concepts also apply to vectors written in column form.

EXAMPLE 4 (a) Show that the following vectors are orthogonal. (b) Compute the norm of each vector.

$$\mathbf{u} = \begin{bmatrix} 1 \\ 2 \\ -5 \end{bmatrix}, \mathbf{v} = \begin{bmatrix} -1 \\ 3 \\ 1 \end{bmatrix}$$

SOLUTION

(a) The dot product of **u** and **v** is obtained by multiplying corresponding components and adding.

$$\mathbf{u} \cdot \mathbf{v} = (1 \times -1) + (2 \times 3) + (-5 \times 1) = 0$$

The dot product is zero, thus the vectors are orthogonal.

(b) The norm of each vector is obtained by taking the square root of the sum of the squares of the components.

$$\|\mathbf{u}\| = \sqrt{1^2 + 2^2 + (-5)^2} = \sqrt{30}$$
$$\|\mathbf{v}\| = \sqrt{(-1)^2 + 3^2 + 1^2} = \sqrt{11}$$

Having defined angles between vectors and magnitudes of vectors, we now complete the geometry of a vector space by defining distances between points.

Distance Between Points

The distance between the points $\mathbf{x} = (x_1, x_2)$ and $\mathbf{y} = (y_1, y_2)$ in $\mathbf{R}^2$ is $d(\mathbf{x}, \mathbf{y}) = \sqrt{(x_1 - y_1)^2 + (x_2 - y_2)^2}$. We define distance in $\mathbf{R}^n$ by generalizing this expression as follows.

> **DEFINITION** Let $\mathbf{x} = (x_1, \ldots, x_n)$ and $\mathbf{y} = (y_1, \ldots, y_n)$ be two points in $\mathbf{R}^n$. The **distance between x and y** is denoted $d(\mathbf{x}, \mathbf{y})$ and is defined by
>
> $$d(\mathbf{x}, \mathbf{y}) = \sqrt{(x_1 - y_1)^2 + \cdots + (x_n - y_n)^2}$$
>
> *Note*: We can also write this distance formula as
>
> $$d(\mathbf{x}, \mathbf{y}) = \|\mathbf{x} - \mathbf{y}\|$$

See Figure 4.12 for distance in $\mathbf{R}^2$, $\mathbf{R}^3$, and $\mathbf{R}^n$.

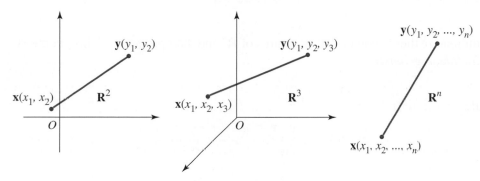

$$\mathbf{R}^2 \colon d(\mathbf{x}, \mathbf{y}) = \sqrt{(x_1 - y_1)^2 + (x_2 - y_2)^2}$$
$$\mathbf{R}^3 \colon d(\mathbf{x}, \mathbf{y}) = \sqrt{(x_1 - y_1)^2 + (x_2 - y_2)^2 + (x_3 - y_3)^2}$$
$$\mathbf{R}^n \colon d(\mathbf{x}, \mathbf{y}) = \sqrt{(x_1 - y_1)^2 + \cdots + (x_n - y_n)^2}$$

Figure 4.12

EXAMPLE 5 Determine the distance between the points $\mathbf{x} = (1, -2, 3, 0)$ and $\mathbf{y} = (4, 0, -3, 5)$ in $\mathbf{R}^4$.

SOLUTION

Applying the above formula for distance we get

$$d(\mathbf{x}, \mathbf{y}) = \sqrt{(1 - 4)^2 + (-2 - 0)^2 + (3 + 3)^2 + (0 - 5)^2}$$
$$= \sqrt{9 + 4 + 36 + 25}$$
$$= \sqrt{74}$$

The norms of vectors and distances between points have certain mathematical properties, which we list below. The following is an important property of distance for example.

EXAMPLE 6 Prove that distance in $\mathbf{R}^n$ has the following **symmetric property**:
$d(\mathbf{x}, \mathbf{y}) = d(\mathbf{y}, \mathbf{x})$.

SOLUTION

Let $\mathbf{x} = (x_1, \ldots, x_n)$ and $\mathbf{y} = (y_1, \ldots, y_n)$. We get

$$d(\mathbf{x}, \mathbf{y}) = \sqrt{(x_1 - y_1)^2 + \cdots + (x_n - y_n)^2}$$
$$= \sqrt{(y_1 - x_1)^2 + \cdots + (y_n - x_n)^2} = d(\mathbf{y}, \mathbf{x})$$

This result tells us that the distance from $\mathbf{x}$ to $\mathbf{y}$ is the same as the distance from $\mathbf{y}$ to $\mathbf{x}$. This property is one we would naturally want a distance function to have.

We now summarize these geometrical structures of $\mathbf{R}^n$ and their properties. This geometry is called *Euclidean geometry*.

Euclidean Geometry of $\mathbf{R}^n$

Dot product of vectors $\mathbf{u}$ *and* $\mathbf{v}$: $\mathbf{u} \cdot \mathbf{v} = u_1 v_1 + \cdots + u_n v_n$

Norm of a vector $\mathbf{u}$: $\|\mathbf{u}\| = \sqrt{(u_1)^2 + \cdots + (u_n)^2}$

Angle between vectors $\mathbf{u}$ *and* $\mathbf{v}$: $\cos \theta = \dfrac{\mathbf{u} \cdot \mathbf{v}}{\|\mathbf{u}\| \|\mathbf{v}\|}$

Distance between points $\mathbf{x}$ *and* $\mathbf{y}$: $d(\mathbf{x}, \mathbf{y}) = \sqrt{(x_1 - y_1)^2 + \cdots + (x_n - y_n)^2}$

Properties of Norm

1. $\|\mathbf{u}\| \geq 0$
   (the length of a vector cannot be negative)

2. $\|\mathbf{u}\| = 0$ if and only if $\mathbf{u} = \mathbf{0}$
   (the length of a vector is zero if and only if the vector is the zero vector)

3. $\|c\mathbf{u}\| = |c| \|\mathbf{u}\|$
   (the length of $c\mathbf{u}$ is $|c|$ times the length of $\mathbf{u}$)

Properties of Distance

1. $d(\mathbf{x}, \mathbf{y}) \geq 0$
   (the distance between two points cannot be negative)

2. $d(\mathbf{x}, \mathbf{y}) = 0$ if and only if $\mathbf{x} = \mathbf{y}$
   (the distance between two points is zero if and only if the points are coincident)

3. $d(\mathbf{x}, \mathbf{y}) = d(\mathbf{y}, \mathbf{x})$ (symmetry property)
   (the distance between $\mathbf{x}$ and $\mathbf{y}$ is the same as the distance between $\mathbf{y}$ and $\mathbf{x}$)

In the exercises that follow you are asked to derive those properties not already discussed. The norms and distances introduced in this section were all derived from the dot product. In numerical work it is often convenient to define other norms and distances using these properties as axioms—basic properties that any norm or distance function should have. The reader will meet some of these ideas in later sections.

We complete this section with three useful theoretical results. The first of these is the *Cauchy-Schwarz inequality*. This inequality enabled us earlier to extend the definition of angle that we have in $\mathbf{R}^2$ to $\mathbf{R}^n$. Here it enables us to extend the Pythagorean Theorem to $\mathbf{R}^n$.

THEOREM 4.3

The Cauchy-Schwarz Inequality*

If **u** and **v** are vectors in $\mathbf{R}^n$, then

$$|\mathbf{u} \cdot \mathbf{v}| \leq \|\mathbf{u}\| \|\mathbf{v}\|$$

Here $|\mathbf{u} \cdot \mathbf{v}|$ denotes the absolute value of the number $\mathbf{u} \cdot \mathbf{v}$.

Proof   We first look at the special case when $\mathbf{u} = \mathbf{0}$. If $\mathbf{u} = \mathbf{0}$, then $|\mathbf{u} \cdot \mathbf{v}| = 0$ and $\|\mathbf{u}\| \|\mathbf{v}\| = 0$. Equality holds.

Let us now consider $\mathbf{u} \neq \mathbf{0}$. The proof of the inequality for this case involves a very interesting application of the quadratic formula. Consider the vector $r\mathbf{u} + \mathbf{v}$, where r is any real number. Using the properties of the dot product we get

$$(r\mathbf{u} + \mathbf{v}) \cdot (r\mathbf{u} + \mathbf{v}) = r^2(\mathbf{u} \cdot \mathbf{u}) + 2r(\mathbf{u} \cdot \mathbf{v}) + \mathbf{v} \cdot \mathbf{v}$$

The properties of the dot product also tell us that

$$(r\mathbf{u} + \mathbf{v}) \cdot (r\mathbf{u} + \mathbf{v}) \geq 0$$

Combining these results we get

$$r^2(\mathbf{u} \cdot \mathbf{u}) + 2r(\mathbf{u} \cdot \mathbf{v}) + \mathbf{v} \cdot \mathbf{v} \geq 0$$

Let $a = \mathbf{u} \cdot \mathbf{u}, b = 2\mathbf{u} \cdot \mathbf{v}, c = \mathbf{v} \cdot \mathbf{v}$. Thus

$$ar^2 + br + c \geq 0$$

This implies that the **quadratic function** $f(r) = ar^2 + br + c$ is never negative. Therefore $f(r)$, whose graph is a parabola, must have either one zero or no zeros. The zeros of $f(r)$ are the roots of the equation

$$ar^2 + br + c = 0$$

The **discriminant** gives information about these zeros. There is one zero if $b^2 - 4ac = 0$ and no zeros if $b^2 - 4ac < 0$. Thus

$$b^2 - 4ac \leq 0$$
$$b^2 \leq 4ac$$
$$(2\mathbf{u} \cdot \mathbf{v})^2 \leq 4(\mathbf{u} \cdot \mathbf{u})(\mathbf{v} \cdot \mathbf{v})$$
$$(\mathbf{u} \cdot \mathbf{v})^2 \leq (\mathbf{u} \cdot \mathbf{u})(\mathbf{v} \cdot \mathbf{v})$$
$$(\mathbf{u} \cdot \mathbf{v})^2 \leq \|\mathbf{u}\|^2 \|\mathbf{v}\|^2$$

Taking the square root of both sides gives the Cauchy-Schwarz inequality.

$$|\mathbf{u} \cdot \mathbf{v}| \leq \|\mathbf{u}\| \|\mathbf{v}\|$$

---

*Augustin Louis Cauchy (1789–1857) was educated in Paris. He worked as an engineer before returning to teach in Paris. He was one of the greatest mathematicians of all time. He was also very prolific, writing seven books and over eight hundred papers. More mathematical concepts and results have been named after him than any other mathematician. Cauchy had a very active life outside mathematics. After the revolution of 1830 he refused to take an oath of allegiance to Louis-Phillipe and went into exile in Turin, but was later able to return to Paris. Cauchy was a devout Catholic and was involved in social work. He was, for example, involved in charities for unwed mothers, aid for the starving in Ireland, and the rehabilitation of criminals. On giving one year's salary to the mayor of Sceaux for charity he commented, "Do not worry, it is only my salary; it is not my money, it is the emperor's."

Herman Amandus Schwarz (1843–1921) was educated in Berlin and taught in Zurich, Göttingen, and Berlin. His greatest asset was his geometric intuition. The methods he developed have turned out to be more significant than the actual problems he worked on. His teaching and his concern for students did not leave him much time for publishing. Another factor that may have contributed to his lack of publishing was a weakness he had for devoting as much thoroughness to the trivial as to the important.

We now extend two familiar geometrical results from $\mathbf{R}^2$ to $\mathbf{R}^n$.

THEOREM 4.4

Let $\mathbf{u}$ and $\mathbf{v}$ be vectors in $\mathbf{R}^n$.

(a) **Triangle inequality:** $\|\mathbf{u} + \mathbf{v}\| \leq \|\mathbf{u}\| + \|\mathbf{v}\|$

This inequality tells us that the length of one side of a triangle cannot exceed the sum of the lengths of the other two sides. See Figure 4.13(a).

(b) **Pythagorean Theorem:** If $\mathbf{u} \cdot \mathbf{v} = 0$, then $\|\mathbf{u} + \mathbf{v}\|^2 = \|\mathbf{u}\|^2 + \|\mathbf{v}\|^2$.

The square of the hypotenuse of a right triangle is equal to the sum of the squares of the other two sides. See Figure 4.13(b).

Proof

(a) By the properties of norm,

$$
\begin{aligned}
\|\mathbf{u} + \mathbf{v}\|^2 &= (\mathbf{u} + \mathbf{v}) \cdot (\mathbf{u} + \mathbf{v}) \\
&= \mathbf{u} \cdot \mathbf{u} + 2\mathbf{u} \cdot \mathbf{v} + \mathbf{v} \cdot \mathbf{v} \\
&= \|\mathbf{u}\|^2 + 2\mathbf{u} \cdot \mathbf{v} + \|\mathbf{v}\|^2 \\
&\leq \|\mathbf{u}\|^2 + 2|\mathbf{u} \cdot \mathbf{v}| + \|\mathbf{v}\|^2
\end{aligned}
$$

The Cauchy-Schwarz inequality implies that

$$
\begin{aligned}
\|\mathbf{u} + \mathbf{v}\|^2 &\leq \|\mathbf{u}\|^2 + 2\|\mathbf{u}\|\|\mathbf{v}\| + \|\mathbf{v}\|^2 \\
&= (\|\mathbf{u}\| + \|\mathbf{v}\|)^2
\end{aligned}
$$

Taking the square root of each side gives the triangle inequality.

(b) By the properties of dot product, and the fact that $\mathbf{u} \cdot \mathbf{v} = 0$, we get

$$
\begin{aligned}
\|\mathbf{u} + \mathbf{v}\|^2 &= (\mathbf{u} + \mathbf{v}) \cdot (\mathbf{u} + \mathbf{v}) \\
&= \mathbf{u} \cdot \mathbf{u} + \mathbf{u} \cdot \mathbf{v} + \mathbf{v} \cdot \mathbf{u} + \mathbf{v} \cdot \mathbf{v} \\
&= \mathbf{u} \cdot \mathbf{u} + 2\mathbf{u} \cdot \mathbf{v} + \mathbf{v} \cdot \mathbf{v} \\
&= \|\mathbf{u}\|^2 + \|\mathbf{v}\|^2
\end{aligned}
$$

proving the Pythagorean Theorem.

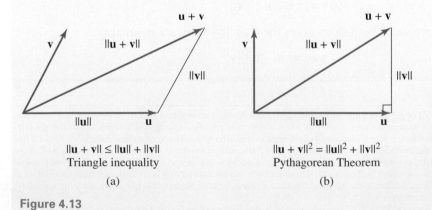

$\|\mathbf{u} + \mathbf{v}\| \leq \|\mathbf{u}\| + \|\mathbf{v}\|$
Triangle inequality
(a)

$\|\mathbf{u} + \mathbf{v}\|^2 = \|\mathbf{u}\|^2 + \|\mathbf{v}\|^2$
Pythagorean Theorem
(b)

Figure 4.13

EXERCISE SET 4.2

Dot Product

1. Determine the dot products of the following pairs of vectors.
   (a) $(2, 1), (3, 4)$ **(b)** $(1, -4), (3, 0)$
   (c) $(2, 0), (0, -1)$ **(d)** $(5, -2), (-3, -4)$

2. Determine the dot products of the following pairs of vectors.
   (a) $(1, 2, 3), (4, 1, 0)$ **(b)** $(3, 4, -2), (5, 1, -1)$
   (c) $(7, 1, -2), (3, -5, 8)$
   (d) $(3, 2, 0), (5, -2, 8)$

3. Determine the dot products of the following pairs of vectors.
   (a) $(5, 1), (2, -3)$ **(b)** $(-3, 1, 5), (2, 0, 4)$
   (c) $(7, 1, 2, -4), (3, 0, -1, 5)$
   (d) $(2, 3, -4, 1, 6), (-3, 1, -4, 5, -1)$
   (e) $(1, 2, 3, 0, 0, 0), (0, 0, 0, -2, -4, 9)$

4. Determine the dot products of the following pairs of column vectors.

   (a) $\begin{bmatrix} 1 \\ 3 \end{bmatrix}, \begin{bmatrix} -2 \\ 5 \end{bmatrix}$ **(b)** $\begin{bmatrix} 5 \\ 0 \end{bmatrix}, \begin{bmatrix} 4 \\ -6 \end{bmatrix}$

   (c) $\begin{bmatrix} 2 \\ 0 \\ -5 \end{bmatrix}, \begin{bmatrix} 3 \\ 6 \\ -4 \end{bmatrix}$ **(d)** $\begin{bmatrix} 1 \\ 3 \\ -7 \end{bmatrix}, \begin{bmatrix} -2 \\ 8 \\ -3 \end{bmatrix}$

Norms of Vectors

5. Find the norms of the following vectors.
   (a) $(1, 2)$ **(b)** $(3, -4)$
   (c) $(4, 0)$ **(d)** $(-3, 1)$
   (e) $(0, 27)$

6. Find the norms of the following vectors.
   (a) $(1, 3, -1)$ **(b)** $(3, 0, 4)$
   (c) $(5, 1, 1)$ **(d)** $(0, 5, 0)$
   (e) $(7, -2, -3)$

7. Find the norms of the following vectors.
   (a) $(5, 2)$ **(b)** $(-4, 2, 3)$
   (c) $(1, 2, 3, 4)$ **(d)** $(4, -2, 1, 3)$
   (e) $(-3, 0, 1, 4, 2)$ **(f)** $(0, 0, 0, 7, 0, 0)$

8. Find the norms of the following column vectors.

   (a) $\begin{bmatrix} 3 \\ 4 \end{bmatrix}$ **(b)** $\begin{bmatrix} 2 \\ -7 \end{bmatrix}$ **(c)** $\begin{bmatrix} 1 \\ 2 \\ 3 \end{bmatrix}$

   (d) $\begin{bmatrix} -2 \\ 0 \\ 5 \end{bmatrix}$ **(e)** $\begin{bmatrix} 2 \\ 3 \\ 5 \\ 9 \end{bmatrix}$

Normalizing Vectors

9. Normalize the following vectors.
   (a) $(1, 3)$ **(b)** $(2, -4)$
   (c) $(1, 2, 3)$ **(d)** $(-2, 4, 0)$
   (e) $(0, 5, 0)$

10. Normalize the following vectors.
   (a) $(4, 2)$ **(b)** $(4, 1, 1)$
   (c) $(7, 2, 0, 1)$ **(d)** $(3, -1, 1, 2)$
   (e) $(0, 0, 0, 7, 0, 0)$

11. Normalize the following column vectors.

   (a) $\begin{bmatrix} 4 \\ 3 \end{bmatrix}$ **(b)** $\begin{bmatrix} 1 \\ -3 \end{bmatrix}$ **(c)** $\begin{bmatrix} 3 \\ 4 \\ 0 \end{bmatrix}$

   (d) $\begin{bmatrix} -1 \\ 2 \\ -5 \end{bmatrix}$ **(e)** $\begin{bmatrix} 3 \\ 0 \\ 1 \\ 8 \end{bmatrix}$

Angles Between Vectors

12. Determine the angles between the following pairs of vectors.
   (a) $(-1, 1), (0, 1)$ **(b)** $(2, 0), (1, \sqrt{3})$
   (c) $(2, 3), (3, -2)$ **(d)** $(5, 2), (-5, -2)$

13. Determine the cosines of the angles between the following pairs of vectors.
   (a) $(4, -1), (2, 3)$ **(b)** $(3, -1, 2), (4, 1, 1)$
   (c) $(2, -1, 0), (5, 3, 1)$ **(d)** $(7, 1, 0, 0), (3, 2, 1, 0)$
   (e) $(1, 2, -1, 3, 1), (2, 0, 1, 0, 4)$

14. Determine the cosines of the angles between the following pairs of column vectors.

   (a) $\begin{bmatrix} 1 \\ 2 \end{bmatrix}, \begin{bmatrix} -1 \\ 4 \end{bmatrix}$ **(b)** $\begin{bmatrix} 5 \\ 1 \end{bmatrix}, \begin{bmatrix} 0 \\ -3 \end{bmatrix}$

   (c) $\begin{bmatrix} 1 \\ -3 \\ 0 \end{bmatrix}, \begin{bmatrix} 2 \\ 5 \\ -1 \end{bmatrix}$ **(d)** $\begin{bmatrix} -2 \\ 3 \\ -4 \end{bmatrix}, \begin{bmatrix} 2 \\ 5 \\ -1 \end{bmatrix}$

Orthogonality

15. Show that the following pairs of vectors are orthogonal.
   (a) $(1, 3), (3, -1)$ **(b)** $(-2, 4), (4, 2)$
   (c) $(3, 0), (0, -2)$ **(d)** $(7, -1), (1, 7)$

16. Show that the following pairs of vectors are orthogonal.
   (a) $(3, -5), (5, 3)$ **(b)** $(1, 2, -3), (4, 1, 2)$
   (c) $(7, 1, 0), (2, -14, 3)$
   (d) $(5, 1, 0, 2), (-3, 7, 9, 4)$
   (e) $(1, -1, 2, -5, 9), (4, 7, 4, 1, 0)$

17. Show that the following pairs of column vectors are orthogonal.

(a) $\begin{bmatrix} 1 \\ 2 \end{bmatrix}, \begin{bmatrix} -6 \\ 3 \end{bmatrix}$ (b) $\begin{bmatrix} 5 \\ -2 \end{bmatrix}, \begin{bmatrix} 4 \\ 10 \end{bmatrix}$

(c) $\begin{bmatrix} 4 \\ -1 \\ 0 \end{bmatrix}, \begin{bmatrix} 2 \\ 8 \\ -1 \end{bmatrix}$ (d) $\begin{bmatrix} -2 \\ 3 \\ 2 \end{bmatrix}, \begin{bmatrix} 2 \\ 6 \\ -7 \end{bmatrix}$

18. Determine nonzero vectors that are orthogonal to the following vectors.

(a) $(1, 3)$ (b) $(7, -1)$

(c) $(-4, -1)$ (d) $(-3, 0)$

19. Determine nonzero vectors that are orthogonal to the following vectors.

(a) $(5, -1)$ (b) $(1, -2, 3)$

(c) $(5, 1, -1)$ (d) $(5, 0, 1, 1)$

(e) $(6, -1, 2, 3)$ (f) $(0, -2, 3, 1, 5)$

20. Determine a vector that is orthogonal to both $(1, 2, -1)$ and $(3, 1, 0)$.

Distances Between Points

21. Find the distances between the following pairs of points.

(a) $(6, 5), (2, 2)$

(b) $(3, 1), (-4, 0)$

(c) $(7, -3), (2, 2)$

(d) $(1, -3), (5, 1)$

22. Find the distances between the following pairs of points.

(a) $(4, 1), (2, -3)$

(b) $(1, 2, 3), (2, 1, 0)$

(c) $(-3, 1, 2), (4, -1, 1)$

(d) $(5, 1, 0, 0), (2, 0, 1, 3)$

(e) $(-3, 1, 1, 0, 2), (2, 1, 4, 1, -1)$

Miscellaneous Results

23. Prove the following two properties of the dot product.

(a) $(\mathbf{u} + \mathbf{v}) \cdot \mathbf{w} = \mathbf{u} \cdot \mathbf{w} + \mathbf{v} \cdot \mathbf{w}$

(b) $c\mathbf{u} \cdot \mathbf{v} = c(\mathbf{u} \cdot \mathbf{v}) = \mathbf{u} \cdot c\mathbf{v}$

24. Prove that if $\mathbf{v}$ is a nonzero vector, then the following vector $\mathbf{u}$ is a unit vector in the direction of $\mathbf{v}$.

$$\mathbf{u} = \frac{1}{\|\mathbf{v}\|} \mathbf{v}$$

25. Prove that two nonzero vectors $\mathbf{u}$ and $\mathbf{v}$ are orthogonal if and only if $\mathbf{u} \cdot \mathbf{v} = 0$.

26. Show that if $\mathbf{v}$ and $\mathbf{w}$ are two vectors in a vector space U and $\mathbf{u} \cdot \mathbf{v} = \mathbf{u} \cdot \mathbf{w}$ for all vectors $\mathbf{u}$ in U, then $\mathbf{v} = \mathbf{w}$.

27. Let $\mathbf{u}, \mathbf{v}_1, \ldots, \mathbf{v}_n$ be vectors in a given vector space. Let $a_1, \ldots, a_n$ be scalars. Prove that

$$\mathbf{u} \cdot (a_1\mathbf{v}_1 + \cdots + a_n\mathbf{v}_n) = a_1\mathbf{u} \cdot \mathbf{v}_1 + \cdots + a_n\mathbf{u} \cdot \mathbf{v}_n$$

28. Let $\mathbf{u}, \mathbf{v}$, and $\mathbf{w}$ be vectors in a given Euclidean space and let c and d be nonzero scalars. Tell whether each of the following expressions is a scalar, a vector, or is not a valid expression.

(a) $(\mathbf{u} \cdot \mathbf{v})\mathbf{w}$ (b) $(\mathbf{u} \cdot \mathbf{v}) \cdot \mathbf{w}$

(c) $\mathbf{u} \cdot \mathbf{v} + c\mathbf{w}$ (d) $\mathbf{u} \cdot \mathbf{v} + c$

(e) $c(\mathbf{u} \cdot \mathbf{v}) + d\mathbf{w}$ (f) $\|\mathbf{u} + c\mathbf{v}\| + d$

(g) $c\mathbf{u} \cdot d\mathbf{v} + \|\mathbf{w}\|\mathbf{v}$ (h) $\|\mathbf{u} \cdot \mathbf{v}\|$

29. Find all the values of c such that $\|c(3, 0, 4)\| = 15$.

30. Prove that $\mathbf{u}$ and $\mathbf{v}$ are orthogonal vectors if and only if

$$\|\mathbf{u} + \mathbf{v}\|^2 = \|\mathbf{u}\|^2 + \|\mathbf{v}\|^2$$

31. Let (a, b) be a vector in $\mathbf{R}^2$. Prove that the vector $(-b, a)$ is orthogonal to (a, b).

32. Let $\mathbf{u}$ and $\mathbf{v}$ be vectors in $\mathbf{R}^n$. Prove that $\|\mathbf{u}\| = \|\mathbf{v}\|$ if and only if $\mathbf{u} + \mathbf{v}$ and $\mathbf{u} - \mathbf{v}$ are orthogonal.

33. Let $\mathbf{u}$ be a vector in $\mathbf{R}^n$ and c a scalar. Prove that the norm of a vector has the following properties.

(a) $\|\mathbf{u}\| \geq \mathbf{0}$

(b) $\|\mathbf{u}\| = 0$ if and only if $\mathbf{u} = \mathbf{0}$

(c) $\|c\mathbf{u}\| = |c| \|\mathbf{u}\|$

34. Consider the vector space $\mathbf{R}^n$. Let $\mathbf{u} = (u_1, \ldots, u_n)$ be a vector in $\mathbf{R}^n$. Prove that both the following satisfy the properties of norm mentioned in Exercise 33. These expressions, even though they do not lead to Euclidean geometry, have all the algebraic properties we expect a norm to have and are used in numerical mathematics.

(a) $\|\mathbf{u}\| = |u_1| + \cdots + |u_n|$ sum of magnitudes norm

(b) $\|\mathbf{u}\| = \max\limits_{i=1\ldots n} |u_n|$ maximum magnitude norm

*Compute these two norms for the vectors $(1, 2), (-3, 4), (1, 2, -5), (0, -2, 7)$.

35. Let $\mathbf{x}, \mathbf{y}$, and $\mathbf{z}$ be points in $\mathbf{R}^n$. Prove that distance has the following properties.

(a) $d(\mathbf{x}, \mathbf{y}) \geq 0$

(b) $d(\mathbf{x}, \mathbf{y}) = 0$ if and only if $\mathbf{x} = \mathbf{y}$

(c) $d(\mathbf{x}, \mathbf{z}) \leq d(\mathbf{x}, \mathbf{y}) + d(\mathbf{y}, \mathbf{z})$

These properties are used as axioms to generalize the concept of distance for certain spaces.

4.3 General Vector Spaces

The vector space $\mathbf{R}^n$ is a set of elements called vectors on which two operations, namely addition and scalar multiplication, have been defined. We have seen that the vector space $\mathbf{R}^n$ is closed under these operations; the sum of two vectors in $\mathbf{R}^n$ lies in $\mathbf{R}^n$ and the scalar multiple of a vector in $\mathbf{R}^n$ also lies in $\mathbf{R}^n$. The vector space $\mathbf{R}^n$ also has other algebraic properties. For example, we have seen that vectors in $\mathbf{R}^n$ are commutative and associative under addition:

$$\mathbf{u} + \mathbf{v} = \mathbf{v} + \mathbf{u}$$
$$\mathbf{u} + (\mathbf{v} + \mathbf{w}) = (\mathbf{u} + \mathbf{v}) + \mathbf{w}$$

Our aim in this section will be to focus on these and other algebraic properties of $\mathbf{R}^n$. We draw up a set of **axioms** based on the properties of $\mathbf{R}^n$. Any set that satisfies these axioms will have similar algebraic properties to the vector space $\mathbf{R}^n$. Such a set will be called a *vector space* and its elements will be called *vectors*.

We now give the formal definition of a vector space, based on the algebraic properties of $\mathbf{R}^n$ that were discussed in Theorem 4.1. Observe that the axioms can be separated into three convenient groups.

DEFINITION

A **vector space** is a set V of elements called **vectors**, having operations of addition and scalar multiplication defined on it that satisfy the following conditions. ($\mathbf{u}$, $\mathbf{v}$, and $\mathbf{w}$ are arbitrary elements of V, and c and d are scalars.)

Closure Axioms

1. The sum $\mathbf{u} + \mathbf{v}$ exists and is an element of V. (V is closed under addition.)
2. $c\mathbf{u}$ is an element of V. (V is closed under scalar multiplication.)

Addition Axioms

3. $\mathbf{u} + \mathbf{v} = \mathbf{v} + \mathbf{u}$ (commutative property)
4. $\mathbf{u} + (\mathbf{v} + \mathbf{w}) = (\mathbf{u} + \mathbf{v}) + \mathbf{w}$ (associative property)
5. There exists an element of V, called the **zero vector**, denoted $\mathbf{0}$, such that $\mathbf{u} + \mathbf{0} = \mathbf{u}$.
6. For every element $\mathbf{u}$ of V there exists an element called the **negative** of $\mathbf{u}$, denoted $-\mathbf{u}$, such that $\mathbf{u} + (-\mathbf{u}) = \mathbf{0}$.

Scalar Multiplication Axioms

7. $c(\mathbf{u} + \mathbf{v}) = c\mathbf{u} + c\mathbf{v}$
8. $(c + d)\mathbf{u} = c\mathbf{u} + d\mathbf{u}$
9. $c(d\mathbf{u}) = (cd)\mathbf{u}$
10. $1\mathbf{u} = \mathbf{u}$

The two most common sets of scalars used in vector spaces are the set of real numbers and the set of complex numbers. The vector spaces are then called **real vector spaces** and **complex vector spaces**. We shall focus primarily on real vector spaces, mentioning briefly, when appropriate, results for complex vector spaces. $\mathbf{R}^n$ satisfies all these conditions. We now give examples of additional vector spaces.

Vector Spaces of Matrices

Consider the set of real 2×2 matrices. Denote this set M_{22}. We have defined operations of addition and scalar multiplication on this set. The set does in fact form a vector space. We shall discuss axioms 1, 3, 4, 5, and 6, leaving the remaining axioms for the reader to check.

Use vector notation for the elements of M_{22}. Let

$$\mathbf{u} = \begin{bmatrix} a & b \\ c & d \end{bmatrix} \quad \text{and} \quad \mathbf{v} = \begin{bmatrix} e & f \\ g & h \end{bmatrix}$$

be two arbitrary 2×2 matrices. We get

Axiom 1:

$$\mathbf{u} + \mathbf{v} = \begin{bmatrix} a & b \\ c & d \end{bmatrix} + \begin{bmatrix} e & f \\ g & h \end{bmatrix} = \begin{bmatrix} a + e & b + f \\ c + g & d + h \end{bmatrix}$$

$\mathbf{u} + \mathbf{v}$ is a 2×2 matrix. Thus M_{22} is closed under addition.

Axioms 3 and 4:

From our previous discussions we know that 2×2 matrices are commutative and associative under addition. (Theorem 2.2.)

Axiom 5:

The 2×2 zero matrix is $\mathbf{0} = \begin{bmatrix} 0 & 0 \\ 0 & 0 \end{bmatrix}$, since

$$\mathbf{u} + \mathbf{0} = \begin{bmatrix} a & b \\ c & d \end{bmatrix} + \begin{bmatrix} 0 & 0 \\ 0 & 0 \end{bmatrix} = \begin{bmatrix} a & b \\ c & d \end{bmatrix} = \mathbf{u}$$

Axiom 6:

If $\mathbf{u} = \begin{bmatrix} a & b \\ c & d \end{bmatrix}$, then $-\mathbf{u} = \begin{bmatrix} -a & -b \\ -c & -d \end{bmatrix}$, since

$$\mathbf{u} + (-\mathbf{u}) = \begin{bmatrix} a & b \\ c & d \end{bmatrix} + \begin{bmatrix} -a & -b \\ -c & -d \end{bmatrix} = \begin{bmatrix} a - a & b - b \\ c - c & d - d \end{bmatrix} = \begin{bmatrix} 0 & 0 \\ 0 & 0 \end{bmatrix} = \mathbf{0}$$

The set M_{22} of real 2×2 matrices is a vector space. The algebraic properties of M_{22} are similar to those of $\mathbf{R}^n$. Similarly, we get

The set of real $m \times n$ matrices, M_{mn}, is a vector space.

Vector Spaces of Functions

Let V be the set of all functions having the real numbers as their domain. Each element of V, such as f, will map the real line into the real line. We shall now introduce operations of addition and scalar multiplication on V that will make it into a vector space.

Let f and g be arbitrary elements of V. Define their sum, $f + g$, to be the function such that

$$(f + g)(x) = f(x) + g(x)$$

This defines $f + g$ to be a function with domain the set of real numbers. To find the value of $f + g$ for any real number x, we add the value of f at x and the value of g at x. This operation is called **pointwise addition**.

We next define scalar multiplication on the elements of V. Let c be an arbitrary scalar. The scalar multiple of f, cf, is the function such that

$$(cf)(x) = c[f(x)]$$

This defines cf to be a function with domain the set of real numbers. To find the value of cf for any real number x, multiply the value of f at x by c. This operation is called **pointwise scalar multiplication**.

To get a feel for these two operations on functions let us consider two specific functions

$$f(x) = x \quad \text{and} \quad g(x) = x^2$$

Then $f + g$ is the function defined by

$$(f + g)(x) = x + x^2$$

The scalar multiple of f by a scalar such as 3 is the function $3f$ defined by

$$(3f)(x) = 3x$$

Having defined operations of addition and scalar multiplication on this function space V let us now check that V is a vector space. We shall look at axioms 1, 2, 5, and 6, leaving the remaining axioms for the reader to check.

Axiom 1:

$f + g$ is defined by $(f + g)(x) = f(x) + g(x)$. Since $f + g$ is a function with domain the set of real numbers it is an element of V. Thus V is closed under addition.

Axiom 2:

cf is defined by $(cf)(x) = c[f(x)]$. Since cf is a function with domain the set of real numbers it is an element of V. Thus V is closed under scalar multiplication.

Axiom 5:

Let $\mathbf{0}$ be the function such that $\mathbf{0}(x) = 0$ for every real number x. $\mathbf{0}$ is called the **zero function**. We get

$$(f + \mathbf{0})(x) = f(x) + \mathbf{0}(x) = f(x) + 0 = f(x) \quad \text{for every real number } x$$

The value of the function $f + \mathbf{0}$ is the same as the value of f at every x. Thus $f + \mathbf{0} = f$. $\mathbf{0}$ is the **zero vector**.

Axiom 6:

Consider the function $-f$ defined by $(-f)(x) = -[f(x)]$. We shall show that $-f$ is the negative of f.

$$\begin{aligned}
[f + (-f)](x) &= f(x) + (-f)(x) \\
&= f(x) - [f(x)] \\
&= 0 \\
&= \mathbf{0}(x)
\end{aligned}$$

The value of the function $[f + (-f)]$ is the same as the value of $\mathbf{0}$ at every x. Thus $[f + (-f)] = \mathbf{0}$. Therefore $-f$ is the negative of f.

> *The set of all functions having the real numbers as their domain, with operations of pointwise addition and scalar multiplication, is a vector space.*

There are many other vector spaces of functions. For example, the set of all functions having domain $[-\pi, \pi]$ with operations of pointwise addition and pointwise scalar multiplication is a vector space. The reader will investigate other function vector spaces in the exercises that follow.

*The Complex Vector Space $\mathbf{C}^n$

We now extend the concept of the real vector space $\mathbf{R}^n$ to a complex vector space $\mathbf{C}^n$. Let $(u_1, \ldots, u_n)$ be a sequence of n complex numbers. The set of all such sequences is denoted $\mathbf{C}^n$. For example, $(2 + 3i, 4 - 6i)$ is an element of $\mathbf{C}^2$, while $(3i, 1 - 5i, 2)$ is an element of $\mathbf{C}^3$.

Let operations of addition and scalar multiplication (by a complex scalar c) be defined on $\mathbf{C}^n$ as follows.

$$(u_1, \ldots, u_n) + (v_1, \ldots, v_n) = (u_1 + v_1, \ldots, u_n + v_n)$$
$$c(u_1, \ldots, u_n) = (cu_1, \ldots, cu_n)$$

For example, consider the two vectors $\mathbf{u} = (2 + i, 3 - 4i)$ and $\mathbf{v} = (1 - 3i, 5 + 3i)$ of $\mathbf{C}^2$ and the scalar $c = 4 + 3i$. Then

$$\mathbf{u} + \mathbf{v} = (2 + i, 3 - 4i) + (1 - 3i, 5 + 3i) = (3 - 2i, 8 - i)$$
$$c\mathbf{u} = (4 + 3i)(2 + i, 3 - 4i) = (5 + 10i, 24 - 7i)$$

It can be shown that $\mathbf{C}^n$ with these two operations is a complex vector space. (See the following exercises.)

We now give a theorem that contains useful properties of vectors. These are properties that were immediately apparent for $\mathbf{R}^n$ and were taken almost for granted. They are not, however, so apparent for all vector spaces.

THEOREM 4.5

Let V be a vector space, $\mathbf{v}$ a vector in V, $\mathbf{0}$ the zero vector of V, c a scalar, and 0 the zero scalar. Then

(a) $0\mathbf{v} = \mathbf{0}$

(b) $c\mathbf{0} = \mathbf{0}$

(c) $(-1)\mathbf{v} = -\mathbf{v}$

(d)   If $c\mathbf{v} = \mathbf{0}$, then either $c = 0$ or $\mathbf{v} = \mathbf{0}$

Proof   We shall prove (a) and (c), leaving the proofs of (b) and (d) for the reader to complete in the exercises that follow.

(a) $0\mathbf{v} + 0\mathbf{v} = (0 + 0)\mathbf{v}$ (axiom 8)

            $= 0\mathbf{v}$ (property of the scalar 0)

Add the negative of $0\mathbf{v}$, namely $-0\mathbf{v}$, to both sides of this equation.

$$(0\mathbf{v} + 0\mathbf{v}) + (-0\mathbf{v}) = 0\mathbf{v} + (-0\mathbf{v})$$
$$0\mathbf{v} + [0\mathbf{v} + (-0\mathbf{v})] = \mathbf{0} \qquad \text{(axioms 4 and 6)}$$
$$0\mathbf{v} + \mathbf{0} = \mathbf{0} \qquad \text{(axiom 6)}$$
$$0\mathbf{v} = \mathbf{0} \qquad \text{(axiom 5)}$$

(c) $\quad (-1)\mathbf{v} + \mathbf{v} = (-1)\mathbf{v} + 1\mathbf{v} \qquad \text{(axiom 10)}$
$$= [(-1) + 1]\mathbf{v} \qquad \text{(axiom 8)}$$
$$= 0\mathbf{v} \qquad \text{(property of scalar 0)}$$
$$= \mathbf{0} \qquad \text{(part (a) above)}$$

Thus $(-1)\mathbf{v}$ is the negative of $\mathbf{v}$ (axiom 6).

EXERCISE SET 4.3

Vector Spaces

1. We have discussed the fact that M_{22}, the set of 2×2 matrices with the usual operations of matrix addition and scalar multiplication satisfies axioms 1, 3, 4, 5, and 6 of a vector space. Prove that M_{22} also satisfies the remaining vector space axioms, completing the proof that M_{22} is a vector space.

2. (a) Let $f(x) = x + 2$ and $g(x) = x^2 - 1$. Compute the functions $f + g$, $2f$, and $3g$.

   (b) Let $f(x) = 2x$ and $g(x) = 4 - 2x$. Compute $f + g$, $3f$, and $-g$.

3. We have shown that the set of functions with operations of pointwise addition and scalar multiplication, with domain the set of real numbers, satisfies axioms 1, 2, 5, and 6 of a vector space. Prove that this set also satisfies the remaining vector space axioms, completing the proof that this set is a vector space.

4. Prove that the set $\mathbf{C}^n$ with the operations of addition and scalar multiplication defined as follows is a vector space.

$$(u_1, \ldots, u_n) + (v_1, \ldots, v_n) = (u_1 + v_1, \ldots, u_n + v_n)$$
$$c(u_1, \ldots, u_n) = (cu_1, \ldots, cu_n)$$

Determine $\mathbf{u} + \mathbf{v}$ and $c\mathbf{u}$ for the following vectors and scalars in $\mathbf{C}^2$.

   (a) $\mathbf{u} = (2 - i, 3 + 4i)$, $\mathbf{v} = (5, 1 + 3i)$, $c = 3 - 2i$.

   (b) $\mathbf{u} = (1 + 5i, -2 - 3i)$, $\mathbf{v} = (2i, 3 - 2i)$, $c = 4 + i$.

5. Prove that the set W of all vectors in $\mathbf{R}^3$ of the form $a(1, 2, 3)$, where a is a real number, is a vector space.

6. Let U be the set of all vectors in $\mathbf{R}^3$ that are perpendicular to a vector $\mathbf{u}$. Prove that U is a vector space.

7. Let W be the set of all 2×2 matrices having every element a positive number. Prove that W is closed under addition, but is not closed under scalar multiplication. Thus it is not a vector space.

8. Is the set of 3×3 symmetric matrices a vector space?

9. Is the set of 2×2 stochastic matrices a vector space? (See Section 2.6 for the definition of stochastic matrix.)

10. (a) Let U be the set of all constant functions with operations of pointwise addition and scalar multiplication, having the real numbers as their domain. Is U a vector space?

    (b) Let V be the set of all nonconstant functions with operations of pointwise addition and scalar multiplication, having the real numbers as their domain. Is V a vector space?

11. (a) Consider the set of all continuous functions with operations of pointwise addition and scalar multiplication, having domain $[0, 1]$. Is this set a vector space?

    (b) Consider the set of all discontinuous functions with operations of pointwise addition and scalar multiplication, having domain $[0, 1]$. Is this set a vector space?

12. Consider the following sets with operations of pointwise addition and scalar multiplication, having the real numbers as their domain. Are they vector spaces? (a) The set of even functions. $(f(-x) = f(x))$. (b) The set of odd functions. $(f(-x) = -f(x))$.

13. Consider the set of polynomial functions of degree 2. Is this set a vector space?

14. Prove that the following sets are not vector spaces. **(a)** The set of all integers. **(b)** The set of all positive numbers.

15. Let V be a vector space, $\mathbf{u}$, $\mathbf{v}$, and $\mathbf{w}$ be vectors in V, and a, b, and c be scalars. Use the axioms of a vector space to prove the following.

(a) $c\mathbf{0} = \mathbf{0}$

(b) If $c\mathbf{v} = \mathbf{0}$, then either $c = 0$ or $\mathbf{v} = \mathbf{0}$.

(c) $-(-\mathbf{v}) = \mathbf{v}$

(d) If $\mathbf{u} + \mathbf{w} = \mathbf{v} + \mathbf{w}$, then $\mathbf{u} = \mathbf{v}$.

(e) If $\mathbf{u} \neq \mathbf{0}$ and $a\mathbf{u} = b\mathbf{u}$, then $a = b$.

16. Let W be the set of all straight lines through the origin in $\mathbf{R}^2$. Let L_1 and L_2 be elements of W. Let $L_1 + L_2$ be the line through the origin whose slope is the sum of the slopes of L_1 and L_2. Let cL_1 be the line through the origin whose slope is c times that of L_1. Is W a vector space under these operations of addition and scalar multiplication?

17. Let U be the set of all circles in $\mathbf{R}^2$ having center the origin. Interpret the origin as being in this set; a circle center the origin with radius zero. Let C_1 and C_2 be elements of W. Let $C_1 + C_2$ be the circle center the origin, whose radius is the sum of the radii of C_1 and C_2. Let kC_1 be the circle center the origin, whose radius is $|k|$ times that of C_1. Determine which vector space axioms hold and which do not.

4.4 Subspaces

In this section we introduce subsets of vector spaces that are vector spaces in their own right. They satisfy all the algebraic conditions of a vector space that we discussed in Section 4.3. Such subsets are called *subspaces*. There are two properties that need to be checked to determine whether a subset is a subspace or not. It must be checked that the sum of two arbitrary vectors of the subset lies in the subset, and that the scalar multiple of an arbitrary vector also lies in the subset. In other words, the subset must be *closed under addition and under scalar multiplication*. All the other vector space properties such as the commutative property $\mathbf{u} + \mathbf{v} = \mathbf{v} + \mathbf{u}$, and the associative property $\mathbf{u} + (\mathbf{v} + \mathbf{w}) = (\mathbf{u} + \mathbf{v}) + \mathbf{w}$ are inherited from the larger space.

For example, consider the subset U of the vector space $\mathbf{R}^3$ consisting of vectors of the form (a, a, b) where the first two components are the same. If we add two such vectors (a, a, b) and (c, c, d) we get $(a + c, a + c, b + d)$, a vector with identical first components. If we multiply (a, a, b) by a scalar k we get (ka, ka, kb), again a vector with identical first components. U is closed under addition and scalar multiplication. Further the vectors of U are commutative and associative—they get these properties from $\mathbf{R}^3$. U contains the zero vector $(0, 0, 0)$ and $\mathbf{u} = (a, a, b)$ has the negative $(-a, -a, b)$. W is a vector space embedded in $\mathbf{R}^3$. It is a subspace of $\mathbf{R}^3$.

Let us look at the geometrical interpretation of U. See Figure 4.14. $\mathbf{R}^3$ is the set of all points in 3-space. U will be the subset of all points that have equal x and y components.

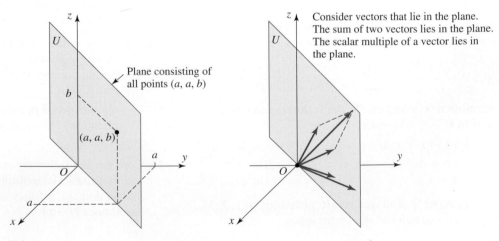

Figure 4.14

These make up a plane perpendicular to the xy plane, through the line $y = x,\ z = 0$. The sum of any two position vectors that lie in this plane will lie in the plane. The scalar multiple of any vector that lies in the plane also lies in the plane.

In general, a subset of a vector space may or may not be closed under addition or scalar multiplication. (We shall see examples of both.) However any subset that is closed under both of these operations satisfies all the other vector space properties. Such a subset is a vector space in its own right.

DEFINITION   Let V be a vector space and U be a nonempty subset of V. If U is a vector space under the operations of addition and scalar multiplication of V it is called a **subspace** of V.

U is a subspace if it is closed under addition and under scalar multiplication. It then inherits the other vector space properties from V.

EXAMPLE 1 Let U be the subset of $\mathbf{R}^3$ consisting of all vectors of the form $(a, 0, 0)$, (with zeros as second and third components). Show that U is a subspace of $\mathbf{R}^3$.

SOLUTION

Let $(a, 0, 0)$ and $(b, 0, 0)$ be two elements of U and k be a scalar. We get

$$(a, 0, 0) + (b, 0, 0) = (a + b, 0, 0) \in U$$

$$k(a, 0, 0) = (ka, 0, 0) \in U$$

The sum and scalar multiple are in U. Thus U is a subspace of $\mathbf{R}^3$.

Geometrically, U is the set of vectors that lies on the x-axis. Note that the sum of two such vectors lies on the x-axis and so does the scalar multiple of any such vector.

EXAMPLE 2 Let W be the set of vectors of the form (a, a^2, b). Show that W is not a subspace of $\mathbf{R}^3$.

SOLUTION

W consists of all elements of $\mathbf{R}^3$ for which the second component is the square of the first. Thus, for example, the vector $(2, 4, 3)$ is in W, whereas the vector $(2, 5, 3)$ is not.

Let (a, a^2, b) and (c, c^2, d) be elements of W. We get

$$(a, a^2, b) + (c, c^2, d) = (a + c, a^2 + c^2, b + d)$$

$$\neq (a + c, (a + c)^2, b + d)$$

Thus $(a, a^2, b) + (c, c^2, d)$ is not an element of W. The set W is not closed under addition. Thus W is not a subspace.

This completes the proof that W is not a subspace. Let us however go on to check for closure under scalar multiplication. Let k be a scalar. We get

$$k(a, a^2, b) = (ka, ka^2, kb)$$

$$\neq (ka, (ka)^2, kb)$$

Thus $k(a, a^2, b)$ is not an element of W. The set W is not closed under scalar multiplication either.

EXAMPLE 3 Prove that the set U of 2×2 diagonal matrices is a subspace of the vector space M_{22} of 2×2 matrices.

SOLUTION

We have to show that U is closed under addition and under scalar multiplication. Consider the following two elements of U.

$$\mathbf{u} = \begin{bmatrix} a & 0 \\ 0 & b \end{bmatrix} \quad \text{and} \quad \mathbf{v} = \begin{bmatrix} p & 0 \\ 0 & q \end{bmatrix}$$

We get

$$\mathbf{u} + \mathbf{v} = \begin{bmatrix} a & 0 \\ 0 & b \end{bmatrix} + \begin{bmatrix} p & 0 \\ 0 & q \end{bmatrix} = \begin{bmatrix} a + p & 0 \\ 0 & b + q \end{bmatrix}$$

Observe that $\mathbf{u} + \mathbf{v}$ is a 2×2 diagonal matrix, and is thus an element of U. U is closed under addition.

Let c be a scalar. We get

$$c\mathbf{u} = c \begin{bmatrix} a & 0 \\ 0 & b \end{bmatrix} = \begin{bmatrix} ca & 0 \\ 0 & cb \end{bmatrix}$$

$c\mathbf{u}$ is a 2×2 diagonal matrix. Thus U is closed under scalar multiplication.

U is a subspace of M_{22}. It is a vector space of matrices, embedded in M_{22}.

EXAMPLE 4 Let P_n denote the set of real polynomial functions of degree $\leq n$. Prove that P_n is a vector space if addition and scalar multiplication are defined on polynomials in a pointwise manner.

SOLUTION

We could prove that P_n, with these two operations, satisfies all the axioms of a vector space. There is, however, an easier way. P_n is a subset of the vector space V of functions having domain the set of real numbers. Let us show that this subset is a subspace of V; it will then be a vector space in its own right. We show that P_n is closed under addition and scalar multiplication.

Let f and g be two elements of P_n defined by

$$f(x) = a_n x^n + a_{n-1} x^{n-1} + \cdots + a_1 x + a_0$$

and

$$g(x) = b_n x^n + b_{n-1} x^{n-1} + \cdots + b_1 x + b_0$$

We first consider addition. $f + g$ is defined by

$$\begin{aligned}
(f + g)(x) &= f(x) + g(x) \\
&= [a_n x^n + a_{n-1} x^{n-1} + \cdots + a_1 x + a_0] \\
&\quad + [b_n x^n + b_{n-1} x^{n-1} + \cdots + b_1 x + b_0] \\
&= (a_n + b_n) x^n + (a_{n-1} + b_{n-1}) x^{n-1} + \cdots \\
&\quad + (a_1 + b_1) x + (a_0 + b_0)
\end{aligned}$$

$(f + g)(x)$ is a polynomial of degree $\leq n$. Thus $f + g$ is an element of P_n. P_n is closed under addition.

We now look at scalar multiplication. cf is defined by

$$(cf)(x) = c[f(x)]$$
$$= c[a_n x^n + a_{n-1} x^{n-1} + \cdots + a_1 x + a_0]$$
$$= ca_n x^n + ca_{n-1} x^{n-1} + \cdots + ca_1 x + ca_0$$

$(cf)(x)$ is a polynomial of degree $\leq n$. Thus cf is an element of P_n. P_n is closed under scalar multiplication.

We have proved that P_n, a subset of the vector space V of functions, is closed under addition and scalar multiplication. It is thus a subspace of V and therefore a vector space.

The following theorem gives an important characteristic of all subspaces.

THEOREM 4.6

Let U be a subspace of a vector space V. U contains the zero vector of V.

Proof Let $\mathbf{u}$ be an arbitrary vector in U and $\mathbf{0}$ be the zero vector of V. Let 0 be the zero scalar. By Theorem 4.5(a) we know that $0\mathbf{u} = \mathbf{0}$. Since U is closed under scalar multiplication this means that $\mathbf{0}$ is in U.

This theorem tells us for example that all subspaces of $\mathbf{R}^3$ contain $(0, 0, 0)$. This means that all subspaces of 3-space pass through the origin. This theorem can sometimes be used as a quick check to show that certain subsets cannot be subspaces. If a given subset does not contain the zero vector it cannot be a subspace. The following example illustrates this approach.

EXAMPLE 5 Let W be the set of vectors of the form $(a, a, a + 2)$. Show that W is not a subspace of $\mathbf{R}^3$.

SOLUTION

We check to see if $(0, 0, 0)$ is in W. Is there a value of a for which $(a, a, a + 2)$ is $(0, 0, 0)$? On equating $(a, a, a + 2)$ to $(0, 0, 0)$ we get

$$(a, a, a + 2) = (0, 0, 0)$$

Equating corresponding components we get

$$a = 0 \quad \text{and} \quad a + 2 = 0$$

This system of equations has no solution. Thus $(0, 0, 0)$ is not an element of W. W is not a subspace.

EXERCISE SET 4.4

Subspaces of R^n

1. Consider the sets of vectors of the following form. Prove that they are subspaces of $\mathbf{R}^3$.

   (a) $(a, 3a, 5a)$ (b) $(a, -a, 2a)$

   (c) $(a, b, a + 2b)$ (d) $(a, b, a - b)$

2. Consider the sets of vectors of the following form. Determine whether the sets are subspaces of $\mathbf{R}^2$ or $\mathbf{R}^3$. Give the geometrical interpretation of each subspace.

   (a) $(a, 0)$ (b) $(a, 2a)$

   (c) $(a, 1)$ (d) $(a, a + 3)$

   (e) $(a, b, 0)$ (f) $(a, b, 2)$

   (g) $(a, b, 2a + 3b)$

3. Are the following sets subspaces of $\mathbf{R}^3$? The set of all vectors of the form (a, b, c), where

   (a) $a + b + c = 0$ (b) $a + b + c = 1$

   (c) $ab = 0$ (d) $ab = 5$

   (e) $ab = ac$ (f) $a = b + c$

4. Which of the following subsets of $\mathbf{R}^3$ are subspaces? The set of all vectors of the form (a, b, c), where a, b, and c are

   (a) integers   (b) nonnegative real numbers

   (c) rational numbers

5. Are the following sets subspaces of $\mathbf{R}^2$? The set of all vectors of the form

   (a) (a, b^2) (b) (a, b^3)

   (c) (a, b) where $a > 0$ (d) (a, b) where $ab < 0$

   (e) (a, b) where a is nonpositive and b is nonnegative

6. Give an example of a subset of $\mathbf{R}^3$, that is

   (a) closed under addition, but not closed under scalar multiplication

   (b) closed under scalar multiplication, but not closed under addition

   Such examples illustrate the **independence** of these two conditions.

7. Prove that the following sets are not subspaces of $\mathbf{R}^3$ by showing that they do not contain the zero vector. The set of all vectors of the form

   (a) $(a, a + 1, b)$ (b) $(a, 3, 2a)$

   (c) $(a, b, a + b - 4)$ (d) (a, b, c) where $a > 0$

8. (a) Give an example of a subset of $\mathbf{R}^2$ that contains the zero vector, but is not a subspace.

   (b) Give an example of a subset of $\mathbf{R}^3$ that contains the zero vector, but is not a subspace.

These examples illustrate that the property of containing the zero vector is a necessary but not sufficient condition for a subset to be a subspace.

9. Let U be a subset of $\mathbf{R}^3$, let $\mathbf{u}_1$ and $\mathbf{u}_2$ be vectors in U, and a and b be scalars. Prove that U is a subspace of $\mathbf{R}^3$ if and only if $a\mathbf{u}_1 + b\mathbf{u}_2$ is a vector in U for all values of a and b.

Subspaces of Matrices

10. Determine which of the following subsets of M_{22} form subspaces.

    (a) The subset having diagonal elements zero.

    (b) The subset consisting of matrices the sum of whose elements is 6. (For example, $\begin{bmatrix} 2 & -1 \\ 0 & 5 \end{bmatrix}$ would be such a matrix.)

    (c) The subset of matrices of the form $\begin{bmatrix} a & a^2 \\ b & b^2 \end{bmatrix}$

    (d) The subset of matrices of the form $\begin{bmatrix} a & a + 2 \\ b & c \end{bmatrix}$

11. Determine which of the following subsets of M_{nn} form subspaces.

    (a) The subset of symmetric matrices.

    (b) The subset of matrices that are not symmetric.

    (c) The subset of antisymmetric matrices. (A matrix A is said to be antisymmetric if $A = -A^t$.)

    (d) The subset of invertible matrices.

12. Which of the following subsets of M_{23} form subspaces?

    (a) The subset of matrices of the form $\begin{bmatrix} a & b & 0 \\ c & d & 0 \end{bmatrix}$

    (b) The subset of matrices of the form $\begin{bmatrix} a & 2a & 3a \\ b & 2b & 3b \end{bmatrix}$

    (c) The subset of matrices of the form $\begin{bmatrix} a & 1 & b \\ c & d & e \end{bmatrix}$

Subspaces of Functions

13. P_3 is the vector space of polynomials of degree ≤ 3, and P_2 is the vector space of polynomials of degree ≤ 2. Prove that P_2 is a subspace of P_3.

14. Let S be the set of all functions of the form $f(x) = ax^2 + bx + 3$, where a and b are real numbers. Is S a subspace of P_2?

15. Is $\mathbf{R}^n$ a subspace of $\mathbf{C}^n$?

16. We have seen that every subspace of a vector space V has to contain the zero vector of V. This is often a quick way of deciding that some subsets cannot be subspaces. Use this criterion to prove the following:

   (a) The set of matrices of the form $\begin{bmatrix} a & 1 \\ b & c \end{bmatrix}$ is not a subspace of M_{22}.

   (b) The set of all functions of the form $f(x) = ax + 2$ is not a subspace of P_2.

17. Consider the vector space of functions defined on the set of real numbers. Which of the following are subspaces of this vector space?

   (a) The subset consisting of all functions f such that $f(0) = 0$.

   (b) The subset consisting of all functions f such that $f(0) = 3$.

   (c) The subset of all constant functions.

Miscellaneous Results

18. If U is a subspace of a vector space V, and if $\mathbf{u}$ and $\mathbf{v}$ are elements of V, but one or both is not in U, can $\mathbf{u} + \mathbf{v}$ be in U? Can $c\mathbf{u}$ be in U for some nonzero scalar c if $\mathbf{u}$ is not in U?

19. Prove that a necessary and sufficient condition for a subset U of a vector space V to be a subspace is that $a\mathbf{u} + b\mathbf{v}$ be in U for all scalars a and b and all vectors $\mathbf{u}$ and $\mathbf{v}$ in U.

20. Prove the following:

   (a) The union of two subspaces need not be a subspace.

   (b) The intersection of two subspaces is a subspace.

21. Prove the following.

   (a) The set of solutions to a homogeneous system of m linear equations in n variables is a subspace of $\mathbf{R}^n$.

   (b) The set of solutions to a nonhomogeneous system of m linear equations in n variables is not a subspace.

4.5 Linear Combinations

This section and the next go together—they form a unit. We introduce many of the concepts that are needed to understand vector spaces. These lead us to the formal definition of dimension in Section 4.7. In Section 4.1 we discussed linear combinations of vectors. We saw for example that the vector $(20, 7, -31)$ is a linear combination of $(2, 5, -3)$, $(-4, 1, 9)$, and $(4, 0, 2)$ since it can be written

$$(20, 7, -31) = 2(2, 5, -3) - 3(-4, 1, 9) + (4, 0, 2)$$

We now give the formal definition of linear combination of vectors. This applies to all vector spaces.

DEFINITION   Let $\mathbf{v}_1, \mathbf{v}_2, \ldots, \mathbf{v}_m$ be vectors in a vector space V. The vector $\mathbf{v}$ in V is a *linear combination* of $\mathbf{v}_1$, $\mathbf{v}_2, \ldots, \mathbf{v}_m$ if there exist scalars $c_1, c_2, \ldots, c_m$ such that $\mathbf{v}$ can be written

$$\mathbf{v} = c_1\mathbf{v}_1 + c_2\mathbf{v}_2 + \ldots + c_m\mathbf{v}_m$$

In general, determining whether a given vector is a linear combination or not of other vectors involves examining a system of linear equations.

EXAMPLE 1 Determine whether the vector $(8, 0, 5)$ is a linear combination of the vectors $(1, 2, 3)$, $(0, 1, 4)$, and $(2, -1, 1)$.

SOLUTION

Examine the following identity for values of c_1, c_2, and c_3.

$$c_1(1, 2, 3) + c_2(0, 1, 4) + c_3(2, -1, 1) = (8, 0, 5)$$

This identity leads to the following system of linear equations.

$$\begin{aligned} c_1 \phantom{{}+{}} &+ 2c_3 = 8 \\ 2c_1 + {}&c_2 - c_3 = 0 \\ 3c_1 + {}&4c_2 + c_3 = 5 \end{aligned}$$

It can be shown that this system of equations has the unique solution,

$$c_1 = 2, c_2 = -1, c_3 = 3$$

Thus, the vector $(8, 0, 5)$ can be written in one way as a linear combination,

$$(8, 0, 5) = 2(1, 2, 3) - 1(0, 1, 4) + 3(2, -1, 1)$$

If the system had no solutions, it would not have been possible to express it as a linear combination of the other vectors.

There is a possibility that a vector can be expressed in many ways as a linear combination of other vectors. The following example illustrates this situation.

EXAMPLE 2 Determine whether the vector $(4, 5, 5)$ is a linear combination of the vectors $(1, 2, 3)$, $(-1, 1, 4)$, and $(3, 3, 2)$.

SOLUTION

Examine the following identity for values of c_1, c_2, and c_3.

$$c_1(1, 2, 3) + c_2(-1, 1, 4) + c_3(3, 3, 2) = (4, 5, 5)$$

This identity leads to the following system of linear equations.

$$\begin{aligned} c_1 - {}&c_2 + 3c_3 = 4 \\ 2c_1 + {}&c_2 + 3c_3 = 5 \\ 3c_1 + {}&4c_2 + 2c_3 = 5 \end{aligned}$$

It can be shown that this system of equations has many solutions,

$$c_1 = -2r + 3, c_2 = r - 1, c_3 = r$$

Thus, the vector $(4, 5, 5)$ can be expressed in many ways as a linear combination of the vectors $(1, 2, 3)$, $(-1, 1, 4)$, and $(3, 3, 2)$,

$$(4, 5, 5) = (-2r + 3)(1, 2, 3) + (r - 1)(-1, 1, 4) + r(3, 3, 2)$$

By letting r take on different values, we get specific combinations. For example,

$$\begin{aligned} r = 3 \text{ gives } (4, 5, 5) &= -3(1, 2, 3) + 2(-1, 1, 4) + 3(3, 3, 2) \\ r = -1 \text{ gives } (4, 5, 5) &= 5(1, 2, 3) - 2(-1, 1, 4) - (3, 3, 2) \end{aligned}$$

The following examples illustrate these concepts for vector spaces of matrices and functions.

EXAMPLE 3 Determine whether the matrix $\begin{bmatrix} -1 & 7 \\ 8 & -1 \end{bmatrix}$ is a linear combination of $\begin{bmatrix} 1 & 0 \\ 2 & 1 \end{bmatrix}$, $\begin{bmatrix} 2 & -3 \\ 0 & 2 \end{bmatrix}$, and $\begin{bmatrix} 0 & 1 \\ 2 & 0 \end{bmatrix}$ in the vector space M_{22} of 2×2 matrices.

SOLUTION

We examine the following identity.

$$c_1 \begin{bmatrix} 1 & 0 \\ 2 & 1 \end{bmatrix} + c_2 \begin{bmatrix} 2 & -3 \\ 0 & 2 \end{bmatrix} + c_3 \begin{bmatrix} 0 & 1 \\ 2 & 0 \end{bmatrix} = \begin{bmatrix} -1 & 7 \\ 8 & -1 \end{bmatrix}$$

Can we find scalars c_1, c_2, c_3 such that this identity holds? Using operations of scalar multiplication and addition of matrices we get

$$\begin{bmatrix} c_1 + 2c_2 & -3c_2 + c_3 \\ 2c_1 + 2c_3 & c_1 + 2c_2 \end{bmatrix} = \begin{bmatrix} -1 & 7 \\ 8 & -1 \end{bmatrix}$$

On equating corresponding elements we get the following system of linear equations.

$$\begin{aligned} c_1 + 2c_2 &= -1 \\ -3c_2 + c_3 &= 7 \\ 2c_1 + 2c_3 &= 8 \\ c_1 + 2c_2 &= -1 \end{aligned}$$

This system has the unique solution $c_1 = 3$, $c_2 = -2$, $c_3 = 1$. The given matrix is thus the following linear combination of the other three matrices.

$$\begin{bmatrix} -1 & 7 \\ 8 & -1 \end{bmatrix} = 3 \begin{bmatrix} 1 & 0 \\ 2 & 1 \end{bmatrix} - 2 \begin{bmatrix} 2 & -3 \\ 0 & 2 \end{bmatrix} + \begin{bmatrix} 0 & 1 \\ 2 & 0 \end{bmatrix}$$

If it had turned out that this system of equations had no solution, then of course the given matrix would not have been a linear combination of the other matrices.

EXAMPLE 4 Determine whether the function $f(x) = 2x^2 + 6x + 7$ is a linear combination of $g(x) = x^2 - 1$ and $h(x) = 2x + 3$.

SOLUTION

We examine the identity $c_1 g(x) + c_2 h(x) = f(x)$ for values of $c_1, c_2,$ and c_3. We get

$$c_1(x^2 - 1) + c_2(2x + 3) = 2x^2 + 6x + 7$$
$$c_1 x^2 + 2c_2 x - c_1 + 3c_2 = 2x^2 + 6x + 7$$

Two polynomials can only be equal for all values of x if the corresponding coefficients are equal. Compare the coefficients of x^2, x, and the constant terms on both sides of the equation. We get the following system of linear equations.

$$\begin{aligned} c_1 &= 2 \\ 2c_2 &= 6 \\ -c_1 + 3c_2 &= 7 \end{aligned}$$

This system has the unique solution $c_1 = 2$, $c_2 = 3$. Thus $f(x) = 2g(x) + 3h(x)$.

If the system of linear equations had no solution then f would not have been a linear combination of g and h. It can easily be seen for example that $f(x) = 2x^2 + 6x + 8$ is not a linear combination of $g(x) = x^2 - 1$ and $h(x) = 2x + 3$.

Spanning a Vector Space

We now introduce finite sets of vectors that represent a whole vector space.

> **DEFINITION** Let $\mathbf{v}_1, \ldots, \mathbf{v}_m$ be vectors in a vector space V. These vectors *span* V if every vector in V can be expressed as a linear combination of them.

For example the vectors $(1, 0, 0)$, $(0, 1, 0)$, $(0, 0, 1)$ span $\mathbf{R}^3$ since we can write an arbitrary vector (x, y, z) of $\mathbf{R}^3$ as the linear combination $(x, y, z) = x(1, 0, 0) + y(0, 1, 0) + z(0, 0, 1)$.

The vectors $(1, 0, 0)$, $(0, 1, 0)$, $(1, 1, 1)$ also span $\mathbf{R}^3$ since we can write $(x, y, z) = (x - z)(1, 0, 0) + (y - z)(0, 1, 0) + z(1, 1, 1)$.

The vectors $(1, 0, 0)$, $(0, 2, 0)$, $(3, 4, 0)$ do not span $\mathbf{R}^3$ since a vector (x, y, z) for which $z \neq 0$ cannot be written as a linear combination of these vectors.

The vectors $(1, 1, 0)$, $(0, 0, 1)$ span the subspace of $\mathbf{R}^3$ consisting of vectors of the form (a, a, b) since we can write

$$(a, a, b) = a(1, 1, 0) + b(0, 0, 1).$$

The following two examples illustrate techniques used to modify vector expressions using operations of scalar multiplication and vector addition.

EXAMPLE 5 Let $\mathbf{v}_1$ and $\mathbf{v}_2$ span a subspace U of a vector space V. Let k_1 and k_2 be nonzero scalars. Show that $k_1\mathbf{v}_1$ and $k_2\mathbf{v}_2$ also span U.

SOLUTION

Let $\mathbf{v}$ be a vector in U. Since $\mathbf{v}_1$ and $\mathbf{v}_2$ span U there exist scalars a and b such that

$$\mathbf{v} = a\mathbf{v}_1 + b\mathbf{v}_2$$

Since k_1 and k_2 are nonzero we can write

$$\mathbf{v} = \frac{a}{k_1}(k_1\mathbf{v}_1) + \frac{b}{k_2}(k_2\mathbf{v}_2)$$

Thus the vectors $k_1\mathbf{v}_1$ and $k_2\mathbf{v}_2$ span U.

EXAMPLE 6 Let $\mathbf{v}_1$, $\mathbf{v}_2$, $\mathbf{v}_3$ span a vector space V. Show that the vectors $\mathbf{v}_1$, $\mathbf{v}_2$, $\mathbf{v}_1 + \mathbf{v}_2 + \mathbf{v}_3$ also span V.

SOLUTION

Let $\mathbf{v}$ be an arbitrary vector in V. Since $\mathbf{v}_1$, $\mathbf{v}_2$, $\mathbf{v}_3$ span V there exist scalars a_1, a_2, a_3 such that $\mathbf{v} = a_1\mathbf{v}_1 + a_2\mathbf{v}_2 + a_3\mathbf{v}_3$. The challenge now is to gradually rewrite this linear combination of $\mathbf{v}_1$, $\mathbf{v}_2$, $\mathbf{v}_3$ to become a linear combination of the vectors $\mathbf{v}_1$, $\mathbf{v}_2$, and $\mathbf{v}_1 + \mathbf{v}_2 + \mathbf{v}_3$. The technique is to introduce appropriate vectors, allowing for this by subtracting them also. We first introduce the vector $(\mathbf{v}_1 + \mathbf{v}_2 + \mathbf{v}_3)$ into the linear combination.

$$
\begin{aligned}
\mathbf{v} &= a_1\mathbf{v}_1 + a_2\mathbf{v}_2 + a_3\mathbf{v}_3 \\
&= a_1\mathbf{v}_1 + a_2\mathbf{v}_2 + a_3(\mathbf{v}_1 + \mathbf{v}_2 + \mathbf{v}_3) - a_3\mathbf{v}_1 - a_3\mathbf{v}_2 \\
&= a_1\mathbf{v}_1 - a_3\mathbf{v}_1 + a_2\mathbf{v}_2 - a_3\mathbf{v}_2 + a_3(\mathbf{v}_1 + \mathbf{v}_2 + \mathbf{v}_3) \\
&= (a_1 - a_3)\mathbf{v}_1 + (a_2 - a_3)\mathbf{v}_2 + a_3(\mathbf{v}_1 + \mathbf{v}_2 + \mathbf{v}_3)
\end{aligned}
$$

Thus $\mathbf{v}$ can be written as a linear combination of $\mathbf{v}_1$, $\mathbf{v}_2$, $\mathbf{v}_1 + \mathbf{v}_2 + \mathbf{v}_3$. These vectors span V.

Generating a Vector Space

We have developed the mathematics for looking at a vector space in terms of a set of vectors that spans the space. It is also useful to be able to do the converse, namely to use a set of vectors to *generate* a vector space.

THEOREM 4.7

Let $\mathbf{v}_1, \ldots, \mathbf{v}_m$ be vectors in a vector space V. Let U be the set consisting of all linear combinations of $\mathbf{v}_1, \ldots, \mathbf{v}_m$. Then U is a subspace of V spanned by the vectors $\mathbf{v}_1, \ldots, \mathbf{v}_m$.

U is said to be the vector space **generated** by $\mathbf{v}_1, \ldots, \mathbf{v}_m$. It is denoted Span $\{\mathbf{v}_1, \ldots, \mathbf{v}_m\}$.

Proof Let

$$
\mathbf{u}_1 = a_1\mathbf{v}_1 + \cdots + a_m\mathbf{v}_m \quad \text{and} \quad \mathbf{u}_2 = b_1\mathbf{v}_1 + \cdots + b_m\mathbf{v}_m
$$

be arbitrary elements of U. Then

$$
\begin{aligned}
\mathbf{u}_1 + \mathbf{u}_2 &= (a_1\mathbf{v}_1 + \cdots + a_m\mathbf{v}_m) + (b_1\mathbf{v}_1 + \cdots + b_m\mathbf{v}_m) \\
&= (a_1 + b_1)\mathbf{v}_1 + \cdots + (a_m + b_m)\mathbf{v}_m
\end{aligned}
$$

$\mathbf{u}_1 + \mathbf{u}_2$ is a linear combination of $\mathbf{v}_1, \ldots, \mathbf{v}_m$. Thus $\mathbf{u}_1 + \mathbf{u}_2$ is in U. U is closed under vector addition.

Let c be an arbitrary scalar. Then

$$
\begin{aligned}
c\mathbf{u}_1 &= c(a_1\mathbf{v}_1 + \cdots + a_m\mathbf{v}_m) \\
&= ca_1\mathbf{v}_1 + \cdots + ca_m\mathbf{v}_m
\end{aligned}
$$

$c\mathbf{u}_1$ is a linear combination of $\mathbf{v}_1, \ldots, \mathbf{v}_m$. Therefore $c\mathbf{u}_1$ is in U. U is closed under scalar multiplication. Thus U is a subspace of V.

By the definition of U every vector in U can be written as a linear combination of $\mathbf{v}_1, \ldots, \mathbf{v}_m$. Thus $\mathbf{v}_1, \ldots, \mathbf{v}_m$ span U.

EXAMPLE 7 Consider the vectors $(-1, 5, 3)$ and $(2, -3, 4)$ in $\mathbf{R}^3$. Let $U = \text{Span}\{(-1, 5, 3), (2, -3, 4)\}$. U will be a subspace of $\mathbf{R}^3$ consisting of all vectors of the form

$$
c_1(-1, 5, 3) + c_2(2, -3, 4)
$$

The following are examples of some of the vectors in U, which are obtained by giving c_1 and c_2 various values.

$$c_1 = 1, c_2 = 1; \text{ vector } (1, 2, 7)$$
$$c_1 = 2, c_2 = 3; \text{ vector } (4, 1, 18)$$

We can visualize U. U is made up of all vectors in the plane defined by the vectors $(-1, 5, 3)$ and $(2, -3, 4)$. See Figure 4.15.

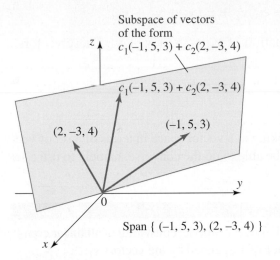

Figure 4.15   Figure 4.16

We can generalize the above result. Let v_1 and v_2 be vectors in the vector space $\mathbf{R}^3$. The subspace Span$\{v_1, v_2\}$ generated by v_1 and v_2 is the set of all vectors of the form $c_1v_1 + c_2v_2$. In general, this space is the plane defined by v_1 and v_2, see Figure 4.16. If v_1 and v_2 are collinear, then the space will be the line defined by these vectors.

EXAMPLE 8 Show that the function $h(x) = 4x^2 + 3x - 7$ lies in the space Span$\{f, g\}$ generated by $f(x) = 2x^2 - 5$ and $g(x) = x + 1$.

SOLUTION

The function h will be in Span$\{f, g\}$ if h is a linear combination of f and g. Let us examine the identity

$$c_1(2x^2 - 5) + c_2(x + 1) = 4x^2 + 3x - 7$$

This gives

$$2c_1x^2 + c_2x - 5c_1 + c_2 = 4x^2 + 3x - 7$$

Two polynomials can only be equal for all values of x if the corresponding coefficients are equal. Compare the coefficients of x^2, x, and the constant terms on both sides of the following equation.

$$\begin{aligned} 2c_1 &= 4 \\ c_2 &= 3 \\ -5c_1 + c_2 &= -7 \end{aligned}$$

This system has the unique solution $c_1 = 2$, $c_2 = 3$. Thus
$$4x^2 + 3x - 7 = 2(2x^2 - 5) + 3(x + 1)$$
The function $h(x) = 4x^2 + 3x - 7$ lies in the space generated by $f(x) = 2x^2 - 5$ and $g(x) = x + 1$.

EXERCISE SET 4.5

Linear Combinations in R^n

1. Determine whether the first vector is a linear combination of the other vectors. If it is, give the combination.
   (a) $(-1, 7); (1, -1), (2, 4)$
   (b) $(8, 13); (1, 2), (2, 3)$
   (c) $(-1, 15); (-1, 4), (2, -8)$
   (d) $(13, 6); (1, 3), (4, 1)$

2. Determine whether the first vector is a linear combination of the other vectors. If it is, give the combination.
   (a) $(-3, 3, 7); (1, -1, 2), (2, 1, 0), (-1, 2, 1)$
   (b) $(-2, 11, 7); (1, -1, 0), (2, 1, 4), (-2, 4, 1)$
   (c) $(2, 7, 13); (1, 2, 3), (-1, 2, 4), (1, 6, 10)$
   (d) $(0, 10, 8); (-1, 2, 3), (1, 3, 1), (1, 8, 5)$
   (e) $(1, 4, -3); (1, 0, 1), (1, 1, 0), (3, 1, 2)$
   (f) $(1, 1, 2); (0, 1, 0), (3, 5, 6), (1, 2, 1)$

3. Give two vectors that are linear combinations of the following vectors.
   (a) $(1, 2), (3, -5)$ (b) $(-1, 0), (3, 1), (2, 4)$
   (c) $(1, -3, 5), (0, 1, 2)$
   (d) $(1, 2, 3), (1, 1, 1), (0, 7, 2), (4, 3, -2)$

Subspaces Generated by Given Sets of Vectors

4. Give three other vectors in the subspace of $\mathbf{R}^3$ generated by the vectors $(1, 2, 3)$ and $(1, 2, 0)$.

5. Give three other vectors in the subspace of $\mathbf{R}^3$ generated by the vectors $(1, 2, 1)$ and $(2, 1, 4)$.

6. Give three other vectors in the subspace of $\mathbf{R}^3$ generated by the vector $(1, 2, 3)$. Sketch the subspace.

7. Give three other vectors in the subspace of $\mathbf{R}^3$ generated by $\{(4, -1, 3)\}$.

8. Give three other vectors in the subspace of $\mathbf{R}^2$ generated by the vector $(1, 2)$. Sketch the subspace.

9. Give three other vectors in the subspace of $\mathbf{R}^4$ generated by the vector $(1, 2, -1, 3)$.

10. Give three other vectors in the subspace of $\mathbf{R}^4$ generated by the vectors $(2, 1, -3, 4), (-3, 0, 1, 5), (4, 1, 2, 0)$.

Spaces of Matrices and Functions

11. In each of the following, determine whether the first matrix is a linear combination of the matrices that follow:
   (a) $\begin{bmatrix} 5 & 7 \\ 5 & -10 \end{bmatrix}$; $\begin{bmatrix} 1 & 2 \\ 3 & -4 \end{bmatrix}$, $\begin{bmatrix} 0 & 3 \\ 1 & 2 \end{bmatrix}$, $\begin{bmatrix} 1 & 2 \\ 0 & 0 \end{bmatrix}$
   (b) $\begin{bmatrix} 7 & 6 \\ -5 & -3 \end{bmatrix}$; $\begin{bmatrix} 3 & 0 \\ 1 & 1 \end{bmatrix}$, $\begin{bmatrix} 0 & 1 \\ 3 & 4 \end{bmatrix}$, $\begin{bmatrix} 1 & 2 \\ 0 & 1 \end{bmatrix}$
   (c) $\begin{bmatrix} 4 & 1 \\ 7 & 10 \end{bmatrix}$; $\begin{bmatrix} 1 & 1 \\ 1 & 1 \end{bmatrix}$, $\begin{bmatrix} 3 & 1 \\ 0 & 0 \end{bmatrix}$, $\begin{bmatrix} -1 & -1 \\ 2 & 3 \end{bmatrix}$

12. In each of the following, determine whether the first function is a linear combination of the functions that follow:
   (a) $f(x) = 3x^2 + 2x + 9$; $g(x) = x^2 + 1$, $h(x) = x + 3$
   (b) $f(x) = 2x^2 + x - 3$; $g(x) = x^2 - x + 1$, $h(x) = x^2 + 2x - 2$
   (c) $f(x) = x^2 + 4x + 5$; $g(x) = x^2 + x - 1$, $h(x) = x^2 + 2x + 1$

13. (a) Is the function $f(x) = x + 5$ in the vector space Span$\{g, h\}$ generated by $g(x) = x + 1$ and $h(x) = x + 3$?
   (b) Is the function $f(x) = 3x^2 + 5x + 1$ in the vector space Span$\{g, h\}$ generated by $g(x) = 2x^2 + 3$ and $h(x) = x^2 + 3x - 1$?
   (c) Give three other functions in the vector space Span$\{g, h\}$ generated by $g(x) = 2x^2 + 3$ and $h(x) = x^2 + 3x - 1$.

General Vector Spaces

14. Let $\mathbf{v}$, $\mathbf{v}_1$, and $\mathbf{v}_2$ be vectors in a vector space V. Let $\mathbf{v}$ be a linear combination of $\mathbf{v}_1$ and $\mathbf{v}_2$. If c_1 and c_2 are nonzero scalars, show that $\mathbf{v}$ is also a linear combination of $c_1\mathbf{v}_1$ and $c_2\mathbf{v}_2$.

15. Let $\mathbf{v}$, $\mathbf{v}_1$, and $\mathbf{v}_2$ be vectors in a vector space V. Let c_1 and c_2 be nonzero scalars. Show that if $\mathbf{v}$ is not a linear combination of $\mathbf{v}_1$ and $\mathbf{v}_2$, then neither is $\mathbf{v}$ a linear combination of $c_1\mathbf{v}_1$ and $c_2\mathbf{v}_2$.

16. Let $\mathbf{v}_1$ and $\mathbf{v}_2$ span a vector space V. Let $\mathbf{v}_3$ be any other vector in V. Show that $\mathbf{v}_1$, $\mathbf{v}_2$, and $\mathbf{v}_3$ also span V.

4.6 Linear Dependence and Independence

We now continue the development of vector space structure from the previous section. We introduce concepts of dependence and independence of vectors.

DEFINITION (a) The set of vectors* $\{v_1, \ldots, v_m\}$ in a vector space V is said to be *linearly dependent* if there exist scalars $c_1, \ldots, c_m$, not all zero, such that

$$c_1 v_1 + \cdots + c_m v_m = 0$$

(b) The set of vectors $\{v_1, \ldots, v_m\}$ is *linearly independent* if $c_1 v_1 + \cdots + c_m v_m = 0$ can only be satisfied when $c_1 = 0, \ldots, c_m = 0$.

EXAMPLE 1 Determine whether the set $\{(1, 2, 0), (0, 1, -1), (1, 1, 2)\}$ is linearly independent in $\mathbf{R}^3$.

SOLUTION

We examine the identity

$$c_1(1, 2, 0) + c_2(0, 1, -1) + c_3(1, 1, 2) = 0$$

The identity leads to the following system of linear equations:

$$
\begin{aligned}
c_1 \quad\quad + c_3 &= 0 \\
2c_1 + c_2 + \;\; c_3 &= 0 \\
- c_2 + 2c_3 &= 0
\end{aligned}
$$

This system has the unique solution $c_1 = 0, c_2 = 0, c_3 = 0$. Thus the set is linearly independent. If the system had other solutions, the vectors would have been linearly dependent.

The following example illustrates how dependency is examined for function vector spaces.

EXAMPLE 2 (a) Show that the set $\{x^2 + 1, 3x - 1, -4x + 1\}$ is linearly independent in P_2. (b) Show that the set $\{x + 1, x - 1, -x + 5\}$ is linearly dependent in P_1.

SOLUTION

(a) Consider the identity $c_1(x^2 + 1) + c_2(3x - 1) + c_3(-4x + 1) = 0$. This gives $c_1 x^2 + (3c_2 - 4c_3)x + c_1 - c_2 + c_3 = 0$. Such a polynomial can only be zero if each coefficient is zero. Thus $c_1 = 0, 3c_2 - 4c_3 = 0, c_1 - c_2 + c_3 = 0$. This system of equations has the unique solution $c_1 = 0, c_2 = 0, c_3 = 0$. The functions are linearly independent.

(b) Consider the identity $c_1(x + 1) + c_2(x - 1) + c_3(-x + 5) = 0$. Rewrite as $(c_1 + c_2 - c_3)x + (c_1 - c_2 + 5c_3) = 0$. Each coefficient must be zero. Thus $c_1 + c_2 - c_3 = 0, c_1 - c_2 + 5c_3 = 0$. This system has many solutions, $c_1 = -2r$, $c_2 = 3r, c_3 = r$. Thus $-2r(x + 1) + 3r(x - 1) + r(-x + 5) = 0$. If $r = 1$ for example, $-2(x + 1) + 3(x - 1) + 1(-x + 5) = 0$. The functions are linearly dependent.

*We shall often just refer to vectors rather than a set of vectors.

The following theorem gives insight into the significance of linear dependence. The vectors are linearly dependent in the sense that it is possible to express one of the vectors linearly in terms of the others.

THEOREM 4.8

A set consisting of two or more vectors in a vector space is linearly dependent if and only if it is possible to express one of the vectors as a linear combination of the other vectors.

Proof   Let the set $\{\mathbf{v}_1, \mathbf{v}_2, \ldots, \mathbf{v}_m\}$ be linearly dependent. Therefore there exist scalars $c_1, c_2, \ldots, c_m$, not all zero, such that

$$c_1\mathbf{v}_1 + c_2\mathbf{v}_2 + \cdots + c_m\mathbf{v}_m = \mathbf{0}$$

Assume that $c_1 \neq 0$. The preceding identity can be rewritten

$$\mathbf{v}_1 = \left(\frac{-c_2}{c_1}\right)\mathbf{v}_2 + \cdots + \left(\frac{-c_m}{c_1}\right)\mathbf{v}_m$$

Thus, $\mathbf{v}_1$ is a linear combination of $\mathbf{v}_2, \ldots, \mathbf{v}_m$.

Conversely, assume that $\mathbf{v}_1$ is a linear combination of $\mathbf{v}_2, \ldots, \mathbf{v}_m$. Therefore there exist scalars $d_2, \ldots, d_m$, such that

$$\mathbf{v}_1 = d_2\mathbf{v}_2 + \cdots + d_m\mathbf{v}_m$$

Rewrite this equation as

$$1\mathbf{v}_1 + (-d_2)\mathbf{v}_2 + \cdots + (-d_m)\mathbf{v}_m = \mathbf{0}$$

Thus the set $\{\mathbf{v}_1, \mathbf{v}_2, \ldots, \mathbf{v}_m\}$ is linearly dependent, completing the proof.

Let us consider the implications of this result for two and three vectors.

Linear Dependence of $\{\mathbf{v}_1, \mathbf{v}_2\}$

The set $\{\mathbf{v}_1, \mathbf{v}_2\}$ is linearly dependent if and only if it is possible to write one vector as a scalar multiple of the other vector. Let $\mathbf{v}_1 = c\mathbf{v}_2$. This means that $\mathbf{v}_1$ and $\mathbf{v}_2$ are collinear. See Figure 4.17 for $\mathbf{R}^3$. On the other hand, $\{\mathbf{v}_1, \mathbf{v}_2\}$ is linearly independent if and only if it is not possible to write one vector as a scalar multiple of the other; $\mathbf{v}_1$ and $\mathbf{v}_2$ are not collinear.

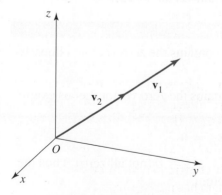

$\{\mathbf{v}_1, \mathbf{v}_2\}$ linearly dependent;
vectors lie on a line

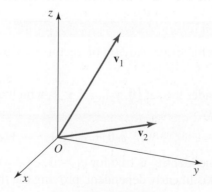

$\{\mathbf{v}_1, \mathbf{v}_2\}$ linearly independent;
vectors do not lie on a line

Linear dependence and independence of $\{\mathbf{v}_1, \mathbf{v}_2\}$ in $\mathbf{R}^3$

Figure 4.17

We shall frequently use this result to quickly see that two vectors are linearly dependent or independent. For example, $(2, -1, 3)$ and $(4, -2, 6)$ are linearly dependent—the second is a scalar multiple of the first. $(1, -2, 4)$ and $(3, -6, 8)$ are linearly independent—the second is not a scalar multiple of the first.

Linear Dependence of $\{v_1, v_2, v_3\}$

The set $\{v_1, v_2, v_3\}$ is linearly dependent if and only if it is possible to write one of the vectors, say v_1, as a linear combination of the other two vectors v_2 and v_3. Let $v_1 = c_2 v_2 + c_3 v_3$. This means that v_1 lies in the plane generated by v_2 and v_3 (or line if they are collinear). See Figure 4.18 for $\mathbf{R}^3$. On the other hand $\{v_1, v_2, v_3\}$ is linearly independent if and only if the vectors do not lie in a plane (or line).

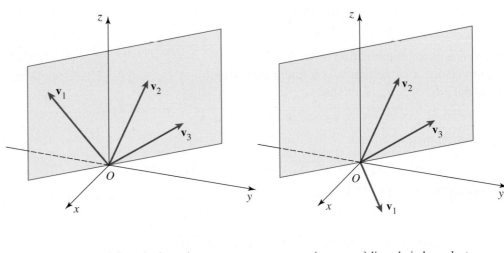

$\{v_1, v_2, v_3\}$ linearly dependent;
vectors lie in a plane (or line)

$\{v_1, v_2, v_3\}$ linearly independent;
vectors do not lie on a plane (or line)

Linear dependence and independence of $\{v_1, v_2, v_3\}$ in $\mathbf{R}^3$

Figure 4.18

We complete this study with two further results about vector spaces.

THEOREM 4.9

Let V be a vector space. Any set of vectors in V that contains the zero vector is linearly dependent.

Proof   Consider the set $\{0, v_2, \dots, v_m\}$, which contains the zero vector. Let us examine the identity

$$c_1 0 + c_2 v_2 + \cdots + c_m v_m = 0$$

We see that the identity is true for $c_1 = 1, c_2 = 0, \dots, c_m = 0$ (not all zero). Thus the set of vectors is linearly dependent, proving the theorem.

THEOREM 4.10

Let the set $\{v_1, \ldots, v_m\}$ be linearly dependent in a vector space V. Any set of vectors in V that contains these vectors will also be linearly dependent.

Proof Since the set $\{v_1, \ldots, v_m\}$ is linearly dependent there exist scalars $c_1, \ldots, c_m$, not all zero, such that

$$c_1 v_1 + \cdots + c_m v_m = 0$$

Consider the set of vectors $\{v_1, \ldots, v_m, v_{m+1}, \ldots, v_n\}$, which contains the given vectors. There are scalars, not all zero, namely $c_1, \ldots, c_m, 0, \ldots, 0$ such that

$$c_1 v_1 + \cdots + c_m v_m + 0 v_{m+1} + \cdots + 0 v_n = 0$$

Thus the set $\{v_1, \ldots, v_m, v_{m+1}, \ldots, v_n\}$ is linearly dependent.

EXAMPLE 3 Let the set $\{v_1, v_2\}$ be linearly independent. Prove that $\{v_1 + v_2, v_1 - v_2\}$ is also linearly independent.

SOLUTION

Let us examine the identity

$$a(v_1 + v_2) + b(v_1 - v_2) = 0 \qquad (1)$$

If we can show that this identity implies $a = 0$ and $b = 0$ then $\{v_1 + v_2, v_1 - v_2\}$ will be linearly independent. We get

$$a v_1 + a v_2 + b v_1 - b v_2 = 0$$
$$(a + b) v_1 + (a - b) v_2 = 0$$

Since $\{v_1, v_2\}$ is linearly independent

$$a + b = 0$$
$$a - b = 0$$

This system has the unique solution $a = 0$, $b = 0$. Returning to identity (1) this means that $\{v_1 + v_2, v_1 - v_2\}$ is linearly independent.

EXERCISE SET 4.6

Linear Dependence in R^n

1. Determine whether the following sets of vectors are linearly dependent or independent.

   (a) $\{(-1, 2), (2, -4)\}$

   (b) $\{(-1, 3), (2, 5)\}$

   (c) $\{(1, -2, 3), (-2, 4, 1), (-4, 8, 9)\}$

   (d) $\{(1, 0, 2), (2, 6, 4), (1, 12, 2)\}$

   (e) $\{(1, 2, 5), (1, -2, 1), (2, 1, 4)\}$

   (f) $\{(1, 1, 1), (-4, 3, 2), (4, 1, 2)\}$

2. Use Theorem 4.10 to show that the following sets of vectors are linearly dependent in $\mathbf{R}^3$.

   (a) $\{(2, -1, 3), (-4, 2, -6), (8, 0, 1)\}$

   (b) $\{(1, -2, 3), (7, 4, -2), (3, -6, 9)\}$

   (c) $\{(5, 2, -3), (3, 0, 4), (-3, 0, -4)\}$

   (d) $\{(1, 1, 1), (2, 2, 2), (0, 1, 5)\}$

3. Find values of t for which the following sets are linearly dependent.

   (a) $\{(-1, 2), (t, -4)\}$

   (b) $\{(3, t), (6, t - 1)\}$

   (c) $\{(2, -t), (2t + 6, 4t)\}$

4. (a) Prove that the set $\{(1, 1), (0, 2)\}$ is linearly independent in $\mathbf{R}^2$.

(b) Prove that the set $\{(1, 1, 2), (0, -1, 3), (0, 0, 5)\}$ is linearly independent in $\mathbf{R}^3$.

(c) Prove that the set $\{(3, -2, 4, 5), (0, 2, 3, -4), (0, 0, 2, 7), (0, 0, 0, 4)\}$ is linearly independent in $\mathbf{R}^4$.

(d) Discuss the pattern of the zero components in the vectors of Parts (a), (b), and (c). How can you use this pattern to construct a set of five linearly independent vectors in $\mathbf{R}^5$?

5. Consider the following matrix, which is in reduced echelon form.

$$\begin{bmatrix} 1 & 0 & 0 & 7 \\ 0 & 1 & 0 & 4 \\ 0 & 0 & 1 & 3 \end{bmatrix}$$

Show that the row vectors form a linearly independent set. Is the set of nonzero row vectors of any matrix in reduced echelon form linearly independent? Discuss.

6. Show that the following sets of vectors are linearly dependent by observing relations between elements of the vectors.

(a) $\{(1, 2, 3), (1, 1, 1), (2, 3, 4)\}$

(b) $\{(1, 2, 5), (3, 6, 15), (-7, 3, 2)\}$

(c) $\{(3, 4, 5), (1, 1, 1), (5, 6, 7)\}$

(d) $\{(1, 1, 1,), (0, 2, -3), (0, 0, 0)\}$

7. The following sets are all linearly independent. Add a vector to each set that will make the new sets linearly dependent.

(a) $\{(1, -1, 0), (2, 1, 3)\}$

(b) $\{(3, -5, 1), (1, 4, 3)\}$

(c) $\{(1, 2, 4), (0, 2, 5)\}$

(d) $\{(3, -2, 4), (2, 5, 7)\}$

Linear Dependence in Matrix and Function Spaces

8. Determine whether the following sets of matrices are linearly dependent in M_{22}.

(a) $\left\{ \begin{bmatrix} 1 & 0 \\ 0 & 0 \end{bmatrix}, \begin{bmatrix} 0 & 2 \\ 0 & 0 \end{bmatrix}, \begin{bmatrix} 0 & 0 \\ 3 & 0 \end{bmatrix}, \begin{bmatrix} 0 & 0 \\ 0 & 4 \end{bmatrix} \right\}$

(b) $\left\{ \begin{bmatrix} 1 & 2 \\ 3 & 1 \end{bmatrix}, \begin{bmatrix} 1 & 1 \\ 1 & 1 \end{bmatrix}, \begin{bmatrix} 2 & 1 \\ 4 & 2 \end{bmatrix} \right\}$

(c) $\left\{ \begin{bmatrix} 1 & 2 \\ -1 & 0 \end{bmatrix}, \begin{bmatrix} 1 & 2 \\ 1 & 1 \end{bmatrix}, \begin{bmatrix} 1 & 2 \\ 5 & 3 \end{bmatrix} \right\}$

(d) $\left\{ \begin{bmatrix} 2 & 4 \\ 0 & 1 \end{bmatrix}, \begin{bmatrix} 0 & -2 \\ 8 & 3 \end{bmatrix}, \begin{bmatrix} -3 & -7 \\ 4 & 0 \end{bmatrix} \right\}$

9. Determine whether the following sets of functions are linearly dependent in P_2.

(a) $\{f, g, h\}$ where $f(x) = 2x^2 + 1$, $g(x) = x^2 + 4x$, $h(x) = x^2 - 4x + 1$

(b) $\{f, g, h\}$ where $f(x) = x^2 + 3$, $g(x) = x + 1$, $h(x) = 2x^2 - 3x + 3$

(c) $\{f, g, h\}$ where $f(x) = x^2 + 3x - 1$, $g(x) = x + 3$, $h(x) = 2x^2 - x + 1$

(d) $\{f, g, h\}$ where $f(x) = -x^2 + 2x - 5$, $g(x) = 5x - 1$, $h(x) = 7$

General Vector Spaces

10. Let $\{\mathbf{v}_1, \mathbf{v}_2\}$ be any vectors in a vector space V. Show that the set $\{\mathbf{v}_1, \mathbf{v}_2, a\mathbf{v}_1 + b\mathbf{v}_2\}$ is linearly dependent for all values of scalars a and b.

11. Let $\{\mathbf{v}_1, \mathbf{v}_2\}$ be linearly independent in a vector space V. Show that if a vector $\mathbf{v}_3$ is not of the form $a\mathbf{v}_1 + b\mathbf{v}_2$, then the set $\{\mathbf{v}_1, \mathbf{v}_2, \mathbf{v}_3\}$ is linearly independent.

12. Let the set $\{\mathbf{v}_1, \mathbf{v}_2\}$ be linearly dependent in a vector space V. Prove that $\{\mathbf{v}_1 + \mathbf{v}_2, \mathbf{v}_1 - \mathbf{v}_2\}$ is also linearly dependent.

13. Prove that every subset of a linearly independent set is linearly independent. Is every subset of a linearly dependent set linearly dependent?

14. Prove that if $\mathbf{u}$ and $\mathbf{v}$ are nonzero orthogonal vectors in $\mathbf{R}^n$ they are linearly independent.

15. Let $\mathbf{v}_1, \mathbf{v}_2$, and $\mathbf{v}_3$ be vectors in $\mathbf{R}^3$. What can you say about the linear dependence or independence of these vectors in the following cases?

(a) Span$\{\mathbf{v}_1, \mathbf{v}_2, \mathbf{v}_3\}$ is $\mathbf{R}^3$.

(b) Span$\{\mathbf{v}_1, \mathbf{v}_2, \mathbf{v}_3\}$ is the same space as Span$\{\mathbf{v}_1, \mathbf{v}_2\}$.

(c) Span$\{\mathbf{v}_1, \mathbf{v}_2, \mathbf{v}_3\}$ is the same space as Span$\{\mathbf{v}_1\}$.

16. State (with a brief explanation) whether the following statements are true or false.

(a) The vector $(1, 4, 2)$ is a linear combination of $(1, 0, 0)$ and $(0, 1, 0)$.

(b) The vectors $(1, 0, 0), (2, 1, 0), (-1, 3, 0)$ span $\mathbf{R}^3$.

(c) The vectors $(1, 1, 0), (0, 0, -1)$ span the subspace of $\mathbf{R}^3$ of vectors of the form (a, a, b).

(d) The set $\{(1, 2, 3), (1, 2, 0), (1, 0, 0)\}$ is linearly dependent.

17. State (with a brief explanation) whether the following statements are true or false.

(a) Let $\mathbf{v}_1, \mathbf{v}_2, \mathbf{v}_3$, be vectors in a vector space V. If $\mathbf{v}$ is a linear combination of $\mathbf{v}_1$ and $\mathbf{v}_2$ it is also a linear combination of $\mathbf{v}_1, \mathbf{v}_2, \mathbf{v}_3$.

(b) It takes at least three vectors to span $\mathbf{R}^3$.

(c) Every vector space has a unique set of vectors that spans it.

(d) Span$\{\mathbf{v}_1, \mathbf{v}_2\}$ is the same space as Span$\{\mathbf{v}_1, 2\mathbf{v}_2\}$

(e) Every set of three vectors in $\mathbf{R}^2$ is linearly dependent.

18. Write down two vectors from $\mathbf{R}^3$ at random. Which is the more likely—that these vectors are linearly dependent or linearly independent?

19. Write down three vectors from $\mathbf{R}^3$ at random. Which is the more likely—that these vectors are linearly dependent or linearly independent?

20. A computer program accepts a number of vectors in $\mathbf{R}^3$ as input and gives the information whether the vectors are linearly dependent or independent as output. Which is the more likely to happen due to round-off error—that the computer states that a given set of linearly independent vectors is linearly dependent, or vice versa?

4.7 Basis and Dimension

We think about a line as being one-dimensional and a plane as being two-dimensional. In this section we introduce the mathematical definition of dimension. It will be compatible with our intuitive ideas. We first bring together the concepts of spanning sets and linear independence.

THEOREM 4.11

Let the vectors $\mathbf{v}_1, \ldots, \mathbf{v}_n$ span a vector space V. Each vector in V can be expressed uniquely as a linear combination of these vectors if and only if the vectors are linearly independent.

Proof

(a) Assume that $\mathbf{v}_1, \ldots, \mathbf{v}_n$ are linearly independent. Let $\mathbf{v}$ be a vector in V. Since $\mathbf{v}_1, \ldots, \mathbf{v}_n$ span V, we can express $\mathbf{v}$ as a linear combination of these vectors. Suppose we can write

$$\mathbf{v} = a_1\mathbf{v}_1 + \cdots + a_n\mathbf{v}_n \quad \text{and} \quad \mathbf{v} = b_1\mathbf{v}_1 + \cdots + b_n\mathbf{v}_n$$

Then

$$a_1\mathbf{v}_1 + \cdots + a_n\mathbf{v}_n = b_1\mathbf{v}_1 + \cdots + b_n\mathbf{v}_n$$

giving

$$(a_1 - b_1)\mathbf{v}_1 + \cdots + (a_n - b_n)\mathbf{v}_n = \mathbf{0}$$

Since $\mathbf{v}_1, \ldots, \mathbf{v}_n$ are linearly independent $(a_1 - b_1) = 0, \ldots, (a_n - b_n) = 0$, implying that $a_1 = b_1, \ldots, a_n = b_n$. There is thus only one way of expressing $\mathbf{v}$ as a linear combination of the $\mathbf{v}_1, \ldots, \mathbf{v}_n$.

(b) We discuss the converse. Let $\mathbf{v}$ be a vector in V. Assume that $\mathbf{v}$ can be written in only one way as a linear combination of $\mathbf{v}_1, \ldots, \mathbf{v}_n$. Note that $0\mathbf{v}_1 + \cdots + 0\mathbf{v}_n = \mathbf{0}$. This must be the only way $\mathbf{0}$ can be written as a linear combination of $\mathbf{v}_1, \ldots, \mathbf{v}_n$. Thus $c_1\mathbf{v}_1 + \cdots + c_n\mathbf{v}_n = \mathbf{0}$ can only be satisfied when $c_1 = 0, \ldots, c_n = 0$. The vectors $\mathbf{v}_1, \ldots, \mathbf{v}_n$ are *linearly independent*.

This result leads naturally to a set of vectors that most suitably represents a vector space—a *basis*.

DEFINITION   A finite set of vectors $\{\mathbf{v}_1, \ldots, \mathbf{v}_n\}$ is called a *basis* for a vector space V if the set spans V and is linearly independent.

Each vector in V can be expressed uniquely as a linear combination of the vectors in a basis (by Theorem 4.11).

In general there are many bases for a vector space. The following theorem leads to a key result that all bases of a given vector space have the same number of vectors.

THEOREM 4.12

Let $\{\mathbf{v}_1, \ldots, \mathbf{v}_n\}$ be a basis for a vector space V. If $\{\mathbf{w}_1, \ldots, \mathbf{w}_m\}$ is a set of more than n vectors in V, then this set is linearly dependent.

Proof   Consider the identity

$$c_1\mathbf{w}_1 + \cdots + c_m\mathbf{w}_m = 0 \tag{1}$$

We shall show that values of $c_1, \ldots, c_m$, not all zero exist, satisfying this identity, thus proving that the vectors are linearly dependent.

The set $\{\mathbf{v}_1, \ldots, \mathbf{v}_n\}$ is a basis for V. Thus each of the vectors $\mathbf{w}_1, \ldots, \mathbf{w}_m$ can be expressed as a linear combination of $\mathbf{v}_1, \ldots, \mathbf{v}_n$. Let

$$\mathbf{w}_1 = a_{11}\mathbf{v}_1 + a_{12}\mathbf{v}_2 + \cdots + a_{1n}\mathbf{v}_n$$
$$\vdots$$
$$\mathbf{w}_m = a_{m1}\mathbf{v}_1 + a_{m2}\mathbf{v}_2 + \cdots + a_{mn}\mathbf{v}_n$$

Substituting for $\mathbf{w}_1, \ldots, \mathbf{w}_m$ into Equation (1) we get,

$$c_1(a_{11}\mathbf{v}_1 + a_{12}\mathbf{v}_2 + \cdots + a_{1n}\mathbf{v}_n) + \cdots + c_m(a_{m1}\mathbf{v}_1 + a_{m2}\mathbf{v}_2 + \cdots + a_{mn}\mathbf{v}_n) = \mathbf{0}$$

Rearranging we get

$$(c_1a_{11} + c_2a_{21} + \cdots + c_ma_{m1})\mathbf{v}_1 + \cdots + (c_1a_{1n} + c_2a_{2n} + \cdots + c_ma_{mn})\mathbf{v}_n = \mathbf{0}$$

Since $\mathbf{v}_1, \ldots, \mathbf{v}_n$ are linearly independent, this identity can be satisfied only if the coefficients are all zero. Thus

$$a_{11}c_1 + a_{21}c_2 + \cdots + a_{m1}c_m = 0$$
$$\vdots$$
$$a_{1n}c_1 + a_{2n}c_2 + \cdots + a_{mn}c_m = 0$$

Thus finding c's that satisfy (1) reduces to finding solutions to this system of n equations in m variables. Since $m > n$, the number of variables is greater than the number of equations. We know that such a system of homogeneous equations has many solutions. There are therefore nonzero values of c's that satisfy (1). Thus the set $\{\mathbf{w}_1, \ldots, \mathbf{w}_m\}$ is linearly dependent.

THEOREM 4.13

All bases for a vector space V have the same number of vectors.

Proof   Let $\{\mathbf{v}_1, \ldots, \mathbf{v}_n\}$ and $\{\mathbf{w}_1, \ldots, \mathbf{w}_m\}$ be two bases for V.

If we interpret $\{\mathbf{v}_1, \ldots, \mathbf{v}_n\}$ as a basis for V and $\{\mathbf{w}_1, \ldots, \mathbf{w}_m\}$ as a set of linearly independent vectors in V, the previous theorem tells us that $m \leq n$. Conversely, if we interpret $\{\mathbf{w}_1, \ldots, \mathbf{w}_m\}$ as a basis for V and $\{\mathbf{v}_1, \ldots, \mathbf{v}_n\}$ as a set of linearly independent vectors in V, then $n \leq m$. Thus $n = m$, proving that both bases consist of the same number of vectors.

We use the number of vectors in a basis for its dimension. Since the standard bases for $\mathbf{R}^2$ and $\mathbf{R}^3$ have two and three vectors, respectively, this fits in with our intuitive understanding of $\mathbf{R}^2$ being two-dimensional and $\mathbf{R}^3$ being three-dimensional.

> **DEFINITION** If a vector space V has a basis consisting of n vectors then the **dimension** of V is said to be n. We write $\dim(V)$ for the dimension of V.

Note that we have defined a basis for a vector space to be a *finite* set of vectors that spans the space and is linearly independent. Such a set does not exist for all vector spaces. When such a finite set exists, we say that the vector space is **finite dimensional**. If such a finite set does not exist we say that the vector space is **infinite dimensional**. We shall meet some function vector spaces that are infinite dimensional in the next chapter. We are primarily interested in finite dimensional vector spaces in this course.

EXAMPLE 1 Consider the set $\{(1, 2, 3), (-2, 4, 1)\}$ of vectors in $\mathbf{R}^3$. These vectors generate a subspace V of $\mathbf{R}^3$ consisting of all vectors of the form

$$\mathbf{v} = c_1(1, 2, 3) + c_2(-2, 4, 1)$$

The vectors $(1, 2, 3)$ and $(-2, 4, 1)$ span this subspace.

Furthermore, since the second vector is not a scalar multiple of the first vector, the vectors are linearly independent.

Therefore $\{(1, 2, 3), (-2, 4, 1)\}$ is a basis for V. Thus $\dim(V) = 2$. We know that V is in fact a plane through the origin.

In this last example we saw that a certain plane through the origin was a two-dimensional subspace of $\mathbf{R}^3$. The following theorem gives us more information about the subspaces of $\mathbf{R}^3$ and their dimensions.

THEOREM 4.14

(a) The origin is a subspace of $\mathbf{R}^3$. The dimension of this subspace is defined to be zero.
(b) The one-dimensional subspaces of $\mathbf{R}^3$ are lines through the origin.
(c) The two-dimensional subspaces of $\mathbf{R}^3$ are planes through the origin. See Figure 4.19.

Proof

(a) Let V be the set $\{(0, 0, 0)\}$, consisting of a single element, the zero vector of $\mathbf{R}^3$. Let c be an arbitrary scalar. Since

$$(0, 0, 0) + (0, 0, 0) = (0, 0, 0) \quad \text{and} \quad c(0, 0, 0) = (0, 0, 0)$$

V is closed under addition and scalar multiplication. It is thus a subspace of $\mathbf{R}^3$. The dimension of this subspace is defined to be zero.
(b) Let $\mathbf{v}$ be a basis for a one-dimensional subspace V of $\mathbf{R}^3$. Every vector in V is thus of the form $c\mathbf{v}$, for some scalar c. We know that these vectors form a line through the origin.
(c) Let $\{\mathbf{v}_1, \mathbf{v}_2\}$ be a basis for a two-dimensional subspace V of $\mathbf{R}^3$. Every vector in V is of the form $c_1\mathbf{v}_1 + c_2\mathbf{v}_2$. V is thus a plane through the origin.

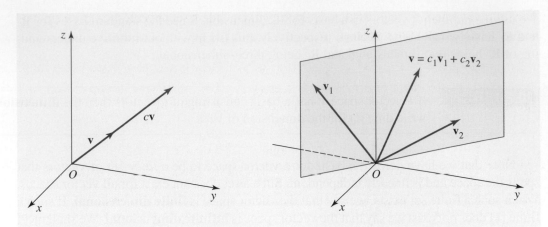

One-dimensional subspace of $\mathbf{R}^3$ with basis $\{\mathbf{v}\}$ is a line through the origin

Two-dimensional subspace of $\mathbf{R}^3$ with basis $\{\mathbf{v}_1, \mathbf{v}_2\}$ is a plane through the origin

One- and two-dimensional subspace of $\mathbf{R}^3$

Figure 4.19

Suppose that a vector space is known to be of dimension n. The following theorem (which we state without proof), tells us that we do not have to check both the linear dependence and spanning conditions to see if a given set is a basis.

THEOREM 4.15

Let V be a vector space of dimension n.
(a) If $S = \{\mathbf{v}_1, \ldots, \mathbf{v}_n\}$ is a set of n linearly independent vectors in V, then S is a basis for V.
(b) If $S = \{\mathbf{v}_1, \ldots, \mathbf{v}_n\}$ is a set of n vectors that spans V, then S is a basis for V.

EXAMPLE 2 Prove that the set $\{(1, 3, -1), (2, 1, 0), (4, 2, 1)\}$ is a basis for $\mathbf{R}^3$.

SOLUTION

The dimension of $\mathbf{R}^3$ is three. Thus a basis for $\mathbf{R}^3$ consists of three vectors. We have the correct number of vectors for a basis. Normally we would have to show that this set is linearly independent and that it spans $\mathbf{R}^3$. Theorem 4.15 tells us that we need to check only one of these two conditions. Let us check for linear independence (this is the easier of the two conditions to check). We get

$$c_1(1, 3, -1) + c_2(2, 1, 0) + c_3(4, 2, 1) = (0, 0, 0)$$

This identity leads to the system of equations

$$c_1 + 2c_2 + 4c_3 = 0$$
$$3c_1 + c_2 + 2c_3 = 0$$
$$-c_1 + c_3 = 0$$

This system has the unique solution $c_1 = 0$, $c_2 = 0$, $c_3 = 0$. Thus the vectors are linearly independent. The set $\{(1, 3, -1), (2, 1, 0), (4, 2, 1)\}$ is therefore a basis for $\mathbf{R}^3$.

The following theorem tells us that any linearly independent set of vectors can be extended to give a basis for a vector space.

THEOREM 4.16

Let V be a vector space of dimension n. Let $\{\mathbf{v}_1, \ldots, \mathbf{v}_m\}$ be a set of m linearly independent vectors in V, where $m < n$. Then there exist vectors $\mathbf{v}_{m+1}, \ldots, \mathbf{v}_n$ such that $\{\mathbf{v}_1, \ldots, \mathbf{v}_m, \mathbf{v}_{m+1}, \ldots, \mathbf{v}_n\}$ is a basis of V.

Proof Since $m < n$, then $\{\mathbf{v}_1, \ldots, \mathbf{v}_m\}$ cannot be a basis. Thus there exists a vector $\mathbf{v}_{m+1}$ in V that does not lie in the subspace generated by $\mathbf{v}_1, \ldots, \mathbf{v}_m$. The set $\{\mathbf{v}_1, \ldots, \mathbf{v}_m, \mathbf{v}_{m+1}\}$ will be linearly independent. If $m + 1 = n$, then $\{\mathbf{v}_1, \ldots, \mathbf{v}_m, \mathbf{v}_{m+1}\}$ is a basis of V. If $m + 1 < n$ there will be a vector $\mathbf{v}_{m+2}$ that does not lie in the subspace generated by $\mathbf{v}_1, \ldots, \mathbf{v}_m, \mathbf{v}_{m+1}$. If $m + 2 = n$, then $\{\mathbf{v}_1, \ldots, \mathbf{v}_m, \mathbf{v}_{m+1}, \mathbf{v}_{m+2}\}$ is a basis for V. One continues adding vectors thus until a basis is found.

In this chapter we have discussed the concepts of vector space, subspace, linear dependence/independence, spanning set, basis, and dimension. The following example brings all these concepts together. The reader should strive for an intuitive "feel" for each of the situations discussed.

EXAMPLE 3 State (with a brief explanation) whether the following statements are true or false.
(a) The vectors $(1, 2)$, $(-1, 3)$, $(5, 2)$ are linearly dependent in $\mathbf{R}^2$.
(b) The vectors $(1, 0, 0)$, $(0, 2, 0)$, $(1, 2, 0)$ span $\mathbf{R}^3$.
(c) $\{(1, 0, 2), (0, 1, -3)\}$ is a basis for the subspace of $\mathbf{R}^3$ consisting of vectors of the form $(a, b, 2a - 3b)$.
(d) Any set of two vectors can be used to generate a two-dimensional subspace of $\mathbf{R}^3$.

SOLUTION

(a) True: The dimension of $\mathbf{R}^2$ is two. Thus any three vectors are linearly dependent.
(b) False: The three vectors are linearly dependent. Thus they cannot span a three-dimensional space: $(1, 0, 0) + (0, 2, 0) - (1, 2, 0) = (0, 0, 0)$.
(c) True: The vectors span the subspace since

$$(a, b, 2a - 3b) = a(1, 0, 2) + b(0, 1, -3)$$

The vectors are also linearly independent since they are not collinear.
(d) False: The two vectors must be linearly independent.

Spaces $\mathbf{R}^n$, Matrices, Polynomials, and $\mathbf{C}^n$

We are now able to talk about bases and dimensions for the spaces $\mathbf{R}^n$, the spaces of matrices, of polynomials, and of the complex vector space $\mathbf{C}^n$. We summarize our conclusions below.

Space $\mathbf{R}^n$ The vectors $(1, 0, \ldots, 0), \ldots (0, 0, \ldots, 1)$ span $\mathbf{R}^n$ since we can write an arbitrary vector

$$(u_1, \ldots, u_n) = u_1(1, 0, \ldots, 0) + \cdots + u_n(0, 0, \ldots, 1)$$

These vectors are linearly independent since

$$a_1(1, 0, \ldots, 0) + a_2(0, 1, \ldots, 0) + \cdots + a_n(0, 0, \ldots, 1) = 0$$

has the unique solution $a_1 = 0$, $a_2 = 0$, $\ldots$, $a_n = 0$. The set of n vectors $\{(1, 0, \ldots, 0), \ldots (0, 0, \ldots, 1)\}$ is the *standard basis* of $\mathbf{R}^n$. The *dimension* of $\mathbf{R}^n$ is n.

Space of Matrices   Consider the vector space M_{22} of 2×2 matrices. The following matrices span M_{22}

$$\begin{bmatrix} 1 & 0 \\ 0 & 0 \end{bmatrix}, \begin{bmatrix} 0 & 1 \\ 0 & 0 \end{bmatrix}, \begin{bmatrix} 0 & 0 \\ 1 & 0 \end{bmatrix}, \begin{bmatrix} 0 & 0 \\ 0 & 1 \end{bmatrix}$$

since an arbitrary matrix in M_{22} can be written

$$\begin{bmatrix} a & b \\ c & d \end{bmatrix} = a\begin{bmatrix} 1 & 0 \\ 0 & 0 \end{bmatrix} + b\begin{bmatrix} 0 & 1 \\ 0 & 0 \end{bmatrix} + c\begin{bmatrix} 0 & 0 \\ 1 & 0 \end{bmatrix} + d\begin{bmatrix} 0 & 0 \\ 0 & 1 \end{bmatrix}.$$

They are also linearly independent since

$$c_1\begin{bmatrix} 1 & 0 \\ 0 & 0 \end{bmatrix} + c_2\begin{bmatrix} 0 & 1 \\ 0 & 0 \end{bmatrix} + c_3\begin{bmatrix} 0 & 0 \\ 1 & 0 \end{bmatrix} + c_4\begin{bmatrix} 0 & 0 \\ 0 & 1 \end{bmatrix} = 0$$

leads to $\begin{bmatrix} c_1 & c_2 \\ c_3 & c_4 \end{bmatrix} = 0$, implying that $c_1 = 0$, $c_2 = 0$, $c_3 = 0$, $c_4 = 0$. This set of matrices thus forms a basis for M_{22}. The dimension of the vector space M_{22} is 4. Similarly, the dimension of the vector space M_{mn} is mn. This type of basis is called the *standard basis* for M_{mn}.

Space of Polynomials   Consider the vector space P_2 of polynomials of degree ≤ 2. The functions x^2, x, and 1 span P_2 since any polynomial $ax^2 + bx + c$ can be written

$$ax^2 + bx + c = a(x^2) + b(x) + c(1)$$

x^2, x, 1 are also linearly independent since

$$c_1(x^2) + c_2(x) + c_3(1) = 0, \qquad \text{for all values of } x$$

implies that $c_1 = 0$, $c_2 = 0$, $c_3 = 0$. $\{x^2, x, 1\}$ is thus a basis for P_2. The dimension of the vector space P_2 is 3. Similarly, the set $\{x^n, x^{n-1}, \ldots, x, 1\}$ is a basis for P_n and the dimension of P_n is $n + 1$. This basis is called the *standard basis* for P_n.

The Space C^n   Consider the complex vector space C^2. The vectors $(1, 0)$ and $(0, 1)$ span C^2 since an arbitrary vector $(a + bi, c + di)$ can be written

$$(a + bi, c + di) = (a + bi)(1, 0) + (c + di)(0, 1).$$

$(1, 0)$ and $(0, 1)$ are also linearly independent in C^2. Thus $\{(1, 0), (0, 1)\}$ is a basis for C^2 and the dimension of C^2 is 2. Similarly $\{(1, \ldots, 0), \ldots, (0, \ldots, 1)\}$ is a basis for C^n and the dimension of C^n is n. This basis is the *standard basis* for C^n.

EXERCISE SET 4.7

Bases for R^n

1. Prove that the following sets are bases for $\mathbf{R}^2$ by showing that they span the space and are linearly independent.

(a) $\{(1, 2), (3, 1)\}$ **(b)** $\{(-1, 4), (2, 5)\}$

(c) $\{(1, 1), (-1, 1)\}$ **(d)** $\{(1, 0), (1, 1)\}$

2. Use Theorem 4.15 to prove that the following sets are bases for $\mathbf{R}^2$.

(a) $\{(1, 3), (-1, 2)\}$ **(b)** $\{(2, 6), (4, 1)\}$

(c) $\left\{ \begin{bmatrix} -1 \\ 2 \end{bmatrix}, \begin{bmatrix} 3 \\ 4 \end{bmatrix} \right\}$ **(d)** $\left\{ \begin{bmatrix} 0 \\ 1 \end{bmatrix}, \begin{bmatrix} 1 \\ 1 \end{bmatrix} \right\}$

3. Which of the following sets of vectors are bases for $\mathbf{R}^2$?

(a) $\{(3, 1), (2, 1)\}$ **(b)** $\{(1, -3), (-2, 6)\}$

(c) $\{(4, 5), (3, 2)\}$ **(d)** $\{(-1, 2), (3, -6)\}$

4. Prove that the following sets are bases for $\mathbf{R}^3$.

(a) $\{(1, 1, 1), (0, 1, 2), (3, 0, 1)\}$

(b) $\{(1, 2, 3), (2, 4, 1), (3, 0, 0)\}$

(c) $\{(0, 0, 1), (2, 3, 1), (4, 1, 2)\}$

(d) $\{(1, 1, 4), (2, 1, 3), (0, 1, 6)\}$

5. Which of the following sets are bases for $\mathbf{R}^3$?

(a) $\{(1, -1, 2), (2, 0, 1), (3, 0, 0)\}$

(b) $\{(2, 1, 0), (-1, 1, 1), (3, 3, 1)\}$

(c) $\left\{ \begin{bmatrix} 3 \\ 1 \\ -1 \end{bmatrix}, \begin{bmatrix} -1 \\ -1 \\ 0 \end{bmatrix}, \begin{bmatrix} 4 \\ 0 \\ -2 \end{bmatrix} \right\}$

(d) $\left\{ \begin{bmatrix} 1 \\ 2 \\ 2 \end{bmatrix}, \begin{bmatrix} -1 \\ 0 \\ 1 \end{bmatrix}, \begin{bmatrix} -3 \\ 1 \\ -1 \end{bmatrix} \right\}$

6. Explain, without performing any computation, why the following sets cannot be bases for the indicated vector spaces.

(a) $\{(3, -2), (6, -4)\}$ for $\mathbf{R}^2$

(b) $\{(1, 3), (4, 1), (1, 1)\}$ for $\mathbf{R}^2$

(c) $\{(0, 0), (1, 3)\}$ for $\mathbf{R}^2$

(d) $\{(1, 0, 1), (2, -1, 3), (-4, 2, -6)\}$ for $\mathbf{R}^3$

(e) $\{(1, 1, 1), (2, 1, 3), (5, 0, 0), (-1, -2, 4)\}$ for $\mathbf{R}^3$

(f) $\{(1, 4), (3, 1, 2), (2, 4, 5)\}$ for $\mathbf{R}^3$

(g) $\{(4, 3, 2), (-1, 0, 5), (2, 7, 1)\}$ for $\mathbf{R}^4$

7. Prove that the subspace of $\mathbf{R}^3$ generated by the vectors $(-1, 2, 1), (2, -1, 0), (1, 4, 3)$ is a two-dimensional subspace of $\mathbf{R}^3$. Give a basis for this subspace.

8. Prove that the vector $(1, 2, -1)$ lies in the two-dimensional subspace of $\mathbf{R}^3$ generated by the vectors $(1, 3, 1)$ and $(1, 4, 3)$.

9. Prove that the vector $(2, 1, 4)$ lies in the two-dimensional subspace of $\mathbf{R}^3$ generated by the vectors $(1, 0, 2)$ and $(1, 1, 2)$.

10. Prove that the vector $(-3, 3, -6)$ lies in the one-dimensional subspace of $\mathbf{R}^3$ generated by the vector $(2, -2, 4)$.

11. Does the vector $(1, 2, -1)$ lie in the subspace of $\mathbf{R}^3$ generated by the vectors $(1, -1, 0)$ and $(3, -1, 2)$?

12. Find a basis for $\mathbf{R}^2$ that includes the vector $(1, 2)$.

13. Find a basis for $\mathbf{R}^3$ that includes the vectors $(1, 1, 1)$ and $(1, 0, -2)$.

14. Find a basis for $\mathbf{R}^3$ that includes the vectors $(-1, 0, 2)$ and $(0, 1, 1)$.

15. Determine a basis for each of the following subspaces of $\mathbf{R}^3$. Give the dimension of each subspace.

(a) the set of vectors of the form (a, a, b)

(b) the set of vectors of the form $(a, a, 2a)$

(c) the set of vectors of the form $(a, b, a + b)$

(d) the set of vectors of the form $(a, 2b, a + 3b)$

(e) the set of vectors of the form (a, b, c) where $a + b + c = 0$

16. Determine a basis for each of the following subspaces of $\mathbf{R}^4$. Give the dimension of each subspace.

(a) the set of vectors of the form $(a, b, a + b, a - b)$

(b) the set of vectors of the form $(a, 2a, b, 0)$

(c) the set of vectors of the form $(2a, b, a + 3b, c)$

(d) the set of vectors of the form (a, a, a, a)

Bases for Matrix and Function Spaces

17. Determine a basis for each of the following vector spaces and give the dimension of the space.

(a) P_3 **(b)** M_{33} **(c)** M_{23}

(d) the subspace of M_{22} consisting of all diagonal matrices

(e) the subspace of M_{22} consisting of all symmetric matrices

18. Consider the vector space M_{22} of real 2×2 matrices. Let V_1 be the set of matrices of the form $\begin{bmatrix} a & b \\ -a & c \end{bmatrix}$ and V_2 be the set of matrices of the form $\begin{bmatrix} p & -p \\ q & r \end{bmatrix}$.

(a) Prove that V_1, V_2, and $V_1 \cap V_2$ are subspaces of M_{22}.

(b) Find bases for V_1, V_2, and $V_1 \cap V_2$. Give the dimensions of these spaces.

19. (a) Is the function $f(x) = x + 5$ in the subspace spanned by $g(x) = x + 1$ and $h(x) = x + 3$?

(b) Is the function $f(x) = 3x^2 + 5x + 1$ in the subspace spanned by $g(x) = 2x^2 + 3$ and $h(x) = x^2 + 3x - 1$?

(c) Give three other functions in the space spanned by $g(x) = 2x^2 + 3$ and $h(x) = x^2 + 3x - 1$.

(d) Give a basis for the space spanned by $f(x) = 2x + 3$, $g(x) = x - 1$, and $h(x) = -x - 4$.

20. Are the following sets bases for the given vector spaces?

(a) $\{f, g, h\}$ where $f(x) = x^2 + 2x - 1$, $g(x) = x + 3, h(x) = x^2 + 3x + 2$, for P_2

(b) $\{f, g, h\}$ where $f(x) = x^2 + x - 3$, $g(x) = x^2 - x + 1, h(x) = x^2 + x - 1$, for P_2

(c) $\left\{ \begin{bmatrix} 1 & 2 \\ 0 & 1 \end{bmatrix}, \begin{bmatrix} 3 & 4 \\ 1 & 1 \end{bmatrix}, \begin{bmatrix} 1 & 2 \\ 1 & 1 \end{bmatrix}, \begin{bmatrix} 0 & 2 \\ 1 & 2 \end{bmatrix} \right\}$, for M_{22}

(d) $\left\{ \begin{bmatrix} 1 & 0 \\ 0 & 0 \end{bmatrix}, \begin{bmatrix} 1 & 2 \\ 0 & 0 \end{bmatrix}, \begin{bmatrix} 1 & 2 \\ 3 & 0 \end{bmatrix}, \begin{bmatrix} 1 & 2 \\ 3 & 4 \end{bmatrix} \right\}$, for M_{22}

(e) $\{(2 - 3i, 1 + 4i), (1 + i, 2)\}$, for C^2

(f) $\{(1 + 2i, 3 - i, 1), (4 + i, 3i, 1 + i),$ $(-2 + 3i, 6 - 5i, 1 - i)\}$, for C^3

General Vector Spaces

21. Let $\{\mathbf{v}_1, \mathbf{v}_2\}$ be a basis for a vector space V. Show that the set of vectors $\{\mathbf{u}_1, \mathbf{u}_2\}$ where $\mathbf{u}_1 = \mathbf{v}_1 + \mathbf{v}_2, \mathbf{u}_2 = \mathbf{v}_1 - \mathbf{v}_2$, is also a basis for V.

22. Let $\{\mathbf{v}_1, \mathbf{v}_2, \mathbf{v}_3\}$ be a basis for a vector space V. Show that the set of vectors $\{\mathbf{u}_1, \mathbf{u}_2, \mathbf{u}_3\}$ where $\mathbf{u}_1 = \mathbf{v}_1, \mathbf{u}_2 = \mathbf{v}_1 + \mathbf{v}_2$, $\mathbf{u}_3 = \mathbf{v}_1 + \mathbf{v}_2 + \mathbf{v}_3$ is also a basis for V.

23. Let $\{\mathbf{v}_1, \mathbf{v}_2, \dots, \mathbf{v}_n\}$ be a basis for a vector space V. Let c be a nonzero scalar. Show that the set $\{c\mathbf{v}_1, c\mathbf{v}_2, \dots, c\mathbf{v}_n\}$ is also a basis for V.

24. Let V be a vector space of dimension n. Prove that no set of $n - 1$ vectors can span V.

25. Let V be a vector space, and let W be a subspace of V. If $\dim(V) = n$ and $\dim(W) = m$, prove that $m \leq n$.

26. Let $\mathbf{u} = (u_1, u_2)$ be a nonzero vector in $\mathbf{R}^2$. Prove that the set of vectors orthogonal to $\mathbf{u}$ forms a one-dimensional subspace of $\mathbf{R}^2$.

27. Let $\mathbf{u} = (u_1, u_2, u_3)$ be a nonzero vector in $\mathbf{R}^3$. Prove that the set of vectors orthogonal to $\mathbf{u}$ forms a two-dimensional subspace of $\mathbf{R}^3$.

28. Let $\mathbf{u} = (u_1, u_2, u_3)$ and $\mathbf{v} = (v_1, v_2, v_3)$ be two nonzero, linearly independent, vectors in $\mathbf{R}^3$. Prove that the set of vectors orthogonal to both $\mathbf{u}$ and $\mathbf{v}$ forms a one-dimensional subspace of $\mathbf{R}^3$.

29. Let U and V be two subspaces of a vector space W. U is said to be orthogonal to V if and only if every vector in U is orthogonal to every vector in V. Give an example of two orthogonal subspaces of $\mathbf{R}^3$.

30. Let V be a vector space of dimension n. Let $S = \{\mathbf{v}_1, \dots, \mathbf{v}_m\}$ span V. Prove that S contains a basis for V.

31. State (with a brief explanation) whether the following statements are true or false.

(a) The set $\{(-1, 2), (3, -6)\}$ is a basis for $\mathbf{R}^2$.

(b) The vectors $(1, 2, 3), (-1, 4, 6)$ span $\mathbf{R}^3$.

(c) The subspace of $\mathbf{R}^3$ generated by the vectors $(1, 2, 3), (0, 1, 4), (1, 3, 7)$ is of dimension two.

(d) Any set of two linearly independent vectors that spans $\mathbf{R}^2$ forms a basis for $\mathbf{R}^2$.

(e) If there are three linearly independent vectors in a vector space the dimension must be greater than or equal to three.

32. State (with a brief explanation) whether the following statements are true or false.

(a) The set $\{(2, 0, 0), (3, 4, 0), (200, 567, 0)\}$ is linearly independent.

(b) A single vector can be added to any two vectors in $\mathbf{R}^3$ to get a basis for $\mathbf{R}^3$.

(c) The maximum number of linearly dependent vectors in $\mathbf{R}^2$ is two.

(d) There exists a set that spans $\mathbf{R}^2$ but which is not linearly independent.

(e) There exists a set that is linearly independent but does not span $\mathbf{R}^3$.

4.8 Rank

In this section the reader is introduced to the concept of the rank of a matrix. Rank enables one to relate matrices to vectors, and vice versa. Rank is a unifying tool that enables us to bring together many of the concepts discussed in the course. Solutions to certain systems of linear equations, singularity of a matrix, and invertibility of a matrix all come together under the umbrella of rank.

DEFINITION   Let A be an $m \times n$ matrix. The rows of A may be viewed as row vectors $\mathbf{r}_1, \ldots, \mathbf{r}_m$, and the columns as column vectors $\mathbf{c}_1, \ldots, \mathbf{c}_n$. Each row vector will have n components, and each column vector will have m components. The row vectors will span a subspace of $\mathbf{R}^n$ called the **row space** of A, and the column vectors will span a subspace of $\mathbf{R}^m$ called the **column space** of A.

EXAMPLE 1   Consider the matrix

$$\begin{bmatrix} 1 & 2 & -1 & 2 \\ 3 & 4 & 1 & 6 \\ 5 & 4 & 1 & 0 \end{bmatrix}$$

The row vectors of A are

$$\mathbf{r}_1 = (1, 2, -1, 2), \mathbf{r}_2 = (3, 4, 1, 6), \mathbf{r}_3 = (5, 4, 1, 0)$$

These vectors span a subspace of $\mathbf{R}^4$ called the row space of A.

   The column vectors of A are

$$\mathbf{c}_1 = \begin{bmatrix} 1 \\ 3 \\ 5 \end{bmatrix}, \mathbf{c}_2 = \begin{bmatrix} 2 \\ 4 \\ 4 \end{bmatrix}, \mathbf{c}_3 = \begin{bmatrix} -1 \\ 1 \\ 1 \end{bmatrix}, \mathbf{c}_4 = \begin{bmatrix} 2 \\ 6 \\ 0 \end{bmatrix}$$

These vectors span a subspace of $\mathbf{R}^3$ called the column space of A.

■

THEOREM 4.17

The row space and the column space of a matrix A have the same dimension.

Proof   Let $\mathbf{r}_1, \ldots, \mathbf{r}_m$ be the row vectors of A. The ith vector is

$$\mathbf{r}_i = (a_{i1}, a_{i2}, \ldots, a_{in})$$

Let the dimension of the row space be s. Let the vectors $\mathbf{v}_1, \ldots, \mathbf{v}_s$ form a basis for the row space. Let the jth vector of this set be

$$\mathbf{v}_j = (b_{j1}, b_{j2}, \ldots, b_{jn})$$

Each of the row vectors of A is a linear combination of $\mathbf{v}_1, \ldots, \mathbf{v}_s$. Let

$$\mathbf{r}_1 = c_{11}\mathbf{v}_1 + c_{12}\mathbf{v}_2 + \cdots + c_{1s}\mathbf{v}_s$$
$$\vdots$$
$$\mathbf{r}_m = c_{m1}\mathbf{v}_1 + c_{m2}\mathbf{v}_2 + \cdots + c_{ms}\mathbf{v}_s$$

Equating the ith components of the vectors on the left and right we get

$$a_{1i} = c_{11}b_{1i} + c_{12}b_{2i} + \cdots + c_{1s}b_{si}$$
$$\vdots$$
$$a_{mi} = c_{m1}b_{1i} + c_{m2}b_{2i} + \cdots + c_{ms}b_{si}$$

This may be written

$$\begin{bmatrix} a_{1i} \\ \vdots \\ a_{mi} \end{bmatrix} = b_{1i}\begin{bmatrix} c_{11} \\ \vdots \\ c_{m1} \end{bmatrix} + b_{2i}\begin{bmatrix} c_{12} \\ \vdots \\ c_{m2} \end{bmatrix} + \cdots + b_{si}\begin{bmatrix} c_{1s} \\ \vdots \\ c_{ms} \end{bmatrix}$$

This implies that each column vector of A lies in a space spanned by a single set of s vectors. Since s is the dimension of the row space of A we get

$$\dim(\text{column space of } A) \leq \dim(\text{row space of } A)$$

By similar reasoning we can show that

$$\dim(\text{row space of } A) \leq \dim(\text{column space of } A)$$

Combining these two results we see that

$$\dim(\text{row space of } A) = \dim(\text{column space of } A),$$

proving the theorem.

DEFINITION   The dimension of the row space and the column space of a matrix A is called the **rank** of A. The rank of A is denoted rank(A).

As we proceed we shall find that the rank of a matrix will be a useful computational and geometrical tool.

EXAMPLE 2 Determine the rank of the matrix

$$A = \begin{bmatrix} 1 & 2 & 3 \\ 0 & 1 & 2 \\ 2 & 5 & 8 \end{bmatrix}$$

SOLUTION

We see by inspection that the third row of A is a linear combination of the first two rows:

$$(2, 5, 8) = 2(1, 2, 3) + (0, 1, 2)$$

Hence the three rows of A are linearly dependent. The rank of A must be less than 3. Since $(1, 2, 3)$ is not a scalar multiple of $(0, 1, 2)$, these two vectors are linearly independent. These vectors form a basis for the row space of A. Thus rank$(A) = 2$.

The above method, based on the definition, is not practical for determining the ranks of larger matrices. We shall give a more systematic method for finding the rank of a matrix. The following theorem, which paves the way for the method, tells us that the rank of a matrix that is in reduced echelon form is immediately known.

THEOREM 4.18

The nonzero row vectors of a matrix A that is in reduced echelon form are a basis for the row space of A. The rank of A is the number of nonzero row vectors.

Proof   Let A be an $m \times n$ matrix with nonzero row vectors of A be $\mathbf{r}_1, \ldots, \mathbf{r}_t$. Consider the identity

$$k_1\mathbf{r}_1 + k_2\mathbf{r}_2 + \cdots + k_t\mathbf{r}_t = \mathbf{0}$$

where $k_1, \ldots, k_t$ are scalars.

The first nonzero element of $\mathbf{r}_1$ is 1. $\mathbf{r}_1$ is the only one of the row vectors to have a nonzero number in this component. Thus, on adding the vectors $k_1\mathbf{r}_1, k_2\mathbf{r}_2, \ldots, k_t\mathbf{r}_t$ we get a vector whose first component is k_1. On equating this vector to zero, we get $k_1 = 0$. The identity then reduces to

$$k_2\mathbf{r}_2 + \cdots + k_t\mathbf{r}_t = \mathbf{0}$$

The first nonzero element of $\mathbf{r}_2$ is 1, and it is the only one of these remaining row vectors with a nonzero number in this component. Thus $k_2 = 0$. Similarly, $k_3, \ldots, k_t$ are all zero. The vectors $\mathbf{r}_1, \ldots, \mathbf{r}_t$ are therefore linearly independent. These vectors span the row space of A. They thus form a basis for the row space of A. The dimension of the row space is t. The rank of A is t, the number of nonzero row vectors in A.

■

EXAMPLE 3 Find the rank of the matrix

$$A = \begin{bmatrix} 1 & 2 & 0 & 0 \\ 0 & 0 & 1 & 0 \\ 0 & 0 & 0 & 1 \\ 0 & 0 & 0 & 0 \end{bmatrix}$$

This matrix is in reduced echelon form. There are three nonzero row vectors, namely $(1, 2, 0, 0)$, $(0, 0, 1, 0)$, and $(0, 0, 0, 1)$. According to the previous theorem these three vectors form a basis for the row space of A. Rank$(A) = 3$.

■

The following theorem relates the row spaces of row equivalent matrices.

THEOREM 4.19

Let A and B be row equivalent matrices. Then A and B have the same row space: rank$(A) = $ rank(B).

Proof Since A and B are row equivalent, the rows of B can be obtained from the rows of A through a sequence of elementary row operations. Therefore each row of B is a linear combination of the rows of A. Thus the row space of B is contained in the row space of A.

In the same way, the rows of A can be obtained from the rows of B through a sequence of elementary row operations, implying that the row space of A is contained in the row space of B.

It follows that the row spaces of A and B are equal. Since their row spaces are equal, their ranks must be equal.

■

The next result brings the last two results together to give a method for finding a basis for the row space of a matrix, and the rank of the matrix.

THEOREM 4.20

Let E be the reduced echelon form of a matrix A. The nonzero row vectors of E form a basis for the row space of A. The rank of A is the number of nonzero row vectors in E.

■

EXAMPLE 4 Find a basis for the row space of the following matrix A and determine its rank.

$$A = \begin{bmatrix} 1 & 2 & 3 \\ 2 & 5 & 4 \\ 1 & 1 & 5 \end{bmatrix}$$

SOLUTION

Use elementary row operations to find the reduced echelon form of the matrix A. We get

$$\begin{bmatrix} 1 & 2 & 3 \\ 2 & 5 & 4 \\ 1 & 1 & 5 \end{bmatrix} \approx \begin{bmatrix} 1 & 2 & 3 \\ 0 & 1 & -2 \\ 0 & -1 & 2 \end{bmatrix} \approx \begin{bmatrix} 1 & 0 & 7 \\ 0 & 1 & -2 \\ 0 & 0 & 0 \end{bmatrix}$$

The two vectors $(1, 0, 7)$, $(0, 1, -2)$ form a basis for the row space of A. Rank$(A) = 2$.

■

The column vectors of a matrix A become the row vectors of the matrix A^t. Thus the column space of A is the row space of A^t. We can find a basis for the column space of a matrix A by using the preceding method to find a basis for the row space of A^t. The following example illustrates the technique.

EXAMPLE 5 Find a basis for the column space of the following matrix A.

$$A = \begin{bmatrix} 1 & 1 & 0 \\ 2 & 3 & -2 \\ -1 & -4 & 6 \end{bmatrix}$$

SOLUTION

The transpose of A is

$$A^t = \begin{bmatrix} 1 & 2 & -1 \\ 1 & 3 & -4 \\ 0 & -2 & 6 \end{bmatrix}$$

The column space of A becomes the row space of A^t. Let us find a basis for the row space of A^t. Compute the reduced echelon form of A^t.

$$\begin{bmatrix} 1 & 2 & 1 \\ 1 & 3 & -4 \\ 0 & -2 & 6 \end{bmatrix} \approx \begin{bmatrix} 1 & 2 & -1 \\ 0 & 1 & -3 \\ 0 & -2 & 6 \end{bmatrix} \approx \begin{bmatrix} 1 & 0 & 5 \\ 0 & 1 & -3 \\ 0 & 0 & 0 \end{bmatrix}$$

The nonzero row vectors of this echelon form, namely $(1, 0, 5)$, $(0, 1, -3)$, are a basis for the row space of A^t. Write these vectors in column form to get a basis for the column space of A. The following vectors are a basis for the column space of A.

$$\begin{bmatrix} 1 \\ 0 \\ 5 \end{bmatrix}, \begin{bmatrix} 0 \\ 1 \\ -3 \end{bmatrix}$$

■

Theorem 4.20 can also be used to determine a basis for the subspace V spanned by a given set of vectors. The vectors are written as the row vectors of a matrix and the reduced echelon form of that matrix computed. The nonzero row vectors of this reduced echelon form give a basis for V. The following example illustrates the method.

EXAMPLE 6 Find a basis for the subspace V of $\mathbf{R}^4$ spanned by the vectors

$$(1, 2, 3, 4), (-1, -1, -4, -2), (3, 4, 11, 8)$$

SOLUTION

We construct a matrix A having these vectors as row vectors.

$$A = \begin{bmatrix} 1 & 2 & 3 & 4 \\ -1 & -1 & -4 & -2 \\ 3 & 4 & 11 & 8 \end{bmatrix}$$

Determine the reduced echelon form of A. We get

$$\begin{bmatrix} 1 & 2 & 3 & 4 \\ -1 & -1 & -4 & -2 \\ 3 & 4 & 11 & 8 \end{bmatrix} \approx \begin{bmatrix} 1 & 2 & 3 & 4 \\ 0 & 1 & -1 & 2 \\ 0 & -2 & 2 & -4 \end{bmatrix} \approx \begin{bmatrix} 1 & 0 & 5 & 0 \\ 0 & 1 & -1 & 2 \\ 0 & 0 & 0 & 0 \end{bmatrix}$$

The nonzero vectors of this reduced echelon form, namely $(1, 0, 5, 0)$ and $(0, 1, -1, 2)$, are a basis for the subspace V. The dimension of this subspace is two.

Systems of Linear Equations

As we proceed in this course we are finding ever more powerful ways of understanding and viewing systems of linear equations. We now use the vector space concepts of linear dependence, basis, and rank to give a geometric understanding of such systems. We have seen how systems of linear equations can have a unique solution, many solutions, or no solution at all. We find that these three situations correspond to types of dependence of vectors and that the three possibilities can be categorized in terms of the ranks of the augmented matrix and the matrix of coefficients.

THEOREM 4.21

Consider a system $AX = B$ of m linear equations in n variables.

(a) If the augmented matrix and the matrix of coefficients have the same rank r and $r = n$, the solution is unique.

(b) If the augmented matrix and the matrix of coefficients have the same rank r and $r < n$, there are many solutions.

(c) If the augmented matrix and the matrix of coefficients do not have the same rank, a solution does not exist.

Proof   The system of equations is

$$a_{11}x_1 + \cdots + a_{1n}x_n = b_1$$
$$\vdots$$
$$a_{m1}x_1 + \cdots + a_{mn}x_n = b_m$$

The system can be written

$$x_1 \begin{bmatrix} a_{11} \\ \vdots \\ a_{m1} \end{bmatrix} + \cdots + x_n \begin{bmatrix} a_{1n} \\ \vdots \\ a_{mn} \end{bmatrix} = \begin{bmatrix} b_1 \\ \vdots \\ b_m \end{bmatrix}$$

That is

$$x_1 \mathbf{a}_1 + \cdots + x_n \mathbf{a}_n = \mathbf{b} \tag{1}$$

Thus the existence and uniqueness of solutions depends upon whether $\mathbf{b}$ can be written as a linear combination of $\mathbf{a}_1, \ldots, \mathbf{a}_n$, and whether this combination is unique or not. Let us now look at three possibilities that can arise.

(a) Since the ranks of the matrix of coefficients and augmented matrix are the same, $\mathbf{b}$ must be linearly dependent on $\mathbf{a}_1, \ldots, \mathbf{a}_n$. Furthermore, since the rank is n, the vectors $\mathbf{a}_1, \ldots, \mathbf{a}_n$ are linearly independent and thus form a basis for the column space of the augmented matrix. Therefore Equation (1) has a unique solution; the solution to the system is unique.

(b) Since the ranks of the matrix of coefficients and augmented matrix are the same, $\mathbf{b}$ must be linearly dependent on $\mathbf{a}_1, \ldots, \mathbf{a}_n$. However since rank $< n$, the vectors $\mathbf{a}_1, \ldots, \mathbf{a}_n$ are linearly dependent. $\mathbf{b}$ can therefore be expressed in more than one way as a linear combination of $\mathbf{a}_1, \ldots, \mathbf{a}_n$. Thus Equation (1) has many solutions; the solution to the system exists but is not unique.

(c) Since the rank of the augmented matrix is not equal to the rank of the matrix of coefficients, $\mathbf{b}$ is linearly independent of $\mathbf{a}_1, \ldots, \mathbf{a}_n$. Thus Equation (1) has no solution; a solution to the system does not exist.

We give the geometric interpretation of the above discussion in Figure 4.20.

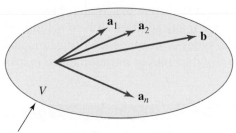

(a) If $\mathbf{a}_1, \ldots, \mathbf{a}_n$ are a basis for V, and $\mathbf{b}$ lies in V, the solution is unique.

(b) If $\mathbf{a}_1, \ldots, \mathbf{a}_n$ are linearly dependent and $\mathbf{b}$ lies in V, there are many solutions.

(c) If $\mathbf{b}$ does not lie in V, there is no solution.

Space spanned by the column vectors of A, $\mathbf{a}_1, \ldots, \mathbf{a}_n$.

Figure 4.20   Geometric interpretation of solutions to $AX = B$

EXAMPLE **7**   Consider the following system of linear equations from Section 1.1.

$$\begin{aligned} x_1 + x_2 + x_3 &= 2 \\ 2x_1 + 3x_2 + x_3 &= 3 \\ x_1 - x_2 - 2x_3 &= -6 \end{aligned}$$

The augmented matrix of this system of equations, and its reduced echelon form are as follows.

$$
\begin{array}{cc}
\text{Augmented} & \text{Reduced echelon} \\
\text{matrix} & \text{form}
\end{array}
$$

$$
\underbrace{\begin{bmatrix} 1 & 1 & 1 & 2 \\ 2 & 3 & 1 & 3 \\ 1 & -1 & -2 & -6 \end{bmatrix}}_{\substack{\text{Matrix of} \\ \text{coefficients}}} \approx \cdots \approx \underbrace{\begin{bmatrix} 1 & 0 & 0 & -1 \\ 0 & 1 & 0 & 1 \\ 0 & 0 & 1 & 2 \end{bmatrix}}_{\substack{\text{Reduced echelon} \\ \text{form}}}
$$

The augmented matrix contains the matrix of coefficients as a submatrix; the reduced echelon form of the augmented matrix contains that of the matrix of coefficients as a submatrix. We see that ranks of the augmented matrix and the matrix of coefficients are equal, both being three. The system thus has a unique solution. The reduced echelon form gives that solution to be $x_1 = -1$, $x_2 = 1$, $x_3 = 2$.

We complete this example by giving the vector space view of things. The system of linear equations can be viewed as the linear combination

$$
x_1 \begin{bmatrix} 1 \\ 2 \\ 1 \end{bmatrix} + x_2 \begin{bmatrix} 1 \\ 3 \\ -1 \end{bmatrix} + x_3 \begin{bmatrix} 1 \\ 1 \\ -2 \end{bmatrix} = \begin{bmatrix} 2 \\ 3 \\ -6 \end{bmatrix}
$$

The existence of solutions depends upon whether $\begin{bmatrix} 2 \\ 3 \\ -6 \end{bmatrix}$ is a linear combination of

$\begin{bmatrix} 1 \\ 2 \\ 1 \end{bmatrix}$, $\begin{bmatrix} 1 \\ 3 \\ -1 \end{bmatrix}$, and $\begin{bmatrix} 1 \\ 1 \\ -2 \end{bmatrix}$. The uniqueness depends upon whether the linear combination is unique. In this particular example we found that there is a unique combination, with $x_1 = -1$, $x_2 = 1$, $x_3 = 2$.

Summary of Results

Our next theorem brings together in a convenient manner a number of results and concepts that have appeared in this course.

THEOREM 4.22

Let A be an $n \times n$ matrix. The following statements are equivalent.

(a) $|A| \neq 0$ (A is nonsingular).
(b) A is invertible.
(c) A is row equivalent to I_n.
(d) The system of equations $AX = B$ has a unique solution.
(e) $\operatorname{rank}(A) = n$.
(f) The column vectors of A form a basis for $\mathbf{R}^n$.

We give the following example to reinforce these results.

> **EXAMPLE 8** Consider the matrix $A = \begin{bmatrix} 1 & -1 & -2 \\ 2 & -3 & -5 \\ -1 & 3 & 5 \end{bmatrix}$. Compute the reduced echelon form of A. Discuss the implications of the answer.
>
> **SOLUTION**
>
> It can be shown that $\begin{bmatrix} 1 & -1 & -2 \\ 2 & -3 & -5 \\ -1 & 3 & 5 \end{bmatrix} \approx \cdots \approx \begin{bmatrix} 1 & 0 & 0 \\ 0 & 1 & 0 \\ 0 & 0 & 1 \end{bmatrix}$. Thus A, a 3×3 matrix, is row equivalent to I_3. The above theorem implies the following results.
>
> **(i)** A is nonsingular. (It can be shown that $|A| = -1$.)
>
> **(ii)** A is invertible. (It can be shown that $A^{-1} = \begin{bmatrix} 0 & 1 & 1 \\ 5 & -3 & -1 \\ -3 & 2 & 1 \end{bmatrix}$.)
>
> **(iii)** The system $\begin{bmatrix} 1 & -1 & -2 \\ 2 & -3 & -5 \\ -1 & 3 & 5 \end{bmatrix} \begin{bmatrix} x_1 \\ x_2 \\ x_3 \end{bmatrix} = \begin{bmatrix} b_1 \\ b_2 \\ b_3 \end{bmatrix}$ has a unique solution for all values of b_1, b_2, and b_3. For example, when $b_1 = 1$, $b_2 = 3$, $b_3 = -2$, it can be shown that $x_1 = 1$, $x_2 = -2$, $x_3 = 1$.
>
> **(iv)** Rank$(A) = 3$. Thus the dimensions of both the row and column spaces of A are 3.
>
> **(v)** The vectors $\begin{bmatrix} 1 \\ 2 \\ -1 \end{bmatrix}$, $\begin{bmatrix} -1 \\ 3 \\ 3 \end{bmatrix}$, $\begin{bmatrix} -2 \\ -5 \\ 5 \end{bmatrix}$ form a basis for $\mathbf{R}^3$.

EXERCISE SET 4.8

Rank

1. Determine the ranks of the following matrices using the definition of rank.

(a) $\begin{bmatrix} 1 & 2 \\ 3 & 4 \end{bmatrix}$ **(b)** $\begin{bmatrix} 1 & -3 \\ -2 & 6 \end{bmatrix}$

(c) $\begin{bmatrix} 4 & 0 \\ 1 & 5 \end{bmatrix}$ **(d)** $\begin{bmatrix} 2 & 4 \\ 3 & 6 \end{bmatrix}$

2. Determine the ranks of the following matrices using the definition of rank.

(a) $\begin{bmatrix} 1 & 2 & -1 \\ 2 & 4 & 2 \\ 1 & 2 & 3 \end{bmatrix}$ **(b)** $\begin{bmatrix} 1 & 0 & 0 \\ 0 & 1 & 0 \\ 0 & 0 & 1 \end{bmatrix}$

(c) $\begin{bmatrix} 2 & 1 & 3 \\ 4 & 2 & 6 \\ 2 & 1 & 3 \end{bmatrix}$ **(d)** $\begin{bmatrix} 1 & 4 & 2 \\ 0 & 1 & 5 \\ 0 & 0 & 1 \end{bmatrix}$

(e) $\begin{bmatrix} 1 & 3 & 4 \\ -1 & 3 & 1 \\ 0 & 6 & 5 \end{bmatrix}$ **(f)** $\begin{bmatrix} 1 & 2 & 3 \\ 4 & 5 & 6 \\ 7 & 8 & 9 \end{bmatrix}$

Basis for Row Space

3. Find the reduced echelon form for each of the following matrices. Use the echelon form to determine a basis for the row space, and the rank of each matrix.

(a) $\begin{bmatrix} 1 & 2 & -1 \\ 2 & 5 & 2 \\ 0 & 2 & 9 \end{bmatrix}$ **(b)** $\begin{bmatrix} 1 & 1 & 8 \\ 0 & 1 & 3 \\ -1 & 1 & -2 \end{bmatrix}$

(c) $\begin{bmatrix} 1 & -3 & 2 \\ -2 & 6 & -4 \\ -1 & 3 & -2 \end{bmatrix}$

4. Find the reduced echelon form for each of the following matrices. Use the echelon form to determine a basis for the row space, and the rank of each matrix.

(a) $\begin{bmatrix} 1 & 4 & 0 \\ -1 & -3 & 3 \\ 2 & 9 & 5 \end{bmatrix}$ (b) $\begin{bmatrix} 1 & 2 & 0 \\ 0 & 1 & 1 \\ -1 & 2 & 3 \end{bmatrix}$

(c) $\begin{bmatrix} 1 & 2 & 3 \\ 0 & -1 & -1 \\ 3 & 4 & 7 \end{bmatrix}$

5. Find the reduced echelon form for each of the following matrices. Use the reduced echelon form to determine a basis for the row space, and the rank of each matrix.

(a) $\begin{bmatrix} 1 & 2 & 3 & 4 \\ -1 & 2 & 0 & 1 \\ 0 & 1 & 0 & 2 \end{bmatrix}$

(b) $\begin{bmatrix} 1 & 2 & -1 & 4 \\ 0 & 1 & -2 & 3 \\ -1 & 0 & -3 & 2 \end{bmatrix}$

(c) $\begin{bmatrix} 1 & 1 & 0 & -1 \\ 2 & 1 & 0 & 0 \\ 3 & 2 & 0 & -1 \\ -1 & 0 & 1 & 1 \end{bmatrix}$

Basis for a Subspace

6. Find bases for the subspaces of $\mathbf{R}^3$ spanned by the following vectors.

(a) $(1, 3, 2), (0, 1, 4), (1, 4, 9)$

(b) $(1, -2, 5), (1, -1, 4), (2, -5, 14)$

(c) $(1, -1, 3), (1, 0, 1), (-2, 1, -4)$

(d) $(2, -6, 4), (-1, 3, -2), (3, -9, 6)$

7. Find bases for the subspaces of $\mathbf{R}^4$ spanned by the following vectors.

(a) $(1, 3, -1, 4), (1, 3, 0, 6), (-1, -3, 0, -8)$

(b) $(1, 2, 0, 1), (-1, -1, 3, 1), (2, 3, -2, 4)$

(c) $(1, 2, 3, 4), (0, -1, 2, 3), (2, 3, 8, 11), (2, 3, 6, 8)$

(d) $(1, -3, -1, 2), (0, 1, -4, 1), (1, -4, 5, 1),$
$(2, -5, -6, 5)$

8. Find bases for both the row and column spaces of the following matrix A. Show that the dimensions of both row space and column space are the same.

$$A = \begin{bmatrix} 1 & 2 & -1 \\ 0 & 1 & 3 \\ 1 & 4 & 6 \end{bmatrix}$$

9. Find bases for both the row and column spaces of the following matrix A. Show that the dimensions of both the row space and the column space are the same.

$$A = \begin{bmatrix} 1 & 3 & 2 \\ 1 & 4 & 1 \\ 2 & 5 & 5 \end{bmatrix}$$

Systems of Linear Equations

10. Consider the following systems of linear equations. The reduced echelon forms (REFs) of the augmented matrices are shown for convenience. (i) Give the reduced echelon form of the matrix of coefficients. (ii) Give the ranks of the matrix of coefficients and augmented matrix. (iii) Give the solution to the system (if it exists). (iv) Express the column vector of constants as a linear combination of the columns of the matrix of coefficients (if possible).

(a) $\begin{aligned} x_1 - 2x_2 + 4x_3 &= 12 \\ 2x_1 - x_2 + 5x_3 &= 18, \\ -x_1 + 3x_2 - 3x_3 &= -8 \end{aligned}$ REF $= \begin{bmatrix} 1 & 0 & 0 & 2 \\ 0 & 1 & 0 & 1 \\ 0 & 0 & 1 & 3 \end{bmatrix}$

(b) $\begin{aligned} 3x_1 - 3x_2 + 3x_3 &= 9 \\ 2x_1 - x_2 + 4x_3 &= 7, \\ 3x_1 - 5x_2 - x_3 &= 7 \end{aligned}$ REF $= \begin{bmatrix} 1 & 0 & 3 & 4 \\ 0 & 1 & 2 & 1 \\ 0 & 0 & 0 & 0 \end{bmatrix}$

(c) $\begin{aligned} x_1 + 4x_2 + x_3 &= 2 \\ x_1 + 2x_2 - x_3 &= 0, \\ 2x_1 + 6x_2 &= 3 \end{aligned}$ REF $= \begin{bmatrix} 1 & 0 & -3 & -2 \\ 0 & 1 & 1 & 1 \\ 0 & 0 & 0 & 1 \end{bmatrix}$

(d) $\begin{aligned} x_1 + x_2 + 3x_3 &= 6 \\ x_1 + 2x_2 + 4x_3 &= 9, \\ 2x_1 + x_2 + 6x_3 &= 11 \end{aligned}$ REF $= \begin{bmatrix} 1 & 0 & 0 & -1 \\ 0 & 1 & 0 & 1 \\ 0 & 0 & 1 & 2 \end{bmatrix}$

11. Consider the systems of linear equations defined by augmented matrices P of the following sizes. The ranks of the augmented matrix and matrix of coefficient Q are given in each case. Will the systems have a single, many, or no solutions?

(a) P is 4×5. Rank$(P) = 4$, rank$(Q) = 4$.

(b) P is 3×4. Rank$(P) = 3$, rank$(Q) = 2$.

(c) P is 4×4. Rank$(P) = 2$, rank$(Q) = 2$.

(d) P is 5×5. Rank$(P) = 4$, rank$(Q) = 3$.

(e) P is 6×6. Rank$(P) = 4$, rank$(Q) = 4$.

(f) P is 7×4. Rank$(P) = 3$, rank$(Q) = 3$.

Miscellaneous Results

12. (a) If A is a 3×5 matrix, what is the largest possible rank of A?

(b) If A is an $m \times n$ matrix, what is the largest possible rank of A?

13. Let A be an $m \times n$ matrix where $m < n$. Prove that

$$\dim(\text{column space of } A) < n.$$

14. (a) Let A be a 3×4 matrix. Prove that the column vectors of A are linearly dependent.

(b) Let A be a 7×3 matrix. Prove that the row vectors of A are linearly dependent.

(c) Let A be an $m \times n$ matrix with $m < n$. Prove that the column vectors of A are linearly dependent.

15. (a) Let A be an $n \times n$ matrix with $|A| \neq 0$. Prove that the row vectors of A form a basis for $\mathbf{R}^n$.

(b) Let A be an $n \times n$ invertible matrix. Prove that the row vectors of A form a basis for $\mathbf{R}^n$.

16. Let A be an $n \times n$ matrix. Prove that the columns of A are linearly independent if and only if $\operatorname{rank}(A) = n$.

17. Let A be an $n \times n$ matrix. Prove that the columns of A span $\mathbf{R}^n$ if and only if the rows of A are linearly independent.

18. A and B are matrices of the same size, prove that

$$\operatorname{rank}(A + B) \leq \operatorname{rank}(A) + \operatorname{rank}(B)$$

19. Let A be an $n \times n$ matrix. The following statements can be divided into two sets, any pair of statements in each set being equivalent. Find the sets.

(a) A is singular.

(b) The system of equations $AX = B$ does not have a unique solution.

(c) The rows of A are linearly independent.

(d) A has an inverse.

(e) The columns of A span $\mathbf{R}^n$.

(f) A is not row equivalent to I_n.

20. State (with a brief explanation) whether the following statements are true or false.

(a) If the rank of A is n, then A cannot have more than n rows.

(b) If matrices A and B are of the same size and k is a scalar, then $\operatorname{rank}(A) + \operatorname{rank}(B) = \operatorname{rank}(A + B)$ and $\operatorname{rank}(kA) = k \operatorname{rank}(A)$.

(c) Consider a system of n equations in n variables. If the matrix of coefficients has rank n, the solution is unique.

(d) If the vectors $\mathbf{a}_1, \mathbf{a}_2, \ldots, \mathbf{a}_n$ are linearly dependent then the system of linear equations $[\mathbf{a}_1 \, \mathbf{a}_2 \ldots \mathbf{a}_n]\mathbf{x} = \mathbf{b}$ has many solutions.

(e) If the system of linear equations $[\mathbf{a}_1 \, \mathbf{a}_2 \ldots \mathbf{a}_n]\mathbf{x} = \mathbf{b}$ has no solution, then $\operatorname{rank}([\mathbf{a}_1 \, \mathbf{a}_2 \ldots \mathbf{a}_n]) < \operatorname{rank}([\mathbf{a}_1 \, \mathbf{a}_2 \ldots \mathbf{a}_n \, \mathbf{b}])$.

4.9 Orthonormal Vectors and Projections

We now focus our attention on sets of unit, orthogonal vectors. We will introduce a method for constructing such sets in $\mathbf{R}^n$. Special properties of these sets will be discussed.

DEFINITION   A set of vectors in a vector space V is said to be an **orthogonal set** if every pair of vectors in the set is orthogonal. The set is said to be an **orthonormal set** if it is orthogonal and each vector is a unit vector.

EXAMPLE 1 Show that the set $\left\{ (1, 0, 0), \left(0, \frac{3}{5}, \frac{4}{5}\right), \left(0, \frac{4}{5}, -\frac{3}{5}\right) \right\}$ is an orthonormal set.

SOLUTION

We first show that each pair of vectors in this set is orthogonal.

$$(1, 0, 0) \cdot \left(0, \tfrac{3}{5}, \tfrac{4}{5}\right) = 0; \qquad (1, 0, 0) \cdot \left(0, \tfrac{4}{5}, -\tfrac{3}{5}\right) = 0; \qquad \left(0, \tfrac{3}{5}, \tfrac{4}{5}\right) \cdot \left(0, \tfrac{4}{5}, -\tfrac{3}{5}\right) = 0$$

Thus the vectors are orthogonal. It remains to show that each vector is a unit vector. We get

$$\|(1, 0, 0)\| = \sqrt{1^2 + 0^2 + 0^2} = 1$$

$$\left\|\left(0, \tfrac{3}{5}, \tfrac{4}{5}\right)\right\| = \sqrt{0^2 + \left(\tfrac{3}{5}\right)^2 + \left(\tfrac{4}{5}\right)^2} = 1$$

$$\left\|\left(0, \tfrac{4}{5}, -\tfrac{3}{5}\right)\right\| = \sqrt{0^2 + \left(\tfrac{4}{5}\right)^2 + \left(-\tfrac{3}{5}\right)^2} = 1$$

The vectors are unit, orthogonal vectors. The set is thus an orthonormal set.

The next theorem tells us that any orthogonal set of nonzero vectors is linearly independent. This result of course also implies that an orthonormal set of nonzero vectors is linearly independent.

THEOREM 4.23

An orthogonal set of nonzero vectors in a vector space is linearly independent.

Proof   Let $\{\mathbf{v}_1, \ldots, \mathbf{v}_m\}$ be an orthogonal set of nonzero vectors in a vector space V. Let us examine the identity

$$c_1\mathbf{v}_1 + c_2\mathbf{v}_2 + \cdots + c_m\mathbf{v}_m = \mathbf{0}$$

We shall show that $c_1 = 0, \ldots, c_m = 0$, proving that the vectors are linearly independent. Let $\mathbf{v}_i$ be the ith vector of the orthogonal set. Take the dot product of each side of this equation with $\mathbf{v}_i$ and use the properties of the dot product. We get

$$(c_1\mathbf{v}_1 + c_2\mathbf{v}_2 + \cdots + c_m\mathbf{v}_m) \cdot \mathbf{v}_i = \mathbf{0} \cdot \mathbf{v}_i$$
$$c_1\mathbf{v}_1 \cdot \mathbf{v}_i + c_2\mathbf{v}_2 \cdot \mathbf{v}_i + \cdots + c_m\mathbf{v}_m \cdot \mathbf{v}_i = 0$$

Since the vectors $\mathbf{v}_1, \ldots, \mathbf{v}_m$ are mutually orthogonal, $\mathbf{v}_j \cdot \mathbf{v}_i = 0$ unless $j = i$. Thus

$$c_i\mathbf{v}_i \cdot \mathbf{v}_i = 0$$

Since $\mathbf{v}_i$ is a nonzero vector then $\mathbf{v}_i \cdot \mathbf{v}_i \neq 0$. Thus $c_i = 0$. Letting $i = 1, \ldots, m$, we get $c_1 = 0, \ldots, c_m = 0$, proving that the vectors are linearly independent.

There are many bases for a vector space. The basis that one uses depends upon the problem under consideration—one selects the basis that best fits the situation. Very often the most suitable basis is either an orthogonal set or an orthonormal set.

> **DEFINITION**   A basis that is an orthogonal set is said to be an **orthogonal basis**. A basis that is an orthonormal set is said to be an **orthonormal basis**.

Consider the standard basis $\{(1, 0, 0), (0, 1, 0), (0, 0, 1)\}$ of $\mathbf{R}^3$. Observe that $\|(1, 0, 0)\| = \sqrt{1^2 + 0^2 + 0^2}$. The vector $(1, 0, 0,)$ is a unit vector. Similarly, $(0, 1, 0)$ and $(0, 0, 1)$ are unit vectors. Further, note that $(1, 0, 0) \cdot (0, 1, 0) = 0$. The vectors $(1, 0, 0)$ and $(0, 1, 0)$ are orthogonal. Similarly the pairs $(1, 0, 0)$, $(0, 0, 1)$, and $(0, 1, 0)$, $(0, 0, 1)$ are orthogonal. Thus $\{(1, 0, 0), (0, 1, 0), (0, 0, 1)\}$ is an orthonormal basis for $\mathbf{R}^3$. This result applies to all the vector spaces $\mathbf{R}^n$.

> **DEFINITION**   The standard basis $\{(1, 0, \ldots, 0), (0, 1, \ldots, 0), \ldots (0, 0, \ldots, 1)\}$ of $\mathbf{R}^n$ is an orthonormal basis.

We know that any vector in a vector space can be expressed as a unique linear combination of basis vectors. The following theorem shows how that linear combination can be easily found for an orthonormal basis.

THEOREM 4.24

Let $\{\mathbf{u}_1, \ldots, \mathbf{u}_n\}$ be an orthonormal basis for a vector space V. Let $\mathbf{v}$ be a vector in V. $\mathbf{v}$ can be written as a linear combination of these basis vectors as follows:

$$\mathbf{v} = (\mathbf{v} \cdot \mathbf{u}_1)\mathbf{u}_1 + (\mathbf{v} \cdot \mathbf{u}_2)\mathbf{u}_2 + \cdots + (\mathbf{v} \cdot \mathbf{u}_n)\mathbf{u}_n$$

Proof Since $\{\mathbf{u}_1, \ldots, \mathbf{u}_n\}$ is a basis there exist scalars $c_1, c_2, \ldots, c_n$ such that

$$\mathbf{v} = c_1\mathbf{u}_1 + c_2\mathbf{u}_2 + \cdots + c_n\mathbf{u}_n$$

We shall show that $c_1 = \mathbf{v} \cdot \mathbf{u}_1, \ldots, c_n = \mathbf{v} \cdot \mathbf{u}_n$.

Let $\mathbf{u}_i$ be the ith base vector. Take the dot product of each side of this equation with $\mathbf{u}_i$ and use the properties of the dot product. We get

$$\begin{aligned}\mathbf{v} \cdot \mathbf{u}_i &= (c_1\mathbf{u}_1 + c_2\mathbf{u}_2 + \cdots + c_n\mathbf{u}_n) \cdot \mathbf{u}_i \\ &- c_1\mathbf{u}_1 \cdot \mathbf{u}_i + c_2\mathbf{u}_2 \cdot \mathbf{u}_i + \cdots + c_n\mathbf{u}_n \cdot \mathbf{u}_i\end{aligned}$$

Since the vectors $\mathbf{u}_1, \ldots, \mathbf{u}_n$ are mutually orthogonal, $\mathbf{u}_j \cdot \mathbf{u}_i = 0$ unless $j = i$. Thus

$$\mathbf{v} \cdot \mathbf{u}_i = c_i\mathbf{u}_i \cdot \mathbf{u}_i$$

Furthermore, since the basis is orthonormal, $\mathbf{u}_i \cdot \mathbf{u}_i = 1$. Therefore

$$\mathbf{v} \cdot \mathbf{u}_i = c_i$$

Thus, letting $i = 1, \ldots, n$, we get $c_1 = \mathbf{v} \cdot \mathbf{u}_1, \ldots, c_n = \mathbf{v} \cdot \mathbf{u}_n$. We can write

$$\mathbf{v} = (\mathbf{v} \cdot \mathbf{u}_1)\mathbf{u}_1 + (\mathbf{v} \cdot \mathbf{u}_2)\mathbf{u}_2 + \cdots + (\mathbf{v} \cdot \mathbf{u}_n)\mathbf{u}_n$$

EXAMPLE 2 The following vectors, $\mathbf{u}_1$, $\mathbf{u}_2$, and $\mathbf{u}_3$, form an orthonormal basis for $\mathbf{R}^3$. Express the vector $\mathbf{v} = (7, -5, 10)$ as a linear combination of these vectors.

$$\mathbf{u}_1 = (1, 0, 0), \mathbf{u}_2 = \left(0, \tfrac{3}{5}, \tfrac{4}{5}\right), \mathbf{u}_3 = \left(0, \tfrac{4}{5}, -\tfrac{3}{5}\right)$$

SOLUTION

We get

$$\begin{aligned}\mathbf{v} \cdot \mathbf{u}_1 &= (7, -5, 10) \cdot (1, 0, 0) = 7 \\ \mathbf{v} \cdot \mathbf{u}_2 &= (7, -5, 10) \cdot \left(0, \tfrac{3}{5}, \tfrac{4}{5}\right) = 5 \\ \mathbf{v} \cdot \mathbf{u}_3 &= (7, -5, 10) \cdot \left(0, \tfrac{4}{5}, -\tfrac{3}{5}\right) = -10\end{aligned}$$

Thus

$$(7, -5, 10) = 7(1, 0, 0) + 5\left(0, \tfrac{3}{5}, \tfrac{4}{5}\right) - 10\left(0, \tfrac{4}{5}, -\tfrac{3}{5}\right)$$

We now introduce a class of matrices that are very useful in computation and in geometry. They are used in constructing suitable coordinate systems for physical situations (Section 5.4), and also have special geometrical properties when used to define "functions" on a vector space. These functions leave the magnitudes of vectors and angles between vectors unchanged. They "preserve the geometry" of the space (Section 6.1).

Orthogonal Matrices

An *orthogonal matrix* is an invertible matrix that has the property

$$A^{-1} = A^t$$

We now see that an orthogonal matrix is a matrix whose columns and rows form orthonormal sets.

THEOREM 4.25

Orthogonal Matrix Theorem
Let A be an $n \times n$ matrix. The following statements are equivalent.
(a) A is orthogonal.
(b) The column vectors of A form an orthonormal set in $\mathbf{R}^n$.
(c) The row vectors of A form an orthonormal set in $\mathbf{R}^n$.

Proof Assume that A is orthogonal. Thus $A^{-1} = A^t$. We have $A^t A = I$ and $AA^t = I$. Let $P = A^t A$. Then p_{ij} is the dot product of row vector i of A^t and column vector j of A. Since row i of A^t is column i of A this means that p_{ij} is the dot product of column vectors i and j of A. Since P is the identity matrix, $p_{ij} = 1$ if $i = j$ and $p_{ij} = 0$ if $i \neq j$. The column vectors of A thus form an orthonormal set.

Similarly, if we start with $P = AA^t$ we can show that the row vectors of A form an orthonormal set.

By using these arguments in the reverse direction it can be shown that the converses are true. Orthonormality of rows or columns imply that A is orthogonal.

The following theorem gives further properties of orthogonal matrices.

THEOREM 4.26

If A is an orthogonal matrix, then
(a) $|A| = \pm 1$.
(b) A^{-1} is an orthogonal matrix.

Proof

(a) Since $A^{-1} = A^t$, we have $AA^t = I$. By the properties of determinants,

$$|AA^t| = 1, \quad |A||A^t| = 1, \quad |A||A| = 1, \quad |A|^2 = 1$$

Thus $|A| = \pm 1$.
(b) Since the row vectors of A form an orthonormal set, the columns of A^t form an orthonormal set. Since $A^{-1} = A^t$, this means that the columns of A^{-1} form an orthonormal set. The previous theorem now implies that A^{-1} is an orthogonal matrix.

We now introduce the concept of projection of one vector onto another. Projection intuitively tells us "how much" of one vector is pointing in the direction of another. It is particularly useful in analyzing forces, where one wants to know the physical effect of a force in a certain direction. We shall use projections to construct sets of orthonormal vectors.

Projection of One Vector onto Another Vector

Let $\mathbf{v}$ and $\mathbf{u}$ be vectors in $\mathbf{R}^n$ with angle α between them. See Figure 4.21(a). The vector $\overrightarrow{OA}$ tells us "how much" of $\mathbf{v}$ is pointing in the direction of $\mathbf{u}$. We call $\overrightarrow{OA}$ the **projection of v onto u**. Let us find an expression for $\overrightarrow{OA}$. We see that

$$
\begin{aligned}
OA &= OB \cos\alpha \\
&= \|\mathbf{v}\| \cos\alpha \\
&= \|\mathbf{v}\| \left(\frac{\mathbf{v}}{\|\mathbf{v}\|} \cdot \frac{\mathbf{u}}{\|\mathbf{u}\|} \right) \\
&= \mathbf{v} \cdot \frac{\mathbf{u}}{\|\mathbf{u}\|}
\end{aligned}
$$

The direction of the vector $\overrightarrow{OA}$ is defined by the unit vector $\mathbf{u}/\|\mathbf{u}\|$. Thus

$$
\overrightarrow{OA} = \left(\mathbf{v} \cdot \frac{\mathbf{u}}{\|\mathbf{u}\|} \right) \frac{\mathbf{u}}{\|\mathbf{u}\|} = \frac{\mathbf{v} \cdot \mathbf{u}}{\mathbf{u} \cdot \mathbf{u}} \mathbf{u}
$$

This expression for projection also holds if $\alpha > 90°$. See Figure 4.21(b). In that case the projection is in the opposite direction to $\mathbf{u}$, and the sign of $(\mathbf{v} \cdot \mathbf{u})/(\mathbf{u} \cdot \mathbf{u})$ is negative.

This discussion leads to the following definition of projection.

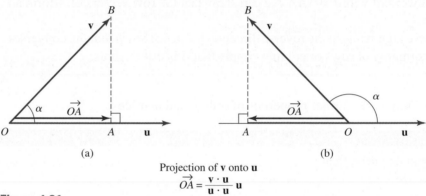

(a)   (b)

Projection of $\mathbf{v}$ onto $\mathbf{u}$

$$
\overrightarrow{OA} = \frac{\mathbf{v} \cdot \mathbf{u}}{\mathbf{u} \cdot \mathbf{u}} \mathbf{u}
$$

Figure 4.21

The **projection** of a vector $\mathbf{v}$ onto a nonzero vector $\mathbf{u}$ in $\mathbf{R}^n$ is denoted

DEFINITION $\mathrm{proj}_{\mathbf{u}} \mathbf{v}$ and is defined by

$$
\mathrm{proj}_{\mathbf{u}} \mathbf{v} = \frac{\mathbf{v} \cdot \mathbf{u}}{\mathbf{u} \cdot \mathbf{u}} \mathbf{u}
$$

See Figure 4.22. The term **component** of $\mathbf{v}$ in the direction of $\mathbf{u}$ is also used for $\mathrm{proj}_{\mathbf{u}} \mathbf{v}$.

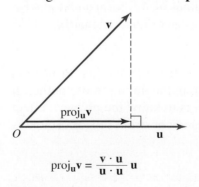

$$
\mathrm{proj}_{\mathbf{u}} \mathbf{v} = \frac{\mathbf{v} \cdot \mathbf{u}}{\mathbf{u} \cdot \mathbf{u}} \mathbf{u}
$$

Figure 4.22

EXAMPLE 3 Determine the projection of the vector $\mathbf{v} = (6, 7)$ onto vector $\mathbf{u} = (1, 4)$.

SOLUTION

For these vectors we get

$$\mathbf{v} \cdot \mathbf{u} = (6, 7) \cdot (1, 4) = 6 + 28 = 34$$
$$\mathbf{u} \cdot \mathbf{u} = (1, 4) \cdot (1, 4) = 1 + 16 = 17$$

Thus

$$\text{proj}_{\mathbf{u}} \, \mathbf{v} = \frac{\mathbf{v} \cdot \mathbf{u}}{\mathbf{u} \cdot \mathbf{u}} \mathbf{u} = \frac{34}{17} (1, 4) = (2, 8)$$

The projection of $\mathbf{v}$ onto $\mathbf{u}$ is $(2, 8)$.

Suppose the vector $\mathbf{v} = (6, 7)$ represents a force acting on a body located at the origin. Then $\text{proj}_{\mathbf{u}} \, \mathbf{v} = (2, 8)$ is the component of the force in the direction of the vector $\mathbf{u} = (1, 4)$. Physically, $(2, 8)$ is the effect of the force in that direction.

We now give a method for constructing an orthonormal basis from a given basis. The method uses vector projections.

THEOREM 4.27

The Gram-Schmidt Orthogonalization Process[*]

Let $\{\mathbf{v}_1, \ldots, \mathbf{v}_n\}$ be a basis for a vector space V. The set of vectors $\{\mathbf{u}_1, \ldots, \mathbf{u}_n\}$ defined as follows is orthogonal. To obtain an orthonormal basis for V, normalize each of the vectors $\mathbf{u}_1, \ldots, \mathbf{u}_n$.

$$\mathbf{u}_1 = \mathbf{v}_1$$
$$\mathbf{u}_2 = \mathbf{v}_2 - \text{proj}_{\mathbf{u}_1} \mathbf{v}_2$$
$$\mathbf{u}_3 = \mathbf{v}_3 - \text{proj}_{\mathbf{u}_1} \mathbf{v}_3 - \text{proj}_{\mathbf{u}_2} \mathbf{v}_3$$
$$\cdots \cdots$$
$$\mathbf{u}_n = \mathbf{v}_n - \text{proj}_{\mathbf{u}_1} \mathbf{v}_n - \text{proj}_{\mathbf{u}_2} \mathbf{v}_n - \cdots - \text{proj}_{\mathbf{u}_{n-1}} \mathbf{v}_n$$

Proof If $\mathbf{v}$ and $\mathbf{u}$ are vectors in $\mathbf{R}^n$, then vector addition tells us that $(\mathbf{v} - \text{proj}_{\mathbf{u}} \, \mathbf{v})$ is orthogonal to $\mathbf{u}$. See Figure 4.23. We apply this result in proving the theorem. Consider the construction of the ith vector, $\mathbf{u}_i$. By subtracting $\text{proj}_{\mathbf{u}_1} \mathbf{v}_i$ from $\mathbf{v}_i$ we get a vector that is orthogonal to $\mathbf{u}_1$. Subtracting $\text{proj}_{\mathbf{u}_2} \mathbf{v}_i$ from the resulting vector gives a vector that is orthogonal to $\mathbf{u}_2$, while still preserving the orthogonality to $\mathbf{u}_1$, and so on. This leads to a vector $\mathbf{u}_i$ that is orthogonal to each of the previous vectors $\mathbf{u}_1, \ldots, \mathbf{u}_{i-1}$. See the following exercises.

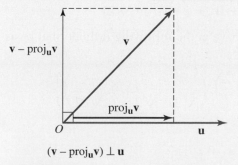

$(\mathbf{v} - \text{proj}_{\mathbf{u}}\mathbf{v}) \perp \mathbf{u}$

Figure 4.23

[*]Jorgen Pederson Gram (1850–1916) was a Danish actuary. Erhard Schmidt (1876–1959) studied at Berlin and Göttingen. He taught at Bonn, Zurich, Erlangen, and Berlin. In 1946 he became the first director of the Research Institute for Mathematics for the German Academy of Sciences. He is considered a founder of the branch of modern mathematics called functional analysis.

EXAMPLE 4 The set $\{(1, 2, 0, 3), (4, 0, 5, 8), (8, 1, 5, 6)\}$ is linearly independent in $\mathbf{R}^4$. The vectors form a basis for a three-dimensional subspace V of $\mathbf{R}^4$. Construct an orthonormal basis for V.

SOLUTION

Let $\mathbf{v}_1 = (1, 2, 0, 3), \mathbf{v}_2 = (4, 0, 5, 8), \mathbf{v}_3 = (8, 1, 5, 6)$. We now use the Gram-Schmidt process to construct an orthogonal set $\{\mathbf{u}_1, \mathbf{u}_2, \mathbf{u}_3\}$ from these vectors.

$$\text{Let } \mathbf{u}_1 = \mathbf{v}_1 = (1, 2, 0, 3)$$

$$\text{Let } \mathbf{u}_2 = \mathbf{v}_2 - \text{proj}_{\mathbf{u}_1} \mathbf{v}_2 = \mathbf{v}_2 - \frac{(\mathbf{v}_2 \cdot \mathbf{u}_1)}{(\mathbf{u}_1 \cdot \mathbf{u}_1)} \mathbf{u}_1$$

$$- (4, 0, 5, 8) - \frac{(4, 0, 5, 8) \cdot (1, 2, 0, 3)}{(1, 2, 0, 3) \cdot (1, 2, 0, 3)} (1, 2, 0, 3)$$

$$= (4, 0, 5, 8) - 2(1, 2, 0, 3)$$

$$= (2, -4, 5, 2)$$

$$\text{Let } \mathbf{u}_3 = \mathbf{v}_3 - \text{proj}_{\mathbf{u}_1} \mathbf{v}_3 - \text{proj}_{\mathbf{u}_2} \mathbf{v}_3 = \mathbf{v}_3 - \frac{\mathbf{v}_3 \cdot \mathbf{u}_1}{\mathbf{u}_1 \cdot \mathbf{u}_1} \mathbf{u}_1 - \frac{\mathbf{v}_3 \cdot \mathbf{u}_2}{\mathbf{u}_2 \cdot \mathbf{u}_2} \mathbf{u}_2$$

$$= (8, 1, 5, 6) - \frac{(8, 1, 5, 6) \cdot (1, 2, 0, 3)}{(1, 2, 0, 3) \cdot (1, 2, 0, 3)} (1, 2, 0, 3)$$

$$- \frac{(8, 1, 5, 6) \cdot (2, -4, 5, 2)}{(2, -4, 5, 2) \cdot (2, -4, 5, 2)} (2, -4, 5, 2)$$

$$= (8, 1, 5, 6) - 2(1, 2, 0, 3) - 1(2, -4, 5, 2)$$

$$= (4, 1, 0, -2)$$

The set $\{(1, 2, 0, 3), (2, -4, 5, 2), (4, 1, 0, -2)\}$ is an orthogonal basis for V. (Check that the dot product of each pair of vectors is indeed zero.)

Let us now compute the norm of each vector and then normalize the vectors to get an orthonormal basis. We get

$$\|(1, 2, 0, 3)\| = \sqrt{1^2 + 2^2 + 0^2 + 3^2} = \sqrt{14}$$

$$\|(2, -4, 5, 2)\| = \sqrt{2^2 + (-4)^2 + 5^2 + 2^2} = 7$$

$$\|(4, 1, 0, -2)\| = \sqrt{4^2 + 1^2 + 0^2 + (-2)^2} = \sqrt{21}$$

Use these values to normalize the vectors to arrive at the following orthonormal basis for V.

$$\left\{ \left(\frac{1}{\sqrt{14}}, \frac{2}{\sqrt{14}}, 0, \frac{3}{\sqrt{14}} \right), \left(\frac{2}{7}, -\frac{4}{7}, \frac{5}{7}, \frac{2}{7} \right), \left(\frac{4}{\sqrt{21}}, \frac{1}{\sqrt{21}}, 0, \frac{-2}{\sqrt{21}} \right) \right\}$$

Projection of a Vector onto a Subspace

We have defined the projection of a vector onto another vector. We now extend this concept to the projection of a vector onto a subspace. The projection of a vector onto a subspace tells us "how much" of the vector lies in the subspace.

Let $\overrightarrow{OA}$ be the projection of a vector $\mathbf{v}$ onto a vector $\mathbf{u}$ in $\mathbf{R}^n$. See Figure 4.24. Let W be the one-dimensional subspace of $\mathbf{R}^n$ consisting of all vectors that lie on the line defined by $\mathbf{u}$. Note that the projection of $\mathbf{v}$ onto any vector $\mathbf{u}$ that lies in this subspace is $\overrightarrow{OA}$. Thus we can define the vector projection of $\mathbf{v}$ onto the subspace W, written $\text{proj}_W \mathbf{v}$, to be $\overrightarrow{OA}$. The simplest expression for $\overrightarrow{OA}$ is obtained by taking $\mathbf{u}$ to be a unit vector. We then get

$$\text{proj}_W \mathbf{v} = \overrightarrow{OA} = (\mathbf{v} \cdot \mathbf{u})\mathbf{u} \quad (\mathbf{u} \text{ a unit vector})$$

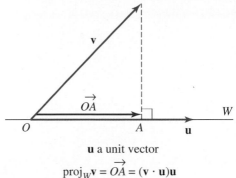

$\mathbf{u}$ a unit vector
$$\text{proj}_W\mathbf{v} = \overrightarrow{OA} = (\mathbf{v} \cdot \mathbf{u})\mathbf{u}$$

Figure 4.24

If W is a subspace of dimension m, then we extend this concept of projection by expressing the projection of $\mathbf{v}$ onto W as a linear combination of the vectors of an orthonormal basis of W as follows.

DEFINITION   Let W be a subspace of $\mathbf{R}^n$. Let $\{\mathbf{u}_1, \ldots, \mathbf{u}_m\}$ be an orthonormal basis for W. If $\mathbf{v}$ is a vector in $\mathbf{R}^n$, the **projection** of $\mathbf{v}$ onto W is denoted $\text{proj}_W \mathbf{v}$ and is defined by

$$\text{proj}_W \mathbf{v} = (\mathbf{v} \cdot \mathbf{u}_1)\mathbf{u}_1 + (\mathbf{v} \cdot \mathbf{u}_2)\mathbf{u}_2 + \cdots + (\mathbf{v} \cdot \mathbf{u}_m)\mathbf{u}_m$$

See Figure 4.25.

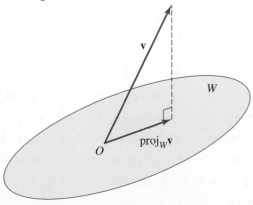

$\{\mathbf{u}_1, \ldots, \mathbf{u}_m\}$ orthonormal basis for W
$$\text{proj}_W\mathbf{v} = (\mathbf{v} \cdot \mathbf{u}_1)\mathbf{u}_1 + \cdots + (\mathbf{v} \cdot \mathbf{u}_m)\mathbf{u}_m$$

Figure 4.25

We say that a vector is **orthogonal to a subspace** W if it is orthogonal to every vector in W. The following theorem tells us that if we have a vector $\mathbf{v}$ in $\mathbf{R}^n$ and if W is a subspace of $\mathbf{R}^n$, then we can **decompose** $\mathbf{v}$ into a vector that lies in the subspace W and a vector that is orthogonal to W.

THEOREM 4.28

Let W be a subspace of $\mathbf{R}^n$. Every vector $\mathbf{v}$ in $\mathbf{R}^n$ can be written uniquely in the form

$$\mathbf{v} = \mathbf{w} + \mathbf{w}_\perp$$

where $\mathbf{w}$ is in W and $\mathbf{w}_\perp$ is orthogonal to W. The vectors $\mathbf{w}$ and $\mathbf{w}_\perp$ are

$$\mathbf{w} = \text{proj}_W \mathbf{v} \quad \text{and} \quad \mathbf{w}_\perp = \mathbf{v} - \text{proj}_W \mathbf{v}$$

See Figure 4.26.

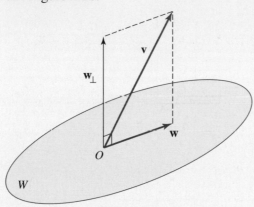

$$\mathbf{v} = \mathbf{w} + \mathbf{w}_\perp \text{ where}$$
$$\mathbf{w} = \text{proj}_W \mathbf{v} \text{ and } \mathbf{w}_\perp = \mathbf{v} - \text{proj}_W \mathbf{v}$$

Figure 4.26

Proof   Observe that

$$\mathbf{w} + \mathbf{w}_\perp = (\text{proj}_W \mathbf{v}) + (\mathbf{v} - \text{proj}_W \mathbf{v})$$
$$= \mathbf{v}$$

Further, $\text{proj}_W \mathbf{v}$ is in W. Let us now show that $(\mathbf{v} - \text{proj}_W \mathbf{v})$ is orthogonal to W. Let $\{\mathbf{u}_1, \ldots, \mathbf{u}_m\}$ be an orthonormal basis for W. We first show that $(\mathbf{v} - \text{proj}_W \mathbf{v})$ is orthogonal to each of these base vectors of W. We will then be able to show that $(\mathbf{v} - \text{proj}_W \mathbf{v})$ is orthogonal to an arbitrary vector in W. We get

$$\mathbf{u}_i \cdot (\mathbf{v} - \text{proj}_W \mathbf{v}) = \mathbf{u}_i \cdot (\mathbf{v} - (\mathbf{v} \cdot \mathbf{u}_1)\mathbf{u}_1 - \cdots - (\mathbf{v} \cdot \mathbf{u}_m)\mathbf{u}_m)$$
$$= \mathbf{u}_i \cdot \mathbf{v} - \mathbf{u}_i \cdot (\mathbf{v} \cdot \mathbf{u}_1)\mathbf{u}_1 - \cdots - \mathbf{u}_i \cdot (\mathbf{v} \cdot \mathbf{u}_m)\mathbf{u}_m$$
$$= \mathbf{u}_i \cdot \mathbf{v} - (\mathbf{v} \cdot \mathbf{u}_1)(\mathbf{u}_i \cdot \mathbf{u}_1) - \cdots - (\mathbf{v} \cdot \mathbf{u}_m)(\mathbf{u}_i \cdot \mathbf{u}_m)$$
$$= \mathbf{u}_i \cdot \mathbf{v} - (\mathbf{v} \cdot \mathbf{u}_i)(\mathbf{u}_i \cdot \mathbf{u}_i), \text{ since } \mathbf{u}_i \cdot \mathbf{u}_j = 0 \text{ unless } i = j$$
$$= \mathbf{u}_i \cdot \mathbf{v} - \mathbf{v} \cdot \mathbf{u}_i, \text{ since } \mathbf{u}_i \cdot \mathbf{u}_i = 1$$
$$= 0$$

Thus $(\mathbf{v} - \text{proj}_W \mathbf{v})$ is orthogonal to each of the base vectors of W.

Let $\mathbf{w}' = c_1\mathbf{u}_1 + \cdots + c_m\mathbf{u}_m$ be an arbitrary vector in W. We get

$$\mathbf{w}' \cdot (\mathbf{v} - \text{proj}_W \mathbf{v}) = (c_1\mathbf{u}_1 + \cdots + c_m\mathbf{u}_m) \cdot (\mathbf{v} - \text{proj}_W \mathbf{v})$$
$$= c_1\mathbf{u}_1 \cdot (\mathbf{v} - \text{proj}_W \mathbf{v}) + \cdots + c_m\mathbf{u}_m \cdot (\mathbf{v} - \text{proj}_W \mathbf{v})$$
$$= 0$$

Thus $(\mathbf{v} - \text{proj}_W \mathbf{v})$ is orthogonal to W.

The proof of uniqueness is omitted.

EXAMPLE 5 Consider the vector $\mathbf{v} = (3, 2, 6)$ in $\mathbf{R}^3$. Let W be the subspace of $\mathbf{R}^3$ consisting of all vectors of the form (a, b, b). Decompose $\mathbf{v}$ into the sum of a vector that lies in W and a vector orthogonal to W.

SOLUTION

We need an orthonormal basis for W. We can write an arbitrary vector of W as follows

$$(a, b, b) = a(1, 0, 0) + b(0, 1, 1)$$

The set $\{(1, 0, 0), (0, 1, 1)\}$ spans W and is linearly independent. It forms a basis for W. The vectors are orthogonal. Normalize each vector to get an orthonormal basis $\{\mathbf{u}_1, \mathbf{u}_2\}$ for W where

$$\mathbf{u}_1 = (1, 0, 0), \quad \mathbf{u}_2 = \left(0, \frac{1}{\sqrt{2}}, \frac{1}{\sqrt{2}}\right)$$

We get

$$\mathbf{w} = \text{proj}_W \mathbf{v} = (\mathbf{v} \cdot \mathbf{u}_1)\mathbf{u}_1 + (\mathbf{v} \cdot \mathbf{u}_2)\mathbf{u}_2$$

$$= ((3, 2, 6) \cdot (1, 0, 0))(1, 0, 0) + \left((3, 2, 6) \cdot \left(0, \frac{1}{\sqrt{2}}, \frac{1}{\sqrt{2}}\right)\right)\left(0, \frac{1}{\sqrt{2}}, \frac{1}{\sqrt{2}}\right)$$

$$= (3, 0, 0) + (0, 4, 4) = (3, 4, 4)$$

and

$$\mathbf{w}_\perp = \mathbf{v} - \text{proj}_W \mathbf{v} = (3, 2, 6) - (3, 4, 4) = (0, -2, 2)$$

Thus the desired decomposition of $\mathbf{v}$ is

$$(3, 2, 6) = (3, 4, 4) + (0, -2, 2)$$

In this decomposition the vector $(3, 4, 4)$ lies in W and the vector $(0, -2, 2)$ is orthogonal to W.

Distance of a Point from a Subspace

We have discussed the distance of a point from another point in $\mathbf{R}^n$. We are now able to talk about the distance of a point from a subspace in $\mathbf{R}^n$. Let $\mathbf{x} = (x_1, \ldots, x_n)$ be a point in $\mathbf{R}^n$, W be a subspace of $\mathbf{R}^n$, and $\mathbf{y} = (y_1, \ldots, y_n)$ be a point in W. It is natural to define the distance of $\mathbf{x}$ from W, denoted $d(\mathbf{x}, W)$ to be the minimum of the distances from $\mathbf{x}$ to the points of W.

$$d(\mathbf{x}, W) = \min\{d(\mathbf{x}, \mathbf{y})\}, \quad \text{for all points } \mathbf{y} \text{ in } W$$

We now find an expression for $d(\mathbf{x}, W)$. We can write

$$\mathbf{x} - \mathbf{y} = (\mathbf{x} - \text{proj}_W \mathbf{x}) + (\text{proj}_W \mathbf{x} - \mathbf{y})$$

This is a decomposition of the vector $\mathbf{x} - \mathbf{y}$ into the sum of a vector $(\mathbf{x} - \text{proj}_W \mathbf{x})$ that is orthogonal to W and a vector $(\text{proj}_W \mathbf{x} - \mathbf{y})$ that lies in W. The Pythagorean Theorem thus gives

$$\|\mathbf{x} - \mathbf{y}\|^2 = \|\mathbf{x} - \text{proj}_W \mathbf{x}\|^2 + \|\text{proj}_W \mathbf{x} - \mathbf{y}\|^2$$

Therefore $\|\mathbf{x} - \mathbf{y}\|$ (which is equal to $d(\mathbf{x}, \mathbf{y})$) has a minimum value of $\|\mathbf{x} - \text{proj}_W \mathbf{x}\|$ when $\text{proj}_W \mathbf{x} - \mathbf{y} = 0$; that is when $\mathbf{y} = \text{proj}_W \mathbf{x}$. Thus

$$d(\mathbf{x}, W) = \|\mathbf{x} - \text{proj}_W \mathbf{x}\|$$

> *The distance of a point from a subspace is the distance of the point from its projection in the subspace. (See Figure 4.27.)*

This discussion also means that $\text{proj}_W \mathbf{x}$ is the closest point to $\mathbf{x}$ in the subspace W.

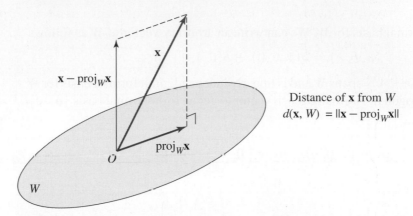

Distance of $\mathbf{x}$ from W
$d(\mathbf{x}, W) = \|\mathbf{x} - \text{proj}_W \mathbf{x}\|$

Figure 4.27

EXAMPLE 6 Find the distance of the point $\mathbf{x} = (4, 1, -7)$ of $\mathbf{R}^3$ from the subspace W consisting of all vectors of the form (a, b, b).

SOLUTION

The previous example tells us that the set $\{\mathbf{u}_1, \mathbf{u}_2\}$, where

$$\mathbf{u}_1 = (1, 0, 0), \quad \mathbf{u}_2 = \left(0, \frac{1}{\sqrt{2}}, \frac{1}{\sqrt{2}}\right),$$

is an orthonormal basis for W. We compute $\text{proj}_W \mathbf{x}$.

$\text{proj}_W \mathbf{x} = (\mathbf{x} \cdot \mathbf{u}_1)\mathbf{u}_1 + (\mathbf{x} \cdot \mathbf{u}_2)\mathbf{u}_2$

$$= ((4, 1, -7) \cdot (1, 0, 0))(1, 0, 0) + \left((4, 1, -7) \cdot \left(0, \frac{1}{\sqrt{2}}, \frac{1}{\sqrt{2}}\right)\right)\left(0, \frac{1}{\sqrt{2}}, \frac{1}{\sqrt{2}}\right)$$

$$= (4, 0, 0) + (0, -3, -3) = (4, -3, -3)$$

Thus

$$\|\mathbf{x} - \text{proj}_W \mathbf{x}\| = \|(4, 1, -7) - (4, -3, -3)\| = \|(0, 4, -4)\| = \sqrt{32}$$

The distance from $\mathbf{x}$ to W is $\sqrt{32}$.

EXERCISE SET 4.9

Orthogonal and Orthonormal Sets

1. Which of the following are orthogonal sets of vectors?

  (a) $\{(1, 2), (2, -1)\}$ **(b)** $\{(3, -1), (0, 5)\}$

  (c) $\{(0, -2), (3, 0)\}$ **(d)** $\{(4, 1), (2, -3)\}$

  (e) $\{(-3, 2), (2, 3)\}$

2. Which of the following are orthogonal sets of vectors?

  (a) $\{(1, 2, 1), (4, -2, 0), (2, 4, -10)\}$

  (b) $\{(3, -1, 1), (2, 0, 1), (1, 1, -2)\}$

  (c) $\{(1, 4, 2), (-2, -1, 3), (6, -1, -1)\}$

  (d) $\{(1, 2, -1, 1), (3, 1, 4, -1), (0, 1, -1, -3)\}$

3. Which of the following are orthonormal sets of vectors?

(a) $\left\{\left(\frac{1}{3}, \frac{2}{3}, \frac{2}{3}\right), \left(\frac{2}{3}, -\frac{2}{3}, \frac{1}{3}\right), \left(\frac{2}{3}, \frac{1}{3}, -\frac{2}{3}\right)\right\}$

(b) $\left\{\left(\frac{1}{\sqrt{10}}, \frac{3}{\sqrt{10}}\right), \left(\frac{-3}{\sqrt{10}}, \frac{1}{\sqrt{10}}\right)\right\}$

(c) $\left\{\left(\frac{1}{\sqrt{2}}, 0, \frac{1}{\sqrt{2}}\right), \left(\frac{1}{\sqrt{2}}, 0, \frac{-1}{\sqrt{2}}\right), (0, 1, 0)\right\}$

(d) $\left\{\left(\frac{1}{\sqrt{6}}, \frac{-1}{\sqrt{6}}, \frac{2}{\sqrt{6}}\right), \left(0, \frac{2}{\sqrt{5}}, \frac{1}{\sqrt{5}}\right),\right.$
$\left.\left(\frac{5}{\sqrt{30}}, \frac{1}{\sqrt{30}}, \frac{-2}{\sqrt{30}}\right)\right\}$

(e) $\left\{\left(\frac{4}{\sqrt{20}}, \frac{2}{\sqrt{20}}, 0\right), \left(\frac{-1}{\sqrt{6}}, \frac{2}{\sqrt{6}}, \frac{1}{\sqrt{6}}\right),\right.$
$\left.\left(\frac{1}{\sqrt{32}}, \frac{-2}{\sqrt{32}}, \frac{5}{\sqrt{32}}\right)\right\}$

Orthonormal Bases

4. The following vectors $\mathbf{u}_1$, $\mathbf{u}_2$, and $\mathbf{u}_3$ form an orthonormal basis for $\mathbf{R}^3$. Use the result of Theorem 4.24 to express the vector $\mathbf{v} = (2, -3, 1)$ as a linear combination of these basis vectors.

$$\mathbf{u}_1 = (0, -1, 0), \quad \mathbf{u}_2 = \left(\tfrac{3}{5}, 0, -\tfrac{4}{5}\right), \quad \mathbf{u}_3 = \left(\tfrac{4}{5}, 0, \tfrac{3}{5}\right)$$

5. The following vectors $\mathbf{u}_1$, $\mathbf{u}_2$, and $\mathbf{u}_3$ form an orthonormal basis for $\mathbf{R}^3$. Use the result of Theorem 4.24 to express the vector $\mathbf{v} = (7, 5, -1)$ as a linear combination of these basis vectors.

$$\mathbf{u}_1 = (1, 0, 0), \mathbf{u}_2 = \left(0, \frac{1}{\sqrt{2}}, \frac{1}{\sqrt{2}}\right), \mathbf{u}_3 = \left(0, \frac{1}{\sqrt{2}}, \frac{-1}{\sqrt{2}}\right)$$

Orthogonal Matrices

6. Prove that each of the following matrices is an orthogonal matrix by showing that the columns form orthonormal sets.

(a) $\begin{bmatrix} 1 & 0 \\ 0 & 1 \end{bmatrix}$ (b) $\begin{bmatrix} 0 & -1 \\ 1 & 0 \end{bmatrix}$

(c) $\begin{bmatrix} \frac{\sqrt{3}}{2} & \frac{1}{2} \\ -\frac{1}{2} & \frac{\sqrt{3}}{2} \end{bmatrix}$ (d) $\begin{bmatrix} 1 & 0 & 0 \\ 0 & 0 & -1 \\ 0 & 1 & 0 \end{bmatrix}$

(e) $\begin{bmatrix} \frac{2}{3} & \frac{2}{3} & \frac{1}{3} \\ -\frac{2}{3} & \frac{1}{3} & \frac{2}{3} \\ \frac{1}{3} & \frac{2}{3} & \frac{2}{3} \end{bmatrix}$

7. The following matrices are orthogonal matrices. Show that they have all the properties indicated in Theorems 4.25 and 4.26.

(a) $\begin{bmatrix} \frac{1}{\sqrt{2}} & \frac{1}{\sqrt{2}} \\ -\frac{1}{\sqrt{2}} & \frac{1}{\sqrt{2}} \end{bmatrix}$ (b) $\begin{bmatrix} \frac{\sqrt{3}}{2} & \frac{1}{2} \\ \frac{1}{2} & -\frac{\sqrt{3}}{2} \end{bmatrix}$

8. Determine the inverse of each of the following matrices.

(a) $\begin{bmatrix} 0 & 1 \\ -1 & 0 \end{bmatrix}$ (b) $\begin{bmatrix} \frac{2}{\sqrt{5}} & \frac{1}{\sqrt{5}} \\ -\frac{1}{\sqrt{5}} & \frac{1}{\sqrt{5}} \end{bmatrix}$

(c) $\begin{bmatrix} \frac{1}{3} & \frac{2}{3} & -\frac{2}{3} \\ \frac{2}{3} & -\frac{2}{3} & \frac{1}{3} \\ -\frac{2}{3} & -\frac{1}{3} & \frac{2}{3} \end{bmatrix}$

9. Let A and B be orthogonal matrices of the same size. Prove that the product matrix AB is also an orthogonal matrix.

10. Let $\mathbf{u}$ be a unit vector. Interpret $\mathbf{u}$ as a column matrix. Show that $A = I - 2\mathbf{u}\mathbf{u}^t$ is an orthogonal matrix.

11. The complex analog of an orthogonal matrix is called a **unitary matrix**. A square matrix A is unitary if $A^{-1} = \overline{A}^t$. If A is unitary prove that A^{-1} is also unitary.

12. Prove that a square matrix A is unitary if and only if its columns form a set of unit, mutually orthogonal vectors in $\mathbf{C}^n$.

Projection and Gram-Schmidt

13. Determine the projection of the vector $\mathbf{v}$ onto the vector $\mathbf{u}$ for the following vectors.

(a) $\mathbf{v} = (7, 4), \mathbf{u} = (1, 2)$

(b) $\mathbf{v} = (-1, 5), \mathbf{u} = (3, -2)$

(c) $\mathbf{v} = (4, 6, 4), \mathbf{u} = (1, 2, 3)$

(d) $\mathbf{v} = (6, -8, 7), \mathbf{u} = (-1, 3, 0)$

(e) $\mathbf{v} = (1, 2, 3, 0), \mathbf{u} = (1, -1, 2, 3)$

14. Determine the projection of the vector $\mathbf{v}$ onto the vector $\mathbf{u}$ for the following vectors.

(a) $\mathbf{v} = (1, 2), \mathbf{u} = (2, 5)$

(b) $\mathbf{v} = (-1, 3), \mathbf{u} = (2, 4)$

(c) $\mathbf{v} = (1, 2, 3), \mathbf{u} = (1, 2, 0)$

(d) $\mathbf{v} = (2, 1, 4), \mathbf{u} = (-1, -3, 2)$

(e) $\mathbf{v} = (2, -1, 3, 1), \mathbf{u} = (-1, 2, 1, 3)$

15. The following vectors form a basis for $\mathbf{R}^2$. Use these vectors in the Gram-Schmidt process to construct an orthonormal basis for $\mathbf{R}^2$.

    (a) $(1, 2), (-1, 3)$ (b) $(1, 1), (6, 2)$
    (c) $(1, -1), (4, -2))$

16. The following vectors form a basis for $\mathbf{R}^3$. Use these vectors in the Gram-Schmidt process to construct an orthonormal basis for $\mathbf{R}^3$.

    (a) $(1, 1, 1), (2, 0, 1), (2, 4, 5)$
    (b) $(3, 2, 0), (1, 5, -1), (5, -1, 2)$

17. Construct an orthonormal basis for the subspace of $\mathbf{R}^3$ spanned by the following vectors.

    (a) $(1, 0, 2), (-1, 0, 1)$ (b) $(1, -1, 1), (1, 2, -1)$

18. Construct an orthonormal basis for the subspace of $\mathbf{R}^4$ spanned by the vectors.

    (a) $(1, 2, 3, 4), (-1, 1, 0, 1)$
    (b) $(3, 0, 0, 0), (0, 1, 2, 1), (0, -1, 3, 2)$

19. Construct a vector in $\mathbf{R}^4$ that is orthogonal to the vector $(1, 2, -1, 1)$.

20. Construct a vector in $\mathbf{R}^4$ that is orthogonal to the vector $(2, 0, 1, 1)$.

Projection onto a Subspace

21. Let W be the subspace of $\mathbf{R}^3$ having basis $\{(1, 1, 2), (0, -1, 3)\}$. Determine the projections of the following vectors onto W.

    (a) $(3, -1, 2)$ (b) $(1, 1, 1)$ (c) $(4, 2, 1)$

22. Let V be the subspace of $\mathbf{R}^4$ having basis $\{(-1, 0, 2, 1), (1, -1, 0, 3)\}$. Determine the projections of the following vectors onto V.

    (a) $(1, -1, 1, -1)$ (b) $(2, 0, 1, -1)$ (c) $(3, 2, 1, 0)$

23. Consider the vector $\mathbf{v} = (1, 2, -1)$ in $\mathbf{R}^3$. Let V be the subspace of $\mathbf{R}^3$ consisting of all vectors of the form (a, a, b). Decompose $\mathbf{v}$ into the sum of a vector that lies in V and a vector orthogonal to V.

24. Consider the vector $\mathbf{v} = (4, 1, -2)$ in $\mathbf{R}^3$. Let V be the subspace of $\mathbf{R}^3$ consisting of all vectors of the form $(a, 2a, b)$. Decompose $\mathbf{v}$ into the sum of a vector that lies in V and a vector orthogonal to V.

25. Consider the vector $\mathbf{v} = (3, 2, 1)$ in $\mathbf{R}^3$. Let W be the subspace of $\mathbf{R}^3$ consisting of all vectors of the form $(a, b, a + b)$. Decompose $\mathbf{v}$ into the sum of a vector that lies in W and a vector orthogonal to W.

Distance from a Subspace

26. Find the distance of the point $\mathbf{x} = (1, 3, -2)$ of $\mathbf{R}^3$ from the subspace W consisting of all vectors of the form (a, a, b).

27. Find the distance of the point $\mathbf{x} = (2, 4, -1)$ of $\mathbf{R}^3$ from the subspace W consisting of all vectors of the form $(a, -2a, b)$.

28. Find the distance of the point $\mathbf{x} = (1, 3, -2)$ of $\mathbf{R}^3$ from the subspace W consisting of all vectors of the form $(a, 2a, 3a)$. Note that this is finding the distance of a point from a line.

Miscellaneous

29. Find an orthogonal basis for $\mathbf{R}^3$ that includes the vectors $(1, 2, -2)$ and $(6, 1, 4)$.

30. Let $\{\mathbf{u}_1, \ldots, \mathbf{u}_n\}$ be an orthonormal basis for a vector space V. Let $\mathbf{v}$ be a vector in V. Show that $\|\mathbf{v}\|$ can be written

$$\|\mathbf{v}\| = \sqrt{(\mathbf{v} \cdot \mathbf{u}_1)^2 + (\mathbf{v} \cdot \mathbf{u}_2)^2 + \cdots + (\mathbf{v} \cdot \mathbf{u}_n)^2}$$

31. Let $\{\mathbf{u}_1 \ldots \mathbf{u}_n\}$ be an orthonormal set in $\mathbf{R}^n$. Prove that this set is a basis for $\mathbf{R}^n$.

32. Let $\mathbf{u}$ and $\mathbf{v}$ be vectors in $\mathbf{R}^n$. We discussed in the proof of the Gram-Schmidt theorem that $(\mathbf{v} - \text{proj}_{\mathbf{u}} \mathbf{v})$ is orthogonal to $\mathbf{u}$. Formally prove this result by showing that $(\mathbf{v} - \text{proj}_{\mathbf{u}} \mathbf{v}) \cdot \mathbf{u} = 0$, using the definition of $\text{proj}_{\mathbf{u}} \mathbf{v}$ and the properties of the dot product.

CHAPTER 4 REVIEW EXERCISES

1. Sketch the following position vectors in $\mathbf{R}^2$.
   (a) $\overrightarrow{OA} = (1, 4)$,   (b) $\overrightarrow{OB} = (-2, 3)$,
   (c) $\overrightarrow{OC} = (4, -1)$

2. Sketch the position vectors $(2, 0, 0)$, $(0, -1, 0)$, $(1, 4, 2)$ in $\mathbf{R}^3$.

3. Compute the following linear combinations for $\mathbf{u} = (3, -1, 5)$, $\mathbf{v} = (2, 3, 7)$, and $\mathbf{w} = (0, 1, -3)$.
   (a) $\mathbf{u} + \mathbf{w}$ (b) $3\mathbf{u} + \mathbf{v}$
   (c) $\mathbf{u} - 2\mathbf{w}$ (d) $4\mathbf{u} - 2\mathbf{v} + 3\mathbf{w}$
   (e) $2\mathbf{u} - 5\mathbf{v} - \mathbf{w}$

4. Determine the dot products of the following pairs of vectors.
   (a) $(1, 2), (3, -4)$ (b) $(1, -2, 3), (4, 2, -7)$
   (c) $(2, 2, -5), (3, 2, -1)$

5. Find the norms of the following vectors.
   (a) $(1, -4)$ (b) $(-2, 1, 3)$
   (c) $(1, -2, 3, 4)$

6. Determine the cosines of the angles between the following vectors.
   (a) $(-1, 1), (2, 3)$ (b) $(1, 2, -3), (4, 1, 2)$

7. Find a vector orthogonal to $(-2, 1, 5)$.

8. Determine the distances between the following pairs of points.
   (a) $(1, -2), (5, 3)$ (b) $(3, 2, 1), (7, 1, 2)$
   (c) $(3, 1, -1, 2), (4, 1, 6, 2)$

9. Find all the values of c such that $\|c(1, 2, 3)\| = 196$.

10. Prove that the set W of all vectors in $\mathbf{R}^3$ of the form $a(1, 3, 7)$, where a is a real number, is a vector space.

11. Consider the set U of 2×2 matrices where all the elements are nonnegative. Is U a vector space?

12. Is the set W of 2×2 matrices whose elements add up to zero a vector space? (For example $\begin{bmatrix} 2 & 1 \\ -7 & 4 \end{bmatrix}$ would be in W.)

13. Let $f(x) = 3x - 1$ and $g(x) = 2x^2 + 3$. Compute the functions $f + g$, $3f$, and $2f - 3g$.

14. (a) Let V be the set of functions having the set of real numbers as domain whose graphs pass through the point $(2, 0)$. Is V a vector space?
    (b) Let W be the set of functions having the set of real numbers as domain whose graphs pass through the point $(2, 1)$. Is W a vector space?

15. Consider the sets of vectors of the following form. Determine which of the sets are subspaces of $\mathbf{R}^3$.
    (a) $(a, b, a - 2)$ (b) $(a, -2a, 3a)$
    (c) $(a, b, 2a - 3b)$ (d) $(a, 2, b)$

16. Which of the following subsets of $\mathbf{R}^3$ are subspaces? The set of all vectors of the form $p(1, 2, 3)$ where p is
    (a) a real number   (b) an integer
    (c) a nonnegative real number (d) a rational number

17. Determine which of the following subsets of M_{22} form subspaces.
    (a) The subset having diagonal elements nonzero.
    (b) The subset of matrices whose $(1, 2)$ element is 0.
       (For example $\begin{bmatrix} 2 & 0 \\ 3 & 4 \end{bmatrix}$ would be such a matrix.)
    (c) The subset of matrices of the form $\begin{bmatrix} a & b \\ -b & c \end{bmatrix}$.
    (d) The subset of 2×2 matrices whose determinants are zero.

18. Let S be the set of all functions of the form $f(x) = 3x^2 + ax - b$, where a and b are real numbers. Is S a subspace of P_2?

19. Is the condition that a subset of a vector space contain the zero vector a necessary and sufficient condition for the subset to be a subspace?

20. In the following sets of vectors, determine whether the first vector is a linear combination of the other vectors.
(a) $(3, 15, -4)$; $(1, 2, -1)$, $(2, 4, 0)$, $(5, 1, 2)$
(b) $(-3, -4, 7)$; $(5, 0, 3)$, $(2, 0, 1)$, $(4, 0, 5)$

21. Determine whether the following matrix D is the linear combination $D = 2A + 3B - C$ of the matrices A, B, and C.

$$D = \begin{bmatrix} 1 & -3 \\ 10 & 18 \end{bmatrix}; A = \begin{bmatrix} 1 & 2 \\ 4 & 3 \end{bmatrix},$$

$$B = \begin{bmatrix} 0 & -2 \\ 1 & 5 \end{bmatrix}, C = \begin{bmatrix} 1 & 1 \\ 1 & 1 \end{bmatrix}$$

22. Explain why the following vectors cannot possibly span $\mathbf{R}^3$.

$$(1, -2, 3), (-2, 4, -6), (0, 6, 4)$$

23. Show that $(10, 9, 8)$ lies in the vector space Span $\{(-1, 3, 1), (4, 1, 2)\}$.

24. Prove that $f(x) = 13x^2 + 8x - 21$ is a linear combination of $g(x) = 2x^2 + x - 3$ and $h(x) = -3x^2 - 2x + 5$.

25. Determine whether the following sets are linearly dependent or independent.
(a) $\{(1, -2, 0), (0, 1, 3), (2, 0, 12)\}$
(b) $\{(-1, 18, 7), (-1, 4, 1), (1, 3, 2)\}$
(c) $\{(5, -1, 3), (2, 1, 0), (3, -2, 2)\}$

26. Prove that the following sets are bases for $\mathbf{R}^2$ or $\mathbf{R}^3$.
(a) $\{(-1, 2), (3, 1)\}$
(b) $\{(4, 2), (1, -1)\}$
(c) $\{(1, 2, 3), (-1, 0, 5), (0, 2, 7)\}$
(d) $\{(8, 1, 0), (4, -1, 3), (5, 2, -3)\}$

27. Show that the functions $f(x) = 2$, $g(x) = 3x$, and $h(x) = -x^2$ span P_2.

28. Find a basis for $\mathbf{R}^3$ that includes the vectors $(1, -2, 3)$ and $(4, 1, -1)$.

29. Determine a basis for the subspace of $\mathbf{R}^4$ that consists of vectors of the form $(a, b, c, a - 2b + 3c)$.

30. Find a basis for the vector space of upper triangular 3×3 matrices.

31. Show that $\{x^2 + 2x - 3, 3x^2 + x - 1, 4x^2 + 3x - 3\}$ is a basis for P_2.

32. Determine the ranks of the following matrices

(a) $\begin{bmatrix} 1 & 2 & -1 \\ -1 & 3 & 4 \\ 0 & 5 & 3 \end{bmatrix}$ **(b)** $\begin{bmatrix} 2 & 1 & 4 \\ -2 & 0 & -1 \\ 3 & 2 & 7 \end{bmatrix}$

(c) $\begin{bmatrix} -2 & 4 & 8 \\ 1 & -2 & 4 \\ 4 & -8 & 16 \end{bmatrix}$

33. Find a basis for the subspace of $\mathbf{R}^4$ spanned by the following set.

$$\{(1, -2, 3, 4), (-1, 3, 1, -2), (2, -3, 10, 10)\}$$

34. Let $\mathbf{v}$ be a linear combination of $\mathbf{v}_1$ and $\mathbf{v}_2$. Let $\mathbf{v}_3$ be any other vector in the same vector space as $\mathbf{v}$, $\mathbf{v}_1$, and $\mathbf{v}_2$. Show that $\mathbf{v}$ is also a linear combination of $\mathbf{v}_1$, $\mathbf{v}_2$, and $\mathbf{v}_3$.

35. Let the set $\{v_1, v_2, v_3\}$ be linearly independent in $\mathbf{R}^3$. Prove that each of the following subsets is also linearly independent. $\{v_1, v_2\}, \{v_1, v_3\}, \{v_2, v_3\}, \{v_1\}, \{v_2\}, \{v_3\}$.

36. Let the set $\{v_1, v_2\}$ be linearly dependent. Prove that $\{v_1 + 2v_2, 3v_1 - v_2\}$ is also linearly dependent.

37. Prove that an $n \times n$ matrix A is row equivalent to I_n if and only if $\text{rank}(A) = n$.

38. Determine the projection of the vector $\mathbf{v}$ onto the vector $\mathbf{u}$ for the following vectors.
  (a) $\mathbf{v} = (1, 3), \mathbf{u} = (2, 4)$
  (b) $\mathbf{v} = (-1, 3, 4), \mathbf{u} = (-1, 2, 4)$

39. Construct an orthonormal basis for the subspace of $\mathbf{R}^4$ spanned by the vectors $(1, 2, 3, -1), (2, 0, -1, 1), (3, 2, 0, 1)$.

40. Determine an orthonormal basis for the subspace of $\mathbf{R}^3$ consisting of vectors of the form $(x, y, x + 2y)$.

41. Let W be the subspace of $\mathbf{R}^3$ having basis $\{(2, 1, 1), (1, -1, 3)\}$. Find the projection of $(3, 1, -2)$ onto W.

42. Find the distance of the point $(1, 2, -4)$ in $\mathbf{R}^3$ from the subspace of vectors of the form $(a, 3a, b)$.

43. Prove that a square matrix A is orthogonal if and only if A^t is orthogonal.

44. Consider the vector $\mathbf{v} = (1, 3, -1)$ in $\mathbf{R}^3$. Let W be the subspace of $\mathbf{R}^3$ consisting of all vectors of the form $(a, b, a - 2b)$. Decompose $\mathbf{v}$ into the sum of a vector that lies in W and a vector orthogonal to W.

45. Prove that if two vectors $\mathbf{u}$ and $\mathbf{v}$ are orthogonal, then they are linearly independent.

46. Prove that if $\mathbf{u}$ is orthogonal to both the vectors $\mathbf{v}$ and $\mathbf{w}$ then $\mathbf{u}$ is orthogonal to every vector in the subspace generated by $\mathbf{v}$ and $\mathbf{w}$.

47. State (with a brief explanation) whether the following statements are true or false.
  (a) $\{(1, 0, 1), (2, 1, 5)\}$ is a basis for the subspace of vectors in $\mathbf{R}^3$ of the form $(a, b, a + 3b)$.
  (b) There are three linearly independent vectors in $\mathbf{R}^2$.
  (c) Any set of three vectors in $\mathbf{R}^3$ that are linearly independent span the space.
  (d) Any two nonzero vectors in $\mathbf{R}^2$ that do not form a basis are collinear.
  (e) Let $v_1, v_2,$ and v_3 be vectors in $\mathbf{R}^3$ written down at random. They are probably linearly dependent.

48. Let V be a vector space of dimension n. State (with a brief explanation) whether the following statements are true or false.
  (a) n is the largest number of linearly dependent vectors in V.
  (b) n is the largest number of linearly independent vectors in V.
  (c) No set of less than n vectors can span V.
  (d) A set of more than n vectors can span V, but cannot be linearly dependent.

Located in Dubai, United Arab Emirates, the Jumeirah
Beach Hotel is a five-star luxury hotel operated by the
Dubai-based hotelier Jumeirah. While the Jumeirah
Beach Hotel was once ranked the 9th tallest building in
Dubai, its rank has fallen lower than the 100th due to the
massive development and expansion of high-rise build-
ings in Dubai. The hotel was designed to resemble the
shape of a wave in order to complement the sail-shaped
Burj Al Arab, which is adjacent to the hotel.

Eigenvalues and Eigenvectors

In this chapter we introduce eigenvalues and eigenvectors for the vector space $\mathbf{R}^n$. Eigenvalues and eigenvectors are special scalars and vectors associated with matrices. Among the many uses of eigenvectors is in selecting coordinate systems. Mathematical models often describe situations relative to a coordinate system. There are usually some preferred systems, where the situation can be viewed in a particularly straightforward manner. Eigenvalues and eigenvectors often lead to these special coordinate systems.

Eigenvalues and eigenvectors are used in many branches of the natural and social sciences and engineering. We discuss the role of eigenvectors to rank pages in the search engine Google. We shall see how eigenvectors are used in demography to predict long term trends. We see their role in meteorology with an example of weather prediction for Tel Aviv. An important application from engineering in the study of oscillating systems is presented.

5.1 Eigenvalues and Eigenvectors

We commence our discussion with the definition of an eigenvalue and eigenvector.

DEFINITION Let A be an $n \times n$ matrix. A scalar λ is called an **eigenvalue** of A if there exists a nonzero vector $\mathbf{x}$ in $\mathbf{R}^n$ such that

$$A\mathbf{x} = \lambda\mathbf{x}.$$

The vector $\mathbf{x}$ is called an **eigenvector** corresponding to λ.

Let us look at the geometrical significance of an eigenvector that corresponds to a nonzero eigenvalue. The vector $A\mathbf{x}$ is in the same or opposite direction as $\mathbf{x}$, depending on the sign of λ. See Figure 5.1. An eigenvector of A is thus a vector whose direction is unchanged or reversed when multiplied by A.

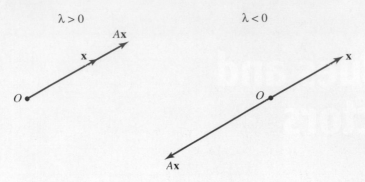

x is an eigenvector of A
$A\mathbf{x}$ is in the same or opposite direction as **x**

Figure 5.1

Computation of Eigenvalues and Eigenvectors

Let A be an $n \times n$ matrix with eigenvalue λ and corresponding eigenvector **x**. Thus $A\mathbf{x} = \lambda\mathbf{x}$. This equation may be rewritten

$$A\mathbf{x} - \lambda\mathbf{x} = \mathbf{0}$$

giving

$$(A - \lambda I_n)\mathbf{x} = \mathbf{0}$$

This matrix equation represents a system of homogeneous linear equations having matrix of coefficients $(A - \lambda I_n)$. $\mathbf{x} = \mathbf{0}$ is a solution to this system. However, eigenvectors have been defined to be nonzero vectors. Further, nonzero solutions to this system of equations can only exist if the matrix of coefficients is singular, $|A - \lambda I_n| = 0$. Hence, solving the equation $|A - \lambda I_n| = 0$ for λ leads to all the eigenvalues of A.

On expanding the determinant $|A - \lambda I_n|$ we get a polynomial in λ. This polynomial is called the **characteristic polynomial** of A. The equation $|A - \lambda I_n| = 0$ is called the **characteristic equation** of A.

The eigenvalues are then substituted back into the equation $(A - \lambda I_n)\mathbf{x} = \mathbf{0}$ to find the corresponding eigenvectors.

We now give a number of examples to illustrate this method.

EXAMPLE 1 Find the eigenvalues and eigenvectors of the matrix

$$A = \begin{bmatrix} -4 & -6 \\ 3 & 5 \end{bmatrix}$$

SOLUTION

Let us first derive the characteristic polynomial of A. We get

$$A - \lambda I_2 = \begin{bmatrix} -4 & -6 \\ 3 & 5 \end{bmatrix} - \lambda \begin{bmatrix} 1 & 0 \\ 0 & 1 \end{bmatrix} = \begin{bmatrix} -4 - \lambda & -6 \\ 3 & 5 - \lambda \end{bmatrix}$$

Note that the matrix $A - \lambda I_2$ is obtained by subtracting λ from the diagonal elements of A. The characteristic polynomial of A is

$$|A - \lambda I_2| = (-4 - \lambda)(5 - \lambda) + 18 = \lambda^2 - \lambda - 2$$

We now solve the characteristic equation of A.

$$\lambda^2 - \lambda - 2 = 0$$
$$(\lambda - 2)(\lambda + 1) = 0$$
$$\lambda = 2 \quad \text{or} \quad -1$$

The eigenvalues of A are 2 and -1.

The corresponding eigenvectors are found by using these values of λ in the equation $(A - \lambda I_2)\mathbf{x} = \mathbf{0}$. There are many eigenvectors corresponding to each eigenvalue.

$\boldsymbol{\lambda = 2}$ We solve the equation $(A - 2I_2)\mathbf{x} = \mathbf{0}$ for $\mathbf{x}$. The matrix $(A - 2I_2)$ is obtained by subtracting 2 from the diagonal elements of A. We get

$$\begin{bmatrix} -6 & -6 \\ 3 & 3 \end{bmatrix} \begin{bmatrix} x_1 \\ x_2 \end{bmatrix} = \mathbf{0}$$

This leads to the system of equations

$$-6x_1 - 6x_2 = 0$$
$$3x_1 + 3x_2 = 0$$

giving $x_1 = -x_2$. The solutions to this system of equations are $x_1 = -r$, $x_2 = r$, where r is a scalar. Thus the eigenvectors of A corresponding to $\lambda = 2$ are nonzero vectors of the form

$$r \begin{bmatrix} -1 \\ 1 \end{bmatrix}$$

$\boldsymbol{\lambda = -1}$ We solve the equation $(A + 1I_2)\mathbf{x} = \mathbf{0}$ for $\mathbf{x}$. The matrix $(A + 1I_2)$ is obtained by adding 1 to the diagonal elements of A. We get

$$\begin{bmatrix} -3 & -6 \\ 3 & 6 \end{bmatrix} \begin{bmatrix} x_1 \\ x_2 \end{bmatrix} = \mathbf{0}$$

This leads to the equations

$$-3x_1 - 6x_2 = 0$$
$$3x_1 + 6x_2 = 0$$

Thus $x_1 = -2x_2$. The solutions to these equations are $x_1 = -2s$, $x_2 = s$, where s is a scalar. The eigenvectors corresponding to $\lambda = -1$ are nonzero vectors of the form

$$s \begin{bmatrix} -2 \\ 1 \end{bmatrix}$$

■

Eigenspaces   The eigenvectors of $\lambda = 2$ are nonzero vectors of the form $r \begin{bmatrix} -1 \\ 1 \end{bmatrix}$. Add the zero vector to this set to get the set of *all* vectors of the form $r \begin{bmatrix} -1 \\ 1 \end{bmatrix}$.

This set is closed under addition and scalar multiplication; it is a subspace.

This subspace $\left\{ r \begin{bmatrix} -1 \\ 1 \end{bmatrix} \right\}$, with a basis $\left\{ \begin{bmatrix} -1 \\ 1 \end{bmatrix} \right\}$, is called the *eigenspace* of $\lambda = 2$. The dimension of the eigenspace is 1. Similarly the eigenspace for $\lambda = -1$ is the one-dimensional space $\left\{ s \begin{bmatrix} -2 \\ 1 \end{bmatrix} \right\}$, with a basis $\left\{ \begin{bmatrix} -2 \\ 1 \end{bmatrix} \right\}$.

These observations lead to the following general result.

THEOREM 5.1

Let A be an $n \times n$ matrix and λ an eigenvalue of A. The set of all eigenvectors corresponding to λ, together with the zero vector, is a subspace of $\mathbf{R}^n$. This subspace is called the **eigenspace** of λ.

Proof   Let V be the set of all eigenvectors corresponding to λ, together with the zero vector. In order to show that V is a subspace we have to show that it is closed under vector addition and scalar multiplication.

Let $\mathbf{x}_1$ and $\mathbf{x}_2$ be vectors in V and c be a scalar. Then $A\mathbf{x}_1 = \lambda\mathbf{x}_1$ and $A\mathbf{x}_2 = \lambda\mathbf{x}_2$. Hence,

$$A\mathbf{x}_1 + A\mathbf{x}_2 = \lambda\mathbf{x}_1 + \lambda\mathbf{x}_2$$
$$A(\mathbf{x}_1 + \mathbf{x}_2) = \lambda(\mathbf{x}_1 + \mathbf{x}_2)$$

Thus $\mathbf{x}_1 + \mathbf{x}_2$ is an eigenvector corresponding to λ. V is closed under addition.

Further, since $A\mathbf{x}_1 = \lambda\mathbf{x}_1$,

$$cA\mathbf{x}_1 = c\lambda\mathbf{x}_1$$
$$A(c\mathbf{x}_1) = \lambda(c\mathbf{x}_1)$$

Therefore $c\mathbf{x}_1$ is an eigenvector corresponding to λ. V is closed under scalar multiplication. Thus V is a subspace of $\mathbf{R}^n$.

EXAMPLE 2   Find the eigenvalues and corresponding eigenspaces of the matrix

$$A = \begin{bmatrix} 5 & 4 & 2 \\ 4 & 5 & 2 \\ 2 & 2 & 2 \end{bmatrix}$$

SOLUTION

The matrix $A - \lambda I_3$ is obtained by subtracting λ from the diagonal elements of A. Thus

$$A - \lambda I_3 = \begin{bmatrix} 5 - \lambda & 4 & 2 \\ 4 & 5 - \lambda & 2 \\ 2 & 2 & 2 - \lambda \end{bmatrix}$$

The characteristic polynomial of A is $|A - \lambda I_3|$. We get, using row and column operations to simplify determinants,

$$|A - \lambda I_3| = \begin{vmatrix} 5 - \lambda & 4 & 2 \\ 4 & 5 - \lambda & 2 \\ 2 & 2 & 2 - \lambda \end{vmatrix} = \begin{vmatrix} 1 - \lambda & -1 + \lambda & 0 \\ 4 & 5 - \lambda & 2 \\ 2 & 2 & 2 - \lambda \end{vmatrix}$$

$$= \begin{vmatrix} 1 - \lambda & 0 & 0 \\ 4 & 9 - \lambda & 2 \\ 2 & 4 & 2 - \lambda \end{vmatrix}$$

$$= (1 - \lambda)[(9 - \lambda)(2 - \lambda) - 8] = (1 - \lambda)[\lambda^2 - 11\lambda + 10]$$

$$= (1 - \lambda)(\lambda - 10)(\lambda - 1) = -(\lambda - 10)(\lambda - 1)^2$$

We now solve the characteristic equation of A

$$-(\lambda - 10)(\lambda - 1)^2 = 0$$

$$\lambda = 10 \quad \text{or} \quad 1$$

The eigenvalues of A are 10 and 1 (repeated).

The corresponding eigenvectors are found by using these values of λ in the equation $(A - \lambda I_3)\mathbf{x} = \mathbf{0}$.

$\lambda = 10$ We get

$$(A - 10I_3)\mathbf{x} = \mathbf{0},$$

$$\begin{bmatrix} -5 & 4 & 2 \\ 4 & -5 & 2 \\ 2 & 2 & -8 \end{bmatrix} \begin{bmatrix} x_1 \\ x_2 \\ x_3 \end{bmatrix} = \mathbf{0}$$

The solutions to this system of equations are $x_1 = 2r$, $x_2 = 2r$, $x_3 = r$, where r is a scalar. Thus the eigenvectors of $\lambda = 10$ are nonzero vectors of the form

$$r \begin{bmatrix} 2 \\ 2 \\ 1 \end{bmatrix}$$

The eigenspace is $\left\{ r \begin{bmatrix} 2 \\ 2 \\ 1 \end{bmatrix} \right\}$. The set $\left\{ \begin{bmatrix} 2 \\ 2 \\ 1 \end{bmatrix} \right\}$ is a basis, and the dimension is 1.

$\lambda = 1$ Let $\lambda = 1$ in $(A - \lambda I_3)\mathbf{x} = \mathbf{0}$. We get

$$(A - 1I_3)\mathbf{x} = \mathbf{0},$$

$$\begin{bmatrix} 4 & 4 & 2 \\ 4 & 4 & 2 \\ 2 & 2 & 1 \end{bmatrix} \begin{bmatrix} x_1 \\ x_2 \\ x_3 \end{bmatrix} = \mathbf{0}$$

The solutions to this system of equations can be shown to be $x_1 = -s - t$, $x_2 = s$, $x_3 = 2t$, where s and t are scalars. Thus the eigenvectors of $\lambda = 1$ are nonzero vectors of the form

$$\begin{bmatrix} -s - t \\ s \\ 2t \end{bmatrix}$$

The eigenspace is $\left\{ \begin{bmatrix} -s - t \\ s \\ 2t \end{bmatrix} \right\}$. To find a basis for this space, separate the parameters s and t as follows. We get linearly independent vectors that span the eigenspace.

$$\begin{bmatrix} -s - t \\ s \\ 2t \end{bmatrix} = s \begin{bmatrix} -1 \\ 1 \\ 0 \end{bmatrix} + t \begin{bmatrix} -1 \\ 0 \\ 2 \end{bmatrix} \text{ leads to the basis } \left\{ \begin{bmatrix} -1 \\ 1 \\ 0 \end{bmatrix}, \begin{bmatrix} -1 \\ 0 \\ 2 \end{bmatrix} \right\}$$

The dimension of this eigenspace is 2.

If an eigenvalue occurs as a k times repeated root of the characteristic equation, we say that it is of **multiplicity** k. Thus $\lambda = 10$ has multiplicity 1, while $\lambda = 1$ has multiplicity 2 in this example.

EXAMPLE 3 Let A be an $n \times n$ matrix A with eigenvalues $\lambda_1, \ldots, \lambda_n$, and corresponding eigenvectors $\mathbf{x}_1, \ldots, \mathbf{x}_n$. Prove that if $c \neq 0$, then the eigenvalues of cA are $c\lambda_1, \ldots, c\lambda_n$ with corresponding eigenvectors $\mathbf{x}_1, \ldots, \mathbf{x}_n$.

SOLUTION

Let λ_i be one of the eigenvalues of A with corresponding eigenvector $\mathbf{x}_i$. Then $A\mathbf{x}_i = \lambda_i \mathbf{x}_i$. Multiply both sides of this equation by c to get

$$cA\mathbf{x}_i = c\lambda_i \mathbf{x}_i$$

Thus $c\lambda_i$ is an eigenvalue of cA with corresponding eigenvector $\mathbf{x}_i$.

Further, since cA is an $n \times n$ matrix, the characteristic polynomial of A is of degree n. The characteristic equation has n roots, implying that cA has n eigenvalues. The eigenvalues of cA are therefore $c\lambda_1, \ldots, c\lambda_n$ with corresponding eigenvectors $\mathbf{x}_1, \ldots, \mathbf{x}_n$.

EXERCISE SET 5.1

Finding Eigenvalues and Eigenvectors

In Exercises 1–8 determine the characteristic polynomials, eigenvalues, and corresponding eigenspaces of the given 2×2 matrices.

1. $\begin{bmatrix} 5 & 4 \\ 1 & 2 \end{bmatrix}$ 2. $\begin{bmatrix} 1 & -2 \\ 1 & 4 \end{bmatrix}$ 3. $\begin{bmatrix} 5 & 6 \\ -2 & -2 \end{bmatrix}$

4. $\begin{bmatrix} 5 & 2 \\ -8 & -3 \end{bmatrix}$ 5. $\begin{bmatrix} 1 & 2 \\ 2 & 1 \end{bmatrix}$ 6. $\begin{bmatrix} 2 & 1 \\ -1 & 4 \end{bmatrix}$

7. $\begin{bmatrix} 3 & -1 \\ 2 & 0 \end{bmatrix}$ 8. $\begin{bmatrix} 2 & -4 \\ -1 & 2 \end{bmatrix}$

In Exercises 9–14 determine the characteristic polynomials, eigenvalues, and corresponding eigenspaces of the given 3×3 matrices.

9. $\begin{bmatrix} 3 & 2 & -2 \\ -3 & -1 & 3 \\ 1 & 2 & 0 \end{bmatrix}$ 10. $\begin{bmatrix} 1 & -2 & 2 \\ -2 & 1 & 2 \\ -2 & 0 & 3 \end{bmatrix}$

11. $\begin{bmatrix} 1 & 0 & 0 \\ -2 & 1 & 2 \\ -2 & 0 & 3 \end{bmatrix}$ 12. $\begin{bmatrix} 1 & 0 & 0 \\ -2 & 5 & -2 \\ -2 & 4 & -1 \end{bmatrix}$

13. $\begin{bmatrix} 15 & 7 & -7 \\ -1 & 1 & 1 \\ 13 & 7 & -5 \end{bmatrix}$ 14. $\begin{bmatrix} 5 & -2 & 2 \\ 4 & -3 & 4 \\ 4 & -6 & 7 \end{bmatrix}$

In Exercises 15 and 16 determine the characteristic polynomials, eigenvalues, and corresponding eigenspaces of the given 4×4 matrices.

15. $\begin{bmatrix} 4 & 2 & -2 & 2 \\ 1 & 3 & 1 & -1 \\ 0 & 0 & 2 & 0 \\ 1 & 1 & -3 & 5 \end{bmatrix}$ 16. $\begin{bmatrix} 3 & 5 & -5 & 5 \\ 3 & 1 & 3 & -3 \\ -2 & 2 & 0 & 2 \\ 0 & 4 & -6 & 8 \end{bmatrix}$

Geometrical Interpretation

In Exercises 17–19 determine the characteristic polynomials, eigenvalues, and corresponding eigenvectors of the given matrices. Interpret your results geometrically.

17. $\begin{bmatrix} 1 & 0 \\ 0 & 1 \end{bmatrix}$ 18. $\begin{bmatrix} 3 & 0 \\ 0 & 3 \end{bmatrix}$ 19. $\begin{bmatrix} -2 & 0 \\ 0 & -2 \end{bmatrix}$

20. Show that the following matrix has no real eigenvalues and thus no eigenvectors. Interpret your result geometrically.

$$\begin{bmatrix} 0 & -1 \\ 1 & 0 \end{bmatrix}$$

21. Show that the following matrix has no real eigenvalues. Interpret your result geometrically. Readers who have a knowledge of complex numbers will observe that it does have complex eigenvalues. All matrices have eigenvalues in the complex number system. The eigenvectors are vectors in $\mathbf{C}^n$.

$$\begin{bmatrix} 1 & 1 \\ -2 & -1 \end{bmatrix}$$

22. Find the eigenvalues and eigenvectors of the identity matrix I_n. Interpret your result geometrically.

Miscellaneous Results

23. Let A be the $n \times n$ matrix having every element 1. Find the eigenvalues and eigenvectors of A.

24. Prove that if A is a diagonal matrix then its eigenvalues are the diagonal elements.

25. Prove that if A is an upper triangular matrix then its eigenvalues are the diagonal elements.

26. Prove that A and A^t have the same eigenvalues.

27. Prove that $\lambda = 0$ is an eigenvalue of a matrix A if and only if A is singular.

28. Prove that if the eigenvalues of a matrix A are $\lambda_1, \ldots, \lambda_n$, with corresponding eigenvectors $\mathbf{x}_1, \ldots, \mathbf{x}_n$ then $\lambda_1^m, \ldots, \lambda_n^m$ are eigenvalues of A^m with corresponding eigenvectors $\mathbf{x}_1, \ldots, \mathbf{x}_n$.

29. Let A be an invertible matrix with eigenvalue λ having corresponding eigenvector $\mathbf{x}$. Prove that λ^{-1} is an eigenvalue of A^{-1} with eigenvector $\mathbf{x}$.

30. Let A be a matrix with eigenvalue λ having corresponding eigenvector $\mathbf{x}$. Let c be a scalar. Prove that $\lambda - c$ is an eigenvalue of $A - cI$ with corresponding eigenvector $\mathbf{x}$.

31. A matrix A is said to be **nilpotent** if $A^k = 0$ for some integer k. Prove that if A is nilpotent then 0 is the only eigenvalue of A.

32. Prove that the constant term of the characteristic polynomial of a matrix A is $|A|$.

33. Determine the eigenvalues and corresponding eigenvectors of the matrices

$$\begin{bmatrix} 1 & 2 \\ 1 & 0 \end{bmatrix}, \begin{bmatrix} 2 & 3 \\ 1 & 0 \end{bmatrix}, \begin{bmatrix} 3 & 4 \\ 1 & 0 \end{bmatrix}, \begin{bmatrix} 4 & 5 \\ 1 & 0 \end{bmatrix}$$

Use these results to make a conjecture about the eigenvalues and eigenvectors of the general matrix, $\begin{bmatrix} a & a+1 \\ 1 & 0 \end{bmatrix}$, where a is any real number. Verify your conjecture.

34. There is a theorem called **The Cayley-Hamilton Theorem**, which states that *a square matrix satisfies its characteristic equation*. That is, if the characteristic equation of a square matrix A is

$$\lambda^n + c_{n-1}\lambda^{n-1} + \cdots + c_0 = 0 \text{ then}$$
$$A^n + c_{n-1}A^{n-1} + \cdots + c_0 I = 0$$

Show that the following matrices satisfy their characteristic equations.

(a) $\begin{bmatrix} 0 & 2 \\ -1 & 3 \end{bmatrix}$ **(b)** $\begin{bmatrix} 8 & -10 \\ 5 & -7 \end{bmatrix}$

(c) $\begin{bmatrix} 6 & -8 \\ 4 & -6 \end{bmatrix}$ **(d)** $\begin{bmatrix} -1 & 5 \\ -10 & 14 \end{bmatrix}$

35. State (with a brief explanation) whether the following statements are true or false.

(a) A 3×3 matrix can have four distinct eigenvalues.

(b) An eigenvector cannot correspond to more than one eigenvalue.

(c) The sum of two eigenvectors is an eigenvector.

(d) Two different matrices can have the same eigenvalues.

*5.2 Google, Demography, and Weather Prediction

We now look at applications of eigenvectors. We discuss the role eigenvalues and eigenvectors play in Google.

The Google Search Engine

Google is a Web search engine that was developed by Larry Page and Sergey Brin when they were graduate students at Stanford University. There are usually a huge number of web pages that correspond to a certain query. Google finds pages that match that query and lists them in the order of their PageRank; PageRank is Google's primary way of deciding a page's importance to a given query. We now see that PageRank is calculated from the eigenvector of a very large matrix A (2.7×2.7 billion in 2002).

Let n be the number of pages that Google examines in a search; number these pages 1 to n. In searching the Web, going from page to page, Google finds that some pages have outgoing links whereas other pages do not (are dead ends). Let p be the fraction of the total number of pages searched that have outgoing links. An $n \times n$ matrix G is defined as follows.

$$g_{ij} = \begin{cases} 1 & \textit{if there is a link from page } i \textit{ to page } j \\ 0 & \textit{otherwise} \end{cases}$$

(G is in fact the adjacency matrix, which is described in Section 2.7, of the digraph having pages as vertices and links as arcs.) Let c_j be the sum of the elements in the jth column of G. Then A is an $n \times n$ matrix defined by

$$a_{ij} = p\left(\frac{g_{ij}}{c_j}\right) + \left(\frac{1-p}{n}\right)$$

Google usually takes $p = 0.85$. A theorem in linear algebra called the Perron-Frobenius theorem guarantees that A is a matrix having largest eigenvalue 1, with corresponding eigenspace of dimension 1. Let $\mathbf{x}$ be an eigenvector corresponding to $\lambda = 1$. Normalize $\mathbf{x}$, such that $\Sigma x_i = 1$. The elements of this normalized vector are Google's PageRank.

The determinant method introduced in this section can be used for finding eigenvectors of small matrices, but is not practical for large matrices, and most certainly not for a matrix of the size encountered in a Google search. The Google company has not revealed how it calculates the eigenvector $\mathbf{x}$ of this very large matrix A. It is generally believed that it is based on the power method discussed in Section 8.5.

More in-depth information on PageRank can be found in:

1. "The World's Largest Matrix Computation," by Cleve Moler, *MATLAB News and Notes*, October 2002, pages 12–13.
2. "Google's PageRank Explained," by Phil Craven, *www.webworkshop.net/pagerank.html*.

Long-Term Predictions

We now discuss how eigenvalues and eigenvectors can be used to predict the long-term behavior of certain Markov chains. Applications in demography and weather prediction are given.

Let us return to the population movement model of Section 2.6. There we found that annual population distributions could be described by a sequence of vectors $\mathbf{x}_0$, $\mathbf{x}_1(=P\mathbf{x}_0)$, $\mathbf{x}_2(=P\mathbf{x}_1)$, $\mathbf{x}_3(=P\mathbf{x}_2)$, ... P is a matrix of transition probabilities that takes us from one vector in the sequence to the following vector. Such a sequence (or chain) of vectors is called a **Markov chain**. Of special interest are Markov chains called *Regular Markov Chains* where the sequence $\mathbf{x}_0$, $\mathbf{x}_1$, $\mathbf{x}_2$, . . . converges to some fixed vector $\mathbf{x}$, where $P\mathbf{x} = \mathbf{x}$. The population movement would then be in a "steady-state" with the total city population and total suburban population remaining constant thereafter. We then write

$$\mathbf{x}_0, \mathbf{x}_1, \mathbf{x}_2, \ldots \to \mathbf{x}$$

Since such a vector $\mathbf{x}$ satisfies $P\mathbf{x} = \mathbf{x}$, it would be an eigenvector of P corresponding to eigenvalue 1. Knowledge of the existence and value of such a vector would give us information about the long-term behavior of the population distribution.

A special class of Markov chains has these desired properties. We now define this class, discuss their properties, and apply the results to our population movement model to make long-term predictions.

> **DEFINITION** The transition matrix P of a Markov chain is said to be **regular** if for some power of P all the components are positive. The chain is then called a **regular Markov chain**.

For example:

$A = \begin{bmatrix} 0.3 & 0.6 \\ 0.7 & 0.4 \end{bmatrix}$ is regular, because all the elements are positive.

$B = \begin{bmatrix} 0.7 & 1 \\ 0.3 & 0 \end{bmatrix}$ is regular, because $B^2 = \begin{bmatrix} 0.79 & 0.7 \\ 0.21 & 0.3 \end{bmatrix}$.

$C = \begin{bmatrix} 0.4 & 0 \\ 0.6 & 1 \end{bmatrix}$ is not regular, because $C^2 = \begin{bmatrix} 0.16 & 0 \\ 0.84 & 1 \end{bmatrix}$, $C^3 = \begin{bmatrix} 0.064 & 0 \\ 0.936 & 1 \end{bmatrix}$, . . . ;

the element in row 1, column 2, will always be zero.

The following theorem, which we do not prove, leads to information about the long-term behavior of regular Markov chains.

THEOREM 5.2

Consider a regular Markov chain having initial vector $\mathbf{x}_0$ and transition matrix P. Then
1. $\mathbf{x}_0$, $\mathbf{x}_1$, $\mathbf{x}_2$, . . . $\to \mathbf{x}$, where $\mathbf{x}$ satisfies $P\mathbf{x} = \mathbf{x}$.
   Thus $\mathbf{x}$ is an eigenvector of P corresponding to $\lambda = 1$.
2. P, P^2, P^3, . . . $\to Q$, where Q is a stochastic matrix. The columns of Q are all identical, each being an eigenvector of P corresponding to $\lambda = 1$.

Let us now apply this theorem to population movements.

Population Predictions

EXAMPLE 1 Determine the long-term trends in population movements between U.S. cities and suburbs.

SOLUTION

We remind the reader of the model that was developed earlier. The populations of U.S. cities and suburbs in 2007 were described by the following vector $\mathbf{x}_0$ (in units of one million), and the populations in the following years are given by a Markov chain with the transition matrix P.

$$
\begin{array}{cc}
\begin{array}{c} \text{Initial} \\ \text{populations} \end{array} &
\end{array}
$$

$$
\mathbf{x}_0 = \begin{bmatrix} 82 \\ 163 \end{bmatrix} \begin{array}{l} \text{city} \\ \text{suburb} \end{array}, \qquad
\begin{array}{c} \quad\;\; \text{(from)} \\ \text{city} \quad \text{suburb} \quad \text{(to)} \end{array}
$$

$$
P = \begin{bmatrix} 0.96 & 0.01 \\ 0.04 & 0.99 \end{bmatrix} \begin{array}{l} \text{city} \\ \text{suburb} \end{array}
$$

Observe that all the elements of P are positive. The chain is therefore regular and the results of the preceding theorem can be applied to give the long-term trends. The theorem tells us that P will have an eigenvalue of 1 and that the steady-state vector $\mathbf{x}$ is a corresponding eigenvector. Thus

$$P\mathbf{x} = \mathbf{x}$$

$$(P - I_2)\mathbf{x} = \mathbf{0}$$

$$\begin{bmatrix} 0.96 - 1 & 0.01 \\ 0.04 & 0.99 - 1 \end{bmatrix}\begin{bmatrix} x_1 \\ x_2 \end{bmatrix} = \mathbf{0}$$

This leads to the system of equations

$$-0.04x_1 + 0.01x_2 = 0$$
$$0.04x_1 - 0.01x_2 = 0$$

giving $x_2 = 4x_1$. The solutions to this system of equations are $x_1 = r$, $x_2 = 4r$, where r is a scalar. Thus the eigenvectors of P corresponding to $\lambda = 1$ are nonzero vectors of the form

$$r\begin{bmatrix} 1 \\ 4 \end{bmatrix}$$

The steady-state vector $\mathbf{x}$ will be a vector of this form. Let us assume that there is no total annual population change over the years. Therefore the sums of the elements of $\mathbf{x}$ and $\mathbf{x}_0$ are equal,

$$r + 4r = 82 + 163$$
$$r = 49$$

The steady-state vector is thus

$$\mathbf{x} = \begin{bmatrix} 49 \\ 196 \end{bmatrix}$$

This implies the following long-term predictions:

$$\text{U.S. city populations} \rightarrow 49 \text{ million}$$

$$\text{U.S. suburban populations} \rightarrow 196 \text{ million}$$

The above theorem gives further information about long-term population trends. Each column of the matrix Q is an eigenvector corresponding to the eigenvalue 1. Let

$$Q = \begin{bmatrix} s & s \\ 4s & 4s \end{bmatrix}$$

Since Q is a stochastic matrix, the sum of the elements in each column is 1. Thus

$$s + 4s = 1$$
$$s = 0.2$$

We get (exhibiting the elements to two decimal places for ease of reading)

$$\overset{P}{\begin{bmatrix} 0.96 & 0.01 \\ 0.04 & 0.99 \end{bmatrix}}, \overset{P^2}{\begin{bmatrix} 0.92 & 0.02 \\ 0.08 & 0.98 \end{bmatrix}}, \overset{P^3}{\begin{bmatrix} 0.89 & 0.03 \\ 0.11 & 0.97 \end{bmatrix}} \cdots \rightarrow \overset{Q}{\begin{bmatrix} 0.2 & 0.2 \\ 0.8 & 0.8 \end{bmatrix}}$$

Let us interpret this result. We focus on the $(2, 1)$ element in each matrix—a similar interpretation will apply to the other elements. We get the sequence

$$0.04, 0.08, 0.11, \ldots \rightarrow 0.8$$

These are the probabilities of moving from the city to suburbia in 1 year, 2 years, 3 years, and so on. The probability gradually increases, approaching 0.8. Q is thus the **long-term transition matrix** of the model. It gives the long-term probabilites of living in the city or suburbia.

$$Q = \begin{matrix} & \overset{\text{(from)}}{\begin{matrix} \text{city} & \text{suburb} \end{matrix}} & \text{(to)} \\ \begin{bmatrix} 0.2 & 0.2 \\ 0.8 & 0.8 \end{bmatrix} & \begin{matrix} \text{city} \\ \text{suburb} \end{matrix} \end{matrix}$$

Long-term probabilities

Observe that the long-term probability of living in the city is 0.2, while the long-term probability of living in suburbia is 0.8. These probabilities are independent of initial location. The long-term probabilities being independent of initial state is a characteristic of regular Markov chains.

We now discuss an interesting application of Markov chains in a model that describes rainfall in Tel Aviv.

Weather in Tel Aviv

EXAMPLE 2 K. R. Gabriel and J. Neumann have developed "A Markov Chain Model for Daily Rainfall Occurrence at Tel Aviv," *Quart J. R. Met. Soc.*, 88(1962), 90–95. The probabilities used were based on data of daily rainfall in Tel Aviv (Nahami Street) for the 27 years 1923 to 1950. Days were classified as wet or dry according to whether or not there had been recorded at least 0.1 mm of precipitation in the 24-hour period from

8 A.M. to 8 A.M. the following day. A Markov chain was constructed for each of the months November through April, these months constituting the rainy season. We discuss the chain developed for November. The model assumes that the probability of rainfall on any day depends only on whether the previous day was wet or dry. The statistics accumulated over the years for November were

A Given Day	Following Day Wet
Wet	117 out of 195
Dry	80 out of 615

Thus the probability of a wet day following a wet day is $\frac{117}{195} = 0.6$. The probability of a wet day following a dry day is $\frac{80}{615} = 0.13$. These probabilities lead to the following transition matrix for the weather pattern in November.

$$\begin{array}{cc} & \text{(A given day)} \\ & \begin{array}{cc} \text{wet} & \text{dry} \end{array} \\ P = & \begin{bmatrix} 0.6 & 0.13 \\ 0.4 & 0.87 \end{bmatrix} \begin{array}{l} \text{wet} \\ \text{dry} \end{array} \quad \text{(following day)} \end{array}$$

On any given day in November, one can use P to predict the weather on a future day in November. For example, if today, a Wednesday, is wet, let us compute the probability that next Saturday will be dry. Saturday is three days hence, thus the various probabilities for the weather on Saturday will be given by the elements of P^3. It can be shown that (exhibiting the elements to two decimal places for ease of reading),

$$\begin{array}{cc} & \text{(Today—Wednesday)} \\ & \begin{array}{cc} \text{wet} & \text{dry} \end{array} \\ P^3 = & \begin{bmatrix} 0.32 & 0.22 \\ 0.68 & 0.78 \end{bmatrix} \begin{array}{l} \text{wet} \\ \text{dry} \end{array} \quad \text{(Saturday)} \end{array}$$

If today is wet, the probability of Saturday being dry is 0.68.

Observe that P, the matrix of transition probabilities is regular. Eigenvectors can thus be used to obtain long-term weather predictions. The eigenvectors of P corresponding to the eigenvalue 1 are found to be nonzero vectors of the form

$$r \begin{bmatrix} 0.325 \\ 1 \end{bmatrix}$$

The column vectors of the long-term transition matrix Q will be eigenvectors of this type, whose components add up to 1 since Q is stochastic. Therefore $0.325r + r = 1$, giving $r = 0.75$ (to 2 decimal places). Thus

$$Q = \begin{bmatrix} 0.25 & 0.25 \\ 0.75 & 0.75 \end{bmatrix}$$

We can interpret this matrix as follows

$$\begin{array}{cc} & \text{(Today)} \\ & \begin{array}{cc} \text{wet} & \text{dry} \end{array} \\ & \begin{bmatrix} 0.25 & 0.25 \\ 0.75 & 0.75 \end{bmatrix} \begin{array}{l} \text{wet} \\ \text{dry} \end{array} \quad \text{(day in the distant future)} \end{array}$$

This implies the following weather forecast for Tel Aviv in November.

Long-Range Forecast for Tel Aviv in November
0.25 probability wet
0.75 probability dry

Weather in Belfast

EXAMPLE 3 A matrix model for weather in Belfast, Northern Ireland, is given by William J. Stuart in *Introduction to the Numerical Solution of Markov Chains*, (Princeton University Press, 1994, 6). Three weather conditions are considered: rainy (R), cloudy (C), and sunny (S). Daily changes are described by the following matrix:

$$
\begin{array}{c}
\text{(A given day)} \\
\begin{array}{ccc} R & C & S \end{array} \\
P = \begin{bmatrix} 0.8 & 0.7 & 0.5 \\ 0.15 & 0.2 & 0.3 \\ 0.05 & 0.1 & 0.2 \end{bmatrix} \begin{array}{c} R \\ C \\ S \end{array} \quad \text{(following day)}
\end{array}
$$

Thus for example, if today is cloudy, the probability that tomorrow is sunny is 0.1. The eigenvectors corresponding to the eigenvalue 1, are found to be of the form

$$
r \begin{bmatrix} 122/11 \\ 27/11 \\ 1 \end{bmatrix}
$$

Using $r = 11/160$ leads to the stochastic matrix that describes the long-term forecast,

$$
\begin{array}{c}
\text{(today)} \\
\begin{array}{ccc} R & C & S \end{array} \\
Q = \begin{bmatrix} 0.7625 & 0.7625 & 0.7625 \\ 0.16875 & 0.16875 & 0.16875 \\ 0.06875 & 0.06875 & 0.06875 \end{bmatrix} \begin{array}{c} R \\ C \\ S \end{array} \quad \text{(day in the distant future)}
\end{array}
$$

According to this model there are probabilities of approximately 0.76, 0.17, and 0.07 of any future day in Belfast being rainy, cloudy, or sunny.

EXERCISE SET 5.2

Applications

1. The populations of U.S. metropolitan and nonmetropolitan areas in 2007 are described by the following vector $\mathbf{x}_0$ (in units of one million), and the populations in the following years are given by a Markov chain with the transition matrix P. (See Exercise 6, Section 2.6.) Determine the long-term predictions for metro and nonmetro populations, assuming no change in their total population.

$$
\begin{array}{c}
\text{Initial populations} \\
\mathbf{x}_0 = \begin{bmatrix} 245 \\ 52 \end{bmatrix} \begin{array}{c} \text{metro} \\ \text{nonmetro} \end{array}
\end{array}
$$

$$
\begin{array}{c}
\text{(from)} \\
\begin{array}{cc} \text{metro} & \text{nonmetro} \end{array} \ \text{(to)} \\
P = \begin{bmatrix} 0.99 & 0.02 \\ 0.01 & 0.98 \end{bmatrix} \begin{array}{c} \text{metro} \\ \text{nonmetro} \end{array}
\end{array}
$$

2. The populations of U.S. cities, suburbs, and nonmetro areas in 2007 are described by the following vector $\mathbf{x}_0$ (in units of one million), and the populations in the following years are given by a Markov chain with the transition matrix P. (See Exercise 7, Section 2.6.) Determine the long-term predictions for the populations of these regions, assuming no change in their total population. (This model is a refinement of the model of the previous exercise in that the metro areas are broken down into city and suburbia.)

$$\mathbf{x}_0 = \begin{bmatrix} 82 \\ 163 \\ 52 \end{bmatrix} \begin{matrix} \text{city} \\ \text{suburb} \\ \text{nonmetro} \end{matrix}$$

Initial populations

$$
\begin{matrix}
 & \text{city} & \text{suburb} & \text{nonmetro} & \text{(to)} \\
P = & \begin{bmatrix} 0.96 & 0.01 & 0.015 \\ 0.03 & 0.98 & 0.005 \\ 0.01 & 0.01 & 0.98 \end{bmatrix} & & & \begin{matrix} \text{city} \\ \text{suburb} \\ \text{nonmetro} \end{matrix}
\end{matrix}
$$

(from)

3. Consider a genetics model in which the offspring of guinea pigs are crossed with hybrids only. (See Exercise 15, Section 2.6.) The transition matrix P for that model is as follows. Prove that P is regular. Let $P, P^2, P^3, \ldots \to Q$. Determine Q. What information about the long-term distribution of guinea pigs does it give?

$$
\begin{matrix}
 & AA & Aa & aa & \text{(to)} \\
P = & \begin{bmatrix} 0.5 & 0.25 & 0 \\ 0.5 & 0.5 & 0.5 \\ 0 & 0.25 & 0.5 \end{bmatrix} & & & \begin{matrix} AA \\ Aa \\ aa \end{matrix}
\end{matrix}
$$

(from)

4. The statistics for rainfall for the month of December in Tel Aviv are as follows:

A Given Day	Following Day Wet
Wet	213 out of 326
Dry	117 out of 511

Use these figures to construct a Markov chain for predicting weather in December in Tel Aviv. (See Example 2.)

(a) If today is a dry Thursday in December in Tel Aviv, what is the probability that next Saturday will also be dry?

(b) Determine the long-range forecast for December in Tel Aviv. What are the probabilities that a distant day will be wet/dry?

5. A psychologist conducts an experiment in which 20 rats are placed at random in a compartment that has been divided into rooms labeled 1, 2, and 3, as shown in Figure 5.2.

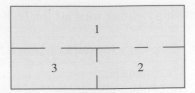

Figure 5.2

Observe that there are four doors in the arrangement. There are three possible states for the rats: they can be in room 1, 2, or 3. Let us assume that the rats move from room to room. A rat in room 1 has the probabilities $p_{11} = 0$, $p_{12} = \frac{2}{3}$, and $p_{13} = \frac{1}{3}$ of moving to the various rooms, based on the distribution of doors. This approach leads to a Markov chain with the following transition matrix P. Predict the long-term distribution of rats. What is the long-term probability that a given marked rat is in room 2?

$$
\begin{matrix}
 & \text{Room 1} & \text{Room 2} & \text{Room 3} & \text{(to)} \\
P = & \begin{bmatrix} 0 & \frac{2}{3} & \frac{1}{2} \\ \frac{2}{3} & 0 & \frac{1}{2} \\ \frac{1}{3} & \frac{1}{3} & 0 \end{bmatrix} & & & \begin{matrix} \text{Room 1} \\ \text{Room 2} \\ \text{Room 3} \end{matrix}
\end{matrix}
$$

(from)

6. 40 rats are placed at random in a compartment having four rooms labeled 1, 2, 3, and 4, as shown in Figure 5.3. Construct a Markov chain to describe the movement of the rats between the rooms. Predict the long-term distribution of rats. What is the long-term probability that a given marked rat is in room 4?

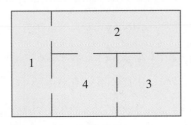

Figure 5.3

7. Two car rental companies A and B are competing for customers at certain airports. A study has been made of customer satisfaction with the various companies. The results are expressed by the following matrix R. The first column of R implies that 75% of those currently using rental company A are satisfied and intend to use A next time, while 25% of those using A are dissatisfied and plan to use B next time. There is a similar interpretation to the second column of R.

$$
\begin{matrix}
 & A & B & \text{(to)} \\
R = & \begin{bmatrix} 75\% & 20\% \\ 25\% & 80\% \end{bmatrix} & & \begin{matrix} A \\ B \end{matrix}
\end{matrix}
$$

(from)

Modify the matrix R to obtain a transition matrix P for a Markov chain that describes the rental patterns. If the current trends continue, how will the rental distribution eventually settle? Express the distribution in percentages that use companies A and B.

8. A market research group has been studying the buying patterns for three competing products I, II, and III. The results of the analysis are described by the following matrix A.

$$
A = \begin{array}{c} \\ \\ \end{array}\overset{\displaystyle \overset{\text{(from)}}{\begin{array}{ccc} \text{I} & \text{II} & \text{III} \end{array}}}{\begin{bmatrix} 80\% & 20\% & 5\% \\ 5\% & 75\% & 5\% \\ 15\% & 5\% & 90\% \end{bmatrix}} \begin{array}{l} \text{I} \\ \text{II} \\ \text{III} \end{array} \text{(to)}
$$

Column 1 implies that of those people currently using product I, 80% plan to continue using it, while 5% plan to switch to product II and 15% to product III. Columns 2 and 3 are to be interpreted similarly. Use the matrix A to construct a Markov chain that describes buying trends. If the current buying patterns continue, determine the likely eventual distribution of sales, in terms of percentages.

9. Prove that 1 is an eigenvalue of every stochastic matrix. (*Hint*: Prove that $|A - 1I| = 0$, using the properties of determinants.)

5.3 Diagonalization of Matrices

We begin this section with a discussion of similarity transformations. Similarity transformations occur frequently in applying linear algebra, very often in the context of relating coordinate systems. We shall use similarity transformations in geometry and in the study of vibrating strings in the next section.

DEFINITION   Let A and B be square matrices of the same size. B is said to be *similar* to A if there exists an invertible matrix C such that $B = C^{-1}AC$. The transformation of the matrix A into the matrix B in this manner is called a *similarity transformation*.

EXAMPLE 1 Consider the following matrices A and C. C is invertible. Use the similarity transformation $C^{-1}AC$ to transform A into a matrix B.

$$
A = \begin{bmatrix} 7 & -10 \\ 3 & -4 \end{bmatrix}, \quad C = \begin{bmatrix} 2 & 5 \\ 1 & 3 \end{bmatrix}
$$

SOLUTION

It can be shown that $C^{-1} = \begin{bmatrix} 3 & -5 \\ -1 & 2 \end{bmatrix}$. We get

$$
\begin{aligned}
B = C^{-1}AC &= \begin{bmatrix} 2 & 5 \\ 1 & 3 \end{bmatrix}^{-1} \begin{bmatrix} 7 & -10 \\ 3 & -4 \end{bmatrix} \begin{bmatrix} 2 & 5 \\ 1 & 3 \end{bmatrix} \\
&= \begin{bmatrix} 3 & -5 \\ -1 & 2 \end{bmatrix} \begin{bmatrix} 7 & -10 \\ 3 & -4 \end{bmatrix} \begin{bmatrix} 2 & 5 \\ 1 & 3 \end{bmatrix} \\
&= \begin{bmatrix} 6 & -10 \\ -1 & 2 \end{bmatrix} \begin{bmatrix} 2 & 5 \\ 1 & 3 \end{bmatrix} = \begin{bmatrix} 2 & 0 \\ 0 & 1 \end{bmatrix}
\end{aligned}
$$

In this example A is transformed into a diagonal matrix B. Not every square matrix can be "diagonalized" in this manner. In this section we shall discuss conditions under which a matrix can be diagonalized, and when it can, ways of constructing an appropriate transforming matrix C. We shall find that eigenvalues and eigenvectors play a key role in this discussion.

THEOREM 5.3

Similar matrices have the same eigenvalues.

Proof   Let A and B be similar matrices. Hence, there exists a matrix C such that $B = C^{-1}AC$. The characteristic polynomial of B is $|B - \lambda I|$. Substituting for B and using the multiplicative properties of determinants, we get

$$
\begin{aligned}
|B - \lambda I| = |C^{-1}AC - \lambda I| &= |C^{-1}(A - \lambda I)C| \\
&= |C^{-1}||A - \lambda I||C| = |A - \lambda I||C^{-1}||C| \\
&= |A - \lambda I||C^{-1}C| = |A - \lambda I||I| \\
&= |A - \lambda I|
\end{aligned}
$$

The characteristic polynomials of A and B are identical. This means that their eigenvalues are the same.

DEFINITION   A square matrix A is said to be **diagonalizable** if there exists a matrix C such that $D = C^{-1}AC$ is a diagonal matrix.

We see in the above example that $A = \begin{bmatrix} 7 & -10 \\ 3 & -4 \end{bmatrix}$ is diagonalizable. The following theorem tells us when it is possible to diagonalize a given matrix A, and if it is possible, how to find the similarity transformation that diagonalizes A.

THEOREM 5.4

Let A be an $n \times n$ matrix.
(a) If A has n linearly independent eigenvectors it is diagonalizable. The matrix C whose columns consist of n linearly independent eigenvectors can be used in a similarity transformation $C^{-1}AC$ to give a diagonal matrix D. The diagonal elements of D will be the eigenvalues of A.
(b) If A is diagonalizable then it has n linearly independent eigenvectors.

Proof
(a) Let A have eigenvalues $\lambda_1, \ldots, \lambda_n$, (which need not be distinct), with corresponding linearly independent eigenvectors $\mathbf{v}_1, \ldots, \mathbf{v}_n$. Let C be the matrix having $\mathbf{v}_1, \ldots, \mathbf{v}_n$ as column vectors.

$$ C = [\mathbf{v}_1 \ldots \mathbf{v}_n] $$

Since $A\mathbf{v}_1 = \lambda_1\mathbf{v}_1, \ldots, A\mathbf{v}_n = \lambda_n\mathbf{v}_n$, matrix multiplication in terms of columns gives

$$
\begin{aligned}
AC &= A[\mathbf{v}_1 \ldots \mathbf{v}_n] \\
&= [A\mathbf{v}_1 \ldots A\mathbf{v}_n] \\
&= [\lambda_1\mathbf{v}_1 \ldots \lambda_n\mathbf{v}_n] \\
&= [\mathbf{v}_1 \ldots \mathbf{v}_n] \begin{bmatrix} \lambda_1 & & 0 \\ & \ddots & \\ 0 & & \lambda_n \end{bmatrix} = C \begin{bmatrix} \lambda_1 & & 0 \\ & \ddots & \\ 0 & & \lambda_n \end{bmatrix}
\end{aligned}
$$

Since the columns of C are linearly independent, C is nonsingular. Thus

$$C^{-1}AC = \begin{bmatrix} \lambda_1 & & 0 \\ & \ddots & \\ 0 & & \lambda_n \end{bmatrix}$$

Therefore if an $n \times n$ matrix A has n linearly independent eigenvectors these eigenvectors can be used as the columns of a matrix C that diagonalizes A. The diagonal matrix has the eigenvalues of A as diagonal elements.

(b) The converse is proved by retracing the above steps. Commence with the assumption that C is a matrix $[\mathbf{v}_1 \dots \mathbf{v}_n]$ that diagonalizes A. Thus, there exist scalars $\gamma_1, \dots, \gamma_n$, such that

$$C^{-1}AC = \begin{bmatrix} \gamma_1 & & 0 \\ & \ddots & \\ 0 & & \gamma_n \end{bmatrix}$$

Retracing these steps we arrive at the conclusion that

$$A\mathbf{v}_1 = \gamma_1\mathbf{v}_1, \dots, A\mathbf{v}_n = \gamma_n\mathbf{v}_n$$

Thus $\mathbf{v}_1, \dots, \mathbf{v}_n$ are eigenvectors of A. Since C is nonsingular these vectors (column vectors of C) are linearly independent. Thus if an $n \times n$ matrix A is diagonalizable it has n linearly independent eigenvectors.

EXAMPLE **2**

(a) Show that the following matrix A is diagonalizable.
(b) Find a diagonal matrix D that is similar to A.
(c) Determine the similarity transformation that diagonalizes A.

$$A = \begin{bmatrix} -4 & -6 \\ 3 & 5 \end{bmatrix}$$

SOLUTION

(a) The eigenvalues and corresponding eigenvectors of this matrix were found in Example 1 of Section 5.1. They are

$$\lambda_1 = 2, \mathbf{v}_1 = r\begin{bmatrix} -1 \\ 1 \end{bmatrix}; \quad \lambda_2 = -1, \quad \mathbf{v}_2 = s\begin{bmatrix} -2 \\ 1 \end{bmatrix}$$

Since A, a 2×2 matrix, has two linearly independent eigenvectors it is diagonalizable.

(b) A is similar to the diagonal matrix D, which has diagonal elements $\lambda_1 = 2$ and $\lambda_2 = -1$. Thus

$$A = \begin{bmatrix} -4 & -6 \\ 3 & 5 \end{bmatrix} \text{ is similar to } D = \begin{bmatrix} 2 & 0 \\ 0 & -1 \end{bmatrix}$$

(c) It is often important to know the transformation that diagonalizes A. We now find the transformation and verify that it does lead to the above matrix D. Select two convenient linearly independent eigenvectors, say

$$\mathbf{v}_1 = \begin{bmatrix} -1 \\ 1 \end{bmatrix} \quad \text{and} \quad \mathbf{v}_2 = \begin{bmatrix} -2 \\ 1 \end{bmatrix}$$

Let these vectors be the column vectors of the diagonalizing matrix C.

$$C = \begin{bmatrix} -1 & -2 \\ 1 & 1 \end{bmatrix}$$

We then get

$$C^{-1}AC = \begin{bmatrix} -1 & -2 \\ 1 & 1 \end{bmatrix}^{-1} \begin{bmatrix} -4 & -6 \\ 3 & 5 \end{bmatrix} \begin{bmatrix} -1 & -2 \\ 1 & 1 \end{bmatrix}$$

$$= \begin{bmatrix} 1 & 2 \\ -1 & -1 \end{bmatrix} \begin{bmatrix} -4 & -6 \\ 3 & 5 \end{bmatrix} \begin{bmatrix} -1 & -2 \\ 1 & 1 \end{bmatrix} = \begin{bmatrix} 2 & 0 \\ 0 & -1 \end{bmatrix} = D$$

If A is similar to a diagonal matrix D under the transformation $C^{-1}AC$ then it can be shown that $A^k = CD^kC^{-1}$. This result can be used to compute A^k. Let us derive this result and then apply it.

$$D^k = (C^{-1}AC)^k = \underbrace{(C^{-1}AC) \dots (C^{-1}AC)}_{k \text{ times}} = C^{-1}A^kC$$

This leads to

$$A^k = CD^kC^{-1}$$

EXAMPLE 3 Compute A^9 for the following matrix A.

$$A = \begin{bmatrix} -4 & -6 \\ 3 & 5 \end{bmatrix}$$

SOLUTION

A is the matrix of the previous example. Use the values of C and D from that example. We get

$$D^9 = \begin{bmatrix} 2 & 0 \\ 0 & -1 \end{bmatrix}^9 = \begin{bmatrix} 2^9 & 0 \\ 0 & (-1)^9 \end{bmatrix} = \begin{bmatrix} 512 & 0 \\ 0 & -1 \end{bmatrix}$$

The transformation now gives

$$A^9 = CD^9C^{-1}$$

$$= \begin{bmatrix} -1 & -2 \\ 1 & 1 \end{bmatrix} \begin{bmatrix} 512 & 0 \\ 0 & -1 \end{bmatrix} \begin{bmatrix} -1 & -2 \\ 1 & 1 \end{bmatrix}^{-1} = \begin{bmatrix} -514 & -1026 \\ 513 & 1025 \end{bmatrix}$$

In practice, powers of matrices such as A^9 can now be quickly found using computers. The importance of the result $A^k = CD^kC^{-1}$ for computing powers is primarily theoretical. It can be used to give expressions for powers of general matrices. The reader will appreciate the technique in the following section, when it is used in solving equations called difference equations.

The following example illustrates that not every matrix is diagonalizable.

EXAMPLE 4 Show that the following matrix A is not diagonalizable.

$$A = \begin{bmatrix} 5 & -3 \\ 3 & -1 \end{bmatrix}$$

SOLUTION

Let us compute the eigenvalues and corresponding eigenvectors of A. We get

$$A - \lambda I_2 = \begin{bmatrix} 5 - \lambda & -3 \\ 3 & -1 - \lambda \end{bmatrix}$$

The characteristic equation is

$$|A - \lambda I_2| = 0,$$
$$(5 - \lambda)(-1 - \lambda) + 9 = 0,$$
$$\lambda^2 - 4\lambda + 4 = 0,$$
$$(\lambda - 2)(\lambda - 2) = 0$$

There is a single repeated eigenvalue, $\lambda = 2$. We now find the corresponding eigenvectors. $(A - 2I_2)\mathbf{x} = \mathbf{0}$ gives

$$\begin{bmatrix} 3 & -3 \\ 3 & -3 \end{bmatrix}\begin{bmatrix} x_1 \\ x_2 \end{bmatrix} = \mathbf{0}$$

This gives $3x_1 - 3x_2 = 0$. Thus $x_1 = r$, $x_2 = r$. The eigenvectors are nonzero vectors of the form

$$r\begin{bmatrix} 1 \\ 1 \end{bmatrix}$$

The eigenspace is a one-dimensional space. A is a 2×2 matrix but it does not have two linearly independent eigenvectors. Thus A is not diagonalizable.

Diagonalization of Symmetric Matrices

We have discussed symmetric matrices and looked at applications of these matrices. Let us now examine the diagonalization of these important matrices. The previous theorem showed us that the diagonalization of a matrix is related to its eigenvectors. The following theorem summarizes the properties of eigenvalues and eigenvectors of symmetric matrices, paving the way for results about their diagonalization.

THEOREM 5.5

Let A be an $n \times n$ symmetric matrix.

(a) All the eigenvalues of A are real numbers.
(b) The dimension of an eigenspace of A is the multiplicity of the eigenvalue as a root of the characteristic equation.
(c) The eigenspaces of A are orthogonal.
(d) A has n linearly independent eigenvectors.

The proof of this theorem is beyond the scope of this class. However the results are extremely important and can be easily understood and used. We illustrate these results for the following symmetric matrix A.

$$A = \begin{bmatrix} 5 & 4 & 2 \\ 4 & 5 & 2 \\ 2 & 2 & 2 \end{bmatrix}$$

The eigenvalues and eigenvectors of this matrix were found in Section 5.1.

The characteristic equation can be written $(\lambda - 10)(\lambda - 1)^2 = 0$. The eigenvalues λ_1, λ_2 and corresponding eigenspaces V_1, V_2 are as follows

$$\lambda_1 = 10, V_1 = \left\{ r \begin{bmatrix} 2 \\ 2 \\ 1 \end{bmatrix} \right\}; \quad \lambda_2 = 1, V_2 = \left\{ s \begin{bmatrix} -1 \\ 1 \\ 0 \end{bmatrix} + t \begin{bmatrix} -1 \\ 0 \\ 2 \end{bmatrix} \right\}$$

We see that the results of the theorem are illustrated as follows:

(a) The eigenvalues of A are real numbers; $\lambda_1 = 10$ and $\lambda_2 = 1$.

(b) $\lambda_1 = 10$ is a root of the characteristic equation of multiplicity 1, and its eigenspace is the one-dimensional space having basis $\left\{ \begin{bmatrix} 2 \\ 2 \\ 1 \end{bmatrix} \right\}$.

$\lambda_2 = 1$ is a root of multiplicity 2, and its eigenspace is the two-dimensional space having basis $\left\{ \begin{bmatrix} -1 \\ 1 \\ 0 \end{bmatrix}, \begin{bmatrix} -1 \\ 0 \\ 2 \end{bmatrix} \right\}$.

(c) Two vector spaces are said to be orthogonal if any vector in the one space is orthogonal to any vector in the other space. Let $\mathbf{v}_1 = r \begin{bmatrix} 2 \\ 2 \\ 1 \end{bmatrix}$ and $\mathbf{v}_2 = s \begin{bmatrix} -1 \\ 1 \\ 0 \end{bmatrix} + t \begin{bmatrix} -1 \\ 0 \\ 2 \end{bmatrix}$

be arbitrary vectors in the two eigenspaces. It can be seen that $\mathbf{v}_1 \cdot \mathbf{v}_2 = 0$, showing that the two eigenspaces are orthogonal.

(d) A is a 3×3 matrix. Observe that it has 3 linearly independent eigenvectors, namely $\begin{bmatrix} 2 \\ 2 \\ 1 \end{bmatrix}, \begin{bmatrix} -1 \\ 1 \\ 0 \end{bmatrix}$, and $\begin{bmatrix} -1 \\ 0 \\ 2 \end{bmatrix}$.

Orthogonal Diagonalization

If a matrix C is orthogonal then we know that $C^{-1} = C^t$. Thus if such a matrix is used in a similarity transformation, the transformation becomes $D = C^t A C$. This type of similarity transformation is of course easier to compute than $D = C^{-1} A C$. This is important if one is performing computations by hand, but not important when using a computer.

There is a deeper significance to **orthogonal diagonalization** related to what we said earlier about the role of similarity transformations in going from one coordinate system to another. Similarity transformations involving an orthogonal matrix are the transformations that are used to relate orthogonal coordinate systems (coordinate systems where the axes are at right angles). The reader will see a geometrical application that relates orthogonal coordinate systems later in the following section.

DEFINITION   A square matrix A is said to be **orthogonally diagonalizable** if there exists an orthogonal matrix C such that $D = C^t A C$ is a diagonal matrix.

The next theorem tells us that the set of orthogonally diagonalizable matrices is in fact the set of symmetric matrices.

THEOREM 5.6

Let A be a square matrix. A is orthogonally diagonalizable if and only if it is a symmetric matrix.

Proof   Assume that A is symmetric. We give the steps that can be taken to construct an orthogonal matrix C such that $D = C^t A C$ is diagonal. The previous theorem ensures us that this algorithm can be carried out.

1. Find a basis for each eigenspace of A.
2. Find an orthonormal basis for each eigenspace. (Use the Gram-Schmidt process if necessary.)
3. Let C be the matrix whose columns are these orthonormal vectors.
4. The matrix $D = C^t A C$ will be a diagonal matrix.

This algorithm will be used to orthogonally diagonalize a given symmetric matrix.

Conversely, assume that A is orthogonally diagonalizable. Thus there exists an orthogonal matrix C such that $D = C^t A C$. Therefore

$$A = CDC^t$$

Use the properties of transpose to get

$$A^t = (CDC^t)^t = (C^t)^t (CD)^t = CDC^t = A$$

Thus A is symmetric.

EXAMPLE **5** Orthogonally diagonalize the following symmetric matrix A.

$$A = \begin{bmatrix} 1 & -2 \\ -2 & 1 \end{bmatrix}$$

SOLUTION

The reader can verify that the eigenvalues and corresponding eigenspaces of this matrix are as follows:

$$\lambda_1 = -1, V_1 = \left\{ s \begin{bmatrix} 1 \\ 1 \end{bmatrix} \right\}; \quad \lambda_2 = 3, V_2 = \left\{ r \begin{bmatrix} -1 \\ 1 \end{bmatrix} \right\}$$

Since A is symmetric we know that it can be diagonalized to give

$$D = \begin{bmatrix} -1 & 0 \\ 0 & 3 \end{bmatrix}$$

Let us determine the transformation. The eigenspaces V_1 and V_2 are, as is to be expected, orthogonal. Use a unit vector in each eigenspace as columns of an orthogonal matrix C. We get

$$C = \begin{bmatrix} \dfrac{1}{\sqrt{2}} & -\dfrac{1}{\sqrt{2}} \\ \dfrac{1}{\sqrt{2}} & \dfrac{1}{\sqrt{2}} \end{bmatrix}$$

The orthogonal transformation that leads to D is

$$C^t A C = \begin{bmatrix} \dfrac{1}{\sqrt{2}} & \dfrac{1}{\sqrt{2}} \\ -\dfrac{1}{\sqrt{2}} & \dfrac{1}{\sqrt{2}} \end{bmatrix} \begin{bmatrix} 1 & -2 \\ -2 & 1 \end{bmatrix} \begin{bmatrix} \dfrac{1}{\sqrt{2}} & -\dfrac{1}{\sqrt{2}} \\ \dfrac{1}{\sqrt{2}} & \dfrac{1}{\sqrt{2}} \end{bmatrix}$$

EXERCISE SET 5.3

Similarity Transformation

1. In each of the following exercises transform the matrix A into a matrix B using the similarity transformation $C^{-1}AC$, with the given matrix C.

(a) $A = \begin{bmatrix} 1 & 2 \\ -1 & 3 \end{bmatrix}, C = \begin{bmatrix} 2 & 5 \\ 1 & 3 \end{bmatrix}$

(b) $A = \begin{bmatrix} -8 & 18 \\ -6 & 13 \end{bmatrix}, C = \begin{bmatrix} 3 & 2 \\ 2 & 1 \end{bmatrix}$

(c) $A = \begin{bmatrix} 0 & 4 \\ 3 & 2 \end{bmatrix}, C = \begin{bmatrix} 2 & 1 \\ 7 & 4 \end{bmatrix}$

2. In each of the following exercises transform the matrix A into a matrix B using the similarity transformation $C^{-1}AC$, with the given matrix C.

(a) $A = \begin{bmatrix} 2 & 0 & 0 \\ -2 & 2 & 1 \\ 2 & 0 & 1 \end{bmatrix}, C = \begin{bmatrix} -1 & 2 & 0 \\ 2 & -3 & 1 \\ -2 & 4 & -1 \end{bmatrix}$

(b) $A = \begin{bmatrix} 1 & 2 & 3 \\ 0 & -1 & 2 \\ 1 & 1 & 0 \end{bmatrix}, C = \begin{bmatrix} 3 & 5 & -1 \\ -2 & -3 & 1 \\ -1 & -2 & 1 \end{bmatrix}$

Diagonalization

3. Diagonalize (if possible) each of the following matrices. Give the similarity transformation.

(a) $\begin{bmatrix} 5 & 4 \\ 1 & 2 \end{bmatrix}$ (b) $\begin{bmatrix} 2 & 1 \\ 2 & 3 \end{bmatrix}$ (c) $\begin{bmatrix} 1 & 1 \\ 0 & 1 \end{bmatrix}$

(d) $\begin{bmatrix} 4 & -1 \\ 2 & 1 \end{bmatrix}$ (e) $\begin{bmatrix} 4 & 1 \\ -1 & 2 \end{bmatrix}$

4. Diagonalize (if possible) each of the following matrices. Give the similarity transformation.

(a) $\begin{bmatrix} -7 & 10 \\ -5 & 8 \end{bmatrix}$ (b) $\begin{bmatrix} 7 & 4 \\ -8 & -5 \end{bmatrix}$

(c) $\begin{bmatrix} 1 & -2 \\ 2 & -3 \end{bmatrix}$ (d) $\begin{bmatrix} 7 & -4 \\ 1 & 3 \end{bmatrix}$

(e) $\begin{bmatrix} a & b \\ 0 & a \end{bmatrix}, b \neq 0$

5. Diagonalize (if possible) each of the following matrices. Give the similarity transformation.

(a) $\begin{bmatrix} 15 & 7 & -7 \\ -1 & 1 & 1 \\ 13 & 7 & -5 \end{bmatrix}$ (b) $\begin{bmatrix} 5 & -2 & 2 \\ 4 & -3 & 4 \\ 4 & -6 & 7 \end{bmatrix}$

(c) $\begin{bmatrix} 1 & 0 & 0 \\ -2 & 1 & 2 \\ -2 & 0 & 3 \end{bmatrix}$ (d) $\begin{bmatrix} 3 & 0 & 0 \\ 1 & 2 & 0 \\ 0 & 0 & -4 \end{bmatrix}$

Orthogonal Diagonalization

6. Orthogonally diagonalize each of the following symmetric matrices. Give the similarity transformation.

(a) $\begin{bmatrix} 1 & 2 \\ 2 & 1 \end{bmatrix}$ (b) $\begin{bmatrix} 11 & 2 \\ 2 & 14 \end{bmatrix}$

(c) $\begin{bmatrix} 3 & 1 \\ 1 & 3 \end{bmatrix}$ (d) $\begin{bmatrix} -1 & -8 \\ -8 & 11 \end{bmatrix}$

7. Orthogonally diagonalize each of the following symmetric matrices. Give the similarity transformation.

(a) $\begin{bmatrix} 1 & 5 \\ 5 & 1 \end{bmatrix}$ (b) $\begin{bmatrix} 9 & 2 \\ 2 & 6 \end{bmatrix}$

(c) $\begin{bmatrix} 1 & 3 \\ 3 & 9 \end{bmatrix}$ (d) $\begin{bmatrix} 1.5 & -0.5 \\ -.5 & 1.5 \end{bmatrix}$

8. Orthogonally diagonalize each of the following symmetric matrices. Give the similarity transformation.

(a) $\begin{bmatrix} 0 & 2 & 0 \\ 2 & 0 & 0 \\ 0 & 0 & 1 \end{bmatrix}$ (b) $\begin{bmatrix} 9 & -3 & 3 \\ -3 & 6 & -6 \\ 3 & -6 & 6 \end{bmatrix}$

(c) $\begin{bmatrix} 1 & 2 & -2 \\ 2 & 4 & -4 \\ -2 & -4 & 4 \end{bmatrix}$

Powers of a Matrix

9. We know that if $D = C^{-1}AC$, then $A^k = CD^kC^{-1}$. Use this result to compute the following powers. (These matrices were diagonalized in Exercise 7.)

(a) $\begin{bmatrix} 1 & 5 \\ 5 & 1 \end{bmatrix}^8$ (b) $\begin{bmatrix} 9 & 2 \\ 2 & 6 \end{bmatrix}^5$

(c) $\begin{bmatrix} 1 & 3 \\ 3 & 9 \end{bmatrix}^6$ (d) $\begin{bmatrix} 1.5 & -0.5 \\ -0.5 & 1.5 \end{bmatrix}^{16}$

10. Compute the following powers. (These matrices were diagonalized in Exercise 8.)

(a) $\begin{bmatrix} 0 & 2 & 0 \\ 2 & 0 & 0 \\ 0 & 0 & 1 \end{bmatrix}^6$ (b) $\begin{bmatrix} 9 & -3 & 3 \\ -3 & 6 & -6 \\ 3 & -6 & 6 \end{bmatrix}^5$

(c) $\begin{bmatrix} 1 & 2 & -2 \\ 2 & 4 & -4 \\ -2 & -4 & 4 \end{bmatrix}^4$

Miscellaneous Results

11. Prove that if A and B are similar matrices, then

(a) $|A| = |B|$.

(b) $\text{rank}(A) = \text{rank}(B)$.

(c) $\text{Tr}(A) = \text{Tr}(B)$.

(d) A^t and B^t are similar.

(e) A is nonsingular if and only if B is nonsingular.

(f) If A is nonsingular then A^{-1} and B^{-1} are similar.

12. Prove that the eigenspaces of a symmetric matrix corresponding to distinct eigenvalues are orthogonal.

13. Let A be diagonalizable and let B be similar to A. Prove that

(a) B is diagonalizable

(b) $B + kI$ is similar to $A + kI$, for any scalar k

14. If A is similar to a diagonal matrix D we say that A is **diagonalizable** to D. Let A be diagonalizable to D. Prove that A^2 is diagonalizable to D^2 and that in general A^n is diagonalizable to D^n.

15. If A is a symmetric matrix we know that it is similar to a diagonal matrix. Is such a diagonal matrix unique? (*Hint*: Does the order of the column vectors in the transforming matrix matter?)

16. Prove that if A is diagonalizable and has eigenvalues $\lambda_1, \ldots, \lambda_n$, then $|A| = \lambda_1 \ldots \lambda_n$.

17. Matrices A and B are said to be **orthogonally similar** if there exists an orthogonal matrix C such that $B = C^{-1}AC$. Show that if A is symmetric and A and B are orthogonally similar then B is also symmetric.

18. Show that if A and B are orthogonally similar and B and C are orthogonally similar, then A and C are orthogonally similar.

*5.4 Quadratic Forms, Difference Equations, and Normal Modes

In this section we see how the techniques of diagonalization are applied in geometry, algebra, and engineering. The methods used in these examples are important in the sciences and in engineering. The reader should master them and be able to use them whenever the need arises. We first introduce the reader to the idea of a coordinate transformation. We determine the relationship between two rectangular coordinate systems with the same origin.

Rotation of Coordinates

Let a rectangular xy-coordinate system be given and a second rectangular coordinate system $x'y'$ be obtained from it through a rotation through an angle θ about the origin. See Figure 5.4.

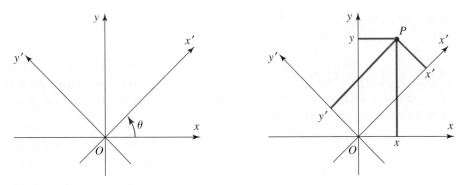

Figure 5.4

An arbitrary point P in the plane has coordinates $\begin{bmatrix} x \\ y \end{bmatrix}$ relative to the xy-coordinate system and coordinates $\begin{bmatrix} x' \\ y' \end{bmatrix}$ relative to the $x'y'$-coordinate system. It can be shown (see Exercise 8 following) that

$$x' = x \cos\theta + y \sin\theta$$
$$y' = -x \sin\theta + y \cos\theta$$

These equations define a **coordinate transformation**, a rotation in this case, between the two coordinate systems. The equations can be written in matrix form.

$$\begin{bmatrix} x' \\ y' \end{bmatrix} = \begin{bmatrix} \cos\theta & \sin\theta \\ -\sin\theta & \cos\theta \end{bmatrix} \begin{bmatrix} x \\ y \end{bmatrix}$$

For example, consider a rotation through an angle of 45°. Suppose P is the point $\begin{bmatrix} 1 \\ 3 \end{bmatrix}$ relative to the xy-coordinate system. Let us find the coordinates of P relative to the $x'y'$-coordinate system. Since $\sin 45° = \dfrac{1}{\sqrt{2}}$ and $\cos 45° = \dfrac{1}{\sqrt{2}}$, we get

$$\begin{bmatrix} x' \\ y' \end{bmatrix} = \begin{bmatrix} \dfrac{1}{\sqrt{2}} & \dfrac{1}{\sqrt{2}} \\ -\dfrac{1}{\sqrt{2}} & \dfrac{1}{\sqrt{2}} \end{bmatrix} \begin{bmatrix} 1 \\ 3 \end{bmatrix} = \begin{bmatrix} 2\sqrt{2} \\ \sqrt{2} \end{bmatrix}$$

Note the similarity between this coordinate transformation and the transformation that defines a rotation in a plane through an angle θ. The similarity arises from the fact that one can interpret a transformation either as moving points in a fixed coordinate system or as changing the coordinate system around fixed points.

Quadratic Forms

The algebraic expression

$$ax^2 + bxy + cy^2$$

where a, b, and c are constants, is called a **quadratic form**. Quadratic forms play an important role in geometry. The reader can verify by multiplying out the matrices that this quadratic form can be written as follows.

$$\begin{bmatrix} x & y \end{bmatrix} \begin{bmatrix} a & \frac{b}{2} \\ \frac{b}{2} & c \end{bmatrix} \begin{bmatrix} x \\ y \end{bmatrix}$$

Let $\mathbf{x} = \begin{bmatrix} x \\ y \end{bmatrix}$ and $A = \begin{bmatrix} a & \frac{b}{2} \\ \frac{b}{2} & c \end{bmatrix}$. We can write the quadratic form

$$\mathbf{x}^t A \mathbf{x}$$

The symmetric matrix A associated with the quadratic from is called the **matrix of the quadratic form**.

EXAMPLE 1 Write the following quadratic form in terms of matrices.

$$5x^2 + 6xy - 4y^2$$

SOLUTION

On comparison with the standard form $ax^2 + bxy + cy^2$, we have

$$a = 5, \quad b = 6, \quad c = -4$$

The matrix form of the quadratic form is thus

$$\begin{bmatrix} x & y \end{bmatrix} \begin{bmatrix} 5 & 3 \\ 3 & -4 \end{bmatrix} \begin{bmatrix} x \\ y \end{bmatrix}$$

Let us develop techniques for understanding and graphing equations of the form

$$ax^2 + bxy + cy^2 + d = 0$$

This equation includes the quadratic form $ax^2 + bxy + cy^2$. We can write the equation in matrix form

$$\mathbf{x}^t A \mathbf{x} + d = 0$$

Since A is a symmetric matrix there exists an orthogonal matrix C such that $C^t A C$ is a diagonal matrix D. Further, since C is orthogonal $C^{-1} = C^t$, and thus $C^t C = I$. This identity enables us to introduce C^tC into the equation at various locations under the guise of the identity matrix I, to diagonalize A, leading to a standard recognizable form of the equation. We get

$$\mathbf{x}^t A\mathbf{x} + d = 0,$$
$$\mathbf{x}^t(CC^t)A(CC^t)\mathbf{x} + d = 0,$$
$$\mathbf{x}^tC(C^tAC)C^t\mathbf{x} + d = 0,$$
$$\mathbf{x}^tCDC^t\mathbf{x} + d = 0,$$
$$(C^t\mathbf{x})^tDC^t\mathbf{x} + d = 0$$

The matrix C^t defines a coordinate transformation $\mathbf{x}' = C^t\mathbf{x}$ from the xy-coordinate system to a new $x'y'$-coordinate system. In the $x'y'$-coordinate system the equation becomes

$$(\mathbf{x}')^tD\mathbf{x}' + d = 0$$

Let $D = \begin{bmatrix} p & 0 \\ 0 & q \end{bmatrix}$ and $x' = \begin{bmatrix} x' \\ y' \end{bmatrix}$. Thus

$$[x' \ \ y']\begin{bmatrix} p & 0 \\ 0 & q \end{bmatrix}\begin{bmatrix} x' \\ y' \end{bmatrix} + d = 0,$$
$$p(x')^2 + q(y')^2 + d = 0$$

If $d \neq 0$, we get

$$\frac{(x')^2}{-d/p} + \frac{(y')^2}{-d/q} = 1$$

This is the equation of a **conic**. Its graph in the $x'y'$-coordinate system can now be sketched, leading to its graph in the original xy-system. The following example illustrates the method.

EXAMPLE 2 Analyze the following equation. Sketch its graph.

$$6x^2 + 4xy + 9y^2 - 20 = 0$$

SOLUTION

This equation includes the quadratic form $6x^2 + 4xy + 9y^2$. Write the equation in matrix form as follows:

$$[x \ \ y]\begin{bmatrix} 6 & 2 \\ 2 & 9 \end{bmatrix}\begin{bmatrix} x \\ y \end{bmatrix} - 20 = 0$$

The eigenvalues and corresponding eigenvectors of the matrix are found. They are

$$\lambda_1 = 10, \mathbf{v}_1 = r\begin{bmatrix} 1 \\ 2 \end{bmatrix}; \quad \lambda_2 = 5, \mathbf{v}_2 = s\begin{bmatrix} -2 \\ 1 \end{bmatrix}$$

Normalize these eigenvectors and write them as the columns of an orthogonal matrix C. We get

$$C = \begin{bmatrix} \dfrac{1}{\sqrt{5}} & \dfrac{-2}{\sqrt{5}} \\ \dfrac{2}{\sqrt{5}} & \dfrac{1}{\sqrt{5}} \end{bmatrix}$$

We now know that the equation can be transformed into the form

$$\begin{bmatrix} x' & y' \end{bmatrix} \begin{bmatrix} 10 & 0 \\ 0 & 5 \end{bmatrix} \begin{bmatrix} x' \\ y' \end{bmatrix} - 20 = 0$$

under the coordinate transformation

$$\begin{bmatrix} x' \\ y' \end{bmatrix} = \begin{bmatrix} \dfrac{1}{\sqrt{5}} & \dfrac{2}{\sqrt{5}} \\ \dfrac{-2}{\sqrt{5}} & \dfrac{1}{\sqrt{5}} \end{bmatrix} \begin{bmatrix} x \\ y \end{bmatrix}$$

On multiplying the matrices and rearranging, the equation in the $x'y'$-coordinate system is

$$\frac{(x')^2}{2} + \frac{(y')^2}{4} = 1$$

We recognize this as being the equation of an ellipse in the $x'y'$-coordinate system. The length of the semimajor axis is 2 and the length of the semiminor axis is $\sqrt{2}$. See Figure 5.5(a). It remains to locate the $x'y'$-coordinate system relative to the xy-coordinate system. The preceding coordinate transformation can be seen to be a rotation of coordinates with $\cos\theta = 1/\sqrt{5}$ and $\sin\theta = 2/\sqrt{5}$; that is, $\theta = 1.107148718$ radians; or $\theta = 63.43494882$ degrees. The coordinate rotation is approximately $63.4°$. The graph is now known in the xy-coordinate system. See Figure 5.5(b).

[Such an orthogonal transformation will in fact always define a rotation.]

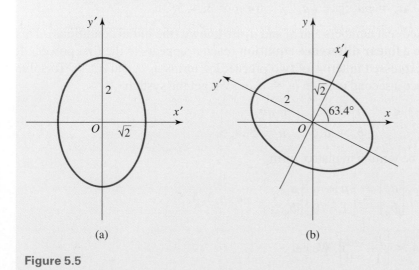

(a)   (b)

Figure 5.5

Difference Equations

Let $a_1, a_2, a_3, \ldots$ be a sequence of real numbers. Such a sequence may be defined by giving its nth term. For example, suppose

$$a_n = n^2 + 1$$

Letting $n = 1, 2, 3, \ldots$ we get the terms of the sequence,

$$2, 5, 10, 17, \ldots$$

Furthermore, any specific term of the sequence can be found. For example, if we want a_{20} then we let $n = 20$ in this expression for a_n and we get

$$a_{20} = (20)^2 + 1 = 401$$

When sequences arise in applications they are often initially defined by a relationship between consecutive terms, with some initial terms known, rather than defined by the nth term. For example a sequence might be defined by the relationship

$$a_n = 2a_{n-1} + 3a_{n-2}, \quad \text{for } n = 3, 4, 5, \ldots$$

with $a_1 = 0$ and $a_2 = 1$. Such an equation is called a **difference equation** (or **recurrence relation**), and the given terms of the sequence are called **initial conditions**. Further terms of the sequence can be found from the difference equation and initial conditions. For example, on letting $n = 3$, 4, and 5 we get

$$a_3 = 2a_2 + 3a_1 = 2(1) + 3(0) = 2$$
$$a_4 = 2a_3 + 3a_2 = 2(2) + 3(1) = 7$$
$$a_5 = 2a_4 + 3a_3 = 2(7) + 3(2) = 20$$

The sequence is 0, 1, 2, 7, 20, ...

However, if one wants to find a specific term such as the 20th term of this sequence, this method of using the difference equation to first find all the preceeding terms is impractical. We need an expression for the nth term of the sequence. The expression for the nth term is called the **solution to the difference equation**. We now illustrate how the tools of linear algebra can be used to solve certain linear difference equations.

Consider the difference equation

$$a_n = pa_{n-1} + qa_{n-2}, \quad \text{for } n = 3, 4, 5, \ldots$$

where p and q are fixed real numbers and a_1 and a_2 are known (the initial conditions). This equation is said to be a **linear difference equation** (each a_i appears to the first power). It is of **order** 2 (a_n is expressed in terms of two preceeding terms a_{n-1} and a_{n-2}). To solve this equation introduce a second relation $b_n = a_{n-1}$. We get the system

$$a_n = pa_{n-1} + qb_{n-1}$$
$$b_n = a_{n-1}, \quad n = 3, 4, 5, \ldots$$

These equations can be written in matrix form

$$\begin{bmatrix} a_n \\ b_n \end{bmatrix} = \begin{bmatrix} p & q \\ 1 & 0 \end{bmatrix} \begin{bmatrix} a_{n-1} \\ b_{n-1} \end{bmatrix}, \quad n = 3, 4, 5, \ldots$$

Let $X_n = \begin{bmatrix} a_n \\ b_n \end{bmatrix}$ and $A = \begin{bmatrix} p & q \\ 1 & 0 \end{bmatrix}$. We get

$$X_n = AX_{n-1}, \quad n = 3, 4, 5, \ldots$$

Thus

$$X_n = AX_{n-1} = A^2 X_{n-2} = A^3 X_{n-3} = \cdots = A^{n-2} X_2, \quad \text{where}$$

$$X_2 = \begin{bmatrix} a_2 \\ b_2 \end{bmatrix} = \begin{bmatrix} a_2 \\ a_1 \end{bmatrix}$$

In most applications A has distinct eigenvalues λ_1 and λ_2. It thus has two linearly independent eigenvectors, so can be diagonalized using a similarity transformation. Let C be a matrix whose columns are linearly independent eigenvectors of A and let

$$D = C^{-1}AC = \begin{bmatrix} \lambda_1 & 0 \\ 0 & \lambda_2 \end{bmatrix}$$

Then

$$A^{n-2} = (CDC^{-1})^{n-2} = (CDC^{-1})(CDC^{-1})\ldots(CDC^{-1}) = CD^{n-2}C^{-1}$$

$$= C\begin{bmatrix} \lambda_1 & 0 \\ 0 & \lambda_2 \end{bmatrix}^{n-2} C^{-1} = C\begin{bmatrix} (\lambda_1)^{n-2} & 0 \\ 0 & (\lambda_2)^{n-2} \end{bmatrix}C^{-1}$$

This gives

$$X_n = C\begin{bmatrix} (\lambda_1)^{n-2} & 0 \\ 0 & (\lambda_2)^{n-2} \end{bmatrix}C^{-1}X_2$$

We now illustrate the application of this result.

EXAMPLE 3 Solve the difference equation

$$a_n = 2a_{n-1} + 3a_{n-2}, \quad \text{for } n = 3, 4, 5, \ldots$$

with initial conditions $a_1 = 0$, $a_2 = 1$. Use the solution to determine a_{15}.

SOLUTION

Construct the system

$$a_n = 2a_{n-1} + 3b_{n-1}$$
$$b_n = a_{n-1}, \quad n = 3, 4, 5, \ldots$$

Write this system in the matrix form

$$\begin{bmatrix} a_n \\ b_n \end{bmatrix} = \begin{bmatrix} 2 & 3 \\ 1 & 0 \end{bmatrix}\begin{bmatrix} a_{n-1} \\ b_{n-1} \end{bmatrix}, \quad n = 3, 4, 5, \ldots$$

Eigenvalues and corresponding eigenvectors of the matrix $\begin{bmatrix} 2 & 3 \\ 1 & 0 \end{bmatrix}$ are

$$\lambda_1 = -1, \mathbf{v}_1 = \begin{bmatrix} 1 \\ -1 \end{bmatrix}; \quad \lambda_2 = 3, \mathbf{v}_2 = \begin{bmatrix} 3 \\ 1 \end{bmatrix}.$$

Let $C = \begin{bmatrix} 1 & 3 \\ -1 & 1 \end{bmatrix}$. Then $C^{-1} = \dfrac{1}{4}\begin{bmatrix} 1 & -3 \\ 1 & 1 \end{bmatrix}$. We get

$$\begin{bmatrix} a_n \\ b_n \end{bmatrix} = C\begin{bmatrix} (\lambda_1)^{n-2} & 0 \\ 0 & (\lambda_2)^{n-2} \end{bmatrix}C^{-1}\begin{bmatrix} a_2 \\ b_2 \end{bmatrix}$$

$$= \frac{1}{4}\begin{bmatrix} 1 & 3 \\ -1 & 1 \end{bmatrix}\begin{bmatrix} (-1)^{n-2} & 0 \\ 0 & (3)^{n-2} \end{bmatrix}\begin{bmatrix} 1 & -3 \\ 1 & 1 \end{bmatrix}\begin{bmatrix} 1 \\ 0 \end{bmatrix}, \text{ since } b_2 = a_1 = 0$$

$$= \frac{1}{4}\begin{bmatrix} (-1)^{n-2} + 3(3)^{n-2} \\ (-1)(-1)^{n-2} + (3)^{n-2} \end{bmatrix} = \frac{1}{4}\begin{bmatrix} (-1)^{n-2} + (3)^{n-1} \\ (-1)^{n-1} + (3)^{n-2} \end{bmatrix}$$

Thus the solution is

$$a_n = \tfrac{1}{4}[(-1)^{n-2} + (3)^{n-1}], \quad n = 3, 4, 5, \ldots$$

This solution can be checked. It gives $a_3 = 2$, $a_4 = 7$, $a_5 = 20$, agreeing with the values obtained by substituting $n = 3, 4, 5$ into the difference equation.

Letting $n = 15$ we get

$$a_{15} = \tfrac{1}{4}[(-1)^{15-2} + (3)^{15-1}] = \tfrac{1}{4}[-1 + 4{,}782{,}969] = 1{,}195{,}742$$

■

Fibonacci Sequence*

The sequence of numbers defined by the difference equation

$$a_n = a_{n-1} + a_{n-2}, \quad \text{for } n = 3, 4, 5, \ldots$$

with initial conditions $a_1 = 1$, $a_2 = 1$ is called a **Fibonacci sequence**. Each term in this sequence is the sum of the two preceding terms.

This sequence was developed by the Italian mathematician Leonardo Fibonacci in the 13th century to describe the growth in population of rabbits. Fibonacci assumed that a pair of rabbits produces a pair of young rabbits every month and that newborn rabbits become adults in two months. At this time they then produce another pair. Starting with an adult pair, how many adult pairs will be in the colony after the first, second, and third months, and so on? Fibonacci showed by counting that the first seven terms of the sequence are

$$1, 1, 2, 3, 5, 8, 13$$

He then made the observation that each term in the sequence was the sum of the two preceding terms, leading to the given difference equation. Fibonacci used the difference equation to compute the number of adult pairs in the colony at any time. We ask the reader to solve the Fibonacci difference equation in the exercises that follow.

Since the time of Fibonacci scientists have used this sequence in fields such as botany, architecture, archaeology, and sociology. It is a puzzle why the sequence arises the way it does in some of these fields. For example, biologists have discovered that the arrangements of leaves around the stems of plants often follow a Fibonacci pattern. The leaves of roses and cabbages, and the scales of spruce trees, also follow a Fibonacci pattern. There are 5 rose leaves and 8 cabbage leaves in a full circle of the stems. There are 21 scales of spruce in a full circle of a tree.

The ratio of consecutive terms of the Fibonacci sequence is interesting. It can be shown that a_n/a_{n-1} approaches the number $(1 + \sqrt{5})/2$ as n increases. (See the following exercises.) This irrational number, $1.6180339885 \ldots$, is called the **golden ratio**. Psychologists have found that rectangles whose length-to-width ratios are near the golden ratio are the most pleasing. Artists such as Leornado da Vinci, Mondrian, and Seurat have made great use of the golden ratio in their works. The dimensions of the Parthenon in Athens are based on the golden ratio.

---

*Leonardo Fibonacci (about 1170–1250) was the first great mathematician of the West. He grew up and lived most of his life in Pisa, Italy. His father was secretary for commerce in Pisa and Fibonacci represented him on business trips to Egypt, Syria, and Greece. He learned the Hindu-Arabic method of numeration and calculation and brought it back to Italy. While Fibonacci was primarily interested in teaching finance and the practical aspects of mathematics, he also taught algebra and geometry to those who wanted to delve into more abstract ideas.

Normal Modes of Oscillating Systems

Consider a horizontal string AB of length $4a$ and negligible mass, loaded with three particles, each of mass m. Let the masses be located at fixed distances a, $2a$, and $3a$ from A. The particles are displaced slightly from their equilibrium position and released. Let us analyze the subsequent motion, assuming that it all takes place in a vertical plane through A and B. See Figure 5.6.

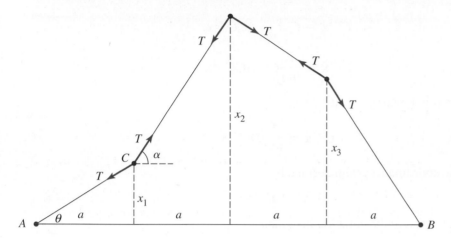

Figure 5.6

Let the vertical displacements of the particles at any instant during the subsequent motion be x_1, x_2, and x_3, as shown in Figure 5.6. Let T be the tension in the string. Consider the motion of the particle at C. The resultant force on this particle in a vertical direction is $T \sin\alpha - T \sin\theta$. If we assume that the displacements are small, the motion of each particle can be assumed to be vertical and the tension can be assumed to be unaltered throughout the motion. Since the angles are small, $\sin\alpha$ can be taken to equal $\tan\alpha$, and $\sin\theta$ to equal $\tan\theta$. Thus the resultant force at C is

$$T \tan\alpha - T \tan\theta = T\frac{(x_2 - x_1)}{a} - T\frac{x_1}{a}$$

$$= \frac{-2Tx_1}{a} + \frac{Tx_2}{a}$$

Applying **Newton's second law of motion** (force = mass × acceleration), we find that the motion of the first particle is described by the equation

$$m\ddot{x}_1 = \frac{-2Tx_1}{a} + \frac{Tx_2}{a}$$

Similarly, the motions of the other two particles are described by the equations

$$m\ddot{x}_2 = \frac{Tx_1}{a} - \frac{2Tx_2}{a} + \frac{Tx_3}{a} \quad \text{and} \quad m\ddot{x}_3 = \frac{Tx_2}{a} - \frac{2Tx_3}{a}$$

These three equations can be combined into the single matrix equation

$$\begin{bmatrix} \ddot{x}_1 \\ \ddot{x}_2 \\ \ddot{x}_3 \end{bmatrix} = \frac{T}{ma} \begin{bmatrix} -2 & 1 & 0 \\ 1 & -2 & 1 \\ 0 & 1 & -2 \end{bmatrix} \begin{bmatrix} x_1 \\ x_2 \\ x_3 \end{bmatrix}$$

Write this equation in matrix form

$$\ddot{\mathbf{x}} = \frac{T}{ma}A\mathbf{x}$$

Observe that A is a symmetric matrix. It can thus be diagonalized. Let C be a matrix consisting of eigenvectors of A as column vectors that can be used to diagonalize A. We can write the preceding equation

$$C^{-1}\ddot{\mathbf{x}} = C^{-1}\frac{T}{ma}ACC^{-1}\mathbf{x},$$

$$C^{-1}\ddot{\mathbf{x}} = \frac{T}{ma}C^{-1}ACC^{-1}\mathbf{x}$$

Let D be the diagonal matrix $C^{-1}AC$. Thus

$$C^{-1}\ddot{\mathbf{x}} = \frac{T}{ma}DC^{-1}\mathbf{x}$$

Introduce a new coordinate system defined by

$$\mathbf{x}' = C^{-1}\mathbf{x}$$

In this coordinate system the equation becomes

$$\ddot{\mathbf{x}}' = \frac{T}{ma}D\mathbf{x}' \tag{1}$$

This matrix equation can be solved for $\mathbf{x}'$. The coordinate relationship

$$\mathbf{x} = C\mathbf{x}'$$

then leads to the solution in terms of the original coordinate system.

Having discussed the general method, let us now implement it for our specific equation. The eigenvalues and corresponding eigenvectors of the symmetric matrix A can be found to be

$$\lambda_1 = -2, \mathbf{v}_1 = r\begin{bmatrix} 1 \\ 0 \\ -1 \end{bmatrix}; \quad \lambda_2 = -2-\sqrt{2}, \mathbf{v}_2 = s\begin{bmatrix} 1 \\ -\sqrt{2} \\ 1 \end{bmatrix};$$

$$\lambda_3 = -2+\sqrt{2}, \mathbf{v}_3 = t\begin{bmatrix} 1 \\ \sqrt{2} \\ 1 \end{bmatrix}$$

Thus let

$$C = \begin{bmatrix} 1 & 1 & 1 \\ 0 & -\sqrt{2} & \sqrt{2} \\ -1 & 1 & 1 \end{bmatrix} \quad \text{and} \quad D = \begin{bmatrix} -2 & 0 & 0 \\ 0 & -2-\sqrt{2} & 0 \\ 0 & 0 & -2+\sqrt{2} \end{bmatrix}$$

The equations of motion (Eq. 1) are

$$\begin{bmatrix} \ddot{x}'_1 \\ \ddot{x}'_2 \\ \ddot{x}'_3 \end{bmatrix} = \frac{T}{ma}\begin{bmatrix} -2 & 0 & 0 \\ 0 & -2-\sqrt{2} & 0 \\ 0 & 0 & -2+\sqrt{2} \end{bmatrix}\begin{bmatrix} x'_1 \\ x'_2 \\ x'_3 \end{bmatrix}$$

This matrix equation represents the following system of three differential equations, each describing a simple harmonic motion.

$$\ddot{x}'_1 = \frac{-2T}{ma}x'_1, \quad \ddot{x}'_2 = \frac{(-2 - \sqrt{2})T}{ma}x'_2, \quad \ddot{x}'_3 = \frac{(-2 + \sqrt{2})T}{ma}x'_3$$

These equations have the following solutions.

$$x'_1 = b_1\cos\left(\sqrt{\frac{2T}{ma}}t + \gamma_1\right),$$

$$x'_2 = b_2\cos\left(\sqrt{\frac{(2 + \sqrt{2})T}{ma}}t + \gamma_2\right),$$

$$x'_3 = b_3\cos\left(\sqrt{\frac{(2 - \sqrt{2})T}{ma}}t + \gamma_3\right)$$

Here $b_1, b_2, b_3, \gamma_1, \gamma_2, \gamma_3$ are constants of integration that depend upon the configuration at time $t = 0$. They are not all independent. In standard interpretation of simple harmonic motion, b_1, b_2, b_3 are amplitudes and $\gamma_1, \gamma_2, \gamma_3$ are phases.

These special coordinates x_1', x_2', and x_3', in which the motion is easiest to describe are called the **normal coordinates** of the motion. The motion in terms of the original coordinates can be obtained by using the relationship

$$\mathbf{x} = C\mathbf{x}'$$

$$\begin{bmatrix} x_1 \\ x_2 \\ x_3 \end{bmatrix} = \begin{bmatrix} 1 & 1 & 1 \\ 0 & -\sqrt{2} & \sqrt{2} \\ -1 & 1 & 1 \end{bmatrix}\begin{bmatrix} x'_1 \\ x'_2 \\ x'_3 \end{bmatrix},$$

$$= x'_1\begin{bmatrix} 1 \\ 0 \\ -1 \end{bmatrix} + x'_2\begin{bmatrix} 1 \\ -\sqrt{2} \\ 1 \end{bmatrix} + x'_3\begin{bmatrix} 1 \\ \sqrt{2} \\ 1 \end{bmatrix}$$

Substituting for x'_1, x'_2, and x'_3 gives

$$\begin{bmatrix} x_1 \\ x_2 \\ x_3 \end{bmatrix} = b_1\cos\left(\sqrt{\frac{2T}{ma}}t + \gamma_1\right)\begin{bmatrix} 1 \\ 0 \\ -1 \end{bmatrix} + b_2\cos\left(\sqrt{\frac{(2 + \sqrt{2})T}{ma}}t + \gamma_2\right)\begin{bmatrix} 1 \\ -\sqrt{2} \\ 1 \end{bmatrix}$$

$$+ b_3\cos\left(\sqrt{\frac{(2 - \sqrt{2})T}{ma}}t + \gamma_3\right)\begin{bmatrix} 1 \\ \sqrt{2} \\ 1 \end{bmatrix}$$

The motion can thus be interpreted as a combination of the following three motions, or **modes**.

Mode 1:

$$\begin{bmatrix} x_1 \\ x_2 \\ x_3 \end{bmatrix} = \cos\left(\sqrt{\frac{2T}{ma}}t + \gamma_1\right)\begin{bmatrix} 1 \\ 0 \\ -1 \end{bmatrix} \text{ (Note that } x_2 = 0 \text{ and } x_3 = -x_1.)$$

Mode 2:

$$\begin{bmatrix} x_1 \\ x_2 \\ x_3 \end{bmatrix} = \cos\left(\sqrt{\frac{(2 + \sqrt{2})T}{ma}}t + \gamma_2\right)\begin{bmatrix} 1 \\ -\sqrt{2} \\ 1 \end{bmatrix} \; (x_2 = -\sqrt{2}x_1 \text{ and } x_3 = x_1)$$

Mode 3:

$$
\begin{bmatrix} x_1 \\ x_2 \\ x_3 \end{bmatrix} = \cos\left(\sqrt{\frac{(2 - \sqrt{2})T}{ma}}\, t + \gamma_3\right) \begin{bmatrix} 1 \\ \sqrt{2} \\ 1 \end{bmatrix} \quad (x_2 = \sqrt{2}x_1 \text{ and } x_3 = x_1)
$$

The actual combination will be a linear combination determined by b_1, b_2, and b_3 that will depend on the initial configuration. Each of these motions is a simple harmonic motion of a particle. These modes, called the **normal modes of oscillation**, are illustrated in Figure 5.7.

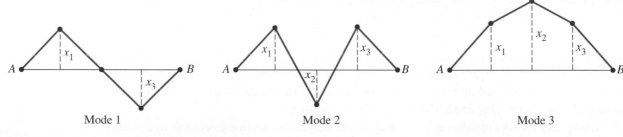

Mode 1   Mode 2   Mode 3

Figure 5.7

EXERCISE SET 5.4

1. Express each of the following quadratic forms in terms of matrices.

(a) $x^2 + 4xy + 2y^2$ (b) $3x^2 + 2xy - 4y^2$
(c) $7x^2 - 6xy - y^2$ (d) $2x^2 + 5xy + 3y^2$
(e) $-3x^2 - 7xy + 4y^2$ (f) $5x^2 + 3xy - 2y^2$

2. Sketch the graph of each of the following equations.

(a) $11x^2 + 4xy + 14y^2 - 60 = 0$
(b) $3x^2 + 2xy + 3y^2 - 12 = 0$
(c) $x^2 - 6xy + y^2 - 8 = 0$
(d) $4x^2 + 4xy + 4y^2 - 5 = 0$

3. Solve the following difference equations and use the solutions to determine the given terms.

(a) $a_n = a_{n-1} + 2a_{n-2}$, $a_1 = 1$, $a_2 = 2$. Find a_{10}.
(b) $a_n = 2a_{n-1} + 3a_{n-2}$, $a_1 = 1$, $a_2 = 3$. Find a_9.
(c) $a_n = 3a_{n-1} + 4a_{n-2}$, $a_1 = 1$, $a_2 = -1$. Find a_{12}.

4. The Fibonacci sequence is defined by the difference equation $a_n = a_{n-1} + a_{n-2}$, with initial conditions $a_1 = 1$, $a_2 = 1$. Solve this difference equation. Show that $\dfrac{a_n}{a_{n-1}}$ approaches the number $\dfrac{1 + \sqrt{5}}{2}$ as n increases, arriving at the *golden ratio*.

5. Determine the normal modes of oscillation of the system in Figure 5.8 when it is displaced slightly from equilibrium.

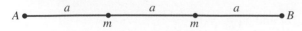

Figure 5.8

6. The motion of a weight attached to a spring is governed by **Hooke's Law**:

$$\text{tension} = k \times \text{extension},$$

where k is a constant of the spring.

Consider the oscillations of the spring described in Figure 5.9. x_1 and x_2 are the displacements of weights of masses m_1 and m_2 at any instant. The extensions of the two springs at that instant are x_1 and $x_2 - x_1$. Application of Hooke's Law gives the following equations of motion

$$m_1\ddot{x}_1 = -k_1 x_1 + k_2(x_2 - x_1) \quad \text{and}$$
$$m_2\ddot{x}_2 = -k_2(x_2 - x_1)$$

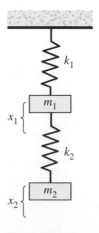

Figure 5.9

The general motion can be analyzed in terms of normal modes as in the example of this section. Analyze the motion when $m_1 = m_2 = M$, $k_1 = 3$, and $k_2 = 2$.

7. Hooke's Law also applies to forces in the extended springs of the system shown in Figure 5.10. x_1 and x_2 are the displacements of balls, each of mass m, at any instant. Application of Hooke's Law gives

$$m\ddot{x}_1 = -k_1 x_1 + k_2(x_2 - x_1) \quad \text{and}$$
$$m\ddot{x}_2 = -k_2(x_2 - x_1)k_3 x_2$$

Analyze the motion in terms of normal modes if the constants of the springs are $k_1 = 1$, $k_2 = 2$, and $k_3 = 3$.

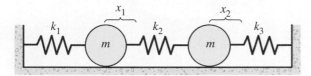

Figure 5.10

8. Let a rectangular xy-coordinate system be given and a second rectangular coordinate system $x'y'$ be obtained from it through a rotation through an angle θ about the origin. An arbitrary point P in the plane has coordinates $\begin{bmatrix} x \\ y \end{bmatrix}$ relative to the xy-coordinate system and coordinates $\begin{bmatrix} x' \\ y' \end{bmatrix}$ relative to the $x'y'$-coordinate system. Use trigonometry to prove that the coordinate transformation is

$$x' = x\cos\theta + y\sin\theta$$
$$y' = x\sin\theta + y\cos\theta$$

CHAPTER 5 REVIEW EXERCISES

1. Determine the characteristic polynomial, eigenvalues, and corresponding eigenspaces of the matrix

$$\begin{bmatrix} 5 & -7 & 7 \\ 4 & -3 & 4 \\ 4 & -1 & 2 \end{bmatrix}$$

2. Let A be an invertible matrix. Show that the eigenvalues of A^{-1} are the inverses of the eigenvalues of A. Prove that A and A^{-1} have the same eigenvectors.

3. Let λ be an eigenvalue of a matrix A with corresponding eigenvector $\mathbf{x}$. If k is a scalar, show that $\lambda - k$ is an eigenvalue of $A - kI$ and that $\mathbf{x}$ is a corresponding eigenvector.

4. Transform the matrix $A = \begin{bmatrix} 4 & -2 \\ 1 & 1 \end{bmatrix}$, into a matrix B using the similarity transformation $C^{-1}AC$, with matrix $C = \begin{bmatrix} 2 & 1 \\ 1 & 1 \end{bmatrix}$.

5. Diagonalize the matrix $\begin{bmatrix} 1 & 1 \\ -2 & 4 \end{bmatrix}$ using a similarity transformation. Give the transformation.

6. Diagonalize the symmetric matrix $\begin{bmatrix} 7 & -2 & 1 \\ -2 & 10 & -2 \\ 1 & -2 & 7 \end{bmatrix}$. Give the similarity transformation.

7. Let the matrix $A = \begin{bmatrix} a & b \\ b & c \end{bmatrix}$ represent an arbitrary 2×2 symmetric matrix. Prove that the characteristic equation of A has real roots for all values of a, b, and c.

8. Show that if A is a symmetric matrix having only one eigenvalue λ then $A = \lambda I$.

9. Sketch the graph of $-x^2 - 16xy + 11y^2 - 30 = 0$.

10. Solve the difference equation $a_n = 4a_{n-1} + 5a_{n-1}$, $a_1 = 3$, $a_2 = 2$ and use the solution to find a_{12}.

The twin bridges of Grand Island, New York. This photo shows the bridges as winter ice floats down the Niagara River toward Niagara Falls. The twin Truss arch bridges were built separately. The North Grand Island Bridge carries Interstate 190 across the Niagara River between Grand Island, New York, and Niagara Falls, New York, while the South Grand Island Bridge carries Interstate 190 between Tonawanda, New York, and Grand Island, New York.

Linear Transformations

A function, or transformation, is a rule that assigns to each element of a set a unique element of another set. Transformations are used in many areas of mathematics and are important in applications for describing the dependency of one variable upon another. We are especially interested in linear transformations. These are transformations that preserve the mathematical structure of a vector space.

Linear transformations are used by scientists and engineers. We shall see the fundamental role of linear transformations in computer graphics. We shall discuss how they are used to produce fractals. Not only are fractals interesting from the graphics point of view—they are spectacular pictures—but they provide ways of transmitting large amounts of data in compressed form. We shall discuss how linear transformations are used to view computer models of aircraft from different angles in aerospace engineering.

6.1 Matrix Transformations, Rotations, and Dilations

The reader will be familiar with functions such as $f(x) = 3x^2 + 4$. The set of allowable x values is called the *domain of the function*. The domain is often the set of real numbers, as here. When $x = 2$, for example, we see that $f(2) = 16$. We say that the image of 2 is 16. We extend these ideas to functions between vector spaces. We usually use the term *transformation* rather than *function* in linear algebra.

For example, consider the transformation T of $\mathbf{R}^3$ into $\mathbf{R}^2$ defined by

$$T(x, y, z) = (2x, y - z)$$

The *domain* of T is $\mathbf{R}^3$ and we say that the *codomain* is $\mathbf{R}^2$. The image of a vector such as $(1, 4, -2)$ can be found by letting $x = 1$, $y = 4$, and $z = -2$ in this equation for T. We get $T(1, 4, -2) = (2, 6)$. The *image* of $(1, 4, -2)$ is $(2, 6)$.

We shall often find it convenient to write vectors in column form when discussing transformations. The preceding transformation can also be written

$$T\left(\begin{bmatrix} x \\ y \\ z \end{bmatrix}\right) = \begin{bmatrix} 2x \\ y - z \end{bmatrix}, \text{ and the image of } \begin{bmatrix} 1 \\ 4 \\ -2 \end{bmatrix} \text{ is } \begin{bmatrix} 2 \\ 6 \end{bmatrix}.$$

> **DEFINITION**   A *transformation* T of $\mathbf{R}^n$ into $\mathbf{R}^m$ is a rule that assigns to each vector $\mathbf{u}$ in $\mathbf{R}^n$ a unique vector $\mathbf{v}$ in $\mathbf{R}^m$. $\mathbf{R}^n$ is called the *domain* of T and $\mathbf{R}^m$ is the *codomain*. We write $T(\mathbf{u}) = \mathbf{v}$; $\mathbf{v}$ is the *image* of $\mathbf{u}$ under T. The term *mapping* is also used for a transformation.

We now introduce a number of useful geometric transformations and find that they can be described by matrices.

Dilation and Contraction

Consider the transformation $T\left(\begin{bmatrix} x \\ y \end{bmatrix}\right) = r\begin{bmatrix} x \\ y \end{bmatrix}$, where r is a positive scalar. T maps every point in $\mathbf{R}^2$ into a point r times as far from the origin. If $r > 1$, T moves points away from the origin and is called a *dilation of factor r*. If $0 < r < 1$, T moves points closer to the origin and is then called a *contraction of factor r*. See Figure 6.1. This equation can be written in the following useful matrix form.

$$T\left(\begin{bmatrix} x \\ y \end{bmatrix}\right) = \begin{bmatrix} r & 0 \\ 0 & r \end{bmatrix}\begin{bmatrix} x \\ y \end{bmatrix}$$

For example, when $r = 3$, we see in the following equation that an image is three times as far from the origin. When $r = 1/2$ an image is half the distance from the origin.

$$T\left(\begin{bmatrix} x \\ y \end{bmatrix}\right) = \begin{bmatrix} 3 & 0 \\ 0 & 3 \end{bmatrix}\begin{bmatrix} x \\ y \end{bmatrix} = 3\begin{bmatrix} x \\ y \end{bmatrix}, \qquad T\left(\begin{bmatrix} x \\ y \end{bmatrix}\right) = \begin{bmatrix} \frac{1}{2} & 0 \\ 0 & \frac{1}{2} \end{bmatrix}\begin{bmatrix} x \\ y \end{bmatrix} = \frac{1}{2}\begin{bmatrix} x \\ y \end{bmatrix}$$

Reflection

Consider the transformation $T\left(\begin{bmatrix} x \\ y \end{bmatrix}\right) = \begin{bmatrix} x \\ -y \end{bmatrix}$. T maps every point in $\mathbf{R}^2$ into its mirror image in the x-axis. T is called a *reflection*. See Figure 6.2. This equation can be written in the following matrix form.

$$T\left(\begin{bmatrix} x \\ y \end{bmatrix}\right) = \begin{bmatrix} 1 & 0 \\ 0 & -1 \end{bmatrix}\begin{bmatrix} x \\ y \end{bmatrix}$$

For example, the image of $\begin{bmatrix} 3 \\ 2 \end{bmatrix}$ under this reflection is $\begin{bmatrix} 1 & 0 \\ 0 & -1 \end{bmatrix}\begin{bmatrix} 3 \\ 2 \end{bmatrix} = \begin{bmatrix} 3 \\ -2 \end{bmatrix}$.

We now find that a rotation about the origin is also a matrix transformation.

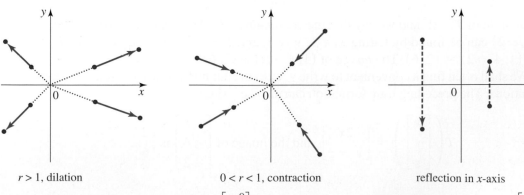

$r > 1$, dilation $0 < r < 1$, contraction   reflection in x-axis

Transformations defined by $\begin{bmatrix} r & 0 \\ 0 & r \end{bmatrix}$ Transformations defined by $\begin{bmatrix} 1 & 0 \\ 0 & -1 \end{bmatrix}$

Figure 6.1   **Figure 6.2**

Rotation about the Origin

Consider a rotation T about the origin through an angle θ, as shown in Figure 6.3. *T maps* the point $A\begin{bmatrix} x \\ y \end{bmatrix}$ into the point $B\begin{bmatrix} x' \\ y' \end{bmatrix}$.

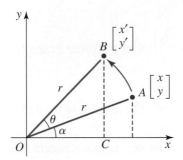

Figure 6.3

The distance OA is equal to OB; let it be r. Let the angle AOC be α. We get

$$x' = OC = r\cos(\alpha + \theta) = r\cos\alpha\cos\theta - r\sin\alpha\sin\theta$$
$$= x\cos\theta - y\sin\theta$$

$$y' = BC = r\sin(\alpha + \theta) = r\sin\alpha\cos\theta + r\cos\alpha\sin\theta$$
$$= y\cos\theta + x\sin\theta$$
$$= x\sin\theta + y\cos\theta$$

These expressions for x' and y' can be combined into a single matrix equation

$$\begin{bmatrix} x' \\ y' \end{bmatrix} = \begin{bmatrix} \cos\theta & -\sin\theta \\ \sin\theta & \cos\theta \end{bmatrix} \begin{bmatrix} x \\ y \end{bmatrix}$$

We thus get the following result.

> A rotation through an angle θ is described by $T\left(\begin{bmatrix} x \\ y \end{bmatrix}\right) = \begin{bmatrix} \cos\theta & -\sin\theta \\ \sin\theta & \cos\theta \end{bmatrix}\begin{bmatrix} x \\ y \end{bmatrix}$

Consider a rotation of $\pi/2$ about the origin. Since $\cos(\pi/2) = 0$ and $\sin(\pi/2) = 1$ the transformation is

$$T\left(\begin{bmatrix} x \\ y \end{bmatrix}\right) = \begin{bmatrix} 0 & -1 \\ 1 & 0 \end{bmatrix}\begin{bmatrix} x \\ y \end{bmatrix}$$

The image of $\begin{bmatrix} 3 \\ 2 \end{bmatrix}$, for example, is $\begin{bmatrix} 0 & -1 \\ 1 & 0 \end{bmatrix}\begin{bmatrix} 3 \\ 2 \end{bmatrix} = \begin{bmatrix} -2 \\ 3 \end{bmatrix}$.

Note that θ is positive for a counterclockwise rotation and negative for a clockwise rotation.

Matrix Transformations

In the previous discussion we found that we could use matrices to define certain transformations. We now see that every matrix in fact defines a transformation. Let A be a matrix and **x** be a column vector such that $A\mathbf{x}$ exists. Then A defines the *matrix transformation* $T(\mathbf{x}) = A\mathbf{x}$. For example,

$$A = \begin{bmatrix} 5 & 3 & -2 \\ 0 & 4 & -1 \end{bmatrix} \text{ defines the transformation } T\left(\begin{bmatrix} x \\ y \\ z \end{bmatrix}\right) = \begin{bmatrix} 5 & 3 & -2 \\ 0 & 4 & -1 \end{bmatrix}\begin{bmatrix} x \\ y \\ z \end{bmatrix}$$

The image of a vector such as $\begin{bmatrix} 1 \\ 3 \\ 4 \end{bmatrix}$ is $\begin{bmatrix} 6 \\ 8 \end{bmatrix}$. We write $\begin{bmatrix} 1 \\ 3 \\ 4 \end{bmatrix} \mapsto \begin{bmatrix} 6 \\ 8 \end{bmatrix}$. Similarly, for example,

$\begin{bmatrix} 3 \\ 1 \\ 2 \end{bmatrix} \mapsto \begin{bmatrix} 14 \\ 2 \end{bmatrix}$. We say that T *maps* $\mathbf{R}^3$ into $\mathbf{R}^2$ and write $T: \mathbf{R}^3 \to \mathbf{R}^2$. The *domain* of the trans-

formation is $\mathbf{R}^3$, the *codomain* is $\mathbf{R}^2$. We can convey this information in a diagram, Figure 6.4.

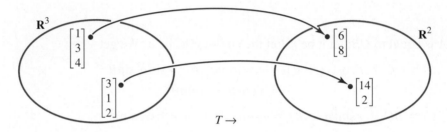

Figure 6.4

DEFINITION   Let A be an $m \times n$ matrix. Let **x** be an element of $\mathbf{R}^n$ written in column matrix form. A defines a *matrix transformation* $T(\mathbf{x}) = A\mathbf{x}$ of $\mathbf{R}^n$ into $\mathbf{R}^m$. The vector $A\mathbf{x}$ is the *image* of **x**. The *domain* of the transformation is $\mathbf{R}^n$ and the *codomain* is $\mathbf{R}^m$.

These transformations have the following geometrical properties (which we do not prove).

Matrix transformations map line segments into line segments (or points). If the matrix is invertible the transformation also maps parallel lines into parallel lines.

The following example illustrates how a square is deformed.

EXAMPLE 1   Consider the transformation $T: \mathbf{R}^2 \to \mathbf{R}^2$ defined by the matrix

$A = \begin{bmatrix} 4 & 2 \\ 2 & 3 \end{bmatrix}$. Determine the image of the unit square under this transformation.

SOLUTION

The unit square is the square whose vertices are the points

$$P\begin{bmatrix} 1 \\ 0 \end{bmatrix}, Q\begin{bmatrix} 1 \\ 1 \end{bmatrix}, R\begin{bmatrix} 0 \\ 1 \end{bmatrix}, O\begin{bmatrix} 0 \\ 0 \end{bmatrix}$$

See Figure 6.5(a). Let us compute the images of these points under the transformation. Multiplying each point by the matrix we get

$$
\begin{array}{cccccccc}
P & P' & Q & Q' & R & R' & O & O
\end{array}
$$

$$
\begin{bmatrix} 1 \\ 0 \end{bmatrix} \mapsto \begin{bmatrix} 4 \\ 2 \end{bmatrix},\ \begin{bmatrix} 1 \\ 1 \end{bmatrix} \mapsto \begin{bmatrix} 6 \\ 5 \end{bmatrix},\ \begin{bmatrix} 0 \\ 1 \end{bmatrix} \mapsto \begin{bmatrix} 2 \\ 3 \end{bmatrix},\ \begin{bmatrix} 0 \\ 0 \end{bmatrix} \mapsto \begin{bmatrix} 0 \\ 0 \end{bmatrix}
$$

The line segments are mapped as follows, Figure 6.5(b),

$$
OP \mapsto OP',\ PQ \mapsto P'Q',\ QR \mapsto Q'R',\ OR \mapsto OR'
$$

This matrix A is invertible ($|A| = 8 \neq 0$). It thus maps parallel lines into parallel lines. The square $PQRO$ is mapped into the parallelogram $P'Q'R'O$.

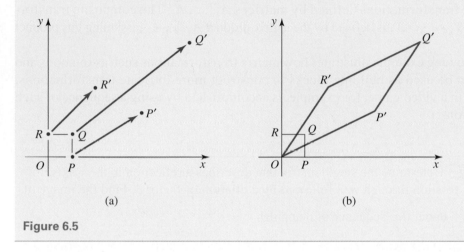

(a)   (b)

Figure 6.5

Solid bodies can be described geometrically. When loads are applied to bodies, changes in shape called *deformations* occur. For example, the square $PQRO$ in Figure 6.5 could represent a physical body that is deformed into the shape $P'Q'R'O$. Such deformations can be modeled and analyzed on computers using these mathematical techniques. The fields of science that investigate such problems are called *elasticity* and *plasticity*.

Composition of Transformations

The reader will be familiar with the concept of combining functions into composite functions, Figure 6.6. Matrix transformations can be combined in a very useful way. Consider the matrix transformations $T_1(\mathbf{x}) = A_1\mathbf{x}$ and $T_2(\mathbf{x}) = A_2\mathbf{x}$. The composite transformation $T = T_2 \circ T_1$ is given by

$$
T(\mathbf{x}) = T_2(T_1(\mathbf{x})) = T_2(A_1\mathbf{x}) = A_2A_1\mathbf{x}
$$

Thus T is defined by the matrix product A_2A_1.

$$
T(\mathbf{x}) = A_2A_1\mathbf{x}
$$

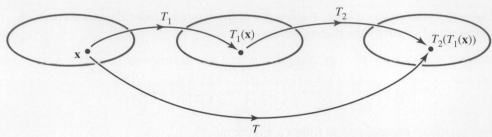

The composite transformation T of T_1 and T_2. $T(\mathbf{x}) = T_2(T_1(\mathbf{x}))$.

Figure 6.6

We can extend the results of this discussion in a natural way. Let $T_1, \ldots T_n$, be a sequence of transformations defined by matrices $A_1, \ldots A_n$. The composite transformation $T = T_n \circ \cdots \circ T_1$ is defined by the matrix product $A_n \ldots A_1$ (assuming this product exists).

The following example illustrates how matrix transformations such as rotations and dilations can be used as building blocks to construct more intricate transformations. (Movement in a video game, for example, is accomplished by using a sequence of such transformations.)

EXAMPLE 2 Determine the single matrix that describes a reflection in the x-axis, followed by a rotation through $\pi/2$ followed by a dilation of factor 3. Find the image of the point $\begin{bmatrix} 1 \\ 2 \end{bmatrix}$ under this sequence of mappings.

SOLUTION

The matrices that define the reflection, rotation, and dilation are

$$\begin{bmatrix} 1 & 0 \\ 0 & -1 \end{bmatrix}, \begin{bmatrix} \cos(\frac{\pi}{2}) & -\sin(\frac{\pi}{2}) \\ \sin(\frac{\pi}{2}) & \cos(\frac{\pi}{2}) \end{bmatrix}, \begin{bmatrix} 3 & 0 \\ 0 & 3 \end{bmatrix}$$

A reflection F followed by a rotation R and then a dilation D is the composite transformation $D \circ R \circ F$. The matrix of this transformation is

$$\begin{bmatrix} 3 & 0 \\ 0 & 3 \end{bmatrix}\begin{bmatrix} \cos(\frac{\pi}{2}) & -\sin(\frac{\pi}{2}) \\ \sin(\frac{\pi}{2}) & \cos(\frac{\pi}{2}) \end{bmatrix}\begin{bmatrix} 1 & 0 \\ 0 & -1 \end{bmatrix} = \begin{bmatrix} 3 & 0 \\ 0 & 3 \end{bmatrix}\begin{bmatrix} 0 & -1 \\ 1 & 0 \end{bmatrix}\begin{bmatrix} 1 & 0 \\ 0 & -1 \end{bmatrix} = \begin{bmatrix} 0 & 3 \\ 3 & 0 \end{bmatrix}$$

The image of the point $\begin{bmatrix} 1 \\ 2 \end{bmatrix}$ is $\begin{bmatrix} 0 & 3 \\ 3 & 0 \end{bmatrix}\begin{bmatrix} 1 \\ 2 \end{bmatrix} = \begin{bmatrix} 6 \\ 3 \end{bmatrix}$.

The reflection followed by the rotation and then dilation maps $A \mapsto B$, $B \mapsto C$, $C \mapsto D$ in Figure 6.7. The composite transformation described by the single matrix maps $A \mapsto D$ directly.

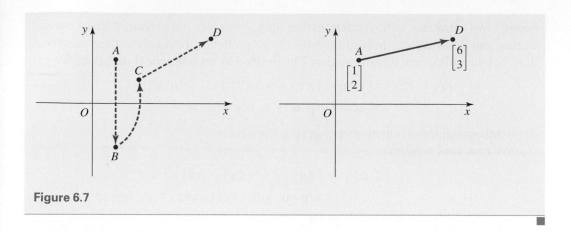

Figure 6.7

The next class of transformations are important in that they *preserve the geometry* of Euclidean Space.

Orthogonal Transformation

An *orthogonal matrix A* (see Section 4.9) is an invertible matrix that has the property

$$A^{-1} = A^t$$

An *orthogonal transformation* is a transformation $T(\mathbf{u}) = A\mathbf{u}$ where A is an orthogonal matrix.

An orthogonal transformation has the following geometrical properties.

THEOREM 6.1

Let T be an orthogonal transformation on $\mathbf{R}^n$. Let $\mathbf{u}$ and $\mathbf{v}$ be elements of $\mathbf{R}^n$. Let P and Q be the points in $\mathbf{R}^n$ defined by $\mathbf{u}$ and $\mathbf{v}$ and let R and S be their images under T. Then

$$\|\mathbf{u}\| = \|T(\mathbf{u})\|$$

$$\text{angle between } \mathbf{u} \text{ and } \mathbf{v} = \text{angle between } T(\mathbf{u}) \text{ and } T(\mathbf{v})$$

$$d(P, Q) = d(R, S)$$

Orthogonal transformations preserve norms, angles, and distances.

See Figure 6.8.

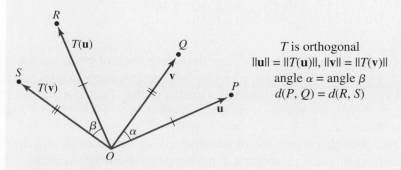

T is orthogonal
$\|\mathbf{u}\| = \|T(\mathbf{u})\|$, $\|\mathbf{v}\| = \|T(\mathbf{v})\|$
angle α = angle β
$d(P, Q) = d(R, S)$

Figure 6.8

Orthogonal transformations preserve the shapes of rigid bodies and are often referred to as *rigid motions*.

Proof   We show that orthogonal transformations preserve dot products. Since norms, angles, and distances are defined in terms of dot products, this leads to their preservation. Let the orthogonal transformation T be defined by an orthogonal matrix A. We get

$$T(\mathbf{u}) \cdot T(\mathbf{v}) = (A\mathbf{u}) \cdot (A\mathbf{v}) = (A\mathbf{u})^t(A\mathbf{v}) = (\mathbf{u}^t A^t)(A\mathbf{v})$$

$$= \mathbf{u}^t A^t A\mathbf{v} = \mathbf{u}^t I\mathbf{v} = \mathbf{u}^t\mathbf{v} = \mathbf{u} \cdot \mathbf{v}$$

Thus orthogonal transformations preserve dot products.

We now look at norms.

$$\|T(\mathbf{u})\| = \|A\mathbf{u}\| = \sqrt{(A\mathbf{u}) \cdot (A\mathbf{u})}$$

$$= \sqrt{\mathbf{u} \cdot \mathbf{u}}, \text{ since dot product is preserved}$$

$$= \|\mathbf{u}\|$$

Thus norm is preserved.

We leave it for the reader to show that orthogonal transformations also preserve angles and distances in Exercise 21.

EXAMPLE 3   Let T be the orthogonal transformation defined by the following orthogonal matrix A. Show that T preserves norms, angles, and distances for the vectors $\mathbf{u}$ and $\mathbf{v}$.

$$A = \begin{bmatrix} \frac{1}{\sqrt{2}} & \frac{1}{\sqrt{2}} \\ -\frac{1}{\sqrt{2}} & \frac{1}{\sqrt{2}} \end{bmatrix}, \mathbf{u} = \begin{bmatrix} 2 \\ 0 \end{bmatrix}, \mathbf{v} = \begin{bmatrix} 3 \\ 4 \end{bmatrix}$$

SOLUTION

We have that

$$T(\mathbf{u}) = A\mathbf{u} = \begin{bmatrix} \sqrt{2} \\ -\sqrt{2} \end{bmatrix} \text{ and } T(\mathbf{v}) = A\mathbf{v} = \begin{bmatrix} \frac{7}{\sqrt{2}} \\ \frac{1}{\sqrt{2}} \end{bmatrix}$$

It can be shown that $\|\mathbf{u}\| = \|T(\mathbf{u})\| = 2$ and $\|\mathbf{v}\| = \|T(\mathbf{v})\| = 5$. Norms of $\mathbf{u}$ and $\mathbf{v}$ are preserved. The angle between $\mathbf{u}$ and $\mathbf{v}$ and the angle between $T(\mathbf{u})$ and $T(\mathbf{v})$ are both found to be $53.13°$. The angle is preserved.

Furthermore,

$$d\left(\begin{bmatrix} 2 \\ 0 \end{bmatrix}, \begin{bmatrix} 3 \\ 4 \end{bmatrix}\right) = d\left(\begin{bmatrix} \sqrt{2} \\ -\sqrt{2} \end{bmatrix}, \begin{bmatrix} \frac{7}{\sqrt{2}} \\ \frac{1}{\sqrt{2}} \end{bmatrix}\right) = \sqrt{17}.$$

Distance is preserved.

Observe that T defines a rotation of points in a plane through an angle of $\pi/4$ in a clockwise direction about the origin. Intuitively, we expect rotations to preserve norms, angles, and distances. Rotation matrices do in fact define orthogonal transformations (see Exercise 20).

We complete this section with a discussion of transformations that, even though they are not truly matrix transformations, are important in mathematics and in applications.

Translation

A *translation* is a transformation $T: \mathbf{R}^n \to \mathbf{R}^n$ defined by

$$T(\mathbf{u}) = \mathbf{u} + \mathbf{v}$$

where $\mathbf{v}$ is a fixed vector.

A translation slides points in a direction and distance defined by the vector $\mathbf{v}$. For example, consider the following translation on $\mathbf{R}^2$:

$$T\left(\begin{bmatrix} x \\ y \end{bmatrix}\right) = \begin{bmatrix} x \\ y \end{bmatrix} + \begin{bmatrix} 4 \\ 2 \end{bmatrix}$$

Let us determine the effect of T on the triangle PQR having vertices $\begin{bmatrix} 1 \\ 2 \end{bmatrix}, \begin{bmatrix} 2 \\ 8 \end{bmatrix}, \begin{bmatrix} 3 \\ 2 \end{bmatrix}$. We see that

$$\begin{array}{cccccc} P & P' & Q & Q' & R & R' \\ \begin{bmatrix} 1 \\ 2 \end{bmatrix} \mapsto \begin{bmatrix} 5 \\ 4 \end{bmatrix}, & & \begin{bmatrix} 2 \\ 8 \end{bmatrix} \mapsto \begin{bmatrix} 6 \\ 10 \end{bmatrix}, & & \begin{bmatrix} 3 \\ 2 \end{bmatrix} \mapsto \begin{bmatrix} 7 \\ 4 \end{bmatrix} \end{array}$$

The triangle PQR is transformed into the triangle $P'Q'R'$ in Figure 6.9.

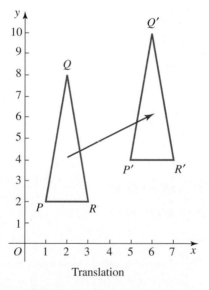

Translation

Figure 6.9

EXAMPLE 4 Find the equation of the image of the line $y = 2x + 3$ under the translation

$$T\left(\begin{bmatrix} x \\ y \end{bmatrix}\right) = \begin{bmatrix} x \\ y \end{bmatrix} + \begin{bmatrix} 2 \\ 1 \end{bmatrix}.$$

SOLUTION

The equation $y = 2x + 3$ describes points on the line of slope 2 and y-intercept 3. T will slide this line into another line. We want to find the equation of this image line. We get

$$T\left(\begin{bmatrix} x \\ y \end{bmatrix}\right) = \begin{bmatrix} x \\ y \end{bmatrix} + \begin{bmatrix} 2 \\ 1 \end{bmatrix} = \begin{bmatrix} x \\ 2x + 3 \end{bmatrix} + \begin{bmatrix} 2 \\ 1 \end{bmatrix} = \begin{bmatrix} x + 2 \\ 2x + 4 \end{bmatrix} = \begin{bmatrix} x' \\ y' \end{bmatrix}$$

We see that $y' = 2x'$ for the image point. Thus the equation of the image line is $y = 2x$.

■

Affine Transformation

An *affine transformation* is a transformation $T: \mathbf{R}^n \rightarrow \mathbf{R}^n$ defined by

$$T(\mathbf{u}) = A\mathbf{u} + \mathbf{v}$$

where A is a matrix and $\mathbf{v}$ is a fixed vector.

An affine transformation can be interpreted as a matrix transformation followed by a translation.

For example, consider the following affine transformation on $\mathbf{R}^2$.

$$T\left(\begin{bmatrix} x \\ y \end{bmatrix}\right) = \begin{bmatrix} 2 & 1 \\ 1 & 1 \end{bmatrix}\begin{bmatrix} x \\ y \end{bmatrix} + \begin{bmatrix} 1 \\ 2 \end{bmatrix}$$

Let us find the image of the unit square in Figure 6.10. We get

$$\begin{array}{cccccccc} P & & P' & Q & & Q' & R & & R' & O & & O' \\ \begin{bmatrix} 1 \\ 0 \end{bmatrix} & \mapsto & \begin{bmatrix} 3 \\ 3 \end{bmatrix}, & \begin{bmatrix} 1 \\ 1 \end{bmatrix} & \mapsto & \begin{bmatrix} 4 \\ 4 \end{bmatrix}, & \begin{bmatrix} 0 \\ 1 \end{bmatrix} & \mapsto & \begin{bmatrix} 2 \\ 3 \end{bmatrix}, & \begin{bmatrix} 0 \\ 0 \end{bmatrix} & \mapsto & \begin{bmatrix} 1 \\ 2 \end{bmatrix} \end{array}$$

Line segments are mapped into line segments. We get

$$OP \rightarrow O'P', PQ \rightarrow P'Q', QR \rightarrow Q'R', OR \rightarrow O'R'$$

The square PQR is transformed into the parallelogram $P'Q'R'O'$.

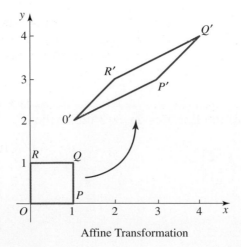

Affine Transformation

Figure 6.10

We have introduced a number of fundamental transformations in this section. The reader will meet other transformations—namely *projection*, *scaling*, and *shear*—in the following section. These basic transformations are the building blocks for creating other transformations using composition.

EXERCISE SET 6.1

Matrix Transformations

1. Consider the transformation T defined by the following matrix A. Find $T(\mathbf{x})$, $T(\mathbf{y})$, and $T(\mathbf{z})$.

$$A = \begin{bmatrix} 0 & -4 \\ 1 & 2 \end{bmatrix}; \mathbf{x} = \begin{bmatrix} 5 \\ -2 \end{bmatrix}, \mathbf{y} = \begin{bmatrix} 4 \\ 0 \end{bmatrix}, \mathbf{z} = \begin{bmatrix} 3 \\ 2 \end{bmatrix}$$

2. Consider the transformation T defined by the following matrix A. Determine the images of the vectors $\mathbf{x}$, $\mathbf{y}$, and $\mathbf{z}$.

$$A = \begin{bmatrix} 3 & -2 & 0 \\ 4 & 2 & 6 \end{bmatrix}; \mathbf{x} = \begin{bmatrix} 1 \\ 2 \\ 3 \end{bmatrix}, \mathbf{y} = \begin{bmatrix} 2 \\ 3 \\ 0 \end{bmatrix}, \mathbf{z} = \begin{bmatrix} -4 \\ 1 \\ 3 \end{bmatrix}$$

3. The following matrix A defines a transformation T. Find the images of the vectors $\mathbf{x}$, $\mathbf{y}$, and $\mathbf{z}$.

$$A = \begin{bmatrix} 1 & 2 \\ -1 & 3 \\ 1 & 2 \end{bmatrix}; \mathbf{x} = \begin{bmatrix} -1 \\ 1 \end{bmatrix}, \mathbf{y} = \begin{bmatrix} 2 \\ 3 \end{bmatrix}, \mathbf{z} = \begin{bmatrix} 5 \\ 2 \end{bmatrix}$$

Dilations, Reflections, Rotations

4. Determine the matrix that defines a reflection in the y-axis. Find the image of $\begin{bmatrix} 3 \\ 2 \end{bmatrix}$ under this transformation.

5. Find the matrix that defines a rotation of a plane about the origin through each of the following angles. Determine the image of the point $\begin{bmatrix} 2 \\ 1 \end{bmatrix}$ under each transformation.

(a) $\dfrac{\pi}{2}$ (b) $-\dfrac{\pi}{2}$ (c) $\dfrac{\pi}{4}$

(d) π (e) $-\dfrac{3\pi}{2}$ (f) $\dfrac{\pi}{6}$

(g) $-\dfrac{\pi}{3}$

6. Find the equation of the image of the unit circle, $x^2 + y^2 = 1$, under a dilation of factor 3.

7. Find the equation of the image of the ellipse $\dfrac{x^2}{4} + \dfrac{y^2}{9} = 1$ under a rotation through an angle of $\dfrac{\pi}{2}$.

Geometry

8. Consider the transformations on $\mathbf{R}^2$ defined by each of the following matrices. Sketch the image of the unit square under each transformation.

(a) $\begin{bmatrix} 0 & -1 \\ 1 & 0 \end{bmatrix}$ (b) $\begin{bmatrix} 2 & 0 \\ 0 & 2 \end{bmatrix}$

(c) $\begin{bmatrix} 3 & 0 \\ 1 & 4 \end{bmatrix}$ (d) $\begin{bmatrix} 4 & -1 \\ 1 & 5 \end{bmatrix}$

9. Sketch the image of the unit square under the transformation defined by each of the following transformations.

(a) $\begin{bmatrix} -2 & -3 \\ 0 & 4 \end{bmatrix}$ (b) $\begin{bmatrix} -2 & -4 \\ -4 & -1 \end{bmatrix}$

(c) $\begin{bmatrix} 0 & -2 \\ 2 & 0 \end{bmatrix}$ (d) $\begin{bmatrix} 0 & 3 \\ -3 & 0 \end{bmatrix}$

Composition of Transformations

10. Let $T_1(\mathbf{x}) = A_1\mathbf{x}$ and $T_2(\mathbf{x}) = A_2\mathbf{x}$ be defined by the following matrices A_1 and A_2. Let $T = T_2 \circ T_1$. Find the matrix that defines T and use it to determine the image of the vector $\mathbf{x}$ under T.

(a) $A_1 = \begin{bmatrix} 1 & 2 \\ 3 & 0 \end{bmatrix}$, $A_2 = \begin{bmatrix} -1 & 0 \\ 1 & 5 \end{bmatrix}$, $\mathbf{x} = \begin{bmatrix} 5 \\ 2 \end{bmatrix}$

(b) $A_1 = \begin{bmatrix} 0 & 1 & 2 \\ 3 & 4 & -1 \end{bmatrix}$, $A_2 = \begin{bmatrix} 2 & 2 \\ 1 & -1 \end{bmatrix}$, $\mathbf{x} = \begin{bmatrix} 0 \\ 1 \\ 3 \end{bmatrix}$

(c) $A_1 = \begin{bmatrix} 3 & -2 \\ 0 & 1 \end{bmatrix}$, $A_2 = \begin{bmatrix} 2 & 2 \\ 1 & -1 \\ 0 & 4 \end{bmatrix}$, $\mathbf{x} = \begin{bmatrix} -3 \\ 2 \end{bmatrix}$

11. Let $T_1(\mathbf{x}) = A_1\mathbf{x}$ and $T_2(\mathbf{x}) = A_2\mathbf{x}$ be defined by the following matrices A_1 and A_2. Let $T = T_2 \circ T_1$. Find the matrix that defines T and use it to determine the image of the vector $\mathbf{x}$ under T.

(a) $A_1 = \begin{bmatrix} 1 & 1 \\ 1 & 1 \end{bmatrix}$, $A_2 = \begin{bmatrix} 2 & 3 \\ 0 & -4 \end{bmatrix}$, $\mathbf{x} = \begin{bmatrix} 1 \\ 3 \end{bmatrix}$

(b) $A_1 = \begin{bmatrix} 2 & 1 & 0 \\ 3 & -2 & 5 \end{bmatrix}$, $A_2 = \begin{bmatrix} 3 & 1 \\ 0 & 2 \end{bmatrix}$, $\mathbf{x} = \begin{bmatrix} 3 \\ -2 \\ 1 \end{bmatrix}$

(c) $A_1 = \begin{bmatrix} 5 & -2 \\ 3 & 6 \end{bmatrix}$, $A_2 = \begin{bmatrix} 3 & 0 \\ 1 & -7 \\ 2 & 5 \end{bmatrix}$, $\mathbf{x} = \begin{bmatrix} 4 \\ -5 \end{bmatrix}$

12. Construct single 2×2 matrices that define the following transformations on $\mathbf{R}^2$. Find the image of the point $\begin{bmatrix} 2 \\ 1 \end{bmatrix}$ under each transformation.

(a) A rotation through $\dfrac{\pi}{2}$ counterclockwise, then a contraction of factor 0.5.

(b) A dilation of factor of 4, then a reflection in the x-axis.

(c) A reflection about the x-axis, a dilation of factor 3, then a rotation through $\dfrac{\pi}{2}$ in a clockwise direction.

13. Find a single matrix that defines a rotation of the plane through an angle of $\dfrac{\pi}{2}$ about the origin, while at the same time moves points to twice their original distance from the origin.

14. Determine a single matrix that defines both a rotation about the origin through an angle θ and a dilation of factor r.

15. Let A be the rotation matrix for $\dfrac{\pi}{4}$. Show that $A^8 = I$, the identity 2×2 matrix. Give a geometrical reason for expecting this result.

16. Show that $A^2 B A^2 = I$ for the following matrices A and B. Give a geometrical reason for expecting this result.

$$A = \begin{bmatrix} \dfrac{1}{\sqrt{2}} & -\dfrac{1}{\sqrt{2}} \\ \dfrac{1}{\sqrt{2}} & \dfrac{1}{\sqrt{2}} \end{bmatrix}, B = \begin{bmatrix} -1 & 0 \\ 0 & -1 \end{bmatrix}$$

17. Determine a rotation through an angle θ followed by a dilation of factor r. Show both algebraically and geometrically that this is equivalent to the dilation followed by the rotation. We say that the rotation and dilation transformations are *commutative*.

18. Let T_1 be a rotation and T_2 be a reflection in the x-axis. Are these transformations commutative? Discuss both algebraically and geometrically.

Orthogonal Transformations

19. Show that the following matrix A is orthogonal. Show that the transformation defined by A preserves the norms of the vectors $\mathbf{u}$ and $\mathbf{v}$, preserves the angle between these vectors,

and also preserves the distance between the points defined by the vectors.

$$A = \begin{bmatrix} 0 & -1 \\ 1 & 0 \end{bmatrix}, \mathbf{u} = \begin{bmatrix} 3 \\ 3 \end{bmatrix}, \mathbf{v} = \begin{bmatrix} 1 \\ 5 \end{bmatrix}$$

20. Prove that a rotation matrix is an orthogonal matrix.

21. Let A be an $n \times n$ orthogonal matrix. A defines a transformation of $\mathbf{R}^n$ into itself. Let $\mathbf{u}$ and $\mathbf{v}$ be elements of $\mathbf{R}^n$. Let P and Q be the points defined by $\mathbf{u}$ and $\mathbf{v}$ and let R and S be the points defined by $A\mathbf{u}$ and $A\mathbf{v}$. See Figure 6.8. Prove that

(a) the angle between $\mathbf{u}$ and $\mathbf{v}$ is equal to the angle between $A\mathbf{u}$ and $A\mathbf{v}$.

(b) the distance between P and Q is equal to the distance between S and T.

Thus the transformation defined by A preserves angles and distances.

Translations and Affine Transformations

22. Find the image of the triangle having vertices $(1, 2)$, $(3, 4)$, and $(4, 6)$ under the translation that takes the point $(1, 2)$ to $(2, -3)$.

23. Find the image of the line $y = 3x + 1$ under the translation

$$T\begin{bmatrix} x \\ y \end{bmatrix} = \begin{bmatrix} x \\ y \end{bmatrix} + \begin{bmatrix} p \\ q \end{bmatrix}$$

where (a) $p = 2, q = 5$ (b) $p = -1, q = 1$.

24. Find and sketch the image of the unit square and the unit circle under the affine transformations $T(\mathbf{u}) = A\mathbf{u} + \mathbf{v}$ defined by the following matrices and vectors.

(a) $A = \begin{bmatrix} 2 & 0 \\ 0 & 2 \end{bmatrix}, \mathbf{v} = \begin{bmatrix} 4 \\ 4 \end{bmatrix}$

(b) $A = \begin{bmatrix} 3 & 0 \\ 0 & 2 \end{bmatrix}, \mathbf{v} = \begin{bmatrix} 4 \\ 1 \end{bmatrix}$

(c) $A = \begin{bmatrix} \dfrac{1}{\sqrt{2}} & -\dfrac{1}{\sqrt{2}} \\ \dfrac{1}{\sqrt{2}} & \dfrac{1}{\sqrt{2}} \end{bmatrix}, \mathbf{v} = \begin{bmatrix} 3 \\ 1 \end{bmatrix}$

(d) $A = \begin{bmatrix} \dfrac{1}{2} & 0 \\ 0 & \dfrac{1}{4} \end{bmatrix}, \mathbf{v} = \begin{bmatrix} 3 \\ 3 \end{bmatrix}$

6.2 Linear Transformations, Graphics, and Fractals

Let us now examine the *properties* of a matrix transformation T. We know that a vector space has two operations, namely addition and scalar multiplication. Let us look at how T interacts with these operations. Consider the matrix transformation $T(\mathbf{u}) = A(\mathbf{u})$. The matrix properties of A imply that

$$T(\mathbf{u} + \mathbf{v}) = A(\mathbf{u} + \mathbf{v}) = A\mathbf{u} + A\mathbf{v}$$
$$= T(\mathbf{u}) + T(\mathbf{v})$$

and

$$T(c\mathbf{u}) = A(c\mathbf{u}) = cA\mathbf{u}$$
$$= cT(\mathbf{u})$$

The implication is that T maps the sum of two vectors into the sum of the images (preserves addition) and maps the scalar multiple of a vector into that same scalar multiple of the image of the vector (preserves scalar multiplication). We say that T *preserves vector space structure*. We call any transformation that has these properties a linear transformation.

DEFINITION

Let $\mathbf{u}$ and $\mathbf{v}$ be vectors in $\mathbf{R}^n$ and let c be a scalar. A transformation $T: \mathbf{R}^n \to \mathbf{R}^m$ is said to be a *linear transformation* if

$$T(\mathbf{u} + \mathbf{v}) = T(\mathbf{u}) + T(\mathbf{v}) \quad \text{(preserves addition)}$$
$$T(c\mathbf{u}) = cT(\mathbf{u}) \qquad \text{(preserves scalar multiplication)}$$

Every matrix transformation is linear. Since dilations, contractions, reflections, and rotations can all be described by matrices, these transformations are linear. Orthogonal transformations are also linear, but translations and affine transformations are not linear (see Exercise 39).

The structure-preserving ideas of a linear transformation are illustrated in Figure 6.11. We now give examples to show how the linearity conditions are in general checked.

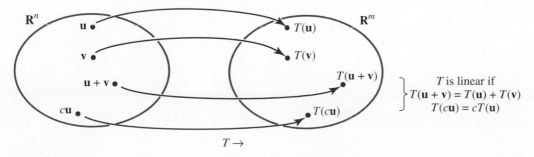

Figure 6.11

EXAMPLE 1 Prove that the following transformation T is linear.

$$T(x, y) = (x - y, 3x)$$

SOLUTION

T maps $\mathbf{R}^2 \to \mathbf{R}^2$. For example $T(5, 1) = (4, 15)$. The image of $(5, 1)$ is $(4, 15)$.

We first show that T preserves addition. Let (x_1, y_1) and (x_2, y_2) be elements of $\mathbf{R}^2$. Then

$$
\begin{aligned}
T((x_1, y_1) + (x_2, y_2)) &= T(x_1 + x_2, y_1 + y_2) && \text{by vector addition} \\
&= (x_1 + x_2 - y_1 - y_2, 3x_1 + 3x_2) && \text{by definition of } T \\
&= (x_1 - y_1, 3x_1) + (x_2 - y_2, 3x_2) && \text{by vector addition} \\
&= T(x_1, y_1) + T(x_2, y_2) && \text{by definition of } T
\end{aligned}
$$

Thus T preserves vector addition.

We now show that T preserves scalar multiplication. Let c be a scalar.

$$
\begin{aligned}
T(c(x_1, y_1)) &= T(cx_1, cy_1) && \text{by scalar multiplication of a vector} \\
&= (cx_1 - cy_1, 3cx_1) && \text{by definition of } T \\
&= c(x_1 - y_1, 3x_1) && \text{by scalar multiplication of a vector} \\
&= cT(x_1, y_1) && \text{by definition of } T
\end{aligned}
$$

Thus T preserves scalar multiplication. T is linear.

Note that T maps $\mathbf{R}^2$ into $\mathbf{R}^2$ in the preceding example. The domain and codomain are the same. A transformation for which the domain and codomain are the same is often referred to as an *operator*.

The following example illustrates a transformation that is not linear.

EXAMPLE 2 Show that the following transformation $T: \mathbf{R}^3 \to \mathbf{R}^2$ is not linear.

$$T(x, y, z) = (xy, z)$$

SOLUTION

Let us first test addition. Let (x_1, y_1, z_1) and (x_2, y_2, z_2) be elements of $\mathbf{R}^3$. Then

$$
\begin{aligned}
T((x_1, y_1, z_1) + (x_2, y_2, z_2)) &= T(x_1 + x_2, y_1 + y_2, z_1 + z_2) \\
&= (x_1 + x_2)(y_1 + y_2), (z_1 + z_2) \\
&= (x_1 y_1 + x_2 y_2 + x_1 y_2 + x_2 y_1, z_1 + z_2)
\end{aligned}
$$

and

$$
\begin{aligned}
T(x_1, y_1, z_1) + T(x_2, y_2, z_2) &= (x_1 y_1, z_1) + (x_2 y_2, z_2) \\
&= (x_1 y_1 + x_2 y_2, z_1 + z_2)
\end{aligned}
$$

Thus, in general,

$$T((x_1, y_1, z_1) + (x_2, y_2, z_2)) \neq T(x_1, y_1, z_1) + T(x_2, y_2, z_2)$$

Since vector addition is not preserved, T is not linear.

(It is not necessary to check the second linearity condition. The fact that one condition is not satisfied is sufficient to prove that T is not linear. It can be shown, in fact, that this particular transformation does not preserve scalar multiplication either.)

In the previous section we used ad hoc ways of arriving at matrices that described certain transformations such as rotations, dilations, and reflections. We now introduce a method for constructing a matrix representation for any linear transformation on $\mathbf{R}^n$. We pave the way with the following example.

EXAMPLE 3 Determine a matrix A that describes the linear transformation

$$T\left(\begin{bmatrix} x \\ y \end{bmatrix}\right) = \begin{bmatrix} 2x + y \\ 3y \end{bmatrix}$$

SOLUTION

It can be shown that T is linear. The domain of T is $\mathbf{R}^2$. We find the effect of T on the standard basis of $\mathbf{R}^2$.

$$T\left(\begin{bmatrix} 1 \\ 0 \end{bmatrix}\right) = \begin{bmatrix} 2 \\ 0 \end{bmatrix} \text{ and } T\left(\begin{bmatrix} 0 \\ 1 \end{bmatrix}\right) = \begin{bmatrix} 1 \\ 3 \end{bmatrix}$$

These vectors will be the columns of the matrix A that describe the transformation. We get

$$A = \begin{bmatrix} 2 & 1 \\ 0 & 3 \end{bmatrix}$$

T can be written as a matrix transformation,

$$T\left(\begin{bmatrix} x \\ y \end{bmatrix}\right) = \begin{bmatrix} 2 & 1 \\ 0 & 3 \end{bmatrix}\begin{bmatrix} x \\ y \end{bmatrix}$$

(We can check that this matrix does work: $\begin{bmatrix} 2 & 1 \\ 0 & 3 \end{bmatrix}\begin{bmatrix} x \\ y \end{bmatrix} = \begin{bmatrix} 2x + y \\ 3y \end{bmatrix}$.) ∎

We now arrive at the general result: We see *why* the above method works.

Matrix Representation

Let T be a linear transformation on $\mathbf{R}^n$. Let $\{\mathbf{e}_1, \mathbf{e}_2, \ldots, \mathbf{e}_n\}$ be the standard basis of $\mathbf{R}^n$ and $\mathbf{u}$ be an arbitrary vector in $\mathbf{R}^n$, written in column form.

$$\mathbf{e}_1 = \begin{bmatrix} 1 \\ 0 \\ \vdots \\ 0 \end{bmatrix}, \mathbf{e}_2 = \begin{bmatrix} 0 \\ 1 \\ \vdots \\ 0 \end{bmatrix}, \ldots, \mathbf{e}_n = \begin{bmatrix} 0 \\ 0 \\ \vdots \\ 1 \end{bmatrix}, \text{ and } \mathbf{u} = \begin{bmatrix} c_1 \\ c_2 \\ \vdots \\ c_n \end{bmatrix}$$

We can express $\mathbf{u}$ in terms of $\{\mathbf{e}_1, \mathbf{e}_2, \ldots, \mathbf{e}_n\}$.

$$\mathbf{u} = c_1\mathbf{e}_1 + \cdots + c_n\mathbf{e}_n$$

Since T is a linear transformation

$$T(\mathbf{u}) = T(c_1\mathbf{e}_1 + \cdots + c_n\mathbf{e}_n)$$

$$= c_1T(\mathbf{e}_1) + \cdots + c_nT(\mathbf{e}_n) \quad \text{(see Exercise 36)}$$

$$= [T(\mathbf{e}_1) \ldots T(\mathbf{e}_n)] \begin{bmatrix} c_1 \\ \vdots \\ c_n \end{bmatrix}$$

where $[T(\mathbf{e}_1) \ldots T(\mathbf{e}_n)]$ is a matrix with columns $T(\mathbf{e}_1), \ldots, T(\mathbf{e}_n)$.

Thus the linear transformation T is defined by the matrix

$$A = [T(\mathbf{e}_1) \ldots T(\mathbf{e}_n)]$$

A is called the *standard matrix* of T.

In this discussion we see the importance of a basis for working with vector spaces. We mentioned earlier how a basis in a sense represents the whole space. Here we see that the effect of a linear transformation on a basis leads to a matrix representation, a representation of the transformation on the whole space.

We derived the matrix for a rotation in the last section using an ad hoc method. You are asked to confirm this matrix in this standard manner, using bases, in the exercises that follow. We now use this method to derive the matrix of a reflection in the line $y = -x$.

EXAMPLE 4   The transformation $T\left(\begin{bmatrix} x \\ y \end{bmatrix}\right) = \begin{bmatrix} -y \\ -x \end{bmatrix}$ defines a reflection in the line $y = -x$, Figure 6.12(a). It can be shown that T is linear. Determine the standard matrix of this transformation. Find the image of $\begin{bmatrix} 4 \\ 1 \end{bmatrix}$.

SOLUTION

We find the effect of T on the standard basis.

$$T\left(\begin{bmatrix} 1 \\ 0 \end{bmatrix}\right) = \begin{bmatrix} 0 \\ -1 \end{bmatrix}, \text{ and } T\left(\begin{bmatrix} 0 \\ 1 \end{bmatrix}\right) = \begin{bmatrix} -1 \\ 0 \end{bmatrix}$$

The standard matrix is thus

$$A = \begin{bmatrix} 0 & -1 \\ -1 & 0 \end{bmatrix}$$

The transformation can be written

$$T\left(\begin{bmatrix} x \\ y \end{bmatrix}\right) = \begin{bmatrix} 0 & -1 \\ -1 & 0 \end{bmatrix}\begin{bmatrix} x \\ y \end{bmatrix}$$

Applying the transformation to the point $\begin{bmatrix} 4 \\ 1 \end{bmatrix}$, we get $T\left(\begin{bmatrix} 4 \\ 1 \end{bmatrix}\right) = \begin{bmatrix} -1 \\ -4 \end{bmatrix}$.

See Figure 6.12(b).

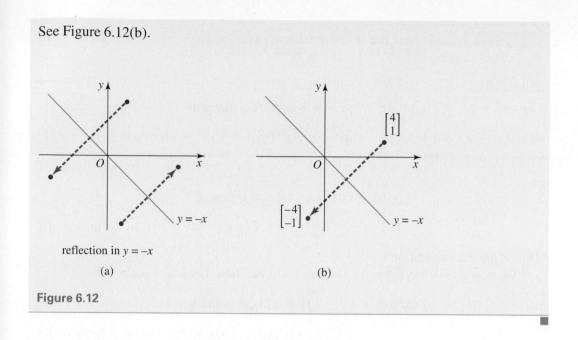

Figure 6.12

Linear Transformations Between Function Spaces

Up to this point we have discussed transformations between vector spaces of the type $\mathbf{R}^n$. The concept of a transformation applies to all vector spaces. Consider the function vector space P_2 of all real polynomial functions of degree ≤ 2. The polynomial $ax^2 + bx + c$ is an arbitrary element of this space. $T(ax^2 + bx + c) = (a + b)x + c$ is an example of a transformation $T: P_2 \rightarrow P_1$. The domain of T is P_2 and the codomain is P_1. Consider the function $2x^2 + 3x + 1$ of P_2. The image of this function is $T(2x^2 + 3x + 1) = (2 + 3)x + 1$. $T(2x^2 + 3x + 1) = 5x + 1$. We now extend the definition of linear transformation to general vector spaces.

DEFINITION Let U and V be vector spaces. Let $\mathbf{u}$ and $\mathbf{v}$ be vectors in U and let c be a scalar. A transformation $T: U \rightarrow V$ is said to be *linear* if

$$T(\mathbf{u} + \mathbf{v}) = T(\mathbf{u}) + T(\mathbf{v})$$

$$T(c\mathbf{u}) = cT(\mathbf{u})$$

If $U = V$, the linear transformation $T: U \rightarrow U$ is often referred to as a *linear operator*.

The first condition implies that T maps the sum of two vectors into the sum of the images of those vectors. The second condition implies that T maps the scalar multiple of a vector into the same scalar multiple of the image of that vector. Thus the operations of addition and scalar multiplication are preserved under a linear transformation.

EXAMPLE 5 Show that the transformation $T(ax^2 + bx + c) = (a + b)x + c$ of $P_2 \rightarrow P_1$ is linear.

SOLUTION

Let $ax^2 + bx + c$ and $px^2 + qx + r$ be arbitrary elements of P_2. Then

$$T((ax^2 + bx + c) + (px^2 + qx + r)) = T((a + p)x^2 + (b + q)x + (c + r))$$

$$= (a + p + b + q)x + (c + r) \qquad \text{by definition of } T$$

$$= (a + b)x + c + (p + q)x + r$$

$$= T(ax^2 + bx + c) + T(px^2 + qx + r) \quad \text{by definition of } T$$

Thus T preserves addition.

We now show that T preserves scalar multiplication. Let k be a scalar. Then

$$T(k(ax^2 + bx + c)) = T(kax^2 + kbx + kc)$$

$$= (ka + kb)x + kc \qquad \text{by definition of } T$$

$$= k((a + b)x + c)$$

$$= kT(ax^2 + bx + c) \qquad \text{by definition of } T$$

T preserves scalar multiplication. Therefore T is a linear transformation.

We now see that the properties of the derivatives that the reader will have met in calculus courses imply that the derivative is a linear transformation.

EXAMPLE 6 Let D be the operation of taking the derivative. (D is the same as $\dfrac{d}{dx}$. It is a more appropriate notation in this context than $\dfrac{d}{dx}$.) D can be interpreted as a mapping of P_n into itself. For example

$$D(4x^3 - 3x^2 + 2x + 1) = 12x^2 - 6x + 2$$

D maps the element $4x^3 - 3x^2 + 2x + 1$ of P_3 into the element $12x^2 - 6x + 2$ of P_3.

Let f and g be elements of P_n and c be a scalar. We know that a derivative has the following properties:

$$D(f + g) = Df + Dg$$

$$D(cf) = cD(f)$$

The derivative thus preserves addition and scalar multiplication of functions. It is a linear transformation.

We now discuss the use of linear transformations in computer graphics.

Transformations in Computer Graphics

Computer graphics is the field that studies the creation and manipulation of pictures with the aid of computers. The impact of computer graphics is felt in many homes through video games; its uses in research, industry, and business are vast and are ever expanding. Architects use computer graphics to explore designs, molecular biologists display and manipulate pictures of molecules to gain insight into their structure, pilots are trained using graphics flight simulators, and transportation engineers use computer-generated transforms in their planning work—to mention a few applications.

The manipulation of pictures in computer graphics is carried out using sequences of transformations. Rotations, reflections, dilations, and contractions are defined by matrices. A sequence of such transformations can be performed by a single transformation defined by the product of the matrices. Unfortunately, translation, as it now stands, uses matrix addition, and any sequence of transformations involving translations cannot be combined in this manner into a single matrix. However if coordinates called **homogeneous coordinates** are used to describe points in a plane, then translations can also be accomplished through matrix multiplication, and any sequence of these transformations can be defined in terms of a single matrix. In homogeneous coordinates a third component of 1 is added to each coordinate, and rotation, reflection, dilation/contraction, and translation R, F, D, and T are defined by the following matrices.

$$
\begin{array}{cc}
\text{X} & R \qquad\qquad\qquad\qquad F \\
\begin{bmatrix} x \\ y \\ 1 \end{bmatrix} \quad A = \begin{bmatrix} \cos\theta & -\sin\theta & 0 \\ \sin\theta & \cos\theta & 0 \\ 0 & 0 & 1 \end{bmatrix} \quad B = \begin{bmatrix} 1 & 0 & 0 \\ 0 & -1 & 0 \\ 0 & 0 & 1 \end{bmatrix} \\
\text{point} \qquad\qquad \text{rotation} \qquad\qquad\qquad \text{reflection}
\end{array}
$$

$$
\begin{array}{cc}
D \qquad\qquad\qquad T \\
C = \begin{bmatrix} r & 0 & 0 \\ 0 & r & 0 \\ 0 & 0 & 1 \end{bmatrix} \quad E = \begin{bmatrix} 1 & 0 & h \\ 0 & 1 & k \\ 0 & 0 & 1 \end{bmatrix} \\
\text{dilation/contraction} \qquad \text{translation} \\
(r > 0)
\end{array}
$$

Thus for example, a dilation D followed by a translation T and then a rotation R would be defined by $R \circ T \circ D(\mathrm{X}) = AEC(\mathrm{X})$. The composite transformation $R \circ T \circ D$ would be described by the single matrix AEC.

Some programming languages provide subroutines for rotation, translation, and dilation/contraction (and also scale and shear, see Exercises 19 and 22) that can be used to move pictures on the screen. To accomplish this movement the subroutines convert screen coordinates into homogeneous coordinates and use the matrices that define these transformations in homogeneous coordinates.

We now illustrate how the transformations are used to rotate a geometrical figure about a point other than the origin.

EXAMPLE 7 Determine the matrix that defines a rotation of points in a plane through an angle θ about a point $P(h, k)$. Use this general result to find the matrix that defines a rotation of the points through an angle of $\pi/2$ about the point $(5, 4)$. Find the image of the triangle having the following vertices $A(1, 2)$, $B(2, 8)$, and $C(3, 2)$ under this rotation. See Figure 6.13.

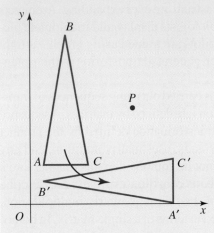

Rotation about P

Figure 6.13

SOLUTION

The rotation about P can be accomplished by a sequence of three of the above transformations:

(a) A translation T_1 that takes P to the origin O.

(b) A rotation R about the origin through an angle θ.

(c) A translation T_2 that takes O back to P.

The matrices that describe these transformations are as follows.

$$
\overset{T_1}{\begin{bmatrix} 1 & 0 & -h \\ 0 & 1 & -k \\ 0 & 0 & 1 \end{bmatrix}}
\overset{R}{\begin{bmatrix} \cos\theta & -\sin\theta & 0 \\ \sin\theta & \cos\theta & 0 \\ 0 & 0 & 1 \end{bmatrix}}
\overset{T_2}{\begin{bmatrix} 1 & 0 & h \\ 0 & 1 & k \\ 0 & 0 & 1 \end{bmatrix}}
$$

The rotation R_p about P can be accomplished as follows.

$$
R_p\left(\begin{bmatrix} x \\ y \\ 1 \end{bmatrix}\right) = T_2 \circ R \circ T_1\left(\begin{bmatrix} x \\ y \\ 1 \end{bmatrix}\right) = \begin{bmatrix} 1 & 0 & h \\ 0 & 1 & k \\ 0 & 0 & 1 \end{bmatrix}\begin{bmatrix} \cos\theta & -\sin\theta & 0 \\ \sin\theta & \cos\theta & 0 \\ 0 & 0 & 1 \end{bmatrix}
$$

$$
\times \begin{bmatrix} 1 & 0 & -h \\ 0 & 1 & -k \\ 0 & 0 & 1 \end{bmatrix}\begin{bmatrix} x \\ y \\ 1 \end{bmatrix}
$$

$$
= \begin{bmatrix} \cos\theta & -\sin\theta & -h\cos\theta + k\sin\theta + h \\ \sin\theta & \cos\theta & -h\sin\theta - k\cos\theta + k \\ 0 & 0 & 1 \end{bmatrix}\begin{bmatrix} x \\ y \\ 1 \end{bmatrix}
$$

To get the specific matrix that defines the rotation of the plane through an angle $\pi/2$ about the point $P(5, 4)$, for example, let $h = 5$, $k = 4$, and $\theta = \pi/2$. The rotation matrix is

$$M = \begin{bmatrix} 0 & -1 & 9 \\ 1 & 0 & -1 \\ 0 & 0 & 1 \end{bmatrix}$$

To find the images of the vertices of the triangle ABC write these vertices in column form as homogeneous coordinates and multiply by M. On performing the matrix multiplications we get

$$\begin{matrix} A & & A' & & B & & B' & & C & & C' \end{matrix}$$

$$\begin{bmatrix} 1 \\ 2 \\ 1 \end{bmatrix} \mapsto \begin{bmatrix} 7 \\ 0 \\ 1 \end{bmatrix} \qquad \begin{bmatrix} 2 \\ 8 \\ 1 \end{bmatrix} \mapsto \begin{bmatrix} 1 \\ 1 \\ 1 \end{bmatrix} \qquad \begin{bmatrix} 3 \\ 2 \\ 1 \end{bmatrix} \mapsto \begin{bmatrix} 7 \\ 2 \\ 1 \end{bmatrix}$$

The triangle with vertices $A(1, 2)$, $B(2, 8)$, $C(3, 2)$ is transformed into the triangle with vertices $A'(7, 0)$, $B'(1, 1)$, $C'(7, 2)$. See Figure 6.13.

■

Fractal Pictures of Nature

Computer graphics systems based on traditional Euclidean geometry are suitable for creating pictures of manmade objects such as machinery, buildings, and airplanes. Images of such objects can be created using lines, circles, and so on. However, these techniques are not appropriate when it comes to constructing images of natural objects such as animals, trees, and landscapes. In the words of mathematician Benoit B. Mandelbrot, "Clouds are not spheres, mountains are not cones, coastlines are not circles, and bark is not smooth, nor does lightning travel in straight lines." However, nature does wear its irregularities in an unexpectedly orderly fashion; it is full of shapes that repeat themselves on different scales within the same object. In 1975 Mandelbrot introduced a new geometry, which he called **fractal geometry**, that can be used to describe natural phenomena. A **fractal** is a convenient label for irregular and fragmented self-similar shapes. Fractal objects contain structures nested within one another. Each smaller structure being a miniature, though not necessarily identical version, of the larger form. The story behind the word fractal is interesting. Mandelbrot came across the Latin adjective *fractus,* from the verb *frangere,* to break, in his son's Latin book. The resonance of the main English cognates fracture and fraction seemed appropriate and he coined the word fractal!

We now discuss methods that have been developed by a research team at the Georgia Institute of Technology for forming images of natural objects using fractals . These fractal images of nature are generated using affine transformations. Figure 6.14 shows a fractal image of a fern being gradually generated. Let us see how this is done.

Figure 6.14

Consider the following four affine transformations $T_1, \ldots, T_4$. Associate probabilities $p_1, \ldots, p_4$ with these transformations.

$$T_1\left(\begin{bmatrix} x \\ y \end{bmatrix}\right) = \begin{bmatrix} 0.86 & 0.03 \\ -0.03 & 0.86 \end{bmatrix} \begin{bmatrix} x \\ y \end{bmatrix} + \begin{bmatrix} 0 \\ 1.5 \end{bmatrix}, \ p_1 = 0.83$$

$$T_2\left(\begin{bmatrix} x \\ y \end{bmatrix}\right) = \begin{bmatrix} 0.2 & -0.25 \\ 0.21 & 0.23 \end{bmatrix} \begin{bmatrix} x \\ y \end{bmatrix} + \begin{bmatrix} 0 \\ 1.5 \end{bmatrix}, \ p_2 = 0.08$$

$$T_3\left(\begin{bmatrix} x \\ y \end{bmatrix}\right) = \begin{bmatrix} -0.15 & 0.27 \\ 0.25 & 0.26 \end{bmatrix} \begin{bmatrix} x \\ y \end{bmatrix} + \begin{bmatrix} 0 \\ 0.45 \end{bmatrix}, \ p_3 = 0.08$$

$$T_4\left(\begin{bmatrix} x \\ y \end{bmatrix}\right) = \begin{bmatrix} 0 & 0 \\ 0 & 0.17 \end{bmatrix} \begin{bmatrix} x \\ y \end{bmatrix} + \begin{bmatrix} 0 \\ 0 \end{bmatrix}, \ p_4 = 0.01$$

The following algorithm is used on a computer to produce the image of the fern.

1. Let $x = 0$, $y = 0$.
2. Use a random generator to select one of the affine transformations T_i according to the given probabilities.
3. Let $(x', y') = T_i(x, y)$.
4. Plot (x', y').
5. Let $(x, y) = (x', y')$.
6. Repeat Steps 2, 3, 4, and 5 twenty thousand times.

As Step 4 is executed, each of twenty thousand times, the image of the fern gradually appears.

Each affine transformation T_i involves six parameters a, b, c, d, e, f, and a probability p_i, as follows

$$T_i\left(\begin{bmatrix} x \\ y \end{bmatrix}\right) = \begin{bmatrix} a & b \\ c & d \end{bmatrix} \begin{bmatrix} x \\ y \end{bmatrix} + \begin{bmatrix} e \\ f \end{bmatrix}, \ p_i$$

The affine transformations and corresponding probabilities that generate a fractal are written as rows of a matrix, called an *iterated function system* (IFS). The IFS for the fern is as follows.

IFS for a fern

T	a	b	c	d	e	f	p
1	0.86	0.03	−0.03	0.86	0	1.5	0.83
2	0.2	−0.25	0.21	0.23	0	1.5	0.08
3	−0.15	0.27	0.25	0.26	0	0.45	0.08
4	0	0	0	0.17	0	0	0.01

The appropriate affine transformations that produce a given fractal object are found by determining transformations that map the object (called the **attractor**) into various disjoint images, the union of which is the whole fractal. A theorem called the **Collage Theorem** then guarantees that the transformations can be grouped into an IFS that produces the fractal.

Different probabilities do not in general lead to different images but they do affect the rate at which the image is produced. Appropriate probabilities are

$$p_i = \frac{\text{area of the image under transformation } T_i}{\text{area of image of object}}$$

These techniques are very valuable because they can be used to produce an image to any desired degree of accuracy using a highly compressed data set. A fractal image containing infinitely many points, whose organization is too complicated to describe directly, can be reproduced using mathematical formulas.

EXERCISE SET 6.2

Check for Linearity

1. Prove that the transformation $T: \mathbf{R}^2 \to \mathbf{R}^2$ defined by $T(x, y) = (2x, x - y)$ is linear. Find the images of the elements $(1, 2)$ and $(-1, 4)$ under this transformation.

2. Prove that $T(x, y) = (3x + y, 2y, x - y)$ defines a linear transformation $T: \mathbf{R}^2 \to \mathbf{R}^3$. Find the images of $(1, 2)$ and $(2, -5)$.

3. Prove that $T: \mathbf{R}^3 \to \mathbf{R}^3$ defined by $T(x, y, z) = (0, y, 0)$ is linear. This transformation is also called a projection. Why is this term appropriate?

4. Prove that the following transformations $T: \mathbf{R}^3 \to \mathbf{R}^2$ are not linear.
   (a) $T(x, y, z) = (3x, y^2)$
   (b) $T(x, y, z) = (x + 2, 4y)$

5. Prove that $T: \mathbf{R}^2 \to \mathbf{R}$ defined by $T(x, y) = x + a$, where a is a nonzero scalar, is not linear.

In Exercises 6–11, determine whether the given transformations are linear.

6. $T(x, y, z) = (2x, y)$ of $\mathbf{R}^3 \to \mathbf{R}^2$.

7. $T(x, y) = x - y$ of $\mathbf{R}^2 \to \mathbf{R}$.

8. $T(x, y) = (x, y, z)$ of $\mathbf{R}^2 \to \mathbf{R}^3$, when (a) $z = 0$,
   (b) $z = 1$.

9. $T(x) = (x, 2x, 3x)$ of $\mathbf{R} \to \mathbf{R}^3$.

10. $T(x, y) = (x^2, y)$ of $\mathbf{R}^2 \to \mathbf{R}^2$.

11. $T(x, y, z) = (x + 2y, x + y + z, 3z)$ of $\mathbf{R}^3 \to \mathbf{R}^3$.

Standard Matrix of a Linear Transformation

12. Find the standard matrix of each of the following linear transformations on $\mathbf{R}^2$.

(a) $T\left(\begin{bmatrix} x \\ y \end{bmatrix}\right) = \begin{bmatrix} 2x \\ x - y \end{bmatrix}$ 　(b) $T\left(\begin{bmatrix} x \\ y \end{bmatrix}\right) = \begin{bmatrix} x - y \\ x + y \end{bmatrix}$

13. Find the standard matrix of each of the following linear transformations on $\mathbf{R}^2$.

(a) $T\left(\begin{bmatrix} x \\ y \end{bmatrix}\right) = \begin{bmatrix} 2x - 5y \\ 3y \end{bmatrix}$

(b) $T\left(\begin{bmatrix} x \\ y \end{bmatrix}\right) = \begin{bmatrix} 2y \\ -3x \end{bmatrix}$

14. Derive the rotation matrix by finding the effect of a rotation on the standard basis of $\mathbf{R}^2$.

15. We have seen examples of rotations, dilations, contractions, and reflections. How would you describe the transformation

$T\left(\begin{bmatrix} x \\ y \end{bmatrix}\right) = \begin{bmatrix} y \\ x \end{bmatrix}$? Show that T is linear. Determine the

standard matrix of T. Find the image of the point $\begin{bmatrix} -3 \\ 5 \end{bmatrix}$.

Projection

16. Find the standard matrix of the operator $T\left(\begin{bmatrix} x \\ y \end{bmatrix}\right) = \begin{bmatrix} x \\ 0 \end{bmatrix}$.

Observe that this transformation projects all points onto the *x*-axis. It is called a *projection operator*.

17. Determine the matrix that defines projection onto the *y*-axis.

18. Determine the matrix that defines projection onto the line $y = x$. Find the image of $\begin{bmatrix} 4 \\ 2 \end{bmatrix}$ under this projection.

Scaling

19. Find the standard matrix of the operator $T\left(\begin{bmatrix} x \\ y \end{bmatrix}\right) = \begin{bmatrix} ax \\ by \end{bmatrix}$,

where *a* and *b* are positive scalars. *T* is called a *scaling of factor a in the x-direction and factor b in the y-direction*. A scaling distorts a figure, since *x* and *y* do not change in the same manner. Sketch the image of the unit square under this transformation when $a = 3$ and $b = 2$.

20. Find the equation of the image of the line $y = 2x$ under a scaling of factor 2 in the *x*-direction and factor 3 in the *y*-direction.

21. Find the equation of the image of the unit circle, $x^2 + y^2 = 1$, under a scaling of factor 4 in the *x*-direction and factor 3 in the *y*-direction.

Shear

22. Find the standard matrix of the operator

$T\left(\begin{bmatrix} x \\ y \end{bmatrix}\right) = \begin{bmatrix} x + cy \\ y \end{bmatrix}$, where *c* is a scalar. *T* is called a *shear*

of factor c in the x-direction. Sketch the image of the unit square under a shear of factor 2 in the *x*-direction. Observe how the *x*-value of each point is increased by a factor of $2y$, causing a shearing of the figure.

23. Sketch the image of the unit square under a shear of factor 0.5 in the *y*-direction.

24. Find the equation of the image of the line $y = 3x$ under a shear of factor 5 in the *x*-direction.

General Matrix Transformations

25. Construct single 2×2 matrices that define the following transformations on $\mathbf{R}^2$. Find the image of the point $\begin{bmatrix} 3 \\ 2 \end{bmatrix}$ under each transformation.

(a) A dilation of factor 3, then a shear of factor 2 in the *x*-direction.

(b) A scaling of factor 3 in the *x*-direction, of factor 2 in the *y*-direction, then a reflection in the line $y = x$.

(c) A dilation of factor 2, then a shear of factor 3 in the *x*-direction, then a rotation through $\pi/2$ counter-clockwise.

26. Find the matrix that maps $\mathbf{R}^2 \to \mathbf{R}^2$ such that $\begin{bmatrix} 1 \\ 2 \end{bmatrix} \mapsto \begin{bmatrix} 7 \\ 3 \end{bmatrix}$ and $\begin{bmatrix} -1 \\ 1 \end{bmatrix} \mapsto \begin{bmatrix} 2 \\ 3 \end{bmatrix}$.

27. Transformations T_1 and T_2 are said to be *commutative* if $T_2 \circ T_1(\mathbf{u}) = T_1 \circ T_2(\mathbf{u})$ for all vectors $\mathbf{u}$. Let R be a rotation, D dilation, F reflection, S scaling, H shear, and A affine transformation. Which pairs of transformations are commutative?

28. Find the matrix that defines a rotation of three-space through an angle of $\pi/2$ about the *z*-axis. (You may consider either direction.)

29. Find the matrix that defines an expansion of three-dimensional space outward from the origin, so that each point moves to three times as far away.

30. Determine the matrix that can be used to define a rotation through $\pi/2$ about the point $(5, 1)$. Find the image of the unit square under this rotation.

31. Consider the general translation defined by the following matrix *T*. Does this transformation have an inverse? If so find it.

$$T = \begin{bmatrix} 1 & 0 & h \\ 0 & 1 & k \\ 0 & 0 & 1 \end{bmatrix}$$

32. Consider the general scaling defined by the following matrix $S(c \neq 0, d \neq 0)$. Does this transformation have an inverse? If so, find it.

$$S = \begin{bmatrix} c & 0 & 0 \\ 0 & d & 0 \\ 0 & 0 & 1 \end{bmatrix}$$

33. Find the image of the triangle having the following vertices A, B, and C (in homogeneous coordinates), under the sequence of transformations T followed by R, followed by S. Sketch the original and final triangle.

$$A \begin{bmatrix} 1 \\ 6 \\ 1 \end{bmatrix}, B \begin{bmatrix} 3 \\ 0 \\ 1 \end{bmatrix}, C \begin{bmatrix} 4 \\ 6 \\ 1 \end{bmatrix};$$

$$T = \begin{bmatrix} 1 & 0 & 4 \\ 0 & 1 & -3 \\ 0 & 0 & 1 \end{bmatrix}, R = \begin{bmatrix} 0 & 1 & 0 \\ -1 & 0 & 0 \\ 0 & 0 & 1 \end{bmatrix},$$

$$S = \begin{bmatrix} 3 & 0 & 0 \\ 0 & 5 & 0 \\ 0 & 0 & 1 \end{bmatrix}$$

Checking Linearity for Function Spaces

34. Show that the transformation $T: P_1 \to P_1$ defined as follows is linear. $T(ax + b) = (a + b)x$.

35. Show that the transformation $T(ax^2 + bx + c) = (a + b + c)x$ of $P_2 \to P_1$ is linear. Find the image of the element $x^2 + 2x + 3$.

36. Show that the transformation $T(ax + b) = ax^2$ of $P_1 \to P_2$ is linear.

37. Is the transformation $T(ax + b) = ax^2 + ax + a$ of $P_1 \to P_2$ linear? Find the image of $2x + 1$. Determine another element of P_1 that has the same image.

38. Show that the transformation $T(ax + b) = ax + 2$ of $P_1 \to P_1$ is not linear.

39. Is the transformation $T(ax + b) = bx + a$ of $P_1 \to P_1$ linear?

40. Is the transformation $T(ax^3 + bx^2 + cx + d) = 2$ of $P_3 \to \mathbf{R}$ linear?

41. Show that the transformation $T: P_2 \to P_2$ defined as follows is linear. $T(ax^2 + bx + c) = cx^2 + a$. Find the image of $3x^2 - x + 2$. Determine another element of P_2 that has the same image.

42. Show that the transformation $T(ax^2 + bx + c) = bx + c$ of $P_2 \to P_1$ is linear. Find the image of $3x^2 - x + 2$. Determine another element of P_2 that has the same image.

43. Prove that the transformation $T(ax + b) = x + a$ of $P_1 \to P_1$ is not linear.

44. Let J be the operation of taking a definite integral from 0 to 1. J can be used to define a mapping of P_n into $\mathbf{R}$. For example

$$J(6x^2 + 2x + 3) = \int_0^1 (6x^2 + 2x + 3)\,dx$$

$$= [2x^3 + x^2 + 3x]_0^1 = 6$$

J maps the element $6x^2 + 2x + 3$ of P_n into the element 6 of $\mathbf{R}$. Show that J is a linear transformation of $P_n \to \mathbf{R}$.

Miscellaneous Results

45. Let T be a linear transformation. Use the fact that T preserves addition and scalar multiplication to show that
(a) $T(-\mathbf{v}) = -T(\mathbf{v})$
(b) $T(\mathbf{v} - \mathbf{w}) = T(\mathbf{v}) - T(\mathbf{w})$

46. Let T be a transformation between vector spaces, $\mathbf{u}$ and $\mathbf{v}$ be vectors in the domain, and a and b be scalars. Prove that T is linear if and only if

$$T(a\mathbf{u} + b\mathbf{v}) = aT(\mathbf{u}) + bT(\mathbf{v})$$

This can be used as an alternative definition of linear transformation.

47. Let T be a linear transformation with domain U. Let $\mathbf{v}_1, \ldots, \mathbf{v}_m$ be vectors in U, and $c_1, \ldots, c_m$ be scalars. Prove that

$$T(c_1\mathbf{v}_1 + \cdots + c_m\mathbf{v}_m) = c_1T(\mathbf{v}_1) + \cdots + c_mT(\mathbf{v}_m)$$

(Can you prove this result by induction?)

48. Prove that the composition of two linear transformations is a linear transformation.

49. State (with a brief explanation) whether the following statements are true or false.
(a) $T(x, y) = (x, y)$ is a linear transformation.
(b) Let $T: U \to V$ be a linear transformation. Distinct vectors in U always have distinct images in V.
(c) If the dimension of V is greater than that of U, there are no linear transformations from U to V.

50. Prove that translations and affine transformations are not linear.

6.3 Kernel, Range, and the Rank/Nullity Theorem

We have seen that a linear transformation is a function from one vector space (called the *domain*) into another vector space called (the *codomain*.) There are two further vector spaces called *kernel* and *range* that are associated with every linear transformation. In this section we introduce and discuss the properties of these spaces.

The following theorem gives an important property of all linear transformations. It paves the way for the introduction of kernel and range.

THEOREM 6.2

Let $T: U \to V$ be a linear transformation. Let $\mathbf{0}_U$ and $\mathbf{0}_V$ be the zero vectors of U and V. Then

$$T(\mathbf{0}_U) = \mathbf{0}_V$$

That is, a linear transformation maps a zero vector into a zero vector.

Proof Let $\mathbf{u}$ be a vector in U and let $T(\mathbf{u}) = \mathbf{v}$. Let 0 be the zero scalar. Since $0\mathbf{u} = \mathbf{0}_U$ and $0\mathbf{v} = \mathbf{0}_V$ and T is linear we get

$$T(\mathbf{0}_U) = T(0\mathbf{u}) = 0T(\mathbf{u}) = 0\mathbf{v} = \mathbf{0}_V$$

For example, $T(x, y, z) = (3x, y + z)$ is a linear transformation of $\mathbf{R}^3 \to \mathbf{R}^2$. The zero vector of $\mathbf{R}^3$ is $(0, 0, 0)$ and the zero vector of $\mathbf{R}^2$ is $(0, 0)$. We see that $T(0, 0, 0) = (0, 0)$.

DEFINITION Let $T: U \to V$ be a linear transformation.

The set of vectors in U that are mapped into the zero vector of V is called the **kernel** of T. The kernel is denoted $\ker(T)$.

The set of vectors in V that are the images of vectors in U is called the **range** of T. The range is denoted $\text{range}(T)$.

We illustrate these sets in Figure 6.15. Whenever we introduce sets in linear algebra we are interested in knowing whether they are vector spaces or not. We now find that the kernel and range are indeed vector spaces.

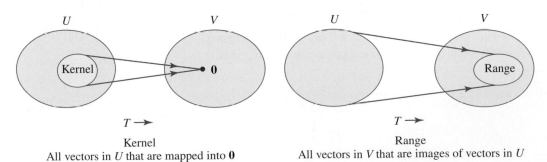

Kernel
All vectors in U that are mapped into $\mathbf{0}$

Range
All vectors in V that are images of vectors in U

Figure 6.15

THEOREM 6.3

Let $T: U \rightarrow V$ be a linear transformation.
(a) The kernel of T is a subspace of U.
(b) The range of T is a subspace of V.

Proof

(a) From the previous theorem we know that the kernel is nonempty since it contains the zero vector of U. To prove that the kernel is a subspace of U it remains to show that it is closed under addition and scalar multiplication.

First we prove closure under addition. Let $\mathbf{u}_1$ and $\mathbf{u}_2$ be elements of $\ker(T)$. Thus $T(\mathbf{u}_1) = \mathbf{0}$ and $T(\mathbf{u}_2) = \mathbf{0}$. Using the linearity of T we get

$$T(\mathbf{u}_1 + \mathbf{u}_2) = T(\mathbf{u}_1) + T(\mathbf{u}_2) = \mathbf{0} + \mathbf{0} = \mathbf{0}$$

The vector $\mathbf{u}_1 + \mathbf{u}_2$ is mapped into $\mathbf{0}$. Thus $\mathbf{u}_1 + \mathbf{u}_2$ is in $\ker(T)$.

Let us now show that $\ker(T)$ is closed under scalar multiplication. Let c be a scalar. Again using the linearity of T we get

$$T(c\mathbf{u}_1) = cT(\mathbf{u}_1) = c\mathbf{0} = \mathbf{0}$$

The vector $c\mathbf{u}_1$ is mapped into $\mathbf{0}$. Thus $c\mathbf{u}_1$ is in $\ker(T)$.

The kernel is closed under addition and under scalar multiplication. It is a subspace of U.

(b) The previous theorem tells us that the range is nonempty since it contains the zero vector of V. To prove that the range is a subspace of V it remains to show that it is closed under addition and scalar multiplication. Let $\mathbf{v}_1$ and $\mathbf{v}_2$ be elements of $\text{range}(T)$. Thus there exist vectors $\mathbf{w}_1$ and $\mathbf{w}_2$ in the domain U such that

$$T(\mathbf{w}_1) = \mathbf{v}_1 \quad \text{and} \quad T(\mathbf{w}_2) = \mathbf{v}_2$$

Using the linearity of T,

$$T(\mathbf{w}_1 + \mathbf{w}_2) = T(\mathbf{w}_1) + T(\mathbf{w}_2) = \mathbf{v}_1 + \mathbf{v}_2$$

The vector $\mathbf{v}_1 + \mathbf{v}_2$ is the image of $\mathbf{w}_1 + \mathbf{w}_2$. Thus $\mathbf{v}_1 + \mathbf{v}_2$ is in the range.

Let c be a scalar. By the linearity of T,

$$T(c\mathbf{w}_1) = cT(\mathbf{w}_1) = c\mathbf{v}_1$$

The vector $c\mathbf{v}_1$ is the image of $c\mathbf{w}_1$. Thus $c\mathbf{v}_1$ is in the range.

The range is closed under addition and under scalar multiplication. It is a subspace of V.

EXAMPLE 1 Find the kernel and range of the linear operator

$$T(x, y, z) = (x, y, 0)$$

SOLUTION

Since the linear operator T maps $\mathbf{R}^3$ into $\mathbf{R}^3$ the kernel and range will both be subspaces of $\mathbf{R}^3$.

Kernel: $\ker(T)$ is the subset that is mapped into $(0, 0, 0)$. We see that

$$T(x, y, z) = (x, y, 0)$$
$$= (0, 0, 0), \text{ if } x = 0, y = 0$$

Thus ker(T) is the set of all vectors of the form $(0, 0, z)$. We express this

$$\text{ker}(T) = \{(0, 0, z)\}$$

Geometrically, ker(T) is the set of all vectors that lie on the z-axis.
Range: The range of T is the set of all vectors of the form $(x, y, 0)$. Thus

$$\text{range}(T) = \{(x, y, 0)\}$$

range(T) is the set of all vectors that lie in the xy-plane.

We illustrate this transformation in Figure 6.16. Observe that T projects the vector (x, y, z) into the vector $(x, y, 0)$ in the xy-plane. T projects all vectors onto the xy-plane. T is an example of a **projection operator**.

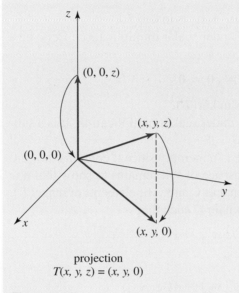

projection
$T(x, y, z) = (x, y, 0)$

Figure 6.16

Projections are important in applications. The world in which we live has three spatial dimensions. When we observe an object, however, we get a two-dimensional impression of that object, the view changing from location to location. Projections can be used to illustrate what three-dimensional objects look like from various locations. Such transformations, for example, are used in architecture, the auto industry, and the aerospace industry. The outline of the object of interest, relative to a suitable coordinate system, is fed into a computer. The computer program contains an appropriate projection transformation that maps the object onto a plane. The output gives a two-dimensional view of the object, the picture being graphed by the computer. In this manner various transformations can be used to lead to various perspectives of an object. Aerospace companies use such graphics systems in designing aircraft. We illustrate these concepts in Figure 6.17, for an aircraft. The projection used is onto the xz-plane. The image represents the view an observer at A has of the aircraft; it would be graphed out by a computer.

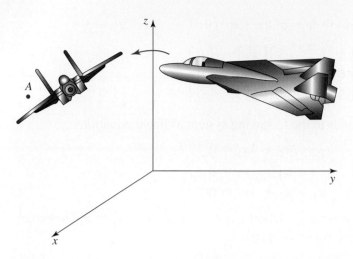

Figure 6.17

We now discuss kernels and ranges of matrix transformations. The following theorem gives us information about the range of a matrix transformation.

THEOREM 6.4

Let $T: \mathbf{R}^n \to \mathbf{R}^m$ be defined by $T(\mathbf{u}) = A\mathbf{u}$. The range of T is spanned by the column vectors of A.

Proof Let $\mathbf{v}$ be a vector in the range. There exists a vector $\mathbf{u}$ such that $T(\mathbf{u}) = \mathbf{v}$. Express $\mathbf{u}$ in terms of the standard basis of $\mathbf{R}^n$.

$$\mathbf{u} = c_1\mathbf{e}_1 + \cdots + c_n\mathbf{e}_n$$

Thus

$$\mathbf{v} = T(c_1\mathbf{e}_1 + \cdots + c_n\mathbf{e}_n)$$
$$= c_1 T(\mathbf{e}_1) + \cdots + c_n T(\mathbf{e}_n)$$

Therefore the column vectors of A, namely, $T(\mathbf{e}_1), \ldots, T(\mathbf{e}_n)$, span the range of T.

EXAMPLE 2 Determine the kernel and the range of the transformation defined by the following matrix.

$$A = \begin{bmatrix} 1 & 2 & 3 \\ 0 & -1 & 1 \\ 1 & 1 & 4 \end{bmatrix}$$

SOLUTION

A is a 3×3 matrix. Thus A defines a linear operator $T: \mathbf{R}^3 \to \mathbf{R}^3$,

$$T(\mathbf{x}) = A\mathbf{x}$$

The elements of $\mathbf{R}^3$ are written in column matrix form for the purpose of matrix multiplication. For convenience, we express the elements of $\mathbf{R}^3$ in row form at all other times.

Kernel: The kernel will consist of all vectors $\mathbf{x} = (x_1, x_2, x_3)$ in $\mathbf{R}^3$ such that

$$T(\mathbf{x}) = \mathbf{0}$$

Thus

$$\begin{bmatrix} 1 & 2 & 3 \\ 0 & -1 & 1 \\ 1 & 1 & 4 \end{bmatrix} \begin{bmatrix} x_1 \\ x_2 \\ x_3 \end{bmatrix} = \begin{bmatrix} 0 \\ 0 \\ 0 \end{bmatrix}$$

This matrix equation corresponds to the following system of linear equations.

$$\begin{aligned} x_1 + 2x_2 + 3x_3 &= 0 \\ -x_2 + x_3 &= 0 \\ x_1 + x_2 + 4x_3 &= 0 \end{aligned}$$

On solving this system we get many solutions, $x_1 = -5r$, $x_2 = r$, $x_3 = r$. The kernel is thus the set of vectors of the form $(-5r, r, r)$.

$$\text{Ker}(T) = \{(-5r, r, r)\}$$

$\text{Ker}(T)$ is a one-dimensional subspace of $\mathbf{R}^3$ with basis $(-5, 1, 1)$.

Range: The range is spanned by the column vectors of A. Write these column vectors as rows of a matrix and compute an echelon form of the matrix. The nonzero row vectors will give a basis for the range. We get

$$\begin{bmatrix} 1 & 0 & 1 \\ 2 & -1 & 1 \\ 3 & 1 & 4 \end{bmatrix} \approx \begin{bmatrix} 1 & 0 & 1 \\ 0 & -1 & -1 \\ 0 & 1 & 1 \end{bmatrix} \approx \begin{bmatrix} 1 & 0 & 1 \\ 0 & 1 & 1 \\ 0 & 1 & 1 \end{bmatrix} \approx \begin{bmatrix} 1 & 0 & 1 \\ 0 & 1 & 1 \\ 0 & 0 & 0 \end{bmatrix}$$

The vectors $(1, 0, 1)$ and $(0, 1, 1)$ span the range of T. An arbitrary vector in the range will be a linear combination of these vectors,

$$s(1, 0, 1) + t(0, 1, 1)$$

Thus the range of T is

$$\text{Range}(T) = \{(s, t, s + t)\}$$

The vectors $(1, 0, 1)$ and $(0, 1, 1)$ are also linearly independent. $\text{Range}(T)$ is a two-dimensional subspace of $\mathbf{R}^3$ with basis $\{(1, 0, 1), (0, 1, 1)\}$.

The following theorem gives an important relationship between the "sizes" of the kernel and the range of a linear transformation.

THEOREM 6.5

Let $T: U \to V$ be a linear transformation. Then

$$\dim \ker(T) + \dim \text{range}(T) = \dim \text{domain}(T)$$

(Observe that this result holds for the linear transformation T of the above example: $\dim \ker(T) = 1$, $\dim \text{range}(T) = 2$, $\dim \text{domain}(T) = 3$.)

Proof Let us assume that the kernel consists of more than the zero vector, and that it is not the whole of U. (The reader is asked to prove the result for these two special cases in Exercise 11.)

Let $\mathbf{u}_1, \ldots, \mathbf{u}_m$ be a basis for $\ker(T)$. Add vectors $\mathbf{u}_{m+1}, \ldots, \mathbf{u}_n$ to this set to get a basis $\mathbf{u}_1, \ldots, \mathbf{u}_n$ for U. We shall show that $T(\mathbf{u}_{m+1}), \ldots, T(\mathbf{u}_n)$ form a basis for the range, thus proving the theorem.

Let $\mathbf{u}$ be a vector in U. $\mathbf{u}$ can be expressed as a linear combination of the basis vectors as follows.

$$\mathbf{u} = a_1\mathbf{u}_1 + \cdots + a_m\mathbf{u}_m + a_{m+1}\mathbf{u}_{m+1} + \cdots + a_n\mathbf{u}_n$$

Thus

$$T(\mathbf{u}) = T(a_1\mathbf{u}_1 + \cdots + a_m\mathbf{u}_m + a_{m+1}\mathbf{u}_{m+1} + \cdots + a_n\mathbf{u}_n)$$

The linearity of T gives

$$T(\mathbf{u}) = a_1T(\mathbf{u}_1) + \cdots + a_mT(\mathbf{u}_m) + a_{m+1}T(\mathbf{u}_{m+1}) + \cdots + a_nT(\mathbf{u}_n)$$

Since $\mathbf{u}_1, \ldots, \mathbf{u}_m$ are in the kernel, this reduces to

$$T(\mathbf{u}) = a_{m+1}T(\mathbf{u}_{m+1}) + \cdots + a_nT(\mathbf{u}_n)$$

$T(\mathbf{u})$ represents an arbitrary vector in the range of T. Thus the vectors $T(\mathbf{u}_{m+1}), \ldots, T(\mathbf{u}_n)$ span the range.

It remains to prove that these vectors are also linearly independent. Consider the identity

$$b_{m+1}T(\mathbf{u}_{m+1}) + \cdots + b_nT(\mathbf{u}_n) = \mathbf{0} \tag{1}$$

where the scalars have been labeled thus for convenience. The linearity of T implies that

$$T(b_{m+1}\mathbf{u}_{m+1} + \cdots + b_n\mathbf{u}_n) = \mathbf{0}$$

This means that the vector $b_{m+1}\mathbf{u}_{m+1} + \cdots + b_n\mathbf{u}_n$ is in the kernel. Thus it can be expressed as a linear combination of the basis of the kernel. Let

$$b_{m+1}\mathbf{u}_{m+1} + \cdots + b_n\mathbf{u}_n = c_1\mathbf{u}_1 + \cdots + c_m\mathbf{u}_m$$

Thus

$$c_1\mathbf{u}_1 + \cdots + c_m\mathbf{u}_m - b_{m+1}\mathbf{u}_{m+1} - \cdots - b_n\mathbf{u}_n = \mathbf{0}$$

Since the vectors $\mathbf{u}_1, \ldots, \mathbf{u}_m, \mathbf{u}_{m+1}, \ldots, \mathbf{u}_n$ are a basis they are linearly independent. Therefore the coefficients are all zero.

$$c_1 = 0, \ldots, c_m = 0, b_{m+1} = 0, \ldots, b_n = 0$$

Returning to identity (1) this implies that $T(\mathbf{u}_{m+1}), \ldots, T(\mathbf{u}_n)$ are linearly independent. The set of vectors $T(\mathbf{u}_{m+1}), \ldots, T(\mathbf{u}_n)$ is a basis for the range. The theorem is proven.

Note that the "bigger" the kernel, the "smaller" the range, and vice versa.

We remind the reader that the subspace spanned by the column vectors of a matrix is called the **column space** of the matrix, and that the dimension of the column space is called the **rank** of the matrix. We now have the following result.

The dimension of the range of a matrix transformation is the rank of the matrix.

The kernel of a linear mapping T is often called the **null space**. Dim $\ker(T)$ is called the **nullity** and dim range(T) is called the **rank** of the transformation. The previous theorem is often referred to as the **rank/nullity theorem** and written in the following form:

$$\text{rank}(T) + \text{nullity}(T) = \dim \text{domain}(T)$$

EXAMPLE 3 Find the dimensions of the kernel and range of the linear transformation T defined by the matrix

$$A = \begin{bmatrix} 1 & 0 & 3 \\ 0 & 1 & 5 \\ 0 & 0 & 0 \end{bmatrix}$$

SOLUTION

Observe that the matrix A is in echelon form. The row vectors are linearly independent. Thus $\text{rank}(A) = 2$, implying that $\dim \text{range}(T) = 2$.
The domain of T is $\mathbf{R}^3$; $\dim \text{domain}(T) = 3$. Therefore $\dim \ker(T) = 1$. ∎

EXERCISE SET 6.3

Kernel and Range

1. Consider the linear transformation T defined by each of the following matrices. Determine the kernel and range of each transformation. Show that $\dim \ker(T) + \dim \text{range}(T) = \dim \text{domain}(T)$ for each transformation.

(a) $\begin{bmatrix} 1 & 2 \\ 3 & 0 \end{bmatrix}$ (b) $\begin{bmatrix} 2 & 0 \\ 3 & 0 \end{bmatrix}$

(c) $\begin{bmatrix} 2 & 4 \\ 4 & 8 \end{bmatrix}$ (d) $\begin{bmatrix} 1 & 2 \\ -1 & 3 \end{bmatrix}$

(e) $\begin{bmatrix} 1 & 2 & 3 \\ 0 & 1 & 2 \end{bmatrix}$ (f) $\begin{bmatrix} 1 & 0 & 0 \\ 0 & 2 & 0 \\ 0 & 0 & 3 \end{bmatrix}$

(g) $\begin{bmatrix} 0 & 1 & 0 \\ 0 & 2 & 0 \\ 0 & 0 & 4 \end{bmatrix}$ (h) $\begin{bmatrix} 1 & 2 & 1 \\ -1 & -2 & 0 \\ 2 & 4 & 1 \end{bmatrix}$

(i) $\begin{bmatrix} 1 & 1 & 5 \\ 0 & 1 & 3 \\ 2 & 1 & 7 \end{bmatrix}$

2. Determine the kernel and range of each of the following transformations. Show that $\dim \ker(T) + \dim \text{range}(T) = \dim \text{domain}(T)$ for each transformation.

(a) $T(x, y, z) = (x, 0, 0)$ of $\mathbf{R}^3 \to \mathbf{R}^3$

(b) $T(x, y, z) = (x + y, z)$ of $\mathbf{R}^3 \to \mathbf{R}^2$

(c) $T(x, y, z) = x + y + z$ of $\mathbf{R}^3 \to \mathbf{R}$

(d) $T(x, y) = (x, 2x, 3x)$ of $\mathbf{R}^2 \to \mathbf{R}^3$

(e) $T(x, y) = (3x, x - y, y)$ of $\mathbf{R}^2 \to \mathbf{R}^3$

3. Let $T: U \to U$. Let $\mathbf{u}$ be an arbitrary vector in U, $\mathbf{v}$ be a fixed vector in U, and $\mathbf{0}$ be the zero vector of U. Determine which

of the following transformations are linear. Find the kernel and range of each linear transformation.

(a) $T(\mathbf{u}) = 5\mathbf{u}$ (b) $T(\mathbf{u}) = 2\mathbf{u} + 3\mathbf{v}$

(c) $T(\mathbf{u}) = \mathbf{u}$ (This is called the *identity transformation* on U.)

(d) $T(\mathbf{u}) = \mathbf{0}$ (This is called the *zero transformation* on U.)

Rank/Nullity Theorem

4. Let $T: U \to V$ be linear. Prove that

$$\dim \ker(T) + \dim \text{range}(T) = \dim \text{domain}(T)$$

(a) when $\ker(T) = \mathbf{0}$ (b) when $\ker(T) = U$

These were the special cases mentioned in the proof of the Rank/Nullity theorem.

5. Let $T: U \to V$ be a linear transformation.

(a) Prove that when $\dim(U) = \dim(V)$, $\text{range}(T) = V$ if and only if $\ker(T) = \mathbf{0}$.

(b) Prove that $\ker(T) = U$ if and only if $\text{range}(T) = \mathbf{0}$.

6. Let $T: U \to V$ be a linear transformation. Prove that the dimension of the range of T can never be greater than the dimension of the domain.

7. Let an $m \times n$ matrix A define a mapping $T: \mathbf{R}^n \to \mathbf{R}^m$. Since A^t is an $n \times m$ matrix, it will define a mapping $T^t: \mathbf{R}^m \to \mathbf{R}^n$. Show that

$$\dim \text{range}(T) = \dim \text{range}(T^t)$$

Linear Transformations on $\mathbf{R}^n$

8. Consider the projection $T(x, y, z) = (x, y, 0)$ of $\mathbf{R}^3 \to \mathbf{R}^3$. Find the set of vectors in $\mathbf{R}^3$ that are mapped by T into the vector $(1, 2, 0)$. Sketch this set.

9. Consider the linear operator $T(x, y) = (x - y, 2y - 2x)$ of $\mathbf{R}^2 \to \mathbf{R}^2$. Find and sketch the set of vectors in $\mathbf{R}^2$ that are mapped by T into the vector $(2, -4)$.

10. Consider the linear operator $T: \mathbf{R}^2 \to \mathbf{R}^2$ defined by $T(x, y) = (2x, 3x)$. Find and sketch the set of vectors that are mapped by T into the vector $(4, 6)$. Is this set a subspace of $\mathbf{R}^2$?

11. Consider the linear transformation $T: \mathbf{R}^3 \to \mathbf{R}^2$ defined by $T(x, y, z) = (x - y, x + z)$. Find and sketch the set of vectors that are mapped by T into the vector $(1, 4)$. Is this set a subspace?

Linear Transformations on Function Spaces

12. Prove that $T: P_2 \to P_2$ defined as follows is linear. Find the kernel and range of T. Give bases for these subspaces.

$$T(a_2 x^2 + a_1 x + a_0) = (a_2 + a_1)x^2 + a_1 x + 2a_0$$

13. Prove that $T: P_3 \to P_2$ defined as follows is linear. Find the kernel and range of T. Give bases for these subspaces.

$$T(a_3 x^3 + a_2 x^2 + a_1 x + a_0) = a_3 x^2 - a_0$$

14. Prove that $g: P_2 \to P_3$ defined as follows is linear. Find the kernel and range of g. Give bases for these subspaces.

$$g(a_2 x^2 + a_1 x + a_0) = 2a_2 x^3 + a_1 x + 3a_0$$

15. Let D be the operation of taking the derivative. Interpret D as a linear operator on P_n. Find the image of $x^3 - 3x^2 + 2x + 1$ under D. Determine the kernel and range of D.

16. Let D^2 be the operation of taking the second derivative.
    (a) Find the image of $2x^2 + 3x^2 - 5x + 4$ under D^2. Prove that D^2 is a linear operator on P_n. Determine the kernel and range of D^2.
    (b) Consider the transformation $D^2 + D + 3$ of P_n into itself. Find the image of $x^3 - 2x^2 + 6x + 1$ under this transformation. Prove that the transformation is linear. Determine the kernel and range of the transformation.

17. Let D be the operation of taking the derivative. Interpret D as an operator on P_n. Find the set of polynomials that are mapped by D into $3x^2 - 4x + 7$.

18. Let D^2 be the operation of taking the second derivative. Interpret D^2 as an operator on P_n. Find the set of polynomials that are mapped by D^2 into $4x^3 + 6x^2 - 2x + 3$.

19. Let p be a polynomial in P_n and let $T(p) = \int_0^1 p(x)\,dx$ be a linear transformation of P_n into $\mathbf{R}$. Find $T(8x^3 + 6x^2 + 4x + 1)$. Determine the kernel and range of T. Find the set of elements of P_n that are mapped into 2.

Miscellaneous

20. Let $\mathbf{x}$ be an arbitrary vector in $\mathbf{R}^n$ and $\mathbf{y}$ a fixed vector. Prove that the transformation $T(\mathbf{x}) = \mathbf{x} \cdot \mathbf{y}$ is a linear transformation of $\mathbf{R}^n \to \mathbf{R}$. ($\mathbf{x} \cdot \mathbf{y}$ is the dot product of $\mathbf{x}$ and $\mathbf{y}$.)

21. Let det be a transformation of $M_{22} \to \mathbf{R}$ defined by $\det(A) = |A|$. (det maps a square matrix into its determinant.) Is det a linear transformation?

22. The equation $T(B) = \begin{bmatrix} 1 & 2 \\ 3 & 4 \end{bmatrix} B$ where B is a 2×2 matrix defines a transformation of M_{22} to M_{22}.

    (a) Find the image of $B = \begin{bmatrix} 2 & 0 \\ 1 & 3 \end{bmatrix}$.

    (b) Prove that this transformation is linear.

23. Let T_1 and T_2 be linear transformations of $U \to V$ and c be a scalar. Let transformations $T_1 + T_2$ and cT_1 be defined by

$$(T_1 + T_2)(\mathbf{u}) = T_1(\mathbf{u}) + T_2(\mathbf{u}) \text{ and } (cT_1)(\mathbf{u}) = cT_1(\mathbf{u})$$

    (a) Prove that $T_1 + T_2$ and cT_1 are both linear transformations.

    (b) Prove that if T_1 and T_2 are matrix transformations defined by the matrices A_1 and A_2, then $(T_1 + T_2)$ is defined by the matrix $(A_1 + A_2)$, and cT_1 is defined by the matrix cA_1.

    (c) Prove that the set of transformations under these operations of addition and scalar multiplication is a vector space.

24. Let U be the vector space of 2×2 matrices. Let A and B be 2×2 matrices. Determine whether the following transformations are linear. Find the kernel and range of each linear transformation.
    (a) $T(A) = A^t$ (b) $T(A) = |A|$
    (c) $T(A) = \text{trace}(A)$ (d) $T(A) = A^2$
    (e) $T(A) = A + B$ (f) $T(A) = a_{11}$
    (g) $T(A) = 0$ (h) $T(A) = I_2$
    (i) $T(A) = A + A^t$ (j) $T(A) =$ reduced echelon form(A)

25. Let $T: U \to V$ be a linear mapping. Let $\mathbf{v}$ be a nonzero vector in V. Let W be the set of vectors in U such that $T(\mathbf{w}) = \mathbf{v}$. Is W a subspace of U?

26. Let A and B be row equivalent matrices. Let T_A and T_B be the linear transformations defined by A and B.
    (a) Show that T_A and T_B have the same kernel.
    (b) Do T_A and T_B have the same range?

6.4 One-to-One Transformations and Inverse Transformations

An element in the range of a transformation may be the image of a single element in the domain or the image of multiple elements in the domain. In this section we study the significance of those transformations for which each element in the range "comes from" a single element in the domain. We shall find that they are the transformations that have an inverse.

> **DEFINITION**   A transformation T is said to be *one-to-one* if each element in the range is the image of just one element in the domain. See Figure 6.18.

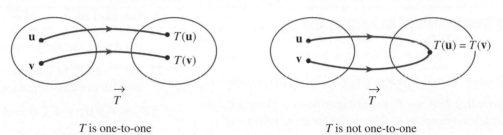

T is one-to-one T is not one-to-one

Figure 6.18

We shall sometimes work with an equivalent form of this definition: The transformation T is one-to-one if $T(\mathbf{u}) = T(\mathbf{v})$ implies that $\mathbf{u} = \mathbf{v}$. For example, the transformation $T(x) = 2x$ is one-to-one. If we take an element in the range such as 6, it is the image of only the single element 3 in the domain. On the other hand, the transformation $T(x) = x^2$ is not one-to-one. For example, if we take the element 9 in the range, it is the image of both -3 and 3 in the domain.

We are of course particularly interested in transformations that are linear. In the following theorems we discuss the requirements for a linear transformation to be one-to-one.

THEOREM 6.6

A linear transformation T is one-to-one if and only if the kernel is the zero vector.

Proof   Let us first assume that T is one-to-one. The kernel consists of all the vectors that are mapped into zero. Since T is one-to-one, the kernel must consist of a single vector. However we know that the zero vector must be in the kernel. Thus the kernel is the zero vector.

Conversely, let us assume that the kernel is the zero vector. Let $\mathbf{u}$ and $\mathbf{v}$ be vectors such that $T(\mathbf{u}) = T(\mathbf{v})$. Using the linear properties of T we get

$$T(\mathbf{u}) - T(\mathbf{v}) = \mathbf{0}$$
$$T(\mathbf{u} - \mathbf{v}) = \mathbf{0}$$

Thus $\mathbf{u} - \mathbf{v}$ is in the kernel. But the kernel is the zero vector. Therefore

$$\mathbf{u} - \mathbf{v} = \mathbf{0}$$
$$\mathbf{u} = \mathbf{v}$$

The transformation is thus one-to-one.

THEOREM 6.7

The linear transformation $T: \mathbf{R}^n \rightarrow \mathbf{R}^n$, defined by $T(\mathbf{x}) = A\mathbf{x}$, is one-to-one if and only if A is nonsingular.

Proof Let T be one-to-one. Thus $\ker(T) = \mathbf{0}$. The rank/nullity theorem implies that $\dim \text{range}(T) = n$. Thus the n columns of A are linearly independent, implying that the rank is n and $|A| \neq 0$.

Conversely, assume that A is nonsingular. This implies that the rank of A is n and that its n column vectors are linearly independent. This means that $\dim \text{range}(T) = n$. The rank nullity theorem now implies that $\ker(T) = \mathbf{0}$. Thus T is one-to-one.

We refer to a linear transformation $T: \mathbf{R}^n \rightarrow \mathbf{R}^n$, defined by $T(\mathbf{x}) = A\mathbf{x}$ with $|A| \neq 0$, as being a *nonsingular transformation*.

EXAMPLE 1 Determine whether the linear transformations T_A and T_B defined by the following matrices are one-to-one or not.

$$\textbf{(a)} \;\; A = \begin{bmatrix} 1 & -2 & 5 & 7 \\ 0 & 1 & 9 & 8 \\ 0 & 0 & 1 & 3 \end{bmatrix} \qquad \textbf{(b)} \;\; B = \begin{bmatrix} 2 & 0 & -1 \\ 3 & 4 & 2 \\ 0 & 7 & 5 \end{bmatrix}$$

SOLUTION

(a) Since the rows of A are linearly independent, $\text{rank}(A) = 3$, implying that dim range$(T_A) = 3$. The domain of T_A is $\mathbf{R}^4$, thus $\dim \text{domain}(T_A) = 4$. By the rank/nullity theorem $\dim \ker(T_A) = 1$. Since $\dim \ker(T_A) \neq 0$ the kernel is not the zero vector. Therefore the mapping is not one-to-one.

(b) It can be shown that $|B| = -9 \neq 0$. The matrix B is nonsingular. Therefore the mapping T_B is one-to-one.

We now find that one of the valuable properties of one-to-one linear transformations is that they preserve linear independence.

THEOREM 6.8

Let $T: U \rightarrow V$ be a one-to-one linear transformation. If the set $\{\mathbf{u}_1, \ldots, \mathbf{u}_n\}$ is linearly independent in U, then $\{T(\mathbf{u}_1), \ldots, T(\mathbf{u}_n)\}$ is linearly independent in V. That is, one-to-one linear transformations preserve linear independence.

Proof Consider the identity

$$a_1 T(\mathbf{u}_1) + \cdots + a_n T(\mathbf{u}_n) = \mathbf{0} \tag{1}$$

for scalars $a_1, \ldots, a_n$. Since T is linear this may be written

$$T(a_1 \mathbf{u}_1 + \cdots + a_n \mathbf{u}_n) = \mathbf{0}$$

T is one-to-one; thus the kernel is the zero vector. Therefore

$$a_1 \mathbf{u}_1 + \cdots + a_n \mathbf{u}_n = \mathbf{0}$$

But the set $\{\mathbf{u}_1, \ldots, \mathbf{u}_n\}$ is linearly independent. Thus

$$a_1 = 0, \ldots, a_n = 0$$

Returning to identity (1), this means that $\{T(\mathbf{u}_1), \ldots, T(\mathbf{u}_n)\}$ is linearly independent.

Invertible Linear Transformation

> **DEFINITION**   Let T be a linear transformation of $\mathbf{R}^n \to \mathbf{R}^n$. T is said to be *invertible* if there exists a transformation S of $\mathbf{R}^n \to \mathbf{R}^n$ such that
>
> $$S(T(\mathbf{u})) = \mathbf{u} \quad \text{and} \quad T(S(\mathbf{u})) = \mathbf{u}$$
>
> for every vector $\mathbf{u}$ in $\mathbf{R}^n$.

The following theorem shows that if such a transformation S exists it is unique and linear. S is called the *inverse* of T and is denoted T^{-1}.

THEOREM 6.9

Let T be a linear transformation of $\mathbf{R}^n \to \mathbf{R}^n$. T is invertible if and only if it is nonsingular. The inverse is unique and linear.

Proof

Assume that T is invertible: By the second part of the preceding definition there is a transformation S such that $T(S(\mathbf{u})) = \mathbf{u}$ for every vector $\mathbf{u}$ in $\mathbf{R}^n$. Let $S(\mathbf{u}) = \mathbf{v}$. This means that there is a vector $\mathbf{v}$ such that $T(\mathbf{v}) = \mathbf{u}$ for every vector $\mathbf{u}$ in $\mathbf{R}^n$. The range of T is thus $\mathbf{R}^n$. Theorems 6.5, 6.6, and 6.7 in sequence then imply that the kernel of T is the zero vector, T is one-to-one, and T is nonsingular.

Assume that T is nonsingular: Let the standard matrix of T be A. Since T is nonsingular, A is invertible. Thus $A^{-1}(A\mathbf{u}) = \mathbf{u}$ and $A(A^{-1}\mathbf{u}) = \mathbf{u}$ for any vector $\mathbf{u}$ in $\mathbf{R}^n$. Let S be the linear transformation with standard matrix A^{-1}. Then $S(T(\mathbf{u})) = \mathbf{u}$ and $T(S(\mathbf{u})) = \mathbf{u}$. S is an inverse of T; T is invertible.

Uniqueness: Let S and P be inverses of an invertible transformation T. Let $\mathbf{v}$ be an arbitrary vector in $\mathbf{R}^n$. Since the range of T is $\mathbf{R}^n$ (see the first part of the Proof), there exists a vector $\mathbf{w}$ such that $T(\mathbf{w}) = \mathbf{v}$. Thus $S(\mathbf{v}) = S(T(\mathbf{w})) = \mathbf{w}$ and $P(\mathbf{v}) = P(T(\mathbf{w})) = \mathbf{w}$. This implies that $S = P$; the inverse is unique. Label it T^{-1}.

Linearity: The preceding inverse S is defined by a matrix, namely A^{-1}. Thus the inverse is linear.

Let us summarize these results.

> **Properties of Invertible Linear Transformations**
>
> Let $T(\mathbf{u}) = A\mathbf{u}$ be a linear transformation of $\mathbf{R}^n \to \mathbf{R}^n$. The following statements are equivalent.
>
> **(a)** T is invertible.
> **(b)** T is nonsingular ($|A| \neq 0$).
> **(c)** T is one-to-one.
> **(d)** The kernel of T is the zero vector.
> **(e)** The range of T is $\mathbf{R}^n$. (We say that T maps $\mathbf{R}^n$ *onto* $\mathbf{R}^n$.)
> **(f)** T^{-1} is linear.
> **(g)** T^{-1} is defined by the matrix A^{-1}.

Note that if T maps $\mathbf{u}$ to $\mathbf{v}$, then T^{-1} "brings back" $\mathbf{v}$ to $\mathbf{u}$. See Figure 6.19. While $\mathbf{v}$ is called the image of $\mathbf{u}$, $\mathbf{u}$ is called the *preimage* of $\mathbf{v}$.

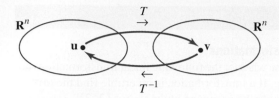

Figure 6.19

EXAMPLE **2** Consider the linear transformation $T(x, y) = (3x + 4y, 5x + 7y)$ of $\mathbf{R}^2 \rightarrow \mathbf{R}^2$. **(a)** Prove that T is invertible and find the inverse of T. **(b)** Determine the preimage of the vector $(1, 2)$.

SOLUTION

(a) Express T in column form to find its standard matrix.

$$T\left(\begin{bmatrix} x \\ y \end{bmatrix}\right) = \begin{bmatrix} 3x + 4y \\ 5x + 7y \end{bmatrix}$$

We find the images of the vectors in the standard basis.

$$T\left(\begin{bmatrix} 1 \\ 0 \end{bmatrix}\right) = \begin{bmatrix} 3 \\ 5 \end{bmatrix}, \quad T\left(\begin{bmatrix} 0 \\ 1 \end{bmatrix}\right) = \begin{bmatrix} 4 \\ 7 \end{bmatrix}$$

These are the columns of the standard matrix A of T. We find that A is invertible proving that T has an inverse.

$$A = \begin{bmatrix} 3 & 4 \\ 5 & 7 \end{bmatrix}, \quad A^{-1} = \begin{bmatrix} 7 & -4 \\ -5 & 3 \end{bmatrix}$$

A^{-1} is the standard matrix of T^{-1}. We get

$$T^{-1}\left(\begin{bmatrix} x \\ y \end{bmatrix}\right) = \begin{bmatrix} 7 & -4 \\ -5 & 3 \end{bmatrix}\begin{bmatrix} x \\ y \end{bmatrix} = \begin{bmatrix} 7x - 4y \\ -5x + 3y \end{bmatrix}$$

Writing in row form for convenience, $T^{-1}(x, y) = (7x - 4y, -5x + 3y)$.

(b) The preimage of $(1, 2)$ will be $T^{-1}(1, 2)$. The preimage of $(1, 2)$ is $(-1, 1)$. See Figure 6.20.

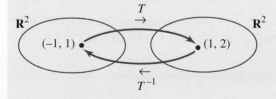

Figure 6.20

EXERCISE SET 6.4

One-to-One Transformations

1. Use the rank/nullity theorem to find the dimensions of the kernels and ranges of the linear transformations defined by the following matrices. State whether the transformations are one-to-one or not.

(a) $A = \begin{bmatrix} 1 & 8 & 2 \\ 0 & 1 & -4 \\ 0 & 0 & 0 \end{bmatrix}$ (b) $B = \begin{bmatrix} 1 & 4 & 2 \\ 0 & 1 & 9 \\ 0 & 0 & 1 \end{bmatrix}$

(c) $C = \begin{bmatrix} 1 & 7 & 3 & 2 \\ 0 & 1 & 5 & 4 \\ 0 & 0 & 1 & 7 \end{bmatrix}$ (d) $D = \begin{bmatrix} 1 & 7 & 3 \\ 0 & 1 & 5 \\ 0 & 0 & 0 \end{bmatrix}$

(e) $E = \begin{bmatrix} 1 & -2 & 3 & 5 \\ 1 & -1 & 8 & 7 \\ 2 & -4 & 6 & 10 \end{bmatrix}$ (f) $F = \begin{bmatrix} 1 & 2 & -3 \\ -2 & 4 & 6 \\ 0 & 2 & -1 \\ 0 & -4 & 2 \end{bmatrix}$

2. Use determinants to decide whether the linear transformations defined by the following matrices are one-to-one or not.

(a) $A = \begin{bmatrix} 1 & 2 & 5 \\ 0 & 3 & 6 \\ 0 & 0 & 4 \end{bmatrix}$ (b) $B = \begin{bmatrix} 3 & 2 & 4 \\ -2 & 0 & 0 \\ 5 & 1 & 2 \end{bmatrix}$

(c) $C = \begin{bmatrix} 1 & 2 & 3 \\ 2 & 4 & 6 \\ 3 & 6 & 9 \end{bmatrix}$ (d) $D = \begin{bmatrix} -1 & 2 & 4 \\ 3 & 1 & 2 \\ 1 & 5 & 10 \end{bmatrix}$

(e) $E = \begin{bmatrix} 2 & 0 & 1 & 3 \\ 0 & 1 & 4 & 7 \\ 0 & 0 & 3 & 8 \\ 0 & 0 & 0 & -4 \end{bmatrix}$

(f) $F = \begin{bmatrix} 2 & 1 & 3 & 0 \\ 4 & 3 & 0 & 2 \\ -1 & 7 & 8 & 1 \\ 0 & -2 & 4 & 0 \end{bmatrix}$

Fixed Points

3. Let $T: U \to U$. A vector $\mathbf{u}$ is said to be a *fixed point* of T if $T(\mathbf{u}) = \mathbf{u}$. Determine the fixed points (if any) of the following transformations.

(a) $T(x, y) = (x, 3y)$

(b) $T(x, y) = (x, 2)$

(c) $T(x, y) = (x, y + 1)$

(d) $T(x, y) = (x, y)$

(e) $T(x, y) = (y, x)$

(f) $T(x, y) = (x + y, x - y)$

(g) Prove that if T is linear the set of fixed points is a subspace.

Inverse Transformations

4. Determine whether the following linear transformations are invertible. If a transformation is invertible find the inverse and compute the preimage of the vector $(2, 3)$.

(a) $T(x, y) = (7x - 3y, 5x - 2y)$

(b) $T(x, y) = (7x + 6y, 8x + 7y)$

(c) $T(x, y) = (2x - y, -4x + 2y)$

(d) $T(x, y) = (x + 4y, 0)$

5. Determine whether the following linear transformations are invertible. If a transformation is invertible, find the inverse and compute the preimage of the vector $(1, -1, 2)$.

(a) $T(x, y, z) = (x + y - z, -3x + 2y - z,$
$\quad 3x - 3y + 2z)$

(b) $T(x, y, z) = (x + z, 4x + 4y + 3z,$
$\quad -4x - 3y - 3z)$

(c) $T(x, y, z) = (x + 2y + 3z, y - z, x + 3y + 2z)$

(d) $T(x, y, z) = (y, z, x)$

Miscellaneous

6. Let $T: U \to V$ be a linear transformation. Prove that

$$\dim \text{range}(T) = \dim \text{domain}(T)$$

if and only if T is one-to-one.

7. Show that no one-to-one linear transformation can exist from $\mathbf{R}^3 \to \mathbf{R}^2$. Generalize this result by proving that no one-to-one linear transformation can exist from $U \to V$ if the dimension of U is greater than the dimension of V.

8. State and prove the converse of Theorem 6.8 about one-to-one linear transformations and linear independence of vectors.

9. Let T be an invertible transformation. Use the linearity of T to prove that if (a) $T(u_1) = v_1$ and $T(u_2) = v_2$, then $T^{-1}(v_1 + v_2) = u_1 + u_2$. (b) $T(u) = v$ and c is a scalar, then $T^{-1}(cv) = cu$. (These confirm that T^{-1} is indeed linear.)

10. Let T_1 and T_2 be invertible transformations of $\mathbf{R}^n \to \mathbf{R}^n$. Prove that $T_2 \circ T_1$ is invertible with inverse $T_1^{-1} \circ T_2^{-1}$.

*6.5 Transformations and Systems of Linear Equations

Linear transformations and the concepts of kernel and range play an important role in the analyses of systems of linear equations. We find that they enable us to "visualize" the sets of solutions.

We have seen that a system of m linear equations in n variables can be written in the matrix form

$$A\mathbf{x} = \mathbf{y}$$

A is an $m \times n$ matrix; it is the matrix of coefficients of the system. The set of solutions is the set of all $\mathbf{x}$'s that satisfy this equation. We now have a very elegant way of looking at this solution set. Let $T: \mathbf{R}^n \rightarrow \mathbf{R}^m$ be the linear transformation defined by A. The system of equations can now be written

$$T(\mathbf{x}) = \mathbf{y}$$

The set of solutions is thus the set of vectors in $\mathbf{R}^n$ that are mapped by T into the vector $\mathbf{y}$. If $\mathbf{y}$ is not in the range of T then the system has no solution. See Figure 6.21.

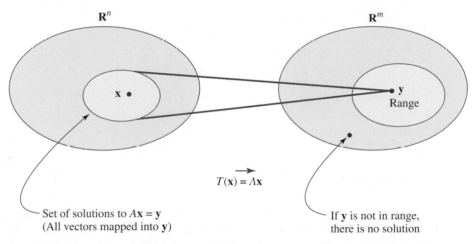

Figure 6.21

Homogeneous Systems

This way of looking at systems of linear equations leads directly to the following result.

THEOREM 6.10

The set of solutions to a homogeneous system of m linear equations in n variables, $A\mathbf{x} = \mathbf{0}$, is a subspace of $\mathbf{R}^n$.

Proof Let T be the linear transformation of $\mathbf{R}^n$ into $\mathbf{R}^m$ defined by A. The set of solutions is the set of vectors in $\mathbf{R}^n$ that are mapped by T into the zero vector. The set of solutions is the kernel of the transformation and is thus a subspace.

EXAMPLE 1 Solve the following homogeneous system of linear equations. Interpret the set of solutions as a subspace. Sketch the subspace of solutions.

$$x_1 + 2x_2 + 3x_3 = 0$$
$$- x_2 + x_3 = 0$$
$$x_1 + x_2 + 4x_3 = 0$$

SOLUTION

Using Gauss-Jordan elimination we get

$$\begin{bmatrix} 1 & 2 & 3 & 0 \\ 0 & -1 & 1 & 0 \\ 1 & 1 & 4 & 0 \end{bmatrix} \approx \begin{bmatrix} 1 & 2 & 3 & 0 \\ 0 & -1 & 1 & 0 \\ 0 & -1 & 1 & 0 \end{bmatrix} \approx \begin{bmatrix} 1 & 2 & 3 & 0 \\ 0 & 1 & -1 & 0 \\ 0 & -1 & 1 & 0 \end{bmatrix}$$

$$\approx \begin{bmatrix} 1 & 0 & 5 & 0 \\ 0 & 1 & -1 & 0 \\ 0 & 0 & 0 & 0 \end{bmatrix}$$

Thus

$$x_1 \quad\quad + 5x_3 = 0$$
$$x_2 - x_3 = 0$$

giving

$$x_1 = -5x_3, \quad x_2 = x_3$$

Assign the value r to x_3. An arbitrary solution is thus

$$x_1 = -5r, \, x_2 = r, \, x_3 = r$$

The solutions are vectors of the form

$$(-5r, r, r)$$

These vectors form a one-dimensional subspace of $\mathbf{R}^3$, with basis $(-5, 1, 1)$. See Figure 6.22. This subspace is the kernel of the transformation defined by the matrix of coefficients of the system, $\begin{bmatrix} 1 & 2 & 3 \\ 0 & -1 & 1 \\ 1 & 1 & 4 \end{bmatrix}$.

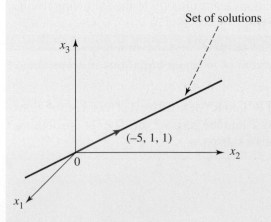

Figure 6.22

Nonhomogeneous Systems

We now find that the set of solutions to a nonhomogeneous system of linear equations does not form a subspace.

Let $A\mathbf{x} = \mathbf{y}$ $(\mathbf{y} \neq 0)$ be a nonhomogeneous system of linear equations. Let $\mathbf{x}_1$ and $\mathbf{x}_2$ be solutions. Thus

$$A\mathbf{x}_1 = \mathbf{y} \quad \text{and} \quad A\mathbf{x}_2 = \mathbf{y}$$

Adding these equations gives

$$A\mathbf{x}_1 + A\mathbf{x}_2 = 2\mathbf{y}$$
$$A(\mathbf{x}_1 + \mathbf{x}_2) = 2\mathbf{y}$$

Therefore $\mathbf{x}_1 + \mathbf{x}_2$ does not satisfy $A\mathbf{x} = \mathbf{y}$. It is not a solution. The set of solutions is not closed under addition; it is not a subspace.

We shall now find that though the set of solutions to a nonhomogeneous system is not itself a subspace, the set can be obtained by sliding a certain subspace. This result will enable us to picture the set of solutions.

THEOREM 6.11

Let $A\mathbf{x} = \mathbf{y}$ be a nonhomogeneous system of m linear equations in n variables. Let $\mathbf{x}_1$ be a particular solution. Every other solution can be written in the form $\mathbf{x} = \mathbf{z} + \mathbf{x}_1$, where $\mathbf{z}$ is an element of the kernel of the transformation T defined by A. The solution is unique if the kernel consists of the zero vector only.

Proof $\mathbf{x}_1$ is a solution. Thus $A\mathbf{x}_1 = \mathbf{y}$. Let $\mathbf{x}$ be an arbitrary solution. Thus $A\mathbf{x} = \mathbf{y}$. Equating $A\mathbf{x}_1$ and $A\mathbf{x}$ we get

$$A\mathbf{x}_1 = A\mathbf{x}$$
$$A\mathbf{x} - A\mathbf{x}_1 = 0$$
$$A(\mathbf{x} - \mathbf{x}_1) = 0$$
$$T(\mathbf{x} - \mathbf{x}_1) = 0$$

Thus $\mathbf{x} - \mathbf{x}_1$ is an element of the kernel of T; call it $\mathbf{z}$.

$$\mathbf{x} - \mathbf{x}_1 = \mathbf{z}$$

We can write

$$\mathbf{x} = \mathbf{z} + \mathbf{x}_1$$

Note that the solution is unique if the only value of $\mathbf{z}$ is 0. That is if the kernel is the zero vector.

This result implies that the set of solutions to a nonhomogeneous system of linear equations $A\mathbf{x} = \mathbf{y}$ can be generated from the kernel of the transformation defined by the matrix of coefficients and a particular solution $\mathbf{x}_1$. If we take any vector $\mathbf{z}$ in the kernel and add $\mathbf{x}_1$ to it we get a solution. Geometrically, this means that the set of solutions is obtained by sliding the kernel in the direction and distance defined by the vector $\mathbf{x}_1$. See Figure 6.23.

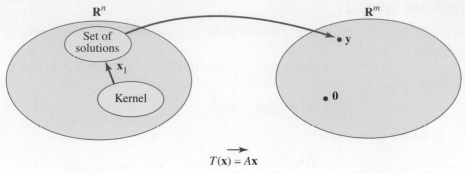

$$T(\mathbf{x}) = A\mathbf{x}$$

Set of solutions to $A\mathbf{x} = \mathbf{y}$

Figure 6.23

EXAMPLE 2 Solve the following system of linear equations. Sketch the set of solutions.

$$\begin{aligned} x_1 + 2x_2 + 3x_3 &= 11 \\ - x_2 + x_3 &= -2 \\ x_1 + x_2 + 4x_3 &= 9 \end{aligned}$$

SOLUTION

Using Gauss-Jordan elimination we get

$$\begin{bmatrix} 1 & 2 & 3 & 11 \\ 0 & -1 & 1 & -2 \\ 1 & 1 & 4 & 9 \end{bmatrix} \approx \begin{bmatrix} 1 & 2 & 3 & 11 \\ 0 & -1 & 1 & -2 \\ 0 & -1 & 1 & -2 \end{bmatrix} \approx \begin{bmatrix} 1 & 2 & 3 & 11 \\ 0 & 1 & -1 & 2 \\ 0 & -1 & 1 & -2 \end{bmatrix}$$

$$\approx \begin{bmatrix} 1 & 0 & 5 & 7 \\ 0 & 1 & -1 & 2 \\ 0 & 0 & 0 & 0 \end{bmatrix}$$

Thus

$$\begin{aligned} x_1 + 5x_3 &= 7 \\ x_2 - x_3 &= 2 \end{aligned}$$

giving

$$x_1 = -5x_3 + 7, \quad x_2 = x_3 + 2$$

Assign the value r to x_3. An arbitrary solution is thus

$$x_1 = -5r + 7, \quad x_2 = r + 2, \quad x_3 = r$$

The solutions are vectors of the form

$$(-5r + 7, r + 2, r)$$

Let us "pull this solution apart." We separate the part involving the parameter r from the constant part. The part involving r will be in the kernel of the transformation defined by A, while the constant part will be a particular solution $\mathbf{x}_1$ to $A\mathbf{x} = \mathbf{y}$.

$$\underbrace{(-5r + 7, r + 2, r)}_{\substack{\text{arbitrary solution} \\ \text{to } A\mathbf{x} = \mathbf{y}}} = \underbrace{r(-5, 1, 1)}_{\text{element of kernel}} + \underbrace{(7, 2, 0)}_{\substack{\text{a particular solution} \\ \text{to } A\mathbf{x} = \mathbf{y}}}$$

We found in Example 1 that the kernel of the mapping defined by this matrix of coefficients is indeed the set of vectors of the form $r(-5, 1, 1)$. The reader can verify by substitution that $(7, 2, 0)$ is indeed a particular solution to the given system.

The set of solutions can be represented geometrically by sliding the kernel, namely the line defined by the vector $(-5, 1, 1)$, in the direction and distance defined by the vector $(7, 2, 0)$. See Figure 6.24.

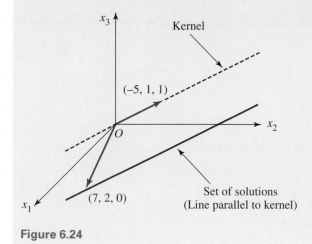

Figure 6.24

Many Systems

Consider a number of linear systems, $A\mathbf{x} = \mathbf{y}_1$, $A\mathbf{x} = \mathbf{y}_2$, $A\mathbf{x} = \mathbf{y}_3, \dots$, all having the same matrix of coefficients A. Let T be the linear transformation defined by A. Let $\mathbf{x}_1, \mathbf{x}_2, \mathbf{x}_3, \dots$, be particular solutions to these systems. Then the sets of solutions to these systems are

$$\ker(T) + \mathbf{x}_1, \quad \ker(T) + \mathbf{x}_2, \quad \ker(T) + \mathbf{x}_3, \dots$$

These sets are "parallel" sets, each being the $\ker(T)$ translated by the amounts $\mathbf{x}_1, \mathbf{x}_2, \mathbf{x}_3, \dots$. Thus, for example, the solutions to the systems

$$x_1 + 2x_2 + 3x_3 = a_1$$
$$- x_2 + x_3 = a_2$$
$$x_1 + x_2 + 4x_3 = a_3$$

will all be straight lines parallel to the line defined by the vector $(-5, 1, 1)$.

EXAMPLE 3 Analyze the solutions to the following system of equations.

$$x_1 - 2x_2 + 3x_3 + x_4 = 1$$
$$2x_1 - 3x_2 + 2x_3 - x_4 = 4$$
$$3x_1 - 5x_2 + 5x_3 = 5$$
$$x_1 - x_2 - x_3 - 2x_4 = 3$$

SOLUTION

Solve using Gauss-Jordan elimination.

$$\begin{bmatrix} 1 & -2 & 3 & 1 & 1 \\ 2 & -3 & 2 & -1 & 4 \\ 3 & -5 & 5 & 0 & 5 \\ 1 & -1 & -1 & -2 & 3 \end{bmatrix} \approx \begin{bmatrix} 1 & -2 & 3 & 1 & 1 \\ 0 & 1 & -4 & -3 & -2 \\ 0 & 1 & -4 & -3 & 2 \\ 0 & 1 & -4 & -3 & 2 \end{bmatrix} \approx \begin{bmatrix} 1 & 0 & -5 & -5 & 5 \\ 0 & 1 & -4 & -3 & 2 \\ 0 & 0 & 0 & 0 & 0 \\ 0 & 0 & 0 & 0 & 0 \end{bmatrix}$$

We get

$$\begin{aligned} x_1 \quad - 5x_3 - 5x_4 &= 5 \\ x_2 - 4x_3 - 3x_4 &= 2 \end{aligned}$$

Express the leading variables in terms of the remaining variables.

$$x_1 = 5x_3 + 5x_4 + 5, \quad x_2 = 4x_3 + 3x_4 + 2$$

The arbitrary solution is

$$(5r + 5s + 5, 4r + 3s + 2, r, s)$$

Separate the parts of this vector as follows:

$$\underbrace{(5r + 5s + 5, 4r + 3s + 2, r, s)}_{\substack{\text{arbitrary solution} \\ \text{to } A\mathbf{x} = \mathbf{y}}} = \underbrace{r(5, 4, 1, 0) + s(5, 3, 0, 1)}_{\text{element of kernel}} + \underbrace{(5, 2, 0, 0)}_{\substack{\text{a particular solution} \\ \text{to } A\mathbf{x} = \mathbf{y}}}$$

Observe that the kernel is a two-dimensional subspace of $\mathbf{R}^4$ with basis $\{(5, 4, 1, 0), (5, 3, 0, 1)\}$. It is a plane through the origin. The set of solutions to the given system is this plane translated in a manner described by the vector $(5, 2, 0, 0)$.

It will be of interest to students who have studied differential equations to now see that these concepts relate to methods used in solving differential equations. The following example illustrates the ideas involved.

EXAMPLE 4 Solve the differential equation $d^2y/dx^2 - 9y = 2x$.

SOLUTION

Let D be the operation of taking the derivative. We can write the differential equation in the form

$$D^2y - 9y = 2x$$
$$(D^2 - 9)y = 2x$$

Let V be the vector space of functions that have second derivatives. $D^2 - 9$ is a linear transformation of V into itself. The solution y to the differential equation is the function that is mapped by $D^2 - 9$ into $2x$. The situation is analogous to solving a system of non-homogeneous linear equations. We get

$$\text{Arbitrary solution} = \begin{pmatrix} \text{element of kernel} \\ \text{of } D^2 - 9 \end{pmatrix} + \begin{pmatrix} \text{particular solution of} \\ D^2 - 9 = 2x \end{pmatrix}$$

Let us first find the kernel of $D^2 - 9$. The general theory tells us that since the operator is of order two, the kernel will be of dimension two, with basis vectors of the form e^{mx}. Since these vectors are in the kernel,

$$(D^2 - 9)e^{mx} = 0$$
$$(m^2 - 9)e^{mx} = 0$$
$$(m - 3)(m + 3)e^{mx} = 0$$
$$m = 3 \quad \text{or} \quad -3$$

Basis vectors of the kernel are thus e^{3x} and e^{-3x}. An arbitrary vector in the kernel can be written $re^{3x} + se^{-3x}$.

It remains to find a particular solution to $D^2y - 9y = 2x$. The simplest solution to find is the one whose second derivative is zero. We get $-9y = 2x$; $y = -2x/9$. Thus an arbitrary solution to the differential equation can be written

$$y = re^{3x} + se^{-3x} - 2x/9$$

■

EXERCISE SET 6.5

Analyzing Solutions

Consider the following systems of linear equations. For convenience, the solutions are given. Express an arbitrary solution as the sum of an element of the kernel of the transformation defined by the matrix of coefficients and a particular solution.

1. $x_1 - x_2 + x_3 = 1$
   $2x_1 - 2x_2 + 3x_3 = 3$
   $x_1 - x_2 - x_3 = -1$
   Solutions: $x_1 = r, x_2 = r, x_3 = 1$

2. $x_1 + x_2 + x_3 = 3$
   $2x_1 + 3x_2 + x_3 = 5$
   $x_1 - x_2 - 2x_3 = -5$
   Solutions: $x_1 = 0, x_2 = 1, x_3 = 2$

3. $x_1 - 2x_2 + 3x_3 = 1$
   $3x_1 - 4x_2 + 5x_3 = 3$
   $2x_1 - 3x_2 + 4x_3 = 2$
   Solutions: $x_1 = r + 1, x_2 = 2r, x_3 = r$

4. $x_1 - x_2 + x_3 = 3$
   $-2x_1 + 2x_2 - 2x_3 = -6$
   Solutions: $x_1 = r - s + 3, x_2 = r, x_3 = s$

5. $x_1 - x_2 - x_3 + 2x_4 = 4$
   $2x_1 - x_2 + 3x_3 - x_4 = 2$
   Solutions: $x_1 = -4r + 3s - 2, x_2 = -5r + 5s - 6,$
   $x_3 = r, x_4 = s$

6. $x_1 + 2x_2 - x_3 + x_4 + 2x_5 = 0$
   $x_2 + 2x_3 - 3x_4 - 3x_5 = 2$
   $x_3 + 5x_4 + 3x_5 = -3$

Solutions: $x_1 = -32r - 23s - 19,$
   $x_2 = 13r + 9s + 8,$
   $x_3 = -5r - 3s - 3,$
   $x_4 = r, x_5 = s$

Miscellaneous Results

7. Consider the transformation $T: \mathbf{R}^3 \rightarrow \mathbf{R}^3$ defined by the following matrix A. The range of this transformation is the set of vectors of the form $(x, y, x + y)$; that is the plane $z = x + y$. Use this information to determine whether solutions exist to the following systems of equations $A\mathbf{x} = \mathbf{y}$, with $\mathbf{y}$ taking on the various values. (You need not determine the solutions if they exist.)

   (a) $\mathbf{y} = (1, 1, 2)$ (b) $\mathbf{y} = (-1, 2, 3)$
   (c) $\mathbf{y} = (3, 2, 5)$ (d) $\mathbf{y} = (2, 4, 5)$

   $$A = \begin{bmatrix} 1 & 2 & 3 \\ 0 & -1 & 1 \\ 1 & 1 & 4 \end{bmatrix}$$

8. In this section we solved nonhomogeneous systems of equations, arriving at an arbitrary solution. The solution was decomposed into the kernel of the transformation defined by the matrix of coefficients and a particular solution to the system. Is the particular solution $\mathbf{x}_1$ unique? If $\mathbf{x}_1$ is not unique, determine a vector other than $(7, 2, 0)$ that can be used in Example 2 of this section.

9. Determine a vector other than $(5, 2, 0, 0)$ that can be used as a particular solution in the analysis of the set of solutions for Example 3 of this section.

10. Construct a nonhomogeneous system of linear equations $A\mathbf{x} = \mathbf{y}$ that has the following matrix of coefficients A and particular solution $\mathbf{x}_1$.

$$A = \begin{bmatrix} 1 & 2 & 1 \\ 0 & 1 & 2 \\ 1 & 1 & -1 \end{bmatrix}, \quad \mathbf{x}_1 = (1, -1, 4)$$

11. Construct a nonhomogeneous system of linear equations $A\mathbf{x} = \mathbf{y}$ that has the following matrix of coefficients A and particular solution $\mathbf{x}_1$.

$$A = \begin{bmatrix} 2 & 1 & 0 \\ 3 & 3 & 1 \\ 0 & 1 & 1 \end{bmatrix}, \quad \mathbf{x}_1 = (2, 0, 3)$$

12. Prove that the solution to a nonhomogeneous system of linear equations is unique if and only if the kernel of the mapping defined by the matrix of coefficients is the zero vector.

Differential Equations

Exercises 13 and 14 are intended for students who have a knowledge of differential equations.

13. Find a basis for the kernel of each of the following operators. Give an arbitrary function in the kernel.

(a) $D^2 + D - 2$ (b) $D^2 + 4D + 3$

(c) $D^2 + 2D - 8$

14. Solve the following differential equations.

(a) $d^2y/dx^2 + 5dy/dx + 6y = 8$
(b) $d^2y/dx^2 - 3dy/dx = 8$
(c) $d^2y/dx^2 - 7dy/dx + 12y = 24$
(d) $d^2y/dx^2 + 8dy/dx = 3e^{2x}$

6.6 Coordinate Vectors

We have discussed various types of vector spaces; the vector spaces $\mathbf{R}^n$, spaces of matrices, and spaces of functions. In this section we shall find how we can use vectors in $\mathbf{R}^n$, called **coordinate vectors**, to describe vectors in any real finite-dimensional vector space. This discussion leads to the conclusion that all finite dimensional vector spaces are in some mathematical sense "the same" as $\mathbf{R}^n$.

DEFINITION   Let U be a vector space with basis $B = \{\mathbf{u}_1, \dots, \mathbf{u}_n\}$ and let $\mathbf{u}$ be a vector in U. We know that there exist unique scalars $a_1, \dots, a_n$, such that

$$\mathbf{u} = a_1\mathbf{u}_1 + \cdots + a_n\mathbf{u}_n$$

The column vector $\mathbf{u}_B = \begin{bmatrix} a_1 \\ \vdots \\ a_n \end{bmatrix}$ is called the **coordinate vector of u** relative to this basis. The scalars

$a_1, \dots, a_n$ are called the **coordinates of u** relative to this basis. Note, we shall use a column vector form for coordinate vectors rather than row vectors. The theory develops most smoothly with this convention.

Notation

Notation is extremely important in mathematics. Much care and attention should be paid to it—it is an intrinsic part of the mathematics. Good notation frees the mind to focus on the concepts involved while poor notation detracts. This section and the following one challenge us to the hilt to find a suitable notation! We have attempted to select the most straightforward notation. The reader should realize that mastering the notation is an important part of the learning experience of this section. We shall be discussing different bases for the vector space U. We have selected the following notation.

We denote two different bases for a vector space U by B and B'. If $\mathbf{u}$ is a vector in U we denote the coordinate vectors of $\mathbf{u}$ relative to these bases by $\mathbf{u}_B$ and $\mathbf{u}_{B'}$.

EXAMPLE 1 Find the coordinate vectors of $\mathbf{u} = (4, 5)$ relative to the following bases B and B' of $\mathbf{R}^2$: **(a)** The standard basis, $B = \{(1, 0), (0, 1)\}$ and **(b)** $B' = \{(2, 1), (-1, 1)\}$.

SOLUTION

(a) By observation we see that

$$(4, 5) = 4(1, 0) + 5(0, 1)$$

Thus $\mathbf{u}_B = \begin{bmatrix} 4 \\ 5 \end{bmatrix}$. The given representation of $\mathbf{u}$ is in fact relative to the standard basis.

(b) Let us now find the coordinate vector of $\mathbf{u}$ relative to B', a basis that is not the standard basis. Let

$$(4, 5) = a_1(2, 1) + a_2(-1, 1)$$

Thus

$$(4, 5) = (2a_1, a_1) + (-a_2, a_2)$$
$$(4, 5) = (2a_1 - a_2, a_1 + a_2)$$

Comparing components leads to the following system of equations.

$$2a_1 - a_2 = 4$$
$$a_1 + a_2 = 5$$

This sytem has the unique solution

$$a_1 = 3, \qquad a_2 = 2$$

Thus $\mathbf{u}_{B'} = \begin{bmatrix} 3 \\ 2 \end{bmatrix}$.

These coordinate vectors have geometrical interpretation. Denote the basis vectors as follows:

$$\mathbf{u}_1 = (1, 0), \mathbf{u}_2 = (0, 1) \quad \text{and} \quad \mathbf{u'}_1 = (2, 1), \mathbf{u'}_2 = (-1, 1)$$

We can write

$$\mathbf{u} = 4\mathbf{u}_1 + 5\mathbf{u}_2, \quad \mathbf{u} = 3\mathbf{u'}_1 + 2\mathbf{u'}_2$$

These vectors are illustrated in Figure 6.25. Observe that each basis defines a coordinate system. B defines the usual rectangular xy-coordinate system. However B' defines a coordinate system $x'y'$ having oblique axes.

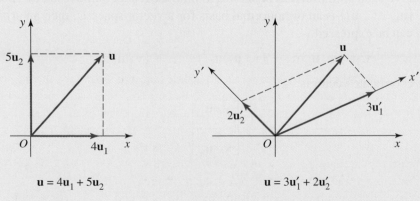

$$\mathbf{u} = 4\mathbf{u}_1 + 5\mathbf{u}_2 \qquad \mathbf{u} = 3\mathbf{u'}_1 + 2\mathbf{u'}_2$$

Figure 6.25

EXAMPLE 2 Find the coordinate vectors of $\mathbf{u} = 5x^2 + x - 3$ relative to the following bases B and B' of P_2.

(a) The standard basis, $B = \{x^2, x, 1\}$

(b) $B' = \{x^2 - x + 1, 3x^2 - 1, 2x^2 + x - 2\}$

SOLUTION

(a) By observation we see that

$$5x^2 + x - 3 = 5(x^2) + (x) - 3(1)$$

Thus $\mathbf{u}_B = \begin{bmatrix} 5 \\ 1 \\ -3 \end{bmatrix}$. The coordinates of $\mathbf{u}$ relative to the standard basis are the

coefficients of $5x^2 + x - 3$.

(b) Once again it is more laborious to find the coordinate vector of $\mathbf{u}$ relative to a basis B' that is not the standard basis. Let

$$5x^2 + x - 3 = a_1(x^2 - x + 1) + a_2(3x^2 - 1) + a_3(2x^2 + x - 2)$$

Thus

$$5x^2 + x - 3 = (a_1 + 3a_2 + 2a_3)x^2 + (-a_1 + a_3)x + (a_1 - a_2 - 2a_3)$$

Comparing coefficients leads to the following system of equations:

$$\begin{aligned} a_1 + 3a_2 + 2a_3 &= 5 \\ -a_1 \quad\quad + a_3 &= 1 \\ a_1 - a_2 - 2a_3 &= -3 \end{aligned}$$

This sytem has the unique solution

$$a_1 = 2, \quad a_2 = -1, \quad a_3 = 3$$

Thus $\mathbf{u}_{B'} = \begin{bmatrix} 2 \\ -1 \\ 3 \end{bmatrix}$.

■

The preceding examples illustrate how the standard bases are the most natural bases to use to express a vector. In general, a system of linear equations has to be solved to find a coordinate vector relative to another basis. However, coordinate vectors can be easily computed relative to an **orthonormal basis**. We remind the reader of the result of Theorem 4.24. If $B = \{\mathbf{u}_1, \ldots, \mathbf{u}_n\}$ is an orthonormal basis for a vector space U, then an arbitrary vector $\mathbf{v}$ in U can be expressed

$$\mathbf{v} = (\mathbf{v} \cdot \mathbf{u}_1)\mathbf{u}_1 + (\mathbf{v} \cdot \mathbf{u}_2)\mathbf{u}_2 + \cdots + (\mathbf{v} \cdot \mathbf{u}_n)\mathbf{u}_n$$

Thus the coordinate vector is

$$\mathbf{v}_B = \begin{bmatrix} \mathbf{v} \cdot \mathbf{u}_1 \\ \vdots \\ \mathbf{v} \cdot \mathbf{u}_n \end{bmatrix}$$

EXAMPLE 3 Find the coordinate vector of $\mathbf{v} = (2, -5, 10)$ relative to the orthonormal basis $B = \{(1, 0, 0), (0, \frac{3}{5}, \frac{4}{5}), (0, \frac{4}{5}, -\frac{3}{5})\}$.

SOLUTION

We get

$$
\begin{aligned}
(2, -5, 10) \cdot (1, 0, 0) &= 2 \\
(2, -5, 10) \cdot (0, \tfrac{3}{5}, \tfrac{4}{5}) &= 5 \\
(2, -5, 10) \cdot (0, \tfrac{4}{5}, -\tfrac{3}{5}) &= -10
\end{aligned}
$$

Thus $\mathbf{v}_B = \begin{bmatrix} 2 \\ 5 \\ -10 \end{bmatrix}$.

There are occasions when bases other than orthonormal bases better fit the situation. It becomes necessary to know how coordinate vectors relative to different bases are related. This is the topic of our next discussion.

Change of Basis

Let $B = \{\mathbf{u}_1, \ldots, \mathbf{u}_n\}$ and $B' = \{\mathbf{u}'_1, \ldots, \mathbf{u}'_n\}$ be bases for a vector space U. A vector $\mathbf{u}$ in U will have coordinate vectors $\mathbf{u}_B$ and $\mathbf{u}_{B'}$ relative to these bases. We now discuss the relationship between $\mathbf{u}_B$ and $\mathbf{u}_{B'}$.

Let the coordinate vectors of $\mathbf{u}_1, \ldots, \mathbf{u}_n$ relative to the basis $B' = \{\mathbf{u}'_1, \ldots, \mathbf{u}'_n\}$ be $(\mathbf{u}_1)_{B'}, \ldots, (\mathbf{u}_n)_{B'}$. The matrix P having these vectors as columns plays a central role in our discussion. It is called the **transition matrix from the basis B to the basis B'**.

$$\text{Transition matrix: } P = [(\mathbf{u}_1)_{B'} \ldots (\mathbf{u}_n)_{B'}]$$

THEOREM 6.12

Let $B = \{\mathbf{u}_1, \ldots, \mathbf{u}_n\}$ and $B' = \{\mathbf{u}'_1, \ldots, \mathbf{u}'_n\}$ be bases for a vector space U. If $\mathbf{u}$ is a vector in U having coordinate vectors $\mathbf{u}_B$ and $\mathbf{u}_{B'}$ relative to these bases then

$$\mathbf{u}_{B'} = P\mathbf{u}_B$$

where P is the transition matrix from B to B',

$$P = [(\mathbf{u}_1)_{B'} \ldots (\mathbf{u}_n)_{B'}]$$

Proof Since $\{\mathbf{u}'_1, \ldots, \mathbf{u}'_n\}$ is a basis for U, each of the vectors $\mathbf{u}_1, \ldots, \mathbf{u}_n$ can be expressed as a linear combination of these vectors. Let

$$\mathbf{u}_1 = c_{11}\mathbf{u}'_1 + \cdots + c_{n1}\mathbf{u}'_n$$
$$\vdots$$
$$\mathbf{u}_1 = c_{1n}\mathbf{u}'_1 + \cdots + c_{nn}\mathbf{u}'_n$$

If $\mathbf{u} = a_1\mathbf{u}_1 + \cdots + a_n\mathbf{u}_n$ we get

$$
\begin{aligned}
\mathbf{u} &= a_1\mathbf{u}_1 + \cdots + a_n\mathbf{u}_n \\
&= a_1(c_{11}\mathbf{u}'_1 + \cdots + c_{n1}\mathbf{u}'_n) + \cdots + a_n(c_{1n}\mathbf{u}'_1 + \cdots + c_{nn}\mathbf{u}'_n) \\
&= (a_1c_{11} + \cdots + a_nc_{1n})\mathbf{u}'_1 + \cdots + (a_1c_{n1} + \cdots + a_nc_{nn})\mathbf{u}'_n
\end{aligned}
$$

The coordinate vector of $\mathbf{u}$ relative to B' can therefore be written

$$\mathbf{u}_{B'} = \begin{bmatrix} (a_1 c_{11} + \cdots + a_n c_{1n}) \\ \vdots \\ (a_1 c_{n1} + \cdots + a_n c_{nn}) \end{bmatrix} = \begin{bmatrix} c_{11} & \cdots & c_{1n} \\ & \vdots & \\ c_{n1} & \cdots & c_{nn} \end{bmatrix} \begin{bmatrix} a_1 \\ \vdots \\ a_n \end{bmatrix}$$

$$= [(\mathbf{u}_1)_{B'} \cdots (\mathbf{u}_n)_{B'}] \mathbf{u}_B$$

proving the theorem.

A change of basis for a vector space from a nonstandard basis to the standard basis is straightforward to accomplish. The following example illustrates the ease with which the transition matrix P is found. The columns of P are in fact the vectors of the first basis.

EXAMPLE 4 Consider the bases $B = \{(1, 2), (3, -1)\}$ and $B' = \{(1, 0), (0, 1)\}$ of $\mathbf{R}^2$. If $\mathbf{u}$ is a vector such that $\mathbf{u}_B = \begin{bmatrix} 3 \\ 4 \end{bmatrix}$, find $\mathbf{u}_{B'}$.

SOLUTION

We express the vectors of B in terms of the vectors of B' to get the transition matrix.

$$(1, 2) = 1(1, 0) + 2(0, 1)$$
$$(3, -1) = 3(1, 0) - 1(0, 1)$$

The coordinate vectors of $(1, 2)$ and $(3, -1)$ are $\begin{bmatrix} 1 \\ 2 \end{bmatrix}$ and $\begin{bmatrix} 3 \\ -1 \end{bmatrix}$. The transition matrix P is thus

$$P = \begin{bmatrix} 1 & 3 \\ 2 & -1 \end{bmatrix}$$

(Observe that the columns of P are the vectors of the basis B.) We get

$$\mathbf{u}_{B'} = \begin{bmatrix} 1 & 3 \\ 2 & -1 \end{bmatrix} \begin{bmatrix} 3 \\ 4 \end{bmatrix} = \begin{bmatrix} 15 \\ 2 \end{bmatrix}$$

Let B and B' be bases for a vector space. We now see that the transition matrices from B to B' and B' to B are related.

THEOREM 6.13

Let B and B' be bases for a vector space U and P be the transition matrix from B to B'. Then P is invertible and the transition matrix from B' to B is P^{-1}.

Proof Let $\mathbf{u}$ be a vector in U having column coordinate vectors $\mathbf{u}_B$ and $\mathbf{u}_{B'}$ relative to the bases B and B'. P is given to be the transition matrix from B to B'; let Q be the transition matrix from B' to B. Then we know that

$$\mathbf{u}_{B'} = P\mathbf{u}_B \quad \text{and} \quad \mathbf{u}_B = Q\mathbf{u}_{B'}$$

We can combine these equations in two ways. Substitute for $\mathbf{u}_B$ from the second equation into the first then substitute for $\mathbf{u}_{B'}$ from the first into the second. We get

$$\mathbf{u}_{B'} = PQ\mathbf{u}_{B'} \quad \text{and} \quad \mathbf{u}_B = QP\mathbf{u}_B$$

Since these results hold for all values of $\mathbf{u}_B$ and $\mathbf{u}_{B'}$, we have that

$$PQ = QP = I$$

Thus $Q = P^{-1}$.

The following example introduces a very useful technique for finding the transition matrix from one basis to another if neither basis is the standard basis. It makes use of the standard basis as an intermediate basis because of the ease with which one can transform to and from the standard basis.

EXAMPLE 5 Consider the bases $B = \{(1, 2), (3, -1)\}$ and $B' = \{(3, 1), (5, 2)\}$ of $\mathbf{R}^2$. Find the transition matrix from B to B'. If $\mathbf{u}$ is a vector such that $\mathbf{u}_B = \begin{bmatrix} 2 \\ 1 \end{bmatrix}$, find $\mathbf{u}_{B'}$.

SOLUTION

Let us use the standard basis $S = \{(1, 0), (0, 1)\}$ as an intermediate basis. The transition matrix P from B to S and the transition matrix P' from B' to S are

$$P = \begin{bmatrix} 1 & 3 \\ 2 & -1 \end{bmatrix}, \quad P' = \begin{bmatrix} 3 & 5 \\ 1 & 2 \end{bmatrix}$$

The transition matrix from S to B' will be $(P')^{-1}$. The transition matrix from B to B' (by way of S) is $(P')^{-1}P$. Thus the transition matrix from B to B' is

$$\begin{bmatrix} 3 & 5 \\ 1 & 2 \end{bmatrix}^{-1} \begin{bmatrix} 1 & 3 \\ 2 & -1 \end{bmatrix} = \begin{bmatrix} 2 & -5 \\ -1 & 3 \end{bmatrix} \begin{bmatrix} 1 & 3 \\ 2 & -1 \end{bmatrix} = \begin{bmatrix} -8 & 11 \\ 5 & -6 \end{bmatrix}$$

This gives

$$\mathbf{u}_{B'} = \begin{bmatrix} -8 & 11 \\ 5 & -6 \end{bmatrix} \begin{bmatrix} 2 \\ 1 \end{bmatrix} = \begin{bmatrix} -5 \\ 4 \end{bmatrix}$$

Isomorphisms

Let $T: U \rightarrow V$. We remind the reader that T is **one-to-one** if each element in the range of T corresponds to just one element in the domain of T. If every element of V is the image of an element of U, the transformation T is said to be **onto**.

The following definition brings the concepts of one-to-one, onto, and linearity together in one transformation.

DEFINITION   Let T be a one-to-one, linear transformation of U onto W. T is called an **isomorphism**. U and W are called **isomorphic** vector spaces.

The significance of isomorphic vector spaces is that they are identical from the mathematical point of view. Since an isomorphism is one-to-one and onto the elements of the two spaces match. Furthermore, an isomorphism is a linear transformation, thus such spaces have similar properties of vector addition and scalar multiplication. The physical appearance of the elements in the two spaces may differ, but mathematically they can be thought of as being the same space.

The act of associating a coordinate vector with every vector in a given real vector space defines a transformation of that vector space to $\mathbf{R}^n$. This transformation is linear and, furthermore, we now see that it is one-to-one and onto.

THEOREM 6.14

Let U be a real vector space with basis $\{\mathbf{u}_1, \ldots, \mathbf{u}_n\}$. Let $\mathbf{u}$ be an arbitrary element of U with coordinate vector $\begin{bmatrix} a_1 \\ \vdots \\ a_n \end{bmatrix}$ relative to this basis. The following transformation T is an isomorphism of U onto $\mathbf{R}^n$.

$$T(\mathbf{u}) = \begin{bmatrix} a_1 \\ \vdots \\ a_n \end{bmatrix}$$

Proof   A transformation T is one-to-one if $T(\mathbf{u}) = T(\mathbf{v})$ implies that $\mathbf{u} = \mathbf{v}$. Let

$$T(\mathbf{u}) = \begin{bmatrix} a_1 \\ \vdots \\ a_n \end{bmatrix} \quad \text{and} \quad T(\mathbf{v}) = \begin{bmatrix} a_1 \\ \vdots \\ a_n \end{bmatrix}$$

Thus

$$\mathbf{u} = a_1\mathbf{u}_1 + \cdots + a_n\mathbf{u}_n \quad \text{and} \quad \mathbf{v} = a_1\mathbf{u}_1 + \cdots + a_n\mathbf{u}_n$$

This implies that $\mathbf{u} = \mathbf{v}$. Thus T is one-to-one.

We now prove that T is onto by showing that every element of $\mathbf{R}^n$ is the image of some element of U. Let $\begin{bmatrix} b_1 \\ \vdots \\ b_n \end{bmatrix}$ be an element of $\mathbf{R}^n$. Then

$$T(b_1\mathbf{u}_1 + \cdots + b_n\mathbf{u}_n) = \begin{bmatrix} b_1 \\ \vdots \\ b_n \end{bmatrix}$$

Thus $\begin{bmatrix} b_1 \\ \vdots \\ b_n \end{bmatrix}$ is the image of the vector $b_1\mathbf{u}_1 + \cdots + b_n\mathbf{u}_n$. Therefore T is onto.

We leave it to the reader to show that T is linear.

An extremely important implication of the last theorem is that every real finite-dimensional vector space U is isomorphic to $\mathbf{R}^n$ for some value of n. Thus every real finite-dimensional vector space is identical from the algebraic viewpoint to $\mathbf{R}^n$. In developing the mathematics of $\mathbf{R}^n$, we have thus in a sense developed the mathematics of all real finite-dimensional vector spaces.

EXERCISE SET 6.6

Coordinate Vectors

In Exercises 1–4, find the coordinate vector of $\mathbf{u}$ relative to the given basis B in $\mathbf{R}^2$.

1. $\mathbf{u} = (2, -3)$; $B = \{(1, 0), (0, 1)\}$

2. $\mathbf{u} = (8, -1)$; $B = \{(3, -1), (2, 1)\}$

3. $\mathbf{u} = (5, 1)$; $B = \{(1, 0), (0, 1)\}$

4. $\mathbf{u} = (-7, -5)$; $B = \{(1, -1), (3, 1)\}$

In Exercises 5–8, find the coordinate vector of $\mathbf{u}$ relative to the given basis B in $\mathbf{R}^3$.

5. $\mathbf{u} = (4, 0, -2)$; $B = \{(1, 0, 0), (0, 1, 0), (0, 0, 1)\}$

6. $\mathbf{u} = (6, -3, 1)$; $B = \{(1, -1, 0), (2, 1, -1), (2, 0, 0)\}$

7. $\mathbf{u} = (3, -1, 7)$; $B = \{(2, 0, -1), (0, 1, 3), (1, 1, 1)\}$

8. $\mathbf{u} = (1, -6, -8)$; $B = \{(1, 2, 3), (1, -1, 0), (0, 1, -2)\}$

In Exercises 9–12, find the coordinate vector of $\mathbf{u}$ relative to the given basis B in P_n.

9. $\mathbf{u} = 7x - 3$; $B = \{x, 1\}$

10. $\mathbf{u} = 2x - 6$; $B = \{3x - 5, x - 1\}$

11. $\mathbf{u} = 4x^2 - 5x + 2$; $B = \{x^2, x, 1\}$

12. $\mathbf{u} = 3x^2 - 6x - 2$; $B = \{x^2, x - 1, 2x\}$

In Exercises 13–15, find the coordinate vector of $\mathbf{u}$ relative to the given orthonormal basis B.

13. $\mathbf{u} = (5, 0, -10)$; $B = \{(0, -1, 0), (\frac{3}{5}, 0, -\frac{4}{5}), (\frac{4}{5}, 0, \frac{3}{5})\}$

14. $\mathbf{u} = (1, 2, 3)$; $B = \{(\frac{1}{3}, \frac{2}{3}, \frac{2}{3}), (\frac{2}{3}, -\frac{2}{3}, \frac{1}{3}), (\frac{2}{3}, \frac{1}{3}, -\frac{2}{3})\}$

15. $\mathbf{u} = (2, 1, 4)$;

$$B = \left\{(1, 0, 0), \left(0, \frac{1}{\sqrt{2}}, \frac{1}{\sqrt{2}}\right), \left(0, \frac{1}{\sqrt{2}}, -\frac{1}{\sqrt{2}}\right)\right\}$$

Transformations from Nonstandard to Standard Bases

In Exercises 16–20, find the transition matrix P from the given basis B to the standard basis B' of $\mathbf{R}^2$. Use this matrix to find the coordinate vectors of $\mathbf{u}$, $\mathbf{v}$, and $\mathbf{w}$ relative to B'.

16. $B = \{(2, 3), (1, 2)\}$ and $B' = \{(1, 0), (0, 1)\}$;

$$\mathbf{u}_B = \begin{bmatrix} 1 \\ 2 \end{bmatrix}, \quad \mathbf{v}_B = \begin{bmatrix} 3 \\ -1 \end{bmatrix}, \quad \mathbf{w}_B = \begin{bmatrix} 2 \\ 4 \end{bmatrix}$$

17. $B = \{(4, 1), (3, -1)\}$ and $B' = \{(1, 0), (0, 1)\}$;

$$\mathbf{u}_B = \begin{bmatrix} 3 \\ 0 \end{bmatrix}, \quad \mathbf{v}_B = \begin{bmatrix} 2 \\ 1 \end{bmatrix}, \quad \mathbf{w}_B = \begin{bmatrix} -2 \\ 1 \end{bmatrix}$$

18. $B = \{(1, 1), (2, -3)\}$ and $B' = \{(1, 0), (0, 1)\}$;

$$\mathbf{u}_B = \begin{bmatrix} 0 \\ 1 \end{bmatrix}, \quad \mathbf{v}_B = \begin{bmatrix} 1 \\ -2 \end{bmatrix}, \quad \mathbf{w}_B = \begin{bmatrix} 3 \\ 1 \end{bmatrix}$$

19. $B = \{(3, 2), (2, 1)\}$ and $B' = \{(1, 0), (0, 1)\}$;

$$\mathbf{u}_B = \begin{bmatrix} 2 \\ -1 \end{bmatrix}, \quad \mathbf{v}_B = \begin{bmatrix} 0 \\ -1 \end{bmatrix}, \quad \mathbf{w}_B = \begin{bmatrix} 5 \\ 3 \end{bmatrix}$$

20. $B = \{(2, -3), (-3, 4)\}$ and $B' = \{(1, 0), (0, 1)\}$;

$$\mathbf{u}_B = \begin{bmatrix} 1 \\ 1 \end{bmatrix}, \quad \mathbf{v}_B = \begin{bmatrix} 3 \\ 0 \end{bmatrix}, \quad \mathbf{w}_B = \begin{bmatrix} 4 \\ 2 \end{bmatrix}$$

Transformations from Standard to Nonstandard Bases

21. Consider the bases $B = \{(1, 0), (0, 1)\}$ and $B' = \{(5, 3), (3, 2)\}$ of $\mathbf{R}^2$. If $\mathbf{u}$ is a vector such that $\mathbf{u}_B = \begin{bmatrix} -2 \\ 5 \end{bmatrix}$, find $\mathbf{u}_{B'}$. (*Hint*: The transition matrix from B to B' is the inverse of that from B' to B.)

22. Consider the bases $B = \{(1, 0), (0, 1)\}$ and $B' = \{(1, 2), (-1, -1)\}$ of $\mathbf{R}^2$. If $\mathbf{u}$ is a vector such that $\mathbf{u}_B = \begin{bmatrix} 8 \\ 3 \end{bmatrix}$, find $\mathbf{u}_{B'}$.

Transformation Between Nonstandard Bases

23. Find the transition matrix P from the basis $B = \{(1, 2), (3, 0)\}$ of $\mathbf{R}^2$ to the basis $B' = \{(2, 1), (3, 2)\}$. If $\mathbf{u}_B = \begin{bmatrix} -3 \\ 2 \end{bmatrix}$, find $\mathbf{u}_{B'}$.

24. Find the transition matrix P from the basis $B = \{(3, -1), (1, 1)\}$ of $\mathbf{R}^2$ to the basis $B' = \{(2, 5), (1, 2)\}$. If $\mathbf{u}_B = \begin{bmatrix} 7 \\ -2 \end{bmatrix}$, find $\mathbf{u}_{B'}$.

Function Spaces

25. Consider the vector space P_2. $B = \{x^2, x, 1\}$ and $B' = \{3x^2, x - 1, 4\}$ are bases for P_2. Find the transition matrix from B to B'. Use this matrix to find the coordinate vectors of $3x^2 + 4x + 8$, $6x^2 + 4$, $8x + 12$, and $3x^2 + 4x + 4$ relative to B'. [*Hint*: The transition matrix from B to B' is the inverse of that from B' to B.]

26. Consider the vector space P_1. $B = \{x, 1\}$ and $B' = \{x + 2, 3\}$ are bases for P_1. Find the transition matrix from B to B'. Use this matrix to find the coordinate vectors of $3x + 3$, $6x$, $6x + 9$, and $12x - 3$ relative to B'.

Isomorphisms

27. Construct an isomorphism from the vector space of diagonal 2×2 matrices onto $\mathbf{R}^2$.

28. Construct an isomorphism from the vector space of symmetric 2×2 matrices onto $\mathbf{R}^3$.

29. Let T be a matrix transformation defined by a square matrix A. Prove that T is an isomorphism if and only if A is nonsingular.

30. Let A be a nonsingular matrix that defines an isomorphism of $\mathbf{R}^n$ onto $\mathbf{R}^n$. (See the previous exercise.) Prove that A^{-1} also defines an isomophism of $\mathbf{R}^n$ onto itself.

6.7 Matrix Representations of Linear Transformations

We have seen that a linear transformation $T: \mathbf{R}^n \to \mathbf{R}^m$ can be defined by a matrix A. In this section we introduce a way of representing a linear transformation between general vector spaces by a matrix. We lead up to this discussion by looking at the information that is necessary to define a linear transformation.

Let f be any function. We know that f is defined if its effect on *every* element of the domain is known. This is usually done by means of an equation that gives the effect of the function on an arbitrary element in the domain. For example, consider the function f defined by

$$f(x) = \sqrt{x - 3}$$

The domain of f is $x \geq 3$. The above equation gives the effect of f on every x in this interval. For example, $f(7) = 2$. Similarly, a linear transformation T is defined if its value at every vector in the domain is known. However, unlike a general function, we shall see that if we know the effect of the linear transformation on a finite subset of the domain (a basis), it will be automatically defined on all elements of the domain.

THEOREM 6.15

Let $T: U \to V$ be a linear transformation. Let $\{\mathbf{u}_1, \ldots, \mathbf{u}_n\}$ be a basis for U. T is defined by its effect on the base vectors, namely by $T(\mathbf{u}_1), \ldots, T(\mathbf{u}_n)$. The range of T is spanned by $\{T(\mathbf{u}_1), \ldots, T(\mathbf{u}_n)\}$.

Thus, defining a linear transformation on a basis defines it on the whole domain.

Proof   Let $\mathbf{u}$ be an element of U. Since $\{\mathbf{u}_1, \ldots, \mathbf{u}_n\}$ is a basis for U, there exist scalars $a_1, \ldots, a_n$ such that

$$\mathbf{u} = a_1\mathbf{u}_1 + \cdots + a_n\mathbf{u}_n$$

The linearity of T gives

$$T(\mathbf{u}) = T(a_1\mathbf{u}_1 + \cdots + a_n\mathbf{u}_n)$$
$$= a_1T(\mathbf{u}_1) + \cdots + a_nT(\mathbf{u}_n)$$

Therefore $T(\mathbf{u})$ is known if $\{T(\mathbf{u}_1), \ldots, T(\mathbf{u}_n)\}$ are known.

Further, $T(\mathbf{u})$ may be interpreted to be an arbitrary element in the range of T. It can be expressed as a linear combination of $\{T(\mathbf{u}_1), \ldots, T(\mathbf{u}_n)\}$. Thus $\{T(\mathbf{u}_1), \ldots, T(\mathbf{u}_n)\}$ spans the range of T.

EXAMPLE 1 Consider the linear transformation $T: \mathbf{R}^3 \to \mathbf{R}^2$ defined as follows on basis vectors of $\mathbf{R}^3$. Find $T(1, -2, 3)$.

$$T(1, 0, 0) = (3, -1), \quad T(0, 1, 0) = (2, 1), \quad T(0, 0, 1) = (3, 0)$$

SOLUTION

Since T is defined on basis vectors of $\mathbf{R}^3$, it is defined on the whole space. To find $T(1, -2, 3)$, express the vector $(1, -2, 3)$ as a linear combination of the basis vectors and use the linearity of T.

$$\begin{aligned}
T(1, -2, 3) &= T(1(1, 0, 0) - 2(0, 1, 0) + 3(0, 0, 1)) \\
&= 1T(1, 0, 0) - 2T(0, 1, 0) + 3T(0, 0, 1) \\
&= 1(3, -1) - 2(2, 1) + 3(3, 0) \\
&= (8, -3)
\end{aligned}$$

We have seen that a linear transformation $T: \mathbf{R}^n \to \mathbf{R}^m$ can be defined by a matrix A:

$$T(\mathbf{u}) = A\mathbf{u}$$

where $A = [T(\mathbf{e}_1) \ldots T(\mathbf{e}_n)]$. Note that A is constructed by finding the effect of T on each of the standard basis vectors of $\mathbf{R}^n$. These ideas can be extended to a linear transformation $T: U \to V$ between general vector spaces. We shall represent the elements of U and V by coordinate vectors and T by a matrix A that defines a transformation of coordinate vectors. As for $\mathbf{R}^n$, the matrix A is constructed by finding the effect of T on basis vectors.

THEOREM 6.16

Let U and V be vector spaces with bases $B = \{\mathbf{u}_1, \ldots, \mathbf{u}_n\}$ and $B' = \{\mathbf{v}_1, \ldots, \mathbf{v}_m\}$ and $T: U \to V$ a linear transformation. If $\mathbf{u}$ is a vector in U with image $T(\mathbf{u})$, having coordinate vectors $\mathbf{a}$ and $\mathbf{b}$ relative to these bases, then

$$\mathbf{b} = A\mathbf{a}$$

where $A = [T(\mathbf{u}_1)_{B'} \ldots T(\mathbf{u}_n)_{B'}]$.

The matrix A thus defines a transformation of coordinate vectors of U in the "same way" as T transforms the vectors of U. See Figure 6.26. A is called the **matrix representation of T** (or **matrix of T**) with respect to the bases B and B'.

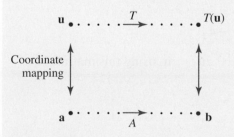

Figure 6.26

Proof Let $\mathbf{u} = a_1\mathbf{u}_1 + \cdots + a_n\mathbf{u}_n$. Using the linearity of T we can write

$$\begin{aligned}
T(\mathbf{u}) &= T(a_1\mathbf{u}_1 + \cdots + a_n\mathbf{u}_n) \\
&= a_1 T(\mathbf{u}_1) + \cdots + a_n T(\mathbf{u}_n)
\end{aligned}$$

Let the effect of T on the basis vectors of U be

$$T(\mathbf{u}_1) = c_{11}\mathbf{v}_1 + \cdots + c_{1m}\mathbf{v}_m$$
$$\vdots$$
$$T(\mathbf{u}_n) = c_{n1}\mathbf{v}_1 + \cdots + c_{nm}\mathbf{v}_m$$

Thus

$$T(\mathbf{u}) = a_1(c_{11}\mathbf{v}_1 + \cdots + c_{1m}\mathbf{v}_m) + \cdots + a_n(c_{n1}\mathbf{v}_1 + \cdots + c_{nm}\mathbf{v}_m)$$
$$= (a_1 c_{11} + \cdots + a_n c_{n1})\mathbf{v}_1 + \cdots + (a_1 c_{1m} + \cdots + a_n c_{nm})\mathbf{v}_m$$

The coordinate vector of $T(\mathbf{u})$ is therefore

$$\mathbf{b} = \begin{bmatrix} (a_1 c_{11} + \cdots + a_n c_{1n}) \\ \vdots \\ (a_1 c_{1m} + \cdots + a_n c_{nm}) \end{bmatrix} = \begin{bmatrix} c_{11} \cdots c_{1n} \\ \vdots \\ c_{1m} \cdots c_{nm} \end{bmatrix} \begin{bmatrix} a_1 \\ \vdots \\ a_n \end{bmatrix}$$
$$= [T(\mathbf{u}_1)_{B'} \ldots T(\mathbf{u}_n)_{B'}]\mathbf{a}$$

proving the theorem.

EXAMPLE **2** Let $T: U \to V$ be a linear transformation. T is defined relative to bases $B = \{\mathbf{u}_1, \mathbf{u}_2, \mathbf{u}_3\}$ and $B' = \{\mathbf{v}_1, \mathbf{v}_2\}$ of U and V as follows.

$$T(\mathbf{u}_1) = 2\mathbf{v}_1 - \mathbf{v}_2$$
$$T(\mathbf{u}_2) = 3\mathbf{v}_1 + 2\mathbf{v}_2$$
$$T(\mathbf{u}_3) = \mathbf{v}_1 - 4\mathbf{v}_2$$

Find the matrix representation of T with respect to these bases and use this matrix to determine the image of the vector $\mathbf{u} = 3\mathbf{u}_1 + 2\mathbf{u}_2 - \mathbf{u}_3$.

SOLUTION

The coordinate vectors of $T(\mathbf{u}_1)$, $T(\mathbf{u}_2)$, and $T(\mathbf{u}_3)$ are

$$\begin{bmatrix} 2 \\ -1 \end{bmatrix}, \begin{bmatrix} 3 \\ 2 \end{bmatrix}, \quad \text{and} \quad \begin{bmatrix} 1 \\ -4 \end{bmatrix}$$

These vectors make up the columns of the matrix of T.

$$A = \begin{bmatrix} 2 & 3 & 1 \\ -1 & 2 & -4 \end{bmatrix}$$

Let us now find the image of the vector $\mathbf{u} = 3\mathbf{u}_1 + 2\mathbf{u}_2 - \mathbf{u}_3$ using this matrix. The coordinate vector of $\mathbf{u}$ is $\mathbf{a} = \begin{bmatrix} 3 \\ 2 \\ -1 \end{bmatrix}$. We get

$$A\mathbf{a} = \begin{bmatrix} 2 & 3 & 1 \\ -1 & 2 & -4 \end{bmatrix} \begin{bmatrix} 3 \\ 2 \\ -1 \end{bmatrix} = \begin{bmatrix} 11 \\ 5 \end{bmatrix}$$

$T(\mathbf{u})$ has coordinate vector $\begin{bmatrix} 11 \\ 5 \end{bmatrix}$. Thus $T(\mathbf{u}) = 11\mathbf{v}_1 + 5\mathbf{v}_2$.

EXAMPLE 3 Consider the linear transformation $T: \mathbf{R}^3 \to \mathbf{R}^2$, defined by $T(x, y, z) = (x + y, 2z)$. Find the matrix of T with respect to the bases $\{\mathbf{u}_1, \mathbf{u}_2, \mathbf{u}_3\}$ and $\{\mathbf{u}'_1, \mathbf{u}'_2\}$ of $\mathbf{R}^3$ and $\mathbf{R}^2$, where

$$\mathbf{u}_1 = (1, 1, 0), \quad \mathbf{u}_2 = (0, 1, 4), \quad \mathbf{u}_3 = (1, 2, 3) \quad \text{and} \quad \mathbf{u}'_1 = (1, 0), \quad \mathbf{u}'_2 = (0, 2)$$

Use this matrix to find the image of the vector $\mathbf{u} = (2, 3, 5)$.

SOLUTION

We find the effect of T on the basis vectors of $\mathbf{R}^3$.

$$T(\mathbf{u}_1) = T(1, 1, 0) = (2, 0) = 2(1, 0) + 0(0, 2) = 2\mathbf{u}'_1 + 0\mathbf{u}'_2$$
$$T(\mathbf{u}_2) = T(0, 1, 4) = (1, 8) = 1(1, 0) + 4(0, 2) = 1\mathbf{u}'_1 + 4\mathbf{u}'_2$$
$$T(\mathbf{u}_3) = T(1, 2, 3) = (3, 6) = 3(1, 0) + 3(0, 2) = 3\mathbf{u}'_1 + 3\mathbf{u}'_2$$

The coordinate vectors of $T(\mathbf{u}_1)$, $T(\mathbf{u}_2)$, and $T(\mathbf{u}_3)$ are thus $\begin{bmatrix} 2 \\ 0 \end{bmatrix}$, $\begin{bmatrix} 1 \\ 4 \end{bmatrix}$, and $\begin{bmatrix} 3 \\ 3 \end{bmatrix}$. These

vectors form the columns of the matrix of T.

$$A = \begin{bmatrix} 2 & 1 & 3 \\ 0 & 4 & 3 \end{bmatrix}$$

Let us now use A to find the image of the vector $\mathbf{u} = (2, 3, 5)$. We determine the coordinate vector of $\mathbf{u}$. It can be shown that

$$\mathbf{u} = (2, 3, 5) = 3(1, 1, 0) + 2(0, 1, 4) - (1, 2, 3) = 3\mathbf{u}_1 + 2\mathbf{u}_2 + (-1)\mathbf{u}_3$$

The coordinate vector of $\mathbf{u}$ is thus $\mathbf{a} = \begin{bmatrix} 3 \\ 2 \\ -1 \end{bmatrix}$. The coordinate vector of $T(\mathbf{u})$ is

$$\mathbf{b} = A\mathbf{a} = \begin{bmatrix} 2 & 1 & 3 \\ 0 & 4 & 3 \end{bmatrix} \begin{bmatrix} 3 \\ 2 \\ -1 \end{bmatrix} = \begin{bmatrix} 5 \\ 5 \end{bmatrix}$$

Therefore $T(\mathbf{u}) = 5\mathbf{u}'_1 + 5\mathbf{u}'_2 = 5(1, 0) + 5(0, 2) = (5, 10)$.

We can check this result directly using the definition $T(x, y, z) = (x + y, 2z)$. For $\mathbf{u} = (2, 3, 5)$ this gives

$$T(\mathbf{u}) = T(2, 3, 5) = (5, 10)$$

While the previous example helped us better understand the idea of matrix representation, the importance of the concept of matrix representation obviously does not lie in its use to find $T(\mathbf{u})$; computing $T(\mathbf{u})$ directly is much easier. We now discuss its relevance.

Importance of Matrix Representations

In the last section we saw that every real finite-dimensional vector space is isomorphic to $\mathbf{R}^n$. This means that any such vector space can be discussed in terms of the appropriate $\mathbf{R}^n$. The fact that every linear transformation can now be represented by a matrix means that all the theoretical mathematics of these vector spaces and their linear transformations can be undertaken in terms of the vector spaces $\mathbf{R}^n$ and matrices.

A second reason is a computational one. The elements of $\mathbf{R}^n$ and matrices can be manipulated on computers. Thus general vector spaces and their linear transformations can be discussed on computers through these representations.

EXAMPLE 4 Consider the linear transformation $T: P_2 \to P_1$ defined by $T(ax^2 + bx + c) = (a + b)x - c$. Find the matrix of T with respect to the bases $\{\mathbf{u}_1, \mathbf{u}_2, \mathbf{u}_3\}$ and $\{\mathbf{u'}_1, \mathbf{u'}_2\}$ of P_2 and P_1, where

$$\mathbf{u}_1 = x^2, \quad \mathbf{u}_2 = x, \quad \mathbf{u}_3 = 1 \quad \text{and} \quad \mathbf{u'}_1 = x, \quad \mathbf{u'}_2 = 1$$

Use this matrix to find the image of $\mathbf{u} = 3x^2 + 2x - 1$.

SOLUTION

Consider the effect of T on each basis vector of P_2.

$$T(\mathbf{u}_1) = T(x^2) = \quad x = 1x + 0(1) \quad = 1\mathbf{u'}_1 + \quad 0\mathbf{u'}_2$$
$$T(\mathbf{u}_2) = \quad T(x) = \quad x = 1x + 0(1) \quad = 1\mathbf{u'}_1 + \quad 0\mathbf{u'}_2$$
$$T(\mathbf{u}_3) = \quad T(1) = -1 = 0x + (-1)(1) = 0\mathbf{u'}_1 + (-1)\mathbf{u'}_2$$

The coordinate vectors of $T(x^2), T(x),$ and $T(1)$ are $\begin{bmatrix} 1 \\ 0 \end{bmatrix}, \begin{bmatrix} 1 \\ 0 \end{bmatrix},$ and $\begin{bmatrix} 0 \\ -1 \end{bmatrix}$. The matrix of T is thus

$$A = \begin{bmatrix} 1 & 1 & 0 \\ 0 & 0 & -1 \end{bmatrix}$$

Let us now use T to find the image of $\mathbf{u} = 3x^2 + 2x - 1$. The coordinate vector of $\mathbf{u}$ relative to the basis $\{x^2, x, 1\}$ is $\mathbf{a} = \begin{bmatrix} 3 \\ 2 \\ -1 \end{bmatrix}$. We get

$$\mathbf{b} = A\mathbf{a} = \begin{bmatrix} 1 & 1 & 0 \\ 0 & 0 & -1 \end{bmatrix} \begin{bmatrix} 3 \\ 2 \\ -1 \end{bmatrix} = \begin{bmatrix} 5 \\ 1 \end{bmatrix}$$

Therefore $T(\mathbf{u}) = 5\mathbf{u'}_1 + 1\mathbf{u'}_2 = 5x + 1$.

We visualize the way this matrix representation works in Figure 6.27. The top half of the figure shows the linear transformation T of P_2 to P_1. The bottom half is analogous to the top half, with A defining a transformation of the coordinate vectors of P_2 into the coordinate vectors of P_1 according to

$$\begin{bmatrix} 1 & 1 & 0 \\ 0 & 0 & -1 \end{bmatrix} \begin{bmatrix} a \\ b \\ c \end{bmatrix} = \begin{bmatrix} a + b \\ -c \end{bmatrix}$$

The bottom half is a **coordinate representation** of the top half.

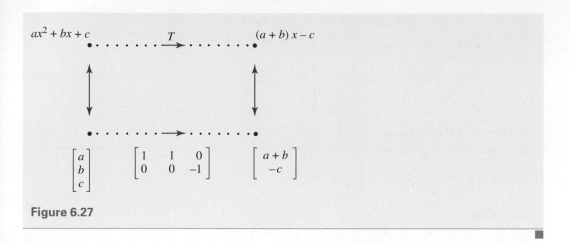

Figure 6.27

When a linear transformation is of a vector space into itself, the matrix representation is usually found relative to a single basis. The following example illustrates this idea.

EXAMPLE 5 Let $D = \frac{d}{dx}$ be the operation of taking the derivative. D is a linear operator on P_2. Find the matrix of D with respect to the basis $\{x^2, x, 1\}$ of P_2.

SOLUTION

We examine the effect of D on the basis vectors.

$$D(x^2) = 2x = 0x^2 + 2x + 0(1)$$
$$D(x) = 1 \;\; = 0x^2 + 0x + 1(1)$$
$$D(1) = 0 \;\; = 0x^2 + 0x + 0(1)$$

The matrix of D is thus

$$A = \begin{bmatrix} 0 & 0 & 0 \\ 2 & 0 & 0 \\ 0 & 1 & 0 \end{bmatrix}$$

The matrix A defines a linear operator on $\mathbf{R}^3$ that is analogous to D on P_2. We have

$$D(ax^2 + bx + c) = 2ax + b \quad \text{and} \quad \begin{bmatrix} 0 & 0 & 0 \\ 2 & 0 & 0 \\ 0 & 1 & 0 \end{bmatrix} \begin{bmatrix} a \\ b \\ c \end{bmatrix} = \begin{bmatrix} 0 \\ 2a \\ b \end{bmatrix}$$

See Figure 6.28.

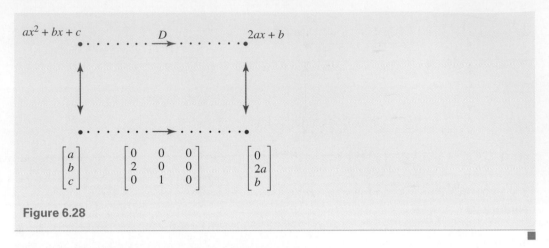

Figure 6.28

Relations Between Matrix Representations

We have seen that the matrix representation of a linear transformation depends upon the bases selected. When linear transformations arise in applications a goal is often to determine a simple matrix representation. Later the reader will, for example, see how bases can be selected that will give diagonal matrix representations of certain linear transformations. At this time we discuss how matrix representations of linear operators relative to different bases are related, paving the way for this work.

We remind the reader that if A and A' are square matrices of the same size, then A' is *similar* to A if there exists an invertible matrix P such that $A' = P^{-1}AP$. The transformation of the matrix A into the matrix A' in this manner is called a similarity *transformation*. We now find that the matrix representations of a linear operator relative to two bases are similar matrices.

We now find that the matrix representation of a linear operator relative to two bases are similar matrices.

THEOREM 6.17

Let U be a vector space with bases B and B'. Let P be the transition matrix from B' to B. If T is a linear operator on U, having matrix A with respect to the first basis and A' with respect to the second basis, then

$$A' = P^{-1}AP$$

Proof   Consider a vector $\mathbf{u}$ in U. Let its coordinate vectors relative to B and B' be $\mathbf{a}$ and $\mathbf{a}'$. The coordinate vectors of $T(\mathbf{u})$ are $A\mathbf{a}$ and $A'\mathbf{a}'$. Since P is the transition matrix from B' to B, we know that

$$\mathbf{a} = P\mathbf{a}' \quad \text{and} \quad A\mathbf{a} = P(A'\mathbf{a}')$$

This second equation may be rewritten

$$P^{-1}A\mathbf{a} = A'\mathbf{a}'$$

Substituting $\mathbf{a} = P\mathbf{a}'$ into this equation gives

$$P^{-1}APa' = A'\mathbf{a}'$$

The effect of the matrices $P^{-1}AP$ and A' as transformations on an arbitrary coordinate vector $\mathbf{a}'$ is the same. Thus these matrices are equal.

EXAMPLE 6 Consider the linear operator $T(x, y) = (2x, x + y)$ on $\mathbf{R}^2$. Find the matrix of T with respect to the standard basis $B = \{(1, 0), (0, 1)\}$ of $\mathbf{R}^2$. Use the transformation $A' = P^{-1}AP$ to determine the matrix A' with respect to the basis $B' = \{(-2, 3), (1, -1)\}$.

SOLUTION

The effect of T on the vectors of the standard basis is

$$T(1, 0) = (2, 1) = 2(1, 0) + 1(0, 1)$$
$$T(0, 1) = (0, 1) = 0(1, 0) + 1(0, 1)$$

The matrix of T relative to the standard basis is

$$A = \begin{bmatrix} 2 & 0 \\ 1 & 1 \end{bmatrix}$$

We now find P the transition matrix from B' to B. Write the vectors of B' in terms of those of B.

$$(-2, 3) = -2(1, 0) + 3(0, 1)$$
$$(1, -1) = 1(1, 0) - 1(0, 1)$$

The transition matrix is

$$P = \begin{bmatrix} -2 & 1 \\ 3 & -1 \end{bmatrix}$$

Therefore

$$A' = P^{-1}AP = \begin{bmatrix} -2 & 1 \\ 3 & -1 \end{bmatrix}^{-1} \begin{bmatrix} 2 & 0 \\ 1 & 1 \end{bmatrix} \begin{bmatrix} -2 & 1 \\ 3 & -1 \end{bmatrix}$$

$$= \begin{bmatrix} 1 & 1 \\ 3 & 2 \end{bmatrix} \begin{bmatrix} 2 & 0 \\ 1 & 1 \end{bmatrix} \begin{bmatrix} -2 & 1 \\ 3 & -1 \end{bmatrix} = \begin{bmatrix} -3 & 2 \\ -10 & 6 \end{bmatrix}$$

■

Diagonal Matrix Representation of a Linear Operator

Let us now see how to find a diagonal matrix representation of a linear operator T, if one exists. A diagonal matrix representation is usually the representation that is most useful.

Let T be a linear operator on a vector space V of dimension n. Let B be a basis for V and let A be the matrix representation of T relative to B. The matrix representation A' of T relative to another basis B' can be obtained using the similarity transformation $A' = P^{-1}AP$, where P is the transition matrix from the basis B' to the basis B. Let us find a basis (if possible) that gives a diagonal form of A'.

Suppose A has eigenvalues $\lambda_1, \ldots, \lambda_n$, with n corresponding, linearly independent eigenvectors $\mathbf{v}_1, \ldots, \mathbf{v}_n$. Let these be the vectors of basis B'.

$$B' = \{\mathbf{v}_1, \ldots, \mathbf{v}_n\}$$

Since P is the transition matrix from B' to B the columns of P will be the column vectors $\mathbf{v}_1, \ldots, \mathbf{v}_n$.

$$P = [\mathbf{v}_1, \ldots, \mathbf{v}_n]$$

The similarity transformation $P^{-1}AP$ now leads to the following diagonal matrix A'. See Theorem 5.6.

$$A' = \begin{bmatrix} \lambda_1 & & 0 \\ & \ddots & \\ 0 & & \lambda_n \end{bmatrix}$$

It is desirable to use an orthogonal matrix for P in this similarity transformation. This is possible if and only if A is symmetric (see Theorem 5.6). If A is symmetric an orthonormal set of eigenvectors can be constructed for the columns of P (using Gram-Schmidt orthogonalization if necessary).

EXAMPLE **7** Consider the linear operator $T(x, y) = (3x + y, x + 3y)$ on $\mathbf{R}^2$. Find a diagonal matrix representation of T. Give the basis for this representation and a geometrical interpretation of T.

SOLUTION

Let us start by finding the matrix representation A relative to the standard basis $B = \{(1, 0), (0, 1)\}$ of $\mathbf{R}^2$. (This is the easiest representation to find.) We get

$$T(1, 0) = (3, 1) = 3(1, 0) + 1(0, 1)$$

$$T(0, 1) = (1, 3) = 1(1, 0) + 3(0, 1)$$

The coordinate vectors of $T(1, 0)$ and $T(0, 1)$ relative to B are $\begin{bmatrix} 3 \\ 1 \end{bmatrix}$ and $\begin{bmatrix} 1 \\ 3 \end{bmatrix}$. The matrix representation of T relative to the standard basis B is thus

$$A = \begin{bmatrix} 3 & 1 \\ 1 & 3 \end{bmatrix}$$

The matrix A has the following eigenvalues and eigenvectors:

$$\lambda_1 = 4, \mathbf{v}_1 = r \begin{bmatrix} 1 \\ 1 \end{bmatrix}; \quad \lambda_2 = 2, \mathbf{v}_2 = s \begin{bmatrix} -1 \\ 1 \end{bmatrix}$$

Since A is symmetric it has an orthonormal set of eigenvectors; we can construct an orthogonal transition matrix P. Normalize the eigenvectors to get the unit vectors

$$\begin{bmatrix} \dfrac{1}{\sqrt{2}} \\ \dfrac{1}{\sqrt{2}} \end{bmatrix}, \begin{bmatrix} \dfrac{-1}{\sqrt{2}} \\ \dfrac{1}{\sqrt{2}} \end{bmatrix}$$

Observe that these vectors, corresponding to distinct eigenvalues, are orthogonal. (This is to be expected since A is symmetric.) Let $B' = \left\{ \left(\dfrac{1}{\sqrt{2}}, \dfrac{1}{\sqrt{2}} \right), \left(-\dfrac{1}{\sqrt{2}}, \dfrac{1}{\sqrt{2}} \right) \right\}$.

This basis leads to the orthogonal transition matrix

$$P = \begin{bmatrix} \dfrac{1}{\sqrt{2}} & \dfrac{-1}{\sqrt{2}} \\ \dfrac{1}{\sqrt{2}} & \dfrac{1}{\sqrt{2}} \end{bmatrix}$$

The similarity transformation $P^{-1}AP$ now gives the following diagonal matrix representation of T relative to this basis B',

$$A' = \begin{bmatrix} \lambda_1 & 0 \\ 0 & \lambda_2 \end{bmatrix} = \begin{bmatrix} 4 & 0 \\ 0 & 2 \end{bmatrix}$$

We now give a geometrical interpretation to the operator T. The standard basis B defines an xy-coordinate system. Let the basis B' define an $x'y'$-coordinate system. Note that the basis B' is obtained from the basis B by rotation through $\pi/4$. See Figure 6.29. The matrix A' tells us that T is a scaling in the $x'y'$ coordinate system, with factor 4 in the x' direction and factor 2 in the y' direction. Thus, for example, T maps the square $PQRO$ into the rectangle $P'Q'R'O$.

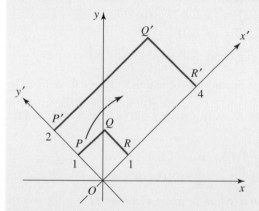

Figure 6.29

This example illustrates a situation that arises frequently in physics and engineering. One is given a mathematical description in an xy-coordinate system. One then searches for a second coordinate system $x'y'$ that better fits the situation and carries out the analysis in that coordinate system. The transformation from the one system to the other is often an orthogonal similarity transformation. We used this approach in the study of oscillating systems in Section 5.4.

EXERCISE SET 6.7

Linear Transformations Defined on Bases

1. Let $T: \mathbf{R}^3 \to \mathbf{R}^2$ be a linear transformation defined as follows on the standard basis of $\mathbf{R}^3$. Find $T(0, 1, -1)$.

$$T(1, 0, 0) = (2, 1), \quad T(0, 1, 0) = (0, -2),$$
$$T(0, 0, 1) = (-1, 1)$$

2. Let $T: \mathbf{R}^2 \to \mathbf{R}$ be a linear transformation defined as follows on the standard basis of $\mathbf{R}^2$. Find $T(3, -2)$.

$$T(1, 0) = 4, \quad T(0, 1) = -3$$

3. Let T be a linear operator on $\mathbf{R}^2$ defined as follows on the standard basis of $\mathbf{R}^2$. Find $T(2, 1)$.

$$T(1, 0) = (2, 5), \quad T(0, 1) = (1, -3)$$

4. Let $T: P_2 \to P_1$ be a linear transformation defined as follows on the standard basis $\{x^2, x, 1\}$ of P_2. Find $T(3x^2 - 2x + 1)$.

$$T(x^2) = 3x + 1, \quad T(x) = 2, \quad T(1) = 2x - 5$$

5. Let T be a linear operator on P_2 defined as follows on the standard basis $\{x^2, x, 1\}$ of P_2. Find $T(x^2 + 3x - 2)$.

$$T(x^2) = x^2 + 3, \quad T(x) = 2x^2 + 4x - 1, \quad T(1) = 3x - 1$$

Matrix of a Linear Transformation

6. Let $T: U \to V$ be a linear transformation. Let T be defined relative to bases $\{\mathbf{u}_1, \mathbf{u}_2\}$ and $\{\mathbf{v}_1, \mathbf{v}_2\}$ of U and V as follows:

$$T(\mathbf{u}_1) = 2\mathbf{v}_1 + 3\mathbf{v}_2, \quad T(\mathbf{u}_2) = 4\mathbf{v}_1 - \mathbf{v}_2$$

Find the matrix of T with respect to these bases. Use this matrix to find the image of the vector $\mathbf{u} = 2\mathbf{u}_1 + 5\mathbf{u}_2$.

7. Let $T: U \to V$ be a linear transformation. Let T be defined relative to bases $\{\mathbf{u}_1, \mathbf{u}_2\}$ and $\{\mathbf{v}_1, \mathbf{v}_2, \mathbf{v}_3\}$ of U and V as follows:

$$T(\mathbf{u}_1) = 2\mathbf{v}_1 + \mathbf{v}_2 - 3\mathbf{v}_3, \quad T(\mathbf{u}_2) = \mathbf{v}_1 - 2\mathbf{v}_2 + \mathbf{v}_3$$

Find the matrix of T with respect to these bases. Use this matrix to find the image of the vector $\mathbf{u} = 4\mathbf{u}_1 - 7\mathbf{u}_2$.

8. Let $T: U \to V$ be a linear transformation. Let T be defined relative to bases $\{\mathbf{u}_1, \mathbf{u}_2, \mathbf{u}_3\}$ and $\{\mathbf{v}_1, \mathbf{v}_2, \mathbf{v}_3\}$ of U and V as follows:

$$T(\mathbf{u}_1) = \mathbf{v}_1 + \mathbf{v}_2 + \mathbf{v}_3, \quad T(\mathbf{u}_2) = 3\mathbf{v}_1 - 2\mathbf{v}_2,$$
$$T(\mathbf{u}_3) = \mathbf{v}_1 + 2\mathbf{v}_2 - \mathbf{v}_3.$$

Find the matrix of T with respect to these bases. Use this matrix to find the image of the vector $\mathbf{u} = 3\mathbf{u}_1 + 2\mathbf{u}_2 - 5\mathbf{u}_3$.

9. Find the matrices of the following linear transformations of $R^3 \to R^2$ with respect to the standard bases of these spaces. Use these matrices to find the images of the vector $(1, 2, 3)$.
   (a) $T(x, y, z) = (x, z)$
   (b) $T(x, y, z) = (3x, y + z)$
   (c) $T(x, y, z) = (x + y, 2x - y)$

10. Find the matrices of the following linear operators on $\mathbf{R}^3$ with respect to the standard basis of $\mathbf{R}^3$. Use these matrices to find the images of the vector $(-1, 5, 2)$.
   *(a)** $T(x, y, z) = (x, 2y, 3z)$ **(b)** $T(x, y, z) = (x, y, z)$
   *(c)** $T(x, y, z) = (x, 0, 0)$
   (d) $T(x, y, z) = (x + y, 3y, x + 2y - 4z)$

11. Consider the linear transformation $T: \mathbf{R}^3 \to \mathbf{R}^2$ defined by $T(x, y, z) = (x - y, x + z)$. Find the matrix of T with respect to the bases $\{\mathbf{u}_1, \mathbf{u}_2, \mathbf{u}_3\}$ and $\{\mathbf{u}'_1, \mathbf{u}'_2\}$ of $\mathbf{R}^3$ and $\mathbf{R}^2$, where

$$\mathbf{u}_1 = (1, -1, 0), \quad \mathbf{u}_2 = (2, 0, 1), \quad \mathbf{u}_3 = (1, 2, 1),$$
$$\text{and} \quad \mathbf{u}'_1 = (-1, 0), \quad \mathbf{u}'_2 = (0, 1)$$

Use this matrix to find the image of the vector $\mathbf{u} = (3, -4, 0)$.

12. Consider the linear transformation $T: \mathbf{R}^2 \to \mathbf{R}^3$ defined by $T(x, y) = (x, x + y, 2y)$. Find the matrix of T with respect to the bases $\{\mathbf{u}_1, \mathbf{u}_2\}$ and $\{\mathbf{u}'_1, \mathbf{u}'_2, \mathbf{u}'_3\}$ of $\mathbf{R}^2$ and $\mathbf{R}^3$, where

$$\mathbf{u}_1 = (1, 3), \quad \mathbf{u}_2 = (4, -1), \quad \text{and}$$
$$\mathbf{u}'_1 = (1, 0, 0), \quad \mathbf{u}'_2 = (0, 2, 0), \quad \mathbf{u}'_3 = (0, 0, -1)$$

Use this matrix to find the image of the vector $\mathbf{u} = (9, 1)$.

13. Consider the linear operator $T: \mathbf{R}^2 \to \mathbf{R}^2$ defined by $T(x, y) = (2x, x + y)$. Find the matrix of T with respect to the basis $\{\mathbf{u}_1, \mathbf{u}_2\}$ of $\mathbf{R}^2$, where

$$\mathbf{u}_1 = (1, 2), \quad \mathbf{u}_2 = (0, -1)$$

Use this matrix to find the image of the vector $\mathbf{u} = (-1, 3)$.

14. Find the matrix of the differential operator D with respect to the basis $\{2x^2, x, -1\}$ of P_2. Use this matrix to find the image of $3x^2 - 2x + 4$.

15. Let V be the vector space of functions having domain $[0, \pi]$ generated by the functions $\sin x$ and $\cos x$. Let D be the operation of taking the derivative with respect to x, and D^2 be the operation of taking the second derivative. Find the matrix of the linear operator $D^2 + 2D + 1$ of V with respect to the basis $\{\sin x, \cos x\}$ of V. Determine the image of the element $3 \sin x + \cos x$ under this transformation.

16. Find the matrix of the following linear transformations with respect to the basis $\{x, 1\}$ of P_1 and $\{x^2, x, 1\}$ of P_2.
   (a) $T(ax^2 + bx + c) = (b + c)x^2 + (b - c)x$ of P_2 into itself.
   (b) $T(ax + b) = bx^2 + ax + b$ of P_1 into P_2.
   (c) $T(ax^2 + bx + c) = 2cx + b - a$ of P_2 into P_1.

17. Find the matrix of the following linear transformation T of P_2 into P_1 with respect to the basis $\{x^2 + x, x, 1\}$ of P_2 and the basis $\{x, 1\}$ of P_1.

$$T(ax^2 + bx + c) = ax + c$$

Determine the image of $3x^2 + 2x - 1$.

18. Find the matrix of the following linear operator T on P_1 with respect to the basis $\{x + 1, 2\}$.

$$T(ax + b) = bx - a$$

Determine the image of $4x - 3$.

Similarity Transformations

19. Find the matrix of the following linear operator T on P_1 with respect to the standard basis $\{x, 1\}$ of P_1.

$$T(ax + b) = (a + b)x - b$$

Use a similarity transformation to then find the matrix of T with respect to the basis $\{x + 1, x - 1\}$ of P_1.

20. Consider the linear operator $T(x, y) = (2x, x + y)$ on $\mathbf{R}^2$. Find the matrix of T with respect to the standard basis for $\mathbf{R}^2$. Use a similarity transformation to then find the matrix with respect to the basis $\{(1, 1), (2, 1)\}$ of $\mathbf{R}^2$.

21. Consider the linear operator $T(x, y) = (x - y, x + y)$ on $\mathbf{R}^2$. Find the matrix of T with respect to the standard basis for $\mathbf{R}^2$. Use a similarity transformation to then find the matrix with respect to the basis $\{(-2, 1), (1, 2)\}$ of $\mathbf{R}^2$.

Miscellaneous Results

22. (a) Let V and W be vector spaces and let U be a subspace of V. Is it always possible to construct a linear transformation of V into W that has U as its kernel?

   (b) Construct a linear transformation of $\mathbf{R}^3$ into $\mathbf{R}^2$ that has the subspace consisting of all vectors of the form $r(1, 3, -1)$ as kernel.

   (c) Construct a linear transformation of $\mathbf{R}^3$ into $\mathbf{R}^2$ that has the subspace consisting of all vectors of the form $(r + s, 2r, -s)$ as kernel.

23. Construct a linear transformation of $\mathbf{R}^2$ into $\mathbf{R}^2$ that has the subspace consisting of all vectors of the form $r(2, -1)$ as kernel.

24. Construct a linear transformation of $\mathbf{R}^2$ into $\mathbf{R}^3$ that has the subspace consisting of all vectors of the form $r(4, 1)$ as kernel.

25. Let $T: U \to U$ for a vector space U be defined by $T(\mathbf{u}) = \mathbf{u}$. Prove that T is linear. T is called the **identity operator** on U. Let B be a basis for U. Show that the matrix of T with respect to B is the identity matrix.

26. Let $T: U \to U$ for a vector space U be defined by $T(\mathbf{u}) = \mathbf{0}$. Prove that T is linear. T is called the **zero operator** on U. Let B be a basis for U. Show that the matrix of T with respect to B is the zero matrix.

27. Let U, V, and W be vector spaces with bases $B = \{\mathbf{u}_1, \ldots, \mathbf{u}_n\}$, $B' = \{\mathbf{v}_1, \ldots, \mathbf{v}_n\}$, and $B'' = \{\mathbf{w}_1, \ldots, \mathbf{w}_n\}$. Let $T: U \to V$ and $L: V \to W$ be linear transformations. Let P be the matrix of T with respect to B and B', and Q be the matrix representation of L with respect to B' and B''. Prove that the matrix of $L \circ T$ with respect to B and B'' is QP.

28. Is it possible for two distinct linear transformations $T: U \to V$ and $L: U \to V$ to have the same matrix with respect to bases B and B' of U and V?

29. Find a diagonal matrix representation for each of the following operators. Determine a basis for each diagonal representation (that involves an orthogonal transformation). Give a geometrical interpretation of T.

(a) $T(x, y) = (4x + 2y, 2x + 4y)$

(b) $T(x, y) = (5x + 3y, 3x + 5y)$

(c) $T(x, y) = (9x + 2y, 2x + 6y)$

(d) $T(x, y) = (14x + 2y, 2x + 11y)$

30. Find a diagonal matrix representation for each of the following operators. Determine a basis for each diagonal

representation. Give a geometrical interpretation of T. (It is not possible to find an orthogonal transformation in these examples.)

(a) $T(x, y) = (8x - 6y, 9x - 7y)$

(b) $T(x, y) = (-2x + 2y, -10x + 7y)$

(c) $T(x, y) = (3x - 4y, 2x - 3y)$

(d) $T(x, y) = (7x + 5y, -10x - 8y)$

CHAPTER 6 REVIEW EXERCISES

1. Consider the transformation T defined by the following matrix A. Determine the images of the vectors **x** and **y**.

$$A = \begin{bmatrix} 1 & 2 & -3 \\ 0 & 4 & 1 \end{bmatrix}; \mathbf{x} = \begin{bmatrix} 3 \\ 0 \\ 1 \end{bmatrix}, \mathbf{y} = \begin{bmatrix} -1 \\ 4 \\ 2 \end{bmatrix}$$

2. Find the matrix that maps $\mathbf{R}^2 \to \mathbf{R}^2$ such that $\begin{bmatrix} 1 \\ 2 \end{bmatrix} \mapsto \begin{bmatrix} 5 \\ 1 \end{bmatrix}$ and $\begin{bmatrix} 3 \\ -2 \end{bmatrix} \mapsto \begin{bmatrix} -1 \\ 11 \end{bmatrix}$.

3. Find a single matrix that defines a rotation of the plane through an angle of $\pi/6$ about the origin, while at the same time moves points to three times their original distance from the origin.

4. Determine the matrix that defines a reflection about the line $y = -x$.

5. Determine the matrix that defines a projection onto the line $y = -x$.

6. Find the equation of the image of the line $y = -5x + 1$ under a scaling of factor 5 in the x-direction and factor 2 in the y-direction.

7. Find the equation of the image of the line $y = 2x + 3$ under a shear of factor 3 in the y-direction.

8. Construct a 2×2 matrix that defines a shear of factor 3 in the y-direction, followed by a scaling of factor 2 in the x-direction, followed by a reflection about the y-axis.

9. Prove that the transformation $T: \mathbf{R}^2 \to \mathbf{R}^2$ defined by $T(x, y) = (2x, x + 3y)$ is linear. Find the image of the vector $(1, 2)$ under this transformation.

10. Determine whether the following transformations are linear.

(a) $T(x, y) = (2x, y, y - x)$ of $\mathbf{R}^2 \to \mathbf{R}^3$

(b) $T(x, y) = (x + y, 2y + 3)$ of $\mathbf{R}^2 \to \mathbf{R}^2$

11. Show that the transformation $T(ax^2 + bx + c) = 2ax + b$ of $P_2 \to P_1$ is linear. Find the image of $3x^2 - 2x + 1$. Determine another element of P_2 that has the same image.

12. Determine the kernel and range of the transformation defined by the matrix $\begin{bmatrix} 6 & 4 \\ 3 & 2 \end{bmatrix}$. Show that

$$\dim \ker(T) + \dim \text{range}(T) - \dim \text{domain}(T).$$

13. Find bases for the kernel and range of the transformation defined by the matrix $\begin{bmatrix} 1 & 1 & 1 \\ 0 & 1 & -1 \\ 2 & 3 & 1 \end{bmatrix}$.

14. Determine whether the linear transformations defined by the following matrices are one-to-one.

(a) $A = \begin{bmatrix} 1 & 3 & -4 & 9 \\ 0 & 1 & 2 & 6 \\ 0 & 0 & 0 & 0 \end{bmatrix}$

(b) $B = \begin{bmatrix} -1 & 2 & 0 & 5 \\ 3 & 7 & 2 & 8 \\ -4 & 2 & 0 & 0 \\ 1 & 3 & 0 & 6 \end{bmatrix}$

15. Determine the kernel and range of the transformation

$$T(x, y, z) = (x, 2x, y - z) \text{ of } \mathbf{R}^3 \to \mathbf{R}^3$$

16. Prove that $g: P_2 \to P_3$ defined as follows is linear. Find the kernel and range of g. Give bases for these subspaces.

$$g(a_2 x^2 + a_1 x + a_0) = (a_2 - a_1)x^3 - a_1 x + 2a_0$$

17. Let D be the operation of taking the derivative. Interpret D as an operator on P_n. Prove that $D^2 - 2D + 1$ is a linear operator. Find the set of polynomials that are mapped by $D^2 - 2D + 1$ into $12x - 4$.

18. Find the coordinate vector of $(-1, 18)$ relative to the basis $\{(1, 3), (-1, 4)\}$.

19. Find the coordinate vector of $3x^2 + 2x - 13$ relative to the basis $\{x^2 + 1, x + 2, x - 3\}$.

20. Find the coordinate vector of $(0, 5, -15)$ relative to the orthonormal basis

$$\{(0, 1, 0), (-\tfrac{3}{5}, 0, \tfrac{4}{5}), (\tfrac{4}{5}, 0, \tfrac{3}{5})\}.$$

21. Find the transition matrix P from the following basis B to the standard basis B' of $\mathbf{R}^2$. Use this matrix to find the coordinate vectors of $\mathbf{u}$, $\mathbf{v}$, and $\mathbf{w}$ relative to B'.

$$B = \{(1, 3), (5, 2)\} \quad \text{and} \quad B' = \{(1, 0), (0, 1)\};$$

$$\mathbf{u}_B = \begin{bmatrix} 3 \\ 1 \end{bmatrix}, \quad \mathbf{v}_B = \begin{bmatrix} 5 \\ -2 \end{bmatrix}, \quad \mathbf{w}_B = \begin{bmatrix} 4 \\ 1 \end{bmatrix}$$

22. Find the transition matrix P from the basis $B = \{(-1, 2), (2, 1)\}$ of $\mathbf{R}^2$ to the basis $B' = \{(4, 3), (-3, 2)\}$. If $\mathbf{u}_B = \begin{bmatrix} 4 \\ 1 \end{bmatrix}$ find $\mathbf{u}_{B'}$.

23. Let T be a linear operator on $\mathbf{R}^2$ defined as follows on the standard basis of $\mathbf{R}^2$. $T(1, 0) = (3, 2)$, $T(0, 1) = (-1, 4)$. Find $T(2, 7)$.

24. Let $T: U \to V$ be a linear transformation. Let T be defined relative to bases $\{\mathbf{u}_1, \mathbf{u}_2\}$ and $\{\mathbf{v}_1, \mathbf{v}_2, \mathbf{v}_3\}$ of U and V as follows:

$$T(\mathbf{u}_1) = \mathbf{v}_1 + 5\mathbf{v}_2 - 2\mathbf{v}_3, \quad T(\mathbf{u}_2) = 3\mathbf{v}_1 - \mathbf{v}_2 + 2\mathbf{v}_3$$

Find the matrix of T with respect to these bases. Use this matrix to find the image of the vector $\mathbf{u} = 2\mathbf{u}_1 - 3\mathbf{u}_2$.

25. Find the matrix of $T(x, y, z) = (2x, -3y)$ relative to the standard bases. Use this matrix to find the image of $(1, 2, 3)$.

26. Find the matrix of $T(ax^2 + bx + c) = (a - b)x^2 + 2cx$ with respect to the basis $\{x^2, x, 1\}$ of P_2. Use the matrix to find the image of $2x^2 - x + 3$.

27. Consider the linear operator $T(x, y) = (3x, x - y)$ on $\mathbf{R}^2$. Find the matrix of T with respect to the standard basis for $\mathbf{R}^2$. Use a similarity transformation to then find the matrix with respect to the basis $\{(1, 2), (2, 3)\}$ of $\mathbf{R}^2$.

Suvarnabhumi Airport, also known as Bangkok
International Airport, is the international airport serving
Southeast Asia and Bangkok, Thailand. The airport has the
world's tallest control tower and the world's third largest
single-building airport terminal. Designed by German-
American architect Helmut Jahn of Murphy/Jahn
Architects, Suvarnabhumi Airport is one of the busiest
airports in Asia.

Inner Product Spaces

In Chapter 4 we generalized the vector space $\mathbf{R}^n$. We drew up a set of axioms based on the properties of $\mathbf{R}^n$, and any set that satisfied those axioms was called a vector space. Such a space had similar algebraic properties to $\mathbf{R}^n$. We now proceed one stage further in this process of generalization; we extend the concepts of dot product of two vectors, norm of a vector, angle between vectors, and distance between points, to general vector spaces. This will enable us, for example, to talk about the magnitudes of functions and orthogonal functions. These concepts are used to approximate functions by polynomials; a technique that is used to implement functions on calculators and computers.

We will no longer be restricted to Euclidean geometry; we will be able to create our own geometries on $\mathbf{R}^n$. We shall, for example, introduce the geometry of special relativity and discuss the implication of this theory for space travel by looking at a voyage to the star Alpha Centauri.

7.1 Inner Product Spaces

The dot product was a key concept on $\mathbf{R}^n$ that led to definitions of norm, angle, and distance. Our approach will be to generalize the dot product of $\mathbf{R}^n$ to a general vector space with a mathematical structure called an **inner product**. This in turn will be used to define norm, angle, and distance for a general vector space. The following definition of inner product is based on the properties of the dot product given in Section 4.2.

DEFINITION   An inner product on a real vector space V is a function that associates a number, denoted $\langle \mathbf{u}, \mathbf{v} \rangle$, with each pair of vectors $\mathbf{u}$ and $\mathbf{v}$ of V. This function satisfies the following conditions for vectors $\mathbf{u}$, $\mathbf{v}$, and $\mathbf{w}$, and scalar c.

1. $\langle \mathbf{u}, \mathbf{v} \rangle = \langle \mathbf{v}, \mathbf{u} \rangle$ (symmetry axiom)
2. $\langle \mathbf{u} + \mathbf{v}, \mathbf{w} \rangle = \langle \mathbf{u}, \mathbf{w} \rangle + \langle \mathbf{v}, \mathbf{w} \rangle$ (additive axiom)
3. $\langle c\mathbf{u}, \mathbf{v} \rangle = c\langle \mathbf{u}, \mathbf{v} \rangle$ (homogeneity axiom)
4. $\langle \mathbf{u}, \mathbf{u} \rangle \geq 0$, and $\langle \mathbf{u}, \mathbf{u} \rangle = 0$ if and only if $\mathbf{u} = \mathbf{0}$ (positive definite axiom)

A vector space V on which an inner product is defined is called an **inner product space**.

Any function on a vector space that satisfies the axioms of an inner product defines an inner product on that space. There can be many inner products on a given vector space. The dot product on $\mathbf{R}^2$ is one inner product on that space. The following example illustrates another inner product on $\mathbf{R}^2$.

EXAMPLE 1 Let $\mathbf{u} = (x_1, x_2)$, $\mathbf{v} = (y_1, y_2)$, and $\mathbf{w} = (z_1, z_2)$ be arbitrary vectors in $\mathbf{R}^2$. Prove that $\langle \mathbf{u}, \mathbf{v} \rangle$ defined as follows is an inner product on $\mathbf{R}^2$.

$$\langle \mathbf{u}, \mathbf{v} \rangle = x_1y_1 + 4x_2y_2$$

Determine the inner product of the vectors $(-2, 5)$, $(3, 1)$ under this inner product.

SOLUTION

We check each of the four inner product axioms. Use the algebraic properties of real numbers and vectors.

Axiom 1:
$$\langle \mathbf{u}, \mathbf{v} \rangle = x_1y_1 + 4x_2y_2$$
$$= y_1x_1 + 4y_2x_2$$
$$= \langle \mathbf{v}, \mathbf{u} \rangle$$

Axiom 2:
$$\langle \mathbf{u} + \mathbf{v}, \mathbf{w} \rangle = \langle (x_1, x_2) + (y_1, y_2), (z_1, z_2) \rangle$$
$$= \langle (x_1 + y_1, x_2 + y_2), (z_1, z_2) \rangle$$
$$= (x_1 + y_1)z_1 + 4(x_2 + y_2)z_2$$
$$= x_1z_1 + y_1z_1 + 4x_2z_2 + 4y_2z_2$$
$$= x_1z_1 + 4x_2z_2 + y_1z_1 + 4y_2z_2$$
$$= \langle (x_1, x_2), (z_1, z_2) \rangle + \langle (y_1, y_2), (z_1, z_2) \rangle$$
$$= \langle \mathbf{u}, \mathbf{v} \rangle + \langle \mathbf{v}, \mathbf{w} \rangle$$

Axiom 3:
$$\langle c\mathbf{u}, \mathbf{v} \rangle = \langle c(x_1, x_2), (y_1, y_2) \rangle = \langle (cx_1, cx_2), (y_1, y_2) \rangle$$
$$= cx_1y_1 + 4cx_2y_2 = c(x_1y_1 + 4x_2y_2)$$
$$= c\langle \mathbf{u}, \mathbf{v} \rangle$$

Axiom 4:
$$\langle \mathbf{u}, \mathbf{u} \rangle = \langle (x_1, x_2), (x_1, x_2) \rangle$$
$$= x_1^2 + 4x_2^2 \geq 0$$

Further, $x_1^2 + 4x_2^2 = 0$ if and only if $x_1 = 0$ and $x_2 = 0$. That is $\mathbf{u} = \mathbf{0}$. Thus $\langle \mathbf{u}, \mathbf{u} \rangle \geq 0$, and $\langle \mathbf{u}, \mathbf{u} \rangle = 0$ if and only if $\mathbf{u} = \mathbf{0}$.

The four inner product axioms are satisfied. $\langle \mathbf{u}, \mathbf{v} \rangle = x_1y_1 + 4x_2y_2$ is an inner product on $\mathbf{R}^2$.

The inner product of the vectors $(-2, 5)$, $(3, 1)$ is

$$\langle (-2, 5), (3, 1) \rangle = (-2 \times 3) + 4(5 \times 1) = 14$$

There are many other inner products on $\mathbf{R}^2$. The dot product is however the most important inner product on $\mathbf{R}^2$ because it leads to the angles and distances of Euclidean geometry. We shall discuss a non-Euclidean geometry based on the inner product of this example in the following section.

We now illustrate inner products on the real vector spaces of matrices and functions.

EXAMPLE 2 Consider the vector space M_{22} of 2×2 matrices. Let **u** and **v**, defined as follows, be arbitrary 2×2 matrices.

$$\mathbf{u} = \begin{bmatrix} a & b \\ c & d \end{bmatrix}, \quad \mathbf{v} = \begin{bmatrix} e & f \\ g & h \end{bmatrix}$$

Prove that the following function is an inner product on M_{22}.

$$\langle \mathbf{u}, \mathbf{v} \rangle = ae + bf + cg + dh$$

(One multiplies corresponding elements of the matrices and adds.)

Determine the inner product of the matrices

$$\begin{bmatrix} 2 & -3 \\ 0 & 1 \end{bmatrix} \quad \text{and} \quad \begin{bmatrix} 5 & 2 \\ 9 & 0 \end{bmatrix}$$

SOLUTION

We shall verify axioms 1 and 3 of an inner product, leaving axioms 2 and 4 for the reader to check in the exercises that follow.

Axiom 1: $\langle \mathbf{u}, \mathbf{v} \rangle = ae + bf + cg + dh = ea + fb + gc + hd$
$$= \langle \mathbf{v}, \mathbf{u} \rangle$$

Axiom 3: Let k be a scalar. Then

$$\langle k\mathbf{u}, \mathbf{v} \rangle = kae + kbf + kcg + kdh = k(ae + bf + cg + dh)$$
$$= k\langle \mathbf{u}, \mathbf{v} \rangle$$

We now compute the inner product of the given matrices. We get

$$\left\langle \begin{bmatrix} 2 & -3 \\ 0 & 1 \end{bmatrix}, \begin{bmatrix} 5 & 2 \\ 9 & 0 \end{bmatrix} \right\rangle = (2 \times 5) + (-3 \times 2) + (0 \times 9) + (1 \times 0) = 4$$

EXAMPLE 3 Consider the vector space P_n, of polynomials of degree $\leq n$. Let f and g be elements of P_n. Prove that the following function defines an inner product on P_n.

$$\langle f, g \rangle = \int_0^1 f(x)g(x)dx$$

Determine the inner product of the polynomials

$$f(x) = x^2 + 2x - 1 \quad \text{and} \quad g(x) = 4x + 1$$

SOLUTION

We shall verify axioms 1 and 2 of an inner product, leaving axioms 3 and 4 for the reader to check in the exercises that follow. We use the properties of integrals.

Axiom 1: $\langle f, g \rangle = \int_0^1 f(x)g(x)dx = \int_0^1 g(x)f(x)dx$

$= \langle g, f \rangle$

Axiom 2: $\langle f + g, h \rangle = \int_0^1 [f(x) + g(x)]h(x)dx$

$= \int_0^1 [f(x)h(x) + g(x)h(x)]dx$

$= \int_0^1 f(x)h(x)dx + \int_0^1 g(x)h(x)dx$

$= \langle f, h \rangle + \langle g, h \rangle$

We now find the inner product of the functions $f(x) = x^2 + 2x - 1$ and $g(x) = 4x + 1$.

$$\langle x^2 + 2x - 1, 4x + 1 \rangle = \int_0^1 (x^2 + 2x - 1)(4x + 1)dx$$

$$= \int_0^1 (4x^3 + 9x^2 - 2x - 1)dx = 2$$

(We shall not include steps of integration in this section. We assume that the reader has the necessary expertise to integrate the functions involved, and to arrive at the given answers.)

Norm of a Vector

The norm of a vector in $\mathbf{R}^n$ can be expressed in terms of the dot product as follows:

$$\|(x_1, \ldots, x_n)\| = \sqrt{(x_1^2 + \cdots + x_n^2)}$$
$$= \sqrt{(x_1, \ldots, x_n) \cdot (x_1, \ldots, x_n)}$$

This definition of norm gave the norms we expected for position vectors in $\mathbf{R}^2$ and $\mathbf{R}^3$ in Euclidean geometry. To get the norm of a vector in a general inner product space we generalize this definition, using the inner product in place of the dot product. The norms that we thus obtain do not necessarily have geometric interpretations, but are often important in numerical work.

DEFINITION   Let V be an inner product space. The norm of a vector $\mathbf{v}$ is denoted $\|\mathbf{v}\|$ and is defined by

$$\|\mathbf{v}\| = \sqrt{\langle \mathbf{v}, \mathbf{v} \rangle}$$

EXAMPLE 4 Consider the vector space P_n of polynomials with inner product

$$\langle f, g \rangle = \int_0^1 f(x)g(x)dx$$

The norm of the function f **generated** by this inner product is

$$\| f \| = \sqrt{\langle f, f \rangle} = \sqrt{\int_0^1 [f(x)]^2 dx}$$

Determine the norm of the function $f(x) = 5x^2 + 1$.

SOLUTION

Using the above definition of norm we get

$$\| 5x^2 + 1 \| = \sqrt{\int_0^1 [5x^2 + 1]^2 dx}$$

$$= \sqrt{\int_0^1 [25x^4 + 10x^2 + 1]dx}$$

$$= \sqrt{\frac{28}{3}}$$

The norm of the function $f(x) = 5x^2 + 1$ is $\sqrt{\frac{28}{3}}$.

Angle Between Two Vectors

The dot product in $\mathbf{R}^n$ was used to define angles between vectors. The angle θ between vectors $\mathbf{u}$ and $\mathbf{v}$ in $\mathbf{R}^n$ is defined by

$$\cos \theta = \frac{\mathbf{u} \cdot \mathbf{v}}{\|\mathbf{u}\| \|\mathbf{v}\|}$$

We generalize this to define the concept of angle for a real vector space, using the inner product in place of the dot product.

DEFINITION Let V be a real inner product space. The angle θ between two nonzero vectors $\mathbf{u}$ and $\mathbf{v}$ in V is given by

$$\cos \theta = \frac{\langle \mathbf{u}, \mathbf{v} \rangle}{\|\mathbf{u}\| \|\mathbf{v}\|}$$

EXAMPLE 5 Consider the inner product space P_n of polynomials with inner product

$$\langle f, g \rangle = \int_0^1 f(x)g(x)dx$$

The angle between two nonzero functions f and g is given by

$$\cos \theta = \frac{\langle f, g \rangle}{\|f\| \|g\|}$$

Determine the cosine of the angle between the functions

$$f(x) = 5x^2 \quad \text{and} \quad g(x) = 3x$$

SOLUTION

We get

$$\langle f, g \rangle = \int_0^1 (5x^2)(3x)dx = \int_0^1 15x^3 = \frac{15}{4}, \quad \|f\| = \|5x^2\| = \sqrt{\int_0^1 (5x^2)^2 dx} = \sqrt{5},$$

$$\|g\| = \|3x\| = \sqrt{\int_0^1 (3x)^2 dx} = \sqrt{3}$$

These give

$$\cos \theta = \frac{15}{4\sqrt{5}\sqrt{3}} = \frac{\sqrt{15}}{4} \quad (\theta = 14.48°)$$

Orthogonal Vectors

Let V be an inner product space. As in $\mathbf{R}^n$, two nonzero vectors in V are said to be orthogonal if the angle between them is a right angle. Thus two nonzero vectors $\mathbf{u}$ and $\mathbf{v}$ in V are orthogonal if

$$\langle \mathbf{u}, \mathbf{v} \rangle = 0$$

EXAMPLE 6 Show that the functions $f(x) = 3x - 2$ and $g(x) = x$ are orthogonal in P_n with inner product $\langle f, g \rangle = \int_0^1 f(x)g(x)dx.$

SOLUTION

We get

$$\langle 3x - 2, x \rangle = \int_0^1 (3x - 2)(x)dx = [x^3 - x^2]_0^1 = 0$$

Thus the functions f and g are orthogonal in this inner product space.

Distance

Our final task is to extend the Euclidean concept of distance to general vector spaces. As for norm, the concept of distance that we arrive at will not in general have direct geometrical interpretation. It can, however, be valuable. For example, it is useful in numerical mathematics to be able to discuss how far apart various functions are.

DEFINITION   Let V be an inner product space with vector norm defined by

$$\|\mathbf{v}\| = \sqrt{\langle \mathbf{v}, \mathbf{v} \rangle}$$

The distance between two vectors (points) $\mathbf{u}$ and $\mathbf{v}$ is denoted $d(\mathbf{u}, \mathbf{v})$ and is defined by

$$d(\mathbf{u}, \mathbf{v}) = \|\mathbf{u} - \mathbf{v}\| \quad (= \sqrt{\langle \mathbf{u} - \mathbf{v}, \mathbf{u} - \mathbf{v} \rangle})$$

EXAMPLE 7 Consider the inner product space P_n of polynomials discussed earlier. Determine which of the functions $g(x) = x^2 - 3x + 5$ or $h(x) = x^2 + 4$ is closest to $f(x) = x^2$.

SOLUTION

We compute the distances between f and g and between f and h. It is easiest to work initially in terms of the squares of the distances.

$$[d(f, g)]^2 = \langle f - g, f - g \rangle = \langle 3x - 5, 3x - 5 \rangle = \int_0^1 (3x - 5)^2 dx = 13$$

$$[d(f, h)]^2 = \langle f - h, f - h \rangle = \langle -4, -4 \rangle = \int_0^1 (-4)^2 dx = 16$$

Thus $d(f, g) = \sqrt{13}$ and $d(f, h) = 4$. The distance between f and h is 4, as we might suspect. g is closer than h to f.

In practice, relative distances between functions are often more significant than absolute distances.

∎

The complex vector space $\mathbf{C}^n$ that was introduced in Section 4.3 is important in many branches of mathematics, physics, and engineering. We now discuss the inner product that is most often used on $\mathbf{C}^n$.

*Inner Product on C^n

For a complex vector space the first axiom of inner product is modified $\langle \mathbf{u}, \mathbf{v} \rangle = \overline{\langle \mathbf{v}, \mathbf{u} \rangle}$. An inner product can then be used to define norm, orthogonality, and distance, as for a real vector space.

Let $\mathbf{u} = (x_1, \ldots, x_n)$ and $\mathbf{v} = (y_1, \ldots, y_n)$ be elements of $\mathbf{C}^n$. The most useful inner product for $\mathbf{C}^n$ is

$$\langle \mathbf{u}, \mathbf{v} \rangle = x_1 \bar{y}_1 + \cdots + x_n \bar{y}_n$$

It can be shown that this definition satisfies the inner product axioms for a complex vector space (see the following exercises). This inner product leads to the following definitions of orthogonality, norm, and distance for $\mathbf{C}^n$.

$$\mathbf{u} \perp \mathbf{v} \text{ if } \langle \mathbf{u}, \mathbf{v} \rangle = 0 \qquad \|\mathbf{u}\| = \sqrt{x_1 \bar{x}_1 + \cdots + x_n \bar{x}_n} \qquad d(\mathbf{u}, \mathbf{v}) = \|\mathbf{u} - \mathbf{v}\|$$

EXAMPLE 8 Consider the vectors $\mathbf{u} = (2 + 3i, -1 + 5i)$, $\mathbf{v} = (1 + i, -i)$ in $\mathbf{C}^2$. Compute

(a) $\langle \mathbf{u}, \mathbf{v} \rangle$, and show that $\mathbf{u}$ and $\mathbf{v}$ are orthogonal

(b) $\|\mathbf{u}\|$ and $\|\mathbf{v}\|$

(c) $d(\mathbf{u}, \mathbf{v})$

SOLUTION

(a) $\langle \mathbf{u}, \mathbf{v} \rangle = (2 + 3i)(1 - i) + (-1 + 5i)(i) = 5 + i - i - 5 = 0$. Thus $\mathbf{u}$ and $\mathbf{v}$ are orthogonal.

(b) $\|\mathbf{u}\| = \sqrt{(2 + 3i)(2 - 3i) + (-1 + 5i)(-1 - 5i)} = \sqrt{13 + 26} = \sqrt{39}$

$\|\mathbf{v}\| = \sqrt{(1 + i)(1 - i) + (-i)(i)} = \sqrt{3}$

$$\text{(c)} \ \ d(\mathbf{u}, \mathbf{v}) = \|\mathbf{u} - \mathbf{v}\| = \|(2 + 3i, -1 + 5i) - (1 + i, -i)\|$$

$$= \|(1 + 2i, -1 + 6i)\|$$

$$= \sqrt{(1 + 2i)(1 - 2i) + (-1 + 6i)(-1 - 6i)} = \sqrt{5 + 37} = \sqrt{42}$$

EXERCISE SET 7.1

Inner Products on R^n

1. Let $\mathbf{u} = (x_1, x_2)$ and $\mathbf{v} = (y_1, y_2)$ be elements of $\mathbf{R}^2$. Prove that the following function defines an inner product on $\mathbf{R}^2$.

$$\langle \mathbf{u}, \mathbf{v} \rangle = 4x_1y_1 + 9x_2y_2$$

2. Let $\mathbf{u} = (x_1, x_2, x_3)$ and $\mathbf{v} = (y_1, y_2, y_3)$ be elements of $\mathbf{R}^3$. Prove that the following function defines an inner product on $\mathbf{R}^3$.

$$\langle \mathbf{u}, \mathbf{v} \rangle = x_1y_1 + 2x_2y_2 + 4x_3y_3$$

3. Let $\mathbf{u} = (x_1, x_2)$ and $\mathbf{v} = (y_1, y_2)$ be elements of $\mathbf{R}^2$. Prove that the following function does not define an inner product on $\mathbf{R}^2$.

$$\langle \mathbf{u}, \mathbf{v} \rangle = 2x_1y_1 - x_2y_2$$

Inner Products on M_{22}

4. Consider the vector space M_{22} of 2×2 matrices. Let $\mathbf{u}$ and $\mathbf{v}$ defined as follows be arbitrary 2×2 matrices.

$$\mathbf{u} = \begin{bmatrix} a & b \\ c & d \end{bmatrix}, \quad \mathbf{v} = \begin{bmatrix} e & f \\ g & h \end{bmatrix}$$

Prove that the following function satisfies axioms 2 and 4 of the inner product.

$$\langle \mathbf{u}, \mathbf{v} \rangle = ae + bf + cg + dh$$

(This exercise completes Example 2 of this section, showing that this function is an inner product.)

5. Let $\mathbf{u}$ and $\mathbf{v}$ defined as follows be arbitrary elements of M_{22}.

$$\mathbf{u} = \begin{bmatrix} a & b \\ c & d \end{bmatrix}, \quad \mathbf{v} = \begin{bmatrix} e & f \\ g & h \end{bmatrix}$$

Prove that the following function is an inner product.

$$\langle \mathbf{u}, \mathbf{v} \rangle = ae + 2bf + 3cg + 4dh$$

Determine the inner products of the following pairs of matrices.

(a) $\mathbf{u} = \begin{bmatrix} 1 & 2 \\ 0 & -3 \end{bmatrix}, \quad \mathbf{v} = \begin{bmatrix} 4 & 1 \\ -3 & 2 \end{bmatrix}$

(b) $\mathbf{u} = \begin{bmatrix} -2 & 4 \\ 1 & 0 \end{bmatrix}, \quad \mathbf{v} = \begin{bmatrix} 5 & -2 \\ 0 & -3 \end{bmatrix}$

Inner Products on P_n

6. Consider the vector space P_n, of polynomials of degree $\leq n$. Let f and g be elements of this space. Prove that the following function satisfies axioms 3 and 4 of the inner product

$$\langle f, g \rangle = \int_0^1 f(x)g(x)dx$$

(This exercise completes Example 3 of this section, showing that this function is an inner product.)

7. Let f and g be arbitrary elements of the vector space P_n. Prove that the following function defines an inner product on P_n if a and b are real numbers with $a < b$. Show that it does not define an inner product if $a \leq b$.

$$\langle f, g \rangle = \int_a^b f(x)g(x)dx$$

In Exercises 8–15, all functions are in the inner product space P_n, with the inner product defined by

$$\langle f, g \rangle = \int_0^1 f(x)g(x)dx$$

8. Determine the inner products of the following functions.
  (a) $f(x) = 2x + 1, g(x) = 3x - 2$
  (b) $f(x) = x^2 + 2, g(x) = 3$
  (c) $f(x) = x^2 + 3x - 2, g(x) = x + 1$
  (d) $f(x) = x^3 + 2x - 1, g(x) = 3x^2 - 4x + 2$

9. Determine the norms of the following functions.
  (a) $f(x) = 4x - 2$ **(b)** $f(x) = 7x^3$
  (c) $f(x) = 3x^2 + 2$ **(d)** $g(x) = x^2 + x + 1$

10. Prove that the functions $f(x) = x^2$ and $g(x) = 4x - 3$ are orthogonal.

11. Prove that the functions $f(x) = 1$ and $g(x) = \frac{1}{2} - x$ are orthogonal.

12. Compute the cosine of the angle between the functions $f(x) = 5x^2$ and $g(x) = 9x$.

13. Determine a function that is orthogonal to $f(x) = 6x + 12$.

14. Compute the distance between the functions $f(x) = x^2 + 3x + 1$ and $g(x) = x^2 + x - 3$.

15. Determine which of the functions $g(x) = x^2 + 2x - 3$ or $h(x) = x^2 - 3x + 4$ is closest to $f(x) = x^2$.

Inner Product Space M_{22}

In Exercises 16–20, all the matrices are elements of the inner product space M_{22} with inner product defined by

$$\left\langle \begin{bmatrix} a & b \\ c & d \end{bmatrix}, \begin{bmatrix} e & f \\ g & h \end{bmatrix} \right\rangle = ae + bf + cg + dh$$

16. Determine the inner products of the following matrices.

(a) $\begin{bmatrix} 1 & 2 \\ 3 & 4 \end{bmatrix}, \begin{bmatrix} -2 & 0 \\ -3 & 5 \end{bmatrix}$

(b) $\begin{bmatrix} 0 & -3 \\ 2 & 5 \end{bmatrix}, \begin{bmatrix} 3 & 6 \\ -2 & -7 \end{bmatrix}$

17. Compute the norms of the following matrices.

(a) $\begin{bmatrix} 1 & 2 \\ 3 & 4 \end{bmatrix}$ (b) $\begin{bmatrix} 0 & 1 \\ -1 & 3 \end{bmatrix}$

(c) $\begin{bmatrix} 5 & -2 \\ -1 & 6 \end{bmatrix}$ (d) $\begin{bmatrix} 4 & -2 \\ -1 & -3 \end{bmatrix}$

18. Prove that the following pairs of matrices are orthogonal.

(a) $\begin{bmatrix} 1 & 2 \\ -1 & 1 \end{bmatrix}, \begin{bmatrix} 2 & 4 \\ 3 & -7 \end{bmatrix}$

(b) $\begin{bmatrix} 5 & 2 \\ -3 & 2 \end{bmatrix}, \begin{bmatrix} -1 & 6 \\ 1 & -2 \end{bmatrix}$

19. Determine a matrix orthogonal to the matrix

$$\begin{bmatrix} 1 & 2 \\ 3 & 4 \end{bmatrix}$$

20. Compute the distances between the following pairs of matrices.

(a) $\begin{bmatrix} 4 & 0 \\ -1 & 3 \end{bmatrix}, \begin{bmatrix} 1 & 1 \\ 1 & 1 \end{bmatrix}$

(b) $\begin{bmatrix} 2 & -3 \\ -1 & 4 \end{bmatrix}, \begin{bmatrix} -3 & 2 \\ 1 & 0 \end{bmatrix}$

Complex Inner Product Space C^n

21. Consider the following vectors $\mathbf{u}$ and $\mathbf{v}$ in the complex inner product space C^2. Compute $\langle \mathbf{u}, \mathbf{v} \rangle$, $\|\mathbf{u}\|$ and $\|\mathbf{v}\|$, $d(\mathbf{u}, \mathbf{v})$. Determine whether $\mathbf{u}$ and $\mathbf{v}$ are orthogonal.

(a) $\mathbf{u} = (2 - i, 3 + 2i)$, $\mathbf{v} = (3 - 2i, 2 + i)$

(b) $\mathbf{u} = (4 + 3i, 1 - i)$, $\mathbf{v} = (2 + i, 4 - 5i)$

(c) $\mathbf{u} = (2 + 3i, -1)$, $\mathbf{v} = (-i, 3 + 2i)$

(d) $\mathbf{u} = (2 - 3i, -2 + 3i)$, $\mathbf{v} = (1, 1)$

22. Consider the following vectors $\mathbf{u}$ and $\mathbf{v}$ in the complex inner product space C^2. Compute $\langle \mathbf{u}, \mathbf{v} \rangle$, $\|\mathbf{u}\|$ and $\|\mathbf{v}\|$, $d(\mathbf{u}, \mathbf{v})$. Determine whether $\mathbf{u}$ and $\mathbf{v}$ are orthogonal.

(a) $\mathbf{u} = (1 + 4i, 1 + i)$, $\mathbf{v} = (1 + i, -4 + i)$

(b) $\mathbf{u} = (2 + 7i, 1 + i)$, $\mathbf{v} = (3 - 4i, 2 + 5i)$

(c) $\mathbf{u} = (1 - 3i, 1 + i)$, $\mathbf{v} = (2 - i, 5i)$

(d) $\mathbf{u} = (3 + i, 2 + 2i)$, $\mathbf{v} = (2 + i, -2 + \frac{2}{3}i)$

23. Let $\mathbf{u} = (x_1, \ldots x_n)$ and $\mathbf{v} = (y_1, \ldots y_n)$ be elements of C_n. Prove that

$$\langle \mathbf{u}, \mathbf{v} \rangle = x_1 \overline{y_1} + \cdots + x_n \overline{y_n}$$

satisfies the inner product axioms for a complex vector space.

24. Prove that if $\mathbf{u}$ and $\mathbf{v}$ are vectors in the complex inner product space C^n and k is a scalar, then

$$\langle \mathbf{u}, k\mathbf{v} \rangle = \overline{k}\langle \mathbf{u}, \mathbf{v} \rangle.$$

Miscellaneous Results

25. Let $\mathbf{u}$ and $\mathbf{v}$ be vectors in a real inner product space and let c be a scalar. Prove that

(a) $\langle \mathbf{0}, \mathbf{v} \rangle = \langle \mathbf{v}, \mathbf{0} \rangle = 0$

(b) $\langle \mathbf{u}, \mathbf{v} + \mathbf{w} \rangle = \langle \mathbf{u}, \mathbf{v} \rangle + \langle \mathbf{u}, \mathbf{w} \rangle$

(c) $\langle \mathbf{u}, c\mathbf{v} \rangle = c\langle \mathbf{u}, \mathbf{v} \rangle$

26. Let $\mathbf{u}$ and $\mathbf{v}$ be vectors in R^n. Let A be an $n \times n$ symmetric matrix such that $\mathbf{w}A\mathbf{w}^t > 0$ for every nonzero vector $\mathbf{w}$ in R^n. Such a matrix is called **positive definite**.

(a) Show that $\langle \mathbf{u}, \mathbf{v} \rangle = \mathbf{u}A\mathbf{v}^t$ satisfies all the axioms of an inner product and thus defines an inner product on R^n. Thus any positive definite matrix can be used to define an inner product on R^n.

(b) Show that the identity matrix I_n defines the dot product on R^n.

(c) Let $\mathbf{u} = (1, 0)$, $\mathbf{v} = (0, 1)$. Find $\langle \mathbf{u}, \mathbf{v} \rangle$, $\|\mathbf{u}\|$ and $\|\mathbf{v}\|$, $d(\mathbf{u}, \mathbf{v})$ for each of the inner products defined by the following positive definite matrices.

$$A = \begin{bmatrix} 2 & 0 \\ 0 & 3 \end{bmatrix}, \quad B = \begin{bmatrix} 1 & 2 \\ 2 & 3 \end{bmatrix},$$

$$C = \begin{bmatrix} 2 & 3 \\ 3 & 5 \end{bmatrix}$$

(d) The inner product of Example 1 of this section can be defined by a positive definite matrix A. Determine A.

27. Let $\mathbf{v}$ be a nonzero vector in an inner product space V. Let W be the set of vectors in V that are orthogonal to $\mathbf{v}$. Prove that W is a subspace of V.

28. State (with a brief explanation) whether the following statements are true or false.

(a) Let **u** and **v** be orthogonal vectors in an inner product space. Then **u** is orthogonal to any vector that is a scalar multiple of **v**.

(b) Let **u** = $(1, 3)$ and **v** = $(-1, 2)$ be vectors in $\mathbf{R}^2$. An inner product exists on $\mathbf{R}^2$ such that **u** and **v** are orthogonal.

(c) Consider the vector space P_2 with inner product $\langle f, g \rangle = \int_0^1 f(x)g(x)dx$. Then $f(x) = 3x$ is a unit vector in this space.

(d) Consider the vector space M_{22} with an inner product defined on it. If a matrix A is orthogonal to matrices B and C it is orthogonal to any linear combination of B and C.

(e) Let **u** and **v** be vectors in an inner product space with $d(\mathbf{u}, \mathbf{v}) = \|\mathbf{u} - \mathbf{v}\|$. The distance between **u** and **v** is equal to the distance between **v** and **u**.

7.2 Non-Euclidean Geometry and Special Relativity

In this section we explore the significance of the new-found mathematical freedom we have in the many inner products that can be placed on $\mathbf{R}^n$. The inner product is a very powerful tool that enables us to control the geometry of $\mathbf{R}^n$. The inner product controls the norms of vectors, angles between vectors, and distances between points, since each of these is defined in terms of the inner product:

$$\|\mathbf{v}\| = \sqrt{\langle \mathbf{v}, \mathbf{v} \rangle}, \quad \cos \theta = \frac{\langle \mathbf{u}, \mathbf{v} \rangle}{\|\mathbf{u}\| \, \|\mathbf{v}\|}, \quad d(\mathbf{u}, \mathbf{v}) = \|\mathbf{u} - \mathbf{v}\| = \sqrt{\langle \mathbf{u} - \mathbf{v}, \mathbf{u} - \mathbf{v} \rangle}$$

Different inner products on $\mathbf{R}^n$ lead to different measures of vector norm, angle, and distance—that is, to different geometries. The dot product is a particular inner product that leads to Euclidean geometry. Let us now look at some of the other geometries that we can get by putting other inner products on $\mathbf{R}^n$. These **non-Euclidean geometries** have been studied by mathematicians and are used by scientists. We shall see how one of these geometries is used to describe space–time in the theory of special relativity.

We first look at a non-Euclidean geometry on $\mathbf{R}^2$. Consider the inner product that was discussed in Example 1 of the previous section. Let (x_1, x_2) and (y_1, y_2) be arbitrary vectors in $\mathbf{R}^2$ and let the inner product be defined by

$$\langle (x_1, x_2), (y_1, y_2) \rangle = x_1 y_1 + 4x_2 y_2$$

This inner product differs from the dot product in the appearance of a 4. Consider the vector $(0, 1)$ in this space. The norm of this vector is

$$\begin{aligned}
\|(0, 1)\| &= \sqrt{\langle (0, 1),(0, 1) \rangle} \\
&= \sqrt{(0 \times 0) + 4(1 \times 1)} \\
&= 2
\end{aligned}$$

We illustrate this vector in Figure 7.1. The norm of this vector in Euclidean geometry is of course 1; in our new geometry, however, the norm is 2.

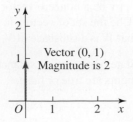

Figure 7.1

Consider the vectors $(1, 1)$ and $(-4, 1)$. The inner product of these vectors is

$$\langle (1, 1), (-4, 1) \rangle = (1 \times -4) + 4(1 \times 1)$$
$$= 0$$

Thus the vectors $(1, 1)$ and $(-4, 1)$ are orthogonal. We illustrate these vectors in Figure 7.2. The vectors do not appear to be at right angles; they are not orthogonal in the Euclidean geometry of our everyday experience. However, the angle between these vectors is a right angle in our new geometry.

Finally, let us use the definition of distance based on this inner product to compute the distance between the points $(1, 0)$ and $(0, 1)$. We have that

$$d((1, 0), (0, 1)) = \| (1, 0) - (0, 1) \|$$
$$= \| (1, -1) \|$$
$$= \sqrt{\langle (1, -1), (1, -1) \rangle}$$
$$= \sqrt{(1 \times 1) + 4(-1 \times -1)}$$
$$= \sqrt{5}$$

This distance is illustrated in Figure 7.3. Note that the distance between these same two points in Euclidean space, on applying the Pythagorean Theorem, is $\sqrt{2}$.

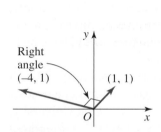

Distance is $\sqrt{5}$

Figure 7.2   **Figure 7.3**

The reader should appreciate that the geometry that we have developed in this example is correct mathematics, even though it does not describe the everyday world in which we live! Mathematics exists as an elegant field in itself, without having to be useful or related to real-world concepts. Even though an important aim of this course is to show how the mathematics that we develop is applied, the reader should be aware that much fine mathematics is not oriented toward applications. Of course, one never knows when such pure mathematics will turn out to be useful. This, for example, happened to be the case for non-Euclidean geometry. Much of non-Euclidean geometry was developed by a German mathematician named Bernhard Riemann in the mid-nineteenth century. In the early part of the twentieth century, Albert Einstein found that this was the type of geometry he needed for describing space–time. In space–time one uses $\mathbf{R}^4$ to represent three spatial dimensions and one time dimension. Euclidean geometry is not the appropriate geometry when space and time are "mixed."

We also mention that matrix inner product spaces and function inner product spaces form the mathematical framework of the theory of quantum mechanics. Quantum mechanics is the mathematical model that was developed in the early part of the twentieth century by such physicists as Niels Bohr, Max Planck, and Werner Heisenberg to describe the behavior of atoms and electrons. Thus the theory of inner product spaces forms the cornerstone of the two foremost physical theories of the twentieth century. These theories have had a great impact on all of our lives; they have, for example, led to the development of both atomic energy and atomic weapons.

Special Relativity

Special relativity was developed by Albert Einstein[*] to describe the physical world in which we live. At the time, Newtonian mechanics was the theory used to describe the motions of bodies. However, experiments had led scientists to believe that large-scale motions of bodies, such as planetary motions, were not accurately described by Newtonian mechanics. One of Einstein's contributions to science was the development of more accurate mathematical models. He first developed special relativity, which did not incorporate gravity; later he included gravity in his general theory of relativity. Here we introduce the mathematical model of special relativity by describing the predictions of the theory and then showing how they arise out of the mathematics.

The nearest star to Earth, other than the sun, is Alpha Centauri; it is about four light-years away. (A light year is the distance light travels in one year, 5.88×10^{12} miles.) Consider a pair of twins, who are separated immediately after birth. Twin 1 remains on Earth, and Twin 2 is flown off to Alpha Centauri in a rocket ship at 0.8 the speed of light. On arriving at Alpha Centauri, he immediately turns around and flies back to Earth at the same high speed. Special relativity predicts that when the twins get together, the twin who remained on Earth will be ten years old while the one who went off to Alpha Centauri will be six years old. We present this scenario in terms of twins, since it clearly displays the phenomenon involved, namely the difference in the aging process. The one twin will be a ten year old child, the other twin a six year old child. The same type of phenomenon is predicted for any two people, one of whom stays on Earth while the other travels to a distant planet and back. The numbers will vary according to how far and how fast the traveler goes. For example, this phenomenon was experienced to a very small degree by the astronauts who went to the moon. On arrival back on Earth they were fractionally younger than they would have been if they had remained on Earth.

Numerous experiments have been constructed to test this hypothesis that time on Earth varies from time recorded on an object moving relative to Earth. We mention one experiment conducted by Professor J. C. Hafele, Department of Physics, Washington University, St. Louis, and Dr. Richard E. Keating, U.S. Naval Observatory, Washington D.C. and reported in the journal *Science*, Volume 177, 1972. During October, 1971, four atomic clocks were flown on regularly scheduled jet flights around the world twice. There were slight differences in the times recorded by the clocks and clocks on Earth. These differences were in accord with the predictions of relativity theory. This aging variation will not however be realized by humans to the extent illustrated by the trip to Alpha Centauri in the near future as the energies required to produce such high speeds in a macroscopic body are prohibitive. This is probably fortunate from a sociological point of view. Thoretically, a man who went off on such a trip could be of an age to marry his great granddaughter when he returned!

We now introduce the mathematical model of special relativity and see how the predictions arise out of this model. Special relativity is a model of space–time and hence involves four coordinates; three space coordinates, x_1, x_2, x_3 and a time coordinate x_4. We use the vector space $\mathbf{R}^4$ to represent space–time. There are many inner products that we can place on $\mathbf{R}^4$; each would lead to a geometry for $\mathbf{R}^4$. None of the inner products, however, leads to a geometry that conforms to experimental results. If one drops the fourth inner product axiom one can come up with a "pseudo" inner product that leads to a geometry that fits experimental

---

[*]Albert Einstein (1879–1955) was one of the greatest scientists of all time. He studied in Zurich, and taught in Zurich, Prague, Berlin, and Princeton. He is best known for his relativity theories, which revolutionized scientific thought with new conceptions of time, space, mass, motion, and gravitation, although he received the Nobel Prize for his work on the photoelectric effect. Einstein moved to the United States when Hitler came to power in Germany. He renounced a pacifist position, being convinced that Hitler's regime could only be put down by force. He wrote a letter with the physicists Teller and Wigner to Roosevelt that helped initiate the efforts that produced the atomic bomb (although Einstein knew nothing of these efforts). Einstein was fond of classical music and played the violin. Although not associated with any formal religion, his nature was deeply religious. He never believed the universe was one of chance or chaos.

results. This geometry of special relativity is called **Minkowski geometry** after Hermann Minkowski[*], who gave this geometrical interpretation of special relativity.

Let $X = (x_1, x_2, x_3, x_4)$ and $Y = (y_1, y_2, y_3, y_4)$ be arbitrary elements of $\mathbf{R}^4$. The following function $\langle X, Y \rangle$ plays the role of an inner product in this geometry.

$$\langle X, Y \rangle = -x_1 y_1 - x_2 y_2 - x_3 y_3 + x_4 y_4$$

Note that $\langle X, Y \rangle$ is not a true inner product; the negative signs cause it to violate axiom 4 of an inner product. See the following exercises.

This pseudo inner product is used as follows to define norms of vectors and distances between points. Observe the use of absolute value signs in these definitions. The justification for these definitions is pure and simple—they are the ones that work for the geometry to describe space–time!

$$\|X\| = \sqrt{|\langle X, X \rangle|}$$

and

$$d(X, Y) = \|X - Y\| = \|(x_1 - y_1, x_2 - y_2, x_3 - y_3, x_4 - y_4)\|$$
$$= \sqrt{|-(x_1 - y_1)^2 - (x_2 - y_2)^2 - (x_3 - y_3)^2 + (x_4 - y_4)^2|}$$

Let us now see how this geometry is used to describe the behavior of time for the voyage to Alpha Centauri. We draw a **space–time diagram**. For convenience, assume that Alpha Centauri lies in the direction of the x_1 axis from Earth. The twin on Earth advances in time, x_4, while the twin on board the rocket advances in time and also moves in the direction of increasing x_1 on the outward voyage and then in a direction of decreasing x_1 on the homeward voyage. The space–time diagram is shown in Figure 7.4. The path of Twin 1 is PQ. The path of Twin 2 is PR to Alpha Centauri and then RQ back to Earth. There is no motion in either the x_2 or x_3 directions; hence we omit these axes in the diagram.

We now look at the situation from the point of view of each twin.

Twin 1 Distance to Alpha Centauri and back = 8 light years; speed of spaceship = 0.8 speed of light.

Thus,

$$\text{Duration of voyage relative to Earth} = \frac{\text{distance}}{\text{speed}} = \frac{8}{0.8} = 10 \text{ years}$$

Geometrically, this means that the length of PQ is 10.

Twin 2 Let us now arrive at coordinates for the various points in the diagram. See Figure 7.5. Let P be the origin, $(0, 0, 0, 0)$, in $\mathbf{R}^4$. Since $PQ = 10$, Q is the point $(0, 0, 0, 10)$ and S is the point $(0, 0, 0, 5)$.

PT is the spatial distance of Alpha Centauri from Earth, namely 4 light years. Thus R is the point $(4, 0, 0, 5)$. We now use Minkowski geometry to compute the distance between R and P.

$$d(P, R) = \|(4, 0, 0, 5) - (0, 0, 0, 0)\|$$
$$= \sqrt{|-(4 - 0)^2 - (0 - 0)^2 - (0 - 0)^2 + (5 - 0)^2|}$$
$$= \sqrt{|-16 + 25|} = \sqrt{9} = 3$$

---

[*]Hermann Minkowski (1864–1909) was educated in Königsberg and taught at Bonn, Königsberg, Zurich, and Göttingen. His mathematical interests were in number theory, geometry, and algebra. He became interested in relativity in his later years. He was the first person to realize that the principle of relativity formulated by Einstein implied the combining of space and time into a four-dimensional space–time continuum. This approach later led Einstein to the general theory of relativity, which incorporated gravity into the mathematical model.

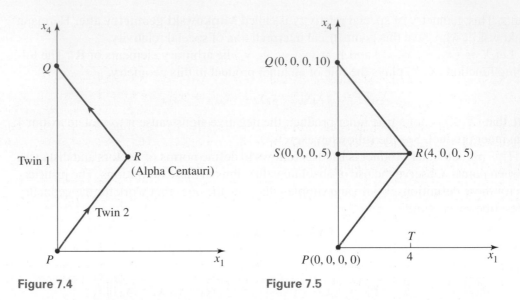

Figure 7.4   Figure 7.5

Similarly, in this geometry,

$$d(R, Q) = 3$$

The Physical Interpretation   We have completed the mathematics. Now comes the very interesting part of relating the geometrical results to the physical situation. The following is a postulate of special relativity that gives the geometry physical meaning.

> *The distance between two points on the path of an observer, such as Twin 2, in Minkowski geometry corresponds to the time recorded by that observer in traveling between the two points.*

Thus $d(P, R) = 3$ implies that Twin 2 ages 3 years in traveling from P to R. Similarly, $d(R, Q) = 3$ implies that Twin 2 ages 3 years in traveling from R to Q. The total duration of the voyage for Twin 2 is thus 6 years.

Therefore, when the twins meet at Q, Twin 1 will have aged 10 years while Twin 2 will have aged 6 years.

Note that this model introduces a new kind of geometry in which the straight line distance is not the shortest distance between two points. In Figure 7.5 the straight line distance PQ from P to Q is 10, whereas the distance $PR + RQ$ from P to Q is 6, which is smaller! Thus, not only is the physical interpretation of the model fascinating, it opens up a new trend in geometrical thinking.

This example illustrates the flexibility that one has in applying mathematics. If the standard body of mathematics, (inner product axioms in this case), does not fit the situation, maybe a slight modification will. Mathematicians have molded and developed mathematics to suit their needs. Mathematics is not as rigid and absolute as it is sometimes made out to be; it is a field that is continually being developed and applied in the spirit presented here.

EXERCISE SET 7.2

Geometries on $\mathbf{R}^2$

1. Consider $\mathbf{R}^2$ with the inner product of this section,

$$\langle (x_1, x_2), (y_1, y_2) \rangle = x_1 y_1 + 4 x_2 y_2$$

Determine the equation of the circle with center the origin and radius one in this space. Sketch the circle.

2. Consider $\mathbf{R}^2$ with the inner product

$$\langle (x_1, x_2), (y_1, y_2) \rangle = 4 x_1 y_1 + 9 x_2 y_2$$

(a) Determine the norms of the following vectors.

$$(1, 0), (0, 1), (1, 1), (2, 3)$$

(b) Prove that the vectors $(2, 1)$ and $(-9, 8)$ are orthogonal in this space. Sketch these vectors. Observe that they are not orthogonal in Euclidean space.

(c) Determine the distance between the points $(1, 0)$ and $(0, 1)$ in this space.

(d) Determine the equation of the circle with center the origin and radius one in this space. Sketch the circle.

3. Consider $\mathbf{R}^2$ with the inner product

$$\langle (x_1, x_2), (y_1, y_2) \rangle = x_1 y_1 + 16 x_2 y_2$$

(a) Determine the norms of the vectors $(1, 0)$, $(0, 1)$, and $(1, 1)$ in this space.

(b) Prove that the vectors $(1, 1)$ and $(-16, 1)$ are orthogonal.

(c) Determine the distances between the points $(5, 0)$ and $(0, 4)$ both in this space and in Euclidean space.

(d) Determine the equation of the circle with center the origin and radius one in this space. Sketch the circle.

4. Determine the inner product that must be placed on $\mathbf{R}^2$ for the equation of the circle center the origin and radius one to be $x^2/4 + y^2/25 = 1$.

Minkowski Geometry

5. Prove that the "pseudo" inner product of Minkowski geometry violates axiom 4 of an inner product. (*Hint*: Find a nonzero element X of $\mathbf{R}^4$ such that $\langle X, X \rangle = 0$.)

6. Use the definition of distance between two points in Minkowski space to show that the distance between the points R and Q in Figure 7.5 is 3.

7. Determine the distance between points $P(0, 0, 0, 0)$ and $M(1, 0, 0, 1)$ in Minkowski geometry. This reveals another interesting aspect of this geometry, namely that noncoincident points can be zero distance apart. The special theory of relativity gives a physical interpretation to such a line PM, all of whose points are zero distance apart; it is the path of a **photon**, a light particle.

8. Prove that the vectors $(2, 0, 0, 1)$ and $(1, 0, 0, 2)$ are orthogonal in Minkowski geometry. Sketch these vectors and observe that they are symmetrical about the "45°" vector, $(1, 0, 0, 1)$. Any pair of vectors that are thus symmetrical about the vector $(1, 0, 0, 1)$ will be orthogonal in Minkowski geometry. Prove this result by demonstrating that the vectors $(a, 0, 0, b)$ and $(b, 0, 0, a)$ are orthogonal.

9. Determine the equations of the circles with center the origin and radii 1, 2, and 3 in the $x_1 x_4$-plane of Minkowski space. Sketch these circles.

Space–Time Travel

10. The star Sirius is 8 light years from Earth. Sirius is the nearest star, other than the sun, that is visible with the naked eye from most parts of North America. It is the brightest appearing of all stars. Light reaches us from the sun in 8 minutes, while it takes 8 years to reach us from Sirius. Suppose a spaceship that leaves for Sirius returns to Earth 20 years later. What is the duration of the voyage for a person on the spaceship? What is the speed of the spaceship?

11. A spaceship makes a round trip to the bright star Capella, which is 45 light years from Earth. If the duration of the voyage relative to Earth is 120 years, what is the duration of the voyage from the traveler's viewpoint?

12. The star cluster Pleiades in the constellation Taurus is 410 light years from Earth. A traveler to the cluster ages 40 years on a round trip. By the time he returns to Earth, how many centuries will have passed on Earth?

13. The star cluster Praesepe in the constellation Cancer is 515 light years from Earth. A traveler flies to Praesepe and back at 0.99999 the speed of light. Determine the durations of the voyage both from the Earth and traveler's viewpoints.

*7.3 Approximation of Functions and Coding Theory

Many problems in the physical sciences and engineering involve approximating given functions by polynomials or trigonometric functions. For example, it may be necessary to approximate $f(x) = e^x$ by a linear function of the form $g(x) = a + bx$, over the interval $[0, 1]$

or by a trigonometric function of the form $h(x) = a + b \sin x + c \cos x$ over the interval $[-\pi, \pi]$. Furthermore, the problem of approximating functions by polynomials is central to software development since most functions are evaluated through polynomial approximations. We now introduce the techniques for approximating functions.

Let $C[a, b]$ be the inner product space of continuous functions over the interval $[a, b]$ with inner product $\langle f, g \rangle = \int_a^b f(x)g(x)dx$, and geometry defined by this inner product. Let W be a subspace of $C[a, b]$. Suppose f is in $C[a, b]$, but outside W, and we want to find the "best" approximation to f that lies in W. We define the "best" approximation to be the function g in W such that the distance $\|f - g\|$ between f and g is a minimum.

> **DEFINITION**   Let $C[a, b]$ be the vector space of continuous functions over the interval $[a, b]$, f be an element of $C[a, b]$, and W be a subspace of $C[a, b]$. The function g in W such that $\int_a^b [f(x) - g(x)]^2\, dx$ is a minimum is called the **least-squares approximation** to f.

This approximation is called "the least-squares approximation," since this distance formula is based on squaring.

We now give a method for finding the least squares approximation $g(x)$. We can really appreciate the approach of extending geometrical structures and results from $\mathbf{R}^n$ to more abstract surroundings. We know from Section 4.9 that if $\mathbf{x}$ is a point in $\mathbf{R}^n$ and U is a subspace of $\mathbf{R}^n$ then the element of U that is closest to $\mathbf{x}$ is $\mathrm{proj}_U \mathbf{x}$. It is very helpful also to imagine functions as geometrical vectors, Figure 7.6. With this picture in mind the analogous result for the function space is as follows.

> *The least-squares approximation to f in the subspace W is $g = \mathrm{proj}_W f$.*

We build on the definition of $\mathrm{proj}_U \mathbf{x}$ to get an expression for $\mathrm{proj}_W f$. If

$$\{\mathbf{u}_1, \ldots, \mathbf{u}_m\}$$

is an orthonormal basis for U then we know that

$$\mathrm{proj}_U \mathbf{x} = (\mathbf{x} \cdot \mathbf{u}_1)\mathbf{u}_1 + (\mathbf{x} \cdot \mathbf{u}_2)\mathbf{u}_2 + \cdots + (\mathbf{x} \cdot \mathbf{u}_m)\mathbf{u}_m$$

Let $\{g_1, \ldots g_n\}$ be an orthonormal basis for W. Replacing the dot product of $\mathbf{R}^n$ by the inner product of the function space we get

$$\mathrm{proj}_W f = \langle f, g_1 \rangle\, g_1 + \cdots + \langle f, g_n \rangle\, g_n$$

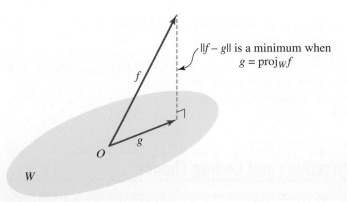

$\|f - g\|$ is a minimum when $g = \mathrm{proj}_W f$

Figure 7.6

EXAMPLE 1 Find the least-squares linear approximation to $f(x) = e^x$ over the interval $[-1, 1]$.

SOLUTION

Let the linear approximation be $g(x) = a + bx$. f is an element of $C[-1, 1]$ and g is an element of the subspace $P_1[-1, 1]$ of polynomials of degree less than or equal to one over $[-1, 1]$. The set $\{1, x\}$ is a basis for $P_1[-1, 1]$. We get

$$\langle 1, x \rangle = \int_{-1}^{1} (1 \cdot x) \, dx = 0$$

Thus the functions are orthogonal. The magnitudes of these vectors are given by

$$\|1\|^2 = \int_{-1}^{1} (1 \cdot 1) \, dx = 2 \quad \text{and} \quad \|x\|^2 = \int_{-1}^{1} (x \cdot x) \, dx = \frac{2}{3}$$

Thus the set $\left\{ \dfrac{1}{\sqrt{2}}, \sqrt{\dfrac{3}{2}}x \right\}$ is an orthonormal basis for $P_1[-1, 1]$. We now get

$$\text{proj}_W f = \langle f, g_1 \rangle g_1 + \langle f, g_2 \rangle g_2$$

$$= \int_{-1}^{1} \left(e^x \sqrt{\frac{1}{2}} \right) dx \frac{1}{\sqrt{2}} + \int_{-1}^{1} \left(e^x \sqrt{\frac{3}{2}}x \right) dx \sqrt{\frac{3}{2}}x$$

$$= \frac{1}{2}(e - e^{-1}) + 3e^{-1}x$$

The least-squares linear approximation to $f(x) = e^x$ over the interval $[-1, 1]$ is $g(x) = \frac{1}{2}(e - e^{-1}) + 3e^{-1}x$. This gives $g(x) = 1.18 + 1.1x$, to two decimal places. See Figure 7.7.

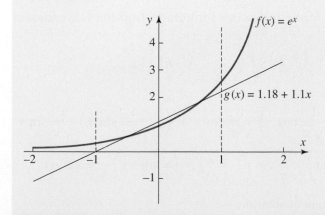

Figure 7.7

In this example we have found the linear approximation to f in $P_1[-1, 1]$. Higher degree polynomial approximations can be found in the space $P_n[-1, 1]$ of polynomials of degree less than or equal to n. To arrive at the approximations an orthogonal basis has to be constructed in $P_n[-1, 1]$ using the Gram-Schmidt orthogonalization process. The orthogonal functions found in this manner are called **Legendre polynomials**. The first six Legendre polynomials are:

$$1, \quad x, \quad x^2 - \tfrac{1}{3}, \quad x^3 - \tfrac{3}{5}x, \quad x^4 - \tfrac{6}{7}x^2 + \tfrac{3}{35}, \quad x^5 - \tfrac{10}{9}x^3 + \tfrac{5}{21}x$$

We next look at approximations of functions in terms of trigonometric functions. Such approximations are widely used in areas such as heat conduction, electromagnetism, electric circuits, and mechanical vibrations. The initial work in this area was undertaken by Jean Baptiste Fourier,[*] a French mathematician who developed the methods to analyze conduction in an insulated bar. The approximations are often used in discussing solutions to partial differential equations that describe physical situations.

Fourier Approximations

Let f be a function in $C[-\pi, \pi]$ with inner product $\langle f, g \rangle = \int_{-\pi}^{\pi} f(x)g(x)\, dx$, and geometry defined by this inner product. Let us find the least-squares approximation of f in the space $T[-\pi, \pi]$ of trigonometric polynomials spanned by the set $\{1, \cos x, \sin x, \ldots, \cos nx, \sin nx\}$, where n is a positive integer.

It can be shown that the vectors $1, \cos x, \sin x, \ldots, \cos nx$, and $\sin nx$ are mutually orthogonal in this space (see the following exercises). The magnitudes of these vectors are given by

$$\|1\|^2 = \int_{-\pi}^{\pi} (1 \cdot 1)\, dx = 2\pi, \quad \|\cos nx\|^2 = \int_{-\pi}^{\pi} (\cos nx \cdot \cos nx)\, dx = \pi,$$

$$\|\sin nx\|^2 = \int_{-\pi}^{\pi} (\sin nx \cdot \sin nx)\, dx = \pi$$

Thus the following set is an orthonormal basis for $T[-\pi, \pi]$.

$$\{g_0, \ldots, g_{2n}\} = \left\{ \frac{1}{\sqrt{2\pi}}, \frac{1}{\sqrt{\pi}}\cos x, \frac{1}{\sqrt{\pi}}\sin x, \ldots, \frac{1}{\sqrt{\pi}}\cos nx, \frac{1}{\sqrt{\pi}}\sin nx \right\}$$

Let us use this orthonormal basis in the following formula to find the least squares approximation g of f.

$$g(x) = \text{proj}_T f = \langle f, g_0 \rangle g_0 + \cdots + \langle f, g_{2n} \rangle g_{2n}$$

We get

$$g(x) = \left\langle f, \frac{1}{\sqrt{2\pi}} \right\rangle \frac{1}{\sqrt{2\pi}} + \left\langle f, \frac{1}{\sqrt{\pi}}\cos x \right\rangle \frac{1}{\sqrt{\pi}}\cos x + \left\langle f, \frac{1}{\sqrt{\pi}}\sin x \right\rangle \frac{1}{\sqrt{\pi}}\sin x$$

$$+ \cdots + \left\langle f, \frac{1}{\sqrt{\pi}}\cos nx \right\rangle \frac{1}{\sqrt{\pi}}\cos nx + \left\langle f, \frac{1}{\sqrt{\pi}}\sin nx \right\rangle \frac{1}{\sqrt{\pi}}\sin nx$$

Let us introduce the following convenient notation.

$$a_0 = \left\langle f, \frac{1}{\sqrt{2\pi}} \right\rangle \frac{1}{\sqrt{2\pi}} = \int_{-\pi}^{\pi} \left(f(x) \cdot \frac{1}{\sqrt{2\pi}} \right) dx \frac{1}{\sqrt{2\pi}} = \frac{1}{2\pi} \int_{-\pi}^{\pi} f(x)\, dx$$

---

[*]Jean Baptiste Fourier (1768–1830) was orphaned at an early age and was placed in a military school. It was at this school that he developed an interest in mathematics. He was jailed during the French Revolution for his defense of the victims of the Terror. He was, however, later made a count by Napoleon and held diplomatic positions in Egypt and France. Fourier is best known for his work on diffusion of heat.

$$a_k = \left\langle f, \frac{1}{\sqrt{\pi}} \cos kx \right\rangle \frac{1}{\sqrt{\pi}} = \int_{-\pi}^{\pi} \left(f(x) \cdot \frac{1}{\sqrt{\pi}} \cos kx \right) dx \frac{1}{\sqrt{\pi}}$$

$$= \frac{1}{\pi} \int_{-\pi}^{\pi} f(x) \cos kx \, dx$$

$$b_k = \left\langle f, \frac{1}{\sqrt{\pi}} \sin kx \right\rangle \frac{1}{\sqrt{\pi}} = \int_{-\pi}^{\pi} \left(f(x) \cdot \frac{1}{\sqrt{\pi}} \sin kx \right) dx \frac{1}{\sqrt{\pi}}$$

$$= \frac{1}{\pi} \int_{-\pi}^{\pi} f(x) \sin kx \, dx$$

The trigonometric approximation of $f(x)$ can now be written

$$g(x) = a_0 + \sum_{k=1}^{n} (a_k \cos kx + b_k \sin kx)$$

$g(x)$ is called the **nth-order Fourier approximation** of $f(x)$. The coefficients a_0, a_1, $b_1, \ldots, a_n, b_n$, are called **Fourier coefficients**.

As n increases this approximation naturally becomes an increasingly better approximation in the sense that $\|f - g\|$ gets smaller. The infinite sum

$$g(x) = a_0 + \sum_{k=1}^{\infty} (a_k \cos kx + b_k \sin kx)$$

is known as the **Fourier series** of f on the interval $[-\pi, \pi]$.

EXAMPLE **2** Find the fourth-order Fourier approximation to $f(x) = x$ over the interval $[-\pi, \pi]$.

SOLUTION

Using the above Fourier coefficients with $f(x) = x$, and using integration by parts, we get

$$a_0 = \frac{1}{2\pi} \int_{-\pi}^{\pi} f(x) dx = \frac{1}{2\pi} \int_{-\pi}^{\pi} x \, dx = \frac{1}{2\pi} \left[\frac{x^2}{2} \right]_{-\pi}^{\pi} = 0$$

$$a_k = \frac{1}{\pi} \int_{-\pi}^{\pi} (f(x) \cdot \cos kx) \, dx = \frac{1}{\pi} \int_{-\pi}^{\pi} (x \cdot \cos kx) \, dx$$

$$= \frac{1}{\pi} \left[\frac{x}{k} \sin kx + \frac{1}{k^2} \cos kx \right]_{-\pi}^{\pi} = 0$$

$$b_k = \frac{1}{\pi} \int_{-\pi}^{\pi} (f(x) \sin kx) \, dx = \frac{1}{\pi} \int_{-\pi}^{\pi} (x \cdot \sin kx) \, dx$$

$$= \frac{1}{\pi} \left[-\frac{x}{k} \cos kx + \frac{1}{k^2} \sin kx \right]_{-\pi}^{\pi} = \frac{2(-1)^{k+1}}{k}$$

The Fourier approximation of f is

$$g(x) = \sum_{k=1}^{n} \frac{2(-1)^{k+1}}{k} \sin kx$$

Taking $k = 1, \ldots 4$, we get the fourth-order approximation

$$g(x) = 2\left(\sin x - \tfrac{1}{2}\sin 2x + \tfrac{1}{3}\sin 3x - \tfrac{1}{4}\sin 4x\right)$$

We have been primarily interested in real vector spaces in this course—that is, vector spaces in which the scalars are real numbers. We have occasionally mentioned complex vector spaces, where the scalars are complex numbers. Our introduction to vector spaces would not be complete without illustrating an application of vector spaces where the scalars arc not rcal or complex numbers. Furthermore, this example is interesting in that there are only a finite number of vectors in the spaces.

Coding Theory

Messages are sent electronically as sequences of 0's and 1's. Errors can occur in messages due to noise or interference. For example, the message $(1, 1, 0, 1, 1, 0, 1)$ could be sent and the message $(1, 1, 0, 1, 0, 0, 1)$ might be received. An error has occured in the fifth entry. We shall now look at methods that are used to detect and correct such errors.

The scalars for a vector space can be sets other than real or complex numbers. The requirement is that the scalars form an algebraic **field**. A field is a set of elements with two operations that satisfy certain axioms. Readers who go on to take a course in modern algebra will study fields. In this example we shall use the field $\{0, 1\}$ of scalars—having only these two elements—with operations of addition and multiplication defined as follows:

$$0 + 0 = 0, \qquad 0 + 1 = 1, \qquad 1 + 0 = 1, \qquad 1 + 1 = 0$$
$$0 \cdot 0 = 0, \qquad 0 \cdot 1 = 1, \qquad 1 \cdot 0 = 0, \qquad 1 \cdot 1 = 1$$

Let V_7 be the vector space of 7-tuples of 0's and 1's over this field of scalars, where addition and scalar multiplication are defined in the usual componentwise manner. For example,

$$(1, 0, 0, 1, 1, 0, 1) + (0, 1, 1, 1, 0, 0, 1)$$
$$= (1 + 0, 0 + 1, 0 + 1, 1 + 1, 1 + 0, 0 + 0, 1 + 1)$$
$$= (1, 1, 1, 0, 1, 0, 0)$$

$$0(1, 0, 0, 1, 1, 0, 1) = (0 \cdot 1, 0 \cdot 0, 0 \cdot 0, 0 \cdot 1, \ 0 \cdot 1, 0 \cdot 0, 0 \cdot 1,) = (0, 0, 0, 0, 0, 0, 0)$$
$$1(1, 0, 0, 1, 1, 0, 1) = (1 \cdot 1, 1 \cdot 0, 1 \cdot 0, 1 \cdot 1, \ 1 \cdot 1, 1 \cdot 0, 1 \cdot 1) = (1, 0, 0, 1, 1, 0, 1)$$

Since each vector in V_7 has seven components, and each of these components can be either 0 or 1, there are 2^7 vectors in this space. The four-dimensional subspace of V_7 having basis

$$\{(1, 0, 0, 0, 0, 1, 1), (0, 1, 0, 0, 1, 0, 1), (0, 0, 1, 0, 1, 1, 0), (0, 0, 0, 1, 1, 1, 1)\}$$

is called a **Hamming Code** and is denoted $C_{7,4}$. The vectors in $C_{7,4}$ can be used as messages. Each vector in $C_{7,4}$ can be written

$$\mathbf{v}_i = a_1(1, 0, 0, 0, 0, 1, 1) + a_2(0, 1, 0, 0, 1, 0, 1)$$
$$+ a_3(0, 0, 1, 0, 1, 1, 0) + a_4(0, 0, 0, 1, 1, 1, 1)$$

Since each of the four scalars a_1, a_2, a_3, a_4 can take one of the values 1 or 0 we have 2^4—that is, 16 vectors—in $C_{7,4}$. The Hamming code can thus be used to send 16 different messages, $\mathbf{v}_1, \ldots \mathbf{v}_{16}$. The reader is asked to list these vectors in the following exercises.

When an error occurs in one location of a transmitted message the resulting incorrect vector lies in V_7, outside the subspace $C_{7,4}$. It can be proved that there is exactly one vector in $C_{7,4}$ that differs from this incorrect vector in one location. Thus the error can be detected and corrected. For example, suppose the received vector is $(1, 0, 1, 1, 0, 1)$. This vector cannot be expressed as a linear combination of the above base vectors; it is not in $C_{7,4}$. There is a single vector in $C_{7,4}$ that differs from this vector in one entry, namely $(1, 0, 1, 0, 1, 0, 1)$. The correct message is $(1, 0, 1, 0, 1, 0, 1)$. The Hamming code is called an **error-correcting code** because of this facility to detect and correct errors.

Let us look at the geometry underlying this code. The distance between two vectors $\mathbf{u}$ and $\mathbf{w}$ in V_7, denoted $d(\mathbf{u}, \mathbf{w})$, is defined to be the number of components that differ in $\mathbf{u}$ and $\mathbf{w}$. Thus for example,

$$d((1, 0, 0, 1, 0, 1, 1), (1, 1, 0, 0, 1, 1, 1)) = 3$$

since the second, fourth, and fifth components of these vectors differ. A sphere of radius 1 about a vector in V_7 contains all those vectors that differ in one component from the vector. It can be shown that the spheres of radii 1, centered at the vectors of $C_{7,4}$, will be disjoint and that every element of V_7 lies in one such sphere. See Figure 7.8. Thus if a vector $\mathbf{u}$ is received that is not in $C_{7,4}$, then it will be in a single sphere having center $\mathbf{v}_i$, the correct message. In practice, electrical circuits called gates are used to test whether the received message is in $C_{7,4}$, and, if it is not, to determine the center of the sphere in which it lies, giving the correct message.

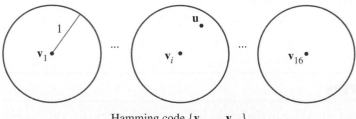

Hamming code $\{\mathbf{v}_1, \ldots, \mathbf{v}_{16}\}$
Incorrect message $\mathbf{u}$; correct message $\mathbf{v}_i$

Figure 7.8

Other error-detecting codes exist. The **Golay code**, for example, is a twelve-dimensional subspace of V_{23} that is denoted $C_{23,12}$. The code space has 2^{12} elements that can be used to represent 4096 messages. This code can detect and correct errors in one, two, or three locations. It can be shown that the spheres of radii 3, centered at the vectors of $C_{23,12}$, are disjoint and that every element of V_{23} lies in one such sphere. If a vector $\mathbf{u}$ that is received has an error in one, two, or three locations it will lie in a sphere centered at the correct message, $\mathbf{v}_i$, and $\mathbf{v}_i$ can be retrieved.

A good introduction to coding theory can be found in *A Commonsense Approach to the Theory of Error Correcting Codes* by Benjamin Arazi, The MIT Press, 1988.

EXERCISE SET 7.3

Least Square Approximations

In Exercises 1–4 find the least squares linear approximations, $g(x) = a + bx$, to the given functions over the stated intervals.

1. $f(x) = x^2$ over the interval $[-1, 1]$

2. $f(x) = e^x$ over $[0, 1]$

3. $f(x) = \sqrt{x}$ over $[0, 1]$

4. $f(x) = \cos x$ over $[0, \pi]$

In Exercises 5–8 find the least-squares quadratic approximations, $g(x) = a + bx + cx^2$, to the given functions over the stated intervals.

5. $f(x) = e^x$ over
   (a) $[-1, 1]$
   (b) $[0, 1]$

6. $f(x) = \sqrt{x}$ over $[0, 1]$

7. $f(x) = x^2$ over $[0, 1]$

8. $f(x) = \sin x$ over $[0, \pi]$

In Exercises 9–11 find the fourth-order Fourier approximations to the given functions over the stated intervals.

9. $f(x) = x$ over $[0, 2\pi]$

10. $f(x) = 1 + x$ over $[-\pi, \pi]$

11. $f(x) = x^2$ over $[0, 2\pi]$

12. Show that the vectors $1, \cos x, \sin x, \ldots, \cos nx$ are mutually orthogonal in the space $C[-\pi, \pi]$

Coding Theory

13. List all the vectors in the Hamming code $C_{7,4}$.

14. List all the vectors of V_7 that lie in the sphere having radius 1 and the following centers.
    (a) $(0, 0, 1, 0, 1, 1, 0)$ (b) $(1, 1, 0, 1, 0, 0, 1)$
    (c) $(1, 1, 1, 1, 1, 1, 1)$ (d) $(0, 0, 0, 0, 0, 0, 0)$

15. The following messages are received using the Hamming code. Find the correct message in each case.
    (a) $(0, 0, 0, 0, 1, 0, 0)$ (b) $(0, 1, 1, 1, 0, 0, 1)$
    (c) $(1, 0, 1, 0, 1, 1, 0)$ (d) $(1, 0, 1, 1, 0, 1, 0)$

16. Consider the vector space V_{23} and the Golay code subspace $C_{23,12}$. Prove that there are
    (a) 2^{23} vectors in V_{23}.   (b) 4096 vectors in $C_{23,12}$.
    (c) 2048 vectors in each sphere of radius 3 about a vector in $C_{23,12}$ (given that each element of V_{23} is in one sphere). How many vectors of distance 1, 2, and 3 are in each sphere of radius 3?

---

*7.4 Least Squares Curves

This section is possibly the most attractive in the book! It illustrates mathematics at its best. Theoretical results developed earlier come together in a very elegant, powerful way to arrive at an important computational tool. We derive the method of finding a polynomial that best fits given data points; a method that is extremely important to the natural sciences, social sciences, and engineering.

We have seen that a system $A\mathbf{x} = \mathbf{y}$ of n equations in n variables, where A is invertible, has the unique solution $\mathbf{x} = A^{-1}\mathbf{y}$. However, if $A\mathbf{x} = \mathbf{y}$ is a system of n equations in m variables where $n > m$, the system does not, in general, have a solution and is said to be **overdetermined**. A is not a square matrix for such a system and A^{-1} does not exist. We shall introduce a matrix called the **pseudoinverse** of A, denoted pinv(A), that leads to a **least squares solution** $\mathbf{x} = \text{pinv}(A)\mathbf{y}$ for an overdetermined system. This is not a true solution, but is in some sense the closest we can get to a true solution for the system. We shall see an application of overdetermined systems in finding curves that "best" fit data.

DEFINITION   Let A be a matrix. The matrix $(A^tA)^{-1}A^t$ is called the **pseudoinverse** of A and is denoted pinv(A).

We have seen that not every matrix has an inverse. Similarly, not every matrix has a pseudoinverse. The matrix A has a pseudoinverse if $(A^tA)^{-1}$ exists.

EXAMPLE 1 Find the pseudoinverse of $A = \begin{bmatrix} 1 & 2 \\ -1 & 3 \\ 2 & 4 \end{bmatrix}$.

SOLUTION

We compute the pseudoinverse of A in stages.

$$A^t A = \begin{bmatrix} 1 & -1 & 2 \\ 2 & 3 & 4 \end{bmatrix} \begin{bmatrix} 1 & 2 \\ -1 & 3 \\ 2 & 4 \end{bmatrix} = \begin{bmatrix} 6 & 7 \\ 7 & 29 \end{bmatrix}$$

$$(A^t A)^{-1} = \frac{1}{|A^t A|} \operatorname{adj}(A^t A) = \frac{1}{125} \begin{bmatrix} 29 & -7 \\ -7 & 6 \end{bmatrix}$$

$$\operatorname{pinv}(A) = (A^t A)^{-1} A^t = \frac{1}{125} \begin{bmatrix} 29 & -7 \\ -7 & 6 \end{bmatrix} \begin{bmatrix} 1 & -1 & 2 \\ 2 & 3 & 4 \end{bmatrix} = \frac{1}{25} \begin{bmatrix} 3 & -10 & 6 \\ 1 & 5 & 2 \end{bmatrix}$$

■

We now use the concept of pseudoinverse to further our understanding of systems of linear equations. Let $A\mathbf{x} = \mathbf{y}$ be a system of n linear equations in m variables with $n > m$, where A is of rank m. Multiply each side of this matrix equation by A^t, to get

$$A^t A \mathbf{x} = A^t \mathbf{y}$$

The matrix $A^t A$ can be shown to be invertible for such a system. Multiply each side of this equation by $(A^t A)^{-1}$ to get

$$\mathbf{x} = [(A^t A)^{-1} A^t] \mathbf{y}$$
$$= \operatorname{pinv}(A) \mathbf{y}$$

This value of $\mathbf{x}$ is called the least squares solution to the system of equations.

$$A\mathbf{x} = \mathbf{y} \qquad \mathbf{x} = \operatorname{pinv}(A)\mathbf{y}$$
system   least squares solution

Let $A\mathbf{x} = \mathbf{y}$ be a system of n linear equations in m variables with $n > m$, where A is of rank m. This system has a least squares solution. If the system has a unique solution, the least squares solution is that unique solution. If the system is overdetermined, the least squares solution is the closest we can get to a true solution. The system cannot have many solutions.

EXAMPLE 2 Find the least squares solution of the following overdetermined system of equations. Sketch the solution.

$$\begin{aligned} x + y &= 6 \\ -x + y &= 3 \\ 2x + 3y &= 9 \end{aligned}$$

SOLUTION

The matrix of coefficients is

$$A = \begin{bmatrix} 1 & 1 \\ -1 & 1 \\ 2 & 3 \end{bmatrix} \quad \text{and} \quad \mathbf{y} = \begin{bmatrix} 6 \\ 3 \\ 9 \end{bmatrix}$$

The column vectors of A are linearly independent. Thus the rank of A is 2. This system has a least squares solution. We compute $\text{pinv}(A)$.

$$A^t A = \begin{bmatrix} 1 & -1 & 2 \\ 1 & 1 & 3 \end{bmatrix} \begin{bmatrix} 1 & 1 \\ -1 & 1 \\ 2 & 3 \end{bmatrix} = \begin{bmatrix} 6 & 6 \\ 6 & 11 \end{bmatrix}$$

$$(A^t A)^{-1} = \frac{1}{|A^t A|} \text{adj}(A^t A) = \frac{1}{30} \begin{bmatrix} 11 & -6 \\ -6 & 6 \end{bmatrix}$$

$$\text{pinv}(A) = (A^t A)^{-1} A^t = \frac{1}{30} \begin{bmatrix} 11 & -6 \\ -6 & 6 \end{bmatrix} \begin{bmatrix} 1 & -1 & 2 \\ 1 & 1 & 3 \end{bmatrix} = \frac{1}{30} \begin{bmatrix} 5 & -17 & 4 \\ 0 & 12 & 6 \end{bmatrix}$$

The least squares solution is

$$\text{pinv}(A)\mathbf{y} = \frac{1}{30} \begin{bmatrix} 5 & -17 & 4 \\ 0 & 12 & 6 \end{bmatrix} \begin{bmatrix} 6 \\ 3 \\ 9 \end{bmatrix} = \begin{bmatrix} \frac{1}{2} \\ 3 \end{bmatrix}$$

The least squares solution is the point $P(\frac{1}{2}, 3)$ in Figure 7.9.

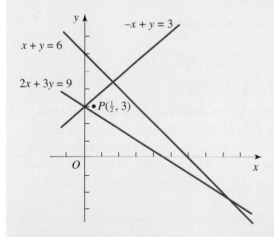

Figure 7.9

Let us now see how least squares solutions can be used to find curves that best fit given data.

Least Squares Curves

Many branches of science and business use equations based on data that has been determined from experimental results. In Section 1.3 we discussed the problem of finding a unique polynomial of degree two that passes through three data points. In many applications, however, there is too much data to lead to an equation that can exactly fit all the data. One then uses the equation of a line or curve that in some sense "best" fits the data. For example, suppose the data consists of the points $(x_1, y_1), \ldots, (x_n, y_n)$, shown in Figure 7.10(a). These points lie approximately on a line. We would want the equation of the line that best fits these points. On the other hand the points might closely fit a parabola, as shown in Figure 7.10(b). We would then want to find the parabola that most closely fits these points.

There are many criteria that can be used for the "best" fit in such cases. The one that has generally been found to be most satisfactory is called the **least squares line or curve**—found by solving an overdetermined system of equations. The least squares line and curve is such that $d_1{}^2 + \cdots + d_n{}^2$ in Figure 7.10 is a minimum. In the previous section we discussed least squares approximations to functions. This discussion is the discrete analogue of that problem. In the function situation we had to find a curve of a certain type that best fit a continuous set of data points over a given interval (the graph of the function); here we want the "best" fit to a discrete set of data points over a given interval. Although the techniques that we develop to arrive at results are different, look for certain similarities in concepts in the two situations. In both situations, the "best" result is the one obtained by minimizing certain squares—hence the term "*least squares.*"

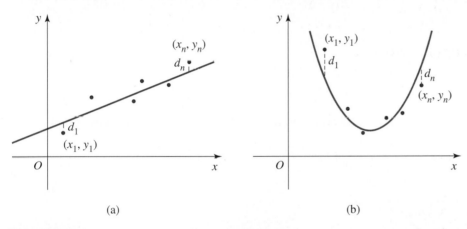

(a)   (b)

Figure 7.10

We now illustrate how to fit a least squares polynomial to given data. The method involves constructing a system of linear equations. The least squares solution to this system of equations gives the coefficients of the polynomial. We shall justify the method after seeing how it works.

EXAMPLE 3 Find the least squares line for the following data points.

$$(1, 1), (2, 2.4), (3, 3.6), (4, 4)$$

SOLUTION

Let the equation of the line be $y = a + bx$. Substituting for these points into the equation of the line we get the overdetermined system

$$
\begin{aligned}
a + b &= 1 \\
a + 2b &= 2.4 \\
a + 3b &= 3.6 \\
a + 4b &= 4
\end{aligned}
$$

We find the least squares solution. The matrix of coefficients A and column vector $\mathbf{y}$ are as follows.

$$A = \begin{bmatrix} 1 & 1 \\ 1 & 2 \\ 1 & 3 \\ 1 & 4 \end{bmatrix} \quad \text{and} \quad \mathbf{y} = \begin{bmatrix} 1 \\ 2.4 \\ 3.6 \\ 4 \end{bmatrix}$$

It can be shown that

$$\text{pinv}(A) = (A^t A)^{-1} A^t = \frac{1}{20} \begin{bmatrix} 20 & 10 & 0 & -10 \\ -6 & -2 & 2 & 6 \end{bmatrix}$$

The least squares solution is

$$[(A^t A)^{-1} A^t]\mathbf{y} = \frac{1}{20} \begin{bmatrix} 20 & 10 & 0 & -10 \\ -6 & -2 & 2 & 6 \end{bmatrix} \begin{bmatrix} 1 \\ 2.4 \\ 3.6 \\ 4 \end{bmatrix} = \begin{bmatrix} 0.2 \\ 1.02 \end{bmatrix}$$

Thus $a = 0.2$, $b = 1.02$.

The equation of the least squares line for this data is

$$y = 0.2 + 1.02x$$

This is the line that is generally considered to be the line of best fit for these points. See Figure 7.11.

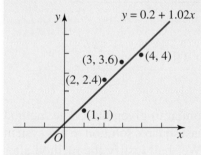

Figure 7.11

EXAMPLE **4** Find the least squares parabola for the following data points:

$$(1, 7), (2, 2), (3, 1), (4, 3)$$

SOLUTION

Let the equation of the parabola be $y = a + bx + cx^2$. Substituting for these points into the equation of the parabola we get the system

$$\begin{aligned} a + b + c &= 7 \\ a + 2b + 4c &= 2 \\ a + 3b + 9c &= 1 \\ a + 4b + 16c &= 3 \end{aligned}$$

We find the least squares solution. The matrix of coefficients A and column vector $\mathbf{y}$ are as follows:

$$A = \begin{bmatrix} 1 & 1 & 1 \\ 1 & 2 & 4 \\ 1 & 3 & 9 \\ 1 & 4 & 16 \end{bmatrix} \quad \text{and} \quad \mathbf{y} = \begin{bmatrix} 7 \\ 2 \\ 1 \\ 3 \end{bmatrix}$$

It can be shown that

$$\text{pinv}(A) = (A^t A)^{-1} A^t = \frac{1}{20} \begin{bmatrix} 45 & -15 & -25 & 15 \\ -31 & 23 & 27 & -19 \\ 5 & -5 & -5 & 5 \end{bmatrix}$$

The least squares solution is

$$[(A^t A)^{-1} A^t]\,\mathbf{y} = \frac{1}{20} \begin{bmatrix} 45 & -15 & -25 & 15 \\ -31 & 23 & 27 & -19 \\ 5 & -5 & -5 & 5 \end{bmatrix} \begin{bmatrix} 7 \\ 2 \\ 1 \\ 3 \end{bmatrix} = \begin{bmatrix} 15.25 \\ -10.05 \\ 1.75 \end{bmatrix}$$

Thus $a = 15.25$, $b = -10.05$, $c = 1.75$.

The equation of the least squares parabola for these data points is

$$y = 15.25 - 10.05x + 1.75x^2$$

We illustrate this parabola in Figure 7.12.

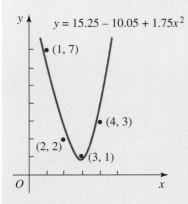

Figure 7.12

We now give the proof of the method used in the above examples to arrive at the least squares polynomials. The proof uses the concept of vector projection onto the range of a matrix mapping. We see geometry at its best in deriving this important result. The proof is challenging and very rewarding!

THEOREM 7.1

Let $(x_1, y_1), \ldots, (x_n, y_n)$ be a set of data points. Let $y = a_0 + \cdots + a_m x^m$ be a polynomial of degree m $(n > m)$ that is to be fitted to these points. Substituting these points into the polynomial leads to a system $A\mathbf{x} = \mathbf{y}$ of n linear equations in the m variables $a_0, \ldots, a_m$, where

$$A = \begin{bmatrix} 1 & x_1 & \cdots & x_1{}^m \\ \vdots & \vdots & \vdots & \vdots \\ 1 & x_n & \cdots & x_n{}^m \end{bmatrix} \quad \text{and} \quad \mathbf{y} = \begin{bmatrix} y_1 \\ \vdots \\ y_n \end{bmatrix}$$

The least squares solution to this system gives the coefficients of the least squares polynomial for these data points.

Proof   The system $A\mathbf{x} = \mathbf{y}$ of n linear equations in m variables $(n > m)$ is in general overdetermined and has no solution. Thus $\mathbf{y}$ is not in the range of A. Let $(x_1, y_1'), \ldots, (x_n, y_n')$ be the "data points" that lie on the least squares curve. Let the corresponding system of equations be $A\mathbf{x} = \mathbf{y}'$, where $\mathbf{y}' = \begin{bmatrix} y_1' \\ \vdots \\ y_n' \end{bmatrix}$. Since the curve passes through these points, this system of n equations in m variables does have a unique solution $\mathbf{x}'$, the components of $\mathbf{x}'$ being the coefficients of the least squares polynomial. Thus $\mathbf{y}'$ is in the range of A. The least squares condition that $d_1{}^2 + \cdots + d_n{}^2$ is a minimum is equivalent to $\|\mathbf{y}' - \mathbf{y}\|$ being a minimum. Thus $\mathbf{y}'$ is the projection of $\mathbf{y}$ onto the range of A. See Figure 7.13. This implies that $\mathbf{y} - \mathbf{y}'$ is orthogonal to the range of A; $\mathbf{y} - \mathbf{y}'$ is orthogonal to the column vectors of A. The dot product of $\mathbf{y} - \mathbf{y}'$ with each column vector of A is zero. Interpreting the column vectors of A as the row vectors of A^t, this orthogonality condition can be expressed as the matrix product $A^t(\mathbf{y} - \mathbf{y}') = \mathbf{0}$. Since $A\mathbf{x}' = \mathbf{y}'$, we get

$$A^t(\mathbf{y} - A\mathbf{x}') = \mathbf{0}$$
$$-A^t A\mathbf{x}' = -A^t\mathbf{y}$$
$$\mathbf{x}' = (A^t A)^{-1} A^t \mathbf{y}$$
$$\mathbf{x}' = \text{pinv}(A)\mathbf{y}$$

The theorem is proved.

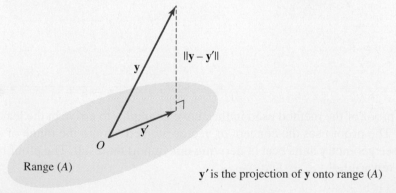

$\mathbf{y}'$ is the projection of $\mathbf{y}$ onto range (A)

Figure 7.13

EXAMPLE 5 **Hooke's Law** states that when a force is applied to a spring the length of the spring will be a linear function of the force. If L is the length of the spring when the force is F, this means that there exist (spring) constants a and b such that

$$L = a + bF$$

We shall now see that the spring constants, and hence the relationship between length and force for a spring, can be determined from experimental results using the method of least squares. Let various weights be suspended from the spring and the length of the spring measured in each case. Let the results be as follows:

Force, F (in ounces)   2   4   6   8
Length, L (in inches)   8.2   11.6   14.3   17.5

Write these statistics as points, where the first component is F and the second component is L. We get

$$(2, 8.2), \ (4, 11.6), \ (6, 14.3), \ (8, 17.5)$$

In theory, these points should all lie on a straight line. In practice, as in most experiments where measurements are taken, the readings are not completely accurate. The least squares line through these points will give the most satisfactory equation for the line. We get the system

$$u + 2b = 8.2$$
$$a + 4b = 11.6$$
$$a + 6b = 14.3$$
$$a + 8b = 17.5$$

The matrix of coefficients A and constant column matrix $\mathbf{y}$ are as follows.

$$A = \begin{bmatrix} 1 & 2 \\ 1 & 4 \\ 1 & 6 \\ 1 & 8 \end{bmatrix} \quad \text{and} \quad \mathbf{y} = \begin{bmatrix} 8.2 \\ 11.6 \\ 14.3 \\ 17.5 \end{bmatrix}$$

We get

$$\text{pinv}(A) = (A^tA)^{-1}A^t = \begin{bmatrix} 1 & 0.5 & 0 & -0.5 \\ -0.15 & -0.05 & 0.05 & 0.15 \end{bmatrix}$$

The least squares solution is

$$[(A^tA)^{-1}A^t]\mathbf{y} = \begin{bmatrix} 1 & 0.5 & 0 & -0.5 \\ -0.15 & -0.05 & 0.05 & 0.15 \end{bmatrix} \begin{bmatrix} 8.2 \\ 11.6 \\ 14.3 \\ 17.5 \end{bmatrix} = \begin{bmatrix} 5.25 \\ 1.53 \end{bmatrix}$$

Thus the spring constants are $a = 5.25$ and $b = 1.53$.

The equation for the spring is

$$L = 5.25 + 1.53F$$

Thus, for example, when a weight of 20 ounces is attached to the spring we can expect the length of the spring to be approximately $5.25 + (1.53 \times 20)$; that is, 35.85 inches.

The following application illustrates a situation that frequently arises in analyses involving a least squares fit. One wants to compare the results from two sets of data points that have the same matrix of coefficients. The techniques of handling many systems having a common matrix of coefficients come in useful.

EXAMPLE 6 Consider the following data for in-state tuition and fees (in dollars) at American 4-year Public and Private Institutions of Higher Education (World Almanac, 2002, page 236). Let us construct the least squares lines for this data and predict tuition costs for the years 2005 and 2010.

	1995	1996	1997	1998	1999	2000
Public	2,848	2,987	3,110	3,229	3,349	3,506
Private	12,243	12,881	13,344	13,973	14,588	15,531

SOLUTION

For convenience let us measure years from 1995, so that 1995 is 0, 1996 is 1, and so on. The preceding statistics give the following data points.

Public   (0, 2848),   (1, 2987),   (2, 3110),   (3, 3229),   (4, 3349),   (5, 3506)

Private   (0, 12243),   (1, 12881),   (2, 13344),   (3, 13973),   (4, 14588),   (5, 15531)

These two sets of data points lead to two systems of equations having the following common matrix of coefficients A and constant vectors $\mathbf{y}_1$ and $\mathbf{y}_2$.

$$A = \begin{bmatrix} 1 & 0 \\ 1 & 1 \\ 1 & 2 \\ 1 & 3 \\ 1 & 4 \\ 1 & 5 \end{bmatrix}, \quad \mathbf{y}_1 = \begin{bmatrix} 2848 \\ 2987 \\ 3110 \\ 3229 \\ 3349 \\ 3506 \end{bmatrix}, \quad \text{and} \quad \mathbf{y}_2 = \begin{bmatrix} 12243 \\ 12881 \\ 13344 \\ 13973 \\ 14588 \\ 15531 \end{bmatrix}$$

The least squares solutions are found,

$$[(A^t A)^{-1} A^t][\mathbf{y}_1 \mathbf{y}_2] = \begin{bmatrix} 2850.428571 & 12175 \\ 128.4285714 & 634 \end{bmatrix}$$

The least squares lines are thus

Public: $y = 2850.4286 + 128.4286x$

Private: $y = 12175 + 634x$

One can use these lines to predict future expenditures. For example, predictions for 2005 $(x = 10)$ are \$4,135 (Public) and \$18,515 (Private). For 2010 $(x = 15)$ we get \$4,777 (Public) and \$21,685 (Private).

We complete this section with a geometrical spinoff from the proof of Theorem 7.1. We find that a pseudoinverse can be used to give the projection of a vector onto a subspace.

THEOREM 7.2

Let W be the subspace of $\mathbf{R}^n$ generated by linearly independent vectors $\mathbf{u}_1, \ldots, \mathbf{u}_m$. Let $A = [\mathbf{u}_1, \ldots, \mathbf{u}_m]$. The projection of a vector $\mathbf{y}$ onto W is given by

$$\text{proj}_W\mathbf{y} = A\,\text{pinv}(A)\mathbf{y}$$

$A\,\text{pinv}(A)$ is called a **projection matrix**.

Proof Since the vectors are linearly independent $\text{pinv}(A)$ exists. The result now follows directly from the proof of the previous theorem. Let $\mathbf{x}'$ be the least squares solution to $A\mathbf{x} = \mathbf{y}$. Thus

$$\mathbf{x}' = \text{pinv}(A)\mathbf{y}$$

Multiply both sides by A.

$$A\mathbf{x}' = A\,\text{pinv}(A)\mathbf{y}$$

But $A\mathbf{x}'$ is the projection of $\mathbf{y}$ onto $\text{range}(A)$. Therefore

$$\text{proj}_W\mathbf{y} = A\,\text{pinv}(A)\mathbf{y}$$

EXAMPLE 7 Find the projection matrix for the plane $x - 2y - z = 0$ in $\mathbf{R}^3$. Use this matrix to find the projection of the vector $(1, 2, 3)$ onto this plane.

SOLUTION

Let W be the subspace of vectors that lie in this plane. W consists of vectors of the form (x, y, z) where $x = 2y + z$. Thus $W = \{(2y + z, y, z)\}$. We can write $W = \{y(2, 1, 0) + z(1, 0, 1)\}$. Therefore W is the space generated by the vectors $(2, 1, 0)$ and $(1, 0, 1)$. Let A be the matrix having these vectors as columns.

$$A = \begin{bmatrix} 2 & 1 \\ 1 & 0 \\ 0 & 1 \end{bmatrix}$$

It can be shown that

$$\text{pinv}(A) = \frac{1}{6}\begin{bmatrix} 2 & 2 & -2 \\ 1 & -2 & 5 \end{bmatrix}$$

The projection matrix is

$$A\,\text{pinv}(A) = \frac{1}{6}\begin{bmatrix} 5 & 2 & 1 \\ 2 & 2 & -2 \\ 1 & -2 & 5 \end{bmatrix}$$

The projection of $(1, 2, 3)$ onto W is computed by multiplying this vector, in column form, by $A\,\text{pinv}(A)$. We get

$$\frac{1}{6}\begin{bmatrix} 5 & 2 & 1 \\ 2 & 2 & -2 \\ 1 & -2 & 5 \end{bmatrix}\begin{bmatrix} 1 \\ 2 \\ 3 \end{bmatrix} = \begin{bmatrix} 2 \\ 0 \\ 2 \end{bmatrix}$$

Thus the projection of $(1, 2, 3)$ onto the plane $x - 2y - z = 0$ is $(2, 0, 2)$.

EXERCISE SET 7.4

Pseudoinverse

Determine the pseudoinverses of the matrices in Exercises 1–4, if they exist.

1. $\begin{bmatrix} 1 & 3 \\ 0 & 1 \\ -1 & 2 \end{bmatrix}$ **2.** $\begin{bmatrix} 1 & 1 \\ 2 & 3 \\ 0 & 1 \end{bmatrix}$

3. $\begin{bmatrix} 1 & 2 \\ 2 & 4 \\ 3 & 6 \end{bmatrix}$ **4.** $\begin{bmatrix} 1 & 2 \\ 3 & 4 \\ 5 & 6 \end{bmatrix}$

Determine the pseudoinverses of the matrices in Exercises 5–8, if they exist.

5. $\begin{bmatrix} 1 & 3 \\ 2 & 0 \\ 1 & -1 \end{bmatrix}$ **6.** $\begin{bmatrix} 4 & 1 & 0 \\ 2 & 3 & 1 \end{bmatrix}$

7. $\begin{bmatrix} 1 & 2 & 3 \\ -2 & 4 & 6 \end{bmatrix}$ **8.** $\begin{bmatrix} 1 & 2 \\ -3 & 5 \end{bmatrix}$

Least Squares Line

Determine the least squares lines for the data points in Exercises 9–14. (*Hint*: Note that the same pseudoinverse is involved for each set of data.)

9. $(1, 0), (2, 4), (3, 7)$ **10.** $(1, 1), (2, 3), (3, 7)$

11. $(1, 1), (2, 5), (3, 9)$ **12.** $(1, 1), (2, 6), (3, 9)$

13. $(1, 7), (2, 2), (3, 0)$ **14.** $(1, 12), (2, 5), (3, 1)$

Determine the least squares line for the data points in Exercises 15–18. (*Hint*: Note that the same pseudoinverse is involved for each set of data.)

15. $(1, 0), (2, 3), (3, 5), (4, 6)$

16. $(1, 1), (2, 3), (3, 5), (4, 9)$

17. $(1, 9), (2, 7), (3, 3), (4, 2)$

18. $(1, 21), (2, 17), (3, 12), (4, 7)$

Determine the least squares line for the data points in Exercises 19–20.

19. $(1, 0), (2, 1), (3, 4), (4, 7), (5, 9)$

20. $(1, 1), (2, 3), (3, 4), (4, 8), (5, 11)$

Least Squares Parabola

Determine the least squares parabola for the data points in Exercises 21–24.

21. $(1, 5), (2, 2), (3, 3), (4, 8)$

22. $(1, 4), (2, 0), (3, 3), (4, 5)$

23. $(1, 2), (2, 5), (3, 7), (4, 1)$

24. $(1, 23), (2, 4), (3, 7), (4, 17)$

Applications of Curve Fitting

25. The following loads (in ounces) are suspended from a spring and the stated lengths (in inches) are obtained. Determine the spring constants and the equation that describes each spring. Predict the length of each spring when a weight of 15 ounces is suspended. (See Example 5.)

(a)

Force F	2	4	6	8
Length L	5.9	8	10.3	12.2

(b)

Force F	2	4	6	8
Length L	7.4	10.3	13.7	16.7

26. The rate of gasoline consumption of a car depends upon its speed. To determine the effect of speed on fuel consumption the Federal Highway Administration tested cars of various weights at various speeds. One car weighing 3980 pounds had the following results. Determine the least squares line corresponding to these data. Use this equation to predict the mileage per gallon of this car at 55 mph.

Speed (miles per hour)	30	40	50	60	70
Gas used (miles per gallon)	18.25	20.00	16.32	15.77	13.61

27. A state highway patrol has the following statistics on age of driver and traffic fatalities. Determine the least squares line and use it to predict the number of driver fatalities at age 22.

Age	20	25	30	35	40	45	50
Number of fatalities	101	115	92	64	60	50	49

28. An advertising company has arrived at the following statistics relating the amount of money spent on advertising a certain product to the realized sale of that product. Find the least squares line and use it to predict the sales when $5000 is spent on advertising.

Dollars spent	1000	1500	2000	2500	3000	3500
Units sold	520	540	582	600	610	615

29. A College of Arts and Sciences has compiled the following data on the verbal SAT scores of ten entering freshmen and their graduating GPA scores four years later. Determine the least squares line and use it to predict the graduating GPA of a student entering with an SAT of 670.

SAT	490	450	640	510	680	610	480	450	600	650
GPA	3.2	2.8	3.9	2.4	4.0	3.5	2.7	3.6	3.5	2.4

30. An insurance company has computed the following percentages of people dying at various ages from 20 to 50. Plot a rough graph to see how the data lies. Predict the percentage of deaths at 23 years of age.

Age in years	20	25	30	35	40	45	50
% deaths	.18	.19	.21	.25	.35	.54	.83

31. The production numbers of a company for the last six-month period were as follows. Determine a least squares line (called the cost-volume formula) and use it to predict total costs for the next month if production was planned to rise sharply to 165 units.

	July	Aug	Sept	Oct	Nov	Dec
Units of production	120	140	155	135	110	105
Total cost (in dollars)	6300	6500	6670	6450	6100	5950

32. The following table gives recent enrollments in public and private schools in the U.S. (World Almanac, 2002, page 235) in units of one million. Plot a rough graph to see that the data seems to lie on a parabola. Construct the least-squares parabola for this data and predict enrollments for the years 2005 and 2010.

	1969	1979	1989	1999	2000
Public	45.550	41.651	40.543	46.857	47.051
Private	5.500	5.000	5.198	6.018	5.851

Projection Matrices

33. Find the projection matrices for the following planes in $\mathbf{R}^3$. Use the matrices to find the projections of the given vectors onto those planes.

(a) plane $x - y - z = 0$; vector $(1, 2, 0)$

(b) plane $x - y + z = 0$; vector $(1, 1, 1)$

(c) plane $x - 2y + z = 0$; vector $(0, 3, 0)$

34. Consider the subspaces W of $\mathbf{R}^3$ generated by the following vectors. Find the projection matrix for each subspace and use it to find the projection of the given vector $\mathbf{y}$ onto W.

(a) W generated by $(1, -1, 1)$ and $(1, 1, -1)$.
$\mathbf{y} = (-2, 1, 3)$.

(b) W generated by $(1, 0, 1)$ and $(2, 1, 0)$. $\mathbf{y} = (6, 0, 12)$.

35. Consider the subspace W generated by the vectors $(1, 1, 1)$ and $(2, 1, 0)$. Find the projection matrix for W. Use this matrix to find the projection of the vector $(3, 0, 1)$ onto W. Show that the projection of $(6, 3, 0)$ onto W is in fact $(6, 3, 0)$. What does this mean?

Miscellaneous Results

36. Trace the steps used in this section to derive the least squares solution to a system of linear equations. These steps follow logically, one after the other—why is the least squares solution not in general an actual solution? Is it ever a solution?

37. Let A be an invertible matrix. Prove that $\mathrm{pinv}(A) = A^{-1}$.

38. Let A and B be two matrices for which the product AB exists.

(a) Is $\mathrm{pinv}(AB) = \mathrm{pinv}(A)\mathrm{pinv}(B)$?

(b) Is $\mathrm{pinv}(AB) = \mathrm{pinv}(B)\mathrm{pinv}(A)$, as for matrix inverse?

39. Prove the following identities.

(a) $\mathrm{pinv}(cA) = (1/c)\mathrm{pinv}(A)$, $c \neq 0$

(b) $\mathrm{pinv}(\mathrm{pinv}(A)) = A$, if A is a square matrix

(c) $\mathrm{pinv}(A^t) = (\mathrm{pinv}(A))^t$, if A is a square matrix

40. Let A be an $m \times n$ matrix. Prove that $\mathrm{pinv}(A)$ exists if and only if A is of rank n. Show that a consequence of this result is that $\mathrm{pinv}(A)$ does not exist if $m < n$. (*Hint*: Make use of the rank/nullity theorem.)

41. Let P be a projection matrix. Prove

(a) P is symmetric

(b) $P^2 = P$ (P is idempotent)

CHAPTER 7 REVIEW EXERCISES

1. Let $\mathbf{u} = (x_1, x_2)$ and $\mathbf{v} = (y_1, y_2)$ be elements of $\mathbf{R}^2$. Prove that the following function defines an inner product on $\mathbf{R}^2$.

$$\langle \mathbf{u}, \mathbf{v} \rangle = 2x_1y_1 + 3x_2y_2$$

2. Determine the inner product, the norms, and the distance between the functions $f(x) = 3x - 1$, $g(x) = 5x + 3$ in the inner product space P_2, with inner product defined by $\langle f, g \rangle = \int_0^1 f(x)g(x)\, dx$.

3. Let $\mathbf{R}^2$ have inner product defined by $\langle (x_1, x_2), (y_1, y_2) \rangle = 2x_1y_1 + 3x_2y_2$. Determine the norms and the distance between $(1, -2)$, $(3, 2)$ in this space.

4. Consider $\mathbf{R}^2$ with the inner product

$$\langle (x_1, x_2), (y_1, y_2) \rangle = 2x_1y_1 + 3x_2y_2$$

Determine the equation of the circle with center the origin and radius one in this space. Sketch the circle.

5. Find the least squares linear approximation to $f(x) = x^2 + 2x - 1$ over the interval $[-1, 1]$.

6. Find the fourth-order Fourier approximation to $f(x) = 2x - 1$ over $[0, 2\pi]$.

7. List all the vectors of V_7 that lie in the sphere having radius 1 and center $(0, 1, 1, 0, 0, 1, 1)$.

8. Find the pseudoinverse of the matrix $\begin{bmatrix} 1 & 2 \\ 3 & 1 \\ 4 & 2 \end{bmatrix}$.

9. Determine the least squares parabola for the data $(1, 6)$, $(2, 2)$, $(3, 5)$, $(4, 9)$.

10. Let A be a matrix with pseudoinverse B. Show that

(a) $ABA = A$

(b) $BAB = B$

(c) AB is symmetric

(d) BA is symmetric

The Emirates Towers complex is located on the Sheikh Zayed Road in Dubai, United Arab Emirates. The two buildings—the Emirates Office Tower and Jumeirah Emirates Towers Hotel—hold the rankings as the 12th- and 29th-tallest buildings in the world, respectively, and are connected by a two-story retail complex known as "The Boulevard."

CHAPTER 8

Numerical Methods

In this chapter we look at numerical techniques for solving systems of linear equations. We add four important methods, namely Gausssian elimination, LU decomposition, the Jacobi method, and the Gauss-Seidel method to our library of techniques of solving systems of linear equations. We discuss the merits of the various methods, including their reliability for solving various types of systems. Certain systems of equations can lead to incorrect results unless great care is taken. We discuss ways of recognizing and solving "delicate" systems.

While the determinant approach of Chapter 5 is useful for finding eigenvalues and eigenvectors of small matrices and for developing the theory, it is not practical for finding eigenvalues and eigenvectors of large matrices that occur in applications. We introduce a numerical technique for finding such eigenvalues and eigenvectors.

Applications discussed include an analysis of networks—ways of describing connectivities of networks and the accessibility of their vertices. Such measures are used to compare connectivities of cities and regions, and to plan where roads should be built.

*8.1 Gaussian Elimination

There are many elimination methods in addition to the method of Gauss-Jordan elimination for solving systems of linear equations. In this section we introduce another elimination method called **Gaussian elimination**. Different methods are suitable for different occasions. It is important to choose the best method for the purpose in mind. We shall discuss the relative merits of Gauss-Jordan elimination and Gaussian elimination. The merits and drawbacks of other methods will be discussed later.

The method of Gaussian elimination involves an *echelon form* of the augmented matrix of the system of equations. An echelon form satisfies the first three of the conditions of the reduced echelon form.

> DEFINITION | A matrix is in **echelon form** if
>
> 1. Any rows consisting entirely of zeros are grouped at the bottom of the matrix.
> 2. The first nonzero element of each row is 1. This element is called a **leading 1**.
> 3. The leading 1 of each row after the first is positioned to the right of the leading 1 of the previous row. (This implies that all the elements below a leading 1 are zero.)

The following matrices are all in echelon form.

$$\begin{bmatrix} 1 & -1 & 2 \\ 0 & 1 & 2 \\ 0 & 0 & 1 \end{bmatrix} \begin{bmatrix} 1 & 3 & -6 & 4 \\ 0 & 0 & 1 & 3 \\ 0 & 0 & 0 & 0 \end{bmatrix} \begin{bmatrix} 1 & 4 & 6 & 2 & 5 & 2 \\ 0 & 0 & 1 & 2 & 3 & 4 \\ 0 & 0 & 0 & 0 & 1 & 6 \end{bmatrix}$$

The difference between a reduced echelon form and an echelon form is that the elements above and below a leading 1 are zero in a reduced echelon form, while only the elements below the leading 1 need be zero in an echelon form.

The Gaussian elimination algorithm is as follows.

Gaussian Elimination

1. Write down the augmented matrix of the system of linear equations.
2. Find an echelon form of the augmented matrix using elementary row operations. This is done by creating leading 1's, then zeros below each leading 1, column by column, starting with the first column.
3. Write down the system of equations corresponding to the echelon form.
4. Use back substitution to arrive at the solution.

We illustrate the method with the following example.

EXAMPLE 1 Solve the following system of linear equations using the method of Gaussian elimination.

$$\begin{aligned} x_1 + 2x_2 + 3x_3 + 2x_4 &= -1 \\ -x_1 - 2x_2 - 2x_3 + x_4 &= 2 \\ 2x_1 + 4x_2 + 8x_3 + 12x_4 &= 4 \end{aligned}$$

SOLUTION

Starting with the augmented matrix, create zeros below the pivot in the first column.

$$\begin{bmatrix} ① & 2 & 3 & 2 & -1 \\ -1 & -2 & -2 & 1 & 2 \\ 2 & 4 & 8 & 12 & 4 \end{bmatrix} \begin{array}{c} \approx \\ R2 + R1 \\ R3 + (-2)R1 \end{array} \begin{bmatrix} 1 & 2 & 3 & 2 & -1 \\ 0 & 0 & ① & 3 & 1 \\ 0 & 0 & 2 & 8 & 6 \end{bmatrix}$$

At this stage we create a zero only *below* the pivot.

$$\begin{array}{c} \approx \\ R3 + (-2)R2 \end{array} \begin{bmatrix} 1 & 2 & 3 & 2 & -1 \\ 0 & 0 & 1 & 3 & 1 \\ 0 & 0 & 0 & ② & 4 \end{bmatrix} \begin{array}{c} \approx \\ (1/2)\,R3 \end{array} \begin{bmatrix} 1 & 2 & 3 & 2 & -1 \\ 0 & 0 & 1 & 3 & 1 \\ 0 & 0 & 0 & 1 & 2 \end{bmatrix}$$

echelon form

We have arrived at the echelon form.
The corresponding system of equations is

$$\begin{aligned} x_1 + 2x_2 + 3x_3 + 2x_4 &= -1 \\ x_3 + 3x_4 &= 1 \\ x_4 &= 2 \end{aligned}$$

Observe that the effect of performing the row operations in this manner to arrive at an echelon form is to eliminate variables from equations. This is called **forward elimination**. This system is now solved by **back substitution**. (The terms **forward pass** and **backward pass** are also used.) The value of x_4 is substituted into the second equation to give x_3. x_3 and x_4 are then substituted into the first equation to get x_1. We get

$$x_3 + 3(2) = 1,$$
$$x_3 = -5$$

Substituting $x_4 = 2$ and $x_3 = -5$ into the first equation,

$$x_1 + 2x_2 + 3(-5) + 2(2) = -1,$$
$$x_1 + 2x_2 = 10,$$
$$x_1 = -2x_2 + 10$$

Let $x_2 = r$. The system has many solutions. The solutions are

$$x_1 = -2r + 10, \quad x_2 = r, \quad x_3 = -5, \quad x_4 = 2$$

The forward elimination of variables in this method was performed using matrices and elementary row operations. The back substitution can also be performed using matrices. The final matrix is then the reduced echelon form of the system. This way of performing the back substitution can be implemented on a computer. We illustrate the method for the system of equations of the previous example.

EXAMPLE 2 Solve the following system of linear equations using the method of Gaussian elimination, performing back substitution using matrices.

$$x_1 + 2x_2 + 3x_3 + 2x_4 = -1$$
$$-x_1 - 2x_2 - 2x_3 + x_4 = 2$$
$$2x_1 + 4x_2 + 8x_3 + 12x_4 = 4$$

SOLUTION

We arrive at the echelon form as in the previous example.

$$\begin{bmatrix} 1 & 2 & 3 & 2 & -1 \\ -1 & -2 & -2 & 1 & 2 \\ 2 & 4 & 8 & 12 & 4 \end{bmatrix} \approx \cdots \approx \begin{bmatrix} 1 & 2 & 3 & 2 & -1 \\ 0 & 0 & 1 & 3 & 1 \\ 0 & 0 & 0 & 1 & 2 \end{bmatrix}$$

echelon form

This marks the end of the forward elimination of variables from equations. We now commence the back substitution using matrices.

$$\begin{bmatrix} 1 & 2 & 3 & 2 & -1 \\ 0 & 0 & 1 & 3 & 1 \\ 0 & 0 & 0 & ① & 2 \end{bmatrix} \begin{array}{c} \\ \text{R1} + (-2)\text{R3} \\ \text{R2} + (-3)\text{R3} \end{array} \approx \begin{bmatrix} 1 & 2 & 3 & 0 & -5 \\ 0 & 0 & ① & 0 & -5 \\ 0 & 0 & 0 & 1 & 2 \end{bmatrix}$$

(Create zeros above the leading 1 in row 3.
This is equivalent to substituting for x_4 from
Equation 3 into Equations 1 and 2.)

$$\approx \\ \text{R1} + (-3)\text{R2} \quad \begin{bmatrix} 1 & 2 & 0 & 0 & 10 \\ 0 & 0 & 1 & 0 & -5 \\ 0 & 0 & 0 & 1 & 2 \end{bmatrix}$$

(Create a zero above the leading 1 in row 2.
This is equivalent to substituting for x_3
from Equation 2 into Equation 1.)

This matrix is the reduced echelon form of the original augmented matrix. The corresponding system of equations is

$$x_1 + 2x_2 \qquad\qquad = \quad 10$$
$$x_3 \qquad = -5$$
$$x_4 = \quad 2$$

Let $x_2 = r$. We get the same solution as previously,

$$x_1 = -2r + 10, \quad x_2 = r, \quad x_3 = -5, \quad x_4 = 2$$

Comparison of Gauss-Jordan and Gaussian Elimination

The method of Gaussian elimination is in general more efficient than Gauss-Jordan elimination in that it involves fewer operations of addition and multiplication. It is during the back substitution that Gaussian elimination picks up this advantage. We now illustrate how Gaussian elimination saves two operations over Gauss-Jordan elimination in the preceding example. Consider the final transformation that brings the matrix to reduced echelon form.

$$\begin{bmatrix} 1 & 2 & 3 & 0 & -5 \\ 0 & 0 & 1 & 0 & -5 \\ 0 & 0 & 0 & 1 & 2 \end{bmatrix} \quad \overset{\approx}{\text{R1} + (-3)\text{R2}} \quad \begin{bmatrix} 1 & 2 & 0 & 0 & 10 \\ 0 & 0 & 1 & 0 & -5 \\ 0 & 0 & 0 & 1 & 2 \end{bmatrix}$$

The aim of this transformation is to create a 0 in location $(1, 3)$. Note that changing the 3 in location $(1, 3)$ to 0 need not in practice involve any arithmetic operations, as one (or the computer) knows in advance that the element is to be zero. The 0 in the $(1, 4)$ location remains unchanged; no arithmetic operations need be performed on it. The row operation **R1** $+ (-3)$**R2** in fact uses only two arithmetic operations—one of multiplication and one of addition—in changing the -5 in the $(1, 5)$ location to 10:

$$-5 + (-3)(-5) = 10$$

$$\qquad\quad \nearrow \qquad \nwarrow$$

addition   multiplication

On the other hand, when the zero is created in the $(1, 3)$ location during Gauss-Jordan elimination, the $(1, 4)$ element and the $(1, 5)$ element are both changed, each involving two operations, one addition and the other multiplication. These two operations for the change in the $(1, 4)$ element are two additional operations involved in Gauss-Jordan elimination.

In larger systems of equations, many more operations are saved in Gaussian elimination during back substitution. The reduction in the number of operations not only saves time on a computer but also increases the accuracy of the final answer. With each arithmetic operation there is a possibility of round-off error on a computer. With large systems, the method of Gauss-Jordan elimination involves approximately 50% more arithmetic operations than does Gaussian elimination (see the following table).

Count of Operations for an $n \times n$ System with Unique Solution

	Number of Multiplications	Number of Additions
Gauss-Jordan elimination	$\dfrac{n^3}{2} + \dfrac{n^2}{2} \approx \dfrac{n^3}{2}$ (for large n)	$\dfrac{n^3}{2} - \dfrac{n}{2} \approx \dfrac{n^3}{2}$
Gaussian elimination	$\dfrac{n^3}{3} + n^2 - \dfrac{n}{3} \approx \dfrac{n^3}{3}$	$\dfrac{n^3}{3} + \dfrac{n^2}{2} - \dfrac{5n}{6} \approx \dfrac{n^3}{3}$

Gauss-Jordan elimination, on the other hand, has the advantage of being more straight-forward for hand computations. It is easier for solving small systems and it is the method that we use in this course when we solve systems of linear equations by hand.

We complete this section with a discussion of the formulas for the total number of operations involved in solving a system of n linear equations in n variables that has a unique solution, using both Gauss-Jordan elimination and Gaussian elimination. We group multiplications and divisions together as multiplications and additions and subtractions together as additions.

If there are many equations, with n large, then the term with the highest power of n dominates the other terms in the preceding formulas. The total number of operations with Gauss-Jordan elimination is approximately n^3 while the total in Gaussian elimination is approximately $\frac{2n^3}{3}$. Gaussian elimination is, thus, approximately 50% more efficient than Gauss-Jordan elimination.

We now derive the above formulas for Gauss-Jordan elimination, leaving it for the reader to arrive at the formulas for Gaussian elimination in the exercises that follow. Let us denote general elements in the matrices by $*$. Assume that there are no row interchanges. Note that when an element is known to become a 1 or a zero there are no arithmetic operations involved; substitution is used. Thus, for example, there are no operations involved in the location where a leading one is created. We get, starting with the $n \times (n + 1)$ augmented matrix of the system,

$n + 1$ columns

$$
n \text{ rows} \left\{
\begin{bmatrix}
* & * & \cdots & * \\
* & * & \cdots & * \\
\vdots & \vdots & & \vdots \\
* & * & \cdots & *
\end{bmatrix}
\right.
\underset{n \text{ mults}}{\approx}
\begin{bmatrix}
1 & * & \cdots & * \\
* & * & \cdots & * \\
\vdots & \vdots & & \vdots \\
* & * & \cdots & *
\end{bmatrix}
\underset{\substack{n \text{ mults} \\ n \text{ adds} \\ (\text{per row})}}{\approx}
\begin{bmatrix}
1 & * & \cdots & * \\
0 & * & \cdots & * \\
\vdots & \vdots & & \vdots \\
0 & * & \cdots & *
\end{bmatrix}
$$

$$
\underset{n-1 \text{ mults}}{\approx}
\begin{bmatrix}
1 & * & \cdots & * \\
0 & 1 & \cdots & * \\
\vdots & \vdots & & \vdots \\
0 & * & \cdots & *
\end{bmatrix}
\underset{\substack{n-1 \text{ mults} \\ n-1 \text{ adds} \\ (\text{per row})}}{\approx}
\begin{bmatrix}
1 & 0 & \cdots & * \\
0 & 1 & \cdots & * \\
\vdots & \vdots & & \vdots \\
0 & 0 & \cdots & *
\end{bmatrix}
$$

$$
\approx \cdots \approx
\underset{1 \text{ mult}}{}
\begin{bmatrix}
0 & 1 & * & * \\
0 & 1 & * & * \\
\vdots & \vdots & \vdots & \vdots \\
0 & 0 & 1 & *
\end{bmatrix}
\underset{\substack{1 \text{ mult} \\ 1 \text{ add} \\ (\text{per row})}}{\approx}
\begin{bmatrix}
1 & 0 & 0 & * \\
0 & 1 & 0 & * \\
\vdots & \vdots & \vdots & \vdots \\
0 & 0 & 1 & *
\end{bmatrix}
$$

We now add up these operations. Remember that there are $(n - 1)$ rows involved every time zeros are created "per row" in the above manner. The following formula is used for the sum of the first n integers

$$[n + (n - 1) + \cdots + 1] = \frac{n(n + 1)}{2}$$

Total number of multiplications

$$= \underbrace{[n + (n + 1) + \cdots + 1]}_{\text{to create leading 1's}} + \underbrace{(n - 1)[n + (n - 1) + \cdots + 1]}_{\text{to create zeros}}$$

$$= n[n + (n - 1) + \cdots + 1] = n\left[\frac{n(n + 1)}{2}\right] = \frac{n^3}{2} + \frac{n^2}{2}$$

Total number of additions

$$= (n - 1)[n + (n + 1) + \cdots + 1] = (n - 1)\left[\frac{n(n + 1)}{2}\right] = \frac{n^3}{2} - \frac{n}{2}$$

EXERCISE SET 8.1

Echelon Form

1. Determine whether or not each of the following matrices is in echelon form.

(a) $\begin{bmatrix} 1 & 2 & 1 \\ 0 & 1 & 3 \\ 0 & 0 & 0 \end{bmatrix}$ (b) $\begin{bmatrix} 1 & 2 & 3 & 4 \\ 0 & 0 & 1 & 0 \\ 0 & 0 & 0 & 1 \end{bmatrix}$

(c) $\begin{bmatrix} 1 & 5 & 6 & 2 \\ 0 & 1 & 0 & 4 \\ 0 & 0 & 1 & 2 \end{bmatrix}$ (d) $\begin{bmatrix} 1 & 7 & 6 & 2 \\ 0 & 0 & 1 & 8 \\ 0 & 0 & 1 & 4 \\ 0 & 0 & 0 & 1 \end{bmatrix}$

2. Determine whether or not each of the following matrices is in echelon form.

(a) $\begin{bmatrix} 1 & 0 & 0 & 3 & 0 \\ 0 & 0 & 1 & 2 & 0 \\ 0 & 0 & 0 & 0 & 1 \end{bmatrix}$ (b) $\begin{bmatrix} 1 & 2 & 4 & 6 \\ 0 & 0 & 1 & 2 \\ 0 & 1 & 3 & 3 \\ 0 & 0 & 0 & 1 \end{bmatrix}$

(c) $\begin{bmatrix} 1 & 3 & 4 & 2 & 3 \\ 0 & 0 & 2 & 5 & 1 \\ 0 & 0 & 0 & 1 & 4 \\ 0 & 0 & 0 & 0 & 6 \end{bmatrix}$ (d) $\begin{bmatrix} 1 & -1 & 4 & 6 \\ 0 & 0 & 1 & 3 \\ 0 & 0 & 0 & 0 \\ 0 & 0 & 0 & 1 \end{bmatrix}$

Gaussian Elimination

In Exercises 3–6, solve the systems of equations using Gaussian elimination.

(a) Perform the back substitution using equations.

(b) Perform the back substitution using matrices.

3. $x_1 + x_2 + x_3 = 6$
$x_1 - x_2 + x_3 = 2$
$x_1 + 2x_2 + 3x_3 = 14$

4. $x_1 - x_2 - x_3 = 2$
$x_1 - x_2 + x_3 = 2$
$3x_1 - 2x_2 + x_3 = 5$

5. $x_1 - x_2 + 2x_3 = 3$
$2x_1 - 2x_2 + 5x_3 = 4$
$x_1 + 2x_2 - x_3 = -3$
$2x_2 + 2x_3 = 1$

6. $x_1 - x_2 + x_3 + 2x_4 - 2x_5 = 1$
$2x_1 - x_2 - x_3 + 3x_4 - x_5 = 3$
$-x_1 - x_2 + 5x_3 - 4x_5 = -3$

Miscellaneous Results

7. Consider a system of four equations in five variables. In general, how many arithmetic operations will be saved by using Gaussian elimination rather than Gauss-Jordan elimination to solve the system? Where are these operations saved?

8. Compare Gaussian elimination to Gauss-Jordan elimination for a system of four equations in six variables. Determine where operations are saved during Gaussian elimination.

9. Can a matrix have more than one echelon form? (*Hint*: Consider a 2×3 matrix and arrive at an echelon form using two distinct sequences of row operations.)

10. Consider a system of n linear equations in n variables that has a unique solution. Show that Gaussian elimination involves $n^3/3 + n^2/2 + n/6$ multiplications and $n^3/3 - n/3$ additions to arrive at an echelon form, and then a further $n^2/2 - n/2$ multiplications and $n^2/2 - n/2$ additions to arrive at the reduced echelon form. Thus a total of $n^3/3 + n^2 - n/3$ multiplications and $n^3/3 + n^2/2 - 5n/6$ additions are involved in solving the system of equations using Gaussian elimination. (The formula for the sum of squares is $n^2 + (n - 1)^2 + \cdots + 1 = [n(n + 1)(2n + 1)]/6$.)

11. Construct a table that gives the number of multiplications and additions in both Gauss-Jordan elimination and Gaussian elimination for systems of n equations in n variables having unique solutions, for $n = 2$ to 10.

*8.2 The Method of *LU* Decomposition

The method of *LU* decomposition is used to solve certain systems of linear equations. The method is widely used on computers. It involves writing the matrix of coefficients as the product of a **lower triangular matrix** *L* and an **upper triangular matrix** *U*. We remind the reader that a lower triangular matrix is a square matrix with zeros in all locations above the main diagonal while an upper triangular matrix is a square matrix with zeros below the main diagonal. The following are examples of lower and upper triangular matrices.

$$\begin{bmatrix} 2 & 0 & 0 & 0 \\ 3 & -1 & 0 & 0 \\ 5 & 2 & 7 & 0 \\ 4 & 0 & -2 & 8 \end{bmatrix} \qquad \begin{bmatrix} 8 & 2 & 5 & -3 \\ 0 & 2 & 9 & 1 \\ 0 & 0 & -4 & 2 \\ 0 & 0 & 0 & 7 \end{bmatrix}$$

lower triangular matrix   upper triangular matrix

A system of equations that has a triangular matrix of coefficients is particularly straightforward to solve. If the matrix is lower triangular, a method called forward substitution is used. If the matrix is upper triangular, the method of back substitution used in Gaussian elimination is applied. The method of *LU decomposition* involves first solving a lower triangular system of equations then an upper triangular system. The following examples illustrate the methods of forward and back substitution. We shall then bring these methods together in *LU* decomposition.

EXAMPLE **1** Solve the following system of equations, which has a lower triangular matrix of coefficients.

$$\begin{aligned} 2x_1 &= 8 \\ x_1 + 3x_2 &= -2 \\ 4x_1 + 5x_2 - x_3 &= 3 \end{aligned}$$

SOLUTION

We solve the system by **forward substitution**. The first equation gives the value of x_1. This value is substituted into the second equation to get x_2. These values of x_1 and x_2 are then substituted into the third equation to get x_3.

1st equation gives

$$2x_1 = 8,$$
$$x_1 = 4$$

2nd equation gives

$$x_1 + 3x_2 = -2,$$
$$4 + 3x_2 = -2,$$
$$x_2 = -2$$

3rd equation gives

$$4x_1 + 5x_2 - x_3 = 3,$$
$$16 - 10 - x_3 = 3,$$
$$x_3 = 3$$

The solution is $x_1 = 4$, $x_2 = -2$, $x_3 = 3$.

EXAMPLE 2 Solve the following system of equations, which has an upper triangular matrix of coefficients.

$$2x_1 - x_2 + 4x_3 = 0$$
$$x_2 - x_3 = 4$$
$$3x_3 = -6$$

SOLUTION

We use the method of **back substitution**. The third equation gives the value of x_3. This value is substituted into the second equation to get x_2. These values of x_2 and x_3 are then substituted into the first equation to get x_1.

3rd equation gives

$$3x_3 = -6,$$
$$x_3 = -2$$

2nd equation gives

$$x_2 - x_3 = 4,$$
$$x_2 + 2 = 4,$$
$$x_2 = 2$$

1st equation gives

$$2x_1 - x_2 + 4x_3 = 0,$$
$$2x_1 - 2 - 8 = 0,$$
$$x_1 = 5$$

The solution is $x_1 = 5$, $x_2 = 2$, $x_3 = -2$.

Forward and back substitution can of course be carried out using row operations on matrices. This is the way they are implemented on a computer.

DEFINITION   Let A be a square matrix that can be factored into the form $A = LU$, where L is a lower triangular matrix and U is an upper triangular matrix. This factoring is called an **LU decomposition** of A.

Not every matrix has an LU decomposition and when it exists, it is not unique. (See the following exercises.) The method that we now introduce can be used to solve a system of linear equations if A has an LU decomposition.

Method of *LU* Decomposition

Let $AX = B$ be a system of n equations in n variables where A has LU decomposition $A = LU$. The system can thus be written

$$LUX = B$$

The method involves writing this system as two subsystems, one of which is lower triangular and the other upper triangular.

$$UX = Y$$
$$LY = B$$

Observe that substituting for Y from the first equation into the second gives the original system $LUX = B$. In practice we first solve $LY = B$ for Y and then solve $UX = Y$ to get the solution X. We now summarize these steps.

Solution of $AX = B$

1. Find the *LU* decomposition of A.

   (If A has no *LU* decomposition the method is not applicable.)
2. Solve $LY = B$ by forward substitution.
3. Solve $UX = Y$ by back substitution.

A key to this method of course is being able to arrive at an *LU* decomposition of the matrix of coefficients A. Suppose it is possible to transform A into an upper triangular form U using a sequence of row operations that involve **adding multiples of rows to rows**. The row operation of adding c times row j to row k for an $n \times n$ matrix A can be performed by the matrix multiplication EA, where E is the matrix obtained by changing the element i_{kj} of I_n to c. E is an **elementary matrix** (Section 2.4). Let the row operations be described by elementary matrices $E_1, \ldots, E_k$. Thus

$$E_k \ldots E_1 A = U$$

We know that each elementary matrix is invertible. Multiply both sides of this equation by $E_1^{-1} \ldots E_k^{-1}$ in succession to get

$$A = E_1^{-1} \ldots E_k^{-1} U$$

Each such elementary matrix is a lower triangular matrix. It can be shown that the inverse of a lower triangular matrix is lower triangular and that the product of lower triangular matrices is also lower triangular. (See the following exercises.) Let $L = E_1^{-1} \ldots E_k^{-1}$. Thus $A = LU$.

EXAMPLE 3 Solve the following system of equations using *LU* decomposition.

$$\begin{aligned} 2x_1 + x_2 + 3x_3 &= -1 \\ 4x_1 + x_2 + 7x_3 &= 5 \\ -6x_1 - 2x_2 - 12x_3 &= -2 \end{aligned}$$

SOLUTION

Let us transform the matrix of coefficients A into upper triangular form U by creating zeros below the main diagonal as follows.

$$\overset{A}{\begin{bmatrix} 2 & 1 & 3 \\ 4 & 1 & 7 \\ -6 & -2 & -12 \end{bmatrix}} \underset{R2 - 2R1}{\approx} \begin{bmatrix} 2 & 1 & 3 \\ 0 & -1 & 1 \\ -6 & -2 & -12 \end{bmatrix}$$

$$\underset{R3 + 3R1}{\approx} \begin{bmatrix} 2 & 1 & 3 \\ 0 & -1 & 1 \\ 0 & 1 & -3 \end{bmatrix} \underset{R3 + R2}{\approx} \overset{U}{\begin{bmatrix} 2 & 1 & 3 \\ 0 & -1 & 1 \\ 0 & 0 & -2 \end{bmatrix}}$$

The elementary matrices that correspond to these row operations are

$$
\begin{array}{ccc}
\text{R2} - \text{2R1} & \text{R3} + \text{3R1} & \text{R3} + \text{R2}
\end{array}
$$

$$
E_1 = \begin{bmatrix} 1 & 0 & 0 \\ -2 & 1 & 0 \\ 0 & 0 & 1 \end{bmatrix}, \quad
E_2 = \begin{bmatrix} 1 & 0 & 0 \\ 0 & 1 & 0 \\ 3 & 0 & 1 \end{bmatrix}, \quad
E_3 = \begin{bmatrix} 1 & 0 & 0 \\ 0 & 1 & 0 \\ 0 & 1 & 1 \end{bmatrix}
$$

The inverses of these matrices are

$$
E_1^{-1} = \begin{bmatrix} 1 & 0 & 0 \\ 2 & 1 & 0 \\ 0 & 0 & 1 \end{bmatrix}, \quad
E_2^{-1} = \begin{bmatrix} 1 & 0 & 0 \\ 0 & 1 & 0 \\ -3 & 0 & 1 \end{bmatrix}, \quad
E_3^{-1} = \begin{bmatrix} 1 & 0 & 0 \\ 0 & 1 & 0 \\ 0 & -1 & 1 \end{bmatrix}
$$

We get

$$
L = E_1^{-1} E_2^{-1} E_3^{-1} = \begin{bmatrix} 1 & 0 & 0 \\ 2 & 1 & 0 \\ -3 & -1 & 1 \end{bmatrix}
$$

Thus

$$
A = \begin{array}{cc} L & U \end{array}
$$

$$
A = \begin{bmatrix} 1 & 0 & 0 \\ 2 & 1 & 0 \\ -3 & -1 & 1 \end{bmatrix} \begin{bmatrix} 2 & 1 & 3 \\ 0 & -1 & 1 \\ 0 & 0 & -2 \end{bmatrix}
$$

We now solve the given system $LUX = B$ by solving the two subsystems $LY = B$ and $UX = Y$. We get

$$
LY = B: \quad \begin{bmatrix} 1 & 0 & 0 \\ 2 & 1 & 0 \\ -3 & -1 & 1 \end{bmatrix} \begin{bmatrix} y_1 \\ y_2 \\ y_3 \end{bmatrix} = \begin{bmatrix} -1 \\ 5 \\ -2 \end{bmatrix}
$$

This lower triangular system has solution $y_1 = -1$, $y_2 = 7$, $y_3 = 2$.

$$
UX = Y: \quad \begin{bmatrix} 2 & 1 & 3 \\ 0 & -1 & 1 \\ 0 & 0 & -12 \end{bmatrix} \begin{bmatrix} x_1 \\ x_2 \\ x_3 \end{bmatrix} = \begin{bmatrix} -1 \\ 7 \\ 2 \end{bmatrix}
$$

This upper triangular system has solution $x_1 = 5$, $x_2 = -8$, $x_3 = -1$.
The solution to the given system is $x_1 = 5$, $x_2 = -8$, $x_3 = -1$.

■

A careful analysis of the above example leads to the following method of arriving at an LU decomposition directly from a knowledge of the row operations that lead to an upper triangular form.

Construction of an *LU* Decomposition of a Matrix

1. Use row operations to arrive at U.

   (The operations must involve adding multiples of rows to rows. In general, if row interchanges are required to arrive at U an LU form does not exist.)

2. The diagonal elements of L are 1's.

   The nonzero elements of L correspond to row operations.

   The row operation $R_k + cR_j$ implies that $l_{kj} = -c$.

We now give a second example to reinforce the understanding of this method of arriving at L and U.

EXAMPLE **4** Solve the following system of equations using *LU* decomposition.

$$\begin{aligned} x_1 - 3x_2 + 4x_3 &= 12 \\ -x_1 + 5x_2 - 3x_3 &= -12 \\ 4x_1 - 8x_2 + 23x_3 &= 58 \end{aligned}$$

SOLUTION

We transform the matrix of coefficients A into upper triangular form U by creating zeros below the main diagonal.

$$\begin{bmatrix} 1 & -3 & 4 \\ -1 & 5 & -3 \\ 4 & -8 & 23 \end{bmatrix} \approx \begin{matrix} \\ R2 + R1 \\ R3 - 4R1 \end{matrix} \begin{bmatrix} 1 & -3 & 4 \\ 0 & 2 & 1 \\ 0 & 4 & 7 \end{bmatrix} \approx R3 - 2R2 \begin{bmatrix} 1 & -3 & 4 \\ 0 & 2 & 1 \\ 0 & 0 & 5 \end{bmatrix}$$

These row operations lead to the following *LU* decomposition of A.

$$A = \begin{bmatrix} 1 & 0 & 0 \\ -1 & 1 & 0 \\ 4 & 2 & 1 \end{bmatrix} \begin{bmatrix} 1 & -3 & 4 \\ 0 & 2 & 1 \\ 0 & 0 & 5 \end{bmatrix}$$

We again solve the given system $LUX = B$ by solving the two subsystems $LY = B$ and $UX = Y$.

$$LY = B: \begin{bmatrix} 1 & 0 & 0 \\ -1 & 1 & 0 \\ 4 & 2 & 1 \end{bmatrix} \begin{bmatrix} y_1 \\ y_2 \\ y_3 \end{bmatrix} = \begin{bmatrix} 12 \\ -12 \\ 58 \end{bmatrix}$$

This lower triangular system has solution $y_1 = 12$, $y_2 = 0$, $y_3 = 10$.

$$UX = Y: \begin{bmatrix} 1 & -3 & 4 \\ 0 & 2 & 1 \\ 0 & 0 & 5 \end{bmatrix} \begin{bmatrix} x_1 \\ x_2 \\ x_3 \end{bmatrix} - \begin{bmatrix} 12 \\ 0 \\ 10 \end{bmatrix}$$

This upper triangular system has solution $x_1 = 1$, $x_2 = -1$, $x_3 = 2$.

The solution to the given system is $x_1 = 1$, $x_2 = -1$, $x_3 = 2$.

■

Numerical Discussion

The method of *LU* decomposition can be applied to any system of n equations in n variables, $AX = B$, which can be transformed into an upper triangular form U using row operations that involve adding multiples of rows to rows. In general, if row interchanges are required to arrive at the upper triangular form then the matrix does not have an *LU* decomposition and the method cannot be used.

The total number of arithmetic operations needed to solve a system of equations using LU decomposition is exactly the same as that needed in Gaussian elimination. The method is the most efficient method available however for solving many systems, all having the same matrix of coefficients when the right-hand sides are not all available in advance. In that situation, the factorization has only to be done once, using $(4n^3 - 3n^2 - n)/6$ arithmetic operations. The solutions of the two triangular systems can then be carried out in $2n^2 - n$ arithmetic operations per system, as compared to $(4n^3 + 9n^2 - 7n)/6$ arithmetic operations for full Gaussian elimination each time.

The fact that LU decomposition is used to solve systems of equations on computers however, where applicable, rather than Gaussian elimination, is a result of the usefulness of LU decomposition of a matrix for many types of computations. If $A = LU$, then

$$A^{-1} = (LU)^{-1} = U^{-1}L^{-1} \quad \text{and} \quad |A| = |LU| = |L||U|$$

The inverse of a triangular matrix can be computed very efficiently and the determinant of a triangular matrix is the product of its diagonal elements. In practice, once one of these matrix functions has been computed using L and U, these matrices are then available for use in other matrix functions. For example, if $|A|$ has been computed then L and U are known, providing a very efficient starting place for solving the system $AX = B$.

The matrix software package MATLAB (The MathWorks, Inc.) discussed in the appendix, for example, uses LU decomposition extensively. MATLAB is one of the most powerful scientific/engineering numerical linear algebra packages available. The name stands for "matrix laboratory."

EXERCISE SET 8.2

Triangular Systems of Equations

In Exercises 1–3 solve the lower triangular systems.

1.
$$\begin{aligned} x_1 &&&= 1 \\ 2x_1 - x_2 &&&= -2 \\ 3x_1 + x_2 - x_3 &&&= 8 \end{aligned}$$

2.
$$\begin{aligned} 2x_1 &&&= 4 \\ -x_1 + x_2 &&&= 1 \\ x_1 + x_2 + x_3 &&&= 5 \end{aligned}$$

3.
$$\begin{aligned} x_1 &&&= -2 \\ 3x_1 + x_2 &&&= -5 \\ x_1 + 4x_2 + 2x_3 &&&= 4 \end{aligned}$$

In Exercises 4–6 solve the upper triangular systems.

4.
$$\begin{aligned} x_1 + x_2 + x_3 &= 3 \\ 2x_2 + x_3 &= 3 \\ 3x_3 &= 6 \end{aligned}$$

5.
$$\begin{aligned} 2x_1 - x_2 + x_3 &= 3 \\ x_2 - 3x_3 &= -7 \\ x_3 &= 3 \end{aligned}$$

6.
$$\begin{aligned} 3x_1 + 2x_2 - x_3 &= -6 \\ 2x_2 + 4x_3 &= 26 \\ x_3 &= 7 \end{aligned}$$

Row Operations

7. Let $A = LU$. The following sequences of transformations were used to arrive at U. Find L in each case.

(a) $R_2 - R_1, R_3 + R_1, R_3 + R_2$

(b) $R_2 + 5R_1, R_3 - 2R_1, R_3 + 7R_2$

(c) $R_2 + \frac{1}{2}R_1, R_3 + \frac{1}{5}R_1, R_3 - R_2$

(d) $R_2 + 4R_1, R_3 - 2R_1, R_4 - 8R_1, R_3 - 2R_2,$
$R_4 + 2R_2, R_4 - 6R_3$

8. Let $A = LU$. Determine the row operations used to arrive at U for each of the following cases of L.

(a) $\begin{bmatrix} 1 & 0 & 0 \\ 2 & 1 & 0 \\ -3 & 5 & 1 \end{bmatrix}$
(b) $\begin{bmatrix} 1 & 0 & 0 \\ -1 & 1 & 0 \\ 4 & 2 & 1 \end{bmatrix}$

(c) $\begin{bmatrix} 1 & 0 & 0 \\ 0 & 1 & 0 \\ \frac{1}{7} & -\frac{3}{4} & 1 \end{bmatrix}$
(d) $\begin{bmatrix} 1 & 0 & 0 & 0 \\ -2 & 1 & 0 & 0 \\ 3 & 5 & 1 & 0 \\ 6 & 0 & -3 & 1 \end{bmatrix}$

Method of *LU* Decomposition

In Exercises 9–16 solve the systems using the method of *LU* decomposition.

9.
$$x_1 + 2x_2 - x_3 = 2$$
$$-2x_1 - x_2 + 3x_3 = 3$$
$$x_1 - x_2 - 4x_3 = -7$$

10.
$$2x_1 + x_2 = 9$$
$$6x_1 + 4x_2 - x_3 = 25$$
$$2x_1 + 4x_3 = 20$$

11.
$$3x_1 - x_2 + x_3 = 10$$
$$-3x_1 + 2x_2 + x_3 = -8$$
$$9x_1 + 5x_2 - 3x_3 = 24$$

12.
$$x_1 + 2x_2 + 3x_3 = -5$$
$$2x_1 + 8x_2 + 7x_3 = -9$$
$$-x_1 + 14x_2 - x_3 = 15$$

13.
$$4x_1 + x_2 + 2x_3 = 18$$
$$-12x_1 - 4x_2 - 3x_3 = -56$$
$$-5x_2 + 16x_3 = -10$$

14.
$$-2x_1 + 3x_3 = 3$$
$$-14x_1 + 3x_2 + 5x_3 = 11$$
$$8x_1 + 9x_2 - 11x_3 = 7$$

15.
$$-2x_1 + 3x_3 = 14$$
$$-4x_1 + 3x_2 + 5x_3 = 30$$
$$8x_1 + 9x_2 - 11x_3 = -34$$

16.
$$2x_1 - 3x_2 + x_3 = -5$$
$$4x_1 - 5x_2 + 6x_3 = 2$$
$$-10x_1 + 19x_2 + 9x_3 = 55$$

In Exercises 17–19 solve the systems using the method of *LU* decomposition.

17.
$$4x_1 + x_2 + 3x_3 = 11$$
$$12x_1 + x_2 + 10x_3 = 28$$
$$-8x_1 - 16x_2 + 6x_3 = -62$$

18.
$$x_1 + 2x_2 - x_3 = 2$$
$$2x_1 + 5x_2 + x_3 = 3$$
$$-x_1 - x_2 + 4x_3 = -3$$

19.
$$4x_1 + x_2 - 2x_3 = 3$$
$$-4x_1 + 2x_2 + 3x_3 = 1$$
$$8x_1 - 7x_2 - 7x_3 = -2$$

In Exercises 20 and 21 solve the systems using the method of *LU* decomposition.

20.
$$x_1 + x_2 - x_3 + 2x_4 = 7$$
$$x_1 + 3x_2 + 2x_3 + 2x_4 = 6$$
$$-x_1 - 3x_2 - 4x_3 + 6x_4 = 12$$
$$4x_2 + 7x_3 - 2x_4 = -7$$

21.
$$2x_1 + x_3 - x_4 = 6$$
$$6x_1 + 3x_2 + 2x_3 - x_4 = 15$$
$$4x_1 + 3x_2 - 2x_3 + 3x_4 = 3$$
$$-2x_1 - 6x_2 + 2x_3 - 14x_4 = 12$$

Miscellaneous Results

22. Prove that the inverse (if it exists) of a lower triangular matrix is also lower triangular.

23. Prove that the product of two lower triangular matrices is lower triangular.

24. Show that an *LU* decomposition of a matrix (if one exists) is not unique by finding two distinct decompositions of the matrix $A = \begin{bmatrix} 6 & -2 \\ 12 & 8 \end{bmatrix}$.

25. Consider the matrix $E = \begin{bmatrix} 1 & 0 & 0 \\ 0 & 0 & 1 \\ 0 & 1 & 0 \end{bmatrix}$ that defines the interchange of rows 2 and 3 of a 3×3 matrix. Show that E has no *LU* decomposition. This example demonstrates that not every matrix can be written in the form *LU*.

26. Let $AX = B$ be a system where $A = LU$ and A is a 3×3 matrix. The formula $(4n^3 - 3n^2 - n)/6$ for the total number of arithmetic operations in arriving at L and U gives 13 operations. How many of these operations are used to compute L and how many to compute U? The formula $2n^2 - n$ for computing the total number of operations then needed to solve $AX = B$ gives 15. How many of these operations are needed to solve $LY = B$ and how many to solve $UX = Y$?

*8.3 Practical Difficulties in Solving Systems of Equations

In this course we have discussed the use of elimination methods such as Gauss-Jordan, Gaussian, and *LU* decomposition for solving systems of linear equations, and we have seen applications of linear systems. In practice, the elements of the matrix of coefficients *A*, and the matrix of constants *B*, in a system of equations $AX = B$, often arise from measurements and are not known exactly. Small errors in the elements of these matrices can cause large errors in the solution, leading to very inaccurate results. In this section, we look into ways of assessing such effects and ways of minimizing them.

The Condition Number of a Matrix

Nonsingular matrices have played a major role in our discussions in this course. The time has now come to look closely at their behavior, especially at the way it affects systems of equations. We introduce the concept of a condition number of a nonsingular matrix *A*. This number is defined in terms of a norm (or magnitude) $\|A\|$ of the matrix.

DEFINITION   Let *A* be a nonsingular matrix. The **condition number** of *A* is denoted $c(A)$ and defined

$$c(A) = \|A\| \, \|A^{-1}\|$$

If $c(A)$ is small then the matrix *A* is said to be **well-conditioned**. If $c(A)$ is large then *A* is **ill-conditioned**.

Let us now see how $c(A)$ can be used to indicate the accuracy of the solution of a system of equations $AX = B$. Suppose $AX = B$ describes a given experiment where the elements of *A* and *B* arise from measurements. Such data depend on the accuracy of instruments and are rarely exact. Let small errors in *A* be represented by a matrix *E* and the corresponding errors in *X* be represented by *e*. Thus $(A + E)(X + e) = B$. If we select appropriate norms for the vectors and matrices, then it can be shown that

$$\frac{\|e\|}{\|X + e\|} \leq c(A) \, \frac{\|E\|}{\|A\|}$$

Thus, if $c(A)$ is small, small errors in *A* can only produce small errors in *X* and the results are accurate. The system of equations is said to be **well-behaved**. On the other hand, if $c(A)$ is large, there is the possibility that small errors in *A* can result in large errors in *X* leading to inaccurate results. Such a system of equations is **ill-conditioned**. (Note that a large value of $c(A)$ is a warning, not a guarantee, of a large error in the solution.)

If *I* is an identity matrix then

$$c(I) = 1 \quad \text{and} \quad c(A) \geq 1$$

(See the following exercises.) Thus 1 is a lower bound for condition numbers. We can intuitively think of the system $IX = B$ as being a best-behaved system. The smaller $c(A)$, then in some sense, the closer *A* is to *I* and the better behaved the system $AX = B$ becomes. On the other end of the scale, the larger $c(A)$, the closer *A* is to being singular, and problems can occur.

The actual value of $c(A)$ will of course depend upon the norm used for *A*. A norm that is commonly used is the so-called 1-norm:

$$\|A\| = \max \{|a_{1j}| + \cdots + |a_{nj}|\} \text{ for } j = 1, \dots, n$$

This norm is the largest number resulting from adding up the absolute values of elements in each column. There are other norms that are more reliable, but less efficient. This norm is a good compromise of reliability and efficiency.

A natural question to ask is, "How large is large for a condition number?" If $c(A)$ is written in the form $c(A) \approx 0.d \times 10^k$, then the above inequality implies that

$$\frac{\|e\|}{\|X + e\|} \leq 10^k \frac{\|E\|}{\|A\|}$$

This inequality gives the following rule of thumb:

If $c(A) \approx 0.d \times 10^k$, then the components of X can usually be expected to have k fewer significant digits of accuracy than the elements of A.

Thus, for example, if $c(A) \approx 100 = (0.1 \times 10^3)$, and the elements of A are known to five-digit precision, the components of X may have only two-digit accuracy. Much therefore depends upon the accuracy of the measurements and the desired accuracy of the solution. $c(A) = 100$ would, however, be considered large by any standard.

A similar relationship involving $c(A)$ exists between errors in the matrix of constants B and the resulting errors in the solution X. We clarify these ideas in the following example.

EXAMPLE 1 The following system of equations describes a certain experiment. Let us show that this system is sensitive to changes in the coefficients and the constants on the right.

$$34.9x_1 + 23.6x_2 = 234$$
$$22.9x_1 + 15.6x_2 = 154$$

The exact solution is $x_1 = 4$, $x_2 = 4$. Let us compute the condition number. The matrix of coefficients and its inverse are

$$A = \begin{bmatrix} 34.9 & 23.6 \\ 22.9 & 15.6 \end{bmatrix}, \quad A^{-1} = \begin{bmatrix} 3.9 & -5.9 \\ -5.725 & 8.725 \end{bmatrix}$$

We get

$$\|A\| = \max\{(34.9 + 22.9), (23.6 + 15.6)\} = \max\{57.8, 39.2\} = 57.8$$
$$\|A^{-1}\| = \max\{(3.9 + 5.725), (5.9 + 8.725)\} = \max\{9.625, 14.625\} = 14.625$$

Thus

$$c(A) = \|A\| \, \|A^{-1}\| = 57.8 \times 14.625 = 845.325$$

This number is very large. The system is ill-conditioned and the solution may not be reliable.

Let us change the coefficient of x_1 in the first equation from 34.9 to 34.8. The new system has solution $x_1 = 6.5574$, $x_2 = 0.2459$! The system is also sensitive to small changes in the constant terms. For example, changing the first constant term from 234 to 235 gives a new solution of $x_1 = 7.9$, $x_2 = -1.725$.

What does one do in a case like this to get meaningful results? If possible, the problem should be reformulated in terms of a well-behaved system of equations. If this is not possible, then one should attempt to derive more accurate data. With more precise data the system should be solved on a computer, using double-precision arithmetic, applying techniques that minimize errors that occur during computation (called round-off errors). One would then have more faith in the results. We shall introduce scaling and pivoting techniques for reducing round-off errors later in this section.

EXAMPLE 2 Find the condition numbers of the following matrices. Decide whether a system of linear equations defined by such a matrix of coefficients is well-behaved.

$$\textbf{(a)} \quad A = \begin{bmatrix} 1 & 1 & -1 \\ 4 & 0 & 1 \\ 0 & 4 & 1 \end{bmatrix} \qquad \textbf{(b)} \quad A = \begin{bmatrix} 1 & 1 & 1 \\ 1 & 2 & 4 \\ 1 & 3 & 9 \end{bmatrix}$$

SOLUTION

(a) The inverse is found to be

$$A^{-1} = \begin{bmatrix} \frac{1}{6} & \frac{5}{24} & -\frac{1}{24} \\ \frac{1}{6} & -\frac{1}{24} & \frac{5}{24} \\ -\frac{2}{3} & \frac{1}{6} & \frac{1}{6} \end{bmatrix}$$

We get

$$\|A\| = \max \{(1 + 4 + 0), (1 + 0 + 4), (1 + 1 + 1)\} = 5$$
$$\|A^{-1}\| = \max \{(\tfrac{1}{6} + \tfrac{1}{6} + \tfrac{2}{3}), (\tfrac{5}{24} + \tfrac{1}{24} + \tfrac{1}{6}), (\tfrac{1}{24} + \tfrac{5}{24} + \tfrac{1}{6})\} = 1$$

Thus

$$c(A) = \|A\| \, \|A^{-1}\| = 5.$$

This value is considered small and the system is thus well-behaved.

This system of equations arose in determining the currents through the electrical network in Example 2, Section 1.3. The values of the currents, namely $I_1 = 1, I_2 = 3, I_3 = 4$ amps can be considered reliable.

(b) This example is of an ill-conditioned system, and illustrates the possibility of constructing an alternative system that is not ill-conditioned. The inverse of A is found.

$$A^{-1} = \begin{bmatrix} 3 & -3 & 1 \\ -2.5 & 4 & -1.5 \\ 0.5 & -1 & 0.5 \end{bmatrix}$$

We get

$$\|A\| = \max \{(1 + 1 + 1), (1 + 2 + 3), (1 + 4 + 9)\} = 14$$
$$\|A^{-1}\| = \max \{(3 + 2.5 + 0.5), (3 + 4 + 1), (1 + 1.5 + 0.5)\} = 8$$

Thus

$$c(A) = \|A\| \, \|A^{-1}\| = 112$$

This is a large number. The system is ill-conditioned.

This matrix of coefficients arose in finding the equation of a polynomial of degree two through three given points. We arrived at the matrix for a system that led to the coefficients of a polynomial of degree two through the points $(1, 6), (2, 3),$ and $(3, 2)$. (See

Example 1 in Section 1.3.) The solution gave the coefficients for the polynomial $y = 11 - 6x + x^2$. How reliable is this equation? In this system the elements of the matrix of coefficients A emerge from the values of $x = 1$, $x = 2$, and $x = 3$; they are often completely accurate. Errors in the solution of $AX = B$ then arise solely from the components of B. Since $c(A) = 112 \approx 0.1 \times 10^3$, the solution could be three digits less accurate than the y values of the points. The equation is not reliable unless the y-data have been measured with a great accuracy.

This example is one in an important class of problems, namely that of finding polynomials through data points. The matrix of coefficients A is a **Vandermonde matrix**. Vandermonde matrices are square matrices of the following form. They are famously ill-conditioned.

$$\begin{bmatrix} 1 & a & a^2 & a^3 & \cdots \\ 1 & b & b^2 & b^3 & \cdots \\ 1 & c & c^2 & c^3 & \cdots \\ \vdots & \vdots & \vdots & \vdots & \vdots \vdots \end{bmatrix}$$

The matrix of coefficients in a curve fitting problem such as this will be a Vandermonde matrix unless one is careful. We can prevent a Vandermonde ill-conditioned system from arising in the first place by using data that is not equispaced. The same applies to least square curve fitting. If the base points are equispaced the resulting least squares equation is extremely sensitive to the y-data. An important rule in curve fitting is to avoid using equispaced data if at all possible, otherwise very accurate data is needed to get reliable equations.

We now come to the second half of this discussion on errors in systems of linear equations. We discuss ways of reducing errors during computation.

Pivoting and Scaling Techniques

When the condition number is small, Gaussian elimination yields accurate answers, even for large systems. However, with larger condition numbers we have to take precautions. When calculations are performed, usually on a computer, only a finite number of digits can be carried—rounding numbers leads to errors that are called **round-off errors**. As computation proceeds in Gaussian elimination, these small inaccuracies produce matrices that would be encountered in using exact arithmetic on a problem with a slightly altered augmented matrix. Thus, ill-conditioned systems are very sensitive to such errors. We now introduce mathematical techniques that can minimize round-off errors during computation.

The method of Gauss-Jordan elimination involves using the first equation in the system to eliminate the first variable from the other equations, and so on. In general, at the rth stage, the rth equation is used to eliminate the rth variable from all other equations. However, we are not restricted to using the rth equation to eliminate the rth variable from the other equations. Any later equation that contains the rth variable can be used to accomplish this elimination. The coefficient of the rth variable in that equation can be taken as **pivot**—we now introduce freedom in selection of pivots. The choice of pivots is very important in obtaining accurate numerical solutions. The following example illustrates how crucial the choice of pivots can be.

EXAMPLE 3 Solve the following system of equations, working to three significant figures.

$$10^{-3}x_1 - x_2 = 1$$
$$2x_1 + x_2 = 0$$

SOLUTION

Let us solve this system in two ways, using two selections of pivots. The augmented matrix of the system is

$$\begin{bmatrix} 10^{-3} & -1 & 1 \\ 2 & 1 & 0 \end{bmatrix}$$

Method 1   Select the $(1, 1)$ element as pivot. (Circle the pivot at each stage.) We get

$$\begin{bmatrix} \boxed{10^{-3}} & -1 & 1 \\ 2 & 1 & 0 \end{bmatrix} \quad \underset{(1^{-3}/10)R1}{\approx} \quad \begin{bmatrix} ① & -1000 & 1000 \\ 2 & 1 & 0 \end{bmatrix}$$

$$\underset{R2 + (-2)R1}{\approx} \quad \begin{bmatrix} 1 & -1000 & 1000 \\ 0 & ⃝2000 & -2000 \end{bmatrix} \quad \underset{(1/2000)R2}{\approx} \quad \begin{bmatrix} 1 & -1000 & 1000 \\ 0 & ① & -1 \end{bmatrix}$$

↑
This element should have been
2001, but has been rounded to
three significant figures.

$$\underset{R1 + (1000)R2}{\approx} \quad \begin{bmatrix} 1 & 0 & 0 \\ 0 & 1 & -1 \end{bmatrix}$$

The solution is $x_1 = 0$, $x_2 = -1$.

Method 2   Select the $(2, 1)$ element as pivot. When a pivot such as this is selected, interchange rows if necessary to bring the pivot into the customary location for creating zeros in that particular column. We get

$$\begin{bmatrix} 10^{-3} & -1 & 1 \\ ② & 1 & 0 \end{bmatrix} \quad \underset{R1 \leftrightarrow R2}{\approx} \quad \begin{bmatrix} ② & 1 & 0 \\ 10^{-3} & -1 & 1 \end{bmatrix}$$

$$\underset{(\frac{1}{2})R1}{\approx} \quad \begin{bmatrix} ① & \frac{1}{2} & 0 \\ 10^{-3} & -1 & 1 \end{bmatrix}$$

$$\underset{R2 + (-10^{-3})R1}{\approx} \quad \begin{bmatrix} 1 & \frac{1}{2} & 0 \\ 0 & ⃝{-1} & 1 \end{bmatrix}$$

$$\underset{(-1)R2}{\approx} \quad \begin{bmatrix} 1 & \frac{1}{2} & 0 \\ 0 & ① & -1 \end{bmatrix}$$

$$\underset{R1 + (-\frac{1}{2})R2}{\approx} \quad \begin{bmatrix} 1 & 0 & \frac{1}{2} \\ 0 & 1 & -1 \end{bmatrix}$$

The solution obtained by pivoting in this manner is $x_1 = \frac{1}{2}$, $x_2 = -1$.

The exact solution to this system can be obtained by applying Gauss-Jordan elimination in the standard way, to full accuracy. It is $x_1 = \frac{1000}{2001}$, $x_2 = -\frac{2000}{2001}$. Observe that the second solution, obtained by pivoting on the $(2, 1)$ element is a reasonably good approximation to the exact solution; the first solution, obtained by pivoting on the $(1, 1)$ element, is not.

Having seen that the choice of pivots is important when solving a system of linear equations, let us now use the preceding example to decide how pivots can be selected to improve accuracy. Why was the selection of the $(2, 1)$ element as pivot a better choice than the selection of the $(1, 1)$ element?

Let us return to look at the matrix prior to the reduced echelon form in *Method 1*. The matrix is

$$\begin{bmatrix} 1 & -1000 & 1000 \\ 0 & 1 & -1 \end{bmatrix}$$

The corresponding system of equations is

$$x_1 - 1000x_2 = 1000$$
$$x_2 = -1$$

Substituting for x_2 from the second equation into the first gives

$$x_1 - 1000(-1) = 1000,$$
$$x_1 + 1000 = 1000$$

Since we are working to three significant figures, observe that $x_1 = 4$ satisfies this equation. $x_1 = -3$ also does; in fact there are many possible values of x_1. When we work to three significant figures, the term x_1 is overpowered by the adjoining 1000 term. The introduction of the large numbers -1000 and 1000 during the first transformation, when pivoting on the $(1, 1)$ element, leads to this inaccuracy in x_1. Observe that in *Method 2*, when pivoting on the $(2, 1)$ element, no such large numbers are introduced and the solution obtained is more accurate.

As further evidence that instability in *Method 1* was caused during the first transformation, note that the condition number of the original system is 3 while the condition number of the system following the transformation is 500.75. The condition numbers of all the systems in *Method 2* are small.

This discussion illustrates that the introduction of large numbers during the elimination process can cause round-off errors. The large numbers in *Method 1* resulted from the choice of a small number, namely 10^{-3}, as pivot; dividing by such a small number introduces a large number. Thus the *rule of thumb in the selection of pivots is to choose the largest number available as pivot.*

Furthermore, this example illustrates that errors can be severe if there are large differences in magnitude between the numbers involved. The final inaccuracy in x_1 resulted from the large difference in magnitude between the coefficient of x_1, namely 1, and the coefficient of x_2, namely 1000. An initial procedure known as scaling can often be used to make all the elements of the augmented matrix of comparable magnitudes, so that one starts off on the best footing possible.

We now give the procedures that are usually used to minimize these causes of errors. It should be stressed that these procedures work most of the time and are the ones most widely adopted. There are, however, exceptions—situations where these procedures do not work. We shall not discuss them here.

Pivoting and Scaling Procedures

Scaling is an attempt to make the elements of the augmented matrix as uniform in magnitude as possible. There are two operations used in scaling a system of equations:

1. Multiplying an equation throughout by a nonzero constant.
2. Replacing any variable by a new variable that is a multiple of the original variable.

In practice, one attempts to scale the system so that the largest element in each row and column of the augmented matrix is of order unity. There is no automatic method of scaling.

Pivots are then selected, eliminating variables in the natural order, $x_1, x_2, \ldots x_n$. At the rth stage, the pivot is generally taken to be the coefficient of x_r in the rth or later equation that has the largest absolute value. In terms of elementary matrix transformations, we scan the relevant column from the rth location down, for the number that has the largest absolute value. Rows are then interchanged if necessary to bring the pivot into the correct location. This procedure ensures that the largest number available is selected as pivot.

We now give an example to illustrate these techniques.

EXAMPLE 4 Solve the following system of equations using pivoting and scaling, working to five significant figures.

$$
\begin{aligned}
0.002x_1 + \quad\quad 4x_2 - 2x_3 &= \quad 1 \\
0.001x_1 + 2.0001x_2 + \quad x_3 &= \quad 2 \\
0.001x_1 + \quad\quad 3x_2 + 3x_3 &= -1
\end{aligned}
$$

SOLUTION

First scale the system by letting

$$y_1 = 0.001x_1$$

The coefficients in the first column will then be of similar magnitude to the other coefficients. To make the notation uniform, let $y_2 = x_2$ and $y_3 = x_3$. We now have the properly scaled system of equations

$$
\begin{aligned}
2y_1 + \quad\quad 4y_2 - 2y_3 &= \quad 1 \\
y_1 + 2.0001y_2 + \quad y_3 &= \quad 2 \\
y_1 + \quad\quad 3y_2 + 3y_3 &= -1
\end{aligned}
$$

The condition number of the original system was 27,096. The condition number of this system is 32.6557. Things have improved!

Proceed now using pivoting, as follows:

$$
\begin{bmatrix}
2 & 4 & -2 & 1 \\
1 & 2.0001 & 1 & 2 \\
1 & 3 & 3 & -1
\end{bmatrix}
\underset{(1/2)\text{R1}}{\approx}
\begin{bmatrix}
① & 2 & -1 & 0.5 \\
1 & 2.0001 & 1 & 2 \\
1 & 3 & 3 & -1
\end{bmatrix}
$$

$$
\underset{\substack{\text{R2} + (-1)\text{R1} \\ \text{R3} + (-1)\text{R1}}}{\approx}
\begin{bmatrix}
1 & 2 & -1 & 0.5 \\
0 & \boxed{0.0001} & 2 & 1.5 \\
0 & \underline{1} & 4 & -1.5
\end{bmatrix}
\underset{\text{R2} \leftrightarrow \text{R3}}{\approx}
\begin{bmatrix}
1 & 2 & -1 & 0.5 \\
0 & ① & 4 & -1.5 \\
0 & 0.0001 & 2 & 1.5
\end{bmatrix}
$$

Scan these elements to determine the one with the largest absolute value. It is 1. This element will be pivot. Swap rows 2 and 3 to bring pivot into the desired location.

$$
\begin{array}{c}
\approx \\
R1 + (-2)R2 \\
R3 + (-0.0001)R2
\end{array}
\begin{bmatrix}
1 & 0 & -9 & 3.5 \\
0 & 1 & 4 & -1.5 \\
0 & 0 & 1.9996 & 1.5002
\end{bmatrix}
\qquad
\begin{array}{c}
\approx \\
(1/1.9996)R3
\end{array}
\begin{bmatrix}
1 & 0 & -9 & 3.5 \\
0 & 1 & 4 & -1.5 \\
0 & 0 & ① & 0.75025
\end{bmatrix}
$$

This element should have been 1.50015, but has been rounded to five significant digits.

$$
\begin{array}{c}
\approx \\
R1 + 9R3 \\
R2 + (-4)R3
\end{array}
\begin{bmatrix}
1 & 0 & 0 & 10.252 \\
0 & 1 & 0 & -4.501 \\
0 & 0 & 1 & 0.75025
\end{bmatrix}
$$

The solution is $y_1 = 10.252$, $y_2 = -4.501$, $y_3 = 0.75025$.

In terms of the original variables, the solution is

$$x_1 = 10{,}252, \quad x_2 = -4.501, \quad x_3 = 0.75025$$

It is interesting to compare this solution with the solution that would have been obtained using the Gauss-Jordan method without the pivoting and scaling refinements. The solution, to five significant digits, would have been

$$x_1 = 11{,}000, \quad x_2 = -5, \quad x_3 = 0.75025$$

On substituting values into the original equations it can be seen that the result using pivoting and scaling is very accurate, while the result obtained without these refinements is unsatisfactory. For example, let us test the first equation, $0.002x_1 + 4x_2 - 2x_3 = 1$, with each set of results. We get

With pivoting: $(0.002 \times 10{,}252) + (4 \times -4.501) - (2 \times 0.75025) = 0.9995 \approx 1$

Without pivoting: $(0.002 \times 11{,}000) + (4 \times -5) - (2 \times 0.75025) = 0.4995 \neq 1$

In this section we have indicated some of the precautions that should be taken when systems of linear equations are used in mathematical models. We have demonstrated some of the shortcomings of elimination methods and techniques that can be used to minimize errors. In the next section we introduce a different approach to solving systems of linear equations, namely iterative methods. Iterative methods give an alternative way of minimizing the effect of round-off errors in solving ill-conditioned systems.

EXERCISE SET 8.3

1-Norm

1. Compute the 1-norms of the following matrices.

(a) $\begin{bmatrix} 1 & 2 \\ 3 & 4 \end{bmatrix}$
(b) $\begin{bmatrix} 5 & 2 \\ -1 & 3 \end{bmatrix}$

(c) $\begin{bmatrix} 1 & 2 & 0 \\ 3 & -5 & 4 \\ 6 & 1 & -2 \end{bmatrix}$
(d) $\begin{bmatrix} 1 & 24 & -3 \\ 6 & 247 & 56 \\ -5 & 7 & -219 \end{bmatrix}$

Condition Number

2. Find the condition numbers of the following matrices.

(a) $\begin{bmatrix} 1 & 2 \\ 3 & 4 \end{bmatrix}$
(b) $\begin{bmatrix} 2 & -2 \\ 3 & 1 \end{bmatrix}$
(c) $\begin{bmatrix} 5 & 2 \\ 8 & 3 \end{bmatrix}$

(d) $\begin{bmatrix} 24 & 21 \\ 6 & 5 \end{bmatrix}$
(e) $\begin{bmatrix} 5.2 & 3.7 \\ 3.8 & 2.6 \end{bmatrix}$

3. Find the condition numbers of the following matrices. Let these be matrices of coefficients for systems of equations. How less accurate than the elements of the matrices can the solutions be?

(a) $\begin{bmatrix} 9 & 2 \\ 4 & 7 \end{bmatrix}$ **(b)** $\begin{bmatrix} 51 & 3 \\ -1 & 27 \end{bmatrix}$ **(c)** $\begin{bmatrix} 300 & 1001 \\ 75 & 250 \end{bmatrix}$

(d) $\begin{bmatrix} 3 & 4 \\ 1 & 2 \end{bmatrix}$ **(e)** $\begin{bmatrix} 5 & 3 \\ 4 & 2 \end{bmatrix}$

4. Find the condition numbers of the following matrices. These matrices are from applications we have discussed. Comment on the reliablity of the solutions.

(a) $\begin{bmatrix} 1 & 1 & -1 \\ 1 & 0 & 2 \\ 1 & -2 & 0 \end{bmatrix}$ (Electrical circuits application, Example 3, Section 1.3)

(b) $\begin{bmatrix} 0.8 & -0.2 & -0.4 \\ -0.6 & 0.4 & 0 \\ 0 & 0 & 0.8 \end{bmatrix}$ (Leontief I/O application, Example 2, Section 2.5)

5. Find the condition number of the matrix $\begin{bmatrix} \frac{1}{3} & \frac{1}{4} & \frac{1}{5} \\ \frac{1}{4} & \frac{1}{5} & \frac{1}{6} \\ \frac{1}{5} & \frac{1}{6} & \frac{1}{7} \end{bmatrix}$.

This is an example of a **Hilbert matrix**. Hilbert matrices are defined by $a_{ij} = \dfrac{1}{i + j + 1}$. These matrices are more ill-conditioned than Vandermonde matrices!

6. Find the condition number of the matrix $A = \begin{bmatrix} 1 & k \\ 1 & 1 \end{bmatrix}$, $k \neq 1$. Observe how the condition number depends on the value of k. For what values of k is $c(A) > 100$?

Properties of Norms and Condition Numbers

7. In this section we have defined the condition number of matrices using 1-norms. Prove that the 1-norm satisfies the properties of norms discussed in Section 4.2, namely:

(a) $\|A\| \geq 0$ **(b)** $\|A\| = 0$ if and only if $A = 0$
(c) $\|cA\| = |c|\|A\|$ **(d)** $\|A + B\| \leq \|A\| + \|B\|$

These conditions are the axioms for a norm on a general vector space. Matrix norms are also required to satisfy the

condition $\|AB\| \leq \|A\| \, \|B\|$. Show that the 1-norm also satisfies this condition.

8. (a) Let I be an identity matrix. Prove that $c(I) = 1$.
   (b) Use the property $\|AB\| \leq \|A\|\|B\|$ of a matrix norm to show that $c(A) \geq 1$.

9. Prove that $c(A) = c(kA)$, for a nonzero constant k. What is the significance of this result in terms of systems of linear equations?

10. Let A be a diagonal matrix. Prove that $c(A) = (\max\{a_{ii}\}) \times (1/\min\{a_{ii}\})$.

Pivoting and Scaling

In Exercises 11–17 solve the systems of equations (if possible), using pivoting and scaling techniques when appropriate.

11. $-x_2 + 2x_3 = 4$
   $x_1 + 2x_2 - x_3 = 1$
   $x_1 + 2x_2 + 2x_3 = 4$

12. $-x_1 + x_2 + 2x_3 = 8$
   $2x_1 + 4x_2 - x_3 = 10$
   $x_1 + 2x_2 + 2x_3 = 2$
   (to 3 significant digits)

13. $-x_1 + 0.002x_2 \qquad\quad = 0$
   $x_1 + \qquad\quad 2x_3 = -2$
   $0.001x_2 + x_3 = 1$

14. $2x_1 + \qquad\qquad\quad x_3 = 1$
   $0.0001x_1 + 0.0002x_2 + 0.0004x_3 = 0.0004$
   $x_1 - \qquad 2x_2 - \qquad 3x_3 = -3$

15. $x_2 - 0.1x_3 = 2$
   $0.02x_1 + 0.01x_2 \qquad\quad = 0.01$
   $x_1 - 4x_2 + 0.1x_3 = 2$
   (to 3 decimal places)

16. $-x_1 + \qquad\qquad 2x_3 = 1$
   $0.1x_1 + 0.001x_2 \qquad = 0.1$
   $0.1x_1 + 0.002x_2 \qquad = 0.2$

17. $x_1 - \qquad 2x_2 - \qquad x_3 = 1$
   $x_1 - 0.001x_2 - \qquad x_3 = 2$
   $x_1 + \qquad 3x_2 - 0.002x_3 = -1$
   (to 3 decimal places)

*8.4 Iterative Methods for Solving Systems of Linear Equations

We have discussed a number of elimination methods for solving systems of linear equations. We now introduce two so-called **iterative methods** for solving systems of n equations in n variables that have a unique solution, called the **Jacobi method** and the **Gauss-Seidel method**. At the close of this section we shall compare the merits of Gaussian elimination and the iterative methods.

An iterative method starts with an initial estimate of the solution, and generates a sequence of values that converges to the solution.

Jacobi Method*

We introduce this method by means of an example. Let us solve the following system of linear equations.

$$\begin{aligned} 6x + 2y - z &= 4 \\ x + 5y + z &= 3 \\ 2x + y + 4z &= 27 \end{aligned} \qquad (1)$$

Rewrite the equations as follows, isolating x in the first equation, y in the second equation, and z in the third equation.

$$\begin{aligned} x &= \frac{4 - 2y + z}{6} \\ y &= \frac{3 - x - z}{5} \\ z &= \frac{27 - 2x - y}{4} \end{aligned} \qquad (2)$$

Now make an estimate of the solution, say $x = 1$, $y = 1$, $z = 1$. The accuracy of the estimate affects only the speed with which we get a good approximation to the solution. Let us label these values $x^{(0)}$, $y^{(0)}$, and $z^{(0)}$. They are called the **initial values** of the iterative process.

$$x^{(0)} = 1, \quad y^{(0)} = 1, \quad z^{(0)} = 1$$

Substitute these values into the right-hand side of (2) to get the next set of values in the iterative process.

$$x^{(1)} = 0.5, \quad y^{(1)} = 0.2, \quad z^{(1)} = 6$$

These values of x, y, and z are now substituted into system (2) again to get

$$x^{(2)} = 1.6, \quad y^{(2)} = -0.7, \quad z^{(2)} = 6.45$$

This process can be repeated to get $x^{(3)}$, $y^{(3)}$, $z^{(3)}$, and so on. Repeating the iteration will give a better approximation to the exact solution at each step. For this straightforward system of equations the solution can easily be seen to be

$$x = 2, \quad y = -1, \quad z = 6$$

Thus after the second iteration one has quite a long way to go.

---

*Carl Gustav Jacobi (1804–1851) was educated in Berlin and taught at Königsberg and Berlin. He was interested in the mathematical fields of number theory, geometry, analysis, and mechanics. He was recognized as being a gifted, enthusiastic teacher. He was primarily interested in mathematics for the sake of the mathematics, rather than having some application in mind. Jacobi also had a deep interest in the history of mathematics.

The following theorem, which we present without proof, gives one set of conditions under which the Jacobi iterative method can be used.

THEOREM 8.1

Consider a system of n linear equations in n variables having a square matrix of coefficients A. If the absolute value of the diagonal element of each row of A is greater than the sum of the absolute values of the other elements in that row, then the system has a unique solution. The Jacobi method will converge to this solution, no matter what the initial values.

Thus the Jacobi method can be used when the diagonal elements **dominate** the rows. Let us return to the previous example to see that the conditions of this theorem are indeed satisfied. The matrix of coefficients is

$$A = \begin{bmatrix} 6 & 2 & -1 \\ 1 & 5 & 1 \\ 2 & 1 & 4 \end{bmatrix}$$

Comparing the magnitude of the diagonal elements with the sums of the magnitudes of the other elements in each row we get the results shown in Table 8.1. Observe that the diagonal elements dominate the rows; the Jacobi method can be used.

Table 8.1

Row	Absolute Value of Diagonal Element	Sum of Absolute Values of Other Elements in that Row
1	6	$\lvert 2 \rvert + \lvert -1 \rvert = 3$
2	5	$\lvert 1 \rvert + \lvert 1 \rvert = 2$
3	4	$\lvert 2 \rvert + \lvert 1 \rvert = 3$

In some systems it may be necessary to rearrange the equations before the above condition is satisfied and this method can be used.

The better the initial estimate, the quicker one gets a result to within the required degree of accuracy. Note that this method has an advantage: if an error is made at any stage, it merely amounts to a new estimate at that point.

There are many thorems similar to the one above that guarantee convergence under varying conditions. Investigations have made available swiftly converging methods that apply to various special systems of equations.

Gauss-Seidel Method[*]

The Gauss-Seidel method is a refinement of the Jacobi method that usually (but not always) gives more rapid convergence. The latest value of each variable is substituted at each step in the iterative process. This method also works if the diagonal elements of the matrix of coefficients dominate the rows. We illustrate the method for the previous systems of equations. As before let us take our initial guess to be

$$x^{(0)} = 1, \quad y^{(0)} = 1, \quad z^{(0)} = 1$$

---

[*]Ludwig von Seidel (1821–1896) studied at Berlin, Königsberg, and Munich and taught in Munich. He worked in probability theory and astronomy. His photometric measurements of fixed stars and planets were the first true measurements of this kind. He investigated the relationship between the frequency of certain diseases and climatic conditions in Munich. It was tragic that Seidel suffered from eye problems and was forced to retire young.

Substituting the latest value of each variable into system (2) gives

$$x^{(1)} = \frac{4 - 2y^{(0)} + z^{(0)}}{6} = 0.5$$

$$y^{(1)} = \frac{3 - x^{(1)} - z^{(0)}}{5} = 0.3$$

$$z^{(1)} = \frac{27 - 2x^{(1)} - y^{(1)}}{4} = 6.4250$$

Observe that we have used $x^{(1)}$, the most up-to-date value of x to get $y^{(1)}$. We have used $x^{(1)}$ and $y^{(1)}$ to get $z^{(1)}$. Continuing, we get

$$x^{(2)} = \frac{4 - 2y^{(1)} + z^{(1)}}{6} = 1.6375$$

$$y^{(2)} = \frac{3 - x^{(2)} - z^{(1)}}{5} = -1.0125$$

$$z^{(2)} = \frac{27 - 2x^{(2)} - y^{(2)}}{4} = 6.1844$$

Tables 8.2 and 8.3 give the results obtained for this system of equations using both methods. They illustrate the more rapid convergence of the Gauss-Seidel method to the exact solution $x = 2, y = -1, z = 6$.

Table 8.2   Jacobi Method

Iteration	x	y	z
Initial Guess	1	1	1
1	0.5	0.2	6
2	1.6	−0.7	6.45
3	1.975	−1.01	6.125
4	2.024167	−1.02	6.015
5	2.009167	−1.007833	5.992917
6	2.001431	−1.000417	5.997375

Table 8.3   Gauss-Seidel Method

Iteration	x	y	z
Initial Guess	1	1	1
1	0.5	0.3	6.425
2	1.6375	−1.01525	6.184375
3	2.034896	−1.043854	5.993516
4	2.013537	−1.001411	5.993584
5	1.999401	−0.998597	5.999949
6	1.999524	−0.9998945	6.000212

Table 8.4 gives the differences between the solutions $x^{(6)}$, $y^{(6)}$, $z^{(6)}$ obtained in the two methods after six iterations and the actual solution $x = 2$, $y = -1$, $z = 6$. The Gauss-Seidel method converges much more rapidly than the Jacobi method.

Table 8.4   Comparison of Methods

	$\lvert x^{(6)} - 2 \rvert$	$\lvert y^{(6)} - (-1) \rvert$	$\lvert z^{(6)} - 6 \rvert$
Jacobi Method	0.001431	0.000417	0.002625
Gauss-Seidel Method	0.000476	0.0001055	0.000212

Comparison of Gaussian Elimination and Gauss-Seidel

Limitations   Gaussian elimination is a finite method and can be used to solve any system of linear equations. The Gauss-Seidel method converges only for special systems of equations; thus it can be used only for those systems.

Efficiency   The efficiency of a method is a function of the number of arithmetic operations (addition, subtraction, multiplication, and division) involved in each method. For a system of n equations in n variables, where the solution is unique, Gaussian elimination involves $(4n^3 + 9n^2 - 7n)/6$ arithmetic operations. The Gauss-Seidel method requires $2n^2 - n$ arithmetic operations per iteration. For large values of n, Gaussian elimination requires approximately $2n^3/3$ arithmetic operations to solve the problem, while Gauss-Seidel requires approximately $2n^2$ arithmetic operations per iteration. Therefore, if the number of iterations is less than or equal to $n/3$, the iterative method requires fewer arithmetic operations.

For example, consider a system of 300 equations in 300 variables. Elimination requires 18,000,000 operations, whereas iteration requires 180,000 operations per iteration. For 100 or fewer iterations, the Gauss-Seidel method involves less arithmetic; it is more efficient. It should be pointed out that Gaussian elimination involves movement of data; for example several rows may need to be interchanged. This is time-consuming and costly on computers. Iterative processes suffer much less from this factor. Thus, even if the number of iterations is more than $n/3$, iteration may require less computer time.

Accuracy   Round-off errors in Gaussian elimination are minimized by using pivoting. However, they can still be sizeable. Errors in Gauss-Seidel are only the errors committed in the final iteration—the result of the next-to-last iteration can be interpreted as a very good initial estimate. Thus in general, when Gauss-Seidel is applicable, it is more accurate than Gaussian elimination. This fact often justifies the use of Gauss-Seidel over Gaussian elimination even when the total amount of computation time involved is greater. For example, when a mathematical model results in an ill-conditioned system and there is no way of constructing an alternative well-behaved description, an iterative method can ensure minimizing bad effects due to computation.

EXERCISE SET 8.4

Determine approximate solutions to the following systems of linear equations using the Gauss-Seidel iterative method. Work to 2 decimal places if you are performing the computation by hand. Use the given initial values. (Initial values are given solely so that answers can be compared.)

1. $4x + y - z = 8$
   $ 5y + 2z = 6$
   $x - y + 4z = 10$
   $x^{(0)} = 1, y^{(0)} = 2, z^{(0)} = 3$

2. $4x - y = 6$
   $2x + 4y - z = 4$
   $x - y + 5z = 10$
   $x^{(0)} = 1, y^{(0)} = 2, z^{(0)} = 3$

3. $5x - y + z = 20$
   $2x + 4y = 30$
   $x - y + 4z = 10$
   $x^{(0)} = 0, y^{(0)} = 0, z^{(0)} = 0$

4. $6x + 2y - z = 30$
   $-x + 8y + 2z = 20$
   $2x - y + 10z = 40$
   $x^{(0)} = 5, y^{(0)} = 6, z^{(0)} = 7$

5. $5x - y + 2z = 40$
   $2x + 4y - z = 10$
   $-2x + 2y + 10z = 8$
   $x^{(0)} = 20, y^{(0)} = 30, z^{(0)} = -40$

6. $6x - y + z + w = 20$
   $x + 8y + 2z = 30$
   $-x + y + 6z + 2w = 40$
   $2x - 3z + 10w = 10$
   $x^{(0)} = 0, y^{(0)} = 0, z^{(0)} = 0, w^{(0)} = 0$

7. Apply the method of Gauss-Seidel to the following system of equations. Observe that the method does not converge. The method does not lead to a solution for this system.

   $5x + 4y - z = 8$
   $x - y + z = 4$
   $2x + y + 2z = 1$
   $x^{(0)} = 0, y^{(0)} = 0, z^{(0)} = 0$

*8.5 Eigenvalues by Iteration and Connectivity of Networks

Numerical techniques exist for evaluating certain eigenvalues and eigenvectors of various types of matrices. Here we present an iterative method called the **power method**. It can be used to determine the eigenvalue with the largest absolute value (if it exists), and a corresponding eigenvector for certain matrices. In many applications one is only interested in the *dominant eigenvalue*. Applications in geography and history that illustrate the importance of the dominant eigenvalue will be given.

DEFINITION Let A be a square matrix with eigenvalues $\lambda_1, \lambda_2, \ldots$. The eigenvalue λ_i is said to be a **dominant eigenvalue** if

$$|\lambda_i| > |\lambda_k|, \, k \neq i$$

The eigenvectors corresponding to the dominant eigenvalue are called the **dominant eigenvectors** of A.

EXAMPLE 1 Let A be a square matrix with eigenvalues $-5, -2, 1$, and 3. Then -5 is the dominant eigenvalue, since

$$|-5| > |-2|, |-5| > |1|, \quad \text{and} \quad |-5| > |3|$$

Let B be a square matrix with eigenvalues $-4, -2, 1$, and 4. There is no dominant eigenvalue since $|-4| = |4|$.

The power method for finding a dominant eigenvector is based on the following theorem.

THEOREM 8.2

Let A be an $n \times n$ matrix having n linearly independent eigenvectors and a dominant eigenvalue. Let $\mathbf{x}_0$ be any nonzero column vector in $\mathbf{R}^n$ having a nonzero component in the direction of a dominant eigenvector. Then the sequence of vectors

$$\mathbf{x}_1 = A\mathbf{x}_0, \quad \mathbf{x}_2 = A\mathbf{x}_1, \ldots, \quad \mathbf{x}_k = A\mathbf{x}_{k-1}, \ldots$$

will approach a dominant eigenvector of A.

Proof   Let the eigenvalues of A be $\lambda_1, \ldots, \lambda_n$, with λ_1 being the dominant eigenvalue. Let $\mathbf{y}_1, \ldots \mathbf{y}_n$, be corresponding linearly independent eigenvectors. These eigenvectors will form a basis for $\mathbf{R}^n$. Thus there exist scalars $a_1, \ldots a_n$, such that

$$\mathbf{x}_0 = a_1\mathbf{y}_1 + \cdots + a_n\mathbf{y}_n \quad \text{where} \quad a_1 \neq 0.$$

We get

$$
\begin{aligned}
\mathbf{x}_k = A\mathbf{x}_{k-1} = A^2\mathbf{x}_{k-2} &= \cdots = A^k\mathbf{x}_0 \\
&= A^k[a_1\mathbf{y}_1 + \cdots + a_n\mathbf{y}_n] \\
&= [a_1 A^k\mathbf{y}_1 + \cdots + a_n A^k\mathbf{y}_n] \\
&= [a_1(\lambda_1)^k\mathbf{y}_1 + \cdots + a_n(\lambda_n)^k\mathbf{y}_n] \\
&= (\lambda_1)^k \left[a_1\mathbf{y}_1 + a_2\left(\frac{\lambda_2}{\lambda_1}\right)^k\mathbf{y}_1 + \cdots + a_n\left(\frac{\lambda_n}{\lambda_1}\right)^k\mathbf{y}_n \right]
\end{aligned}
$$

Since $\left|\dfrac{\lambda_i}{\lambda_1}\right| < 1$, for $i = 2, 3, \ldots$ then $\left(\dfrac{\lambda_i}{\lambda_1}\right)^k$ will approach 0 as k increases, and the vector $\mathbf{x}_k$ will approach $(\lambda_1)^k a_1\mathbf{y}_1$, a dominant eigenvector.

The following theorem tells us how to determine the eigenvalue corresponding to a given eigenvector. Once a dominant eigenvector has been found, the dominant eigenvalue can be determined by applying this result.

THEOREM 8.3

Let $\mathbf{x}$ be an eigenvector of a matrix A. The corresponding eigenvalue is given by

$$\lambda = \frac{A\mathbf{x} \cdot \mathbf{x}}{\mathbf{x} \cdot \mathbf{x}}$$

This quotient is called the **Rayleigh quotient**.

Proof   Since λ is the eigenvalue corresponding to $\mathbf{x}$, $A\mathbf{x} = \lambda\mathbf{x}$. Thus

$$\frac{A\mathbf{x} \cdot \mathbf{x}}{\mathbf{x} \cdot \mathbf{x}} = \frac{\lambda\mathbf{x} \cdot \mathbf{x}}{\mathbf{x} \cdot \mathbf{x}} = \frac{\lambda(\mathbf{x} \cdot \mathbf{x})}{\mathbf{x} \cdot \mathbf{x}} = \lambda$$

If Theorem 8.2 as it now stands is used to compute a dominant eigenvector, the components of the vectors $\mathbf{x}_1, \mathbf{x}_2, \ldots$ may become very large, causing significant round-off errors to occur. This problem is overcome by dividing each component of $\mathbf{x}_i$ by the absolute value of its largest component and then using this vector (a vector in the same direction as

$\mathbf{x}_i$) in the following iteration. We refer to this procedure as **scaling** the vector. We now summarize the power method.

The Power Method for an $n \times n$ matrix A

Select an arbitrary nonzero column vector $\mathbf{x}_0$ having n components.

Iteration 1: Compute $A\mathbf{x}_0$.

      Scale $A\mathbf{x}_0$ to get $\mathbf{x}_1$.

      Compute $\dfrac{A\mathbf{x}_1 \cdot \mathbf{x}_1}{\mathbf{x}_1 \cdot \mathbf{x}_1}$

Iteration 2: Compute $A\mathbf{x}_1$.

      Scale $A\mathbf{x}_1$ to get $\mathbf{x}_2$.

      Compute $\dfrac{A\mathbf{x}_2 \cdot \mathbf{x}_2}{\mathbf{x}_2 \cdot \mathbf{x}_2}$, and so on.

Then

$$\mathbf{x}_0, \mathbf{x}_1, \mathbf{x}_2, \ldots \text{ converges to a dominant eigenvector}$$

and

$$\frac{A\mathbf{x}_1 \cdot \mathbf{x}_1}{\mathbf{x}_1 \cdot \mathbf{x}_1}, \frac{A\mathbf{x}_2 \cdot \mathbf{x}_2}{\mathbf{x}_2 \cdot \mathbf{x}_2}, \cdots \quad \begin{array}{l}\text{converges to the}\\ \text{dominant eigenvalue}\end{array}$$

if

A has n linearly independent eigenvectors and $\mathbf{x}_0$ has a nonzero component in the direction of a dominant eigenvector.

These conditions for the Power Method come from Theorem 8.2. We know in advance that some $n \times n$ matrices, such as **symmetric matrices**, have n linearly independent eigenvectors. The method can thus be applied to find the dominant eigenvalue of a symmetric matrix. One does not know in advance that the selected vector $\mathbf{x}_0$ has a nonzero component in the direction of a dominant eigenvector. There is a high probability that it will have. Further ramifications of this condition are discussed in the exercises that follow.

EXAMPLE 2 Find the dominant eigenvalue and a dominant eigenvector of the following symmetric matrix.

$$\begin{bmatrix} 5 & 4 & 2 \\ 4 & 5 & 2 \\ 2 & 2 & 2 \end{bmatrix}$$

SOLUTION

Let

$$\mathbf{x}_0 = \begin{bmatrix} 1 \\ -2 \\ 3 \end{bmatrix},$$

be an arbitrary column vector with 3 components. Computations were carried out to nine decimal places. We display the results, rounded to 3 decimal places for clarity of viewing. See Table 8.5.

Table 8.5

Iteration	Ax	Scaled Vector	$(x \cdot Ax)/(x \cdot x)$
1	$\begin{bmatrix} 3 \\ 0 \\ 4 \end{bmatrix}$	$x_1 = \begin{bmatrix} 0.75 \\ 0 \\ 1 \end{bmatrix}$	5
2	$\begin{bmatrix} 5.75 \\ 5 \\ 3.5 \end{bmatrix}$	$x_2 = \begin{bmatrix} 1 \\ 0.870 \\ 0.609 \end{bmatrix}$	9.889
3	$\begin{bmatrix} 9.696 \\ 9.565 \\ 4.957 \end{bmatrix}$	$x_3 = \begin{bmatrix} 1 \\ 0.987 \\ 0.511 \end{bmatrix}$	9.999
4	$\begin{bmatrix} 0.997 \\ 9.955 \\ 4.996 \end{bmatrix}$	$x_4 = \begin{bmatrix} 1 \\ 0.999 \\ 0.501 \end{bmatrix}$	10
5	$\begin{bmatrix} 0.997 \\ 9.996 \\ 5 \end{bmatrix}$	$x_5 = \begin{bmatrix} 1 \\ 1 \\ 0.5 \end{bmatrix}$	10

Thus, after 5 iterations we have arrived at the dominant eigenvalue of 10, with a corresponding eigenvector of $\begin{bmatrix} 1 \\ 1 \\ 0.5 \end{bmatrix}$. These results are in agreement with the discussion of Section 5.1, where we computed the eigenvalues and eigenvectors of this matrix using determinants. ∎

This method has the same advantage as the Gauss-Seidel iterative method discussed in the previous section in that any error in computation only means that a new arbitrary vector has been introduced at that stage. The method is very accurate; the only round-off errors that occur are those that arise in the final iteration. The method has the disadvantage that it may converge only very slowly for large matrices.

Deflation

If the matrix A is symmetric, a technique called **deflation** can be used for determining further eigenvalues and eigenvectors. The method is based on the following theorem, which we state without proof.

THEOREM 8.4

Let A be a symmetric matrix with eigenvalues $\lambda_1, \ldots, \lambda_n$, labeled according to absolute value, λ_1 being the dominant eigenvalue. Let $\mathbf{x}$ be a dominant unit eigenvector. Then

1. The matrix $B = A - \lambda_1 \mathbf{x}\mathbf{x}^t$ has eigenvalues $0, \lambda_2, \ldots, \lambda_n$. B is symmetric.
2. If $\mathbf{y}$ is an eigenvector of B corresponding to one of the eigenvalues $\lambda_2, \ldots, \lambda_n$, it is also an eigenvector of A for the same eigenvalue.

One determines λ_1 and $\mathbf{x}$ of the matrix A using the power method. The matrix B is then constructed. Observe that λ_2 is the dominant eigenvalue of B (if one exists). Apply the power method to B to get λ_2 and a corresponding eigenvector, which is also an eigenvector of A. The process is then repeated. The disadvantage of this method is that the eigenvalues become increasingly inaccurate due to compounding of errors. Furthermore, if at any stage there is no dominant eigenvalue, the method terminates at that point; no further eigenvalues can be computed.

Readers who are interested in further discussion of the power method and deflation are referred to *Numerical Analysis* by Richard L. Burden and J. Douglas Faires, Eighth Edition, Brooks-Cole Publishers, 2004.

Accessibility Index of a Network

Geographers use eigenvectors within the field of graph theory to analyze networks. The most common types of networks investigated are roads, airline routes, sea routes, navigable rivers, and canals. Here we illustrate the concepts in terms of a road network.

Consider a network of n cities linked together by two-way roads. The cities can be represented by the vertices of a digraph and the roads as edges.

Let us assume that it is possible to travel along roads from any city to another; the digraph is said to be **connected**. Geographers are interested in the **accessibility** of each city. The **Gould index** is used as a measure of accessibility. This index is obtained from an eigenvector corresponding to the dominant eigenvalue of an $n \times n$ matrix that describes the network.

The network consists of two-way roads. Thus, if there is an edge from vertex i to vertex j, there will also be an edge from vertex j to vertex i. The adjacency matrix A will therefore be symmetric. Construct the *augmented adjacency matrix* $B = A + I_n$. B is also symmetric. This implies that B has n linearly independent eigenvectors. There is a theorem, called the **Perron-Frobenius theorem**, which implies that this augmented adjacency matrix B of a connected digraph does have a dominant eigenvalue and that the corresponding eigenspace is one-dimensional with a basis vector having all positive components. The power method is used in practice to compute the dominant eigenvalue and a corresponding eigenvector. The components of this eigenvector give the accessibilities of the cities corresponding to the vertices of the digraph.

We now justify this use of the components of the dominant eigenvector. A highly accessible vertex should have a large number of paths to other vertices. The elements of B^m give the number of paths of length m between vertices, including possible stopovers at vertices along the way. The sum of the elements of row i of B^m gives the number of m-paths out of vertex i. Let $\mathbf{e}$ be the column vector having n components, all of which are 1. Then $B^m \mathbf{e}$ will be a column vector, the ith component of which is the sum of the elements of row i of B^m. Thus the components of the vector $B^m \mathbf{e}$ will give the number of m-paths out of the various vertices. Since all the components of $\mathbf{e}$ are 1, and there is a dominant eigenvector of B having all positive components, $\mathbf{e}$ has a nonzero component in the direction of a dominant eigenvector. Theorem 8.2 tells us that as m increases, $B^m \mathbf{e}$ will approach a dominant eigenvector. As m increases, $B^m\mathbf{e}$ accounts for longer and longer paths out of each vertex. Thus the ratios of the components of the dominant eigenvector will give the relative accessibilities of the vertices.

This justification of the accessibility index is adapted from "Linear Algebra in Geography: Eigenvectors of Networks," by Philip D. Straffin, Jr., *Mathematics Magazine*, Vol. 53, No. 5, November 1980, pages 269–276. This interesting article has a discussion of the Perron-Frobenius Theorem and references to applications ranging from growth of the São Paolo economy to an analysis of the U.S. interstate road system to an analysis of trade routes in medieval Russia (which we include later in this section).

We now illustrate the concepts for a simple transportation network.

EXAMPLE 3 **Connectivity of a Transportation Network**

Consider the transportation network described in Figure 8.1. The adjacency matrix A and the augmented matrix B are given.

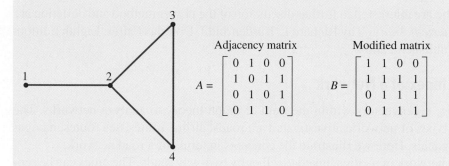

Figure 8.1

The dominant eigenvalue and corresponding eigenvector of B are found (to two decimal places) using the power method to be $\lambda = 3.17$ and $\mathbf{x} = (0.46, 1, 0.85, 0.85)$.

> *The eigenvalue gives a measure of the overall accessibility of the network. The components of the eigenvector give the relative accessibilities of the vertices.*

Note that vertex 2 has the highest connectivity, vertex 1 has the lowest connectivity, and vertices 3 and 4 have equal connectivity, as is to be expected on looking at the the graph. The convention adopted by geographers is to divide this eigenvector by the sum of the components. This is a form of "normalizing" the vector so that the sum of the components is then 1. The components of the resulting vector are called the **Gould accessibility indices** of the vertices. We get

$$\frac{1}{(0.46 + 1 + 0.85 + 0.85)}(0.46, 1, 0.85, 0.85) = (0.15, 0.32, 0.27, 0.27)$$

This vector gives the indices of the various vertices, as shown in Table 8.6.

Table 8.6

Vertex	Gould Accessibility Index
1	0.15
2	0.32
3	0.27
4	0.27

EXAMPLE **4** **Connectivity of Cuba**

Consider the map of Cuba with its principal cities and highways, Figure 8.2.
Let us determine its overall connectivity index, and the connectivity of each city.

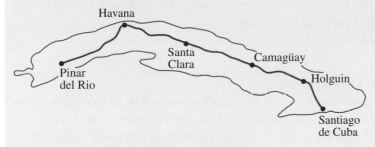

Figure 8.2   Cuba

The adjacency matrix and the augmented matrix are found. We use letters to denote the
cities, P for Pinar del Rio, H for Havana, and so on.

$$
A = \begin{array}{c} \\ P \\ H \\ SC \\ C \\ H \\ Sd \end{array}
\begin{array}{cccccc} P & H & SC & C & H & Sd \\ \end{array}
\begin{bmatrix}
0 & 1 & 0 & 0 & 0 & 0 \\
1 & 0 & 1 & 0 & 0 & 0 \\
0 & 1 & 0 & 1 & 0 & 0 \\
0 & 0 & 1 & 0 & 1 & 0 \\
0 & 0 & 0 & 1 & 0 & 1 \\
0 & 0 & 0 & 0 & 1 & 0
\end{bmatrix}, B =
\begin{bmatrix}
1 & 1 & 0 & 0 & 0 & 0 \\
1 & 1 & 1 & 0 & 0 & 0 \\
0 & 1 & 1 & 1 & 0 & 0 \\
0 & 0 & 1 & 1 & 1 & 0 \\
0 & 0 & 0 & 1 & 1 & 1 \\
0 & 0 & 0 & 0 & 1 & 1
\end{bmatrix}
$$

The dominant eigenvalue and corresponding eigenvector of B are found (to two decimal
places) using the power method to be

$$\lambda = 2.80 \quad \text{and} \quad \mathbf{x} = (0.45, 0.80, 1.00, 1.00, 0.80, 0.45)$$

We normalize $\mathbf{x}$ to get the Gould indices:

$$\frac{1}{(0.45 + 0.80 + 1.00 + 1.00 + 0.80 + 0.45)} (0.45, 0.80, 1.00, 1.00, 0.80, 0.45)$$

$$= (0.10, 0.18, 0.22, 0.22, 0.18, 0.10)$$

The connectivities of the cities are shown in Table 8.7.

Table 8.7

City	Accessibility Index
Pinar del Rio	0.10
Havana	0.18
Santa Clara	0.22
Camagüay	0.22
Holguin	0.18
Santiago de Cuba	0.10

The overall connectivity of the network is 2.80.

Observe that Havana is not the most connected city on the island. Note also the symmetries in the table, which fit in with the network of roads. We would expect Pinar del Rio and Santiago de Cuba to have the same connectivity. The same applies for Havana and Holguin, and for Santa Clara and Camagüay.

EXAMPLE 5   Moscow as Capital of Russia

Geographers have maintained that Moscow eventually assumed its dominant position in central Russia because of its strategic location on the medieval trade routes. Techniques of graph theory have been used in a paper "A Graph Theoretical Approach to Historical Geography" by F. R. Pitts, *Professional Geographer*, 17, 1965, 15–20 to question this theory.

The vertices of the graph in Figure 8.3 here represent centers of population in Russia in the twelfth and thirteenth centuries and the edges represent major trade routes. Some of the centers are Moscow (35), Kiev (4), Novgorod (1), and Smolensk (3). The Gould index of accessibility of each vertex in the network was computed. This involved finding the dominant eigenvector of a 39 × 39 adjacency matrix using a computer. We list the seven highest indices in Table 8.8. Observe that Moscow ranks sixth in accessibility. The conclusion arrived at in this analysis was that sociological and political factors other than its location on the trade routes must have been important in Moscow's rise.

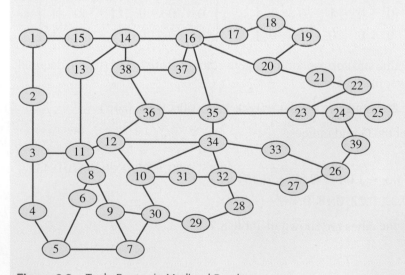

Figure 8.3   Trade Routes in Medieval Russia

Table 8.8

City	Accessibility Index
Kozelsk (10)	0.0837
Kolomna (34)	0.0788
Vyazma (12)	0.0722
Bryansk (8)	0.0547
Mtsensk (30)	0.0493
Moscow (35)	0.0490
Dorogobusch (11)	0.0477

EXAMPLE 6 **Comparison of Air Routes**

While the components of the dominant eigenvector are used to give internal information about a digraph, namely a comparison of the accessibility of its vertices, the dominant eigenvalues of various digraphs are used to compare the connectedness of distinct digraphs. The higher the dominant eigenvalue of a digraph, the greater its overall connectedness. Table 8.9 gives a comparison of the connectedness of internal airline networks of eight countries based on their dominant eigenvalues. It is surprising to find Sweden low on this list. Readers who are interested in reading more about this application and others like it are referred to *Applications of Graph Theory*, edited by Robin J. Wilson and Lowell W. Beineke, Academic Press, Chapter 10, 1979.

Table 8.9

Country	Dominant Eigenvalue
US	6
France	5.243
United Kingdom	4.610
India	4.590
Canada	4.511
USSR	3.855
Sweden	3.301
Turkey	2.903

EXERCISE SET 8.5

Dominant Eigenvalue and Eigenvector

In Exercises 1–4 use the power method to determine the dominant eigenvalue and a dominant eigenvector of the given matrices.

1. $\begin{bmatrix} 1 & 7 & -7 \\ -1 & 3 & -1 \\ -1 & -5 & 7 \end{bmatrix}$
2. $\begin{bmatrix} 9 & 4 & -4 \\ -1 & 1 & 1 \\ 7 & 4 & -2 \end{bmatrix}$

3. $\begin{bmatrix} 13 & 7 & -7 \\ -2 & -1 & 2 \\ 12 & 7 & -6 \end{bmatrix}$

4. $\begin{bmatrix} 17 & 8 & -8 \\ 0 & 1 & 0 \\ 16 & 8 & -7 \end{bmatrix}$

In Exercises 5–8 use the power method to determine the dominant eigenvalue and a dominant eigenvector of the given matrices.

5. $\begin{bmatrix} 1 & 1 & -1 & 1 \\ 1 & 1 & 1 & -1 \\ -3 & 3 & 3 & 3 \\ -3 & 3 & 1 & 5 \end{bmatrix}$ 6. $\begin{bmatrix} 3 & -1 & 1 & -1 \\ -3 & 1 & -3 & 3 \\ -3 & 3 & 7 & 3 \\ -1 & 5 & 7 & 3 \end{bmatrix}$

7. $\begin{bmatrix} 84 & 5 & -5 & 5 \\ 1 & 0 & 1 & -1 \\ -1 & 1 & 0 & 1 \\ 3 & 5 & -5 & 6 \end{bmatrix}$ 8. $\begin{bmatrix} 4.5 & 5.5 & -5.5 & 5.5 \\ 1.5 & 0.5 & 1.5 & -1.5 \\ -1 & 1 & 0 & 1 \\ 3 & 5 & -6 & 7 \end{bmatrix}$

In Exercises 9–13 determine *all* eigenvalues and corresponding eigenvectors of the given *symmetric* matrices using the power method with deflation.

9. $\begin{bmatrix} 5 & 4 & 2 \\ 4 & 5 & 2 \\ 2 & 2 & 2 \end{bmatrix}$ 10. $\begin{bmatrix} 4 & 0 & 2 \\ 0 & 4 & 0 \\ 2 & 0 & 4 \end{bmatrix}$

11. $\begin{bmatrix} 7 & 0 & 5 \\ 0 & 2 & 0 \\ 5 & 0 & 7 \end{bmatrix}$ 12. $\begin{bmatrix} 4 & 0 & 0 & 6 \\ 0 & 2 & -4 & 0 \\ 0 & -4 & 2 & 0 \\ 6 & 0 & 0 & 4 \end{bmatrix}$

13. $\begin{bmatrix} 5 & 4 & 1 & 1 \\ 4 & 5 & 1 & 1 \\ 1 & 1 & 4 & 2 \\ 1 & 1 & 2 & 4 \end{bmatrix}$

Gould Accessibility Index

14. Use the power method to compute the Gould accessibility index of the vertices of the graphs in Figure 8.4. Use three iterations in each case with the given initial vectors. (Initial vectors have been given so that answers can be compared.) Comment on the choice of initial vectors.

(a) Initial vector $\begin{bmatrix} 1 \\ 2 \\ 2 \\ 1 \end{bmatrix}$

(b) Initial vector $\begin{bmatrix} 1 \\ 1 \\ 3 \\ 3 \\ 1 \\ 1 \end{bmatrix}$

(c) Initial vector $\begin{bmatrix} 2 \\ 1 \\ 1 \\ 1 \\ 1 \end{bmatrix}$

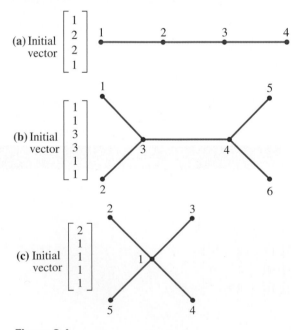

Figure 8.4

15. Consider the road network described in Figure 8.5.

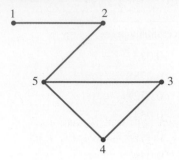

Figure 8.5

(a) Find the connectivities of the vertices of the network to two decimal places.

(b) Between which two vertices should a road be built to most increase the connectivity?

16. Determine the Gould indices of the North and South Islands of New Zealand, Figure 8.6. Which city has the highest connectivity on each island? Compare the connectivities of the two islands. Which is most connected?

17. Construct digraphs of the major road systems of the states of Alabama and Georgia. Determine the Gould indices of each state. Which city has the highest connectivity in each state? Compare the connectivities of the two states. Which is most connected? Where should an intercity road be built to most effectively increase the connectivity of each state?

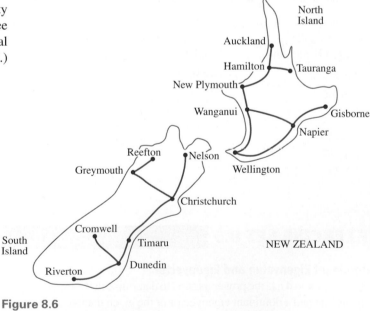

Figure 8.6

CHAPTER 8 REVIEW EXERCISES

1. Solve the following lower triangular system of equations.

$$2x_1 \qquad\qquad = 4$$
$$3x_1 - x_2 \qquad = 7$$
$$x_1 + 3x_2 + 2x_3 = 1$$

2. Let $A = LU$. The sequence of transformations $R_2 - \frac{3}{2}R_1$, $R_3 + \frac{2}{7}R_1$, $R_3 - 4R_2$ was used to arrive at U. Find L.

3. Solve the following system of equations using LU decomposition.

$$6x_1 + x_2 - x_3 = 5$$
$$-6x_1 + x_2 + x_3 = 1$$
$$12x_1 + 12x_2 + x_3 = 52$$

4. Let $A = LU$. If you are given L and U how many arithmetic operations are needed to compute A, knowing the locations of the zeros above and below the diagonals in L and U?

5. Find the condition number of the matrix $\begin{bmatrix} 250 & 401 \\ 125 & 201 \end{bmatrix}$.

Let this matrix be the matrix of coefficients of a system of equations. How less accurate than the elements of the matrix can the solutions be?

6. Find the condition numbers of the matrices

(a) $\begin{bmatrix} 1 & 1 & -1 \\ 1 & 0 & 2 \\ 0 & 3 & 2 \end{bmatrix}$

(b) $\begin{bmatrix} 1 & -1 & -1 & 0 & 0 \\ 0 & 0 & 1 & -1 & -1 \\ 1 & 1 & 0 & 0 & 0 \\ 1 & 0 & 0 & 2 & 0 \\ 0 & 0 & 0 & 2 & 2 \end{bmatrix}$

These are the matrices of coefficients of the systems of equations that describe the currents in the electrical circuits of Exercises 8 and 13 of Section 1.3. Comment on the reliability of the solutions.

7. Prove that $c(A) = c(A^{-1})$.

8. Does $c(A)$ define a linear mapping of $M_{nn} \to \mathbf{R}$?

9. Solve the following systems of equations using pivoting and scaling techniques.

(a)
$$-0.4x_2 + 0.002x_3 = 0.2$$
$$100x_1 + x_2 - 0.01x_3 = 1$$
$$200x_1 + 2x_2 - 0.03x_3 = 1$$

(b)
$$- x_2 + 0.001x_3 = 6$$
$$0.01x_1 + 0.0002x_3 = -0.2$$
$$0.01x_1 + 0.01x_2 + 0.00003x_3 = 0.02$$

10. Solve the following system of equations using the Gauss-Seidel method with initial value $x^{(0)} = 1$, $y^{(0)} = 2$, $z^{(0)} = -3$.

$$6x + y - z = 27$$
$$2x + 7y - 3z = 9$$
$$x + y + 4z = 18$$

11. Use the power method to determine the dominant eigenvalue and corresponding eigenvector of the matrix

$$\begin{bmatrix} 1 & 3 & 4 \\ -2 & 4 & 9 \\ -5 & 8 & 7 \end{bmatrix}$$

12. Use the power method with inflation to find all eigenvalues and eigenvectors for the symmetric matrix

$$\begin{bmatrix} 3 & 2 & 1 \\ 2 & 4 & -2 \\ 1 & -2 & 3 \end{bmatrix}$$

The Sage Gateshead Music and Arts Centre, designed by British architect Norman Foster, is located on the River Tyne in Gateshead, in the northeast of England. The Arts Centre is made up of three different performance halls specifically designed for superior acoustics.

Linear Programming

W e complete this course with an introduction to linear programming. Linear programming is concerned with the efficient allocation of limited resources to meet desired objectives. It is widely used in industry and in government.

Historically, linear programming was first developed and applied in 1947 by George Dantzig, Marshall Wood, and their associates at the U.S. Department of the Air Force; the early applications of linear programming were thus in the military field. Today linear programming is also used in business and industry. To illustrate the importance of the field, we point out that the 1975 Nobel Prize in Economic Science was awarded to two scientists, Professors Leonid Kantorovich of the Soviet Union and Tjalling C. Koopmans of the United States, for their "contributions to the theory of optimum allocation of resources."

Today, the Exxon Corporation, for example, uses linear programming to determine the optimal blending of gasoline. The Armour Company uses linear programming in determining specifications for a processed cheese spread. The H.J. Heinz Company uses linear programming to determine shipment schedules for its products between its factories and warehouses.

*9.1 A Geometrical Introduction to Linear Programming

Systems of linear equations have been an important component of this course. We now turn our attention to a related topic, namely systems of linear inequalities. Linear programming involves analyzing systems of linear inequalities to arrive at optimal solutions to problems.

An expression such as $x + 2y = 6$ is called a **linear equation**. The graph of this equation is a straight line that represents all points in the plane that satisfy the equation. An expression such as $x + 2y \leq 6$ is called a **linear inequality**. It will also have a graph, namely the set of points that satisfy this condition. For example, the point $x = 2$, $y = 1$ satisfies the inequality while $x = 3$, $y = 4$ does not satisfy the inequality. The point $(2, 1)$ is on the graph of the inequality while the point $(3, 4)$ is not on the graph.

It can be proved that the graph of an inequality such as $x + 2y \leq 6$ will consist of all the points that lie on the line $x + 2y = 6$, together with all the points on one side of the line. A suitable test point, not on the line, can be used to determine which of the two sides is on the graph. For example, let us select $x = 0$, $y = 0$ as a test point for this inequality. It satisfies the inequality. Thus the graph consists of all points on the line $x + 2y = 6$ together with all points on the side of the line containing the origin. See Figure 9.1.

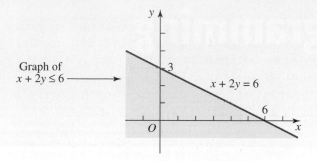

Graph of
$x + 2y \le 6$

$x + 2y = 6$

Figure 9.1

The inequalities in a linear programming problem arise as mathematical descriptions of constraints. We now give examples to illustrate how limitations on resources are described by inequalities. The first example shows how a monetary constraint can be represented mathematically by an inequality. The second example shows how a time constraint can be described by an inequality.

EXAMPLE 1 A company manufactures two types of cameras, the Pronto I and the Pronto II. The Pronto I costs $10 to manufacture, while the Pronto II costs $15. The total funds available for the production are $23,500. We can describe this monetary constraint on the production of cameras by a linear inequality. Let the company manufacture x of Pronto I and y of Pronto II.

Total cost of producing x Pronto I cameras at $10 per camera = $10x$

Total cost of producing y Pronto II cameras at $15 per camera = $15y$

Thus,

$$\text{Total manufacturing cost} = \$(10x + 15y)$$

Since the funds available are $23,500 we get the constraint

$$10x + 15y \le 23{,}500$$

EXAMPLE 2 A company makes two types of microcomputers, the Jupiter and the Cosmos. It takes 27 hours to assemble a Jupiter computer and it takes 34 hours to assemble a Cosmos computer. The total labor time available for this work is 800 hours. Let us describe this time constraint by means of an inequality. Let the company assemble x Jupiter and y Cosmos computers.

Total time to assemble x Jupiters at 27 hours per micro = $27x$ hours

Total time to assemble y Cosmos at 34 hours per micro = $34y$ hours

Thus,

$$\text{Total time to assemble computers} = (27x + 34y) \text{ hours}$$

Since the available time is 800 hours we get the constraint

$$27x + 34y \le 800$$

Let us now look at a linear programming problem that involves determining maximum profit under both monetary and time constraints.

A Linear Programming Problem

A company manufactures two types of handheld calculators, model Calc1 and model Calc2. It takes 5 hours and 2 hours to manufacture a Calc1 and Calc2, respectively. The company has 900 hours available per week for the production of calculators. The manufacturing cost of each Calc1 is $8 and the manufacturing cost of a Calc2 is $10. The total funds available per week for production are $2,800. The profit on each Calc1 is $3 and the profit on each Calc2 is $2. How many of each type of calculator should be manufactured weekly to obtain maximum profit?

In this problem there are two constraints, namely time and money. The aim of the company is to maximize profit under these constraints. We can solve the problem in three stages:

1. Construct the mathematical model—that is, the mathematical description of the problem.
2. Illustrate the mathematics by means of a graph.
3. Use the graph to determine the solution.

Step 1   The Mathematical Model   We first find the inequalities that describe the time and monetary constraints. Let the company manufacture x of Calc1 and y of Calc2.

Time Constraint: time to manufacture x Calc1 at 5 hours each $= 5x$ hours

time to manufacture y Calc2 at 2 hours each $= 2y$ hours

Thus,   total manufacturing time $= (5x + 2y)$ hours

There are 900 hours available. Therefore

$$5x + 2y \leq 900$$

Monetary Constraint: total cost of manufacturing x Calc1 at $8 each $= \$8x$

total cost of manufacturing y Calc2 at $10 each $= \$10y$

Thus,   total production costs $= \$(8x + 10y)$

There is $2,800 available for the production of calculators. Therefore

$$8x + 10y \leq 2800$$

Furthermore, x and y represent numbers of calculators manufactured. These numbers cannot be negative. We therefore get two more constraints,

$$x \geq 0, \quad y \geq 0$$

The constraints on the production process are thus described by the following **system of linear inequalites**.

$$5x + 2y \leq 900$$
$$8x + 10y \leq 2800$$
$$x \geq 0$$
$$y \geq 0$$

We next find a mathematical expression for profit.

Profit: weekly profit on x calculators at $3 per calculator $= \$3x$

weekly profit on y calculators at $2 per calculator $= \$2y$

Thus, total weekly profit $= \$(3x + 2y)$.

Let us introduce a **profit function**

$$f = 3x + 2y$$

The problem thus reduces mathematically to finding values of x and y that give the maximum value of f under the above constraints. Such a function f is called the **objective function** of the linear programming problem.

Step 2 Graphical Representation of the Constraints We now find the points (x, y) that satisfy all the inequalities in this system; these are called the **solutions** to the system. It can be shown that the graph of $5x + 2y \leq 900$ consists of points on and below the line $5x + 2y = 900$. Similarly, the points that satisfy the inequality $8x + 10y \leq 2800$ are on and below the line $8x + 10y = 2800$. The condition $x \geq 0$ is satisfied by all points on and to the right of the y-axis. The condition $y \geq 0$ is satisfied by all points on and above the x-axis. The set of solutions to the system of inequalities is the region that is common to all these regions; the shaded region in Figure 9.2. Such a region, which satisfies all the constraints, is called the **feasible region** of the linear programming problem and the points in this region are called **feasible solutions**. Any such point corresponds to a possible schedule for production. Among these points is one (or possibly more than one) that gives a maximum value to the objective function. Such a point is called an **optimal solution** to the linear programming problem. The next step will be to find this point.

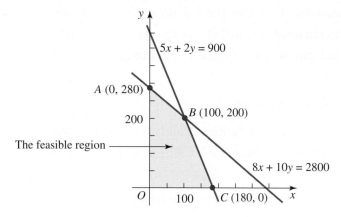

Figure 9.2

Step 3 Determining the Optimal Solution It would be an endless task to examine the value of the objective function $f = 3x + 2y$ at all points in the feasible region. It has been proved that the maximum value of f will occur at a vertex of the feasible region, namely at one of the points A, B, C, or O; or, if there is more than one point at which the maximum occurs, it will be along one edge of the region, such as AB or BC. Hence we have only to examine the values of $f = 3x + 2y$ at the vertices A, B, C, and O. These vertices are found by determining the points of intersection of the lines. We get

$$A(0, 280): \quad f_A = 3(0) \quad + 2(280) = 560$$
$$B(100, 200): \quad f_B = 3(100) + 2(200) = 700$$
$$C(180, 0): \quad f_C = 3(180) + 2(0) \quad = 540$$
$$O(0, 0): \quad f_O = 3(0) \quad + 2(0) \quad = 0$$

Thus the maximum value of f is 700 and it occurs at B, namely when $x = 100$ and $y = 200$. The interpretation of these results is that the maximum weekly profit is \$700 and this occurs when 100 Calc1 calculators and 200 Calc2 calculators are manufactured.

We now look at a linear programming problem that has many solutions.

EXAMPLE **3** Find the maximum value of $f = 8x + 2y$ under the constraints

$$4x + y \leq 32$$
$$4x + 3y \leq 48$$
$$x \geq 0$$
$$y \geq 0$$

SOLUTION

The constraints are represented graphically by the shaded region in Figure 9.3. The vertices of the feasible region are found and the value of the objective function calculated at each vertex.

$$A(0, 16): f_A = 8(0) + 2(16) = 32$$
$$B(6, 8):\ \ f_B = 8(6) + 2(8)\ \ = 64$$
$$C(8, 0):\ \ f_C = 8(8) + 2(0)\ \ = 64$$
$$O(0, 0):\ \ f_O = 8(0) + 2(0)\ \ =\ \ 0$$

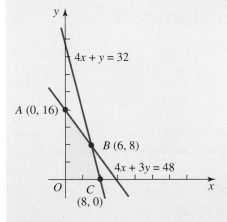

Figure 9.3

The maximum value of f is 64 and this occurs at two adjacent vertices B and C. When this happens in a linear programming problem the objective function will have the same value at every point on the edge joining the adjacent vertices. Thus f has a maximum value of 64 at all points along BC. The equation of BC is $y = -4x + 32$. Thus any point along the line $y = -4x + 32$ from $(6, 8)$ to $(8, 0)$ is an optimal solution. At $x = 7$, $y = 4$, for example, we have $f = 8(7) + 2(4) = 64$.

If x and y correspond to numbers of items manufactured and f corresponds to profit, as in the previous example, then there is a certain flexibility in the production schedule. Any schedule corresonding to a point along BC will lead to maximum profit.

Minimum Value of a Function

A problem that involves determining the minimum value of an objective function f can be solved by looking for the maximum value of $-f$, the negative of f, over the same feasible region. The general result is as follows.

> *The minimum value of a function f over a region S occurs at the point(s) of maximum value of $-f$, and is the negative of that maximum value.*

Let us derive this result. Let f have minimum value f_A at the point A in the region. Then if B is any other point in the region

$$f_A \leq f_B$$

Multiply both sides of this inequality by -1 to get

$$-f_A \geq -f_B$$

This implies that A is a point of maximum value of $-f$. Furthermore, the minimum value is f_A, the negative of the maximum value $-f_A$. These steps can be reversed, proving that the converse holds, thus verifying the result.

EXAMPLE 4 Find the minimum value of $f = 2x - 3y$ under the constraints

$$x + 2y \leq 10$$
$$2x + y \leq 11$$
$$x \geq 0$$
$$y \geq 0$$

The feasible region is shown in Figure 9.4. To find the minimum value of f over the region let us determine the maximum value of $-f$. The vertices are found and the value of $-f$ computed at each vertex.

$$
\begin{aligned}
A(0,5): &\ -f_A = -2(0) + 3(5) &=&\ \ 15 \\
B(4,3): &\ -f_B = -2(4) + 3(3) &=&\ \ \ 1 \\
C(\tfrac{11}{2},0): &\ -f_C = -2(\tfrac{11}{2}) + 3(0) &=&\ -11 \\
O(0,0): &\ -f_O = -2(0) + 3(0) &=&\ \ \ 0
\end{aligned}
$$

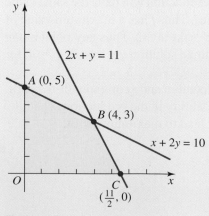

Figure 9.4

The maximum value of $-f$ is 15 at A. Thus the minimum value of $f = 2x - 3y$ is -15 when $x = 0$, $y = 5$.

Discussion of the Method

Notice that each feasible region we have discussed is such that the whole of the segment of a straight line joining any two points within the region lies within that region. Such a region is called **convex**. See Figure 9.5. A theorem states that *the feasible region in a linear programming problem is convex.*

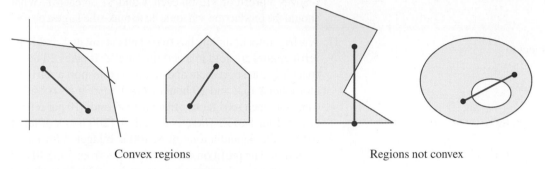

Convex regions   Regions not convex

Figure 9.5

    We now give a geometrical explanation of why we can expect the maximum value of the objective function to occur at a vertex, or along one side of the feasible region. Let $S = ABCO$ in Figure 9.6(a) be a feasible region and let the objective function be $f = ax + by$. f has a value at each point within S. Let $P(x_p, y_p)$ be a point in S. Then $f_p = ax_p + by_p$. This implies that P lies on the line $ax + by = f_p$. The y-intercept of this line is f_p/b. See Figure 9.6(a). Therefore f_p will have a maximum value when the y-intercept is a maximum. It can be seen from the figure that this will occur when P is the vertex B.

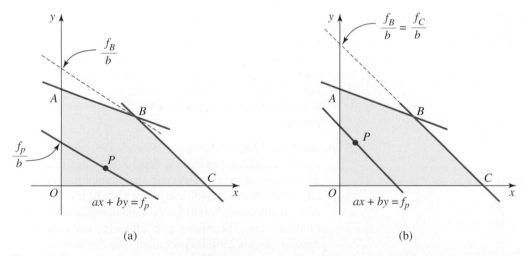

(a)   (b)

Figure 9.6

Some lines $ax + by = f_p$ will have maximum y-intercept when P is at B, as shown, others will have a maximum y-intercept when P is at A or when P is at C. The vertex will depend upon the slope of the line. If the line is parallel to an edge of the region S then the y-intercept will be a maximum when P is any point along that edge. See Figure 9.6(b).

EXERCISE SET 9.1

Linear Programming and Geometry

In Exercises 1–5 solve the linear programming problems.

1. Maximize $f = 2x + y$ subject to

   $4x + y \leq 36$
   $4x + 3y \leq 60$
   $x \geq 0, y \geq 0$

2. Maximize $f = x - 4y$ subject to

   $x + 2y \leq 4$
   $x + 6y \leq 8$
   $x \geq 0, y \geq 0$

3. Maximize $f = 4x + 2y$ subject to

   $x + 3y \leq 15$
   $2x + y \leq 10$
   $x \geq 0, y \geq 0$

4. Maximize $f = 2x + y$ subject to

   $4x + y \leq 16$
   $x + y \leq 7$
   $x \geq 0, y \geq 0$

5. Maximize $f = 4x + y$ subject to

   $2x + y \leq 4$
   $6x + y \leq 8$
   $x \geq 0, y \geq 0$

In Exercises 6–8 solve the linear programming problems.

6. Minimize $f = -3x - y$ subject to

   $x + y \leq 150$
   $4x + y \leq 450$
   $x \geq 0, y \geq 0$

7. Minimize $f = -2x + y$ subject to

   $2x + y \leq 440$
   $4x + y \leq 680$
   $x \geq 0, y \geq 0$

8. Minimize $f = -x + 2y$ subject to

   $x + 2y \leq 4$
   $x + 4y \leq 6$
   $x \geq 0, y \geq 0$

Applications of Linear Programming

9. A company manufactures two types of hand calculators, model $C1$ and model $C2$. It takes 1 hour and 4 hours in labor time to manufacture each $C1$ and $C2$, respectively. The cost of manufacturing the $C1$ is $30 and that of manufacturing a $C2$ is $20. The company has 1600 hours of labor time available and $18,000 in running costs. The profit on each $C1$ is $10 and on each $C2$ is $8. What should the production schedule be to ensure maximum profit?

10. A company manufactures two products, X and Y. Two machines, I and II are needed to manufacture each product.

It takes 3 minutes on each machine to produce an X. To produce a Y takes 1 minute on machine I and 2 minutes on machine II. The total time available on machine I is 3000 minutes and on machine II the time available is 4500 minutes. The company realizes a profit of $15 on each X and $7 on each Y. What should the production schedule be to make the largest profit?

11. A refrigerator company has two plants at towns X and Y. Its refrigerators are sold in a certain town Z. It takes 20 hours (packing, transportation, and so on) to transport a refrigerator from X to Z and 10 hours from Y to Z. It costs $60 to transport each refrigerator from X to Z, and $10 per refrigerator from Y to Z. There is a total of 1200 hours of labor time available and a total of $2400 is budgeted for transportation. The profit on each refrigerator from X is $40 and the profit on each refrigerator from Y is $20. How should the company schedule the transportation of refrigerators so as to maximize profit?

12. The maximum daily production of an oil refinery is 1400 barrels. The refinery can produce two types of oil; gasoline for automobiles and heating oil for domestic purposes. The production costs per barrel are $6 for gasoline and $8 for heating oil. The daily production budget is $9600. The profit per barrel is $3.50 on gasoline and $4 on heating oil. What is the maximum profit that can be realized daily and what quantities of each type of oil are then produced?

13. A tailor has 80 square yards of cotton material and 120 square yards of woolen material. A suit requires 2 square yards of cotton and 1 square yard of wool. A dress requires 1 square yard of cotton and 3 square yards of wool. How many of each garment should the tailor make to maximize income, if a suit and a dress each sell for $90? What is the maximum income?

14. A city has $1,800,000 to purchase cars. Two cars, the Arrow and the Gazelle, are under consideration, costing $12,000 and $15,000, respectively. The estimated annual maintainance cost on the Arrow is $400 and on the Gazelle it is $300. The city will allocate $40,000 for the total annual maintainance of these cars. The Arrow gets 28 miles per gallon and the Gazelle gets 25 miles per gallon. The city wants to maximize the "gasoline efficiency number" of this group of cars. For x Arrows and y Gazelles this number would be $28x + 25y$. How many of each model should be purchased?

15. A car dealer imports foreign cars by way of two ports of entry, A and B. A total of 120 cars are needed in city C and 180 in city D. There are 100 cars available at A and 200 at B. It takes 2 hours to transport each car from A to C and 6 hours from B to C. It takes 4 hours and 3 hours to transport each car from A and B to D. The dealer has 1030 hours in driver time available to move the cars. As many cars as possible should be moved from B to C, as drivers will not

be available for this route in the future. What schedule will achieve this?

16. A school district is buying new busses. It has a choice of two types. The Torro costs $18,000 and holds 25 passengers. The Sprite costs $22,000 and holds 30 passengers. $572,000 has been budgeted for the new busses. A maximum of 30 drivers will be available to drive the busses. At least 17 Sprite busses must be ordered because of the desirability of having a certain number of large capacity busses. How many of each type should be purchased to carry a maximum number of students?

17. A manufacturer makes two types of fertilizer, X and Y, using chemicals A and B. Fertilizer X is made up of 80% chemical A and 20% chemical B. Fertilizer Y is made up of 60% chemical A and 40% chemical B. The manufacturer requires at least 30 tons of fertilizer X and at least 50 tons of fertilizer Y. He has available 100 tons of chemical A and 50 tons of B. He wants to make as much fertilizer as possible. What quantities of X and Y should be produced?

18. A hospital wants to design a dinner menu containing two items, M and N. Each ounce of M provides 1 unit of vitamin A and 2 units of vitamin B. Each ounce of N provides 1 unit of vitamin A and 1 unit of vitamin B. The two dishes must provide at least 7 units of vitamin A and at least 10 units of vitamin B. If each ounce of M costs 8 cents and each ounce of N costs 12 cents, how many ounces of each item should the hospital serve to minimize cost?

19. A clothes manufacturer has 10 square yards of cotton material, 10 square yards of wool material, and 6 square yards of silk material. A pair of slacks requires 1 square yard of cotton, 2 square yards of wool, and 1 square yard of silk. A skirt requires 2 square yards of cotton, 1 square yard of wool, and 1 square yard of silk. The net profit on a pair of slacks is $3 and the net profit on a skirt is $4. How many skirts and how many slacks should be made to maximize profit?

20. (a) A shipper has trucks that can carry a maximum of 12,000 pounds of cargo with a maximum volume of 9000 cubic feet. He ships for two companies: Pringle Co. has packages weighing 5 pounds each and a volume of 5 cubic feet, and Williams Co. packages weigh 6 pounds with a volume of 3 cubic feet. By contract, the shipper makes 30 cents on each package from Pringle and 40 cents on each package from Williams. How many packages from each company should the shipper carry?

(b) A lawyer points out the fine print at the bottom of the Pringle contact that says that the shipper must carry at least 240 packages from Pringle. How should the shipper now divide the work? How much did the clause cost him in profit?

---

*9.2 The Simplex Method

The graphical method of solving a linear programming problem introduced in the last section has its limitations. The method demonstrated for two variables can be extended to linear programming problems involving three variables; the feasible region then being a convex subset of three-dimensional space. However, for problems involving more than two variables the geometrical approach becomes impractical. We now introduce the **simplex method**, an algebraic method that can be used for any number of variables. This method was developed by George B. Dantzig of the University of California, Berkeley, in 1947. The simplex method has the advantage of being readily programmable for a computer. Today, "packages" of computer programs based on the simplex method are offered commercially by software companies. Industrial customers often pay sizable monthly fees for the use of these programs. Exxon, for example, currently uses linear programming in the scheduling of its drilling operations, in the allocation of crude oil among its refineries, in the control of refinery operations, in the distribution of its products, and in the planning of business strategy. It has been estimated that linear programming accounts for between 5% and 10% of the company's total computing time. The simplex method involves reformulating constraints in terms of linear equations and then uses elementary row operations in a manner very similar to that of Gauss-Jordan elimination, to arrive at the solution.

We introduce the simplex method by using it to arrive at the solution of the first linear programming problem discussed in the previous section. Let us find the maximum value of $f = 3x + 2y$ subject to the constraints

$$5x + 2y \leq 900$$
$$8x + 10y \leq 2800$$
$$x \geq 0$$
$$y \geq 0$$

Consider the first inequality, $5x + 2y \leq 900$. For each x and y that satisfy this condition there will exist a value for a nonnegative variable u such that

$$5x + 2y + u = 900$$

The value of u will be the number that must be added to $5x + 2y$ to bring it up to 900. Similarly, for each x and y that satisfy $8x + 10y \leq 2800$, there will be a value for a nonnegative variable v such that $8x + 10y + v = 2800$. Such variables u and v are called **slack variables**—they make up the slack in the original variables.

Thus the constraints in the linear programming problem may be written

$$5x + 2y + u \quad\quad = 900$$
$$8x + 10y \quad\quad + v = 2800$$
$$x \geq 0, y \geq 0, u \geq 0, v \geq 0$$

Finally, the objective function $f = 3x + 2y$ is rewritten in the form

$$-3x - 2y + f = 0$$

The entire problem now becomes that of determining the solution to the following system of equations

$$5x + 2y + u \quad\quad\quad = 900$$
$$8x + 10y \quad + v \quad\quad = 2800$$
$$-3x - 2y \quad\quad + f = 0$$

such that f is as large as possible, under the restrictions $x \geq 0, y \geq 0, u \geq 0, v \geq 0$.

We have thus reformulated the problem in terms of a system of linear equations under certain constraints. The system of equations, consisting of three equations in the five variables x, y, u, v, and f will have many solutions in the region defined by $x \geq 0, y \geq 0, u \geq 0, v \geq 0$. Any such solution is called a **feasible solution**. A solution that maximizes f is called an **optimal solution**; this is the solution we are interested in.

We determine the optimal solution by using elementary row operations in an algorithm called the **simplex algorithm**. The method of Gauss-Jordan elimination involved selecting pivots and using them to create zeros in columns in a systematic manner. In the simplex algorithm pivots are again selected and used to create zeros in columns, but using different criteria from Gauss-Jordan elimination.

The Simplex Algorithm

1. Write down the augmented matrix of the system of equations. This is called the **initial simplex tableau**.
2. Locate the negative element in the last row, other than the last element, that is largest in magnitude. (If two or more entries share this property, any one of these can be selected.) If all such entries are nonnegative the tableau is in final form.

3. Divide each positive element in the column defined by this negative entry into the corresponding element of the last column.
4. Select the divisor that yields the smallest quotient. This element is called a **pivot element**. (If two or more elements share this property, any one of these can be selected as pivot.)
5. Use row operations to create a 1 in the pivot location and zeros elsewhere in the pivot column.
6. Repeat Steps 2–5 until all such negative elements have been eliminated from the last row. The final matrix is called the **final simplex tableau**. It leads to the optimal solution.

Let us apply the simplex algorithm to our linear programming problem. Starting with the initial simplex tableau we get

$$
\begin{array}{c}
\text{pivot since} \rightarrow \\
\frac{900}{5} < \frac{2800}{8}
\end{array}
\begin{bmatrix}
⑤ & 2 & 1 & 0 & 0 & 900 \\
8 & 10 & 0 & 1 & 0 & 2800 \\
-3 & -2 & 0 & 0 & 1 & 0
\end{bmatrix}
\begin{array}{c}
\approx \\
(1/5)R1
\end{array}
\begin{bmatrix}
① & \frac{2}{5} & \frac{1}{5} & 0 & 0 & 180 \\
8 & 10 & 0 & 1 & 0 & 2800 \\
-3 & -2 & 0 & 0 & 1 & 0
\end{bmatrix}
$$

$$
\begin{array}{c}
\approx \\
R2 - (8)R1 \\
R3 + (3)R1
\end{array}
\begin{bmatrix}
1 & \frac{2}{5} & \frac{1}{5} & 0 & 0 & 180 \\
0 & ⑭/5 & -\frac{8}{5} & 1 & 0 & 1360 \\
0 & -\frac{4}{5} & \frac{3}{5} & 0 & 1 & 540
\end{bmatrix}
\begin{array}{c}
\approx \\
(5/34)R2
\end{array}
\begin{bmatrix}
1 & \frac{2}{5} & \frac{1}{5} & 0 & 0 & 180 \\
0 & ① & -\frac{4}{17} & \frac{5}{34} & 0 & 200 \\
0 & -\frac{4}{5} & \frac{3}{5} & 0 & 1 & 540
\end{bmatrix}
$$

$$
\begin{array}{c}
\approx \\
R1 - (2/5)R2 \\
R3 + (4/5)R1
\end{array}
\begin{bmatrix}
1 & 0 & \frac{25}{85} & -\frac{1}{17} & 0 & 100 \\
0 & 1 & -\frac{4}{17} & \frac{5}{34} & 0 & 200 \\
0 & 0 & \frac{7}{17} & \frac{2}{17} & 1 & 700
\end{bmatrix}
$$

We have arrived at the final tableau; all negative elements have been eliminated from the last row. This tableau corresponds to the following system of equations.

$$
\begin{aligned}
x + \frac{25}{85}u - \frac{1}{17}v &= 100 \\
y - \frac{4}{17}u + \frac{5}{34}v &= 200 \\
\frac{7}{17}u + \frac{2}{17}v + f &= 700
\end{aligned}
$$

The solutions to this system are of course identical to those of the original system, since it has been derived from the original system using elementary row operations. Since $u \geq 0$, $v \geq 0$, the last equation tells us that f has a maximum value of 700 that takes place when $u = 0$ and $v = 0$. On substituting these values of u and v back into the system we get $x = 100$, $y = 200$. Thus the maximum value of $f = 3x + 2y$ is $f = 700$, and this takes place when $x = 100$, $y = 200$. This result agrees with that obtained in the previous section, using geometry.

Note that the reasoning used in arriving at this maximum value of f implies that the element in the last row and last column of the final tableau will always correspond to the maximum value of f.

The next example illustrates the application of the simplex method for a function of three variables.

EXAMPLE 1 Determine the maximum value of the function $f = 3x + 5y + 8z$ subject to the following constraints.

$$x + y + z \leq 100$$
$$3x + 2y + 4z \leq 200$$
$$x + 2y \leq 150$$
$$x \geq 0, y \geq 0, z \geq 0$$

The corresponding system of equations, with slack variables u, v, and w is

$$x + y + z + u = 100$$
$$3x + 2y + 4z + v = 200$$
$$x + 2y + w = 150$$
$$-3x - 5y - 8z + f = 0$$

with $x \geq 0, y \geq 0, z \geq 0, u \geq 0, v \geq 0, w \geq 0$.

The simplex tableaux are as follows:

$$\begin{bmatrix} 1 & 1 & 1 & 1 & 0 & 0 & 0 & 100 \\ 3 & 2 & ④ & 0 & 1 & 0 & 0 & 200 \\ 1 & 2 & 0 & 0 & 0 & 1 & 0 & 150 \\ -3 & -5 & -8 & 0 & 0 & 0 & 1 & 0 \end{bmatrix} \underset{(1/4)R2}{\approx} \begin{bmatrix} 1 & 1 & 1 & 1 & 0 & 0 & 0 & 100 \\ \frac{3}{4} & \frac{1}{2} & ① & 0 & \frac{1}{4} & 0 & 0 & 50 \\ 1 & 2 & 0 & 0 & 0 & 1 & 0 & 150 \\ -3 & -5 & -8 & 0 & 0 & 0 & 1 & 0 \end{bmatrix}$$

$$\underset{\substack{R1 - R2 \\ R4 + (8)R2}}{\approx} \begin{bmatrix} \frac{1}{4} & \frac{1}{2} & 0 & 1 & -\frac{1}{4} & 0 & 0 & 50 \\ \frac{3}{4} & \frac{1}{2} & 1 & 0 & \frac{1}{4} & 0 & 0 & 50 \\ 1 & ② & 0 & 0 & 0 & 1 & 0 & 150 \\ 3 & -1 & 0 & 0 & 2 & 0 & 1 & 400 \end{bmatrix} \underset{(1/2)R3}{\approx} \begin{bmatrix} \frac{1}{4} & \frac{1}{2} & 0 & 1 & -\frac{1}{4} & 0 & 0 & 50 \\ \frac{3}{4} & \frac{1}{2} & 1 & 0 & \frac{1}{4} & 0 & 0 & 50 \\ \frac{1}{2} & ① & 0 & 0 & 0 & \frac{1}{2} & 0 & 75 \\ 3 & -1 & 0 & 0 & 2 & 0 & 1 & 400 \end{bmatrix}$$

$$\underset{\substack{R1 - (1/2)R3 \\ R2 - (1/2)R3 \\ R4 + R3}}{\approx} \begin{bmatrix} 0 & 0 & 0 & 1 & -\frac{1}{4} & -\frac{1}{4} & 0 & \frac{25}{2} \\ \frac{1}{2} & 0 & 1 & 0 & \frac{1}{4} & -\frac{1}{4} & 0 & \frac{25}{2} \\ \frac{1}{2} & 1 & 0 & 0 & 0 & \frac{1}{2} & 0 & 75 \\ \frac{7}{2} & 0 & 0 & 0 & 2 & \frac{1}{2} & 1 & 475 \end{bmatrix}$$

maximum value of the function ←

This final tableau gives the following system of equations:

$$u - \tfrac{1}{4}v - \tfrac{1}{4}w = 12\tfrac{1}{2}$$
$$\tfrac{1}{2}x + z + \tfrac{1}{4}v - \tfrac{1}{4}w = 12\tfrac{1}{2}$$
$$\tfrac{1}{2}x + y + \tfrac{1}{2}w = 75$$
$$3\tfrac{1}{2}x \qquad 2v + \tfrac{1}{2}w + f = 475$$

The constraints are

$$x \geq 0, \quad y \geq 0, \quad z \geq 0, \quad u \geq 0, \quad v \geq 0, \quad w \geq 0$$

The final equation, under these constraints, implies that f has a maximum of 475, when $x = 0, v = 0, w = 0$. On substituting these values back into the equations we get $y = 75, z = 12\tfrac{1}{2}, u = 12\tfrac{1}{2}$.

Thus $f = 3x + 5y + 8z$ has a maximum value of 475 at $x = 0, y = 75, z = 12\tfrac{1}{2}$.

The next example illustrates the application of the simplex method when there are many optimal solutions. (We use Example 3 of the previous section.)

EXAMPLE 2   Find the maximum value of $f = 8x + 2y$ subject to the constraints

$$4x + y \leq 32$$
$$4x + 3y \leq 48$$
$$x \geq 0$$
$$y \geq 0$$

The simplex tableaux are as follows:

$$
\begin{bmatrix}
④ & 1 & 1 & 0 & 0 & 32 \\
4 & 3 & 0 & 1 & 0 & 48 \\
-8 & -2 & 0 & 0 & 1 & 0
\end{bmatrix}
\approx
\begin{bmatrix}
① & \frac{1}{4} & \frac{1}{4} & 0 & 0 & 8 \\
4 & 3 & 0 & 1 & 0 & 48 \\
-8 & -2 & 0 & 0 & 1 & 0
\end{bmatrix}
\approx
\begin{bmatrix}
1 & \frac{1}{4} & \frac{1}{4} & 0 & 0 & 8 \\
0 & 2 & -1 & 1 & 0 & 16 \\
0 & 0 & 2 & 0 & 1 & 64
\end{bmatrix}
$$

The final tableau gives the following system of equations:

$$x + \tfrac{1}{4}y + \tfrac{1}{4}u \qquad\quad = 8$$
$$2y - u + v \qquad = 16$$
$$2u \qquad + f = 64$$

with $x \geq 0, y \geq 0, u \geq 0, v \geq 0$.

The last equation implies that f has a maximum value of 64, when $u = 0$. Substituting $u = 0$ into the other equations gives

$$x + \tfrac{1}{4}y \qquad = 8$$
$$2y + v = 16$$

with $x \geq 0, y \geq 0, v \geq 0$.

Any point (x, y) that satisfies these conditions is an optimal solution. The second equation tells us that v can be any number in the interval $[0, 16]$ and that y can be any number in the interval $[0, 8]$. The first equation then tells us that $x = 8$ when $y = 0$ and $x = 6$ when $y = 8$.

Thus the maximum value of $f = 8x + 2y$ is 64. This is achieved at a point on the line $x + \tfrac{1}{4}y = 8$ between $(6, 8)$ and $(8, 0)$. This result agrees with that of Example 3 of the previous section when the geometric approach was used.   ■

In this section we have introduced the simplex algorithm for maximizing a function f when all constraints are of the type $ax + by + \cdots \leq k$, where k is nonnegative, and all the variables are nonnegative. This is called a **standard linear programming problem**. Constraints in linear programming problems can involve $=$, $\leq$, and $\geq$, and some of the variables can be negative. The way slack variables are used in setting up an initial tableau varies, depending upon the types of constraints. Once the initial tableau has been constructed however, the above algorithm is used to arrive at a final tableau that leads to the solution. Students who are interested in reading more about linear programming are referred to *Applied Linear Programming* by Michael R. Greenberg, Academic Press, 1978. Dantzig's discovery of the simplex method ranks high among the achievments of twentieth-century mathematics.

EXERCISE SET 9.2

Simplex Method in Maximizing Functions

In Exercises 1–9 use the simplex method to maximize the functions under the given constraints. (Exercises 1 and 2 were solved using geometry in the previous section.)

1. Maximize $f = 2x + y$ subject to

$$4x + y \leq 36$$
$$4x + 3y \leq 60$$
$$x \geq 0, y \geq 0$$

2. Maximize $f = x - 4y$ subject to

$$x + 2y \leq 4$$
$$x + 6y \leq 8$$
$$x \geq 0, y \geq 0$$

3. Maximize $f = 4x + 6y$ subject to

$$x + 3y \leq 6$$
$$3x + y \leq 8$$
$$x \geq 0, y \geq 0$$

4. Maximize $f = 10x + 5y$ subject to

$$x + y \leq 180$$
$$3x + 2y \leq 480$$
$$x \geq 0, y \geq 0$$

5. Maximize $f = x + 2y + z$ subject to

$$3x + y + z \leq 35$$
$$x - 10y - 4z \leq 20$$
$$x \geq 0, y \geq 0, z \geq 0$$

6. Maximize $f = 100x + 200y + 50z$ subject to

$$5x + 5y + 10z \leq 1000$$
$$10x + 8y + 5z \leq 2000$$
$$10x + 5y \leq 500$$
$$x \geq 0, y \geq 0, z \geq 0$$

7. Maximize $f = 2x + 4y + z$ subject to

$$-x + 2y + 3z \leq 6$$
$$-x + 4y + 5z \leq 5$$
$$-x + 5y + 7z \leq 7$$
$$x \geq 0, y \geq 0, z \geq 0$$

8. Maximize $f = x + 2y + 4z - w$ subject to

$$5x + 4z + 6w \leq 20$$
$$4x + 2y + 2z + 8w \leq 40$$
$$x \geq 0, y \geq 0, z \geq 0, w \geq 0$$

9. Maximize $f = x + 2y - z + 3w$ subject to

$$2x + 4y + 5z + 6w \leq 24$$
$$4x + 4y + 2z + 2w \leq 4$$
$$x \geq 0, y \geq 0, z \geq 0, w \geq 0$$

Applications of the Simplex Method

10. A company uses three machines, I, II, and III to produce items X, Y, and Z. The production of each item involves the use of more than one machine. It takes 2 minutes on I and 4 on II to manufacture a single X. It takes 3 minutes on I and 6 minutes on III to manufacture a Y. It takes 1 minute on I, 2 minutes on II, and 3 minutes on III to manufacture a Z. The total time available on each machine per day is 6 hours. The profits are $10, $8, and $12 on each of X, Y, and Z, respectively. How should the company allocate the production times on the machines in order to maximize total profit?

11. An industrial furniture company manufactures desks, cabinets, and chairs. These items involve metal, wood, and plastic. The following table gives the amounts that go into each product (in convenient units) and the profit on each item.

	Metal	Wood	Plastic	Profit
Desk	3	4	2	$16
Cabinet	6	1	1	$12
Chair	1	2	2	$ 6

If the company has 800 units of metal, 400 units of wood, and 100 units of plastic available, how should it allocate these resources in order to maximize total profit?

12. A company produces washing machines at three factories, A, B, and C. The washing machines are sold in a certain city P. It costs $10, $20, and $40 to transport each washing machine from A, B, and C, respectively, to P. It involves 6, 4, and 2 hours of packing and transportation time to get a washing machine from A, B, and C, respectively, to P. There is $6000 budgeted weekly for transportation of the washing machines to P and a total of 4000 hours of labor available. The profit on each machine from A is $12, on each from B is $20, and on each from C is $16. How should the company schedule the transportation of washing machines from A, B, and C to P in order to maximize total profit?

13. A manufacturer makes three lines of tents, all from the same material. The Aspen, Alpine, and Cub tents require 60, 30, and 15 square yards of material, respectively. The manufacturing costs of the Aspen, Alpine, and Cub are $32, $20, and $12. The material is available in amounts of 7800 square yards weekly, and the weekly working budget is $8320. The profits on the Aspen, Alpine, and Cub are $12, $8, and $4. What should the weekly production schedule of these tents be to maximize total profit?

Simplex Method in Minimizing Functions

In Exercises 14–16 use the simplex method to minimize the functions under the given constraints. (Exercise 14 was solved using geometry in the previous section.)

14. Minimize $f = -x + 2y$ subject to

$$x + 2y \leq 4$$
$$x + 4y \leq 6$$
$$x \geq 0, y \geq 0$$

15. Minimize $f = -2x + y$ subject to

$$2x + 2y \leq 8$$
$$x - y \leq 2$$
$$x \geq 0, y \geq 0$$

16. Minimize $f = 2x + y - z$ subject to

$$x + 2y - 2z \leq 20$$
$$2x + y + \qquad \leq 10$$
$$x + 3y + 4z \leq 15$$
$$x \geq 0, y \geq 0, z \geq 0$$

---

*9.3 Geometrical Explanation of the Simplex Method

We now explain, by means of an example, the sequence of transformations used in the simplex method. Let us maximize the function $f = 2x + 3y$ subject to the constraints

$$x + 2y \leq 8$$
$$3x + 2y \leq 12 \tag{1}$$
$$x \geq 0$$
$$y \geq 0$$

The region that satisfies these constraints is $ABCO$ in Figure 9.7. Recall that any point in this region is called a **feasible solution**. A feasible solution that gives a maximum value of f is called an **optimal solution**.

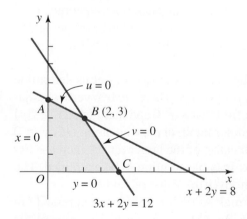

Figure 9.7

Let us reformulate the constraints using slack variables u and v:

$$x + 2y + u \qquad\qquad = 8$$
$$3x + 2y \qquad + v \qquad = 12 \tag{2}$$
$$-2x - 3y \qquad\qquad + f = 0$$

with $x \geq 0, y \geq 0, u \geq 0, v \geq 0$.

Observe that $u = 0$ along AB, since AB is a segment of the line $x + 2y = 8$. $v = 0$ along BC, since BC is a segment of the line $3x + 2y = 12$. Furthermore, $x = 0$ along OA and $y = 0$ along OC. Thus the boundaries of the feasible region $ABCO$ are such that one of the variables is zero along each boundary.

Let us now look at the vertices in terms of the four variables x, y, u, and v. Vertex O is the point $x = 0$, $y = 0$, $u = 8$, $v = 12$. (One obtains the u and v values at O by letting

$x = 0$ and $y = 0$ in the constraints.) Vertex A lies on OA and AB. Thus $x = 0$ and $u = 0$ at A. Substituting these values into the constraints gives $y = 4$ and $v = 4$. Thus A is the point $x = 0, y = 4, u = 0, v = 4$. In a similar manner we find that B is the point $x = 2, y = 3, u = 0, v = 0$ and C is the point $x = 4, y = 0, u = 4, v = 0$.

Observe that certain variables are zero while others are nonzero at each vertex. The variables that are not zero are called **basic variables** and the remaining variables are called **nonbasic variables**. We summarize the results thus far in Table 9.1.

Table 9.1

Vertex	Coordinates	Basic Variables ($\neq 0$)	Nonbasic Variables ($= 0$)
O	$x = 0, \ y = 0, \ u = 8, \ v = 12$	u, v	x, y
A	$x = 0, \ y = 4, \ u = 0, \ v = 4$	y, v	x, u
B	$x = 2, \ y = 3, \ u = 0, \ v = 0$	x, y	u, v
C	$x = 4, \ y = 0, \ u = 4, \ v = 0$	x, u	y, v

The method that we shall now develop starts at a vertex feasible solution, O in our case, and then proceeds through a sequence of adjacent vertices, each one giving an increased value of f, until an optimal solution is reached. The initial simplex tableau corresponds (in a way to be explained later) to the situation at the initial feasible solution; further tableaux represent the pictures at other vertices. We must examine each vertex in turn to see if it is an optimal solution. If it is not an optimal solution, we have to decide which neighboring vertex to move to next. Our aims are to first develop the tools to carry out the procedure geometrically, and then to translate the geometrical concepts into analogous algebraic ones. The advantage of carrying out the procedure algebraically is that it can then be implemented on the computer. We shall find that the algebraic procedure that results is the simplex algorithm.

Let us start at O, a vertex feasible solution. The value of f at O is 0. Let us determine whether it is necessary to move from O to a neighboring vertex. There are two vertices adjacent to O, namely A and C. In moving along OA we find that $x = 0$ and that $f(= 2x + 3y)$ increases by 3 for every unit increase in y. On the other hand, in moving along OC, $y = 0$ and f increases by 2 for every unit increase in x. Because of the larger rate of increase in f along OA this path is selected; A becomes the next feasible solution to be examined. The value of f at A is 12. This is indeed an improvement over the value of f at O.

Let us now determine whether or not A is optimal. At the vertex O, we expressed f in terms of the nonbasic variables x and y; $f = 2x + 3y$. Let us now write f in terms of the nonbasic variables of the point A, namely x and u. To do this we substitute the value of y from the first constraint in system (2) into $f = 2x + 3y$. We get $y = 4 - x/2 - u/2$. The substitution gives, $f = 12 + x/2 - 3u/2$. If we move from A to B along AB then $u = 0$ along this route, and f increases by $\frac{1}{2}$ for every unit increase in x. Since f increases we move to B. The value of f at B is 13, an increase over its value at A.

We now determine whether B is an optimal solution. The nonbasic variables at B are u and v. We express f in terms of these variables. From the original restrictions (2) we get $x = 2 + u/2 - v/2$ and $y = 3 - 3u/4 + v/4$. Thus $f = 13 - 5u/4 - v/4$. In moving from B to C along BC, $v = 0$ and f decreases by $\frac{5}{4}$ for every unit increase in u. Thus f has a maximum value at B. The maximum value is 13.

We have developed the geometrical ideas. Let us now translate them into algebraic form. The initial feasible solution is O. The basic variables at O are u and v, with values 8 and 12, respectively. The value of f at O is 0. The initial tableau, corresponding to system of equations (2), reflects this information.

Tableau for Vertex O
coefficients of

$$\begin{array}{ccccc} x & y & u & v & f \\ \left[\begin{array}{ccccc} 1 & 2 & 1 & 0 & 0 \\ 3 & 2 & 0 & 1 & 0 \\ -2 & -3 & 0 & 0 & 1 \end{array}\right. & & & & \left.\begin{array}{c} 8 \\ 12 \\ 0 \end{array}\right] \begin{array}{l} \leftarrow u \\ \leftarrow v \\ \leftarrow f \end{array} \begin{array}{l} \text{basic} \\ \text{variables} \end{array} \end{array}$$

The last column gives the values of the basic variables and f at O. Knowing the basic variables, we also know the nonbasic variables. This is the tableau associated with vertex O.

The geometric discussion told us that O was not an optimal solution, and led us to A. The basic variables at O are u and v; at A the basic variables are y and v. In going from O to A, y replaces u as a basic variable. u is called the **departing basic variable** and y is called the **entering basic variable**. The entering basic variable y corresponds to the largest rate of increase of f, namely 3. In terms of the given tableau, this is reflected in y being the column corresponding to the negative entry, -3, in the last row, having largest magnitude. Determining the negative entry in the last row having largest magnitude enables us to select the entering variable. If no such entry exists then there is no appropriate entering variable and we have arrived at the final tableau.

The next step is to determine the departing variable for the tableau. In going from O to A the departing variable is u. The first two rows of the tableau correspond to the equations

$$x + 2y + u = 8$$
$$3x + 2y + v = 12$$

In going along OA, $x = 0$. The equations may be written

$$u = 8 - 2y$$
$$v = 12 - 2y$$

Since $u \geq 0$ and $v \geq 0$, the maximum value to which y can increase is 4, when $u = 0$ and $v = 4$ (at the point A). Thus u becomes a nonbasic variable. Looking at the above equations we see that the reason u arrives at the value 0 before v is that $\frac{8}{2} < \frac{12}{2}$. In terms of the tableau, this corresponds to dividing the elements in the entering variable column into the corresponding elements of the last column. The row containing the numbers that give the smallest result gives the departing variable. Thus,

$$\begin{array}{ccccc} x & y & u & v & f \\ \left[\begin{array}{ccccc} 1 & 2 & 1 & 0 & 0 \\ 3 & 2 & 0 & 1 & 0 \\ -2 & -3 & 0 & 0 & 1 \end{array}\right. & & & & \left.\begin{array}{c} 8 \\ 12 \\ 0 \end{array}\right] \begin{array}{l} u \leftarrow \text{departing variable} \\ v \\ f \end{array} \end{array}$$

$$\uparrow$$
entering variable

We now have a method for selecting the entering and departing variables in any tableau. Let us now transform the preceding tableau into the tableau having the new basic variables y and v. The new tableau should have the values of the new basic variables and the value of f at the new vertex as its last column. The initial constraints, from which the initial tableau was derived, were as follows

$$x + 2y + u \qquad\qquad = 8$$
$$3x + 2y \qquad + v \qquad = 12$$
$$-2x - 3y \qquad\qquad + f = 0$$

Observe that the basic variables u and v appear in a single equation each, the coefficient being 1 in each case. It is this fact that causes u and v to assume the values 8 and 12 at O when the nonbasic variables are 0. Furthermore, the fact that the last equation involves f and only nonbasic variables cause f to assume the value 0 on the right-hand side of the equation. These are the characteristics that we must attempt to obtain for the tableau that represents the vertex A, in terms of the basic variables y and v of A. This is achieved by selecting the element that lies in the entering variable column and departing variable row as pivot, creating a 1 in its location and 0's elsewhere in this column. Thus the sequence of tableaux becomes as follows:

$$
\begin{array}{ccccc}
x & y & u & v & f \\
\end{array}
\begin{bmatrix}
1 & ② & 1 & 0 & 0 & 8 \\
3 & 2 & 0 & 1 & 0 & 12 \\
-2 & -3 & 0 & 0 & 1 & 0
\end{bmatrix}
\begin{array}{l} u \leftarrow \\ v \\ f \end{array}
\approx
\begin{array}{ccccc}
x & y & u & v & f \\
\end{array}
\begin{bmatrix}
\frac{1}{2} & ① & \frac{1}{2} & 0 & 0 & 4 \\
3 & 2 & 0 & 1 & 0 & 12 \\
-2 & -3 & 0 & 0 & 1 & 0
\end{bmatrix}
\begin{array}{l} u \leftarrow \\ v \\ f \end{array}
$$

$$
\approx
\begin{array}{ccccc}
x & y & u & v & f \\
\end{array}
\begin{bmatrix}
\frac{1}{2} & 1 & \frac{1}{2} & 0 & 0 & 4 \\
2 & 0 & -1 & 1 & 0 & 4 \\
-\frac{1}{2} & 0 & \frac{3}{2} & 0 & 1 & 12
\end{bmatrix}
\begin{array}{l} y \\ u \\ f \end{array}
\begin{array}{l} \text{new basic} \\ \text{variables} \\ \\ \text{new value of } f \end{array}
$$

Let us verify that this tableau does indeed correspond to a system of equations that gives $y = 4$, $v = 4$, and $f = 12$ in the preceding order, if we take y and v as basic variables. The corresponding equations are as follows:

$$\frac{1}{2}x + y + \frac{1}{2}u = 4$$

$$2x - u + v = 4$$

$$-\frac{1}{2}x + \frac{3}{2}u + f = 12$$

Taking x and u as nonbasic variables, we make them zero, and we do indeed get $y = 4$, $v = 4$, $f = 12$. This is the tableau for the vertex A.

The analysis is now repeated for this tableau. Because of the negative entry $-\frac{1}{2}$ in the last row, this value of $f = 12$ is not the optimal solution. The new entering variable is x because of the negative sign in this column. The departing variable is v, since $\frac{4}{2} < 4/\frac{1}{2}$. We get

$$
\begin{array}{ccccc}
x & y & u & v & f \\
\end{array}
\begin{bmatrix}
\frac{1}{2} & 1 & \frac{1}{2} & 0 & 0 & 4 \\
② & 0 & -1 & 1 & 0 & 4 \\
-\frac{1}{2} & 0 & \frac{3}{2} & 0 & 1 & 12
\end{bmatrix}
\begin{array}{l} y \\ v \leftarrow \text{departing variable} \\ f \end{array}
$$

$\uparrow$ entering variable

The entering variable x will replace the v in the second row. The tableaux are as follows.

$$
\approx
\begin{array}{ccccc}
x & y & u & v & f \\
\end{array}
\begin{bmatrix}
\frac{1}{2} & 1 & \frac{1}{2} & 0 & 0 & 4 \\
① & 0 & -\frac{1}{2} & \frac{1}{2} & 0 & 2 \\
-\frac{1}{2} & 0 & \frac{3}{2} & 0 & 1 & 12
\end{bmatrix}
\begin{array}{l} y \\ v \leftarrow \\ f \end{array}
\approx
\begin{array}{ccccc}
x & y & u & v & f \\
\end{array}
\begin{bmatrix}
0 & 1 & \frac{3}{4} & -\frac{3}{4} & 0 & 3 \\
1 & 0 & -\frac{1}{2} & \frac{1}{2} & 0 & 2 \\
0 & 0 & \frac{5}{4} & \frac{1}{4} & 1 & 13
\end{bmatrix}
\begin{array}{l} y \\ x \\ f \end{array}
\begin{array}{l} \text{new basic} \\ \text{variables} \\ \\ \text{new value of } f \end{array}
$$

$\uparrow$

This is the final tableau. The basic variables are x and y, assuming the values 2 and 3 respectively. This is the tableau for the vertex B. The value of f is 13; it is the maximum value possible under the given constraints.

Let us now analyze a linear programming problem that has many solutions, to see how the final tableau is interpreted.

EXAMPLE 1 Let us return to Example 2 of the previous section, a linear programming problem that has many solutions.

The objective function is $f = 8x + 2y$. The constraints are

$$4x + \ y \le 32$$
$$4x + 3y \le 48$$
$$x \ge 0$$
$$y \ge 0$$

Express the constraints using slack variables u and v.

$$4x + \ y + u \qquad\quad = 32$$
$$4x + 3y \qquad + v \qquad = 48$$
$$-8x - 2y \qquad\qquad + f = \ 0$$

with $x \ge 0, y \ge 0, u \ge 0, v \ge 0$.

The simplex tableau are

$$
\begin{array}{ccccc}
x & y & u & v & f \\
\end{array}
$$

$$
\begin{bmatrix}
④ & 1 & 1 & 0 & 0 & 32 \\
4 & 3 & 0 & 1 & 0 & 48 \\
-8 & -2 & 0 & 0 & 1 & 0
\end{bmatrix}
\begin{array}{l}
u \;\leftarrow \text{departing variable} \\
v \\
f
\end{array}
$$

$$\uparrow$$
$$\text{entering variable}$$

$$
\approx
\begin{bmatrix}
① & \frac{1}{4} & \frac{1}{4} & 0 & 0 & 8 \\
4 & 3 & 0 & 1 & 0 & 48 \\
-8 & -2 & 0 & 0 & 1 & 0
\end{bmatrix}
\begin{array}{l}
u \leftarrow \\
v \\
f
\end{array}
\qquad
\approx
\begin{bmatrix}
1 & \frac{1}{4} & \frac{1}{4} & 0 & 0 & 8 \\
0 & 2 & -1 & 1 & 0 & 16 \\
0 & 0 & 2 & 0 & 1 & 64
\end{bmatrix}
\begin{array}{l}
x \\
v \\
f
\end{array}
$$

$$\uparrow$$

This is the final tableau. It leads to a maximum value of $f = 64$. This occurs at $x = 8, v = 16$. Since y is not a basic variable, $y = 0$. Thus we have found the optimal solution $x = 8, y = 0$ (the point C in Figure 9.3). When the simplex method is applied to a problem that has many solutions, it stops as soon as it has found one optimal solution, as shown here. We now show how to extend the algorithm to find other solutions.

Observe that in the final tableau, the coefficient of y, a nonbasic variable is 0 in the last row. Each coefficient of a nonbasic variable in this row indicates the rate at which f increases as that variable is increased. Thus, making y an entering or departing variable neither increases or decreases f. Let us use y as an entering variable and v as a departing variable. We get the following tableaux.

$$
\begin{array}{ccccc}
 & x & y & u & v & f \\
\end{array}
$$

$$
\begin{bmatrix}
1 & \frac{1}{4} & \frac{1}{4} & 0 & 0 & 8 \\
0 & \boxed{2} & -1 & 1 & 0 & 16 \\
0 & 0 & 2 & 0 & 1 & 64
\end{bmatrix}
\begin{array}{l}
x \\
v \leftarrow \text{departing variable} \\
f
\end{array}
$$

$$\underset{\text{entering variable}}{\uparrow}$$

$$
\approx
\begin{bmatrix}
1 & \frac{1}{4} & \frac{1}{4} & 0 & 0 & 8 \\
0 & \boxed{1} & -\frac{1}{2} & \frac{1}{2} & 0 & 8 \\
0 & 0 & 2 & 0 & 1 & 64
\end{bmatrix}
\begin{array}{l}
x \leftarrow \\
v \\
f
\end{array}
\approx
\begin{bmatrix}
1 & 0 & \frac{3}{8} & -\frac{1}{8} & 0 & 6 \\
0 & 1 & -\frac{1}{2} & \frac{1}{2} & 0 & 8 \\
0 & 0 & 2 & 0 & 1 & 64
\end{bmatrix}
\begin{array}{l}
x \\
y \\
f
\end{array}
$$

$$\uparrow$$

This tableau leads to the optimal solution $x = 6$, $y = 8$ (the point B in Figure 9.3). All points on the line between these two optimal solutions of $(8, 0)$ and $(6, 8)$, which were obtained using the simplex tableaux, are also optimal solutions.

EXERCISE SET 9.3

In Exercises 1–6 use the simplex method to maximize the functions under the given constraints. Determine the basic and nonbasic variables and the entering variables and departing variables for each tableau. Determine the optimal solution and the maximum value of the objective function directly from the final tableau. (These problems were given in the previous set of exercises. Use the tableaux that you have already derived, and check your previous answers.)

1. Maximize $f = 2x + y$ subject to
   $$4x + y \le 36$$
   $$4x + 3y \le 60$$
   $$x \ge 0, y \ge 0$$
   (Exercise 1, Section 9.2)

2. Maximize $f = x - 4y$ subject to
   $$x + 2y \le 4$$
   $$x + 6y \le 8$$
   $$x \ge 0, y \ge 0$$
   (Exercise 2, Section 9.2)

3. Maximize $f = x + 2y + z$ subject to
   $$3x + y + z \le 3$$
   $$x - 10y - 4z \le 20$$
   $$x \ge 0, y \ge 0, z \ge 0$$
   (Exercise 5, Section 9.2)

4. Maximize $f = 100x + 200y + 50z$ subject to
   $$5x + 5y + 10z \le 1000$$
   $$10x + 8y + 5z \le 2000$$
   $$10x + 5y \le 500$$
   $$x \ge 0, y \ge 0, z \ge 0$$
   (Exercise 6, Section 9.2)

5. Maximize $f = 2x + 4y + z$ subject to
   $$-x + 2y + 3z \le 6$$
   $$-x + 4y + 5z \le 5$$
   $$-x + 5y + 7z \le 7$$
   $$x \ge 0, y \ge 0, z \ge 0, w \ge 0$$
   (Exercise 7, Section 9.2)

6. Maximize $f = x + 2y + 4z - w$ subject to
   $$5x + 4z + 6w \le 20$$
   $$4x + 2y + 2z + 8w \le 40$$
   $$x \ge 0, y \ge 0, z \ge 0, w \ge 0$$
   (Exercise 8, Section 9.2)

CHAPTER 9 REVIEW EXERCISES

1. Maximize $f = 2x + 3y$ subject to

$2x + 4y \leq 16$
$3x + 2y \leq 12$
$x \geq 0, y \geq 0$

2. Maximize $f = 6x + 4y$ subject to

$x + 2y \leq 16$
$3x + 2y \leq 24$
$x \geq 0, y \geq 0$

3. Minimize $f = 4x + y$ subject to

$3x + 2y \leq 21$
$x + 5y \leq 20$
$x \geq 0, y \leq 0$

4. A farmer has to decide how many acres of a 40-acre plot are to be devoted to growing strawberries and how many to growing tomatoes. There will be 300 hours of labor available for the picking. It takes 8 hours to pick an acre of strawberries and 6 hours to pick an acre of tomatoes. The profit per acre is $700 on the strawberries compared to $600 on the tomatoes. How many acres of each should be grown to maximize profit?

5. A company is buying lockers. It has narrowed the choice down to two kinds, X and Y. X has a volume of 36 cubic feet, while Y has a volume of 44 cubic feet. X occupies an area of 6 square feet and costs $54, while Y occupies 8 square feet and costs $60. A total of 256 square feet of floor space is available and $2100 in funds are available for the purchase of the lockers. At least 20 of the larger lockers are needed. How many of each type of locker should be purchased in order to maximize volume?

6. Use the simplex method to maximize $f = 2x + y + z$ under the constraints

$x + 2y + 4z \leq 20$
$2x + 4y + 4z \leq 60$
$3x + 4y + z \leq 90$
$x \geq 0, y \geq 0, z \geq 0$

7. A furniture company finishes two kinds of tables, X and Y. There are three stages in the finishing process, namely sanding, staining, and varnishing. The times in minutes involved for each of these processes are as follows:

	Sanding	Staining	Varnishing
Table X	10	8	4
Table Y	5	4	8

The three types of equipment needed for sanding, staining, and varnishing are each available for 5 hours per day. Each type of equipment can handle only one table at a time. The profit on each X table is $8 and on each Y table is $4. How many of each type of table should be finished daily to maximize total profit?

8. Minimize $f = x - 2y + 4z$ subject to

$x - y + 3z \leq 4$
$2x + 2y - 3z \leq 6$
$-x + 2y + 3z \leq 2$
$x \geq 0, y \geq 0, z \geq 0$

This bronze arch, made in the shape of a whale, was designed by Canadian architect Frank Gehry for the Olympic Port of Barcelona for the 1992 Summer Olympics.

Cross Product*

n this section we introduce the cross product of two vectors in $\mathbf{R}^3$. The cross product is an important tool in many areas of science and engineering.

DEFINITION Let $\mathbf{u} = (u_1, u_2, u_3)$ and $\mathbf{v} = (v_1, v_2, v_3)$ be two vectors in $\mathbf{R}^3$. The cross product of $\mathbf{u}$ and $\mathbf{v}$ is denoted $\mathbf{u} \times \mathbf{v}$ and is the vector

$$\mathbf{u} \times \mathbf{v} = (u_2v_3 - u_3v_2, u_3v_1 - u_1v_3, u_1v_2 - u_2v_1)$$

EXAMPLE 1 Determine the cross product $\mathbf{u} \times \mathbf{v}$ of the vectors

$$\mathbf{u} = (1, -2, 3) \text{ and } \mathbf{v} = (0, 1, 4)$$

SOLUTION

Using this definition and letting $(u_1, u_2, u_3) = (1, -2, 3)$ and $(v_1, v_2, v_3) = (0, 1, 4)$, we get

$$\mathbf{u} \times \mathbf{v} = ((-2 \times 4) - (3 \times 1), (3 \times 0) - (1 \times 4), (1 \times 1) - (-2 \times 0))$$
$$= (-11, -4, 1)$$

We shall use the notation $\mathbf{i} = (1, 0, 0)$, $\mathbf{j} = (0, 1, 0)$, $\mathbf{k} = (0, 0, 1)$ for these **standard basis vectors** for $\mathbf{R}^3$. See Figure A1. This notation is commonly used in the sciences.

We can write $\mathbf{u} \times \mathbf{v}$ in the convenient form

$$\mathbf{u} \times \mathbf{v} = (u_2v_3 - u_3v_2)\mathbf{i} + (u_3v_1 - u_1v_3)\mathbf{j} + (u_1v_2 - u_2v_1)\mathbf{k}$$

It can be easily seen (by expanding the determinant) that

$$\mathbf{u} \times \mathbf{v} = \begin{vmatrix} \mathbf{i} & \mathbf{j} & \mathbf{k} \\ u_1 & u_2 & u_3 \\ v_1 & v_2 & v_3 \end{vmatrix}$$

*Optional—some schools cover this material in the calculus sequence.

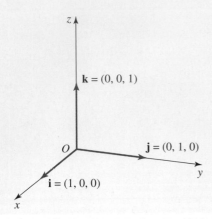

Figure A1

Note that this is not an ordinary determinant, since the elements of the first row are vectors, not scalars. The expansion of this determinant in terms of the first row gives a useful way of remembering the components of the vector $\mathbf{u} \times \mathbf{v}$. Furthermore, the algebraic properties of a determinant conveniently describe the algebraic properties of the cross product.

We shall find it convenient to use all three of the above forms of the cross product at various times.

EXAMPLE **2** Use the determinant form of the cross product to compute $\mathbf{u} \times \mathbf{v}$ for the vectors

$$\mathbf{u} = (-2, 4, 1) \text{ and } \mathbf{v} = (3, 1, 5)$$

SOLUTION

We get, expanding the determinant in terms of the first row,

$$\mathbf{u} \times \mathbf{v} = \begin{vmatrix} \mathbf{i} & \mathbf{j} & \mathbf{k} \\ -2 & 4 & 1 \\ 3 & 1 & 5 \end{vmatrix}$$

$$= ((4 \times 5) - (1 \times 1))\mathbf{i} - ((-2 \times 5) - (1 \times 3))\mathbf{j} + ((-2 \times 1) - (4 \times 3))\mathbf{k}$$

$$= 19\mathbf{i} + 13\mathbf{j} - 14\mathbf{k}$$

$$= (19, 13, -14)$$

We now use this determinant form of the cross product to derive some of the properties of the cross product.

THEOREM A.1

Let $\mathbf{u}$ and $\mathbf{v}$ be vectors in $\mathbf{R}^3$. Then

$$\mathbf{u} \times \mathbf{u} = \mathbf{0}$$

$$\mathbf{u} \times \mathbf{v} = -(\mathbf{v} \times \mathbf{u})$$

Proof Let $\mathbf{u} = (u_1, u_2, u_3)$ and $\mathbf{v} = (v_1, v_2, v_3)$. We get

$$\mathbf{u} \times \mathbf{u} = \begin{vmatrix} \mathbf{i} & \mathbf{j} & \mathbf{k} \\ u_1 & u_2 & u_3 \\ u_1 & u_2 & u_3 \end{vmatrix} = \mathbf{0}$$

since a determinant with two equal rows is zero. Note that since $\mathbf{u} \times \mathbf{u}$ is a vector, the zero is a vector, not a scalar.

Further, using the properties of determinants we get

$$\mathbf{u} \times \mathbf{v} = \begin{vmatrix} \mathbf{i} & \mathbf{j} & \mathbf{k} \\ u_1 & u_2 & u_3 \\ v_1 & v_2 & v_3 \end{vmatrix} \underset{R1 \leftrightarrow R2}{=} -\begin{vmatrix} \mathbf{i} & \mathbf{j} & \mathbf{k} \\ v_1 & v_2 & v_3 \\ u_1 & u_2 & u_3 \end{vmatrix} = -(\mathbf{v} \times \mathbf{u})$$

Thus, in particular,

$$\mathbf{i} \times \mathbf{i} = \mathbf{0}, \quad \mathbf{j} \times \mathbf{j} = \mathbf{0}, \quad \mathbf{k} \times \mathbf{k} = \mathbf{0}$$

$$\mathbf{i} \times \mathbf{j} = -(\mathbf{j} \times \mathbf{i}), \quad \mathbf{i} \times \mathbf{k} = -(\mathbf{k} \times \mathbf{i}), \quad \mathbf{j} \times \mathbf{k} = -(\mathbf{k} \times \mathbf{j})$$

THEOREM A.2

If $\mathbf{i} = (1, 0, 0)$, $\mathbf{j} = (0, 1, 0)$, $\mathbf{k} = (0, 0, 1)$, then

$$\mathbf{i} \times \mathbf{j} = \mathbf{k}, \quad \mathbf{j} \times \mathbf{k} = \mathbf{i}, \quad \mathbf{k} \times \mathbf{i} = \mathbf{j}$$

See Figure A2.

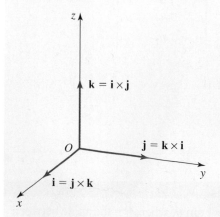

Figure A2

Proof To arrive at $\mathbf{i} \times \mathbf{j} = \mathbf{k}$ use $\mathbf{i} = (1, 0, 0)$ and $\mathbf{j} = (0, 1, 0)$ in the determinant form of the cross product as follows.

$$\mathbf{i} \times \mathbf{j} = \begin{vmatrix} \mathbf{i} & \mathbf{j} & \mathbf{k} \\ 1 & 0 & 0 \\ 0 & 1 & 0 \end{vmatrix}$$

$$= ((0 \times 0) - (0 \times 1))\mathbf{i} + ((1 \times 0) - (0 \times 0))\mathbf{j} + ((1 \times)1 - (0 \times 0))\mathbf{k}$$

$$= \mathbf{k}$$

$\mathbf{j} \times \mathbf{k} = \mathbf{i}$ and $\mathbf{k} \times \mathbf{i} = \mathbf{j}$ are derived similarly.

The Direction and Magnitude of the Vector u × v

THEOREM A.3

Let **u** and **v** be vectors in $\mathbf{R}^3$. Then

$$\mathbf{u} \cdot (\mathbf{u} \times \mathbf{v}) = 0$$

$$\mathbf{v} \cdot (\mathbf{u} \times \mathbf{v}) = 0$$

This result tells us that the vector $(\mathbf{u} \times \mathbf{v})$ is orthogonal to both **u** and **v**. Therefore it is orthogonal to the space spanned by **u** and **v**; namely the plane defined by **u** and **v**. See Figure A3.

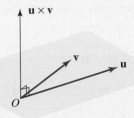

$\mathbf{u} \times \mathbf{v} \perp$ plane spanned by **u** and **v**

Figure A3

Proof

$$\mathbf{u} \cdot (\mathbf{u} \times \mathbf{v}) = [u_1\mathbf{i} + u_2\mathbf{j} + u_3\mathbf{k}] \cdot \begin{vmatrix} \mathbf{i} & \mathbf{j} & \mathbf{k} \\ u_1 & u_2 & u_3 \\ v_1 & v_2 & v_3 \end{vmatrix}$$

$$= \begin{vmatrix} u_1 & u_2 & u_3 \\ u_1 & u_2 & u_3 \\ v_1 & v_2 & v_3 \end{vmatrix} = 0$$

It can be proved similarly that $\mathbf{v} \cdot (\mathbf{u} \times \mathbf{v}) = 0$.

THEOREM A.4

Let **u** and **v** be vectors in $\mathbf{R}^3$. Then $\|\mathbf{u} \times \mathbf{v}\| = \|\mathbf{u}\| \, \|\mathbf{v}\| \sin \theta$, where θ is the angle between **u** and **v**.

Proof   Using the definition of magnitude of a vector we get

$$\|\mathbf{u} \times \mathbf{v}\|^2 = (u_2v_3 - u_3v_2)^2 + (u_3v_1 - u_1v_3)^2 + (u_1v_2 - u_2v_1)^2$$

On expanding the squares this can be rewritten

$$= (u_1^2 + u_2^2 + u_3^2)(v_1^2 + v_2^2 + v_3^2) - (u_1v_1 + u_2v_2 + u_3v_3)^2$$
$$= \|\mathbf{u}\|^2 \|\mathbf{v}\|^2 - (\mathbf{u} \cdot \mathbf{v})^2$$
$$= \|\mathbf{u}\|^2 \|\mathbf{v}\|^2 - (\|\mathbf{u}\| \|\mathbf{v}\| \cos\theta)^2$$
$$= \|\mathbf{u}\|^2 \|\mathbf{v}\|^2 - \|\mathbf{u}\|^2 \|\mathbf{v}\|^2 \cos^2\theta$$
$$= \|\mathbf{u}\|^2 \|\mathbf{v}\|^2 (1 - \cos^2\theta)$$
$$= \|\mathbf{u}\|^2 \|\mathbf{v}\|^2 \sin^2\theta$$

Thus

$$\|\mathbf{u} \times \mathbf{v}\| = \|\mathbf{u}\| \|\mathbf{v}\| \sin\theta$$

Area of a Triangle

The preceding result leads to the area of a triangle that is defined by two vectors. Consider the triangle whose edges are the vectors **u** and **v**. See Figure A4. We get

$$\text{Area of a triangle} = \left(\frac{1}{2}\right) \text{base} \times \text{height}$$

$$= \left(\frac{1}{2}\right) \|\mathbf{u}\| \|\mathbf{v}\| \sin\theta$$

$$= \left(\frac{1}{2}\right) \|\mathbf{u} \times \mathbf{v}\|$$

$$\text{Area of a triangle with edges } \mathbf{u} \text{ and } \mathbf{v} = \left(\tfrac{1}{2}\right) \|\mathbf{u} \times \mathbf{v}\|$$

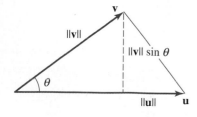

Length of base = ‖u‖
Height = ‖v‖ sin θ

Figure A4

EXAMPLE 3 Determine the area of the triangle having vertices $A(1, 1, 1)$, $B(-2, 3, 5)$, $C(1, 7, 2)$.

SOLUTION

We sketch this triangle in Figure A5. The points B and C define the following two edge vectors, starting from the point A.

$$\vec{AB} = (-2, 3, 5) - (1, 1, 1) = (-3, 2, 4)$$

$$\vec{AC} = (1, 7, 2) - (1, 1, 1) = (0, 6, 1)$$

$$\text{Thus, the area of a triangle} = \left(\frac{1}{2}\right) \| \vec{AB} \times \vec{AC} \|$$

$$= \left(\frac{1}{2}\right) \|(-3, 2, 4) \times (0, 6, 1)\|$$

$$= \left(\frac{1}{2}\right) \|(-22, 3, -18)\|$$

$$= \left(\frac{1}{2}\right) \sqrt{22^2 + 3^2 + 18^2}$$

$$= \left(\frac{1}{2}\right) \sqrt{817}$$

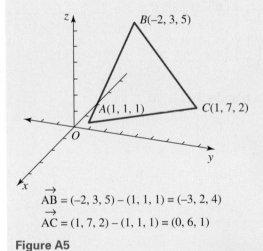

$$\vec{AB} = (-2, 3, 5) - (1, 1, 1) = (-3, 2, 4)$$
$$\vec{AC} = (1, 7, 2) - (1, 1, 1) = (0, 6, 1)$$

Figure A5

Volume of a Parallelepiped

Consider the parallelepiped whose adjacent edges are defined by the vectors **u**, **v**, and **w**. See Figure A6.

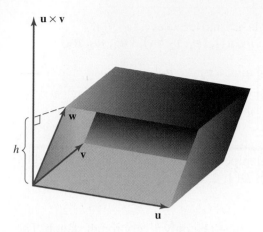

Figure A6

The area of the base is twice the area of the triangle defined by **u** and **v**. Thus the area of the base $= \|\mathbf{u} \times \mathbf{v}\|$. Further, volume $= \|\mathbf{u} \times \mathbf{v}\| \times h$, where h is the height.

Observe that:

$$h = \text{magnitude of projection of } \mathbf{w} \text{ onto } \mathbf{u} \times \mathbf{v}$$

$$= \text{magnitude of } \left(\frac{\mathbf{w} \cdot (\mathbf{u} \times \mathbf{v})}{(\mathbf{u} \times \mathbf{v})(\mathbf{u} \times \mathbf{v})}(\mathbf{u} \times \mathbf{v}) \right)$$

$$= \text{magnitude of } \left(\frac{\mathbf{w} \cdot (\mathbf{u} \times \mathbf{v})}{\|\mathbf{u} \times \mathbf{v}\|^2}(\mathbf{u} \times \mathbf{v}) \right)$$

$$= \frac{|\mathbf{w} \cdot (\mathbf{u} \times \mathbf{v})|}{\|\mathbf{u} \times \mathbf{v}\|}$$

Thus:

Volume of a parallelepiped with adjacent edges **u**, **v**, and **w**

$$= |\mathbf{w} \cdot (\mathbf{u} \times \mathbf{v})|$$

The expression $\mathbf{w} \cdot (\mathbf{u} \times \mathbf{v})$ is called the **triple scalar product** of **u**, **v**, and **w**. It can be conveniently written as a determinant. Let

$$\mathbf{u} = (u_1, u_2, u_3), \mathbf{v} = (v_1, v_2, v_3), \text{ and } \mathbf{w} = (w_1, w_2, w_3)$$

Then,

$$\mathbf{w} \cdot (\mathbf{u} \times \mathbf{v}) = [w_1\mathbf{i} + w_2\mathbf{j} + w_3\mathbf{k}] \cdot \begin{vmatrix} \mathbf{i} & \mathbf{j} & \mathbf{k} \\ u_1 & u_2 & u_3 \\ v_1 & v_2 & v_3 \end{vmatrix}$$

$$= \begin{vmatrix} w_1 & w_2 & w_3 \\ u_1 & u_2 & u_3 \\ v_1 & v_2 & v_3 \end{vmatrix} = \begin{vmatrix} u_1 & u_2 & u_3 \\ v_1 & v_2 & v_3 \\ w_1 & w_2 & w_3 \end{vmatrix}$$

Thus:

Volume of a parallelepiped with edges $(u_1, u_2, u_3), (v_1, v_2, v_3), (w_1, w_2, w_3)$

$$= \text{absolute value of} \begin{vmatrix} u_1 & u_2 & u_3 \\ v_1 & v_2 & v_3 \\ w_1 & w_2 & w_3 \end{vmatrix}$$

EXAMPLE 4 Find the volume of the parallelepiped having adjacent edges defined by the points $A(1, 1, 3)$, $B(3, 7, 1)$, $C(-2, 3, 3)$, $D(1, 2, 8)$.

SOLUTION

We sketch the parallelepiped in Figure A7. This need be only a very rough sketch. The points A, B, C, and D define the following three adjacent edge vectors,

$$\overrightarrow{AB} = (3, 7, 1) - (1, 1, 3) = (2, 6, -2)$$

$$\overrightarrow{AC} = (-2, 3, 3) - (1, 1, 3) = (-3, 2, 0)$$

$$\overrightarrow{AD} = (1, 2, 8) - (1, 1, 3) = (0, 1, 5)$$

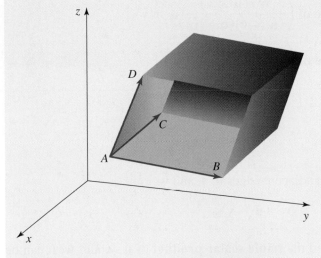

Figure A7

The volume of the parallelepiped is thus

$$= \text{absolute value of} \begin{vmatrix} 2 & 6 & -2 \\ -3 & 2 & 0 \\ 0 & 1 & 5 \end{vmatrix}$$

$$= \text{absolute value of } (116)$$

$$= 116$$

We have discussed numerous operations on vectors in this course. Addition, scalar multiplication, and dot product have been defined. In this section we have introduced yet another operation, namely the cross product. We now conveniently summarize the properties of the cross product. The summary includes properties already derived and further properties that tell us how the cross product interacts with addition, scalar multiplication, and dot product. The reader is asked to derive the new properties in the exercises that follow.

Properties of the Cross Product

Let $\mathbf{u}$, $\mathbf{v}$, and $\mathbf{w}$ be vectors in $\mathbf{R}^3$ and let c be a scalar. Then

(a) $\mathbf{u} \times \mathbf{u} = 0$

(b) $\mathbf{u} \times \mathbf{v} = -(\mathbf{u} \times \mathbf{v})$

(c) $\mathbf{u} \cdot (\mathbf{u} \times \mathbf{v}) = 0$

(d) $\mathbf{v} \cdot (\mathbf{u} \times \mathbf{v}) = 0$

(e) $\mathbf{u} \times 0 = 0 \times \mathbf{u} = 0$

(f) $\mathbf{u} \cdot (\mathbf{v} \times \mathbf{w}) = (\mathbf{u} \times \mathbf{v}) \cdot \mathbf{w}$

(g) $\mathbf{u} \times (\mathbf{v} + \mathbf{w}) = (\mathbf{u} \times \mathbf{v}) + (\mathbf{u} \times \mathbf{w})$

(h) $c(\mathbf{u} \times \mathbf{v}) = c\mathbf{u} \times \mathbf{v} = \mathbf{u} \times c\mathbf{v}$

If $\mathbf{i} = (1, 0, 0), \mathbf{j} = (0, 1, 0), \mathbf{k} = (0, 0, 1)$, then,

(i) $\mathbf{i} \times \mathbf{i} = 0, \mathbf{j} \times \mathbf{j} = 0, \mathbf{k} \times \mathbf{k} = 0$

(j) $\mathbf{i} \times \mathbf{j} = \mathbf{k}, \mathbf{j} \times \mathbf{k} = \mathbf{i}, \mathbf{k} \times \mathbf{i} = \mathbf{j}$

(k) $\mathbf{i} \times \mathbf{j} = -(\mathbf{j} \times \mathbf{i}), \mathbf{i} \times \mathbf{k} = -(\mathbf{k} \times \mathbf{i}), \mathbf{j} \times \mathbf{k} = -(\mathbf{k} \times \mathbf{j})$

EXERCISE SET APPENDIX A

Cross Product

1. If $\mathbf{u} = (1, 2, 3), \mathbf{v} = (-1, 0, 4),$ and $\mathbf{w} = (1, 2, -1),$ compute the following using the definition of the cross product.

(a) $\mathbf{u} \times \mathbf{v}$　　　　(b) $\mathbf{v} \times \mathbf{u}$

(c) $\mathbf{u} \times \mathbf{w}$　　　　(d) $\mathbf{v} \times \mathbf{w}$

(e) $(\mathbf{u} \times \mathbf{v}) \times \mathbf{w}$

2. If $\mathbf{u} = (-2, 2, 4), \mathbf{v} = (3, 0, 5),$ and $\mathbf{w} = (4, -2, 1),$ compute the following using the determinant form of the cross product.

(a) $\mathbf{u} \times \mathbf{v}$　　　　(b) $\mathbf{u} \times \mathbf{w}$

(c) $\mathbf{w} \times \mathbf{v}$　　　　(d) $\mathbf{v} \times \mathbf{w}$

(e) $(\mathbf{w} \times \mathbf{v}) \times \mathbf{u}$

3. If $\mathbf{u} = 2\mathbf{i} + 3\mathbf{j} + \mathbf{k}, \mathbf{v} = -\mathbf{i} + 2\mathbf{j} + 4\mathbf{k},$ and $\mathbf{w} = 3\mathbf{i} - 7\mathbf{k},$ compute the following cross products.

(a) $\mathbf{u} \times \mathbf{v}$　　　　(b) $\mathbf{u} \times \mathbf{w}$

(c) $\mathbf{w} \times \mathbf{u}$　　　　(d) $\mathbf{v} \times \mathbf{w}$

(e) $(\mathbf{w} \times \mathbf{u}) \times \mathbf{v}$

4. If $\mathbf{u} = (3, 1, -2), \mathbf{v} = (4, -1, 2),$ and $\mathbf{w} = (0, 3, -2),$ compute the following.

(a) $\mathbf{u} \times \mathbf{v}$　　　　(b) $3\mathbf{v} \times 2\mathbf{w}$

(c) $(\mathbf{w} \times \mathbf{u}) \cdot \mathbf{v}$　　　　(d) $(\mathbf{w} + 2\mathbf{u}) \times \mathbf{v}$

(e) $\mathbf{u} \cdot (\mathbf{v} \times \mathbf{w})$　　　　(f) $(\mathbf{v} \times \mathbf{u}) \cdot (\mathbf{w} \times \mathbf{v})$

Areas and Volumes

5. Determine the areas of the triangles having the following vertices.

(a) $A(1, 2, 1), B(-3, 4, 6), C(1, 8, 3)$

(b) $A(3, -1, 2), B(0, 2, 6), C(7, 1, 5)$

(c) $A(1, 0, 0), B(0, 5, 2), C(3, -4, 8)$

6. Find the volumes of the parallelepipeds having adjacent edges defined by the following points.

(a) $A(1, 2, 5), B(4, 8, 1), C(-3, 2, 3), D(0, 3, 9)$

(b) $A(3, 1, 6), B(-2, 3, 4,), C(0, 2, -5), D(3, -1, 4)$

(c) $A(0, 1, 2), B(-3, 1, 4), C(5, 2, 3), D(-3, -2, 1)$

Miscellaneous Results

7. Prove the following.

   (a) $\mathbf{i} \cdot \mathbf{i} = 1, \mathbf{j} \cdot \mathbf{j} = 1, \mathbf{k} \cdot \mathbf{k} = 1$

   (b) $\mathbf{i} \cdot \mathbf{j} = 0, \mathbf{i} \cdot \mathbf{k} = 0, \mathbf{j} \cdot \mathbf{k} = 0$

8. Prove that $(\mathbf{i} \times \mathbf{j}) \cdot \mathbf{k} = 1$.

9. Let $\mathbf{u}$ be a vector in $\mathbf{R}^3$. Prove that $\mathbf{u} \times \mathbf{0} = \mathbf{0} \times \mathbf{u} = \mathbf{0}$.

10. Let $\mathbf{u}$, $\mathbf{v}$, and $\mathbf{w}$ be vectors in $\mathbf{R}^3$. Prove that

$$\mathbf{u} \cdot (\mathbf{v} \times \mathbf{w}) = (\mathbf{u} \times \mathbf{v}) \cdot \mathbf{w}$$

11. Let $\mathbf{u}$, $\mathbf{v}$, and $\mathbf{w}$ be vectors in $\mathbf{R}^3$. Prove that

$$\mathbf{u} \times (\mathbf{v} \times \mathbf{w}) = (\mathbf{u} \cdot \mathbf{w})\mathbf{v} - (\mathbf{u} \cdot \mathbf{v})\mathbf{w}$$

12. Let $\mathbf{u}$ and $\mathbf{v}$ be vectors in $\mathbf{R}^3$ and let c be a scalar. Prove that

$$c(\mathbf{u} \times \mathbf{v}) = c\mathbf{u} \times \mathbf{v} = \mathbf{u} \times c\mathbf{v}$$

13. Let $\mathbf{u}$ and $\mathbf{v}$ be nonzero vectors in $\mathbf{R}^3$. Prove that $\mathbf{u}$ and $\mathbf{v}$ are parallel if and only if $\mathbf{u} \times \mathbf{v} = \mathbf{0}$.

14. Let $\mathbf{u}$, $\mathbf{v}$, and $\mathbf{w}$ be vectors in $\mathbf{R}^3$. Prove that

$$\mathbf{u} \times (\mathbf{v} \times \mathbf{w}) + \mathbf{v} \times (\mathbf{w} \times \mathbf{u}) + \mathbf{w} \times (\mathbf{u} \times \mathbf{v}) = \mathbf{0}$$

15. Prove that the position vectors $\mathbf{u}$, $\mathbf{v}$, and $\mathbf{w}$ of $\mathbf{R}^3$ all lie in a plane if and only if $\mathbf{u} \cdot (\mathbf{v} \times \mathbf{w}) = \mathbf{0}$.

16. Let $\mathbf{t}$, $\mathbf{u}$, $\mathbf{v}$, and $\mathbf{w}$ be vectors in $\mathbf{R}^3$. Prove that

$$(\mathbf{t} \times \mathbf{u}) \cdot (\mathbf{v} \times \mathbf{w}) = \begin{vmatrix} \mathbf{t} \cdot \mathbf{v} & \mathbf{u} \cdot \mathbf{v} \\ \mathbf{t} \cdot \mathbf{w} & \mathbf{u} \cdot \mathbf{w} \end{vmatrix}$$

Equations of Planes and Lines in Three-Space*

The tools that we develop in this section are useful for discussing planes and lines in three-space. We first look at planes.

Planes in $\mathbf{R}^3$

Let $P_0(x_0, y_0, z_0)$ be a point in a plane. Let (a, b, c) be a vector perpendicular to the plane, called a **normal** to the plane. See Figure B1. These two quantities, namely a point in a plane and a normal vector to the plane, characterize the plane. There is only one plane through a given point having a given normal. We shall now derive the equation of a plane passing through the point $P_0(x_0, y_0, z_0)$ having normal (a, b, c).

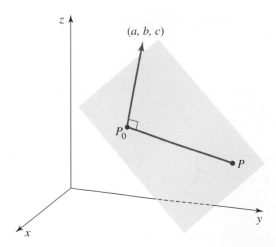

Figure B1

Let $P(x, y, z)$ be an arbitrary point in the plane. We get

$$\overrightarrow{P_0P} = (x, y, z) - (x_0, y_0, z_0)$$
$$= (x - x_0, y - y_0, z - z_0)$$

---

*Some courses cover this material in the calculus sequence.

The vector $\overrightarrow{P_0P}$ lies in the plane. Thus the vectors (a, b, c) and $\overrightarrow{P_0P}$ are orthogonal. Their dot product is zero. This observation leads to an equation for the plane.

$$(a, b, c) \cdot \overrightarrow{P_0P} = 0$$
$$(a, b, c) \cdot (x - x_0, y - y_0, z - z_0) = 0$$
$$a(x - x_0) + b(y - y_0) + c(z - z_0) = 0$$

This is called the **point-normal form** of the equation of the plane.

Let us now rewrite the equation. Multiplying out we get

$$ax - ax_0 + by - by_0 + cz - cz_0 = 0$$
$$ax + by + cz - ax_0 - by_0 - cz_0 = 0$$

The last three terms are constant. Combine them into a single constant d. We get

$$ax + by + cz + d = 0$$

This is called the **general form** of the equation of the plane.

Point-normal equation of a plane: $a(x - x_0) + b(y - y_0) + c(z - z_0) = 0$

General form equation of a plane: $ax + by + cz + d = 0$

EXAMPLE 1 Find the point-normal and general forms of the equations of the plane passing through the point $(1, 2, 3)$ having normal $(-1, 4, 6)$.

SOLUTION

Let $(x_0, y_0, z_0) = (1, 2, 3)$ and $(a, b, c) = (-1, 4, 6)$. The point-normal form is

$$-1(x - 1) + 4(y - 2) + 6(z - 3) = 0$$

Multiplying out and simplifying,

$$-x + 1 + 4y - 8 + 6z - 18 = 0$$

The general form is

$$-x + 4y + 6z - 25 = 0$$

EXAMPLE 2 Determine the equation of the plane through the three points $P_1(2, -1, 1)$, $P_2(-1, 1, 3)$, and $P_3(2, 0, -3)$.

SOLUTION

The vectors $\overrightarrow{P_1P_2}$ and $\overrightarrow{P_1P_3}$ lie in the plane. Thus $\overrightarrow{P_1P_2} \times \overrightarrow{P_1P_3}$ will be normal to the plane. We get

$$\overrightarrow{P_1P_2} = (-1, 1, 3) - (2, -1, 1) = (-3, 2, 2)$$
$$\overrightarrow{P_1P_3} = (2, 0, -3) - (2, -1, 1) = (0, 1, -4)$$

Thus

$$\overrightarrow{P_1P_3} \times \overrightarrow{P_1P_3} = \begin{vmatrix} \mathbf{i} & \mathbf{j} & \mathbf{k} \\ -3 & 2 & 2 \\ 0 & 1 & -4 \end{vmatrix}$$
$$= (-8 - 2)\mathbf{i} - (12 - 0)\mathbf{j} + (-3 - 0)\mathbf{k}$$
$$= -10\mathbf{i} - 12\mathbf{j} - 3\mathbf{k}$$

Let $(x_0, y_0, z_0) = (2, -1, 1)$ and $(a, b, c) = (-10, -12, -3)$. The point-normal form gives

$$-10(x - 2) - 12(y + 1) - 3(z - 1) = 0$$
$$-10x + 20 - 12y - 12 - 3z + 3 = 0$$

The equation of the plane is

$$-10x - 12y - 3z + 11 = 0$$

Observe that each of the given points $(2, -1, 1)$, $(-1, 1, 3)$, and $(2, 0, -3)$ satisfies this equation.

■

EXAMPLE 3 Prove that the following system of equations has no solution.

$$2x - y + 3z = 6$$
$$-4x + 2y - 6z = 2$$
$$x + y + 4z = 5$$

SOLUTION

Consider the general form of the equation of a plane

$$ax + by + cz + d = 0$$

The vector (a, b, c) is normal to this plane. Interpret each of the given equations as defining a plane in $\mathbf{R}^3$. On comparison with the general form it is seen that the following vectors are normals to these three planes.

$$(2, -1, 3), (-4, 2, -6), \text{ and } (1, 1, 4)$$

Observe that

$$(-4, 2, -6) = -2(2, -1, 3)$$

The normals to the first two planes are parallel. Thus these two planes are parallel, and they are distinct. The three planes can therefore have no points in common. The system of equations has no solution.

■

Lines in $\mathbf{R}^3$

Consider a line through the point $P_0(x_0, y_0, z_0)$ in the direction defined by the vector (a, b, c). See Figure B2. Let $P(x, y, z)$ be any other point on the line. We get

$$\overrightarrow{P_0P} = (x - x_0, y - y_0, z - z_0)$$

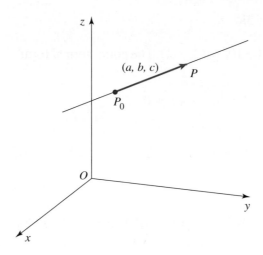

Figure B2

The vectors $\overrightarrow{P_0P}$ and (a, b, c) are parallel. Thus there exists a scalar t such that

$$\overrightarrow{P_0P} = t(a, b, c)$$

Equating the two expressions for $\overrightarrow{P_0P}$ we get

$$(x - x_0, y - y_0, z - z_0) = t(a, b, c)$$

This is called the **vector equation** of the line. Comparing the components of the vectors on the left and right of this equation gives

$$x - x_0 = ta, y - y_0 = tb, z - z_0 = tc.$$

Rearranging these equations as follows gives the **parametric equations** of a line in $\mathbf{R}^3$.

Parametric equations of a line

$$x = x_0 + ta, y = y_0 + tb, z = z_0 + tc, \qquad -\infty < t < \infty$$

As t varies we get the points on the line.

EXAMPLE 4 Find the parametric equations of the line through the point $(1, 2, 5)$ in the direction $(4, 3, 1)$. Determine any two points on the line.

SOLUTION

Let $(x_0, y_0, z_0) = (1, 2, 5)$ and $(a, b, c) = (4, 3, 1)$. The parametric equations of the line are

$$x = 1 + 4t, \, y = 2 + 3t, \, z = 5 + t, \, -\infty < t < \infty$$

By letting t take on two convenient values we can determine two other points on the line. For example, $t = 1$ leads to the point $(5, 5, 6)$ and $t = -2$ leads to the point $(-7, -4, 3)$.

∎

EXAMPLE 5 Find the parametric equations of the line through the points $(-1, 2, 6)$ and $(1, 5, 4)$.

SOLUTION

Let (x_0, y_0, z_0) be the point $(-1, 2, 6)$. The direction of the line is given by the vector

$$(a, b, c) = (1, 5, 4) - (-1, 2, 6) = (2, 3, -2)$$

The parametric equations of the line are

$$x = -1 + 2t, \, y = 2 + 3t, \, z = 6 - 2t, \, -\infty < t < \infty$$

∎

The parametric equations of a line for which a, b, and c are all nonzero lead to another useful way of expressing the line, called the **symmetric equations** of the line. Isolating the variable t in each of the three parametric equations we get

$$t = \frac{x - x_0}{a} \quad t = \frac{y - y_0}{b} \quad t = \frac{z - z_0}{c}$$

Equating these expressions for t, we get the symmetric equations.

Symmetric equations of a line

$$\frac{x - x_0}{a} = \frac{y - y_0}{b} = \frac{z - z_0}{c}$$

In this form the line can be interpreted as the intersection of the planes

$$\frac{x - x_0}{a} = \frac{y - y_0}{b} \quad \text{and} \quad \frac{y - y_0}{b} = \frac{z - z_0}{c}$$

EXAMPLE 6 Determine the symmetric equations of the line through the points $(-4, 2, -6)$ and $(1, 4, 3)$.

SOLUTION

The direction of the line is given by the vector

$$(a, b, c) = (1, 4, 3) - (-4, 2, -6) = (5, 2, 9)$$

Let (x_0, y_0, z_0) be the point $(-4, 2, -6)$. (Either point can be used.) The symmetric equations of the line are

$$\frac{x - (-4)}{5} = \frac{y - 2}{2} = \frac{z - (-6)}{9}$$

Write these equations in the form

$$\frac{x + 4}{5} = \frac{y - 2}{2} \quad \text{and} \quad \frac{y - 2}{2} = \frac{z + 6}{9}$$

Cross multiply both of these equations and simplify.

$$2(x + 4) = 5(y - 2) \text{ and } 9(y - 2) = 2(z + 6)$$

$$2x - 5y + 18 = 0 \text{ and } 9y - 2z - 30 = 0$$

The line through the points $(-4, 2, -6)$ and $(1, 4, 3)$ is thus the intersection of the planes

$$2x - 5y + 18 = 0 \text{ and } 9y - 2z - 30 = 0$$

EXERCISE SET APPENDIX B

Planes in R³

1. Determine the point-normal form and the general form of the equations of the plane through each of the following points, having the given normals.

   (a) point $(1, -2, 4)$; normal $(1, 1, 1)$

   (b) point $(-3, 5, 6)$; normal $(-2, 4, 5)$

   (c) point $(0, 0, 0)$; normal $(1, 2, 3)$

   (d) point $(4, 5, -2)$; normal $(-1, 4, 3)$

2. Determine a general form of the equation of the plane through each of the following sets of three points.

   (a) $P_1(1, -2, 3)$, $P_2(1, 0, 2)$, and $P_3(-1, 4, 6)$

   (b) $P_1(0, 0, 0)$, $P_2(1, 2, 4)$, and $P_3(-3, 5, 1)$

   (c) $P_1(-1, -1, 2)$, $P_2(3, 5, 4)$, and $P_3(1, 2, 5)$

   (d) $P_1(7, 1, 3)$, $P_2(-2, 4, -3)$, and $P_3(5, 4, 1)$

3. Prove that the planes $3x - 2y + 4z - 3 = 0$ and $-6x + 4y - 8z + 7 = 0$ are parallel.

4. Use the general form of the equation of a plane to prove that the following system of equations has no solutions.

$$x + y - 3z = 7$$
$$3x - 6y + 9z = 6$$
$$-x + 2y - 3z = 2$$

5. Find the equation of the plane parallel to the plane $2x - 3y + z + 4 = 0$, passing through the point $(1, 2, -3)$.

Lines in R³

6. Determine parametric equations and symmetric equations of the lines through the following points in the given directions.

   (a) point $(1, 2, 3)$; direction $(-1, 2, 4)$

   (b) point $(-3, 1, 2)$; direction $(1, 1, 1)$

   (c) point $(0, 0, 0)$; direction $(-2, -3, 5)$

   (d) point $(-2, -4, 1)$; direction $(2, -2, 4)$

7. Find the equation of the line through the point $(1, 2, -4)$, parallel to the line $x = 4 + 2t$, $y = -1 + 3t$, $z = 2 + t$, where $-\infty < t < \infty$.

8. Find the equation of the line through the point $(2, -3, 1)$ in a direction orthogonal to the line

$$\frac{x + 1}{3} = \frac{y - 1}{2} = \frac{z + 2}{5}$$

9. Determine the equation of the line through the point $(4, -1, 3)$ perpendicular to the plane

$$2x - y + 4z + 7 = 0.$$

Miscellaneous Results

10. Give general forms for the equations of the xy-plane, xz-plane, and yz-plane.

11. Show that there are many planes that contain the three points $P_1(3, -5, 5)$, $P_2(-1, 1, 3)$, and $P_3(5, -8, 6)$. Interpret your conclusion geometrically.

12. Find an equation for the plane through the point $(4, -1, 5)$, in a direction perpendicular to the line $x = 1 - t$, $y = 3 + 2t$, $z = 5 - 4t$, where $-\infty < t < \infty$.

13. Show that the line $x = 1 + t$, $y = 14 - t$, $z = 2 - t$, where $-\infty < t < \infty$, lies in the plane

$$2x - y + 3z + 6 = 0.$$

14. Prove that the line $x = 4 + 2t$, $y = 5 + t$, $z = 7 + 2t$, where $-\infty < t < \infty$, never intersects the plane $3x + 2y - 4z + 7 = 0$.

15. Find an equation of the line through the point $(5, -1, 2)$ in a direction perpendicular to the line $x = 5 - 2t$, $y = 2 + 3t$, $z = 2t$, where $-\infty < t < \infty$.

16. Prove that the lines $x = -1 - 4t$, $y = 4 + 4t$, $z = 7 + 8t$, where $-\infty < t < \infty$, and $x = 4 + 3h$, $y = 1 - h$, $z = 5 + 2h$, where $-\infty < h < \infty$, intersect at right angles. Find the point of intersection.

17. Two lines are said to be **skew** if they are not parallel and do not intersect. Show that the following two lines are skew.

$$x = 1 + 4t, \ y = 2 + 5t, \ z = 3 + 3t,$$

where $-\infty < t < \infty$

$$x = 2 + h, \ y = 1 - 3h, \ z = -1 - 2h,$$

where $-\infty < h < \infty$

18.(a) Show that the set of all points on the plane $2x + 3y - 4z = 0$ forms a subspace of $\mathbf{R}^3$. Find a basis for this subspace.

   (b) Show that the set of all points on the plane $3x - 4y + 2z - 6 = 0$ does not form a subspace of $\mathbf{R}^3$.

   (c) Prove that the set of all points on the plane $ax + by + cz + d = 0$ forms a subspace of $\mathbf{R}^3$ if and only if $d = 0$.

19. Prove that a line can be written in symmetric form if and only if it is not orthogonal to the x-axis, y-axis, or z-axis.

Graphing Calculator Manual

Introduction

This graphing calculator discussion is included for readers who plan to use a calculator to perform matrix operations. Keystrokes are given for the TI-83 and TI-83 Plus calculators. The primary aim, however, is to convey concepts—to illustrate when and how to use a calculator in linear algebra. The reader should find out how to perform each operation on his/her own calculator. Practice problems have been included to reinforce ideas.

C1. Reduced Echelon Form of a Matrix

Find the reduced echelon form of

$$A = \begin{bmatrix} 1 & -2 & 4 & 12 \\ 2 & -1 & 5 & 18 \\ -1 & 3 & -3 & -8 \end{bmatrix}$$

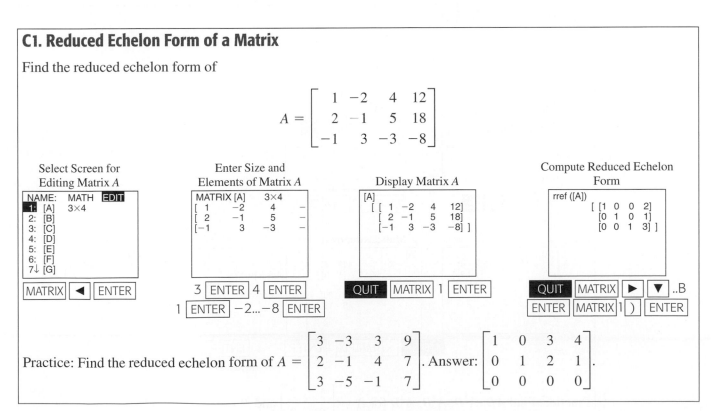

Select Screen for Editing Matrix A

Enter Size and Elements of Matrix A

Display Matrix A

Compute Reduced Echelon Form

Practice: Find the reduced echelon form of $A = \begin{bmatrix} 3 & -3 & 3 & 9 \\ 2 & -1 & 4 & 7 \\ 3 & -5 & -1 & 7 \end{bmatrix}$. Answer: $\begin{bmatrix} 1 & 0 & 3 & 4 \\ 0 & 1 & 2 & 1 \\ 0 & 0 & 0 & 0 \end{bmatrix}$.

C2. Matrix Operations

Find $A + B$, AB, and $6A$ for $A = \begin{bmatrix} 2 & 4 \\ 1 & 3 \end{bmatrix}$ and $B = \begin{bmatrix} 5 & -2 \\ 0 & 3 \end{bmatrix}$.

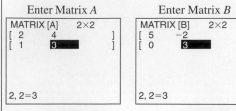

Enter Matrix A	Enter Matrix B	$A + B$ and AB	$6A$

Edit A QUIT	Edit B QUIT	MATRIX 1 + MATRIX 2 ENTER	6 MATRIX 1 ENTER
		MATRIX 1 × MATRIX 2 ENTER	

Practice: Find $A(B + 3C)$ for $A = \begin{bmatrix} 0 & -1 \\ 2 & 5 \end{bmatrix}$, $B = \begin{bmatrix} 2 & 4 & 1 \\ 3 & 0 & 2 \end{bmatrix}$, and $C = \begin{bmatrix} -7 & 3 & -2 \\ 0 & -4 & 1 \end{bmatrix}$.

Answer: $\begin{bmatrix} -3 & 12 & -5 \\ -23 & -34 & 15 \end{bmatrix}$.

C3. Powers of a Matrix

Find (i) A^4 (ii) A^2, A^3, A^4 for $A = \begin{bmatrix} 2 & 4 \\ 1 & 3 \end{bmatrix}$.

(i)

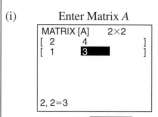   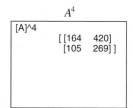

Edit A QUIT	MATRIX 1 ^ 4 ENTER

(ii)

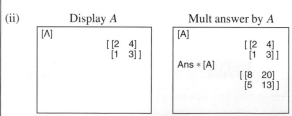

Display A	Mult answer by A	A^3	A^4

	ANS × MATRIX 1 ENTER	ENTER	ENTER

(Repeated use of ENTER gives further powers.)

Practice: Find AB, A^2B, A^3B, A^4B for $A = \begin{bmatrix} 1 & 3 \\ 0 & -2 \end{bmatrix}$ and $B = \begin{bmatrix} 4 \\ 1 \end{bmatrix}$.
(Use A*Ans repeatedly)

Answer: $AB = \begin{bmatrix} 7 \\ -2 \end{bmatrix}$, $A^2B = \begin{bmatrix} 1 \\ 4 \end{bmatrix}$, $A^3B = \begin{bmatrix} 13 \\ -8 \end{bmatrix}$, $A^4B = \begin{bmatrix} -11 \\ 16 \end{bmatrix}$

C4. Transpose of a Matrix

Find the transpose of $A = \begin{bmatrix} 1 & -1 & -2 \\ 2 & -3 & -5 \\ -1 & 3 & 5 \end{bmatrix}$. Show that AA^t is symmetric.

Enter Matrix A A^t Compute AA^t

```
MATRIX [A]    3×3
[ 1      -1    -2      ]
[ 2      -3    -5      ]
[-1       3     5      ]



3, 3=5
```

```
[A]ᵗ
   [[ 1     2    -1]
    [-1    -3     3]
    [-2    -5     5]]
```

```
[A] * [A]ᵗ
   [[  6    15   -14]
    [ 15    38   -36]
    [-14   -36    35]]
```

Edit A QUIT   MATRIX 1 MATRIX ▶ 2 ENTER   Symmetric

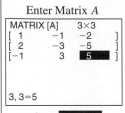

C5. Inverse of a Matrix

Find the inverse of $A = \begin{bmatrix} 1 & -1 & -2 \\ 2 & -3 & -5 \\ -1 & 3 & 5 \end{bmatrix}$. Let $B = \begin{bmatrix} 5 \\ 3 \\ 7 \end{bmatrix}$.

Solve the system $AX = B$.

Enter Matrix A A^{-1} Enter B and compute $A^{-1}B$

```
MATRIX [A]    3×3
[ 1      -1    -2      ]
[ 2      -3    -5      ]
[-1       3     5      ]



3, 3=5
```

```
[A]⁻¹
   [[  0     1     1]
    [  5    -3    -1]
    [ -3     2     1]]
```

```
[A]⁻¹ * [B]
                [[ 10]
                 [  9]
                 [ -2]]
```

Edit A QUIT   MATRIX 1 x⁻¹ ENTER   Get x = 10, y = 9, z = −2.

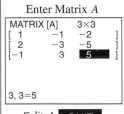

C6. Determinant of a Matrix

Find the determinant of $A = \begin{bmatrix} 1 & 2 & -1 \\ 3 & 0 & 1 \\ 4 & 2 & 1 \end{bmatrix}$.

Enter Matrix A $|A|$

```
MATRIX [A]    3×3
[1       2    -1       ]
[3       0     1       ]
[4       2     1       ]



3, 3=1
```

```
det ([A])
                      -6
```

Edit A QUIT   MATRIX ▶ 1 MATRIX 1) ENTER

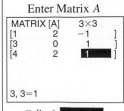
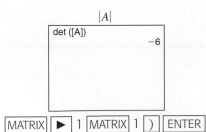

Practice: Show that $A = \begin{bmatrix} 1 & 3 & -3 \\ 3 & 2 & 0 \\ 4 & 5 & -3 \end{bmatrix}$ is singular.

C7. Summary of Formats for Row Operations

Press $\boxed{\text{MATRIX}}$ $\boxed{\blacktriangleright}$ to get the Matrix Math Menu.

Scroll and select C, D, E, F for the operation.

C: Interchange two rows
   rowSwap([A], 1st row, 2nd row)

D: Add a row to another row
   row+([A], row to be added, row to be added to)

E: Multiply a row by a number
   *row(number,[A], row to be multiplied)

F: Add a multiple of a row to another row
   *row(number,[A], row to be multiplied, row to be added to)

Example: Perform the operation $\begin{bmatrix} 1 & -2 & 4 & 12 \\ 2 & -1 & 5 & 18 \\ -1 & 3 & -3 & -8 \end{bmatrix}$ $\underset{R2+(-2)R1}{\approx}$ $\begin{bmatrix} 1 & -2 & 4 & 12 \\ 0 & 3 & -3 & -6 \\ -1 & 3 & -3 & -8 \end{bmatrix}$

(Note: The calculator essentially uses the convention $-2R1 + R2$.)

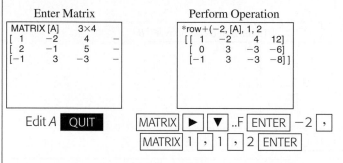

Enter Matrix

```
MATRIX [A]      3×4
[ 1      -2      4      -
[ 2      -1      5      -
[-1       3     -3      -
```

          Edit A ▮QUIT▮

Perform Operation

```
*row+(-2, [A], 1, 2
[[ 1     -2      4     12]
 [ 0      3     -3     -6]
 [-1      3     -3     -8]]
```

$\boxed{\text{MATRIX}}$ $\boxed{\blacktriangleright}$ $\boxed{\blacktriangledown}$ ..F $\boxed{\text{ENTER}}$ -2 $\boxed{,}$
$\boxed{\text{MATRIX}}$ 1 $\boxed{,}$ 1 $\boxed{,}$ 2 $\boxed{\text{ENTER}}$

A sequence of row operations can be performed by using ANS for the matrix A.

MATLAB Manual

Contents

Introduction

A computer package can be used in the introductory linear algebra course to

- relieve the student of the drudgery of doing arithmetic, so that he or she is free to concentrate on concepts.
- provide the student with a tool for learning mathematics by investigating ideas and exploring patterns.
- introduce the student to some of the benefits and pitfalls of working with computers.
- provide a tool for implementing mathematical models.

MATLAB is the most widely used software for working with matrices. Many schools are now using this software and numerous summer workshops have been given on its integration into the introductory linear algebra course, hence its selection for this text. The sections are presented here in the order that the topics appear in the text. The instructor can integrate the material into the course in any way he or she desires.

The MATLAB discussions are written in computer manual style. Explanations are brief and to the point. A brief summary of each topic is included, and it is shown how to use MATLAB for that topic. The discussions include exercises. *The exercises should be of general interest as many can be implemented on any matrix algebra software package or on a graphing calculator.* Some exercises are open-ended and encourage exploration on the part of the student. Many of the assignments can be projects and students can work on them in groups.

MATLAB has built-in functions such as inv(A) for finding the inverse of a matrix. It is also possible to write programs (called M-files) in the MATLAB language to tap these built-in functions and make use of MATLAB matrix input/output. The programs that are included with this text were written by Lisa Coulter and myself.

My co-workers in software development in linear algebra have been Lisa Coulter, my wife Donna, and my son Jeff. I wish to thank them all for their important contributions. I thank Cleve Moler of MATLAB for the support that he has given me, and the National Science Foundation for a curriculum development grant that supported the development of the original MATLAB package.

MATLAB can be obtained from The MATH WORKS Inc., 3 Apple Hill Drive, Natick, MA 01760-2098. Phone: (508) 647-7000. Fax: (508) 647-7001. E-mail: info@mathworks.com. A student version is available from Prentice Hall. Programs written in the MATLAB language to accompany this book (*The Linear Algebra with Applications Toolbox M-Files*) can be downloaded from www.stetson.edu/~gwilliam/mfiles.htm.

D.1 Entering and Displaying a Matrix (Section 1.1)

The MATLAB prompt is »
To enter a matrix:

   (i) start with [

  (ii) separate elements of the matrix with a space (or a comma)

 (iii) use ; to mark the end of each row

 (iv) end the matrix with]

EXAMPLE 1 Enter the matrix $A = \begin{bmatrix} 3 & 5 \\ 1 & 2 \end{bmatrix}$ into the workspace.

»A = [3 5;1 2]   {Type thus, followed by Enter (or return) key}
A =
3 5   {Output}
1 2   {$A = [3\ 5;1\ 2]$; i.e., with semicolon suppresses this output}

■

EXAMPLE 2 A matrix that is in the workspace can be displayed at any time. For example, the above matrix A is in the workspace.

»A   {Type A , followed by Enter (or return) key to display the matrix}
A =
3 5
1 2

■

EXERCISE 1

Enter and display the following matrices in MATLAB.

(a) $\begin{bmatrix} 5 & 4 \\ 2 & 9 \end{bmatrix}$ (b) $\begin{bmatrix} 2 & 4 & 1 \\ 5 & 0 & -3 \end{bmatrix}$ (c) $\begin{bmatrix} 9 \\ 3 \\ 2 \\ 0 \end{bmatrix}$ (d) $\begin{bmatrix} -2 & 3 & 7 & 1 \end{bmatrix}$

D.2 Solving Systems of Linear Equations (Sections 1.1–1.3)

Gauss-Jordan Elimination

MATLAB contains functions for working with matrices. New functions can also be written in the MATLAB language. The function **Gjelim** has been written for Gauss-Jordan elimination. It includes a "rational number" option (the default mode is decimal number), a "count of arithmetic operations" option, and an "all steps" option that displays all the steps. **Gjelim** and other MATLAB programs are found in The Linear Algebra Toolbox available from http://www.stetson.edu/~gwilliam/mfiles.htm.

EXAMPLE 1 Solve the following system using Gauss-Jordan elimination in all steps mode.

$$x - 2y + 4z = 12$$
$$2x - y + 5z = 18$$
$$-x + 3y - 3z = 8$$

»A = [1 −2 4 12;2 −1 5 18;   {enter the augmented matrix}
−1 3 −3 −8];
»Gjelim(A)   {author-defined function for GJelim for A}
Rational numbers? y/n: n   {option of rational arithmetic}

Count of operations? y/n: n   {option of keeping count of arith ops}
All steps? y/n: y   {option of displaying all the steps}
initial matrix   {start of output}
   1 −2   4 12
   2 −1   5 18
−1   3 −3 −8
[press return at each step to continue]
create zero   **- reduced echelon form -**
   1 −2   4 12   1 0 0 2
   0   3 −3 −6   ...   0 1 0 1   {end of output}
−1   3 −3 −8   0 0 1 3

We see that the system has the unique solution $x = 2, y = 1, z = 3$.

Selecting Your Own Elementary Transformations

The function **Transf** (for transform) can be used to perform your own sequence of elementary row transformations on matrices—the computer does the arithmetic for you.

EXAMPLE **2** Solve the following system of equations using Gauss-Jordan elimination using **transf**—you are given the opportunity to select the transformations at each step.

$$x - 2y + 4z = 12$$
$$2x - y + 5z = 18$$
$$-x + 3y - 3z = -8$$

»**A = [1 −2 4 12;2 −1 5 18;−1 3 −3 −8]**   {enter augmented matrix}
A =
     1 −2   4 12
     2 −1   5 18
   −1   3 −3 −8
»**Transf(A)**   {author defined function}
Format: (r)ational numbers or (d)ecimal numbers? d   {choice of format}
Operation: (a)dd (m)ultiply (d)ivide (s)wap (e)xit a   {select operation, addition}

add the multiple: −2   {enter the operations}
     of row: 1   {   on   }
     to row: 2   { these lines }
 1 −2   4 12
 0   3 −3 −6   {the new matrix}
−1   3 −3 −8
⋮   {perform intermediate steps}

Operation: (a)dd (m)ultiply (d)ivide (s)wap (e)xit e   {exit at appropriate time}

ans =
  1 0 0 2
  0 1 0 1
  0 0 1 3

The system has the unique solution $x = 2$, $y = 1$, $z = 3$.

[A» = transf(A) can be useful to construct a new matrix A obtained using elementary transformations. Explore this format.]

Help A help facility is available to get online information on MATLAB topics. To get information about a function enter "help *function*." For example, to get information about the function **Transf** we would enter "help transf" as follows:

»**help Transf**
Elementary row transformations for row reduction   {the information}
Add multiples of rows, multiply, divide, swap
Gives choice of format—rational numbers or decimal numbers
Calling format: Transf(A)

rref   MATLAB has a function **rref** to arrive directly at the reduced echelon form of a matrix. For example,
»**A = [1 −2 4 12;2 −1 5 18;−1 3 −3 −8];**
»**rref(A)**
ans =
  1 0 0 2   {the reduced echelon form of A}
  0 1 0 1
  0 0 1 3

EXERCISE 1

1. Solve (if possible) each of the following systems of three equations in three variables using **Gjelim** with "all steps" option. Look closely at the purpose of each operation.

   (a) $x + 4y + 3z = 1$
   $2x + 8y + 11z = 7$
   $x + 6y + 7z = 3$

   (b) $x + 2y - z = 3$
   $2x + 4y - 2z = 6$
   $3x + 6y + 2z = -1$

   (c) $x + y + 2z = 2.4$
   $x + 2y + 2z = 2.8$
   $2x + 2y + 3z = 4.05$
   (use the rational number option)

   (d) $x + 2y - 4z = -1.5$
   $x + 3y - 7z = -2.95$
   $-2x - 4y + 9z = 3.35$
   (use the rational number option)

2. Use the function **Transf** to solve the following systems of equations using Gauss-Jordan elimination. Use **rref** to check your answers. The author-defined function **Frac**(A) can be used to express the elements of A as rational numbers ("frac" for fraction form). Express the solution to **(d)** in terms of rational numbers.

   (a) $x - 2y = -8$
   $3x - 5y = -19$

   (b) $3x - 6y = 1$
   $x - 2y = 2$

(c) $\quad x + 2y = 4$
$\quad\quad 2x + 4y = 8$

(d) $\quad 3x - 4y = 2$
$\quad\quad x + y = 1$

3. Use **Transf** to solve the following systems using Gauss-Jordan elimination.

(a) $\quad x \quad\quad + z = 3$
$\quad\quad\quad 2y - 2z = -4$
$\quad\quad\quad -y - 2z = 5$

(b) $\quad x + y + 3z = 6$
$\quad\quad x + 2y + 4z = 9$
$\quad\quad 2x + y + 6z = 11$

(c) $\quad 4x + 8y - 12z = 44$
$\quad\quad 3x + 6y - 8z = 32$
$\quad\quad -2x - y \quad\quad = -7$

4. Construct a system of linear equations having solution $x = 1, y = 2, z = 3$, in which all three types of elementary row operations are used to arrive at this solution using Gauss-Jordan elimination.

5. The function **Gjpic** has been written to display the lines at each step in Gauss-Jordan elimination for a system of linear equations in two variables. Use **Gjpic** to display the lines for the system

$$x + 3y = 9$$
$$-2x + y = -4$$

6. Compute the reduced echelon forms of a number of 3×3 matrices entered at random, using **rref**. Why do you get these reduced echelon forms? How can you get another type of reduced echelon form?

(Use the function **Echtest** to test REFs of 5000 random 3×3 matrices—we got 97 not I_3, in 5000 matrices.)

7. Use the Gauss-Jordan Elimination function **Gjelim** to find the reduced echelon forms of

$$\begin{bmatrix} 1 & 2 & 3 \\ 4 & 5 & 6 \\ 7 & 8 & 9 \end{bmatrix} \text{ and } \begin{bmatrix} 7 & 8 & 9 \\ 4 & 5 & 6 \\ 1 & 2 & 3 \end{bmatrix}$$

Do you expect these two reduced forms to differ? Why do they differ?

8. There are many ways of solving systems of equations. In this exercise we introduce the method of *Gaussian elimination*.

The functions **Gjalg** and **Galg** have been written to give the reader practice at mastering the Gauss-Jordan and Gaussian algorithms by looking at patterns. Use the help facility to get information about these functions. Use these functions to see the difference in the algorithms in solving a system of three equations in three variables.

Solve the following system of equations using the function **Gelim** that performs Gaussian elimination with the "all steps" option. You get the same REF as you would with **Gjelim**, with a matrix called the *echelon* form of A appearing along the way. Describe the characteristics of a matrix in echelon form. Describe the algorithm used in Gaussian elimination.

$$x + y + z = 3$$
$$2x + 3y + z = 2$$
$$x - y - 2z = -5$$

9. Determine the number of operations needed to solve the system of equations of Exercise 8 by using

(a) Gauss-Jordan elimination

(b) Gaussian elimination

Explain, by investigating in "all steps" mode, how the arithmetic operations are counted. Where *exactly* does Gaussian elimination gain over Gauss-Jordan elimination?

10. Consider the algorithms for Gauss-Jordan elimination and Gaussian elimination for a system of n linear equations in n variables. Show, by analyzing these algorithms, that the counts of arithmetic operations (flop counts) for these methods are given by polynomials of degree 3 in n. Use the help facility to get information about **Gjinfo** and **Ginfo**. Use **Gjinfo** and **Ginfo** to determine the specific counts for such systems with $n = 2, 3, 4, 5$. Use this data in a system of linear equations to arrive at the specific polynomials that give the flop counts for these methods. Use the function **graph** to graph both these polynomials for the interval [2, 100] in steps of 10. Observe how the graphs diverge as n increases—what does this mean?

Determine the total number of arithmetic operations needed to solve a system of linear equations having a 20×20 matrix of coefficients using each of the two methods.

11. Use Gauss-Jordan elimination to determine the equation of the unique polynomial of degree two that passes through the following points (Section 1.3):

(a) $(1, 8), (3, 26), (5, 60)$

(b) $(1, -3), (2, -1), (3, 9), (4, 33)$

D.3 Matrix Operations (Sections 2.1, 2.2)

MATLAB can be used to perform matrix algebra. The symbols for standard operations are:

$$\text{Addition: } + \quad \text{Subtraction: } - \quad \text{Multiplication: } * \quad \text{Power: } \wedge \quad \text{Transpose: } \text{'}$$

Let us enter the matrices $A = \begin{bmatrix} 3 & 5 \\ 1 & 2 \end{bmatrix}$, $B = \begin{bmatrix} 1 & 0 \\ 2 & -1 \end{bmatrix}$, and $C = \begin{bmatrix} 3 & -2 \\ 0 & -1 \end{bmatrix}$ into the workspace. We shall work with these matrices in this section.

$$\text{»A = [3 5;1 2]; »B = [1 0;2 -1]; »C = [3 -2;0 -1];}$$

EXAMPLE **1** Compute AB.

```
»A*B
ans =
   13 -5
    5 -2
```

EXAMPLE **2** Compute A^3.

```
»A^3
ans =
   67 120
   24  43
```

EXERCISE 1

(a) Compute $A + B$ for matrices A and B.

(b) Compute B^{64}. (Explore various powers of matrix B.)

(c) Use your answer to (b) to predict the value of B^{129}. Use MATLAB to check your answer.

(d) Compute $(B*A)^t$.

(e) Compute $(P + P^t)$ for various matrices. Make a conjecture and prove it.

(f) Enter various square matrices P and compute PP^t for each matrix. Examine the outputs and make a conjecture. Prove this conjecture.

Algebraic Expressions

Algebraic expressions involving matrices can be computed. The hierarchy of operations is $\wedge$, $*$, $+$, and $-$.

EXAMPLE 3 Compute $2A - B^3 + 4C$ for the matrices A, B, and C in the workspace.

```
»2*A−B^3+4*C
ans =
    17  2
     0  1
```

EXERCISE 2

Compute the following expressions, using the following matrices A, B, and C.

(a) $A^2 + 3B - 2AB$ (b) $4A^3 - BC^t$

$$A = \begin{bmatrix} 3 & 5 \\ 1 & 2 \end{bmatrix}, B = \begin{bmatrix} 1 & 0 \\ 2 & -1 \end{bmatrix}, C = \begin{bmatrix} 3 & -2 \\ 0 & -1 \end{bmatrix}$$

Matrix Elements

Individual matrix elements can be listed and manipulated.

EXAMPLE 4 List the element in row 2, col 1 of B. Change this element to 3.

```
»B(2,1)
ans =
    2

»B(2,1) = 3
B =
    1   0
    3  -1
```

EXERCISE 3

Enter the matrix $P = \begin{bmatrix} 1 & 2 & -4 \\ 5 & 7 & 0 \end{bmatrix}$. Change the element p_{23} to -3. This technique can be used to correct input in the elements of a matrix, without having to re-enter the whole matrix.

†EXERCISE 4

Let $P = \begin{bmatrix} 0 & -2 & 3 \\ 1 & 4 & 5 \end{bmatrix}$ and $Q = \begin{bmatrix} 5 & 2 & -1 \\ 0 & 6 & 3 \end{bmatrix}$.

Compute $4p_{13} - 5q_{23} + 2(p_{22})^2$.

---

† indicates exercises with answers provided in the back of the book.

Submatrices, Rows, and Columns of a Matrix

In work involving matrices it is often important to be able to manipulate rows and columns of matrices, and also work with submatrices. MATLAB has a special function for selecting a submatrix of a given matrix. The functions for selecting a row or column become special cases of this function.

A(i:j, p:q)—select the submatrix of A lying from row i to row j, and col p to col q.
A(i, :)—select row i of A
A(:, p)—select column p of A

EXAMPLE 5 Consider the following matrix A.
(a) Let X be the 2nd row of A, and Y be the 3rd column. Compute XAY.
(b) Let B be the matrix lying from row 2 to row 4, column 1 to column 3 of A. Compute B^5.
(c) Let C be the matrix consisting of columns 1, 2, and 4 of A. Construct C.

$$A = \begin{bmatrix} 1 & 3 & 2 & 5 \\ 0 & 3 & 5 & 1 \\ -6 & 3 & 5 & 9 \\ -8 & -2 & 3 & 4 \end{bmatrix}$$

(a) »A = [1 3 2 5;0 3 5 1;−6 3 5 9;−8 −2 3 4];
»X = A(2, :); »Y = A(:, 3);
»X*A*Y
ans =
   405

(b) »B = A(2:4, 1:3);
»B^5
ans =
    29778 12117   1385
    11406 18135 19205
  −23600 13702 31203

(c) »C = [A(:, 1) A(:, 2) A(:, 4)]
C =
   1   3   5
   0   3   1
  −6   3   9
  −8 −2   4

EXERCISE 5

Consider the following matrix A. Use submatrices as in Example 5 to compute the following.

(a) Let X be the 3rd row of A, and Y be the 4th column. Compute XA^tY.

(b) Let B be the matrix lying from row 1 to row 3, column 2 to column 4 of A. Compute B^7.

(c) Let C be the matrix consisting of rows 1, 3, and 4 of A. Compute CA.

$$A = \begin{bmatrix} 4 & -2 & 5 & 8 \\ 0 & 1 & 2 & 7 \\ -3 & 2 & 5 & 4 \\ 1 & 0 & -2 & 3 \end{bmatrix}$$

D.4 Computational Considerations (Section 2.2)

Most matrix computation is now done on computers. Efficient algorithms are important—they mean time and accuracy. In computing a product such as ABC we can use $(AB)C$ or $A(BC)$, since matrix multiplication is *associative*. This property is useful because the number of scalar multiplications involved one way is usually less than the number involved the other way. (Scalar multiplication is more time consuming than addition; thus analyses of efficiency are done in terms of multiplications.)

> **EXAMPLE 1** A is 2×2, B is 2×3, and C is 3×1. Compute the number of multiplications involved in performing $(AB)C$ and $A(BC)$.
>
> »**Ops**   {author-defined function for computing # of multiplications}
> **Give # matrices (3 or 4): 3**
> **rows in A: 2...cols in C: 1**
> **ways =**
>   **AB×C   A×BC**
> **numbers =**
>   **18   10**
>
> Thus the number of scalar multiplications in $(AB)C$ is 18, while in $A(BC)$ it is 10. When products of many large matrices are involved the savings can be substantial.

EXERCISE 1

Verify by hand that the above number of multiplications are indeed 18 and 10.

EXERCISE 2

Let A be an $m \times r$ matrix, B an $r \times n$, and C an $n \times s$. Determine a formula that gives the number of scalar multiplications in computing AB. Use this result to derive a formula for the number of scalar multiplications in computing $(AB)C$ and $A(BC)$. Check your formula with your MATLAB results in the following exercise.

EXERCISE 3

Use MATLAB to find the most and least efficient ways of performing the following products. A is 5×14, B is 14×87, C is 87×3, and D is 3×42.

(a) ABC

(b) $ABCD$

EXERCISE 4

Use MATLAB to find the most and least efficient ways of performing the following products. A is 2×45, B is 45×45, and C is 45×3.

(a) ABC

(b) AB^2

(c) AB^3

(d) AB^2C

EXERCISE 5

Let A be an $m \times r$ matrix, B an $r \times n$, and C an $n \times s$. Determine a formula that gives the number of scalar additions in computing AB. Use this result to derive a formula for the number of scalar additions in computing $(AB)C$ and $A(BC)$. Copy the function **ops** onto your own disc. Modify it to take into account additions as well as multiplications. Test your formula and function on a 2×2 matrix A, a 2×3 matrix B, and a 3×1 matrix C.

D.5 Inverse of a Matrix (Section 2.4)

MATLAB has a matrix inverse function **inv**(A).

EXAMPLE 1 Find the inverse of the matrix $A = \begin{bmatrix} 3 & 5 \\ 1 & 2 \end{bmatrix}$.

```
»A = [3 5;1 2];
»inv(A)
ans =
    2.0000  -5.0000
   -1.0000   3.0000
```

[MATLAB uses a default "short format" of 5 decimal digits. Computations use IEEE arithmetic, and are to 2^{-52} (approx 2.22×10^{-16}).] Try some matrix outputs using other formats. »**help format** gives a list of formats available in MATLAB.

EXERCISE 1

Use MATLAB to prove that $C = \begin{bmatrix} 2 & -5 \\ 1 & 3 \end{bmatrix}$ is the inverse of

$A = \begin{bmatrix} 3 & 5 \\ 1 & 2 \end{bmatrix}$ by showing that $AC = CA = I_2$.

EXERCISE 2

Enter the matrix $B = \begin{bmatrix} 1 & 0 \\ 2 & -1 \end{bmatrix}$. Compute the first four powers of B.

(a) By looking at these results why do you expect $(B^2)^{-1} = B^2$? Compute $(B^2)^{-1}$ using MATLAB to verify that this is indeed the case.

(b) Predict the value of $(B^{194})^{-1}$. Use MATLAB to check your answer.

EXAMPLE 2 It is often preferable to use rational numbers rather than decimal numbers. The author-defined function **Frac**(x) returns the rational approximation to x. For example,

$$\text{Frac}(3.375) = 27/8$$

Let us compute the inverse of the matrix $A = \begin{bmatrix} 1/2 & 3 \\ -3/5 & 6 \end{bmatrix}$ and express the answer with rational elements.

A = [1/2 3;−3/5 6];   {the matrix can be entered
»inv(A)   with rational elements}
ans =
   1.25000000000000   −0.62500000000001   {using "long format," 15 digits}
   0.12500000000000   0.10416666666667
»Frac(inv(A))   {author-defined Frac for "fraction"}
ans =
   5/4   −5/8   {inverse in terms
   1/8   5/48   of rational numbers}

†EXERCISE 3

Compute the inverse of each of the following matrices. Express your answers in terms of both decimal and rational numbers.

(a) $A = \begin{bmatrix} 1 & 4 \\ 3 & 5 \end{bmatrix}$ (b) $B = \begin{bmatrix} 1 & 9 \\ 11 & 13 \end{bmatrix}$

Similar Matrices

Two square matrices of the same size, A and B, are said to be similar if there exists an invertible matrix C such that $B = C^{-1}AC$. The transformation of A into B is called a **similarity transformation**.

EXERCISE 4

Use MATLAB to perform the similarity transformation $B = C^{-1}AC$, where

(a) $A = \begin{bmatrix} 7 & -10 \\ 3 & -4 \end{bmatrix}$ and $C = \begin{bmatrix} 2 & 5 \\ 1 & 3 \end{bmatrix}$

(b) $A = \begin{bmatrix} 1 & 2 \\ -1 & 3 \end{bmatrix}$ and $C = \begin{bmatrix} 2 & 5 \\ 1 & 3 \end{bmatrix}$

(c) $A = \begin{bmatrix} 1 & 2 & 3 \\ 0 & -1 & 2 \\ 1 & 1 & 0 \end{bmatrix}$ and $C = \begin{bmatrix} 3 & 5 & -1 \\ -2 & -3 & 1 \\ -1 & -2 & 1 \end{bmatrix}$

(d) $A = \begin{bmatrix} 5 & 2 & 3 \\ 8 & 2 & 1 \\ -3 & 2 & 5 \end{bmatrix}$ and $C = \begin{bmatrix} 7 & -2 & 3 \\ 2 & 0 & 1 \\ 9 & -3 & 5 \end{bmatrix}$

EXERCISE 5

Write your own function **sim**(A, C) for performing a similarity transformation $B = C^{-1}AC$. The function should be on your own disc, call matrices A and C from the matrix command window, and return B to the command window. Check your function out on the similarity transformations of Exercise 4.

D.6 Solving Systems of Equations Using Matrix Inverse (Section 2.4)

A system of linear equations can be written in matrix form $AX = B$. If A is invertible there is a unique solution $X = A^{-1}B$.

EXAMPLE **1** Solve the system

$$3x + 5y = 1$$

$$x + 2y = 2$$

We can write this system and solution in the following matrix form

$$
\begin{array}{ccccccc}
A & X & B & & X & A^{-1} & B
\end{array}
$$

$$
\begin{bmatrix} 3 & 5 \\ 1 & 2 \end{bmatrix} \begin{bmatrix} x \\ y \end{bmatrix} = \begin{bmatrix} 1 \\ 2 \end{bmatrix}, \qquad \begin{bmatrix} x \\ y \end{bmatrix} = \begin{bmatrix} 3 & 5 \\ 1 & 2 \end{bmatrix}^{-1} \begin{bmatrix} 1 \\ 2 \end{bmatrix}
$$

Enter A and B and compute the solution X.

```
»A = [3 5;1 2];  B = [1;2];
»X = inv(A)*B
X =
  -8.0000
   5.0000        {The solution is x = −8, y = 5.}
```

EXERCISES 1 AND 2

Solve the following systems of equations using this matrix inverse method. Check your answers using **rref**.

1. $x + 2y = 6$
   $x + 4y = 2$

2. $x + 2y \qquad = 4$
   $2x + y - z = 2$
   $3x + y + z = -2$

Left Division in MATLAB

Again consider the system of equations $AX = B$ where A is nonsingular. The unique solution can be conveniently found in MATLAB using left division, $X = A \backslash B$.

EXAMPLE **2** Solve the system of equations

$$x + y + z = 2$$

$$2x + 3y + z = 3$$

$$x - y - 2z = -6$$

We can write this system in the matrix form

$$
\begin{array}{ccc}
A & X & B
\end{array}
$$

$$
\begin{bmatrix} 1 & 1 & 1 \\ 2 & 3 & 1 \\ 1 & -1 & -2 \end{bmatrix} \begin{bmatrix} x \\ y \\ z \end{bmatrix} = \begin{bmatrix} 2 \\ 3 \\ -6 \end{bmatrix}
$$

Enter A and B and compute the solution X.

```
»A = [1 1 1;2 3 1;1 −1 −2]; B = [2;3; −6];
»X = A\B
```

$$X =$$
$$-1.0000$$
$$1.0000$$
$$2.0000 \quad \text{\{The solution is } x = -1, y = 1, z = 2.\}}$$

EXERCISES 3 AND 4

Solve the following systems using left division. Check your answers using **rref**.

3. $2x + y = 4$
$4x + 3y = -6$

4. $x + 2y - z = 2$
$x + y + 2z = 0$
$x - y - z = 1$

Left division can be conveniently used to solve many systems, all having the same invertible matrix of coefficients.

EXAMPLE **3** Solve the systems

$$
\begin{aligned}
x - y + 3z &= b_1 \\
2x - y + 4z &= b_2 \quad \text{where} \\
-x + 2y - 4z &= b_3
\end{aligned}
\qquad
\begin{bmatrix} b_1 \\ b_2 \\ b_3 \end{bmatrix}
=
\begin{bmatrix} 8 \\ 11 \\ -11 \end{bmatrix},
\begin{bmatrix} 0 \\ 1 \\ 2 \end{bmatrix},
\begin{bmatrix} 3 \\ 3 \\ -4 \end{bmatrix}
\text{ in turn.}
$$

$$
\text{Let } A = \begin{bmatrix} 1 & -1 & 3 \\ 2 & -1 & 4 \\ -1 & 2 & -4 \end{bmatrix}
\text{ and } B = \begin{bmatrix} 8 & 0 & 3 \\ 11 & 1 & 3 \\ -11 & 2 & -4 \end{bmatrix}
$$

Letting X equal $A \backslash B$ in MATLAB, the columns of X give the three solutions.

```
»A = [1 -1 3;2 -1 4;-1 2 -4];
»B = [8 0 3;11 1 3;-11 2 -4];
»X = A\B
X =
       1.0000  0.0000  -2.0000
      -1.0000  3.0000   1.0000
       2.0000  1.0000   2.0000
```

The solutions to the three systems are

$$
\begin{aligned}
x &= 1, & y &= -1, & z &= 2 \\
x &= 0, & y &= 3, & z &= 1 \\
x &= -2, & y &= 1, & z &= 2
\end{aligned}
$$

EXERCISE 5

5. Solve the systems

$$
\begin{aligned}
x - 2y + 3z &= b_1 \\
x - y + 2z &= b_2 \quad \text{where} \\
2x - 3y + 6z &= b_3
\end{aligned}
\qquad
\begin{bmatrix} b_1 \\ b_2 \\ b_3 \end{bmatrix}
=
\begin{bmatrix} 6 \\ 5 \\ 14 \end{bmatrix},
\begin{bmatrix} -5 \\ -3 \\ -8 \end{bmatrix},
\begin{bmatrix} 4 \\ 3 \\ 9 \end{bmatrix}
\text{ in turn.}
$$

D.7 Cryptography (Section 2.4)

Matrix algebra is used in cryptography to code and decode messages. Let A be an invertible $n \times n$ matrix—the secret, encoding, matrix. Assign a number to each letter in the alphabet and space between words. Write the message in numerical form. Break the enumerated message into a sequence of $n \times 1$ columns and write them as columns of a matrix X, padding any remaining locations with the "space number." The multiplication AX encodes the message. The receiver uses A^{-1}, the decoding matrix, to decode the message.

EXAMPLE **1** Let us send the message "BUY IBM STOCK" using the encoding matrix

$$A = \begin{bmatrix} -3 & -3 & -4 \\ 0 & 1 & 1 \\ 4 & 3 & 4 \end{bmatrix}$$

Let us associate each letter with its position in the alphabet. A is 1, B is 2, and so on. Let a space between words be denoted by the number 27. Thus the message becomes

B	U	Y	*	I	B	M	*	S	T	O	C	K
2	21	25	27	9	2	13	27	19	20	15	3	11

Enter this form of the message as columns of a matrix X, which has three rows, as follows.

$$X = \begin{bmatrix} 2 & 27 & 13 & 20 & 11 \\ 21 & 9 & 27 & 15 & 27 \\ 25 & 2 & 19 & 3 & 27 \end{bmatrix}.$$

Multiply by A to get the message in encoded matrix form Y.

```
»A = [-3 -3 -4;0 1 1;4 3 4];
»X = [2 27 13 20 11;21 9 27 15 27;25 2 19 3 27];
»Y = A*X
Y =
  -169 -116 -196 -117 -222
    46   11   46   18   54
   171  143  209  137  233
```

The receiver gets this encoded message and uses inv(A) as follows to reproduce X and thus the original message.

```
»A = [-3 -3 -4;0 1 1;4 3 4];
»Y = [-169 -116 -196 -117 -222;46 11 46 18 54;
       171 143 209 137 233];
»X = inv(A)*Y
X =
  2 27 13 20 11
 21  9 27 15 27
 25  2 19  3 27
```

The columns of this matrix, written in linear form, give the original message.

2	21	25	27	9	2	13	27	19	20	15	3	11
B	U	Y	*	I	B	M	*	S	T	O	C	K

Readers who are interested in an introduction to cryptography are referred to *Coding Theory and Cryptography* edited by David Joyner, Springer-Verlag, 2000. This is an excellent collection of articles that contains historical, elementary, and advanced discussions.

EXERCISE

Decode the following messages, which were sent using the preceding encoding matrix A.

(a) {398, 25, −104, 359, 411, 12, −124, 409, 280, 41, −67, 243, 122, −9, −50, 152, 346, 19, −89, 309}.

(b) {392, 17, −108, 367, 106, 9, −25, 90, 181, −27, −69, 210, 238, −11, −86, 268, 83, −43, −57, 152, 191, 75, −11, 92}.

D.8 Leontief I/O Model (Section 2.5)

Let A be the I/O matrix that describes the interdependence of industries in an economy. Let X describe industrial production and D the demand of the industries and the open sector. Then

$$X = AX + D$$

total output — interindustry portion — open sector portion

The output level required to meet the demand is

$$X = (I - A)^{-1} D$$

EXAMPLE 1 An economy of three industries has the following I/O matrix A and demand matrix D. Find the output levels necessary to meet these demands. Units are in millions of dollars.

$$A = \begin{bmatrix} 0.2 & 0.2 & 0.3 \\ 0.5 & 0.5 & 0 \\ 0 & 0 & 0.2 \end{bmatrix}, D = \begin{bmatrix} 9 \\ 12 \\ 16 \end{bmatrix}$$

The identity $n \times n$ matrix in MATLAB is **eye**(n). Thus

$$X = (I - A)^{-1} D \text{ is written } X = \text{inv}(\text{eye}(3) - A)*D$$

Enter the following sequence

```
»A = [.2 .2 .3;.5 .5 0;0 0 .2]; D = [9;12;16];
»X = inv(eye(3)−A)*D
X =
    33.0000
    57.0000
    20.0000
```

The necessary output levels in dollars are 33, 57, 20 millions.

†EXERCISE 1

An economy of three industries has the following I/O matrix A and demand matrix D. Find the output levels necessary to meet these demands. Units are in millions of dollars.

$$A = \begin{bmatrix} 0.2 & 0.2 & 0 \\ 0.4 & 0.4 & 0.6 \\ 0.4 & 0.1 & 0.4 \end{bmatrix}, D = \begin{bmatrix} 36 \\ 72 \\ 36 \end{bmatrix}$$

EXERCISE 2

An economy of two industries has the following I/O matrix A. Find the output levels necessary to meet the demands D_1, D_2, and D_3. Units are in millions of dollars. (*Hint*: Combine D_1, D_2, and D_3 into one matrix D.)

$$A = \begin{bmatrix} 0.2 & 0.6 \\ 0.4 & 0.1 \end{bmatrix}, D_1 = \begin{bmatrix} 24 \\ 12 \end{bmatrix}, D_2 = \begin{bmatrix} 8 \\ 6 \end{bmatrix}$$

$$D_3 = \begin{bmatrix} 0 \\ 12 \end{bmatrix}$$

†EXERCISE 3

Consider the following economy consisting of three industries. The output levels of the industries are given. Determine the amounts available for the open sector from each industry.

$$A = \begin{bmatrix} 0.10 & 0.10 & 0.20 \\ 0.20 & 0.10 & 0.30 \\ 0.40 & 0.30 & 0.15 \end{bmatrix}, X = \begin{bmatrix} 6 \\ 4 \\ 5 \end{bmatrix}$$

EXERCISE 4

An economy of two industries has the following I/O matrix A and demand matrix D. Apply the Leontief model to this situation. Why does it not work? Explain the situation from both a mathematical and practical viewpoint. Extend your results to the general model.

$$A = \begin{bmatrix} 0.8 & 0.4 \\ 0.2 & 0.6 \end{bmatrix}, D = \begin{bmatrix} 20 \\ 32 \end{bmatrix}$$

D.9 Markov Chains (Sections 2.6, 5.2)

Let P be the matrix of transition probabilities for a Markov chain. Let X_0 be the initial distribution. Then

$$\text{distribution after } n \text{ steps } X_n = P^n X$$

probability of starting in state i and being in state j after n steps is the (j, i)th element of P^n, denoted $p_{ji}^{(n)}$

If P is **regular** (some power of P has all positive elements), then long-term trends are

$$X_n \rightarrow X \text{ and } P^n \rightarrow Q$$

EXAMPLE 1 Annual population movement between U.S. cities and suburbs in 2007 is described by the following stochastic matrix P. The population distribution in 2007 is given by the matrix X_0, in units of one million.

$$P = \begin{array}{c} \text{(from)} \\ \begin{array}{cc} \text{city} & \text{suburb} \end{array} \\ \begin{bmatrix} 0.96 & 0.01 \\ 0.04 & 0.99 \end{bmatrix} \begin{array}{c} \text{city} \\ \text{suburb} \end{array} \end{array} \qquad X_0 = \begin{array}{c} \text{(to)} \\ \begin{bmatrix} 82 \\ 163 \end{bmatrix} \begin{array}{c} \text{city} \\ \text{suburb} \end{array} \end{array}$$

(a) Predict the population distributions for the next five years.

(b) What is the probability that a person who was in the city in 2007 is living in the suburbs in the year 2012?

(a) »P = [.96 .01;.04 .99]; X0 = [82;163];
»Markov(P,X0)   {author-defined function to compute distributions}
Give number of steps: 5
Display all steps? y/n: y   {option is available of displaying all steps or of going immediately to the final answer. Latter is useful when long-term predictions are desired.}

step =   **. . . step =**
  1   **5**
distribution =   **distribution =**
  80.3500 {City in 2008}   **74.5348** {City in 2012}
  164.65 {Suburb 2008}   **170.4652** {Suburb in 2012}
Bar graphs? y/n: y   {option of drawing bar graphs is available for two states}

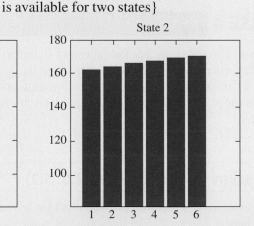

The predicted population distributions for 2012 are $\begin{bmatrix} 74.5348 \\ 170.4652 \end{bmatrix} \begin{array}{c} \text{city} \\ \text{suburb} \end{array}$.

City populations are steadily decreasing while suburban populations are rising.

(b) 2012 is 5 years after 2007. We compute P^5.

```
»P = [.96 .01;.04 .99];
»P^5
ans =
  0.8190  0.0452
  0.1810  0.9548
```

The probability that a person who was in the city (State 1) in 2007 is living in the suburbs (State 2) in 2012 (5 years later) is $p_{21}^{(5)} = 0.1810$. ∎

†EXERCISE 1

Construct a model of population flow between metropolitan and nonmetropolitan areas of the United States, given that their respective populations in 2007 were 245 million and 52 million. The probabilities are given by the matrix

$$\begin{array}{cc} \text{(from)} & \text{(to)} \\ \text{metro} \quad \text{nonmetro} \end{array}$$
$$\begin{bmatrix} 0.99 & 0.02 \\ 0.01 & 0.98 \end{bmatrix} \begin{array}{l} \text{metro} \\ \text{nonmetro} \end{array}$$

(a) Predict the population distributions of metropolitan and nonmetropolitan areas for the years 2008 through 2012.

(b) If a person was living in a metropolitan area in 2007, what is the probability that the person is still living in a metropolitan area in 2012?

EXERCISE 2

Consider the city/suburb model of this section.

(a) By computing $P^n X$ using MATLAB, for various increasing values of n (e.g., $n = 50, 100, 150$, and so forth.) show that the population distributions of the cities and suburbs will approach 49 and 196 million, unless conditions change. Arrive at these same numbers using mathematics.

(b) Let $P^n \to Q$. Determine Q and use it to show that the long-term probabilities of living in the city and suburbia are 0.2 and 0.8.

EXERCISE 3

Consider the city/suburb model of this section.

(a) The population of the United States increased by 1% per annum during the period 2000 to 2007. Allow for this increase in your model, and use MATLAB to predict the populations for the years 2008 to 2012.

(b) Assume that the populations of cities due to births, deaths, and immigration increased by 1.2% during the period 2000 to 2007, and that the population of the suburbs increased by .8% due to these same factors. Allow for these increases in your model and predict the population for the years 2008 to 2012.

EXERCISE 4

Consider the population movement model of this section.

(a) Interchange the columns of P to get a matrix P'. Look at the bar graphs for the behavior of city and suburban populations for five years, and compare them with those obtained with transition matrix P. Which model is most realistic, the P or P' model?

(b) Consider other such matrices P and corresponding matrices P'. Examine the bar graphs and make a hypothesis. Prove or disprove your hypothesis.

EXERCISE 5

Consider the above population movement model for flow between cities and suburbs. Determine the population distributions for 2002 to 2006—prior to 2007. Is the chain going from 2007 into the past a Markov chain? What are the characteristics of the matrix that takes one from distribution to distribution into the past?

†EXERCISE 6

Construct a model of population flows between cities, suburbs, and nonmetropolitan areas of the United States. Their respective populations in 2007 were 82 million, 163 million, and 52 million. The stochastic matrix giving the probabilities of the moves is

$$
\begin{array}{cc}
 & \begin{array}{ccc} \text{(from)} & & \text{(to)} \\ \text{city} & \text{suburb} & \text{nonmetro} \end{array} \\
\begin{bmatrix} 0.96 & 0.01 & 0.015 \\ 0.03 & 0.98 & 0.005 \\ 0.01 & 0.01 & 0.98 \end{bmatrix} & \begin{array}{l} \text{city} \\ \text{suburb} \\ \text{nonmetro} \end{array}
\end{array}
$$

(a) Predict the population distributions for the years 2008 to 2012.

(b) Determine the long-term population distributions.

(c) Determine the long-term probabilities of living in the city, suburbs, and nonmetropolitan areas.

D.10 Digraphs (Section 2.7)

Let A be the adjacency matrix of a digraph. Let $a_{ij}^{(n)}$ be the element in row i, column j of A^n. Then

$$ a_{ij}^{(n)} = \text{number of paths of length } n \text{ from vertex } i \text{ to vertex } j $$

The *distance* from one vertex to another is the length of the shortest path from that vertex to the other. The function **Digraph**(A) sketches a digraph and uses powers of the adjacency matrix A to give the distance from one vertex to the other. It also gives the number of distinct shortest paths.

EXAMPLE 1 Consider the digraph defined by the following adjacency matrix A. Sketch the digraph. Determine the distance from vertex 4 to vertex 1. Find the number of paths from vertex 4 to vertex 1 that give this distance.

$$
A = \begin{bmatrix}
0 & 1 & 0 & 0 & 0 \\
1 & 0 & 0 & 0 & 1 \\
0 & 1 & 0 & 0 & 1 \\
0 & 0 & 1 & 0 & 1 \\
0 & 1 & 0 & 1 & 0
\end{bmatrix}
$$

»A = [0 1 0 0 0;1 0 0 0 1;0 1 0 0 1;0 0 1 0 1;0 1 0 1 0];
　　»**Digraph(A)**,　　　{author-defined function to sketch digraph and find distance}
Want distances between vertices? y/n: y
give 1st vertex: 4
give 2nd vertex: 1
distance =
　　3
numberpaths =
　　2
Again? y/n: n

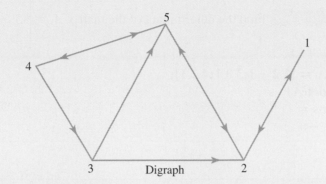

Digraph

The distance from vertex 4 to vertex 1 is 3. There are two distinct paths that give this distance. From the sketch we see that they are

$$4 \to 3 \to 2 \to 1 \text{ and } 4 \to 5 \to 2 \to 1$$

The user has the opportunity to repeat for other vertices, thus finding the distance matrix. ■

†EXERCISE 1

Consider the digraph defined by the following adjacency matrix A. Sketch the digraph. Determine the distance from vertex 2 to vertex 1. Find the number of paths from vertex 2 to vertex 1 that give this distance.

$$A = \begin{bmatrix} 0 & 1 & 0 & 0 & 0 \\ 0 & 0 & 0 & 1 & 0 \\ 1 & 0 & 0 & 0 & 0 \\ 0 & 0 & 1 & 0 & 1 \\ 1 & 0 & 0 & 1 & 0 \end{bmatrix}$$

EXERCISE 2

The following tables represent information obtained from questionaires given to groups of people. Construct the digraphs that describe the leadership structures within the groups. Compute the distance matrix and rank the members according to their influence on each group. Who should M_3 strive to influence directly in order to become the most influencial person in the group?

(a)

Group member	Person whose opinion valued
M_1	M_4
M_2	M_1 and M_5
M_3	M_2
M_4	M_3
M_5	M_1

(b)

Group member	Person whose opinion valued
M_1	M_5
M_2	M_1
M_3	M_1 and M_4
M_4	M_5
M_5	M_3

D.11 Determinants (Sections 3.1–3.3)

MATLAB has a determinant function **det**(A).

EXAMPLE 1 Find the determinant of the matrix $A = \begin{bmatrix} 1 & 2 & -1 \\ 3 & 0 & 1 \\ 4 & 2 & 1 \end{bmatrix}$.

»**A = [1 2 −1;3 0 1;4 2 1];**
»**det(A)**
ans =
    −6

EXAMPLE 2 Determine the minor of the element a_{32} of the matrix $A = \begin{bmatrix} 1 & 0 & 3 \\ 4 & -1 & 2 \\ 0 & -2 & 1 \end{bmatrix}$.

»**A = [1 0 3;4 −1 2;0 −2 1];**
»**Minor(A, 3, 2)**   {author-defined function for finding minors}
The submatrix is   {The submatrix obtained on deleting
  1 3   row 3 and column 2.}
  4 2
The minor is   {The minor is the determinant of the submatrix.}
  −10

EXERCISE 1

Use the determinant function to compute the determinants of the following matrices. Use the minor function to determine the minors of the indicated elements.

(a) $A = \begin{bmatrix} -1 & 2 & 3 \\ 0 & 1 & 4 \\ 1 & 1 & 2 \end{bmatrix}$.

Find minors of a_{22} and a_{31}.

(b) $B = \begin{bmatrix} 1 & -7 & 6 \\ 2 & 1 & 0 \\ 3 & 1 & -5 \end{bmatrix}$.

Find minors of b_{21} and b_{33}.

EXERCISE 2

Compute the determinants of the following matrices and explain the results.

(a) $\begin{bmatrix} 1 & -2 & 3 \\ 0 & 4 & 1 \\ 3 & -6 & 9 \end{bmatrix}$

(b) $\begin{bmatrix} 2 & 3 & -3 \\ -4 & 1 & 6 \\ 6 & 2 & -9 \end{bmatrix}$

(c) $\begin{bmatrix} 1 & 5 & 5 \\ 0 & -2 & -2 \\ 3 & 1 & 1 \end{bmatrix}$

(d) $\begin{bmatrix} 1 & -2 & 4 \\ 0 & -1 & 3 \\ -3 & 6 & -12 \end{bmatrix}$

EXERCISE 3

Section 3.2 gives an elimination method for computing the determinant of a matrix. Use the function **transf** to perform the row operations to compute the determinants of the following matrices. Check your answers with **det**(A).

(a) $\begin{vmatrix} 1 & -1 & 0 & 2 \\ -1 & 1 & 0 & 0 \\ 2 & -2 & 0 & 1 \\ 3 & 1 & 5 & -1 \end{vmatrix}$ (b) $\begin{vmatrix} 2 & 1 & 3 & 1 \\ -2 & 3 & -1 & 2 \\ 2 & 1 & 2 & 3 \\ -4 & -2 & 0 & -1 \end{vmatrix}$

D.12 Cramer's Rule (Section 3.3)

Inverse of a Matrix Using its Adjoint

If A is a square matrix with $|A| \neq 0$, then A is invertible with

$$A^{-1} = \frac{1}{|A|} \text{adj}(A)$$

There is an author-defined function **Adj**(A) for computing the adjoint of a matrix.

EXERCISE 1

Use MATLAB to compute the inverses of the following matrices using both **inv**(A) and **Adj**(A)/**det**(A). Compare the results using the two methods.

(a) $\begin{bmatrix} 1 & 2 & 3 \\ 0 & 1 & 2 \\ 4 & 5 & 3 \end{bmatrix}$ (b) $\begin{bmatrix} 0 & 3 & 3 \\ 1 & 2 & 3 \\ 1 & 4 & 6 \end{bmatrix}$ (c) $\begin{bmatrix} 1 & 2 & -1 \\ 2 & 4 & -3 \\ 1 & -2 & 0 \end{bmatrix}$

Cramer's Rule

Let $AX = B$ be a system of n linear equations in n variables such that $|A| \neq 0$. The system has a unique solution

$$x_1 = \frac{|A_1|}{|A|}, \; x_2 = \frac{|A_2|}{|A|}, \ldots, x_n = \frac{|A_n|}{|A|}$$

where A_i is the matrix obtained by replacing column i of A with B.

EXAMPLE 1 Solve the following system of equations using Cramer's rule.

$$x_1 + 3x_2 + x_3 = -2$$
$$2x_1 + 5x_2 + x_3 = -5$$
$$x_1 + 2x_2 + 3x_3 = 6$$

The matrix of coefficients A and column matrix of constants B, are

$$A = \begin{bmatrix} 1 & 3 & 1 \\ 2 & 5 & 1 \\ 1 & 2 & 3 \end{bmatrix} \text{ and } B = \begin{bmatrix} -2 \\ -5 \\ 6 \end{bmatrix}$$

We get, using MATLAB

»**A = [1 3 1;2 5 1;1 2 3]; »B = [−2;−5;6];**
»**Cramer(A, B)**   {author-defined function for Cramer's rule}
All information? y/n: y   {option of getting all determinants, or just the solution}
det_matrix_coeffs =
  −3
variable =   {x_1 variable for this system}
  1

Matrix =	**determinant =**	**... determinant =**	**ans =**
−2 3 1	**−3**	**−9**	**1 −2 3**
−5 5 1	{$|A_1|$}	{$|A_3|$}	{unique solution,
6 2 3			$x_1 = 1,\ x_2 = -2,\ x_3 = 3$}
{A_1}			

EXERCISE 2

Solve the following systems of equations using Cramer's rule.

(a) $\quad x_1 + 3x_2 + 4x_3 = 3$
$\quad\ 2x_1 + 6x_2 + 9x_3 = 5$
$\quad\ 3x_1 + \ x_2 - 2x_3 = 7$

(b) $x_1 + 2x_2 + x_3 = 9$
$\quad x_1 + 3x_2 - x_3 = 4$
$\quad x_1 + 4x_2 - x_3 = 7$

(c) $\ 2x_1 + \ x_2 + 3x_3 = 2$
$\quad 3x_1 - 2x_2 + 4x_3 = 2$
$\quad\ x_1 + 4x_2 - 2x_3 = 1$

D.13 Dot Product, Norm, Angle, Distance (Section 4.2)

The dot product on n-dimensional Euclidean space is

$$\mathbf{u} \cdot \mathbf{v} = u_1 v_1 + \cdots + u_n v_n$$

The geometry is defined in terms of the dot product.

norm	*angle*	*distance X to Y*
$\|\mathbf{u}\| = \sqrt{\mathbf{u} \cdot \mathbf{u}}$	$\cos\theta = \dfrac{\mathbf{u} \cdot \mathbf{v}}{\|\mathbf{u}\|\|\mathbf{v}\|}$	$d(X, Y) = \|X - Y\|$

The following functions have been written to compute these quantities.

Dot(u, v), **Mag**(u), **Angle**(u, v), **Dist**(X, Y)

EXAMPLE 1 Let $\mathbf{u} = (1, -2, 4)$, $\mathbf{v} = (3, 1, 2)$, $X = (1, 2, 3)$, $Y = (-4, 2, 3)$. Find $\mathbf{u} \cdot \mathbf{v}$, $\|\mathbf{u}\|$, the angle between $\mathbf{u}$ and $\mathbf{v}$, and $d(X, Y)$.
　　We enter the data and use the functions, one at a time.

»u = [1 2 4]; »v = [3 1 2]; »X = [1 2 3]; »Y = [−4 2 3];

dot prod	*magnitude*	*angle*	*distance*
»**Dot(u,v)**	»**Mag(u)**	»**Angle(u,v)**	»**Dist(X,Y)**
ans =	ans =	ans =	ans =
13	4.5826	40.6964	5
		{degrees}	

†EXERCISE 1

Let $\mathbf{u} = (2, 0, 6)$, $\mathbf{v} = (4, -5, -1)$, $X = (3, 3, 3)$, $Y = (-2, 5, 1)$. Find $\mathbf{u} \cdot \mathbf{v}$, $\|\mathbf{u}\|$, the angle between $\mathbf{u}$ and $\mathbf{v}$, and $d(X, Y)$.

EXERCISE 2

Let $\mathbf{u} = (1, 2, -5, 3)$, $\mathbf{v} = (0, 4, 2, -8)$, $X = (6, 3, 2, 7)$, $Y = (9, -4, -3, 2)$. Find $\mathbf{u} \cdot \mathbf{v}$, $\|\mathbf{u}\|$, the angle between $\mathbf{u}$ and $\mathbf{v}$, and $d(X, Y)$.

†EXERCISE 3

Let $\mathbf{u} = (3, 2, -1)$ and $\mathbf{v} = (-5, 9, 4)$. Compute $\|\mathbf{u}\|$, $\|\mathbf{v}\|$, and $\|\mathbf{u} + \mathbf{v}\|$, using MATLAB, and verify that the triangle inequality holds.

EXERCISE 4

Use MATLAB to investigate the following matrix A. Determine as many properties as you can. (*Hint*: Look at the vectors that make up the rows of A, det(A), etc.) Construct two other matrices, one 2×2, the other 3×3, that have the same properties as A.

$$A = \begin{bmatrix} \dfrac{1}{\sqrt{2}} & \dfrac{1}{\sqrt{2}} \\ -\dfrac{1}{\sqrt{2}} & \dfrac{1}{\sqrt{2}} \end{bmatrix}$$

D.14 Linear Combinations, Dependence, Basis, Rank (Sections 4.5–4.8)

Discussions of linear combination, dependence, and basis often reduce to solving systems of linear equations.

EXAMPLE 1 Determine whether or not $(-1, 1, 5)$ is a linear combination of $(1, 2, 3)$, $(0, 1, 4)$, $(2, 3, 6)$.

Examine the identity

$$a(1, 2, 3) + b(0, 1, 4) + c(2, 3, 6) = (-1, 1, 5)$$

This reduces to the system

$$a \qquad\quad 2c = -1$$
$$2a + b + 3c = \;\;\; 1$$
$$3a + 4b + 6c = \;\;\; 5$$

Observe that the given vectors become the *columns* of the augmented matrix A. Using MATLAB we get

```
»A = [1 0 2 −1;2 1 3 1;3 4 6 5]
»rref(A)
ans =
1 0 0   1
0 1 0   2
0 0 1 −1                    {the reduced echelon form}
```

Thus $a = 1, b = 2, c = -1$. The linear combination is unique

$$(-1, 1, 5) = (1, 2, 3) + 2(0, 1, 4) - (2, 3, 6)$$

EXAMPLE 2 Prove that the vectors $(1, 2, 3)$, $(-2, 1, 1)$, $(8, 6, 10)$ are linearly dependent. We examine the identity

$$a(1, 2, 3) + b(-2, 1, 1) + c(8, 6, 10) = 0$$

This reduces to the system

$$a - 2b + \;\;\; 8c = 0$$
$$2a + \;\; b + \;\;\; 6c = 0$$
$$3a + \;\; b + 10c = 0$$

Observe that again the given vectors become the columns of the augmented matrix A. Using MATLAB we get

```
»A = [1 −2 8 0;2 1 6 0;3 1 10 0];
»rref(A)
ans =
1 0   4 0
0 1 −2 0
0 0   0 0                {the reduced echelon form}
```

Thus $a = -4r, b = 2r, c = r$. The vectors are linearly dependent

$$-4r(1, 2, 3) + 2r(-2, 1, 1) + r(8, 6, 10) = 0$$

EXAMPLE 3 Determine a basis for the subspace of $\mathbf{R}^4$ spanned by the vectors $(1, 2, 3, 4), (0, -1, 2, 3), (2, 3, 8, 11), (2, 3, 6, 8)$. Write these vectors as the row vectors of a matrix A and compute the reduced echelon form of A.

```
»A = [1 2 3 4;0 −1 2 3;2 3 8 11;2 3 6 8];
»rrcf(A)
ans =
1 0 0 −.5
0 1 0   0
0 0 1 1.5
0 0 0   0                {the reduced echelon form}
```

The vectors form a 3-dimensional subspace of $\mathbf{R}^4$ having basis $\{(1, 0, 0, -.5), (0, 1, 0, 0), (0, 0, 1, 1.5)\}$.

EXERCISE 1

Determine whether the first vector is a linear combination of the other vectors.

(a) $(-3, 3, 7); (1, -1, 2), (2, 1, 0), (-1, 2, 1)$

(b) $(2, 7, 13); (1, 2, 3), (-1, 2, 4), (1, 6, 10)$

(c) $(0, 10, 8); (-1, 2, 3), (1, 3, 1), (1, 8, 5)$

EXERCISE 2

Determine whether the following sets of vectors are linearly dependent.

(a) $\{(-1, 3, 2), (1, -1, -3), (-5, 9, 13)\}$

(b) $\{(1, -1, 2, 1), (4, -1, 6, 2), (-2, -1, 1, -2)\}$

EXERCISE 3

Use the determinant function **det**(A) and rank function **rank**(A) to determine whether or not the following vectors are linearly dependent.

(a) $(2, 4, 5, -3), (-3, 2, 7, 0), (9, 2, 4, 6), (3, 2, -6, 5)$

(b) $(1, -1, 3, 2), (-1, 2, 4, -5), (3, -4, 2, 9), (7, 2, 1, 3)$

EXERCISE 4

Use the computer to determine the rank of the matrix

$$\begin{bmatrix} 1 & 2 & 3 & 4 & 5 \\ 6 & 7 & 8 & 9 & 10 \\ 11 & 12 & 13 & 14 & 15 \\ 16 & 17 & 18 & 19 & 20 \\ 21 & 22 & 23 & 24 & 25 \end{bmatrix}$$

(a) Explain the result.

(b) Generalize the result.

EXERCISE 5

Determine bases for the subspaces spanned by the following sets of vectors.

(a) $\{(2, 4, 6), (3, -2, 1), (-7, 14, 7)\}$

(b) $\{(2, 1, 3, 0, 4), (-1, 2, 3, 1, 0), (3, -1, 0, -1, 4)\}$

(c) $\{(1, 2, -1, 0, 4, 1), (3, 4, -2, 0, 8, 2),$
$(2, -1, 3, 2, 1, 4), (3, 1, 2, 2, 5, 5),$
$(2, 2, -1, 0, 4, 1)\}$

D.15 Projection, Gram-Schmidt Orthogonalization (Section 4.9)

The projection of a vector **u** onto a vector **v** is written proj$_v$**u** and is defined by

$$\text{proj}_v \mathbf{u} = \frac{\mathbf{u} \cdot \mathbf{v}}{\mathbf{v} \cdot \mathbf{v}} \mathbf{v}$$

The function **Proj(u,v)** has been written to compute projections.

> EXAMPLE 1 Let **u** = (1, 2, 4), **v** = (3, 1, 2). Find the projection of **u** onto **v**. We enter the data and use the function **Proj(u,v)**
>
> ```
> »u = [1 2 4]; »v = [3 1 2]; ;
> »Proj(u, v)              {author-defined function}
> 2.7857   0.9286   1.8571
> ```
>
> The projection of **u** onto **v** is the vector (2.7857, 0.9286, 1.8571).

†EXERCISE 1

Let **u** = (2, 0, 6), **v** = (4, -5, -1). Find the projection of **u** onto **v**.

EXERCISE 2

Let **u** = (1, 2, -5, 3), **v** = (0, 4, 2, -8). Find the projection of **u** onto **v**.

The Gram-Schmidt orthogonalization process provides a way in which an orthonormal basis for a vector space or subspace can be constructed from a given set of vectors.

EXAMPLE **2** Find an orthonormal basis for the subspace of $\mathbf{R}^4$ spanned by the three vectors $\{(1, 3, -1, 2), (0, -4, 5, 1),$ and $(-7, 2, 1, 0)\}$.

»Gramschm   {author-defined function for G/S orthogonalization}
Gram-Schmidt Orthogonalization
Number of vectors to be orthonormalized: 3
Number of elements in each vector: 4
Vector 1:   {enter the vectors}
[1   3 −1   2]
⋮

All steps? y/n: y   {select "All steps" option.}
[press return at each step to continue]

Initial vectors:
Vector 1   {displays original vectors}
3   3 −1   2
⋮

Perform orthogonalization

Orthogonalizing vector 1
Orthogonal basis vector   {gives orthogonal vectors}
3   3 −1   2
⋮

Normalizing vectors
Normalize vector 1   {gives orthonormal set}
0.2582   0.7746 −0.2582   0.5164
⋮

Orthonormal basis
  0.2582   0.7746   −0.2582   0.5164
  0.1925   −0.1925   0.7698   0.5774
 −0.9194   0.3048   0.2212   0.1131

The three vectors $(0.2582, 0.7746, -0.2582, 0.5164)$, $(0.1925, -0.1925, 0.7698, 0.5774)$, $(-0.9194, 0.3048, 0.2212, 0.1131)$ form an orthonormal basis to the subspace of $\mathbf{R}^4$.

When the given vectors are not linearly independent, the routine **Gram** will warn you of this fact and find an orthonormal basis for the subspace that is spanned by the vectors. ∎

EXERCISE 3

Find orthonormal bases for the subspaces spanned by the following sets.

(a) $\{(2, 3, 1), (5, -1, 7)\}$

(b) $\{(4, 2, 1, 0), (1, 5, 2, -8), (3, 10, 6, 3)\}$

(c) $\{(1, 2, 3, 4), (5, -2, 3, 8), (-13, 10, -3, -16)\}$

Explain in detail why the subspace (c) is 2-dimensional.

D.16 Eigenvalues and Eigenvectors (Sections 5.1, 5.2)

Let A be a square matrix. Let X be a non-zero vector such that

$$AX = \lambda X$$

λ is said to be an **eigenvalue** of A with corresponding **eigenvector X**.

EXAMPLE 1 Find the eigenvalues and corresponding eigenvectors of $A = \begin{bmatrix} 1 & 6 \\ 5 & 2 \end{bmatrix}$.

```
»A = [1 6;5 2];
»[X, D] = eig(A)          {gives a diagonal matrix D of eigenvalues of A.}
                          {columns of X will be corresponding normalized
                          eigenvectors.}
X =                       D =
 -0.7682   -0.7071        -4   0
  0.6402   -0.7071         0   7
```

Thus $\lambda_1 = -4$, $\mathbf{v}_1 = \begin{bmatrix} -.7682 \\ .6402 \end{bmatrix}$, and $\lambda_2 = 7$, $\mathbf{v}_2 = \begin{bmatrix} -.7071 \\ -.7071 \end{bmatrix}$. Note that MATLAB scales each eigenvector so that the norm of each is 1.

---

†EXERCISE 1

Compute the eigenvalues and corresponding eigenvectors of

$$A = \begin{bmatrix} 5 & 4 & 2 \\ 4 & 5 & 2 \\ 2 & 2 & 2 \end{bmatrix}$$

EXERCISE 2

Use MATLAB to compute the eigenvalues and eigenvectors of the matrices $\begin{bmatrix} 1 & 2 \\ 1 & 0 \end{bmatrix}$, $\begin{bmatrix} 2 & 3 \\ 1 & 0 \end{bmatrix}$, $\begin{bmatrix} 3 & 4 \\ 1 & 0 \end{bmatrix}$. Use these results to make a conjecture about the eigenvalues and eigenvectors of $\begin{bmatrix} a & a+1 \\ 1 & 0 \end{bmatrix}$. Verify your conjecture.

EXERCISE 3

Construct a symmetric matrix A. Use $[\mathbf{X}, \mathbf{D}] = \mathbf{eig}(\mathbf{A})$ to find eigenvalues and eigenvectors of A. Use MATLAB to compute $B = X^{-1}AX$. Investigate such similarity transformations on other symmetric matrices. Make some conjectures.

Long-Term Behavior

Eigenvalues and eigenvectors are used in understanding the long-term behavior of phenomena. Consider a regular Markov chain having initial vector $\mathbf{X}_0$ and transition matrix P. Suppose P is regular (some power of P has all positive elements). Then the following sequences are convergent,

$$\mathbf{X}_1 \, (= P\mathbf{X}_0), \; \mathbf{X}_2 \, (= P\mathbf{X}_1), \ldots, \mathbf{X}_n \, (= P\mathbf{X}_{n-1}), \ldots \to X$$

and

$$P, P^2, P^3 \ldots \to Q \; \text{(a stochastic matrix)}$$

Further,

(a) X is an eigenvector of P corresponding to the eigenvalue 1.

(b) The column vectors of Q are all identical, each being an eigenvector of P corresponding to the eigenvalue 1.

EXAMPLE 2 Let us return to the population movement model of D.9. Annual population movement between U.S. cities and suburbs in 2007 is described by the following stochastic matrix P. Observe that P is regular. The population distribution in 2007 is given by the matrix X_0, in units of one million.

$$P = \begin{bmatrix} 0.96 & 0.01 \\ .04 & .99 \end{bmatrix} \begin{matrix} \text{city} \\ \text{suburb} \end{matrix} \qquad X_0 = \begin{bmatrix} 82 \\ 163 \end{bmatrix} \begin{matrix} \text{city} \\ \text{suburb} \end{matrix}$$

$$\begin{matrix} \text{(from)} & \text{(to)} \\ \text{city} \quad \text{suburb} & \end{matrix}$$

Let us use eigen concepts to find X and Q and interpret these results.

The eigenvalues and eigenvectors of P are computed.

```
»P = [.96 .01;.04 .99];
»[V, D] = eig(P)
V =                        D =
 -0.7071   -0.2425          0.9500   0
  0.7071   -0.9701          0        1.0000
```

Thus $\lambda_1 = 0.95$, $\mathbf{v}_1 = \begin{bmatrix} -.7071 \\ .7071 \end{bmatrix}$, and $\lambda_2 = 1$, $\mathbf{v}_2 = \begin{bmatrix} -0.2425 \\ -0.9701 \end{bmatrix}$.

We use the eigenvector corresponding to the eigenvalue 1. Each step in the Markov process redistributes the initial total of 245, eventually reaching an eigenvector, a vector whose components are in the same ratio as $\mathbf{v}_2$ above. Thus

$$X = \begin{bmatrix} 0.2425 \times 245/1.2126 \\ 0.9701 \times 245/1.2126 \end{bmatrix} = \begin{bmatrix} 49 \\ 196 \end{bmatrix} \begin{matrix} \text{city} \\ \text{suburb} \end{matrix}$$

City and suburban populations will gradually approach 49 and 196 million, respectively.

The columns of Q are an eigenvector. Since Q is stochastic the components of each column will be in the same ratio as those of $\mathbf{v}_2$, but now adding up to 1. We get

$$
Q = \begin{bmatrix} 0.2425/1.2126 & 0.2425/1.2126 \\ 0.9701/1.2126 & 0.9701/1.2126 \end{bmatrix} = \begin{array}{c} \text{(from)} \\ \begin{array}{cc} \text{city} & \text{suburb} \end{array} \\ \begin{bmatrix} 0.2 & 0.2 \\ 0.8 & 0.8 \end{bmatrix} \end{array} \begin{array}{l} \text{(to)} \\ \text{city} \\ \text{suburb} \end{array}
$$

The long-term probability of living in the city is 0.2, while the long-term probability of living in suburbia is 0.8—independent of initial location.

†EXERCISE 4

Construct a model of population flow between metropolitan and nonmetropolitan areas of the United States, given that their respective populations in 2007 were 245 million and 52 million. The probabilities are given by the matrix

$$
\begin{array}{c} \text{(from)} \\ \begin{array}{ccc} \text{metro} & \text{nonmetro} & \text{(to)} \end{array} \\ \begin{bmatrix} 0.99 & 0.02 \\ 0.01 & 0.98 \end{bmatrix} \begin{array}{l} \text{metro} \\ \text{nonmetro} \end{array} \end{array}
$$

Determine the long-term predictions for metro and nonmetro populations, assuming no change in their total populations.

EXERCISE 5

Forty rats are placed at random in a compartment having four rooms labeled 1, 2, 3, and 4, as shown in the figure. Construct a Markov chain to describe the movement of the rats between the rooms. Predict the long-term distribution of rats. What is the long-term probability that a given marked rat is in room 4?

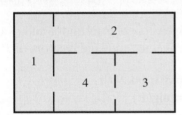

D.17 Transformations Defined by Matrices (Sections 6.1, 6.2)

Let A be an $m \times n$ matrix. Let $\mathbf{x}$ be a vector in $\mathbf{R}^n$. Write $\mathbf{x}$ as a column matrix. The transformation $T: \mathbf{R}^n \to \mathbf{R}^m$ defined by $T(\mathbf{x}) = A\mathbf{x}$ is called a matrix transformation. Such a transformation is linear.

The function **Map**(A) sketches and finds the image of the unit square, $O(0, 0)$, $P(1, 0)$, $Q(1, 1)$, $R(0, 1)$, under the transformation defined by the matrix A. The image is denoted $O*P*Q*R*$. Let us rotate the unit square through an angle of $\pi/4$. The matrix that defines a rotation through an angle α about the origin is

$$
\begin{bmatrix} \cos\alpha & -\sin\alpha \\ \sin\alpha & \cos\alpha \end{bmatrix}
$$

Use this matrix with $\alpha = \pi/4$. (π is entered as pi in MATLAB)

»A = [cos(pi/4) −sin(pi/4); sin(pi/4) cos(pi/4)];
»Map(A);   {author-defined function}
O* =
    0.0000   0.0000
P* =
    0.5253   0.8509
Q* =
    −0.3256   1.3762
R* =
    −0.8509   1.5253

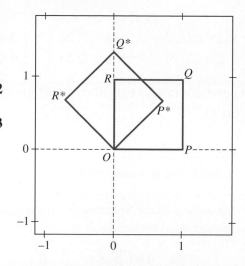

EXERCISES 1–9

1. Sketch and find the vertices of the image of the unit square under a rotation of $2\pi/3$ about the origin.

2. Sketch and find the vertices of the image of the unit square under a rotation of $-\pi/6$.

3. Sketch and find the vertices of the image of the unit square under a dilation of factor 3.

4. Determine the matrix that defines a reflection in the y-axis. Find and sketch the image of the unit square under this transformation.

5. Consider the transformations on $\mathbf{R}^2$ defined by each of the following matrices. Find the image of the unit square under each transformation.

   (a) $\begin{bmatrix} 3 & 0 \\ 1 & 4 \end{bmatrix}$ (b) $\begin{bmatrix} 4 & -1 \\ 1 & 5 \end{bmatrix}$

   (c) $\begin{bmatrix} 0 & 3 \\ -3 & 0 \end{bmatrix}$

6. The function **Compmap**(A, B) gives the image of the unit square under the composite transformation $B \circ A$. Find the image of the unit square under each of the following transformations.

   (a) A rotation through $\pi/2$ counterclockwise, then a dilation of factor 2.

   (b) A dilation of factor 4, then a reflection in the x-axis.

   (c) A rotation through $\pi/3$ then a reflection about the line $y = x$.

7. Construct single 2×2 matrices that define the following transformations on $\mathbf{R}^2$. Find the image of the unit square under each transformation.

   (a) A dilation of factor 3, then a shear of factor 2 in the x-direction.

   (b) A scaling of factor 3 in the x-direction, of factor 2 in the y-direction, then a reflection in the line $y = x$.

   (c) A dilation of factor 2, then a shear of factor 3 in the y-direction, then a rotation through $\pi/3$ counterclockwise.

8. Use **Affine**(A, **v**) to sketch the image of the unit square under the affine transformation

   $$A = \begin{bmatrix} 3 & 2 \\ 1 & 1 \end{bmatrix}, \mathbf{v} = \begin{bmatrix} 3 \\ 1 \end{bmatrix}$$

9. Use **Affine**(A, **v**) to draw the images of the unit square under the affine transformations $T_n(\mathbf{u}) = A^n\mathbf{u} + \mathbf{v}$, for $n = 1, 2, 3$, defined by the following matrix A and vector **v**. What does the image approach as $n \to \infty$?

   (a) $A = \begin{bmatrix} 0.5 & 0 \\ 0 & 0.5 \end{bmatrix}, \mathbf{v} = \begin{bmatrix} 2 \\ 2 \end{bmatrix}$

   (b) $B = \begin{bmatrix} \cos(45) & -\sin(45) \\ \sin(45) & \cos(45) \end{bmatrix}$

   $C = \begin{bmatrix} 0.5 & 0 \\ 0 & 0.5 \end{bmatrix}, A = C \circ B, \mathbf{v} = \begin{bmatrix} 3 \\ 2 \end{bmatrix}$

D.18 Fractals (Section 6.2)

Fractals can be generated using affine transformations on $\mathbf{R}^2$

$$T_i(u) = A_i u + v_i$$

The algorithm is
1. Let $x = 0$, $y = 0$.
2. Use the random # generator to select an affine transformation T_i.
3. Let $(x', y') = T_i(x, y)$.
4. Plot (x', y').
5. Let $(x, y) = (x', y')$.
6. Repeat Steps 2, 3, 4, and 5 five thousand times.

EXAMPLE **1** Let us generate a Black Spleenwort fern using the following four affine transformations T_i with associated probabilities p_i.

$$T_1 = \begin{bmatrix} 0.86 & 0.03 \\ -0.03 & 0.86 \end{bmatrix} \begin{bmatrix} x \\ y \end{bmatrix} + \begin{bmatrix} 0 \\ 1.5 \end{bmatrix}, p_1 = 0.83$$

$$T_2 = \begin{bmatrix} 0.2 & -0.25 \\ 0.21 & 0.23 \end{bmatrix} \begin{bmatrix} x \\ y \end{bmatrix} + \begin{bmatrix} 0 \\ 1.5 \end{bmatrix}, p_2 = 0.08$$

$$T_3 = \begin{bmatrix} -0.15 & 0.27 \\ 0.25 & 0.26 \end{bmatrix} \begin{bmatrix} x \\ y \end{bmatrix} + \begin{bmatrix} 0 \\ 0.45 \end{bmatrix}, p_3 = 0.08$$

$$T_4 = \begin{bmatrix} 0 & 0 \\ 0 & 0.17 \end{bmatrix} \begin{bmatrix} x \\ y \end{bmatrix} + \begin{bmatrix} 0 \\ 0 \end{bmatrix}, p_4 = 0.01$$

The function **Fern**(n) implements this algorithm for n iterations.

»**Fern(5000)**   {author-defined function}
use "command period (Mac)", "Ctrl C (PC)" to stop at any time
- press Enter to get picture -

EXERCISE 1

Modify the function **Fern** to zoom in on **(a)** the tip of the fern, and **(b)** the stem of the bottom right-hand leaf. Observe the self-similar fractal characteristics of the fern. Investigate the effects of varying the parameters of the affine transformations. Determine which parameters lead to squatter ferns and which to stunted ferns. Determine the role of each transformation in the set.

IFS Code

The affine transformations and probabilities that describe fractals can be written as matrices (IFS codes), each row corresponding to an affine transformation and probability.

$$\begin{bmatrix} x' \\ y' \end{bmatrix} = \begin{bmatrix} a & b \\ c & d \end{bmatrix} \begin{bmatrix} x \\ y \end{bmatrix} + \begin{bmatrix} e \\ f \end{bmatrix}, p$$

EXERCISE 2

The IFS code for a fractal called the "Sierpinski triangle" is

T	a	b	c	d	e	f	p
1	0.5	0	0	0.5	0.5	0.5	0.34
2	0.5	0	0	0.5	1	0	0.33
3	0.5	0	0	0.5	0	0	0.33

Run the fractal function **Sierpinski**. How many different sized triangles have you been able to produce in your image?

EXERCISE 3

The IFS codes for a fractal tree and "dragon" are as follows. Run the fractal function **tree**. Modify the function to get the **dragon** fractal.

Tree

T	a	b	c	d	e	f	p
1	0.42	0.42	−0.42	0.42	0	0.2	0.4
2	0.42	−0.42	0.42	0.42	0	0.2	0.4
3	0.1	0	0	0.1	0	0.2	0.15
4	0	0	0	0.5	0	0	0.05

Dragon

T	a	b	c	d	e	f	p
1	0.5	0.5	−0.5	0.5	0.125	0.625	0.5
2	0.5	0.5	−0.5	0.5	−0.125	0.375	0.5

EXERCISE 4

The first of the fern affine transformations can be written in the form

$$T_1 = AB \begin{bmatrix} x \\ y \end{bmatrix} + \begin{bmatrix} 0 \\ 1.5 \end{bmatrix}$$

where A is dilation and B is a rotation.

(a) Find A and B, and hence the dilation factor and the angle of rotation of this transformation.

(b) If the angle of rotation is α, determine and use the transformation that "rotates the fern" through an angle 1.5α, the dilation factor being unchanged.

(c) How do you make "controlled changes" in the first row of the IFS of the fern to rotate it and vary its size? Experiment with different angles—45° is interesting!

(d) Find an IFS that makes the fern (a) bend to the left (b) become vertical.

†EXERCISE 5

Use mathematics to find the coordinates of the tip of the fern.
Use the picture to check your answer.

D.19 Kernel and Range (Section 6.3)

Let the $m \times n$ matrix A define a linear transformation T of $\mathbf{R}^n$ to $\mathbf{R}^m$. The kernel of A is the set of all vectors $\mathbf{u}$ such that $A\mathbf{u} = 0$. The range of A is the set of all vectors $\mathbf{w}$ for which there is a vector $\mathbf{v}$ such that $A\mathbf{v} = \mathbf{w}$. The kernel and range of T are subspaces found by solving systems of linear equations.

EXAMPLE **1** Find the kernel and range of the linear transformation T defined by the following matrix A.

$$A = \begin{bmatrix} 1 & 2 & 3 \\ 0 & -1 & 1 \\ 1 & 1 & 4 \end{bmatrix}$$

The kernel will consist of the elements of $\mathbf{R}^3$ such that

$$\begin{bmatrix} 1 & 2 & 3 \\ 0 & -1 & 1 \\ 1 & 1 & 4 \end{bmatrix} \begin{bmatrix} x \\ y \\ z \end{bmatrix} = \begin{bmatrix} 0 \\ 0 \\ 0 \end{bmatrix}$$

Solve this system of linear equations using MATLAB.

```
»B = [1  2  3  0;0 −1  1  0;1  1  4  0];
»rref(B)
ans =
1 0    5 0
0 1 −1 0
0 0    0 0                    {the reduced echelon form}
```

The system has many solutions $x = -5r$, $y = r$, $z = r$. The kernel of T is the one-dimensional vector space with basis $\{(-5, 1, 1)^t\}$.

The range of T will be the subspace of $\mathbf{R}^3$ spanned by the column vectors of A. A basis for the range can be found by finding the reduced echelon form of A^t using MATLAB. The transposes of the row vectors of this form give a basis for the range.

```
»rref(A')
ans =
1  0  1
0  1  1
0  0  0                       {the reduced echelon form}
```

The range of T is the two-dimensional space with basis $\{(1, 0, 1)^t, (0, 1, 1)^t\}$
Observe that the rank/nullity theorem holds:

$$\text{dim domain}(3) = \text{dim ker}(1) + \text{dim range}(2)$$

EXERCISE 1

1. Find bases for the kernels and ranges of the linear transformations defined by the following matrices. Show that the rank/nullity holds in each case.

(a) $\begin{bmatrix} 1 & 1 & 5 \\ 0 & 1 & 3 \\ 2 & 1 & 7 \end{bmatrix}$　　(b) $\begin{bmatrix} 1 & 2 & 3 \\ -1 & 2 & 4 \\ 1 & 6 & 10 \end{bmatrix}$　　(c) $\begin{bmatrix} 6 & 1 & 3 \\ -1 & 2 & 4 \\ 0 & 1 & 5 \end{bmatrix}$　　(d) $\begin{bmatrix} -1 & 3 & 6 \\ 2 & -6 & -12 \\ 1 & -3 & -6 \end{bmatrix}$

D.20 Inner Product, Non-Euclidean Geometry (Sections 7.1, 7.2)

Let **u** and **v** be row vectors in $\mathbf{R}^n$. Let A be an $n \times n$ real symmetric matrix. Suppose A is such that $\mathbf{u}A\mathbf{u}^t > 0$ for all nonzero vectors **u**. Such a matrix is said to be **positive definite**. An inner product on $\mathbf{R}^n$ is defined by $\langle \mathbf{u}, \mathbf{v} \rangle = \mathbf{u}A\mathbf{v}^t$. This inner product leads to non-Euclidean geometries, where

$$\|\mathbf{u}\| = \sqrt{\langle \mathbf{u}, \mathbf{u} \rangle}, \qquad \cos \theta = \frac{\langle \mathbf{u}, \mathbf{v} \rangle}{\|\mathbf{u}\|\,\|\mathbf{v}\|}, \qquad d(X, Y) = \|X - Y\|$$

MATLAB functions have been written to explore these non-Euclidean spaces where

$$\langle \mathbf{u}, \mathbf{v} \rangle = \mathbf{u}A\mathbf{v}^t$$

The "g" in each of these functions stands for "generalized."

　Ginner(u,v,A): inner product of **u** and **v**.
　Gnorm(u,A): norm of **u**.
　Gangle(u,v,A): angle between **u** and **v** in degrees.
　Gdist(X,Y,A): distance between the points X and Y.
　Circles(A): sketches the circles of radii 1, 2, and 3 in $\mathbf{R}^2$.

EXAMPLE 1 Consider $\mathbf{R}^2$ with inner product defined by the matrix $A = \begin{bmatrix} 1 & 0 \\ 0 & 4 \end{bmatrix}$. This inner product is positive definite. Let us use MATLAB to determine

(a) The norm of the vector $(0, 1)$.

(b) The angle between the vectors $(1, 1)$ and $(-4, 1)$.

(c) The distance between the points $(1, 0)$ and $(0, 1)$.

(d) Graph the circles having centers the origin and radii 1, 2, and 3 in this space.

Using the author-defined functions we get

>>A = [1 0;0 4]; >>u = [0 1];
>>Gnorm(u,A)
ans =
  2   {the magnitude of $(0, 1)$ is 2 in this space; it is 1 in Euclidean space}

>>u = [1 1]; >>v = [−4 1];
>>Gangle(u,v,A)
ans =
  90   {units are degrees—the vectors are at right angles in this space}

>>X = [1 0]; >>Y = [0 1];
>>Gdist(X,Y,A)
ans =
  2.2361   {the distance between the points $(1, 0)$ and $(0, 1)$ is $\sqrt{5}$ in this space}
                                      {it is $\sqrt{2}$ in Euclidean space}

>>Circles(A)
★★computing pts for circle radius 1★★
★★computing pts for circle radius 2★★
★★computing pts for circle radius 3★★

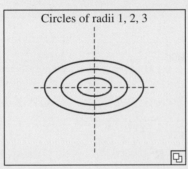

Circles of radii 1, 2, 3

Observe that the circles appear as ellipses in this geometry. The reader is asked to determine the equations of these circles in the exercises that follow.

---

†EXERCISE 1

Consider the above non-Euclidean geometry on $\mathbf{R}^2$, having inner product defined by the matrix $A = \begin{bmatrix} 1 & 0 \\ 0 & 4 \end{bmatrix}$. Determine the following using mathematics, and compare your results with the above computer output.

  (a) The norm of the vector $(0, 1)$.

  (b) The angle between the vectors $(1, 1)$ and $(−4, 1)$.

  (c) The distance between the points $(1, 0)$ and $(0, 1)$.

  (d) The equations of the circles having centers the origin and radii 1, 2, and 3.

  (e) Prove that the matrix A is positive definite.

EXERCISE 2

Find the matrix that defines the dot product, and hence Euclidean geometry on $\mathbf{R}^2$. Use this matrix to sketch the circles of radius 1, 2, and 3, having centers at the origin, in Euclidean space using MATLAB.

EXERCISE 3

Show that the following matrix A is positive definite and hence can be used to define a non-Euclidean geometry on $\mathbf{R}^2$.

$$A = \begin{bmatrix} 6 & 2 \\ 2 & 9 \end{bmatrix}$$

Use the function **Circles**(A) to sketch the circles having centers at the origin and radii 1, 2, and 3 in this space. Find the equations of these circles.

†EXERCISE 4

The following matrix A is not positive definite. It can however be used to define a pseudo inner product $\langle \mathbf{u}, \mathbf{v} \rangle = \mathbf{u}A\mathbf{v}^t$ on $\mathbf{R}^2$ (the positive definite axiom is violated), with $\|\mathbf{u}\| = \sqrt{|\langle \mathbf{u}, \mathbf{u} \rangle|}$.

$$A = \begin{bmatrix} -1 & 0 \\ 0 & 1 \end{bmatrix}$$

(a) Find a vector $\mathbf{u}$ such that $\|\mathbf{u}\| = 0$, but $\mathbf{u} \neq \mathbf{0}$. (This proves that A is not positive definite.)

(b) Find all vectors $\mathbf{u}$ in $\mathbf{R}^2$ such that $\|\mathbf{u}\| = 0$, $\mathbf{u} \neq \mathbf{0}$. Sketch these vectors.

(c) Find two distinct points in this space that are zero distance apart. Can you generalize? Find the relationship between points that are zero distance apart?

(d) Use **circles**(A) to sketch the circles center the origin and radii 1, 2, and 3 in this space. Find the equations of these circles.

This space is two-dimensional **Minkowski space**. Four dimensional Minkowski space is the space of special relativity. It is discussed in Section 6.2 and in the following section, D.21.

EXERCISE 5

Use **circles**(A) to sketch circles having centers the origin and radii 1, 2, and 3 in the spaces with inner products defined by the following matrices.

(a) $A = \begin{bmatrix} 1 & 0 \\ 0 & 0 \end{bmatrix}$

(b) $B = \begin{bmatrix} 0 & 0 \\ 0 & 1 \end{bmatrix}$

(c) $C = \begin{bmatrix} 1 & 1 \\ 1 & 1 \end{bmatrix}$

Determine the equations of these circles.

EXERCISE 6

Minkowski geometry on $\mathbf{R}^3$ is defined using the pseudo inner product $\langle \mathbf{u}, \mathbf{v} \rangle = \mathbf{u}A\mathbf{v}^t$ where

$$A = \begin{bmatrix} -1 & 0 & 0 \\ 0 & -1 & 0 \\ 0 & 0 & 1 \end{bmatrix}$$

(a) Find all vectors $\mathbf{u}$ in $\mathbf{R}^3$ such that $\langle \mathbf{u}, \mathbf{u} \rangle = 0, \mathbf{u} \neq \mathbf{0}$.

(b) Determine the spheres center the origin and radii 1, 2, and 4.

D.21 Space–Time Travel (Section 7.2)

The geometry of Minkowski space can be used to describe travel through space-time. An astronaut goes to Alpha Centauri, which is 4 light years from Earth, with .8 speed of light and returns to earth. We can use the function **space**, based on the pseudo inner product of Minkowski space, to simulate a space voyage and predict the duration of the voyage, both from the point of view of Earth and the astronaut.

```
»Space                          {author-defined function}
Give the speed: .8
  the distance: 4
  name of astronaut: Geraint
Nationality
English(E), French(F), German(G), Spanish(S), Turkish(T), Welsh(W): W
        Times
Earth       Spaceship
E-time      S-time
  1.0000    0.6000
  2.0000    1.2000
  3.0000    1.8000
  4.0000    2.4000
  5.0000    3.0000
  6.0000    3.6000
  7.0000    4.2000
  8.0000    4.8000
  9.0000    5.4000
 10.0000    6.0000
times =
 10.0000    6.0000
```

Graphics window gives picture of voyage

The time lapse on earth is 10 years; on the spaceship it is 6 years.

†EXERCISE 1

A rocket goes off to Capella, 45 light years from Earth, at .75 speed of light and returns to Earth. How long will the trip take both from Earth's point of view and from the point of view of a person on the spaceship?

EXERCISE 2

A spaceship goes on a round trip to Pleiades, 410 light years from Earth, at .9999 speed of light. How many years will have passed on Earth, and how many will have passed on the spaceship when the ship returns?

D.22 Pseudoinverse and Least Squares Curves (Section 7.4)

The pseudoinverse of a matrix A is $\text{pinv}(A) = (A^t A)^{-1} A^t$. It can be used to give the "best solution" possible for an overdetermined system of linear equations (more equations than variables). It can also be used to compute the coefficients in a least squares polynomial.

EXAMPLE **1** Find the pseudoinverse of $A = \begin{bmatrix} 1 & 2 \\ 2 & 4 \\ 3 & 0 \end{bmatrix}$.

```
»A = [1 2;2 4;3 0];
»pinv(A)                    {MATLAB pseudoinverse function}
ans =
 -0.0000  -0.0000   0.3333
  0.1000   0.2000  -0.1667
```

EXERCISE 1

Verify the answer to the previous example by computing $(A^t A)^{-1} A^t$.

EXERCISE 2

Find the pseudoinverse of each of the following matrices. Check that pinv(A) and $(A^t A)^{-1} A^t$ agree.

(a) $\begin{bmatrix} 1 & 2 \\ 1 & 5 \\ 1 & -1 \end{bmatrix}$ (b) $\begin{bmatrix} 1 & 0 \\ 2 & 4 \\ -3 & 1 \end{bmatrix}$

EXERCISE 3

Use MATLAB and various invertible matrices to investigate the pseudoinverse of an invertible matrix. Make a conjecture and prove the conjecture.

EXERCISE 4

Compute, if possible, the pseudoinverse of the matrix

$A = \begin{bmatrix} 1 & 2 & 3 \\ 4 & 5 & 6 \end{bmatrix}$ using $(A^t A)^{-1} A^t$. Investigate pseudo-

inverses of matrices of various sizes. Arrive at a result (with proof) about the existence of a pseudoinverse of a matrix based on the definition $(A^t A)^{-1} A^t$.

Least Squares Solution

If $AX = Y$ is an overdetermined system then the least squares solution is $X = \text{pinv}(A)$. This is the "closest" we can get to a true solution.

EXAMPLE **2** Find the least squares solution to the overdetermined system

$$x + y = 6$$
$$-x + y = 3$$
$$2x + 3y = 9$$

```
»A = [1 1;-1 1;2 3]; »Y = [6;3;9];
»X = pinv(A)*Y,
X =
0.5000
3.0000
```

The least squares solution is $x = \frac{1}{2}, y = 3$.

EXERCISE 5

Find the least squares solution to the following systems.

(a) $x + y = 3$
     $x - y = 1$
     $2x + 3y = 5$

(b) $2x - y = 0$
     $x + 2y = 3$
     $x + y = 1$

(c) $3x + 2y = 5$
     $x - y = 1$
     $2x + y = 4$

Least Squares Curve

Let $AX = Y$ be an overdetermined system of equations obtained by substituting n data points into a polynomial of degree m. The coefficients of the least squares polynomial of degree m are given by $X = \text{pinv}(A)*Y$.

EXAMPLE **3** Find the least squares line for the data points $(1, 1), (2, 4), (3, 2), (4, 4)$.

»Lsq	{author-defined function to compute least squares curve, based on the preceding result.}
give degree of polynomial: 1	{input information about the polynomial and the data.}
give # of data points: 4	
x values of data points: 1	
: 2	
: 3	
: 4	
y values of data points: 1	
: 4	
: 2	
: 4	
computing	

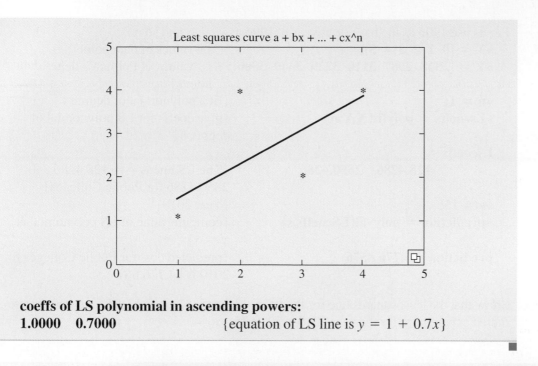

Least squares curve a + bx + ... + cx^n

coeffs of LS polynomial in ascending powers:
1.0000   0.7000   {equation of LS line is $y = 1 + 0.7x$}

†EXERCISE 6

Find and graph the least squares line for the following data points.
$(1, 1), (2, 5), (3, 9)$

EXERCISE 7

Find and graph the least squares polynomial of degree two for
the data point $(1, 7), (2, 2), (3, 1), (4, 3)$.

We now give a sequence of MATLAB functions that can be used to compute least squares curves and make predictions when a graph is not needed. We introduce the functions **polyfit** and **polyval**. **polyfit** returns the coefficients of a least squares polynomial in *descending order*. **polyval** gives the value of a polynomial at a point.

EXAMPLE 4 Consider the following data for in-state tuition and fees ($) at American 4-year Public and Private Institutions of Higher Education (World Almanac, 2002, page 236). Let us construct the least squares lines for this data and predict tuition costs for the year 2010.

	1995	1996	1997	1998	1999	2000
Public	2,848	2,987	3,110	3,229	3,349	3,506
Private	12,243	12,881	13,344	13,973	14,588	15,531

Let us use 1995 as the base year, year 0, 1996 as year 1, and so on.

»X = [0 1 2 3 4 5];	{x-values of data points}
»Y = [2848 2987 3110 3229 3349 3506];	{y-values of Public Colleges data points}
»n = 1;	{fit a polynomial of degree 1}
»Lscoeffs = polyfit(X,Y,n)	{give coeffs for LS polynomial of degree n}
Lscoeffs =	
128.4286 2850.4286	{The LS line is $y = 128.4286x + 2850.4286$ for Public Colleges}
»s = 15;	{Year 2010}
»prediction = polyval(LScoeffs,s)	{compute value of LS polynomial at $x = s$}
prediction = 4776.8576	{predicted cost for Public Colleges for 2010 is \$4,776.86}

Show that the least squares line for Private Colleges is $634x + 12175$, and that the predicted expenditure for the year 2010 is \$21,685.

†EXERCISE 8

A College of Arts and Sciences has compiled the following data on the verbal SAT scores of ten entering freshmen and their graduating GPA scores four years later. Determine the least squares line and use it to predict the graduating GPA of a student entering with an SAT score of 670.

SAT	490	450	640	510	680	610	480	450	600	650
GPA	3.2	2.8	3.9	2.4	4.0	3.5	2.7	3.6	3.5	2.4

EXERCISE 9

The following table gives the percentage of the U.S. population that was foreign-born for the years 1900 through 2000. (*Source*: Bureau of the Census, U.S. Dept. of Commerce.) The author is one of these immigrants, having come to the United States from Great Britain in 1962. Find a least squares polynomial of degree three that best fits these data and use it to predict the number of immigrants for the year 2005.

Year	1900	1910	1920	1930	1940	1950	1960	1970	1980	1990	2000
% Foreign Born	13.6	14.7	13.2	11.6	8.8	6.9	5.4	4.7	6.2	8.0	10.4

EXERCISE 10

Researchers who have studied the sense of time in humans have concluded that a person's subjective estimate of time is not equal to the elapsed time, but is a linear function $s(t) = mt + b$ of the elapsed time, for some constants m and b. With a coworker who has a watch, estimate intervals of 10 minutes, for a given period while you are reading. You estimate times 10 minutes, 20 minutes, . . . , and the coworker records the actual times t_1, t_2, ... Use this data to find the least squares line that describes the behavior of your subjective time. What would your estimate be of 3 hours. Does time go slowly or fast for you?

EXERCISE 11

Consider the following data. What must the *y*-value of the last point be to ensure that the least squares line has equation $y = mx + 0.3763$ with $m > 1.4458$?

$(1, 2.1), (2, 3.2), (4, 5.6), (6, y)$

Check your answer.

D.23 *LU* Decomposition (Section 8.2)

A nonsingular matrix may be expressed as the product of lower and upper triangular matrices. Perform Gaussian elimination on the matrix to get the upper triangular form; the lower triangular form consists of the product of the inverses of the row exchange operations.

EXAMPLE 1 Find the *LU* factorization of $A = \begin{bmatrix} 1 & 3 & 0 \\ 2 & 0 & 2 \\ 1 & 2 & 1 \end{bmatrix}$.

```
»A = [1 3 0;2 0 2;1 2 1];          {Enter the matrix A}
»Lufact(A)                          {author-defined function for LU factorization}
```

LU factorization of a matrix

All steps? y/n: n   {all steps option}

initial matrix A
```
1 3 0
2 0 2
1 2 1
```

computing

final L matrix:
```
1.0000    0        0
2.0000  1.0000     0
1.0000  0.1667  1.0000
```

final U matrix:
```
1.0000    3.0000  0
   0     -6.0000  2.0000
   0        0      0.6667
```

One of the major advantages of *LU* factorization is that, once it is found, we can solve $AX = B$ very easily. Observe that $Ax = LUX = B$ so that $X = U^{-1}L^{-1}B$. Since *L* and *U* are triangular, their inverses are easy to compute.

EXAMPLE 2 Find the solution of $AX = B$ using *LU* factorization, where

$$A = \begin{bmatrix} 1 & 3 & 0 \\ 2 & 0 & 2 \\ 1 & 2 & 1 \end{bmatrix}, \quad B = \begin{bmatrix} -2 & 4 \\ 2 & 0 \\ -1 & 2 \end{bmatrix}$$

```
»A = [1  3  0;2  0  2;1  2  1];        {Enter the matrices for A and B.}
»B = [−2  4;2  0;−1  2];
»LUsolve(A,B)                          {author-defined function to solve AX = B using
                                        the LU factorization}
```

Routine to solve the system $AX = B$ using LU factorization.

All steps? y/n: n

initial matrix A
```
 1  3  0
 2  0  2
 1  2  1
```

computing

final L matrix:
```
 1.0000     0        0
 2.0000  1.0000      0
 1.0000  0.1667  1.0000
```                                    {display the LU factorization}

**final U matrix:**
```
 1.0000   3.0000   0
      0   −6.0000 2.0000
      0   0   0.6667
```

[press return to continue]

solving AX = LUX = B:
B:
```
−2  4
 2  0
−1  2
```

solution:
```
 1.0000    1.0000
−1.0000    1.0000
 0.0000  −1.0000
```

Solution for $B = (-2, 2, -1)^t$ is $X = (1, -1, 0)^t$; for $B = (4, 0, 2)^t$ is $X - (1, 1, -1)^t$.

EXERCISE 1

Find the LU decomposition of the following matrices.

(a) $\begin{bmatrix} 3 & 5 \\ 1 & 2 \end{bmatrix}$ 　　(b) $\begin{bmatrix} 4 & 1 & 3 \\ -2 & 6 & 0 \\ 1 & 5 & 1 \end{bmatrix}$ 　　(c) $\begin{bmatrix} 2 & 0 & 2 \\ 1 & 4 & 1 \\ 2 & 5 & 5 \end{bmatrix}$

EXERCISE 2

Solve the following systems using the *LU* decomposition.

(a) $A = \begin{bmatrix} 4 & 1 & 3 \\ -2 & 6 & 0 \\ 1 & 5 & 1 \end{bmatrix}, B = \begin{bmatrix} 11 & 6 \\ 4 & -8 \\ 8 & -3 \end{bmatrix}$

(b) $A = \begin{bmatrix} 3 & 5 \\ 1 & 2 \end{bmatrix}, B = \begin{bmatrix} 8 & 1 \\ 3 & 0 \end{bmatrix}$

D.24 Condition Number of a Matrix (Section 8.3)

The condition number of a matrix is

$$c(A) = \|A\| \, \|A^{-1}\|$$

If $c(A)$ is large the matrix is ill-conditioned. The system $AX = Y$ would then be very sensitive to small changes in A and Y and the solution could be unreliable.

We use the 1-norm for the matrix,

$$\|A\| = \max\{|a_{i1}| + \cdots + |a_{in}|\}, \text{for } i = 1, \ldots n$$

This norm is a good compromise in reliability and efficiency.

EXAMPLE **1** Find the equation of the polynomial of degree 2 through the points $(1, 6)$, $(2, 3)$, $(3, 2)$.

SOLUTION

Let the polynomial be $y = a + bx + cx^2$. Thus

$$a + b + c = 6$$
$$a + 2b + 4c = 3$$
$$a + 3b + 9c = 2$$

The solution is $a = 11, b = -6, c = 1$. Thus the polynomial is $y = 11 - 6x + x^2$.

The matrix of coefficients is

$$\begin{bmatrix} 1 & 1 & 1 \\ 1 & 2 & 4 \\ 1 & 3 & 9 \end{bmatrix}$$

Let us find its condition number.

```
»A = [1 1 1;1 2 4;1 3 9];    {author-defined function that gives matrix inverse,
»Cn(A),                       determinant, 1-norm, and condition number}

inverse_matrix =
    3.0000  -3.0000   1.0000
   -2.5000   4.0000  -1.5000
    0.5000  -1.0000   0.5000

determinant_matrix =    2
onenorm_matrix =       14
onenorm_invmatrix =     8
condition_number =    112
```

The system is ill-conditioned. The y-values usually correspond to measurements—they better be very accurate! A is an example of a Vandermonde matrix—such matrices are famously ill-conditioned. In a problem such as this, if data is not equispaced, a Vandermonde matrix will not arise.

$$\text{Vandermonde matrices:} \quad \begin{bmatrix} 1 & a & a^2 & a^3 & \cdots \\ 1 & b & b^2 & b^3 & \cdots \\ 1 & c & c^2 & c^3 & \cdots \\ \vdots & \vdots & \vdots & \vdots & \cdots \end{bmatrix}$$

EXERCISE 1

Find the condition numbers of the following matrices.

(a) $\begin{bmatrix} 1 & 2 \\ 3 & 4 \end{bmatrix}$ (b) $\begin{bmatrix} 24 & 21 \\ 6 & 5 \end{bmatrix}$ (c) $\begin{bmatrix} 1 & -1 & 2 \\ 4 & 3 & 0 \\ 5 & 1 & 4 \end{bmatrix}$ (d) $\begin{bmatrix} 2 & 3 & -1 \\ 4 & 5 & 3 \\ 10 & 13 & 4 \end{bmatrix}$

EXERCISE 2

Find the condition number of the matrix $\begin{bmatrix} \frac{1}{3} & \frac{1}{4} & \frac{1}{5} \\ \frac{1}{4} & \frac{1}{5} & \frac{1}{6} \\ \frac{1}{5} & \frac{1}{6} & \frac{1}{7} \end{bmatrix}$.

This is an example of a Hilbert matrix. Hilbert matrices are defined by $a_{ij} = \dfrac{1}{i + j + 1}$. These matrices are more ill-conditioned than Vandermonde matrices!

EXERCISE 3

Find a 3×3 matrix other than a Hilbert matrix that has a condition number over 900. (*Hint*: Look for a relationship between determinants and condition numbers.) Why would the determi- nant of a matrix however be an unsatisfactory definition of condition number? Discuss.

†EXERCISE 4

Find the equation of the polynomial of degree three through the points $(1, 7)$, $(2, 3)$, $(3, 9)$, $(4, 1)$. Compute the condition number of the matrix of coefficients in the system of equations that arises.

D.25 Jacobi and Gauss-Seidel Iterative Methods (Section 8.4)

The Jacobi and Gauss-Seidel methods are iterative methods for solving certain linear systems of equations. These methods differ from the direct methods such as Gaussian elimination in that one starts with an initial guess for the solution and then computes successive approximations. Under certain conditions these approximations get closer and closer to the true solution. The primary use of these methods is in the solution of large sparse systems of equations for which Gaussian elimination is extremely expensive. The major drawback is that the methods do not always converge to a solution. Two functions **Jacobi** and **Gseidel** have been written for MATLAB that implement these methods.

EXAMPLE 1 Solve the following system of equations by the Jacobi method.

$$4x + y \quad\quad = 1$$
$$x + 4y \quad\quad = 1$$
$$y + 4z = 1$$

»A = [4 1 0 1;   {augmented matrix}
      1 4 0 1;
      0 1 4 1];
»Jacobi(A)   {author-defined function for Jacobi iteration}
Jacobi iteration
Number of iterations: {maximum # of iterations before stopping process}
  10
Tolerance: .0001   {how close to true solution before stopping process}
Initial estimate: {initial guess for the solution}
  [0 0 0]
All steps? y/n: n   {"all steps" option displays every iteration}

initial matrix
  4 1 0 1
  1 4 0 1
  0 1 4 1

****computing****

Successful run   {If solution is approximated to desired tolerance within
                       specified # of iterations the run is considered successful.}

Number of iterations
7
Solution
  0.2000
  0.2000
  0.2000

The Gauss-Seidel method would have given the same solution. The options are the same as for the Jacobi method. In general, the Gauss-Seidel method will converge faster than the Jacobi method.

As stated earlier, these iterative methods will not always converge to the solution. One condition for convergence is that the matrix of coefficients for the system of equations be strictly diagonally dominant. If it is not, then the approximations may get further away from the true solution with each successive step. We now illustrate this situation.

■

EXAMPLE 2 Let us apply the Gauss-Seidel method to the following system of equations.

$$x + 2y = \quad 3$$
$$5x + 7y = 12$$

This system has the unique solution $x_1 = 1$, $y_2 = 1$. Note that the matrix of coefficients is not strictly diagonally dominant.

»A = [1 2 3;5 7 12];   {augmented matrix of the system}
»Gseidel(A)   {author-defined function}
Gauss-Seidel iteration
Number of iterations: 4
Tolerance: 0.01
Initial estimate: [0 0]
All steps? y/n: y   {Select the "all steps" option to examine the estimate
                                 after each iteration.}

initial matrix
  1 2 3
  5 7 12

[press return at each step to continue]

iteration ...	iteration	**Desired tolerance not achieved**
1	4	**Value of X at end of run:**
X =	X =	6.8309
3.0000	6.8309	−3.1649
−0.4286	−3.1649	

Note that instead of getting closer to the true solution of $x = 1_1$, $y_2 = 1$ the approximations stray.

EXERCISE 1

Find the solution (if possible) to the following systems using both iterative methods.

(a) $10x - y + 2z = 9.5$
$\quad 5x + 10y \quad\quad = 10$
$\quad 3x + 4y + 8z = 5$

(b) $3x + \quad\quad 5z = 10$
$\quad 10x + y + 2z = 5$
$\quad\quad 5y + z = 0$

EXERCISE 2

What effect does each of the following have on the solution of systems using these iterative methods: changing the tolerance, changing the number of iterations, and changing the initial guess for the solution? Experiment with each of these inputs and draw some general conclusions.

D.26 The Simplex Method in Linear Programming (Section 9.2)

Linear programming is a mathematical method for finding optimal solutions to problems. It is widely used in business, industry, and government. In the simplex method of linear programming, constraints are written in matrix form, and row operations applied, to arrive at the optimal solution.

EXAMPLE 1 Let us find the maximum value of $f = 3x + 2y$ subject to the constraints

$$5x + 2y \leq 900$$
$$8x + 10y \leq 2800$$
$$x \geq 0$$
$$y \geq 0$$

Introduce slack variables u and v, and rewrite the system of inequalities as a system of equations that includes f, in the following manner.

$$
\begin{array}{rcrcrcrcr}
5x & + & 2y & + & u & & & & = & 900 \\
8x & + & 10y & & & + & v & & = & 2800 \\
-3x & - & 2y & & & & & + f = & & 0
\end{array}
$$

with $x \geq 0$, $y \geq 0$, $u \geq 0$, $v \geq 0$.

The problem becomes that of finding the maximum value of f under these constraints. The simplex method is an algorithm that uses row operations on the augmented matrix of this system to lead to the maximum value of f.

A MATLAB function called **Simplex** has been written for the simplex method. Let us apply it to this problem.

```
»A = [5 2 1 0 0 900;        {Enter the augmented matrix}
      8 10 0 1 0 2800;
      -3 -2 0 0 1 0];
»Simplex(A)                  {author-defined function}
Rational numbers? y/n: n    {Option of rational numbers format}
All steps? y/n: n            {Option of displaying all the steps}
initial simplex tableau
   5   2 1 0 0   900
   8  10 0 1 0  2800
  -3  -2 0 0 1     0          {displays initial tableau to check}
```

—final simplex tableau—

1.0000	0	0.2941	−0.0588	0	100.0000
0	1.0000	−0.2353	0.1471	0	200.0000
0	0	0.4118	0.1176	1.0000	700.0000

The final tableau is displayed, to be interpreted by the user. The maximum value of f is 700, and it takes place at $x = 100$, $y = 200$ (see Section 9.2).

EXERCISES

Use the **simplex** function to maximize the following functions under the given constraints.

1. Maximize $f = 2x + y$ subject to

$4x + y \leq 36$
$4x + 3y \leq 60$
$x \geq 0, y \geq 0$

2. Maximize $f = x - 4y$ subject to

$x + 2y \leq 4$
$x + 6y \leq 8$
$x \geq 0, y \geq 0$

3. Maximize $f = 4x + 6y$ subject to

$x + 3y \leq 6$
$3x + y \leq 8$
$x \geq 0, y \geq 0$

D.27 Cross Product (Appendix A)

Let $A = (a_1, a_2, a_3)$ and $B = (b_1, b_2, b_3)$. The cross product of A and B is

$$A \times B = (a_2 b_3 - b_2 a_3, b_1 a_3 - a_1 b_3, a_1 b_2 - b_1 a_2)$$

The cross product of A and B is also given by the following determinant.

$$A \times B = \begin{vmatrix} i & j & k \\ a_1 & a_2 & a_3 \\ b_1 & b_2 & b_3 \end{vmatrix}$$

Here i, j, and k are the unit vectors in the x-, y-, and z- directions.

EXAMPLE 1 Find the cross product of $(1, 2, 1)$ and $(1, -2, 1)$.

```
»a = [1  2  1];  »b = [1  −2  1];     {enter the vectors}
»c = Cross(a,b)                        {author-defined function}

c =
4  0  −4
```
The cross product of the vectors is $(4, 0, -4)$.

EXERCISE 1

Find the cross product of

(a) $(3, 4, 0), (5, -2, 9)$

(b) $(3, -1, 2), (6, 8, -3)$.

Triple Scalar Product

Let $A = (a_1, a_2, a_3)$, $B = (b_1, b_2, b_3)$, and $C = (c_1, c_2, c_3)$. The triple scalar product of A, B, and C is

$$A \cdot (B \times C) = \begin{vmatrix} a_1 & a_2 & a_3 \\ b_1 & b_2 & b_3 \\ c_1 & c_2 & c_3 \end{vmatrix}$$

EXAMPLE 2 Find the triple scalar product of $(1, 2, 1)$, $(1, -2, 1)$, and $(3, 0, 6)$.

```
»a = [1  2  1];  b = [1  −2  1];  c = [3  0  6];    {enter the vectors}
»d = Tdot(a,b,c)                                     {author-defined function}

d =
12
```
The triple scalar product is 12.

EXERCISE 2

Let $A = (3, 6, 0)$, $B = (0, 1, 5)$, and $C = (2, -2, 6)$. Find

(a) $A \cdot (B \times C)$ (b) $A \cdot (A \times B)$ (c) $A \cdot (B \times B)$

D.28 MATLAB Commands, Functions, and M-Files

Information

help	on-screen help facility
who	lists current variables in memory
whos	lists current variables and their sizes

Interrupting and Ending

exit	exits MATLAB
quit	exits MATLAB
Ctrl C	stops execution of current command

Matrix Operators

*	multiplication
^	raise to power
'	conjugate transpose
\	left division

Matrix and Vector Functions

A(i:j,p:q)	select submatrix of A lying from row i to row j, col p to col q
A(i,:)	select row I of A
A(:,p)	select column p of A
det(A)	determinant
eig(A)	eigenvalues
[X,D] = eig(A)	gives norm eigenvectors as columns of X and eigenvalues as diagonal elements of D

eye(m,n)	generates identity matrix
hilb(n)	generates a Hilbert matrix
inv(A)	matrix inverse
ones(m,n)	generates matrix of all ones
pinv(A)	pseudoinverse
rand(m,n)	generates a random matrix
rank(A)	rank
rref(A)	reduced echelon form
zeros(m,n)	generates matrix of all zeros

Special Functions

clock	returns vector [year month day hour minute seconds] t1 = clock records this vector
etime(t2,t1)	returns the time in seconds between t2 and t1
flops	returns cumulative number of floating point operations
flop(0)	resets the count to zero
polyfit(X,Y,n)	gives the least squares polynomial of degree n that fits data described by vectors **X** and **Y**
polyvalue(p,s)	evaluates the polynomial p at s

D.29 The Linear Algebra with Applications Toolbox M-Files

These programs can be downloaded from www.stetson.edu/~gwilliam/mfiles.htm.

To get information about a file:
enter help . . . (e.g., help **Space**)

Add	elementary row operation
Adj	adjoint of a matrix
Affine	affine transformation of unit square
Angle	angle between two vectors (in degrees)
Circle	circles in an inner prod space
Cminor	compute a minor of a matrix
Cn	condition number and matrix stats
Compmap	compositon of two mappings
Cramer	Cramer's rule
Cross	cross product of vectors
Digraph	picture and info on digraph
Dist	distance between two points in $\mathbf{R}^n$
Div	elementary row operation

Dot	dot product of two vectors in $\mathbf{R}^n$
Echtest	test random 3×3 matrices for REF
Fern	fractal fern
Frac	rational form for matrix elements
Galg	pattern of algorithm in Gaussian elimination
Gangle	angle in inner product space
Gdist	distance in inner product space
Gelim	Gaussian elimination
Ginfo	stats for Gaussian elimination
Ginner	inner prod in inner product space
Gjalg	pattern of alg in G-J elimination
Gjdet	determinant using G-J elimination
Gjelim	Gauss-Jordan elimination
Gjinfo	stats for G-J elimination

Gjpic	lines for G-J elimination	**Permprod**	product of permutations
Gjinv	matrix inv using G-J elimination	**Pic**	function for drawing graphs
Gnorm	norm in inner product space	**Picture**	graphs of lines
Gpic	lines for Gaussian elimination	**Pivot**	complete pivoting
Gramschm	Gram-Schmidt orthogonalization	**Pow**	powers of a matrix
Graph	graph of polynomial function	**Proj**	vector projection in $\mathbf{R}^n$
Gseidel	Gauss-Seidel method	**Sierp**	Sierpinski triangle
Jacobi	Jacobi method	**Siml**	similarity transformation
Leoinv	matrix inv in Leontief model	**Simplex**	simplex method
Leontief	predictions using Leontief	**Space**	space–time simulation
Lindep	check linear independence	**Square**	fractal square
Lsq	least squares curve	**Swap**	elementary row operation
Lufact	*LU* decomposition of matrix	**Tdot**	triple scalar product
Lusolve	solution using **Lufact**	**Transf**	elementary row operations
Mag	magnitude of vector in $\mathbf{R}^n$	**Tree**	fractal tree
Map	matrix mapping of unit square		
Markov	Markov chain model		
Minor	minors of a matrix		
Mult	elementary row operation		
Ops	# ops in matrix mult		
Perminv	inv of a permutation		

Note: The names of our functions all start with a capital letter to distinguish them from MATLAB functions.

Some versions of MATLAB are case sensitive. If a case sensitive warning appears it can be ignored.

Answers to Selected Exercises

Exercise Set 1.1

1. (a) 3×3 (c) 2×4 (e) 3×5

2. $1, 4, 9, -1, 3, 8$

5. (a) $\begin{bmatrix} 1 & 3 \\ 2 & -5 \end{bmatrix}$ and $\begin{bmatrix} 1 & 3 & 7 \\ 2 & -5 & -3 \end{bmatrix}$

(c) $\begin{bmatrix} -1 & 3 & -5 \\ 2 & -2 & 4 \\ 1 & 3 & 0 \end{bmatrix}$ and $\begin{bmatrix} -1 & 3 & -5 & -3 \\ 2 & -2 & 4 & 8 \\ 1 & 3 & 0 & 6 \end{bmatrix}$

(e) $\begin{bmatrix} 5 & 2 & -4 \\ 0 & 4 & 3 \\ 1 & 0 & -1 \end{bmatrix}$ and $\begin{bmatrix} 5 & 2 & -4 & 8 \\ 0 & 4 & 3 & 0 \\ 1 & 0 & -1 & 7 \end{bmatrix}$

6. (a) $\begin{aligned} x_1 + 2x_2 &= 3 \\ 4x_1 + 5x_2 &= 6 \end{aligned}$ (d) $\begin{aligned} 8x_1 + 7x_2 + 5x_3 &= -1 \\ 4x_1 + 6x_2 + 2x_3 &= 4 \\ 9x_1 + 3x_2 + 7x_3 &= 6 \end{aligned}$

(f) $\begin{aligned} -2x_2 &= 4 \\ 5x_1 + 7x_2 &= -3 \\ 6x_1 &= 8 \end{aligned}$ (h) $\begin{aligned} x_1 + 2x_2 - x_3 &= 6 \\ x_2 + 4x_3 &= 5 \\ x_3 &= -2 \end{aligned}$

7. (a) $\begin{bmatrix} 1 & 3 & -2 & 0 \\ 1 & 2 & -3 & 6 \\ 8 & 3 & 2 & 5 \end{bmatrix}$ (c) $\begin{bmatrix} 1 & 2 & 3 & -1 \\ 0 & 3 & 10 & 0 \\ 0 & -8 & -1 & -1 \end{bmatrix}$

(e) $\begin{bmatrix} 1 & 0 & 0 & -23 \\ 0 & 1 & 0 & 17 \\ 0 & 0 & 1 & 5 \end{bmatrix}$

8. (a) Elements in the first column, except for the leading 1, become zero. x_1 is eliminated from all equations except the first.

(c) The leading 1 in row 2 is moved to the left of the leading nonzero term in row 3. The second equation now contains x_2 with leading coefficient 1.

9. (a) Elements in the third column, except for the leading 1, become zero. x_3 is eliminated from all equations except the third.

(c) The leading nonzero element in row 3 becomes 1. Equation 3 is now solved for x_3.

10. (a) $x_1 = 2$ and $x_2 = 5$

(c) $x_1 = 10, x_2 = -9, x_3 = -7$

(e) $x_1 = 1, x_2 = 4, x_3 = 2$

11. (a) $x_1 = 0, x_2 = 4, x_3 = 2$

(c) $x_1 = 2, x_2 = 3, x_3 = -1$

(d) $x_1 = 3, x_2 = 0, x_3 = 2$

12. (a) $x_1 = 6, x_2 = -3, x_3 = 2$

(b) $x_1 = \frac{1}{2}, x_2 = \frac{3}{2}, x_3 = -\frac{1}{2}$

(d) $x_1 = -2, x_2 = -5, x_3 = -1, x_4 = 5$

13. (a) The solutions are in turn, $x_1 = 1, x_2 = 1$; $x_1 = -2$, $x_2 = 3$; and $x_1 = -1, x_2 = 2$.

(c) The solutions are in turn, $x_1 = 1, x_2 = 2, x_3 = 3$; $x_1 = -1, x_2 = 2, x_3 = 0$; and $x_1 = 0, x_2 = 1, x_3 = 2$.

Exercise Set 1.2

1. (a) Yes

(c) No. The second column contains a leading 1, so other elements in that column should be zero.

(e) Yes

(h) No. The second row does not have 1 as the first nonzero number.

2. (a) No. The leading 1 in row 3 is not to the right of the leading 1 in row 2.

(c) Yes

(e) No. The row containing all zeros should be at the bottom of the matrix.

(g) No. The leading 1 in row 3 is not to the right of the leading 1 in row 2. Also, since column 3 contains a leading 1, all other numbers in that column should be zero.

(i) Yes

3. (a) $x_1 = 2, x_2 = 4, x_3 = -3$

   (c) $x_1 = -3r + 6, x_2 = r, x_3 = -2$

   (e) $x_1 = -5r + 3, x_2 = -6r - 2, x_3 = -2r - 4, x_4 = r$

4. (a) $x_1 = -2r - 4s + 1, x_2 = 3r - 5s - 6, x_3 = r, x_4 = s$

   (c) $x_1 = 2r - 3s + 4, x_2 = r, x_3 = -2s + 9, x_4 = s, x_5 = 8$

5. (a) $x_1 = 2, x_2 = -1, x_3 = 1$

   (c) $x_1 = 3 - 2r, x_2 = 4 + r, x_3 = r$

   (e) $x_1 = 4 - 3r, x_2 = 1 - 2r, x_3 = r$

6. (a) No solution

   (c) $x_1 = 1 - 2r, x_2 = r, x_3 = -2$

   (e) $x_1 = 2, x_2 = 3, x_3 = 1$

7. (a) $x_1 = 4 - 2r, x_2 = 6 + 5r, x_3 = r$

   (c) $x_1 = 3 - 2r, x_2 = r, x_3 = 2,$ and $x_4 = 1$

   (e) $x_1 = -2r - 3s, x_2 = 3r - s, x_3 = r, x_4 = s$

8. (a) $x_1 = -2r - 4, x_2 = r + 3, x_3 = r, x_4 = 2$

   (c) $x_1 = 7 - 2r - s, x_2 = 1 + 3r - 4s, x_3 = r, x_4 = s$

   (d) No solution

   (g) $x_1 = 3, x_2 = -1$

9. (a) The following system of equations has no solution, since the equations are inconsistent.

$$3x_1 + 2x_2 - x_3 + x_4 = 4$$

$$3x_1 + 2x_2 - x_3 + x_4 = 1$$

   (b) Choose a solution, e.g., $x_1 = 1, x_2 = 2$. Now make up equations giving x_1 and x_2 the values 1 and 2:

$$x_1 + x_2 = 3$$

$$x_1 + 2x_2 = 5$$

$$x_1 - 2x_2 = -3$$

13. (a) No   (b) No

14. (a) 1st system: $x_1 = -1 - 2r, x_2 = 3 - 3r, x_3 = r,$

   2nd system: $x_1 = 4 - 2r, x_2 = -1 - 3r, x_3 = r.$

16. $[I_3{:}X]$

Exercise Set 1.3

1. $4 - 3x + x^2 = y$

3. $3 + 2x = y$

4. $5 + x + 2x^2 = y$. When $x = 2, y = 5 + 2 + 8 = 15$.

6. $-3 + x - 2x^2 + x^3 = y$

7. $I_1 = 5, I_2 = 1, I_3 = 6$

9. $I_1 = 2, I_2 = 1, I_3 = 3$

11. $I_1 = 10, I_2 = 7, I_3 = 3$

13. $I_1 = \frac{7}{3}, I_2 = \frac{5}{3}, I_3 = \frac{2}{3}, I_4 = \frac{5}{6}, I_5 = \frac{1}{6}$

17. A: $x_1 - x_4 = 100$ B: $x_1 - x_2 = 200$

   C: $-x_2 + x_3 = 150$ D: $x_3 - x_4 = 50$

Solving these equations simultaneously gives

$$x_1 = x_4 + 100, x_2 = x_4 - 100, x_3 = x_4 + 50$$

$x_2 = 0$ is theoretically possible. In that case, $x_4 = 100$, $x_1 = 200, x_3 = 150$. This flow is not likely to be realized in practice unless branch BC is completely closed.

20. Let $y = a_0 + a_1 x + a_2 x^2$. These polynomials must pass through $(1, 2)$ and $(3, 4)$. Thus

$$a_0 + a_1 + a_2 = 2$$

$$a_0 + 3a_1 + 9a_2 = 4.$$

$$\begin{bmatrix} 1 & 1 & 1 & 2 \\ 1 & 3 & 9 & 4 \end{bmatrix} \approx \begin{bmatrix} 1 & 0 & -3 & 1 \\ 0 & 1 & 4 & 1 \end{bmatrix}.$$

$$a_0 = 3a_2 + 1, a_1 = -4a_2 + 1.$$

Let $a_2 = r$. The family of polynomials is $y = (3r + 1) + (-4r + 1)x + rx^2$. $r = 0$ gives the line $y = 1 + x$ that passes through these points. When $r > 0$, the polynomials open up and when $r < 0$, the polynomials open down.

Chapter 1 Review Exercises

1. (a) 2×3 (b) 2×2

   (c) 1×4 (d) 3×1 (e) 4×6

2. $0, 6, 5, 1, 9$

3. $I_5 = \begin{bmatrix} 1 & 0 & 0 & 0 & 0 \\ 0 & 1 & 0 & 0 & 0 \\ 0 & 0 & 1 & 0 & 0 \\ 0 & 0 & 0 & 1 & 0 \\ 0 & 0 & 0 & 0 & 1 \end{bmatrix}$

4. (a) $\begin{bmatrix} 1 & 2 \\ 4 & -3 \end{bmatrix}, \begin{bmatrix} 1 & 2 & 6 \\ 4 & -3 & -1 \end{bmatrix}$

(b) $\begin{bmatrix} 2 & 1 & -4 \\ 1 & -2 & 8 \\ 3 & 5 & -7 \end{bmatrix}, \begin{bmatrix} 2 & 1 & -4 & 1 \\ 1 & -2 & 8 & 0 \\ 3 & 5 & -7 & -3 \end{bmatrix}$

(c) $\begin{bmatrix} -1 & 2 & -7 & -2 \\ 3 & -1 & 5 & 3 \\ 4 & 3 & 0 & 5 \end{bmatrix}, \begin{bmatrix} -1 & 2 & -7 \\ 3 & -1 & 5 \\ 4 & 3 & 0 \end{bmatrix}$

(d) $\begin{bmatrix} 1 & 0 & 0 \\ 0 & 1 & 0 \\ 0 & 0 & 1 \end{bmatrix}, \begin{bmatrix} 1 & 0 & 0 & 1 \\ 0 & 1 & 0 & 5 \\ 0 & 0 & 1 & -3 \end{bmatrix}$

(e) $\begin{bmatrix} -2 & 3 & -8 & 5 \\ 1 & 5 & 0 & -6 \\ 0 & -1 & 2 & 3 \end{bmatrix}, \begin{bmatrix} -2 & 3 & -8 & 5 & -2 \\ 1 & 5 & 0 & -6 & 0 \\ 0 & -1 & 2 & 3 & 5 \end{bmatrix}$

5. (a) $\begin{aligned} 4x_1 + 2x_2 &= 0 \\ -3x_1 + 7x_2 &= 8 \end{aligned}$

(b) $\begin{aligned} x_1 + 9x_2 &= -3 \\ 3x_2 &= 2 \end{aligned}$

(c) $\begin{aligned} x_1 + 2x_2 + 3x_3 &= 4 \\ 5x_1 \qquad\quad -3x_3 &= 6 \end{aligned}$

(d) $\begin{aligned} x_1 \qquad\quad &= 5 \\ x_2 \qquad &= -8 \\ x_3 &= 2 \end{aligned}$

(e) $\begin{aligned} x_1 + 4x_2 - x_3 &= 7 \\ x_2 + 3x_3 &= 8 \\ x_3 &= -5 \end{aligned}$

6. (a) Yes   **(b)** Yes

(c) No. There is a 2 (a nonzero element) above the leading 1 of row 2.

(d) Yes

(e) No. The leading 1 in row 3 is not positioned to the right of the leading 1 in row 2.

7. (a) $x_1 = 3$ and $x_2 = -1$

(b) $x_1 = 3, x_2 = -2, x_3 = 4$

(c) $x_1 = \frac{56}{9}, x_2 = \frac{17}{3}, x_3 = 2, x_4 = \frac{17}{9}$

8. (a) No solution

(b) $x_1 = 2, x_2 = -1 - 2r, x_3 = r, x_4 = 3$

9. If a matrix A is in reduced echelon form, it is clear from the definition that the leading 1 in any row cannot be to the left of the diagonal element in that row. Therefore if

$A \neq I_n$, there must be some row that has its leading 1 to the right of the diagonal element in that row. Suppose row j is such a row and the leading 1 is in position (j, k), where $j < k \le n$. Then if rows $j + 1, j + 2, \ldots,$ $j + (n - k) < n$ all contain nonzero terms, the leading 1 in these rows must be at least as far to the right as columns $k + 1, k + 2, \ldots, k + (n - k) = n$, respectively. The leading 1 in row $j + (n - k) + 1$ must then be to the right of column n. But there is no column to the right of column n, so row $j + (n - k) + 1$ must consist of all zeros.

10. Let E be the reduced echelon form of A. Since B is row equivalent to A, B is also row equivalent to E. But since E is in reduced echelon form, it must be the reduced echelon form of B.

11. $4 - 3x + 2x^2 = y$

12. $I_1 = 3, I_2 = 2, I_3 = 1$

13. $x_1 = x_8 + 160, x_1 = x_2 + 100, x_3 = x_2 + 80,$ $x_3 = x_4 + 120, x_5 = x_4 + 90, x_5 = x_6 + 130,$ $x_7 = x_6 + 100, x_7 = x_8 + 80$. Minimum flow allowable along x_8 is 20. Then $x_1 = 180, x_2 = 80, x_3 = 160,$ $x_4 = 40, x_5 = 130, x_6 = 0, x_7 = 100$.

CHAPTER 2

Exercise Set 2.1

1. (a) $\begin{bmatrix} 2 & 4 \\ 3 & 9 \\ 14 & -10 \end{bmatrix}$ **(c)** $\begin{bmatrix} -9 & 5 \\ -3 & 0 \end{bmatrix}$

(e) does not exist   **(g)** $\begin{bmatrix} 8 & 4 \\ -5 & 5 \\ 4 & 4 \end{bmatrix}$

2. (a) does not exist   **(d)** $\begin{bmatrix} -3 & -7 & 19 \\ 27 & -35 & -7 \\ -19 & 17 & 9 \end{bmatrix}$

(f) $\begin{bmatrix} 21 \\ 6 \\ 1 \end{bmatrix}$

3. (a) B **(d)** does not exist

(e) D **(g)** $\begin{bmatrix} 5 & 7 & 2 \\ 27 & 35 & 4 \end{bmatrix}$

4. (a) $\begin{bmatrix} 27 \\ 23 \\ 9 \end{bmatrix}$ **(d)** $[27]$

(f) does not exist   **(h)** $\begin{bmatrix} 13 & 8 & 13 \\ -5 & 76 & -33 \\ 11 & -16 & 35 \end{bmatrix}$

5. (a) $\begin{bmatrix} -12 & 2 \\ -9 & -24 \\ -20 & -24 \end{bmatrix}$ **(c)** $\begin{bmatrix} 8 & 2 \\ 4 & 1 \\ 0 & 6 \end{bmatrix}$

(e) does not exist   **(g)** $\begin{bmatrix} -14 & 0 \\ 12 & 10 \end{bmatrix}$

7. (a) The sum of the elements in each row is 1.

8. (a) 3×2 **(c)** does not exist

(e) 4×2 **(g)** does not exist

9. (a) 2×2 **(b)** 2×3

(d) does not exist   **(e)** 3×2

(g) does not exist

10. (a) -1 **(c)** 9

11. (a) 0 **(c)** 1

12. (a) -6

13. (a) -14

14. (a) $AB_1 = \begin{bmatrix} 1 & 2 \\ 3 & 0 \end{bmatrix}\begin{bmatrix} -2 \\ 4 \end{bmatrix} = \begin{bmatrix} 6 \\ -6 \end{bmatrix}$,

$AB_2 = \begin{bmatrix} 1 & 2 \\ 3 & 0 \end{bmatrix}\begin{bmatrix} 3 \\ 1 \end{bmatrix} = \begin{bmatrix} 5 \\ 9 \end{bmatrix}$,

$AB = \begin{bmatrix} 6 & 5 \\ -6 & 9 \end{bmatrix}$.

15. (a) $AB = 4\begin{bmatrix} 3 \\ 4 \\ 8 \end{bmatrix} + 3\begin{bmatrix} -2 \\ 2 \\ 5 \end{bmatrix} - 5\begin{bmatrix} 0 \\ 7 \\ 6 \end{bmatrix}$

16. row2 of $AB = ($row2 of $A) \times B$

$= \begin{bmatrix} 4 & 0 & 3 \end{bmatrix}\begin{bmatrix} 8 & 1 & 3 \\ 2 & 1 & 0 \\ 4 & 6 & 3 \end{bmatrix}$

$= \begin{bmatrix} 44 & 22 & 21 \end{bmatrix}$

17. The third row of AB is the third row of A times each of the columns of B in turn. Since the third row of A is all zeros, each of the products is zero.

21. (a) Submatrix products:

$\begin{bmatrix} 2 \\ -1 \end{bmatrix}\begin{bmatrix} 3 & 0 \end{bmatrix} + \begin{bmatrix} 1 \\ 0 \end{bmatrix}\begin{bmatrix} 2 & 1 \end{bmatrix}$

$= \begin{bmatrix} 6 & 0 \\ -3 & 0 \end{bmatrix} + \begin{bmatrix} 2 & 1 \\ 0 & 0 \end{bmatrix} = \begin{bmatrix} 8 & 1 \\ -3 & 0 \end{bmatrix}$,

$[3][3 \quad 0] + [1][2 \quad 1] = [9 \quad 0] + [2 \quad 1] = [11 \quad 1]$.

$AB = \begin{bmatrix} 8 & 1 \\ -3 & 0 \\ 11 & 1 \end{bmatrix}$

22. (a) $B = \begin{bmatrix} -1 & -2 \\ 0 & 3 \\ 4 & 1 \end{bmatrix}$ or $\begin{bmatrix} -1 & -2 \\ 0 & 3 \\ 4 & 1 \end{bmatrix}$

24. (a) $\begin{bmatrix} 2 & 3 & 2 & 1 \\ 4 & 0 & 0 & 0 \\ 1 & 0 & 0 & 0 \\ 5 & 0 & 0 & 0 \end{bmatrix}\begin{bmatrix} 1 & 2 \\ -1 & 3 \\ 4 & 0 \\ 2 & 5 \end{bmatrix} = \cdots = \begin{bmatrix} 9 & 18 \\ 4 & 8 \\ 1 & 2 \\ 5 & 10 \end{bmatrix}$

25. (a) True: If $A + B$ exists, then A and B are the same size. If $B + C$ exists, B and C are the same size. Thus A and C are the same size. $A + C$ exists.

(c) False: For example, let A and B be square, with A being the identity matrix I. Then $AB = B$ and $BA = B$. This is called a counterexample.

Exercise Set 2.2

1. (a) $AB = \begin{bmatrix} 4 & 7 & 10 \\ 0 & -5 & -4 \end{bmatrix}$ and BA does not exist.

(c) $AD = DA = \begin{bmatrix} 4 & 4 \\ -2 & 2 \end{bmatrix}$

2. $ABC = \begin{bmatrix} 18 \\ -10 \\ 14 \end{bmatrix}$

4. (a) $\begin{bmatrix} 8 & -5 \\ 6 & 11 \end{bmatrix}$ **(c)** $\begin{bmatrix} 9 & 3 \\ 3 & 5 \end{bmatrix}$

5. (a) $\begin{bmatrix} 26 & 34 \\ 187 & 383 \end{bmatrix}$ **(c)** $\begin{bmatrix} 50 & 156 \\ 57 & 109 \end{bmatrix}$

6. (a) does not exist   **(c)** 3×6

7. (a) 3×3 **(c)** does not exist

(e) 2×3

9. (a) The (i, j)th element of $A + (B + C)$ is $a_{ij} + (b_{ij} + c_{ij})$. The (i, j)th element of $(A + B) + C$ is $(a_{ij} + b_{ij}) + c_{ij} = a_{ij} + (b_{ij} + c_{ij})$. Since their elements are the same, $A + (B + C) = (A + B) + C$.

11. The (i, j)th element of cA is ca_{ij}. If $cA = O_{mn}$, then $ca_{ij} = 0$ for all i and j. So either $c = 0$ or all $a_{ij} = 0$, in which case $A = O_{mn}$.

12. (a) $9B^2 - AB + 2BA$　**(c)** $-2BA - 2B^2$

13. (a) $A^2 + AB - BA - B^2$

　　(c) $-A^3 + A^2B + BA^2 + BAB - 2ABA - 2AB^2 + B^2A$

14. (a) The matrices that commute with the given matrix are all matrices of the form $\begin{bmatrix} a & 0 \\ c & c + a \end{bmatrix}$, where a and c can take any real values.

15. $AX_1 = AX_2$ does not imply that $X_1 = X_2$.

16. (a) $A^2 = AA$, which is the product of two $n \times n$ matrices.

17. If $AB = BA$.

21. (a) The (i, j)th element of $A + B$ is $a_{ij} + b_{ij} = 0 + 0$ if $i \ne j$, so $A + B$ is a diagonal matrix.

22. The nondiagonal elements of AB and of BA are zero. (See Exercise 21c.) The diagonal elements of AB are the elements $a_{i1}b_{1i} + a_{i2}b_{2i} + \cdots + a_{in}b_{ni} = a_{ii}b_{ii}$ and the diagonal elements of BA are $b_{i1}a_{1i} + b_{i2}a_{2i} + \cdots + b_{in}a_{ni} = b_{ii}a_{ii} = a_{ii}b_{ii}$. Thus, $AB = BA$.

24. (a) yes　　　**(c)** no　　　**(e)** yes

25. The idempotent matrices of the given form are

$$\begin{bmatrix} 1 & 0 \\ 0 & 0 \end{bmatrix}, \begin{bmatrix} 1 & 0 \\ 0 & 1 \end{bmatrix}, \begin{bmatrix} 1 & b \\ 0 & 0 \end{bmatrix}, \text{ and } \begin{bmatrix} 1 & 0 \\ c & 0 \end{bmatrix},$$

where b and c are any nonzero real numbers.

27. If $AB = BA$, then $(AB)^2 = (AB)(AB) = A(BA)B = A(AB)B = A^2B^2 = AB$.

31. (a) $\begin{bmatrix} 2 & 3 \\ 3 & -8 \end{bmatrix} \begin{bmatrix} x_1 \\ x_2 \end{bmatrix} = \begin{bmatrix} 4 \\ -1 \end{bmatrix}$

32. (a) $\begin{bmatrix} 1 & 8 & -2 \\ 4 & -7 & 1 \\ -2 & -5 & -2 \end{bmatrix} \begin{bmatrix} x_1 \\ x_2 \\ x_3 \end{bmatrix} = \begin{bmatrix} 3 \\ -3 \\ 1 \end{bmatrix}$

　　(b) $\begin{bmatrix} 5 & 2 \\ 4 & -3 \\ 3 & 1 \end{bmatrix} \begin{bmatrix} x_1 \\ x_2 \end{bmatrix} = \begin{bmatrix} 6 \\ -2 \\ 9 \end{bmatrix}$

33. Since X_1 is a solution then the scalar multiple of X_1 by a is a solution. aX_1 is a solution. Similarly, bX_2 is a solution. Since the sum of two solutions is a solution, $aX_1 + bX_2$ is a solution.

34. e.g., $X_3 = X_1 + X_2 = \begin{bmatrix} -4 \\ 1 \\ -6 \\ 2 \end{bmatrix} + \begin{bmatrix} -17 \\ 3 \\ -3 \\ 1 \end{bmatrix} = \begin{bmatrix} -21 \\ 4 \\ -9 \\ 3 \end{bmatrix}$.

$$X_4 = 2X_1 = \begin{bmatrix} -4 \\ 1 \\ -6 \\ 2 \end{bmatrix} = \begin{bmatrix} -8 \\ 2 \\ -12 \\ 4 \end{bmatrix}.$$

$$X_5 = X_3 + X_4 = \begin{bmatrix} -29 \\ 6 \\ -21 \\ 7 \end{bmatrix}, \ X_4 = -1X_5 = \begin{bmatrix} 29 \\ -6 \\ 21 \\ -7 \end{bmatrix}.$$

By Exercise 33, the linear combinations $a\begin{bmatrix} -4 \\ 1 \\ -6 \\ 2 \end{bmatrix} + b\begin{bmatrix} -17 \\ 3 \\ -3 \\ 1 \end{bmatrix}$ are solutions for all values of a and b. We want a solution for which $x_1 = -3$, $x_2 = 2$. Thus $-4a - 17b = -3$. $a + 3b = 2$. Gives $a = 5$, $b = -1$.

There is a unique solution $5\begin{bmatrix} -4 \\ 1 \\ -6 \\ 2 \end{bmatrix} - 1\begin{bmatrix} -17 \\ 3 \\ -3 \\ 1 \end{bmatrix} = \begin{bmatrix} -3 \\ 2 \\ -27 \\ 9 \end{bmatrix}$.

39. (a) False: $(A + B)(A - B) = A^2 + BA - AB - B^2 = A^2 - B^2$ only if $BA - AB = 0$; that is $AB = BA$.

　　(c) False: ABC is mq matrix; m rows, q columns. Thus mq elements.

40. (a) Transmission matrix $\begin{bmatrix} 1 & 0 \\ \dfrac{-1}{R} & 1 \end{bmatrix}$.

41. (a) Transmission matrix $\begin{bmatrix} 1 & -1 \\ 2 & 0 \end{bmatrix}$.

Exercise Set 2.3

1. (a) $A^t = \begin{bmatrix} -1 & 2 \\ 2 & -3 \end{bmatrix}$ symmetric

(c) $C^t = \begin{bmatrix} 3 & 2 \\ -1 & 4 \end{bmatrix}$ not symmetric

(d) $D^t = \begin{bmatrix} 4 & -2 & 7 \\ 5 & 3 & 0 \end{bmatrix}$ not symmetric

(f) $F^t = \begin{bmatrix} 1 & -1 & 3 \\ -1 & 2 & 0 \\ 3 & 0 & 4 \end{bmatrix}$ symmetric

(h) $H^t = \begin{bmatrix} 1 & 4 & -2 \\ 2 & 5 & 6 \\ 3 & 6 & 7 \end{bmatrix}$ not symmetric

2. (a) $\begin{bmatrix} 1 & 2 & 4 \\ 2 & 6 & 5 \\ 4 & 5 & 2 \end{bmatrix}$

3. (a) 4×2 **(c)** does not exist   **(e)** 4×3

5. (a) $(A + B + C)^t = (A + (B + C))^t =$
$A^t + (B + C)^t = A^t + (B^t + C^t) = A^t + B^t + C^t$

9. If A is symmetric, then $A = A^t$. By Theorem 2.4, this means $A^t = A = (A^t)^t$. So A^t is symmetric.

11. (a) $\begin{bmatrix} 0 & -1 \\ 1 & 0 \end{bmatrix}$

(c) If A and B are antisymmetric, then $A + B = (-A^t) + (-B^t) = -(A^t + B^t) = -(A + B)^t$.

13. $B = \frac{1}{2}(A + A^t)$ is symmetric and $C = \frac{1}{2}(A - A^t)$ is antisymmetric. $A = B + C$.

15. (a) -2 **(c)** -1

17. $tr(A + B + C) = tr(A + (B + C))$
$\qquad = tr(A) + tr(B + C)$
$\qquad = tr(A) + (tr(B) + tr(C))$
$\qquad = tr(A) + tr(B) + tr(C)$

21. $A + B = \begin{bmatrix} 3 + i & 8 + i \\ 5 + 2i & 4 - 2i \end{bmatrix}$,

$AB = \begin{bmatrix} -2 - i & 40 + 15i \\ -12 - 19i & 19 - i \end{bmatrix}$,

$BA = \begin{bmatrix} -6 + 24i & 5 - 20i \\ 14 + 13i & 23 - 26i \end{bmatrix}$,

23. $A^* = \overline{A}^t = \begin{bmatrix} 2 + 3i & 2 \\ -5i & 5 + 4i \end{bmatrix} \neq A$, not hermitian

$D^* = \overline{D}^t = \begin{bmatrix} -2 & 3 - 5i \\ 3 + 5i & 9 \end{bmatrix} = D$, hermitian

25. (a) The (i, j)th element of $A^* + B^*$ is $\overline{a_{ji}} + \overline{b_{ji}} = \overline{a_{ji} + b_{ji}}$, the (i, j)th element of $(A + B)^*$.

27. (a) $g_{12} = 0$, $g_{13} = 1$, $g_{23} = 1$, so $1 \to 3 \to 2$ or $2 \to 3 \to 1$
$p_{12} = 1$, so $1 \to 2$ or $2 \to 1$, gives no information

(c) $g_{12} = 1$, $g_{13} = 2$, $g_{23} = 3$, so $1 \to 3 \to 2$ or $2 \to 3 \to 1$
$p_{12} = 1$, $p_{13} = 2$, $p_{14} = 1$, $p_{23} = 2$, $p_{24} = 2$ $p_{34} = 2$, so $1 \to 3 \to \{2 \leftrightarrow 4\}$ or $\{2 \leftrightarrow 4\} \to 3 \to 1$

(e) $g_{12} = 1$, $g_{13} = 0$, $g_{14} = 1$, $g_{23} = 0$, $g_{24} = 0$, $g_{34} = 1$, so $2 \to 1 \to 4 \to 3$ or $3 \to 4 \to 1 \to 2$
$p_{12} = 0$, $p_{13} = 1$, $p_{14} = 0$, $p_{23} = 1$, $p_{24} = 1$, $p_{34} = 0$, so $1 \to 3 \to 2 \to 4$ or $4 \to 2 \to 3 \to 1$

28. (a) $G^t = (AA^t)^t = (A^t)^t A^t = AA^t = G$, and
$P^t = (A^t A)^t = A^t (A^t)^t = A^t A = P$.

Exercise Set 2.4

1. (a) B is the inverse of A.

(c) B is not the inverse of A.

2. (a) B is the inverse of A.

(c) B is not the inverse of A.

3. (a) $\begin{bmatrix} 1 & 0 \\ -2 & 1 \end{bmatrix}$ **(c)** $\begin{bmatrix} \frac{3}{2} & -\frac{1}{2} \\ -2 & 1 \end{bmatrix}$

(e) The inverse does not exist.

4. (a) $\begin{bmatrix} \frac{7}{3} & -3 & -\frac{1}{3} \\ -\frac{8}{3} & 3 & \frac{2}{3} \\ \frac{4}{3} & -1 & -\frac{1}{3} \end{bmatrix}$ **(c)** The inverse does not exist.

5. (a) $\begin{bmatrix} -\frac{3}{7} & \frac{5}{7} & -\frac{11}{7} \\ \frac{2}{7} & -\frac{1}{7} & -\frac{2}{7} \\ \frac{2}{7} & -\frac{1}{7} & \frac{5}{7} \end{bmatrix}$ **(c)** The inverse does not exist.

6. (a) $\begin{bmatrix} -\frac{1}{5} & -\frac{1}{5} & \frac{1}{5} & 0 \\ -\frac{1}{5} & \frac{1}{5} & 0 & \frac{1}{5} \\ \frac{1}{5} & 0 & \frac{1}{5} & -\frac{1}{5} \\ 0 & \frac{1}{5} & -\frac{1}{5} & -\frac{1}{5} \end{bmatrix}$ **(c)** $\begin{bmatrix} 1 & 0 & 1 & -1 \\ 0 & -1 & -3 & 4 \\ 1 & 0 & -1 & 2 \\ -3 & 0 & 0 & -1 \end{bmatrix}$

7. (a) $\begin{bmatrix} 3 & -8 \\ -1 & 3 \end{bmatrix}$ **(c)** $\begin{bmatrix} 2 & -3 \\ -\frac{3}{2} & \frac{5}{2} \end{bmatrix}$

8. (a) $\begin{bmatrix} x_1 \\ x_2 \end{bmatrix} = \begin{bmatrix} -5 & 2 \\ 3 & -1 \end{bmatrix} \begin{bmatrix} 2 \\ 4 \end{bmatrix} = \begin{bmatrix} -2 \\ 2 \end{bmatrix}$

(c) $\begin{bmatrix} x_1 \\ x_2 \end{bmatrix} = \begin{bmatrix} -\frac{1}{5} & \frac{3}{5} \\ \frac{2}{5} & -\frac{1}{5} \end{bmatrix} \begin{bmatrix} 5 \\ 10 \end{bmatrix} = \begin{bmatrix} 5 \\ 0 \end{bmatrix}$

(e) $\begin{bmatrix} x_1 \\ x_2 \end{bmatrix} = \begin{bmatrix} 2 & -1 \\ -\frac{3}{4} & \frac{1}{2} \end{bmatrix} \begin{bmatrix} 6 \\ 1 \end{bmatrix} = \begin{bmatrix} 11 \\ -4 \end{bmatrix}$

9. (a) $\begin{bmatrix} x_1 \\ x_2 \\ x_3 \end{bmatrix} = \begin{bmatrix} \frac{1}{9} & \frac{3}{9} & \frac{5}{9} \\ \frac{3}{9} & 0 & -\frac{3}{9} \\ -\frac{2}{9} & \frac{3}{9} & -\frac{1}{9} \end{bmatrix} \begin{bmatrix} 2 \\ 0 \\ 1 \end{bmatrix} = \begin{bmatrix} \frac{7}{9} \\ \frac{3}{9} \\ -\frac{5}{9} \end{bmatrix}$

(c) $\begin{bmatrix} x_1 \\ x_2 \\ x_3 \end{bmatrix} = \begin{bmatrix} -40 & 16 & 9 \\ 13 & -5 & -3 \\ 5 & -2 & -1 \end{bmatrix} \begin{bmatrix} 1 \\ 3 \\ 15 \end{bmatrix} = \begin{bmatrix} 143 \\ -47 \\ -16 \end{bmatrix}$

(e) $\begin{bmatrix} x_1 \\ x_2 \\ x_3 \end{bmatrix} = \begin{bmatrix} 2 & 3 & 1 \\ 3 & 3 & 1 \\ 2 & 4 & 1 \end{bmatrix} \begin{bmatrix} 5 \\ -2 \\ 1 \end{bmatrix} = \begin{bmatrix} 5 \\ 10 \\ 3 \end{bmatrix}$

10. $\begin{bmatrix} x_1 \\ x_2 \\ x_3 \\ x_4 \end{bmatrix} = \begin{bmatrix} -\frac{14}{17} & \frac{8}{17} & \frac{4}{17} & \frac{5}{17} \\ \frac{3}{17} & -\frac{9}{17} & \frac{4}{17} & \frac{5}{17} \\ \frac{5}{17} & \frac{2}{17} & \frac{1}{17} & -\frac{3}{17} \\ \frac{18}{17} & -\frac{3}{17} & -\frac{10}{17} & -\frac{4}{17} \end{bmatrix} \begin{bmatrix} 5 \\ 6 \\ 1 \\ 7 \end{bmatrix} = \begin{bmatrix} 1 \\ 0 \\ 1 \\ 2 \end{bmatrix}$

12. (a) $(cA)\left(\frac{1}{c}A^{-1}\right) = \left(c\frac{1}{c}\right)(AA^{-1}) = I$, and

$$\left(\frac{1}{c}A^{-1}\right)(cA) = \left(\frac{1}{c}c\right)(A^{-1}A) = I.$$

Thus $(cA)^{-1} = \frac{1}{c}A^{-1}$.

13. $A = \begin{bmatrix} \frac{3}{2} & -\frac{1}{2} \\ -2 & 1 \end{bmatrix}$

15. (a) $(3A)^{-1} = \begin{bmatrix} \frac{2}{3} & -\frac{1}{3} \\ -\frac{5}{3} & 1 \end{bmatrix}$

(c) $A^{-2} = (A^{-1})^2 = \begin{bmatrix} 9 & -5 \\ -25 & 14 \end{bmatrix}$

16. (a) $(2A^t)^{-1} = \frac{1}{2}\begin{bmatrix} 2 & -9 \\ -1 & 5 \end{bmatrix}$

(c) $(AA^t)^{-1} = \begin{bmatrix} 85 & -47 \\ -47 & 26 \end{bmatrix}$

17. $x = 2$

19. $A = \begin{bmatrix} -\frac{1}{4} & \frac{1}{4} \\ -\frac{3}{16} & \frac{1}{8} \end{bmatrix}$

21. $(A^tB^t)^{-1} = (B^t)^{-1}(A^t)^{-1} = (B^{-1})^t(A^{-1})^t = (A^{-1}B^{-1})^t$

24. (a) $AB = AC$ so $A^{-1}AB = A^{-1}AC$. Thus $B = C$.

28. (a) True: A is invertible. Thus there exists a matrix denoted A^{-1}, such that $AA^{-1} = A^{-1}A = I$. This also shows that the inverse of A^{-1} is A.

(c) False: For example, let

$$A = \begin{bmatrix} 1 & 0 & 0 \\ 0 & 0 & 1 \\ 0 & 1 & 0 \end{bmatrix}, A^{-1} = \begin{bmatrix} 1 & 0 & 0 \\ 0 & 0 & 1 \\ 0 & 1 & 0 \end{bmatrix}$$

29. (a) 8,125 multiplications and 7,800 additions

(b) 16,250 multiplications and 15,000 additions

31. $\begin{bmatrix} 0 & 1 & 0 \\ 1 & 0 & 0 \\ 0 & 0 & 1 \end{bmatrix} \begin{bmatrix} a & b & c \\ d & e & f \\ g & h & i \end{bmatrix} = \begin{bmatrix} d & e & f \\ a & b & c \\ g & h & i \end{bmatrix}.$

$\begin{bmatrix} 1 & 0 & 0 \\ 0 & 1 & 0 \\ 0 & 0 & -2 \end{bmatrix} \begin{bmatrix} a & b & c \\ d & e & f \\ g & h & i \end{bmatrix} = \begin{bmatrix} a & b & c \\ d & e & f \\ -2g & -2h & -2i \end{bmatrix}.$

$\begin{bmatrix} 1 & 0 & 0 \\ 0 & 1 & 0 \\ 4 & 0 & 1 \end{bmatrix} \begin{bmatrix} a & b & c \\ d & e & f \\ g & h & i \end{bmatrix} =$

$\begin{bmatrix} a & b & c \\ d & e & f \\ g + 4a & h + 4b & i + 4c \end{bmatrix}.$

33. (a) Interchange rows 1 and 3.

(c) Multiply row 3 by 4.

(e) Add -4 times row 1 to row 3.

34. $0 \leq Y \leq 255, -151.98 \leq I \leq 151.89,$
$-133.365 \leq Q \leq 133.365.$

37. 57, 44, 26, 24, 17, 13, -1, 6

39. PEACE

Exercise Set 2.5

1. (a) $a_{32} = 0.25$

(c) electrical industry $(a_{43} = 0.30)$

(e) steel industry $(a_{21} = 0.40)$

2. $\begin{bmatrix} 60 \\ 40 \end{bmatrix}, \begin{bmatrix} \frac{45}{2} \\ \frac{50}{3} \end{bmatrix}, \begin{bmatrix} 15 \\ 20 \end{bmatrix}$

4. $\begin{bmatrix} 210 \\ 175 \end{bmatrix}, \begin{bmatrix} \frac{100}{7} \\ \frac{50}{3} \end{bmatrix}, \begin{bmatrix} 40 \\ \frac{70}{3} \end{bmatrix}, \begin{bmatrix} 150 \\ 105 \end{bmatrix}$

5. $\begin{bmatrix} 15 \\ 24 \\ 32 \end{bmatrix}, \begin{bmatrix} 15 \\ 32 \\ 56 \end{bmatrix}, \begin{bmatrix} 30 \\ 56 \\ 48 \end{bmatrix}$

7. $D = \begin{bmatrix} 2.4 \\ 5 \end{bmatrix}$

9. $D = \begin{bmatrix} 4 \\ 0.9 \\ 0.65 \end{bmatrix}$

Exercise Set 2.6

1. (a) stochastic

(c) not stochastic

(e) stochastic

3. If A and B are doubly stochastic matrices,
then AB is stochastic, and A^t and B^t are stochastic,
so $B^t A^t = (AB)^t$ is stochastic. Thus AB is doubly
stochastic.

4. (a) $p_{21} = 0.1$

(c) Vacant land in 2000 has the highest probability (0.35)
of becoming office land in 2005.

5. (a) 0.0780

7. 2008: City: 81.13 million, Suburb: 162.46 million,
Nonmetro: 53.41 million

2009: City: 80.3106 million, Suburb: 161.9118 million,
Nonmetro: 54.7777 million

2010: City: 79.5389 million, Suburb: 161.3567 million,
Nonmetro: 56.1044 million

Probability of living in the city in 2007 then living in a
nonmetropolitan area in 2009 is .0197.

8. 2010: City population 79.6354, suburban population
172.7883.

11. (a) 0.2

(b) White-collar 14,000, Manual 16,000

13. Small 59,488, Large 30,512

15. $AA \frac{1}{4}, Aa \frac{1}{2}, aa \frac{1}{4}$

Exercise Set 2.7

1. (a)

adjacency matrix

$$\begin{bmatrix} 0 & 1 & 0 \\ 1 & 0 & 1 \\ 1 & 0 & 0 \end{bmatrix},$$

distance matrix

$$\begin{bmatrix} 0 & 1 & 2 \\ 1 & 0 & 1 \\ 1 & 2 & 0 \end{bmatrix}$$

(c)

adjacency matrix

$$\begin{bmatrix} 0 & 1 & 1 & 1 & 1 \\ 0 & 0 & 0 & 0 & 0 \\ 0 & 0 & 0 & 0 & 0 \\ 0 & 0 & 0 & 0 & 0 \\ 0 & 0 & 0 & 0 & 0 \end{bmatrix},$$

distance matrix

$$\begin{bmatrix} 0 & 1 & 1 & 1 & 1 \\ x & 0 & x & x & x \\ x & x & 0 & x & x \\ x & x & x & 0 & x \\ x & x & x & x & 0 \end{bmatrix}$$

2. (a) 2   **(c)** undefined

3. (a) P_1 P_2 **(d)** P_2 P_3

          P_4 P_3 P_5 P_4

5.

	R	B	D	G	I
Raccoon	0	0	0	0	1
Bird	0	0	0	1	1
Deer	0	0	0	1	0
Grass	0	0	0	0	0
Insect	0	0	0	1	0

7. P_2 P_3 **(a)** $P_2 \to P_3 \to P_4 \to P_5$
                              length $= 3$

          P_1 **(b)** $P_3 \to P_4 \to P_5 \to P_1 \to P_2$
                              length $= 4$

    P_5 P_4

distance matrix

$$\begin{bmatrix} 0 & 1 & 1 & 2 & 3 \\ 1 & 0 & 1 & 2 & 3 \\ 3 & 4 & 0 & 1 & 2 \\ 2 & 3 & 3 & 0 & 1 \\ 1 & 2 & 2 & 3 & 0 \end{bmatrix}$$

8. (a) $(a_{22})^2 = 1$. One 2-path from P_2 to P_2.

$(a_{24})^2 = 0$. No 2-path from P_2 to P_4.

$(a_{31})^2 = 1$. One 2-path from P_3 to P_1.

$(a_{42})^2 = 1$. One 2-path from P_4 to P_2.

$(a_{12})^3 = 1$. One 3-path from P_1 to P_2.

$(a_{24})^3 = 0$. No 3-path from P_2 to P_4.

$(a_{32})^3 = 1$. One 3-path from P_3 to P_2.

$(a_{41})^3 = 1$. One 3-path from P_4 to P_1.

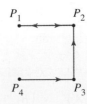

9. (a) No arcs from P_3 to any other vertex.

(c) Three arcs from P_5.

(f) No 4-paths lead to P_3.

10. (a) No arcs from P_2 to any other vertex.

(c) There is an arc from P_4 to every other vertex.

(e) Five arcs from P_3.

(g) Seven arcs in the digraph.

(i) Four 5-paths lead to P_3.

11.

Digraph

13. (a)

Digraph

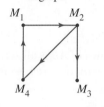

$$D = \begin{bmatrix} 0 & 1 & 2 & 2 \\ 2 & 0 & 1 & 1 \\ x & x & 0 & x \\ 1 & 2 & 3 & 0 \end{bmatrix} \begin{matrix} 5 \\ 4 \\ 3x \\ 6 \end{matrix}$$

Most to least influential: M_2, M_1, M_4, M_3

15.

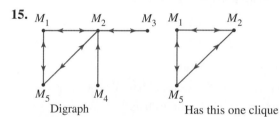

Digraph   Has this one clique

17. If the digraph of A contains an arc from P_i to P_j, then the digraph of A^t contains an arc from P_j to P_i. If the digraph of A does not contain an arc from P_i to P_j, then the digraph of A^t does not contain an arc from P_j to P_i

19. If a path from P_i to P_j contains P_k twice, $P_i \rightarrow \cdots \rightarrow P_k \rightarrow \cdots \rightarrow P_k \rightarrow \cdots \rightarrow P_j$, then there is a shorter path from P_i to P_j obtained by omitting the path from P_k to P_k

21. $a_{ik}a_{jk} = 1$ if station k can receive messages directly from both station i and station j and $a_{ik}a_{jk} = 0$, otherwise. Thus $c_{ij} = a_{i1}a_{j1} + a_{i2}a_{j2} + \cdots + a_{in}a_{jn}$ is the number of stations that can receive messages directly from both station i and station j.

22. (a) $\begin{bmatrix} 1 & 1 & 0 & 0 \\ 1 & 1 & 0 & 0 \\ 1 & 1 & 1 & 0 \\ 1 & 1 & 1 & 1 \end{bmatrix}$ **(c)** $\begin{bmatrix} 1 & 1 & 1 & 1 \\ 1 & 1 & 1 & 1 \\ 1 & 1 & 1 & 1 \\ 1 & 1 & 1 & 1 \end{bmatrix}$

23. $n - 1$

24. (a) Yes   **(b)** No

28. (a) No

Chapter 2 Review Exercises

1. (a) $\begin{bmatrix} 28 & 0 \\ 100 & -6 \end{bmatrix}$ **(b)** does not exist

(c) $\begin{bmatrix} 28 & 0 \\ 69 & -6 \end{bmatrix}$ **(d)** $\begin{bmatrix} -6 \\ 26 \end{bmatrix}$

(e) $\begin{bmatrix} 54 & -9 & 27 \\ 46 & -6 & 14 \end{bmatrix}$ **(f)** does not exist

2. (a) 2×2 **(b)** 2×3

(c) 2×2 **(d)** does not exist

(e) 3×2 **(f)** does not exist

(g) 2×2

3. (a) 16   **(b)** -8

4. (a) $AB_1 = \begin{bmatrix} 3 & 1 \\ 7 & 2 \end{bmatrix}\begin{bmatrix} 1 \\ 2 \end{bmatrix} = \begin{bmatrix} 5 \\ 11 \end{bmatrix},$

$AB_2 = \begin{bmatrix} 3 & 1 \\ 7 & 2 \end{bmatrix}\begin{bmatrix} 3 \\ 0 \end{bmatrix} = \begin{bmatrix} 9 \\ 21 \end{bmatrix},$

$AB_3 = \begin{bmatrix} 3 & 1 \\ 7 & 2 \end{bmatrix}\begin{bmatrix} 6 \\ -1 \end{bmatrix} = \begin{bmatrix} 17 \\ 40 \end{bmatrix}.$

$AC = \begin{bmatrix} 5 & 9 & 17 \\ 11 & 21 & 40 \end{bmatrix}.$

(b) $PQ = 2\begin{bmatrix} 1 \\ 5 \end{bmatrix} - 3\begin{bmatrix} 2 \\ -1 \end{bmatrix} + 5\begin{bmatrix} 3 \\ 4 \end{bmatrix}$

(c) $B = \begin{bmatrix} 2 & 4 & 2 \\ -1 & 3 & 7 \\ 0 & 1 & -2 \end{bmatrix}, \begin{bmatrix} 2 & 4 & 2 \\ -1 & 3 & 7 \\ 0 & 1 & -2 \end{bmatrix},$

$\begin{bmatrix} 2 & 4 & 2 \\ -1 & 3 & 7 \\ 0 & 1 & -2 \end{bmatrix}, \text{ or } \begin{bmatrix} 2 & 4 & 2 \\ -1 & 3 & 7 \\ 0 & 1 & -2 \end{bmatrix}.$

5. (a) $\begin{bmatrix} 9 & 0 \\ 5 & 4 \end{bmatrix}$ **(b)** $\begin{bmatrix} -4 & 1 \\ 4 & -2 \end{bmatrix}$

(c) $\begin{bmatrix} -39 & 15 \\ 24 & -12 \end{bmatrix}$ **(d)** $\begin{bmatrix} 4 & 2 \\ 0 & 2 \end{bmatrix}$

6. e.g., $X_3 = \begin{bmatrix} 5 \\ 3 \\ -1 \\ 2 \end{bmatrix} + \begin{bmatrix} 3 \\ -1 \\ 1 \\ 2 \end{bmatrix} = \begin{bmatrix} 8 \\ 2 \\ 0 \\ 4 \end{bmatrix}$

$X_4 = 2\begin{bmatrix} 5 \\ 3 \\ -1 \\ 2 \end{bmatrix} = \begin{bmatrix} 10 \\ 6 \\ -2 \\ 4 \end{bmatrix}$

$X_5 = 3\begin{bmatrix} 3 \\ -1 \\ 1 \\ 2 \end{bmatrix} = \begin{bmatrix} 9 \\ -3 \\ 3 \\ 6 \end{bmatrix}$

$X_6 = X_4 + X_5 = \begin{bmatrix} 10 \\ 6 \\ -2 \\ 4 \end{bmatrix} + \begin{bmatrix} 9 \\ -3 \\ 3 \\ 6 \end{bmatrix} = \begin{bmatrix} 19 \\ 3 \\ 1 \\ 10 \end{bmatrix}$

Solution when $x = 1$, $y = 9$: $\begin{bmatrix} 1 \\ 9 \\ -5 \\ -2 \end{bmatrix}$

7. (a) $\begin{bmatrix} \frac{1}{9} & \frac{4}{9} \\ \frac{2}{9} & -\frac{1}{9} \end{bmatrix}$

(b) $\begin{bmatrix} 0 & 2 & -1 \\ 1 & 1 & -1 \\ -\frac{2}{3} & -1 & 1 \end{bmatrix}$

(c) $\begin{bmatrix} -40 & 16 & 9 \\ 13 & -5 & -3 \\ 5 & -2 & -1 \end{bmatrix}$

8. $\begin{bmatrix} x_1 \\ x_2 \\ x_3 \end{bmatrix} = \begin{bmatrix} 14 & -8 & -1 \\ -17 & 10 & 1 \\ -19 & 11 & 1 \end{bmatrix}\begin{bmatrix} 1 \\ 5 \\ 7 \end{bmatrix} = \begin{bmatrix} -33 \\ 40 \\ 43 \end{bmatrix}$

9. $A = \begin{bmatrix} 3 & 6 \\ 2 & 5 \end{bmatrix}$

11. (a) $R1 \leftrightarrow R3:\begin{bmatrix} 0 & 0 & 1 \\ 0 & 1 & 0 \\ 1 & 0 & 0 \end{bmatrix}$

$-4R2 + R1:\begin{bmatrix} 1 & -4 & 0 \\ 0 & 1 & 0 \\ 0 & 0 & 1 \end{bmatrix}$

(b) Add 2 times row 1 to row 3. Interchange rows 2 and 3 of I_3.

12. $(cA)^n = (cA)(cA) \cdots (cA) = c^n A^n.$

13. The ith diagonal element of AA^t is

$$a_{i1}a_{i1} + a_{i2}a_{i2} + \cdots + a_{in}a_{in}$$
$$= (a_{i1})^2 + (a_{i2})^2 + \cdots + (a_{in})^2.$$

Since each term is a square, the sum can equal zero only if each term is zero, i.e., if each $a_{ij} = 0$. Thus if $AA^t = 0$, then $A = 0$.

14. If A is a symmetric matrix, then $A = A^t$, so $AA^t = A^2 = A^tA$ and A is normal.

15. If $A = A^2$, then $A^t = (A^2)^t = (A^t)^2$, so A^t is idempotent.

16. If $n < p$, $A^n \neq 0$, so $(A^n)^t \neq 0$, so $(A^t)^n \neq 0$ (Exercise 7, Section 2.3). $A^p = 0$, so $(A^p)^t = (A^t)^p = 0$. Thus, A^t is nilpotent with degree of nilpotency $= p$.

17. $A = A^t$, so $A^{-1} = (A^t)^{-1} = (A^{-1})^t$, so A^{-1} is symmetric.

18. If row i of A is all zeros then row i of AB is all zeros for any matrix B. The ith diagonal term of I_n is 1, so there is no matrix B for which $AB = I_n$.

19. $A + B = \begin{bmatrix} 5 + i & 5 - 5i \\ 6 + 10i & -3 + i \end{bmatrix}$

$AB = \begin{bmatrix} 35 + 24i & -3 + 6i \\ 7 + 6i & 12 - 6i \end{bmatrix}$

$A^* = A$, so A is hermitian.

20. If A is a real symmetric matrix, then $\overline{A} = A = A^t$ so $A^* = A^t = A$. Thus A is hermitian.

21. $g_{12} = 1$, $g_{13} = 0$, $g_{14} = 0$, $g_{15} = 0$, $g_{23} = 0$, $g_{24} = 0$, $g_{25} = 1$, $g_{34} = 1$, $g_{35} = 1$, $g_{45} = 0$, so $1 \to 2 \to 5 \to 3 \to 4$ or $4 \to 3 \to 5 \to 2 \to 1$.

$p_{12} = 1$, $p_{13} = 0$, $p_{14} = 0$, $p_{23} = 1$, $p_{24} = 0$, $p_{34} = 1$, so $1 \to 2 \to 3 \to 4$ or $4 \to 3 \to 2 \to 1$.

22. 559,875 college-educated, 490,125 noncollege-educated. Probability 0.41.

23. (a) No arcs lead to vertex 4.

(b) There are two arcs from vertex 3.

(c) There are four 3-paths from vertex 2.

(d) No 2-paths lead to vertex 3.

(e) There are two 3-paths from vertex 4 to vertex 4.

(f) There are three pairs of vertices joined by 4-paths.

CHAPTER 3

Exercise Set 3.1

1. (a) 7   (c) 14
2. (a) 3   (c) 13
3. (a) $M_{11} = C_{11} = -6$ (c) $M_{23} = -13, C_{23} = 13$
4. (a) $M_{11} = C_{13} = 12$ (c) $M_{31} = C_{31} = -3$
5. (a) $M_{12} = -59, C_{12} = 59$

   (c) $M_{33} = -177, C_{33} = -177$
6. (a) -15 (c) -7
7. (a) 393   (c) 0
8. (a) -31 (c) -45
9. (a) -31 (c) 12
10. (a) -27 (c) -72
11. (a) -248 (c) 21
12. $x = 5$ or $x = -1$
14. $x = \sqrt{3}$ or $x = -\sqrt{3}$
16. They have the same cofactor expansion using the third column.
18. (a) even   (c) odd

    (e) even
19. (a) odd   (c) odd

    (e) odd

Exercise Set 3.2

1. (a) -5 (c) 0
2. (a) 20   (c) 0
3. (a) -4 (c) -2
4. (a) -5 (c) 5
5. The second answer is correct.
6. (a) Row 3 is all zeros.

   (c) Row 3 is -3 times row 1.
7. (a) Column 3 is all zeros.

   (c) Row 3 is 3 times row 1.
8. (a) 12   (c) 9

   (e) 9
9. (a) -6 (c) -6

   (e) -12

10. (a) 3   (c) 3
12. (a) -12 (b) -60
13. (a) -4 (c) 0
14. (a) -104
15. Expand by row (or column) 1 at each stage.
18. Suppose the sum of the elements in each column is 0. Add each row after the first to row 1 to get a matrix B, which is row equivalent to A and has the same determinant as A. The first row of B is all zeros, so $|A| = |B| = 0$.
20. $|AB| = |A||B| = |B||A| = |BA|$.
23. If B is obtained from A using an elementary row operation, then $|B| = |A|$ or $|B| = -|A|$ or $|B| = c|A|$. Since $c \neq 0$ by definition, $|B| \neq 0$ if and only if $|A| \neq 0$.

Exercise Set 3.3

1. (a) invertible   (c) not invertible
2. (a) invertible   (c) invertible
3. (a) invertible   (c) invertible
4. (a) not invertible   (c) invertible
5. (a) $\dfrac{-1}{10}\begin{bmatrix} 2 & -4 \\ -3 & 1 \end{bmatrix}$ (c) not invertible

6. (a) $\dfrac{-1}{3}\begin{bmatrix} -7 & 9 & 1 \\ 8 & -9 & -2 \\ -4 & 3 & 1 \end{bmatrix}$

   (c) $\dfrac{-1}{4}\begin{bmatrix} -6 & 2 & -2 \\ -3 & 1 & 1 \\ -8 & 4 & 0 \end{bmatrix}$

7. (a) $-\begin{bmatrix} -1 & 2 & 0 \\ 2 & -1 & -2 \\ 0 & -2 & 1 \end{bmatrix}$

   (c) The inverse does not exist.
8. (a) $x_1 = 2, x_2 = 3$

   (c) $x_1 = 2, x_2 = 3$
9. (a) $x_1 = -2, x_2 = 5$

   (b) $x_1 = 1, x_2 = 4$
10. (a) $x_1 = 1, x_2 = 2, x_3 = -1$

    (c) $x_1 = \dfrac{1}{2}, x_2 = \dfrac{1}{4}, x_3 = \dfrac{1}{4}$

11. (a) $x_1 = 3, x_2 = 0, x_3 = 1$

(c) $x_1 = \dfrac{1}{8}, x_2 = \dfrac{1}{4}, x_3 = \dfrac{1}{2}$

12. (a) $|A| = 0$, so this system of equations cannot be solved using Cramer's rule.

(c) $x_1 = \dfrac{1}{2}, x_2 = \dfrac{3}{8}, x_3 = \dfrac{3}{4}$

13. (a) not a unique solution

(c) not a unique solution

14. (a) unique solution

(c) not a unique solution

15. $\lambda = 7, x_1 = x_2 = r; \lambda = -4, x_1 = -\dfrac{6r}{5}, x_2 = r.$

17. $\lambda = 1, x_1 = -s - \dfrac{r}{2}, x_2 = s, x_3 = r;$

$\lambda = 10, x_1 = x_2 = 2r, x_3 = r.$

18. $AX = \lambda X = \lambda I_n X$, so $AX - \lambda I_n X = 0$. Thus $(A - \lambda I_n)X = 0$, and there is a nontrivial solution if and only if $|A - \lambda I_n| = 0$.

21. If A is invertible then $\dfrac{1}{|A|} \mathrm{adj}(A) = A^{-1}$, so that

$A\dfrac{1}{|A|}\mathrm{adj}(A) = AA^{-1} = I_n$. Thus $\dfrac{1}{|A|}A = [\mathrm{adj}(A)]^{-1}$.

25. If $|A| = \pm 1$, then $A^{-1} = \dfrac{1}{|A|}\mathrm{adj}(A) = \pm\mathrm{adj}(A)$, and since all elements of A are integers, all elements of $\mathrm{adj}(A)$ are integers.

27. $AX = B_2$ has a unique solution if and only if $|A| \neq 0$ if and only if $AX = B_1$ has a unique solution.

28. (a) True: $|A^2| = |AA| = |A||A| = (|A|)^2$.

(c) True: $A^{-1} = \dfrac{\mathrm{adj}(A)}{|A|}$ (Thm 3.5). Thus if $|A| = 1$,

$A^{-1} = \mathrm{adj}(A)$.

Chapter 3 Review Exercises

1. (a) -7 **(b)** -18 **(c)** 29

2. (a) $M_{12} = -13, C_{12} = 13$

(b) $M_{31} = 1, C_{31} = 1$

(c) $M_{22} = 4, C_{22} = 4$

3. (a) 63 **(b)** 40

4. $x = 3$ or $x = 2$

5. (a) -15 **(b)** 51 **(c)** -65

6. (a) 6 **(b)** 2 **(c)** -12

7. (a) 10 **(b)** -20 **(c)** 82

8. (a) -54 **(b)** 32 **(c)** -8

(d) 16 **(e)** -8 **(f)** -4

9. $|B| \neq 0$ so B^{-1} exists, and A and B^{-1} can be multiplied. Let $C = AB^{-1}$. Then $CB = AB^{-1}B = A$.

10. $|C^{-1}AC| = |C^{-1}||AC| = |C^{-1}||A||C|$
$= |C^{-1}||C||A| = |C^{-1}C||A| = |A|$

12. If $A^2 = A$ then $|A||A| = |A|$ so $|A| = 1$ or 0. If A is also invertible then $|A| \neq 0$ so $|A| = 1$.

13. (a) $\begin{bmatrix} 2 & -5 \\ -1 & 3 \end{bmatrix}$ **(b)** $\dfrac{1}{17}\begin{bmatrix} 5 & -2 \\ 1 & 3 \end{bmatrix}$

(c) no inverse **(d)** $\dfrac{1}{20}\begin{bmatrix} 4 & -5 & 3 \\ 36 & 10 & -18 \\ -8 & 0 & 4 \end{bmatrix}$

14. (a) $x_1 = 1, x_2 = -3$ **(b)** $x_1 = 2, x_2 = -1, x_3 = 0$

16. $|A|$ is the product of the diagonal elements, and since $|A| \neq 0$ all diagonal elements must be nonzero.

17. If $|A| = \pm 1$ then $A^{-1} = \pm\mathrm{adj}(A)$, so $X = A^{-1}AX = A^{-1}B = \pm\mathrm{adj}(A)B$, which has all integer components.

CHAPTER ④

Exercise Set 4.1

1.   **2.**

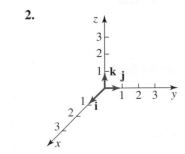

3. (a)

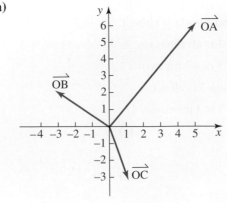

(c)

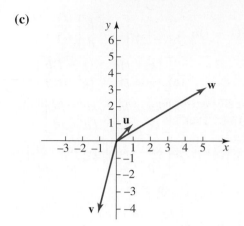

5. (a) $(3, 12)$ **(c)** $(1, 3)$

   (d) $(-1, -2, -1)$ **(g)** $(-5, 20, -15, 10, -25)$

6. (a) $(-2, 7)$ **(b)** $(13, -1)$

   (d) $(17, -4)$

7. (a) $(4, 5, 1)$ **(b)** $(3, 5, 8)$

   (d) $(24, 23, -1)$

8. (a) $\begin{bmatrix} 1 \\ -1 \end{bmatrix}$ **(b)** $\begin{bmatrix} -14 \\ 10 \end{bmatrix}$

9. (a) $\begin{bmatrix} 7 \\ 2 \\ 1 \end{bmatrix}$ **(c)** $\begin{bmatrix} -7 \\ 6 \\ 15 \end{bmatrix}$

Exercise Set 4.2

1. (a) 10 **(c)** 0

2. (a) 6 **(c)** 0

3. (a) 7 **(c)** -1 **(e)** 0

4. (a) 13 **(c)** 26

5. (a) $\sqrt{5}$ **(c)** 4 **(e)** 27

6. (a) $\sqrt{11}$ **(c)** $3\sqrt{3}$ **(e)** $\sqrt{62}$

7. (a) $\sqrt{29}$ **(c)** $\sqrt{30}$ **(e)** $\sqrt{30}$

8. (a) 5 **(c)** $\sqrt{14}$ **(e)** $\sqrt{119}$

9. (a) $\left(\dfrac{1}{\sqrt{10}}, \dfrac{3}{\sqrt{10}} \right)$ **(c)** $\left(\dfrac{1}{\sqrt{14}}, \dfrac{2}{\sqrt{14}}, \dfrac{3}{\sqrt{14}} \right)$

   (d) $\left(\dfrac{-1}{\sqrt{5}}, \dfrac{2}{\sqrt{5}}, 0 \right)$

10. (a) $\left(\dfrac{2}{\sqrt{5}}, \dfrac{1}{\sqrt{5}} \right)$ **(c)** $\left(\dfrac{7}{3\sqrt{6}}, \dfrac{2}{3\sqrt{6}}, 0, \dfrac{1}{3\sqrt{6}} \right)$

   (e) $(0, 0, 0, 1, 0, 0)$

11. (a) $\begin{bmatrix} \dfrac{4}{5} \\ \dfrac{3}{5} \end{bmatrix}$ **(c)** $\begin{bmatrix} \dfrac{3}{5} \\ \dfrac{4}{5} \\ 0 \end{bmatrix}$ **(e)** $\begin{bmatrix} \dfrac{3}{\sqrt{74}} \\ 0 \\ \dfrac{1}{\sqrt{74}} \\ \dfrac{8}{\sqrt{74}} \end{bmatrix}$

12. (a) $45°$ **(b)** $60°$

13. (a) $\dfrac{5}{\sqrt{17}\sqrt{13}}$ $(\theta = 70.3462°)$

   (b) $\dfrac{13}{6\sqrt{7}}$ $(\theta = 35.0229°)$

   (d) $\dfrac{23}{10\sqrt{7}}$ $(\theta = 29.6205°)$

14. (a) $\dfrac{7}{\sqrt{5}\sqrt{17}}$ $(\theta = 40.6013°)$

   (c) $\dfrac{-13}{10\sqrt{3}}$ $(\theta = 138.6385°)$

15. (a) $(1, 3) \cdot (3, -1) = (1 \times 3) + (3 \times -1) = 0$, so the vectors are orthogonal.

16. (a) $(3, -5) \cdot (5, 3) = (3 \times 5) + (-5 \times 3) = 0$.

   (c) $(7, 1, 0) \cdot (2, -14, 3)$
$= (7 \times 2) + (1 \times -14) + (0 \times 3) = 0$.

17. (a) $\begin{bmatrix} 1 \\ 2 \end{bmatrix} \cdot \begin{bmatrix} -6 \\ 3 \end{bmatrix} = (1 \times -6) + (2 \times 3) = 0$.

   (c) $\begin{bmatrix} 4 \\ -1 \\ 0 \end{bmatrix} \cdot \begin{bmatrix} 2 \\ 8 \\ -1 \end{bmatrix} = (4 \times 2) + (-1 \times 8) + (0 \times -1) = 0$.

18. (a) any vector of the form $(-3b, b)$

   (c) any vector of the form $(a, -4a)$

19. (a) any vector of the form $(a, 5a)$

   (c) any vector of the form $(a, b, 5a + b)$

   (e) any vector of the form $(a, 6a + 2c + 3d, c, d)$

20. any vector of the form $(a, -3a, -5a)$

21. (a) 5 **(c)** $5\sqrt{2}$

22. (a) $2\sqrt{5}$ **(c)** $3\sqrt{6}$

   (e) $2\sqrt{11}$

24. $\mathbf{u}$ is a scalar multiple of $\mathbf{v}$, so it has the same direction as $\mathbf{v}$. The magnitude of $\mathbf{u}$ is

$$\|\mathbf{u}\| = \frac{\|1\|}{\|\mathbf{v}\|} \sqrt{(\mathbf{v}_1)^2 + (\mathbf{v}_2)^2 + \cdots + (\mathbf{v}_n)^2} = \frac{\|\mathbf{v}\|}{\|\mathbf{v}\|} = 1,$$

so $\mathbf{u}$ is a unit vector.

26. If $\mathbf{u} \cdot \mathbf{v} = \mathbf{u} \cdot \mathbf{w}$, then $\mathbf{u} \cdot (\mathbf{v} - \mathbf{w}) = 0$ for all vectors $\mathbf{u}$ in U. Since $\mathbf{v} - \mathbf{w}$ is a vector in U, this means that $(\mathbf{v} - \mathbf{w}) \cdot (\mathbf{v} - \mathbf{w}) = 0$. Therefore $\mathbf{v} - \mathbf{w} = \mathbf{0}$, so $\mathbf{v} = \mathbf{w}$.

28. (a) vector   **(c)** not valid

  (f) scalar   **(h)** not valid

29. $c = \pm 3$

31. $(a, b) \cdot (-b, a) = (a \times (-b)) + (b \times a) = 0$, so $(-b, a)$ is orthogonal to (a, b).

32. It can be shown that $(\mathbf{u} + \mathbf{v}) \cdot (\mathbf{u} - \mathbf{v}) = \|\mathbf{u}\| - \|\mathbf{v}\|$, so $\|\mathbf{u}\| = \|\mathbf{v}\|$ if and only if $(\mathbf{u} + \mathbf{v}) \cdot (\mathbf{u} - \mathbf{v}) = 0$ if and only if $\mathbf{u} + \mathbf{v}$ and $\mathbf{u} - \mathbf{v}$ are orthogonal.

34. (a) $\|(1, 2)\| = |1| + |2| = 3$,
  $\|(-3, 4)\| = |-3| + |4| = 7$,
  $\|(1, 2, -5)\| = |1| + |2| + |-5| = 8$, and
  $\|(0, -2, 7)\| = |0| + |-2| + |7| = 9$.

  (b) $\|(1, 2)\| = |2| = 2$, $\|(-3, 4)\| = |4| = 4$,
  $\|(1, 2, -5)\| = |-5| = 5$, and
  $\|(0, -2, 7)\| = |7| = 7$

35. (a) $d(\mathbf{x}, \mathbf{y}) = \|\mathbf{x} - \mathbf{y}\| \geq 0$

  (c) $d(\mathbf{x}, \mathbf{z}) = \|\mathbf{x} - \mathbf{y} + \mathbf{y} - \mathbf{z}\| \leq \|\mathbf{x} - \mathbf{y}\| + \|\mathbf{y} - \mathbf{z}\| = d(\mathbf{x}, \mathbf{y}) + d(\mathbf{y}, \mathbf{z})$, from the triangle inequality

Exercise Set 4.3

1. $\mathbf{u} = \begin{bmatrix} a & b \\ c & d \end{bmatrix}$ and $\mathbf{v} = \begin{bmatrix} e & f \\ g & h \end{bmatrix}$ in M_{22};
k and l are scalars.

axiom 2 $k\mathbf{u} = k\begin{bmatrix} a & b \\ c & d \end{bmatrix} = \begin{bmatrix} ka & kb \\ kc & kd \end{bmatrix}$ is in M_{22}.

axiom 7 $k(\mathbf{u} + \mathbf{v}) = k\begin{bmatrix} a + e & b + f \\ c + g & d + h \end{bmatrix}$

$= \begin{bmatrix} k(a + e) & k(b + f) \\ k(c + g) & k(d + h) \end{bmatrix}$

$= \begin{bmatrix} ka + ke & kb + kf \\ kc + kg & kd + kh \end{bmatrix}$

$= \begin{bmatrix} ka & kb \\ kc & kd \end{bmatrix} + \begin{bmatrix} ke & kf \\ kg & kh \end{bmatrix}$

$= k\mathbf{u} + k\mathbf{v}.$

axiom 8 $(k + l)\mathbf{u} = (k + l)\begin{bmatrix} a & b \\ c & d \end{bmatrix}$

$= \begin{bmatrix} (k + l)a & (k + l)b \\ (k + l)c & (k + l)d \end{bmatrix}$

$= \begin{bmatrix} ka + la & kb + lb \\ kc + lc & kd + ld \end{bmatrix}$

$= \begin{bmatrix} ka & kb \\ kc & kd \end{bmatrix} + \begin{bmatrix} la & lb \\ lc & ld \end{bmatrix}$

$= k\mathbf{u} + l\mathbf{u}.$

axiom 9 $k(l\mathbf{u}) = k\begin{bmatrix} la & lb \\ lc & ld \end{bmatrix}$

$= \begin{bmatrix} kla & klb \\ klc & kld \end{bmatrix}$

$= \begin{bmatrix} (kl)a & (kl)b \\ (kl)c & (kl)d \end{bmatrix}$

$= (kl)\begin{bmatrix} a & b \\ c & d \end{bmatrix}$

$= (kl)\mathbf{u}.$

axiom 10 $1\mathbf{u} = 1\begin{bmatrix} a & b \\ c & d \end{bmatrix} = \begin{bmatrix} a & b \\ c & d \end{bmatrix} = \mathbf{u}.$

2. (a) $(f + g)(x) = x^2 + x + 1$, $(2f)(x) = 2x + 4$, and $(3g)(x) = 3x^2 - 3$.

3. f, g, and h are functions and c and d are scalars.

axiom 3 $(f + g)(x) = f(x) + g(x) = g(x) + f(x)$
$= (g + f)(x),$
so $f + g = g + f.$

axiom 4 $((f + g) + h)(x) = (f + g)(x) + h(x)$
$= (f(x) + g(x)) + h(x)$
$= f(x) + (g(x) + h(x))$
$= f(x) + (g + h)(x)$
$= (f + (g + h))(x),$
so $(f + g) + h = f + (g + h).$

axiom 7 $(c(f + g))(x) = c(f(x) + g(x))$
$= c(f(x)) + c(g(x))$
$= (cf)(x) + (cg)(x)$
$= (cf + cg)(x),$
so $c(f + g) = cf + cg.$

axiom 8 $((c + d)f)(x) = (c + d)(f(x))$
$= c(f(x)) + d(f(x))$
$= (cf)(x) + (df)(x)$
$= (cf + df)(x),$
so $(c + d)f = cf + df.$

axiom 9 $c((df)(x)) = c(d(f(x)))$
$$= (cd)f(x)$$
$$= ((cd)f)(x),$$
so $(cd)f = c(df).$

axiom 10 $(1f)(x) = 1(f(x))$
$$= f(x), \quad \text{so}$$
$$1f = f.$$

4. (a) $(7 - i, 4 + 7i), (4 - 7i, 17 + 6i).$

5. $W = \{a(1, 2, 3)\}.$

axiom 1: Let $a(1, 2, 3)$ and $b(1, 2, 3)$ be elements of W. $a(1, 2, 3) + b(1, 2, 3) = (a + b)(1, 2, 3)$. This is an element of W, since it is a scalar multiple of $(1, 2, 3)$. W is closed under addition. Axiom 2: Let c be a scalar. $c(a(1, 2, 3)) = (ca)(1, 2, 3)$. This is an element of W since it is a scalar multiple of $(1, 2, 3)$. W is closed under scalar multiplication. The elements of W are all in $\mathbf{R}^3$. They inherit all other vector space properties such as $\mathbf{u} + \mathbf{v} = \mathbf{v} + \mathbf{u}$ and the zero vector $(0, 0, 0)$ from $\mathbf{R}^3$.

7. Let A be a 2×2 matrix with all its elements positive. $(-1)A$ will have all negative elements. Thus $(-1)A$ is not in W. W is not closed under scalar multiplication, thus it is not a vector space.

10. (a) Vector space

(b) Not a vector space. Not closed under addition.

15. (a) $c\mathbf{0} \underset{\text{axiom 5}}{=} c\mathbf{0} + \mathbf{0} \underset{\text{axiom 6}}{=} c\mathbf{0} + (c\mathbf{0} + -(c\mathbf{0}))$

$\underset{\text{axiom 4}}{=} (c\mathbf{0} + c\mathbf{0}) + -(c\mathbf{0})$

$\underset{\text{axiom 7}}{=} c(\mathbf{0} + \mathbf{0}) + -(c\mathbf{0})$

$\underset{\text{axiom 5}}{=} c\mathbf{0} + -(c\mathbf{0}) \underset{\text{axiom 6}}{=} \mathbf{0}.$

(c) By axiom 6 there is a vector $-(-\mathbf{v})$ such that $(-\mathbf{v}) + (-(-\mathbf{v})) = \mathbf{0}$. However, $(-\mathbf{v}) + \mathbf{v} \underset{\text{axiom 3}}{=} \mathbf{v} + (-\mathbf{v}) \underset{\text{axiom 6}}{=} \mathbf{0}$. Thus the vector $-(-\mathbf{v})$ is $\mathbf{v}$.

(e) $\mathbf{0} \underset{\text{axiom 6}}{=} a\mathbf{u} + (-a\mathbf{u}) \underset{\text{given}}{=} b\mathbf{u} + (-a\mathbf{u})$

$\underset{\text{axiom 8}}{=} (b + -a)\mathbf{u},$

so from Part (b), $b + -a = 0$, and therefore $b = a$.

Exercise Set 4.4

1. (a) $(a, 3a, 5a) + (b, 3b, 5b) =$
$(a + b, 3(a + b), 5(a + b))$ and
$c(a, 3a, 5a) = (ca, 3ca, 5ca)$; thus the sum and scalar product of vectors in the set are also in the set, and so the set is a subspace of $\mathbf{R}^2$. The set is a line defined by the vector $(1, 3, 5)$.

(c) $(a, b, a + 2b) + (c, d, c + 2d) =$
$(a + c, b + d, (a + c) + 2(b + d))$ and
$k(a, b, a + 2b) = (ka, kb, ka + 2kb)$; thus the sum and scalar product of vectors in the set are also in the

set, and so the set is a subspace of $\mathbf{R}^2$. The set is the plane $z = x + 2y$.

2. (a) $(a, 0) + (b, 0) = (a + b, 0)$ and $c(a, 0) = (ca, 0)$; the sum and scalar product of vectors in the set are also in the set, and so the set is a subspace of $\mathbf{R}^2$. The set is a line, the x-axis.

(c) $(a, 1) + (b, 1) = (a + b, 2)$. Last component is 2. $(a + b, 2)$ is not in the set. Set is not closed under addition. Thus not a subspace. Let us check scalar multiplication. $k(a, 1) = (ka, k)$. (ka, k) is not in subset unless $k = 1$. Thus not closed under scalar multiplication either.

(e) $(a, b, 0) + (d, e, 0) = (a + d, b + e, 0)$ and $c(a, b, 0) = (ca, cb, 0)$; the sum and scalar product of vectors in the set are also in the set, and so the set is a subspace of $\mathbf{R}^3$. The set is the xy-plane.

(f) $(a, b, 2) + (c, d, 2) = (a + c, b + d, 4)$. Last component is 4. $(a + c, b + d, 4)$ is not in the set. Not closed under addition. Thus not a subspace. Let us check scalar multiplication. $k(a, b, 2) = (ka, kb, 2k)$. Not in subset unless $k = 1$. Thus not closed under scalar multiplication either.

3. (a) Subspace   **(c)** Not a subspace

(e) Not a subspace

4. (a) Not a subspace   **(c)** Not a subspace

5. (a) Not a subspace   **(c)** Not a subspace

7. (a) If $(a, a + 1, b) = (0, 0, 0)$, then $a = a + 1 = b = 0$. $a = a + 1$ is impossible. Therefore $(0, 0, 0)$ is not in the set.

(c) If $(a, b, a + b - 4) = (0, 0, 0)$, then $a = b = a + b - 4 = 0$, but if $a = b = 0$ then $a + b - 4 = -4$, not zero, so $(0, 0, 0)$ is not in the set.

10. (a) Subspace   **(c)** Not a subspace

11 (a) Subspace

12. (a) Subspace   **(c)** Not a subspace

13. Every element of P_2 is an element of P_3. Both P_2 and P_3 are vector spaces with the same operations and the same set of scalars, so P_2 is a subspace of P_3.

14. Not a subspace

15. Not a subspace

17. (a) $(f + g)(x) = f(x) + g(x)$ so $(f + g)(0) = 0 + 0 = 0$, and $(cf)(x) = c(f(x))$ so $(cf)(0) = c0 = 0$. Thus this subset is a subspace.

Exercise Set 4.5

1. (a) $(-1, 7) = -3(1, -1) + (2, 4)$.

(c) Not a linear combination

2. (a) Unique combination, $(-3, 3, 7) = 2(1, -1, 2) - (2, 1, 0) + 3(-1, 2, 1)$.

(c) $(2, 7, 13)$ is not a linear combination of $(1, 2, 3)$, $(-1, 2, 4)$, and $(1, 6, 10)$.

(d) Many combinations, $(0, 10, 8) = (2 - c)(-1, 2, 3) + (2 - 2c)(1, 3, 1) + c(1, 8, 5)$, where c is any real number.

3. (a) Any vector of the form $a(1, 2) + b(3, -5)$, e.g., $1(1, 2) + 2(3, -5) = (7, -8)$. $3(1, 2) - 2(3, -5) = (-3, 16)$.

(c) Any vector of the form $a(1, -3, 5) + b(0, 1, 2)$, e.g., $1(1, -3, 5) + 1(0, 1, 2) = (1, -2, 7)$. $2(1, -3, 5) - 3(0, 1, 2) = (2, -9, 4)$.

4. e.g., $(1, 2, 3) + (1, 2, 0) = (2, 4, 3)$, $(1, 2, 3) - (1, 2, 0) = (0, 0, 3)$, $2(1, 2, 3) = (2, 4, 6)$.

6. e.g., $-(1, 2, 3) = (-1, -2, -3)$, $2(1, 2, 3) = (2, 4, 6)$, $(\frac{1}{2})(1, 2, 3) = (\frac{1}{2}, 1, \frac{3}{2})$.

9. e.g., $-(1, 2, -1, 3) = (-1, -2, 1, -3)$, $2(1, 2, -1, 3) = (2, 4, -2, 6)$, $(0.1)(1, 2, -1, 3) = (0.1, 0.2, -0.1, 0.3)$.

10. e.g., $-(2, 1, -3, 4) = (-2, -1, 3, -4)$, $(2, 1, -3, 4) + (-3, 0, 1, 5) = (-1, 1, -2, 9)$, $5(4, 1, 2, 0) = (20, 5, 10, 0)$.

11. (a) $\begin{bmatrix} 5 & 7 \\ 5 & -10 \end{bmatrix} = 2\begin{bmatrix} 1 & 2 \\ 3 & -4 \end{bmatrix} - \begin{bmatrix} 0 & 3 \\ 1 & 2 \end{bmatrix} + 3\begin{bmatrix} 1 & 2 \\ 0 & 0 \end{bmatrix}$

(c) Not a linear combination.

12. (a) $3x^2 + 2x + 9 = 3(x^2 + 1) + 2(x + 3)$.

(c) $x^2 + 4x + 5 = -2(x^2 + x - 1) + 3(x^2 + 2x + 1)$.

13. (a) Yes; $f(x) = 2h(x) - g(x)$.

14. If $\mathbf{v} = a\mathbf{v}_1 + b\mathbf{v}_2$, then $\mathbf{v} = \dfrac{a}{c_1}c_1\mathbf{v}_1 + \dfrac{b}{c_2}c_2\mathbf{v}_2$

Exercise Set 4.6

1. (a) $2(-1, 2) + (2, -4) = \mathbf{0}$. Vectors are linearly dependent.

(c) $-2(1, -2, 3) - 3(-2, 4, 1) + (-4, 8, 9) = \mathbf{0}$. The vectors are linearly dependent.

(e) Thus the vectors $a(1, 2, 5) + b(1, -2, 1) + c(2, 1, 4) = 0 \Rightarrow a = b = c = 0$. Thus the vectors are linearly independent.

2. (a) $2(2, -1, 3) + (-4, 2, -6) = \mathbf{0}$, so the vectors $(2, -1, 3)$ and $(-4, 2, -6)$ are linearly dependent, and any set of vectors containing these vectors is linearly dependent.

(c) $(3, 0, 4) + (-3, 0, -4) = \mathbf{0}$, so the vectors $(3, 0, 4)$ and $(-3, 0, -4)$ are linearly dependent, and any set of vectors containing these vectors is linearly dependent.

3. (a) $t = 2$ **(c)** $t = -7$

6. (a) Observe that $(2, 3, 4) = (1, 2, 3) + (1, 1, 1)$. Thus $1(1, 2, 3) + 1(1, 1, 1) - 1(2, 3, 4) = \mathbf{0}$.

(c) $(5, 6, 7) = (3, 4, 5) + 2(1, 1, 1)$. Thus $1(3, 4, 5) + 2(1, 1, 1) - 1(5, 6, 7) = \mathbf{0}$.

7. (a) Add any vector of the form $\mathbf{v} = a(1, -1, 0) + b(2, 1, 3)$, since $a(1, -1, 0) + b(2, 1, 3) - \mathbf{v} = \mathbf{0}$; e.g., $2(1, -1, 0) + 3(2, 1, 3) = (8, 1, 9)$. Set $\{(1, -1, 0), (2, 1, 3), (8, 1, 9)\}$ is linearly dependent.

(c) Add any vector of the form $a(1, 2, 4) + b(0, 2, 5)$; e.g., $2(1, 2, 4) - 1(0, 2, 5) = (2, 2, 3)$. Set $\{(1, 2, 4), (0, 2, 5), (2, 2, 3)\}$ is linearly dependent.

8. (a) Linearly independent:

$$a\begin{bmatrix} 1 & 0 \\ 0 & 0 \end{bmatrix} + b\begin{bmatrix} 0 & 2 \\ 0 & 0 \end{bmatrix} + c\begin{bmatrix} 0 & 0 \\ 3 & 0 \end{bmatrix}$$
$$+ d\begin{bmatrix} 0 & 0 \\ 0 & 4 \end{bmatrix} = \begin{bmatrix} 0 & 0 \\ 0 & 0 \end{bmatrix}$$

implies that $a = b = c = d = 0$.

(c) Linearly dependent:

$$2\begin{bmatrix} 1 & 2 \\ -1 & 0 \end{bmatrix} + (-3)\begin{bmatrix} 1 & 2 \\ 1 & 1 \end{bmatrix}$$
$$+ \begin{bmatrix} 1 & 2 \\ 5 & 3 \end{bmatrix} = \begin{bmatrix} 0 & 0 \\ 0 & 0 \end{bmatrix}$$

9. (a) $1(2x^2 + 1) + (-1)(x^2 + 4x) + (-1)(x^2 - 4x + 1) = 0$. Linearly dependent.

(c) $(x^2 + 3x - 1) + b(x + 3) + c(2x^2 - x + 1) = 0$ implies that $a = b = c = 0$, so the functions are linearly independent.

10. $a\mathbf{v}_1 + b\mathbf{v}_2 + (-1)(a\mathbf{v}_1 + b\mathbf{v}_2) = \mathbf{0}$ for all values of a and b.

15. (a) $\mathbf{v}_1, \mathbf{v}_2, \mathbf{v}_3$ span $\mathbf{R}^3$. They do not lie on a line or in a plane. $\mathbf{v}_1, \mathbf{v}_2$, and $\mathbf{v}_3$ are linearly independent.

(b) $\mathbf{v}_3$ lies in the space spanned by $\mathbf{v}_2, \mathbf{v}_1$. This space may be a plane or a line, depending on whether or not $\mathbf{v}_1$ and $\mathbf{v}_2$ are independent. $\mathbf{v}_1, \mathbf{v}_2$, and $\mathbf{v}_3$ are linearly dependent.

16. (a) False: Consider $(1, 4, 2) = a(1, 0, 0) + b(0, 1, 0)$. Then $(1, 4, 2) = (a, b, 0)$. System $1 = a, 4 = b$, $2 = 0$ has no solution. Geometry: $(1, 0, 0)$ and $(0, 1, 0)$ have no component in the z direction while $(1, 4, 2)$ does. Thus $(1, 4, 2)$ cannot be a linear combination of $(1, 0, 0)$ and $(0, 1, 0)$.

(b) False: A vector can be expressed as a linear combination of $(1, 0, 0), (2, 1, 0), (-1, 3, 0)$ only if the last component is zero. For example, $(0, 0, 1)$ cannot be expressed as a linear combination of these vectors.

17. (a) True: $\mathbf{v} = a\mathbf{v}_1 + b\mathbf{v}_2 + 0\mathbf{v}_3$.

(b) True: One vector spans the line defined by it. Two vectors span the plane defined by them. Need at least three vectors that do not lie in a plane to span $\mathbf{R}^3$.

18. It is more likely that the vectors will be independent. For them to be dependent, one has to be a multiple of the other.

Exercise Set 4.7

1. In each case below neither vector is a multiple of the other, so the vectors are linearly independent.

(a) $(x_2, x_1) = \dfrac{3x_2 - x_1}{5}(1, 2) + \dfrac{2x_1 - x_2}{5}(3, 1)$

(c) $(x_1, x_2) = \dfrac{x_1 + x_2}{2}(1, 1) + \dfrac{-x_1 + x_2}{2}(-1, 1)$

2. (a) and **(c)**. In each case neither vector is a multiple of the other, so the vectors are linearly independent and therefore a basis for $\mathbf{R}^2$. (Theorem 4.15)

3. (a) These two vectors are linearly independent since neither is a multiple of the other. Thus they are a basis for $\mathbf{R}^2$. (Theorem 4.15)

(b) $(-2, 6) = -2(1, -3)$, so these vectors are not linearly independent and therefore not a basis for $\mathbf{R}^2$.

4. (a) $a(1, 1, 1) + b(0, 1, 2) + c(3, 0, 1) = (0, 0, 0)$ if and only if $a + 3c = 0, a + b = 0$, and $a + 2b + c = 0$. System has the unique solution $a = b = c = 0$, so the three vectors are linearly independent. Thus they are a basis for $\mathbf{R}^3$. (Theorem 4.15)

(c) $a(0, 0, 1) + b(2, 3, 1) + c(4, 1, 2) = (0, 0, 0)$ if and only if $2b + 4c = 0, 3b + c = 0$, and $a + b + 2c = 0$. The system has the unique solution $a = b = c = 0$, so the three vectors are linearly independent. Thus they are a basis for $\mathbf{R}^3$. (Theorem 4.15)

5. (a) basis

(c) not a basis

6. (a) These vectors are linearly dependent.

(b) A basis for $\mathbf{R}^2$ can contain only two vectors.

(d) The third vector is a multiple of the second, so the set is not linearly independent.

(f) $(1, 4)$ is not in $\mathbf{R}^3$.

7. $(1, 4, 3) = 3(-1, 2, 1) + 2(2, -1, 0)$, and $(-1, 2, 1)$ and $(2, -1, 0)$ are linearly independent, so the dimension is 2 and $(-1, 2, 1)$ and $(2, -1, 0)$ are a basis for the subspace.

8. $(1, 2, -1) = 2(1, 3, 1) - (1, 4, 3)$

10. $(-3, 3, -6) = -\dfrac{3}{2}(2, -2, 4)$

12. $\{(1, 2), (0, 1)\}$

13. $\{(1, 1, 1), (1, 0, -2), (1, 0, 0)\}$

15. (a) basis $= \{(1, 1, 0), (0, 0, 1)\}$, dimension $= 2$

(c) basis $= \{(1, 0, 1), (0, 1, 1)\}$, dimension $= 2$

(e) basis $= \{(1, 0, -1), (0, 1, -1)\}$, dimension $= 2$

16. (a) basis $= \{(1, 0, 1, 1), (0, 1, 1, -1)\}$, dimension $= 2$

(c) basis $= \{(2, 0, 1, 0), (0, 1, 3, 0), (0, 0, 0, 1)\}$, dimension $= 3$

17. (a) The set $\{x^3, x^2, x, 1\}$ is a basis. The dimension is 4.

(d) $\begin{bmatrix} 1 & 0 \\ 0 & 0 \end{bmatrix}$ and $\begin{bmatrix} 0 & 0 \\ 0 & 1 \end{bmatrix}$ are a basis. The dimension is 2.

19. (a) Yes $f(x) = 2h(x) - g(x)$.

(c) $g(x) + h(x) = (2x^2 + 3) + (x^2 + 3x - 1)$
$$= 3x^2 + 3x + 2,$$
$$g(x) - h(x) = (2x^2 + 3) - (x^2 + 3x - 1)$$
$$= x^2 - 3x + 4,$$
and $2g(x) = 2(2x^2 + 3) = 4x^2 + 6$ are functions in the space spanned by $g(x)$ and $h(x)$.

20. (a) $f(x) + g(x) - h(x) = 0$, so the three given functions are not linearly independent and therefore not a basis for P_2.

(c) $\begin{bmatrix} 1 & 2 \\ 0 & 1 \end{bmatrix} - \begin{bmatrix} 3 & 4 \\ 1 & 1 \end{bmatrix} + 2\begin{bmatrix} 1 & 2 \\ 1 & 1 \end{bmatrix} - \begin{bmatrix} 0 & 2 \\ 1 & 2 \end{bmatrix}$
$$= \begin{bmatrix} 0 & 0 \\ 0 & 0 \end{bmatrix},$$
so these matrices are linearly dependent and therefore not a basis for M_{22}.

(e) The set $\{(1, 0), (0, 1)\}$ is a basis for $\mathbf{C}^2$, so the dimension of $\mathbf{C}^2$ is 2 and any set of two linearly independent vectors in $\mathbf{C}^2$ is a basis. The given vectors are linearly independent since neither is a multiple of the other, so they are a basis.

21. Since $\dim(V) = 2$, any basis for V must contain 2 vectors. In Example 3 in Section 4.6, it was shown that the vectors $\mathbf{u}_1 = \mathbf{v}_1 + \mathbf{v}_2$ and $\mathbf{u}_2 = \mathbf{v}_1 - \mathbf{v}_2$ are linearly independent. Thus from Theorem 4.15 the set $\{\mathbf{u}_1, \mathbf{u}_2\}$ is a basis for V.

23. Since V is n-dimensional, any basis for V must contain n vectors. We need only show that the vectors $c\mathbf{v}_1, c\mathbf{v}_2, \ldots, c\mathbf{v}_n$ are linearly independent.

   If $a_1 c\mathbf{v}_1 + a_2 c\mathbf{v}_2 + \cdots + a_n c\mathbf{v}_n = \mathbf{0}$, then $a_1 c = a_2 c = \cdots = a_n c = 0$ and since $c \neq 0$, $a_1 = a_2 = \cdots = a_n = 0$. Thus the vectors $c\mathbf{v}_1, c\mathbf{v}_2, \ldots, c\mathbf{v}_n$ are linearly independent and a basis for V.

25. Let $\{w_1, \ldots, w_m\}$ be a basis for W. If $m > n$ these vectors would have to be linearly dependent by Theorem 4.12. Being base vectors, they are linearly independent. Thus $m \leq n$.

26. Suppose (a, b) is orthogonal to (u_1, u_2). Then $(a, b) \cdot (u_1, u_2) = au_1 + bu_2 = 0$, so $au_1 = -bu_2$. Thus if $u_1 \neq 0$ (a, b) is orthogonal to (u_1, u_2) if and only if (a, b) is of the form $\left(\dfrac{-bu_2}{u_1}, b \right) = \dfrac{b}{u_1}(-u_2, u_1)$; that is, (a, b) is orthogonal to (u_1, u_2) if and only if (a, b) is in the vector space with basis $(-u_2, u_1)$. If $u_1 = 0$, then $b = 0$ and (a, b) is orthogonal to (u_1, u_2) if and only if (a, b) is of the form $a(1, 0)$; that is, (a, b) is in the vector space with basis $(1, 0)$.

31. (a) False. The vectors are linearly dependent.

   (c) True. $(1, 2, 3) + (0, 1, 4) = (1, 3, 7)$, and $(1, 2, 3)$ and $(0, 1, 4)$ are linearly independent, so the dimension is 2.

32. (a) False. The three vectors lie in the 2D subspace of $\mathbf{R}^3$ with basis $\{(1, 0, 0), (0, 1, 0)\}$. Thus they are linearly dependent.

   (c) False. In fact any set of more than two vectors in $\mathbf{R}^2$ is a linearly dependent set.

Exercise Set 4.8

1. (a) 2

   (c) 2

2. (a) 2

   (c) 1

   (e) 2

3. (a) $\begin{bmatrix} 1 & 0 & 0 \\ 0 & 1 & 0 \\ 0 & 0 & 1 \end{bmatrix}$. Basis $(1, 0, 0)$, $(0, 1, 0)$, and $(0, 0, 1)$. Rank 3.

   (c) $\begin{bmatrix} 1 & -3 & 2 \\ 0 & 0 & 0 \\ 0 & 0 & 0 \end{bmatrix}$. Basis $(1, -3, 2)$. Rank 1.

4. (a) $\begin{bmatrix} 1 & 0 & 0 \\ 0 & 1 & 0 \\ 0 & 0 & 1 \end{bmatrix}$. Basis $(1, 0, 0)$, $(0, 1, 0)$, and $(0, 0, 1)$. Rank 3.

   (c) $\begin{bmatrix} 1 & 0 & 1 \\ 0 & 1 & 1 \\ 0 & 0 & 0 \end{bmatrix}$. Basis $(1, 0, 1)$ and $(0, 1, 1)$. Rank 2.

5. (a) $\begin{bmatrix} 1 & 0 & 0 & 3 \\ 0 & 1 & 0 & 2 \\ 0 & 0 & 1 & -1 \end{bmatrix}$. Basis $(1, 0, 0, 3)$, $(0, 1, 0, 2)$, $(0, 0, 1, -1)$. Rank 3.

   (b) $\begin{bmatrix} 1 & 0 & 3 & -2 \\ 0 & 1 & -2 & 3 \\ 0 & 0 & 0 & 0 \end{bmatrix}$. Basis $(1, 0, 3, -2)$ and $(0, 1, -2, 3)$. Rank 2.

6. (a) The given vectors are linearly independent so they are a basis for the space they span, which is $\mathbf{R}^3$.

   (c) $\begin{bmatrix} 1 & -1 & 3 \\ 1 & 0 & 1 \\ -2 & 1 & -4 \end{bmatrix} \approx \cdots \approx \begin{bmatrix} 1 & 0 & 1 \\ 0 & 1 & -2 \\ 0 & 0 & 0 \end{bmatrix}$.

   Vectors $(1, 0, 1)$ and $(0, 1, -2)$ are a basis for the space spanned by the given vectors.

7. (a) $\begin{bmatrix} 1 & 3 & -1 & 4 \\ 1 & 3 & 0 & 6 \\ -1 & -3 & 0 & -8 \end{bmatrix} \approx \cdots \approx \begin{bmatrix} 1 & 3 & 0 & 0 \\ 0 & 0 & 1 & 0 \\ 0 & 0 & 0 & 1 \end{bmatrix}$.

   Vectors $(1, 3, 0, 0)$, $(0, 0, 1, 0)$, and $(0, 0, 0, 1)$ are a basis for the subspace. The given vectors are linearly independent so they are a basis also.

   (c) $\begin{bmatrix} 1 & 2 & 3 & 4 \\ 0 & -1 & 2 & 3 \\ 2 & 3 & 8 & 11 \\ 2 & 3 & 6 & 8 \end{bmatrix} \approx \cdots \approx \begin{bmatrix} 1 & 0 & 0 & -.5 \\ 0 & 1 & 0 & 0 \\ 0 & 0 & 1 & 1.5 \\ 0 & 0 & 0 & 0 \end{bmatrix}$.

   Vectors $(1, 0, 0, -0.5)$, $(0, 1, 0, 0)$, and $(0, 0, 1, 1.5)$ are a basis for the subspace.

8. $A = \begin{bmatrix} 1 & 2 & -1 \\ 0 & 1 & 3 \\ 1 & 4 & 6 \end{bmatrix} \approx \cdots \approx \begin{bmatrix} 1 & 0 & 0 \\ 0 & 1 & 0 \\ 0 & 0 & 1 \end{bmatrix}$.

Vectors $(1, 0, 0)$, $(0, 1, 0)$, and $(0, 0, 1)$ are a basis for the row space of A.

$A^t = \begin{bmatrix} 1 & 0 & 1 \\ 2 & 1 & 4 \\ -1 & 3 & 6 \end{bmatrix} \approx \cdots \approx \begin{bmatrix} 1 & 0 & 0 \\ 0 & 1 & 0 \\ 0 & 0 & 1 \end{bmatrix}$.

Vectors $(1, 0, 0)$, $(0, 1, 0)$, and $(0, 0, 1)$ are a basis for the row space of A^t. Therefore the column vectors

$$\begin{bmatrix} 1 \\ 0 \\ 0 \end{bmatrix}, \begin{bmatrix} 0 \\ 1 \\ 0 \end{bmatrix}, \text{ and } \begin{bmatrix} 0 \\ 0 \\ 1 \end{bmatrix}$$

are a basis for the column space of A. Both the row space and the column space of A have dimension 3.

10. (a) (i) $\begin{bmatrix} 1 & 0 & 0 \\ 0 & 1 & 0 \\ 0 & 0 & 1 \end{bmatrix}$.

(ii) Ranks 3 and 3. Thus unique solution.

(iii) $x_1 = 2$, $x_2 = 1$, $x_3 = 3$.

(iv) $\begin{bmatrix} 12 \\ 18 \\ -8 \end{bmatrix} = 2\begin{bmatrix} 1 \\ 2 \\ -1 \end{bmatrix} + 1\begin{bmatrix} -2 \\ -1 \\ 3 \end{bmatrix} + 3\begin{bmatrix} 4 \\ 5 \\ -3 \end{bmatrix}$.

(b) (i) $\begin{bmatrix} 1 & 0 & 3 \\ 0 & 1 & 2 \\ 0 & 0 & 0 \end{bmatrix}$.

(ii) Ranks 2 and 2. Thus many solutions.

(iii) $x_1 = -3r + 4$, $x_2 = -2r + 1$, $x_3 = r$.

(iv) $\begin{bmatrix} 9 \\ 7 \\ 7 \end{bmatrix} = (-3r + 4)\begin{bmatrix} 3 \\ 2 \\ 3 \end{bmatrix} + (-2r + 1)\begin{bmatrix} -3 \\ -1 \\ -5 \end{bmatrix}$

$+ r\begin{bmatrix} 3 \\ 4 \\ -1 \end{bmatrix}$.

11. Use Theorem 4.21 **(a)** Unique solution **(b)** No solution **(c)** Many solutions

12. (a) There are three rows, so the row space must have dimension ≤ 3. Five columns, column space must have dimension ≤ 5. Since dim row space = dim col space, this dim ≤ 3. Rank $(A) \leq 3$.

13. dim(column space of A) = dim(row space of A) $\leq$ number of rows of $A = m < n$.

14. (a) The rank of A cannot be greater than 3, so a basis for the column space of A can contain no more than three vectors. The four column vectors in A must therefore be linearly dependent by Theorem 4.12.

16. If the n columns of A are linearly independent, they are a basis for the column space of A and dim(column space of A) is n, so rank$(A) = n$.

   If rank$(A) = n$, then the n column vectors of A are basis for the column space of A because they span the column space of A (Theorem 4.15), but this means they are linearly independent.

20. (a) False

   (c) True

Exercise Set 4.9

1. (a) orthogonal

   (b) not orthogonal

   (d) not orthogonal

2. (a) orthogonal

   (b) not orthogonal

3. (a) orthonormal

   (c) orthonormal

   (e) not orthonormal

4. $\mathbf{v} = 3\mathbf{u}_1 + \dfrac{2}{5}\mathbf{u}_2 + \dfrac{11}{5}\mathbf{u}_3$

6. We show that the column vectors are unit vectors and that they are mutually orthogonal.

(a) $\mathbf{a}_1 = \begin{bmatrix} 1 \\ 0 \end{bmatrix}$ and $\mathbf{a}_2 = \begin{bmatrix} 0 \\ 1 \end{bmatrix}$. $\|\mathbf{a}_1\| = \|\mathbf{a}_2\| = 1$ and

$\mathbf{a}_1 \cdot \mathbf{a}_2 = 0$

(c) $\mathbf{a}_1 = \begin{bmatrix} \frac{\sqrt{3}}{2} \\ -\frac{1}{2} \end{bmatrix}$ and $\mathbf{a}_2 = \begin{bmatrix} \frac{1}{2} \\ \frac{\sqrt{3}}{2} \end{bmatrix}$. $\|\mathbf{a}_1\|^2 = \dfrac{3}{4} + \dfrac{1}{4} = 1$

and $\|\mathbf{a}_2\|^2 = \dfrac{1}{4} + \dfrac{3}{4} = 1$, so $\|\mathbf{a}_1\| = \|\mathbf{a}_2\| = 1$, and

$\mathbf{a}_1 \cdot \mathbf{a}_2 = \left(\dfrac{\sqrt{3}}{2} \times \dfrac{1}{2}\right) - \left(\dfrac{1}{2} \times \dfrac{\sqrt{3}}{2}\right) = 0$.

8. (a) $\begin{bmatrix} 0 & -1 \\ 1 & 0 \end{bmatrix}$ **(c)** $\begin{bmatrix} \frac{1}{3} & \frac{2}{3} & -\frac{2}{3} \\ \frac{2}{3} & -\frac{2}{3} & -\frac{1}{3} \\ -\frac{2}{3} & \frac{1}{3} & \frac{2}{3} \end{bmatrix}$

11. To show that A^{-1} is unitary, it is necessary to show that $(A^{-1})^{-1} = (\overline{A^{-1}})^t$. Since A is unitary $A^{-1} = \overline{A}^t$. Thus $(A^{-1})^t = (\overline{A}^t)^t = \overline{A}$, so that $(\overline{A^{-1}})^t = A = (A^{-1})^{-1}$.

13. (a) $(3, 6)$ **(c)** $(2, 4, 6)$

    (e) $\left(\frac{1}{3}, -\frac{1}{3}, \frac{2}{3}, 1 \right).$

14. (a) $\left(\frac{24}{29}, \frac{60}{29} \right)$ **(c)** $(1, 2, 0)$

    (e) $\left(-\frac{2}{15}, \frac{4}{15}, \frac{2}{15}, \frac{2}{5} \right)$

15. (a) $\left\{ \left(\frac{1}{\sqrt{5}}, \frac{2}{\sqrt{5}} \right), \left(-\frac{2}{\sqrt{5}}, \frac{1}{\sqrt{5}} \right) \right\}$

    (c) $\left\{ \left(\frac{1}{\sqrt{2}}, -\frac{1}{\sqrt{2}} \right), \left(\frac{1}{\sqrt{2}}, \frac{1}{\sqrt{2}} \right) \right\}$

16. (a) $\left\{ \left(\frac{1}{\sqrt{3}}, \frac{1}{\sqrt{3}}, \frac{1}{\sqrt{3}} \right), \left(\frac{1}{\sqrt{2}}, -\frac{1}{2}, 0 \right), \right.$

    $\left. \left(-\frac{1}{\sqrt{6}}, -\frac{1}{\sqrt{6}}, \frac{2}{\sqrt{6}} \right) \right\}$

17. (a) $\left\{ \left(\frac{1}{\sqrt{5}}, 0, \frac{2}{\sqrt{5}} \right), \left(-\frac{2}{\sqrt{5}}, 0, \frac{1}{\sqrt{5}} \right) \right\}$

19. Start with any vector that is not a multiple of $(1, 2, -1, -1)$ and use the Gram-Schmidt process to find a vector orthogonal to $(1, 2, -1, -1)$.

21. (a) $\left(\frac{5}{7}, \frac{13}{35}, \frac{86}{35} \right)$ **(c)** $\left(\frac{15}{7}, \frac{109}{35}, \frac{48}{35} \right)$

22. (a) $\left(-\frac{4}{31}, \frac{3}{31}, \frac{2}{31}, -\frac{8}{31} \right)$**(c)** $\left(\frac{21}{62}, -\frac{4}{31}, -\frac{13}{31}, \frac{11}{62} \right)$

23. $\mathbf{v} = \mathbf{w} + \mathbf{w}_\perp = \left(\frac{3}{2}, \frac{3}{2}, -1 \right) + \left(-\frac{1}{2}, \frac{1}{2}, 0 \right)$

25. $\mathbf{v} = \mathbf{w} + \mathbf{w}_\perp = \left(\frac{5}{3}, \frac{2}{3}, \frac{7}{3} \right) + \left(\frac{4}{3}, \frac{4}{3}, -\frac{4}{3} \right)$

26. $\sqrt{2}$

28. $\dfrac{\sqrt{2730}}{14}$

29. Any set $\{(1, 2, -2), (6, 1, 4), (a, b, c)\}$ where $(1, 2, -2) \cdot (a, b, c) = 0$ and $(6, 1, 4) \cdot (a, b, c) = 0$; i.e., $a + 2b - 2c = 0$ and $6a + b + 4c = 0$. One solution is $a = -10, b = 16$, and $c = 11$. Thus $\{(1, 2, -2), (6, 1, 4), (-10, 16, 11)\}$ is an orthogonal basis.

31. $\{\mathbf{u}_1, \mathbf{u}_2, \ldots, \mathbf{u}_n\}$ is a set of n linearly independent vectors in a vector space with dimension n. Therefore $\{\mathbf{u}_1, \mathbf{u}_2, \ldots, \mathbf{u}_n\}$ is a basis for the vector space.

Chapter 4 Review Exercises

1.

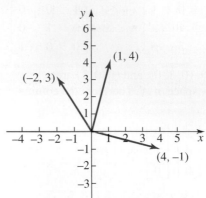

2.

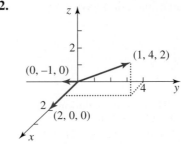

3. (a) $(3, 0, 2)$ **(b)** $(11, 0, 22)$

    (c) $(3, -3, 11)$ **(d)** $(8, -7, -3)$

    (e) $(-4, -18, -22)$

4. (a) -5 **(b)** -21 **(c)** 15

5. (a) $\sqrt{17}$ **(b)** $\sqrt{14}$ **(c)** $\sqrt{30}$

6. (a) $\dfrac{1}{\sqrt{26}}$ **(b)** 0

7. For example, $(1, 2, 0)$.

8. (a) $\sqrt{41}$ **(b)** $3\sqrt{2}$ **(c)** $5\sqrt{2}$

9. $\pm 14\sqrt{14}$

10. Let $a(1, 3, 7)$ and $b(1, 3, 7)$ be elements of W. Then $a(1, 3, 7) + b(1, 3, 7) = (a + b)(1, 3, 7)$, an element of V. Let k be a scalar. Then $k(a(1, 3, 7)) = (ka)(1, 3, 7)$, an element of V. V is closed under addition and scalar multiplication—it is a subspace.

11. U is not a vector space.

12. W is a vector space.

13. $(f + g)(x) = 3x - 1 + 2x^2 + 3 = 2x^2 + 3x + 2$, $3f(x) = 3(3x - 1) = 9x - 3$. $(2f - 3g)(x) = 2(3x - 1) - 3(2x^2 + 3) = -6x^2 + 6x - 11$.

14. (a) Let f and g be in V. Then $f(2) = 0$ and $g(2) = 0$. $(f + g)(2) = f(2) + g(2) = 0$. Thus $f + g$ is in V. Let k be a scalar. Then $kf(2) = k(f(2)) = k(0) = 0$. kf is in V. V is closed under addition and scalar multiplication. V is a subspace.

   (b) Let f and g be in W. Then $f(2) = 1$ and $g(2) = 1$. $(f + g)(2) = 2$. Thus $f + g$ is not in W. W is not closed under addition. It is not a vector space.

15. (a) Not a subspace

   (b) Subspace

   (c) Subspace

   (d) Not a subspace

16. Only the subset (a) is a subspace.

17. (a) Not a subspace

   (b) Subspace

   (c) Subspace

   (d) Not a subspace

18. Not a subspace

19. Necessary: Let $\mathbf{v}$ be in the subspace and 0 be the zero scalar. Then $0\mathbf{v} = \mathbf{0}$, the zero vector (Theorem 4.5(a)). The subspace is closed under scalar multiplication. Thus $\mathbf{0}$ is in the subspace.
   Not sufficient: The subset of $\mathbf{R}^2$ consisting of vectors of the form (a, a^2) contains the zero vector. It is not closed under addition, thus not a subspace.

20. (a) $(3, 15, -4) = 2(1, 2, -1) + 3(2, 4, 0) - (5, 1, 2)$.

   (b) Not a linear combination. The three vectors lie in the xz plane. $(-3, -4, 7)$ does not lie in this plane.

21. No: $2A + 3B - C = 2\begin{bmatrix} 1 & 2 \\ 4 & 3 \end{bmatrix} + 3\begin{bmatrix} 0 & -2 \\ 1 & 5 \end{bmatrix} - \begin{bmatrix} 1 & 1 \\ 1 & 1 \end{bmatrix} = \begin{bmatrix} 1 & -3 \\ 10 & 20 \end{bmatrix}$.

22. $(1, -2, 3)$ and $(-2, 4, -6)$ are linearly dependent. Need at least three linearly independent vectors to span $\mathbf{R}^3$. Thus vectors do not span $\mathbf{R}^3$.

23. $(10, 9, 8) = 2(-1, 3, 1) + 3(4, 1, 2)$.

24. $13x^2 + 8x - 21 = 2(2x^2 + x - 3) - 3(-3x^2 - 2x + 5)$.

25. (a) Linearly dependent

   (b) Linearly dependent

   (c) Linearly independent

26. (a) and (b). In each case the set consists of two linearly independent vectors. By Theorem 4.15 they are therefore a basis for $\mathbf{R}^2$.

   (c) and (d). In each case the set consists of three linearly independent vectors. By Theorem 4.15 they are therefore a basis for $\mathbf{R}^3$.

27. $ax^2 + bx + c = -a(-x^2) + \left(\frac{b}{3}\right)(3x) + \left(\frac{c}{2}\right)2$.

28. Any of the sets $\{(1, -2, 3), (4, 1, -1), (1, 0, 0)\}$, $\{(1, -2, 3), (4, 1, -1), (0, 1, 0)\}$, $\{(1, -2, 3), (4, 1, -1), (0, 0, 1)\}$ is a basis for $\mathbf{R}^3$.

29. $(a, b, c, a - 2b + 3c) = a(1, 0, 0, 1) + b(0, 1, 0, -2) + c(0, 0, 1, 3)$. The linearly independent set $\{(1, 0, 0, 1), (0, 1, 0, -2), (0, 0, 1, 3)\}$ is a basis for the subspace.

30. $\left\{ \begin{bmatrix} 1 & 0 & 0 \\ 0 & 0 & 0 \\ 0 & 0 & 0 \end{bmatrix}, \begin{bmatrix} 0 & 1 & 0 \\ 0 & 0 & 0 \\ 0 & 0 & 0 \end{bmatrix}, \begin{bmatrix} 0 & 0 & 1 \\ 0 & 0 & 0 \\ 0 & 0 & 0 \end{bmatrix}, \begin{bmatrix} 0 & 0 & 0 \\ 0 & 1 & 0 \\ 0 & 0 & 0 \end{bmatrix}, \begin{bmatrix} 0 & 0 & 0 \\ 0 & 0 & 1 \\ 0 & 0 & 0 \end{bmatrix}, \begin{bmatrix} 0 & 0 & 0 \\ 0 & 0 & 0 \\ 0 & 0 & 1 \end{bmatrix} \right\}$.

31. $a(x^2 + 2x - 3) + b(3x^2 + x - 1) + c(4x^2 + 3x - 3) = 0$ if and only if $a + 3b + 4c = 0$, $2a + b + 3c = 0$, and $-3a - b - 3c = 0$. This system of homogeneous equations has the unique solution $a = b = c = 0$. Thus the given functions are linearly independent. The dimension of P_2 is 3, so the three functions are a basis.

32. (a) $\begin{bmatrix} 1 & 2 & -1 \\ -1 & 3 & 4 \\ 0 & 5 & 3 \end{bmatrix} \approx \cdots \approx \begin{bmatrix} 1 & 2 & -1 \\ 0 & 1 & \frac{3}{5} \\ 0 & 0 & 0 \end{bmatrix}$, thus rank is 2.

   (b) $\begin{bmatrix} 2 & 1 & 4 \\ -2 & 0 & -1 \\ 3 & 2 & 7 \end{bmatrix} \approx \cdots \approx \begin{bmatrix} 1 & \frac{1}{2} & 2 \\ 0 & 1 & 3 \\ 0 & 0 & 1 \end{bmatrix}$, thus rank is 3.

   (c) $\begin{bmatrix} -2 & 4 & 8 \\ 1 & -2 & 4 \\ 4 & -8 & 16 \end{bmatrix} \approx \begin{bmatrix} 1 & -2 & -4 \\ 0 & 0 & 1 \\ 0 & 0 & 0 \end{bmatrix}$, thus rank is 2.

33. $\{(1, -2, 3, 4), (0, 1, 4, 2)\}$

34. $\mathbf{v} = a\mathbf{v}_1 + b\mathbf{v}_2 = a\mathbf{v}_1 + b\mathbf{v}_2 + 0\mathbf{v}_3$

35. If $a\mathbf{v}_1 + b\mathbf{v}_2 = \mathbf{0}$ then $a\mathbf{v}_1 + b\mathbf{v}_2 + 0\mathbf{v}_3 = \mathbf{0}$, and since the set $\{\mathbf{v}_1, \mathbf{v}_2, \mathbf{v}_3\}$ is linearly independent this means $a = b = 0$, so $\{\mathbf{v}_1, \mathbf{v}_2\}$ must be linearly independent.

If $a\mathbf{v}_1 + c\mathbf{v}_3 = \mathbf{0}$ then $a\mathbf{v}_1 + 0\mathbf{v}_2 + c\mathbf{v}_3 = \mathbf{0}$, and since the set $\{\mathbf{v}_1, \mathbf{v}_2, \mathbf{v}_3\}$ is linearly independent this means $a = c = 0$, so $\{\mathbf{v}_1, \mathbf{v}_3\}$ must be linearly independent.

If $b\mathbf{v}_2 + c\mathbf{v}_3 = 0$ then $0\mathbf{v}_1 + b\mathbf{v}_2 + c\mathbf{v}_3 = 0$, and since the set $\{\mathbf{v}_1, \mathbf{v}_2, \mathbf{v}_3\}$ is linearly independent this means $b = c = 0$, so $\{\mathbf{v}_2, \mathbf{v}_3\}$ must be linearly independent. Since $\{\mathbf{v}_1, \mathbf{v}_2, \mathbf{v}_3\}$ is linearly independent, none of the three vectors can be the zero vector. Thus $a\mathbf{v}_1 = \mathbf{0}$ (or $b\mathbf{v}_2 = \mathbf{0}$ or $c\mathbf{v}_3 = \mathbf{0}$) only if $a = 0$ (or $b = 0$ or $c = 0$).

36. $a(\mathbf{v}_1 + 2\mathbf{v}_2) + b(3\mathbf{v}_1 - \mathbf{v}_2) = (a + 3b)\mathbf{v}_1 + (2a - b)\mathbf{v}_2$. There are scalars c and d, not both zero, with $c\mathbf{v}_1 + d\mathbf{v}_2 = \mathbf{0}$. Solve the system

$$c = a + 3b, d = 2a - b: a = \frac{3d + c}{7}, b = \frac{2c - d}{7}.$$

$3d + c$ and $2c - d$ cannot both be zero unless both c and d are zero, so at least one of a and b is nonzero and $a(\mathbf{v}_1 + 2\mathbf{v}_2) + b(3\mathbf{v}_1 - \mathbf{v}_2) = c\mathbf{v}_1 + d\mathbf{v}_2 = \mathbf{0}$. So $\mathbf{v}_1 + 2\mathbf{v}_2$ and $3\mathbf{v}_1 - \mathbf{v}_2$ are linearly dependent.

37. If $\text{rank}(A) = n$, the row space of A is $\mathbf{R}^n$, so the reduced echelon form of A must have n linearly independent rows; i.e., it must be I_n. If A is row equivalent to I_n, the rows of I_n are linear combinations of the rows of A. But the rows of I_n span $\mathbf{R}^n$, so the rows of A span $\mathbf{R}^n$ and $\text{rank}(A) = n$.

38. (a) $\left(\dfrac{7}{5}, \dfrac{14}{5}\right)$ **(b)** $\left(-\dfrac{23}{21}, \dfrac{46}{21}, \dfrac{92}{21}\right)$

39. Orthogonal: $\mathbf{u}_1 = (1, 2, 3, -1)$,

$$\mathbf{u}_2 = \left(\frac{32}{15}, \frac{4}{15}, -\frac{9}{15}, \frac{13}{15}\right),$$

$$\mathbf{u}_3 = \left(-\frac{26}{86}, \frac{72}{86}, -\frac{33}{86}, \frac{19}{86}\right).$$

Orthonormal: $\left\{\left(\dfrac{1}{\sqrt{15}}, \dfrac{2}{\sqrt{15}}, \dfrac{3}{\sqrt{15}}, -\dfrac{1}{\sqrt{15}}\right),\right.$

$$\left(\frac{32}{\sqrt{1290}}, \frac{4}{\sqrt{1290}}, -\frac{9}{\sqrt{1290}}, \frac{13}{\sqrt{1290}}\right),$$

$$\left.\left(-\frac{26}{\sqrt{7310}}, \frac{72}{\sqrt{7310}}, -\frac{33}{\sqrt{7310}}, \frac{19}{\sqrt{7310}}\right)\right\}$$

40. $\left\{\left(\dfrac{1}{\sqrt{2}}, 0, \dfrac{1}{\sqrt{2}}\right), \left(-\dfrac{1}{\sqrt{3}}, \dfrac{1}{\sqrt{3}}, \dfrac{1}{\sqrt{3}}\right)\right\}$

41. $\left(\dfrac{294}{150}, \dfrac{345}{150}, -\dfrac{183}{150}\right)$

42. $\dfrac{1}{10}\sqrt{10}$

43. If A is orthogonal, the rows of A form an orthonormal set. The rows of A are the columns of A^t, so from the definition of orthogonal matrix, A^t is orthogonal. Interchange A and A^t in the argument above to show that if A^t is orthogonal then A is orthogonal.

44. $W = \{a(1, 0, 1) + b(0, 1, -2)\}$. $\mathbf{u}_1 = (1, 0, 1)$ and

$$\mathbf{u}_2 = (0, 1, -2) - \frac{(0, 1, -2) \cdot (1, 0, 1)}{(1, 0, 1) \cdot (1, 0, 1)}(1, 0, 1) =$$

$(0, 1, -2) - (-1)(1, 0, 1) = (1, 1, -1)$ are an

orthogonal basis. The vectors $\left(\dfrac{1}{\sqrt{2}}, 0, \dfrac{1}{\sqrt{2}}\right)$ and

$\left(\dfrac{1}{\sqrt{3}}, \dfrac{1}{\sqrt{3}}, -\dfrac{1}{\sqrt{3}}\right)$ are therefore an orthonormal basis.

$\mathbf{v} = (1, 3, -1) = \mathbf{w} + \mathbf{w}_\perp$ where $\mathbf{w} = \left(\dfrac{5}{3}, \dfrac{5}{3}, -\dfrac{5}{3}\right)$

and $\mathbf{w}_\perp = \left(-\dfrac{2}{3}, \dfrac{4}{3}, \dfrac{2}{3}\right)$.

45. Suppose $\mathbf{u}$ and $\mathbf{v}$ are orthogonal and that $a\mathbf{u} + b\mathbf{v} = 0$, so that $a\mathbf{u} = -b\mathbf{v}$. Then $a\mathbf{u} \cdot a\mathbf{u} = a\mathbf{u} \cdot (-b\mathbf{v}) = (-ab)\mathbf{u} \cdot \mathbf{v} = 0$, so $a\mathbf{u} = \mathbf{0}$. Thus $a = 0$ since $\mathbf{u} \neq \mathbf{0}$. In the same way $b = 0$, and so $\mathbf{u}$ and $\mathbf{v}$ are linearly independent.

46. If $\mathbf{u} \cdot \mathbf{v} = 0$ and $\mathbf{u} \cdot \mathbf{w} = 0$, then $\mathbf{u} \cdot (a\mathbf{v} + b\mathbf{w}) = \mathbf{u} \cdot (a\mathbf{v}) + \mathbf{u} \cdot (b\mathbf{w}) = a(\mathbf{u} \cdot \mathbf{v}) + b(\mathbf{u} \cdot \mathbf{w}) = 0$.

47. (a) True

(b) False

(c) True

(d) True

(e) False

48. (a) False

(b) True

(c) True

(d) False

CHAPTER 5

Exercise Set 5.1

1. $\lambda^2 - 7\lambda + 6; 6, 1; r\begin{bmatrix} 4 \\ 1 \end{bmatrix}, s\begin{bmatrix} -1 \\ 1 \end{bmatrix}$

2. $\lambda^2 - 5\lambda + 6; 2, 3; r\begin{bmatrix} -2 \\ 1 \end{bmatrix}, s\begin{bmatrix} -1 \\ 1 \end{bmatrix}$

4. $\lambda^2 - 2\lambda + 1; 1; r\begin{bmatrix} 1 \\ -2 \end{bmatrix}$

6. $\lambda^2 - 6\lambda + 9; 3; r\begin{bmatrix} 1 \\ 1 \end{bmatrix}$

8. $\lambda^2 - 4\lambda$; 0, 4; $r\begin{bmatrix} 2 \\ 1 \end{bmatrix}$, $s\begin{bmatrix} -2 \\ 1 \end{bmatrix}$

9. $-\lambda^3 + 2\lambda^2 + \lambda - 2$; 1, −1, 2; $r\begin{bmatrix} 1 \\ 0 \\ 1 \end{bmatrix}$, $s\begin{bmatrix} 1 \\ -1 \\ 1 \end{bmatrix}$, $t\begin{bmatrix} 0 \\ 1 \\ 1 \end{bmatrix}$

10. $(1 - \lambda)^2(3 - \lambda)$; 1, 3; $r\begin{bmatrix} 1 \\ 1 \\ 1 \end{bmatrix}$, $s\begin{bmatrix} 0 \\ 1 \\ 1 \end{bmatrix}$

13. $(1 - \lambda)(2 - \lambda)(8 - \lambda)$; 1, 2, 8; $r\begin{bmatrix} 1 \\ -1 \\ 1 \end{bmatrix}$, $s\begin{bmatrix} 0 \\ 1 \\ 1 \end{bmatrix}$, $t\begin{bmatrix} 1 \\ 0 \\ 1 \end{bmatrix}$

15. $(2 - \lambda)(2 - \lambda)(4 - \lambda)(6 - \lambda)$; 2, 4, 6;

$$r\begin{bmatrix} 1 \\ -1 \\ 0 \\ 0 \end{bmatrix} + s\begin{bmatrix} 0 \\ 0 \\ 1 \\ 1 \end{bmatrix}, t\begin{bmatrix} 0 \\ 1 \\ 0 \\ 1 \end{bmatrix}, p\begin{bmatrix} 1 \\ 0 \\ 0 \\ 1 \end{bmatrix}$$

17. Characteristic polynomial $(1 - \lambda)^2$; eigenvalue 1. Eigenvectors are all the vectors in $\mathbf{R}^2$. Matrix is the identity transformation that maps each vector in $\mathbf{R}^2$ into itself.

19. Characteristic polynomial $(-2 - \lambda)^2$; eigenvalue −2. Eigenvectors are all the vectors in $\mathbf{R}^2$. Matrix maps each vector $\mathbf{v}$ in $\mathbf{R}^2$ into the vector $-2\mathbf{v}$.

20. $\begin{vmatrix} -\lambda & -1 \\ 1 & -\lambda \end{vmatrix} = \lambda^2 + 1 \neq 0$ for any real value of λ, so there are no real eigenvalues. The given matrix is a rotation matrix that rotates each vector in $\mathbf{R}^2$ through a 90° angle. Thus no vector has the same or opposite direction as its image.

24. If A is a diagonal matrix with diagonal elements a_{ii}, then $A - \lambda I_n$ is also a diagonal matrix with diagonal elements $a_{ii} - \lambda$. Thus $|A - \lambda I_n|$ is the product of the terms $a_{ii} - \lambda$, and the solutions of the equation $|A - \lambda I_n| = 0$ are the values $\lambda = a_{ii}$, the diagonal elements of A.

26. $(A - \lambda I_n)^t = A^t - (\lambda I_n)^t = A^t - \lambda I_n$, so $|A - \lambda I_n| = |(A - \lambda I_n)^t| = |A^t - \lambda I_n|$, that is, A and A^t have the same characteristic polynomial and therefore the same eigenvalues.

29. Have that $A\mathbf{x} = \lambda\mathbf{x}$. Thus $A^{-1}(A\mathbf{x}) = A^{-1}(\lambda\mathbf{x})$. $(A^{-1}A)\mathbf{x} = \lambda A^{-1}\mathbf{x}$. $(I)\mathbf{x} = \lambda A^{-1}\mathbf{x}$. $\mathbf{x} = \lambda A^{-1}\mathbf{x}$. $\lambda^{-1}\mathbf{x} = A^{-1}\mathbf{x}$.

32. The characteristic polynomial of A is $|A - \lambda I_n| = \lambda^n + c_{n-1}\lambda^{n-1} + \cdots + c_1\lambda + c_0$. Substituting $\lambda = 0$, this equation becomes $|A| = c_0$.

34. (a) $\begin{vmatrix} -\lambda & 2 \\ -1 & 3 - \lambda \end{vmatrix} = (-\lambda)(3 - \lambda) + 2$
$$= \lambda^2 - 3\lambda + 2.$$

$$\begin{bmatrix} 0 & 2 \\ -1 & 3 \end{bmatrix}^2 - 3\begin{bmatrix} 0 & 2 \\ -1 & 3 \end{bmatrix} + 2\begin{bmatrix} 1 & 0 \\ 0 & 1 \end{bmatrix}$$

$$= \begin{bmatrix} -2 & 6 \\ -3 & 7 \end{bmatrix} + \begin{bmatrix} 2 & -6 \\ 3 & -7 \end{bmatrix}$$

$$= \begin{bmatrix} 0 & 0 \\ 0 & 0 \end{bmatrix}.$$

(c) $\begin{vmatrix} 6 - \lambda & -8 \\ 4 & -6 - \lambda \end{vmatrix} = (6 - \lambda)(-6 - \lambda) + 32$
$$= \lambda^2 - 4.$$

$$\begin{bmatrix} 6 & -8 \\ 4 & -6 \end{bmatrix}^2 - 4\begin{bmatrix} 1 & 0 \\ 0 & 1 \end{bmatrix}$$

$$= \begin{bmatrix} 4 & 0 \\ 0 & 4 \end{bmatrix} - \begin{bmatrix} 4 & 0 \\ 0 & 4 \end{bmatrix}$$

$$= \begin{bmatrix} 0 & 0 \\ 0 & 0 \end{bmatrix}.$$

35. (a) False. Let A be a 3 × 3 matrix. The characteristic equation of A is $|A - \lambda I_n| = 0$. This will be a polynomial of degree 3 in λ. Thus 3, 2, or 1 distinct roots. 3, 2, or 1 distinct eigenvalues. In general an $n \times n$ matrix has $n, n - 1, \ldots, 2$, or 1 distinct eigenvalues.

(c) False. Show that the sum of two eigenvectors from different eigenspaces is not an eigenvector.

Exercise Set 5.2

1. Metropolitan areas 198 million; nonmetropolitan areas 99 million.

3. $Q = \begin{bmatrix} 0.25 & 0.25 & 0.25 \\ 0.5 & 0.5 & 0.5 \\ 0.25 & 0.25 & 0.25 \end{bmatrix}$.

Long-term probabilities of Types AA, Aa, and aa are 0.25, 0.5, and 0.25.

4. $P = \begin{array}{c} \\ \text{Wet} \\ \text{Dry} \end{array} \begin{bmatrix} \text{Wet} & \text{Dry} \\ 0.65 & 0.23 \\ 0.35 & 0.77 \end{bmatrix}$

(a) 0.67   (b) Wet day: 0.4; Dry day: 0.6.

6. Room $\quad$ 1 $\quad$ 2 $\quad$ 3 $\quad$ 4

$$P = \begin{bmatrix} 0 & \frac{1}{3} & 0 & \frac{1}{4} \\ \frac{1}{2} & 0 & \frac{1}{3} & \frac{1}{4} \\ 0 & \frac{1}{3} & 0 & \frac{1}{2} \\ \frac{1}{2} & \frac{1}{3} & \frac{2}{3} & 0 \end{bmatrix} \begin{matrix} 1 \\ 2 \\ 3 \\ 4 \end{matrix}.$$

Distribution of rats in rooms 1, 2, 3, and 4 is 2:3:3:4. Long-term probability that a given rat will be in room 4 is $\frac{1}{3}$.

7. $P = \begin{bmatrix} 0.75 & 0.20 \\ 0.25 & 0.80 \end{bmatrix}$. Eventual distribution will be 44.4% using company A and 55.6% using company B.

9. The sum of the terms in each column of $A - I$ is zero. Thus $|A - 1I| = |A - I| = 0$, and 1 is an eigenvalue of A.

Exercise Set 5.3

1. (a) $\begin{bmatrix} 3 & -5 \\ -1 & 2 \end{bmatrix}\begin{bmatrix} 1 & 2 \\ -1 & 3 \end{bmatrix}\begin{bmatrix} 2 & 5 \\ 1 & 3 \end{bmatrix} = \begin{bmatrix} 7 & 13 \\ -2 & -3 \end{bmatrix}$

(c) $\begin{bmatrix} 4 & -1 \\ -7 & 2 \end{bmatrix}\begin{bmatrix} 0 & 4 \\ 3 & 2 \end{bmatrix}\begin{bmatrix} 2 & 1 \\ 7 & 4 \end{bmatrix} = \begin{bmatrix} 92 & 53 \\ -156 & -90 \end{bmatrix}$

2. (a) $\begin{bmatrix} -1 & 2 & 2 \\ 0 & 1 & 1 \\ 2 & 0 & -1 \end{bmatrix}\begin{bmatrix} 2 & 0 & 0 \\ -2 & 2 & 1 \\ 2 & 0 & 1 \end{bmatrix}\begin{bmatrix} -1 & 2 & 0 \\ 2 & -3 & 1 \\ -2 & 4 & -1 \end{bmatrix}$

$$= \begin{bmatrix} 2 & 0 & 0 \\ 0 & 2 & 0 \\ 0 & 0 & 1 \end{bmatrix}$$

3. (a) $\frac{1}{5}\begin{bmatrix} 1 & 1 \\ -1 & 4 \end{bmatrix}\begin{bmatrix} 5 & 4 \\ 1 & 2 \end{bmatrix}\begin{bmatrix} 4 & -1 \\ 1 & 1 \end{bmatrix} = \begin{bmatrix} 6 & 0 \\ 0 & 1 \end{bmatrix}$

(c) Cannot be diagonalized.

(d) $\begin{bmatrix} -1 & 1 \\ 2 & -1 \end{bmatrix}\begin{bmatrix} 4 & -1 \\ 2 & 1 \end{bmatrix}\begin{bmatrix} 1 & 1 \\ 2 & 1 \end{bmatrix} = \begin{bmatrix} 2 & 0 \\ 0 & 3 \end{bmatrix}$

4. (a) $\begin{bmatrix} -2 & 0 \\ 0 & 3 \end{bmatrix}$, transformation matrix $\begin{bmatrix} 2 & 1 \\ 1 & 1 \end{bmatrix}$

(c) Cannot be diagonalized.

(e) Cannot be diagonalized.

5. (a) $\begin{bmatrix} -1 & -1 & 1 \\ -1 & 0 & 1 \\ 2 & 1 & -1 \end{bmatrix}\begin{bmatrix} 15 & 7 & -7 \\ -1 & 1 & 1 \\ 13 & 7 & -5 \end{bmatrix}\begin{bmatrix} 1 & 0 & 1 \\ -1 & 1 & 0 \\ 1 & 1 & 1 \end{bmatrix}$

$$= \begin{bmatrix} 1 & 0 & 0 \\ 0 & 2 & 0 \\ 0 & 0 & 8 \end{bmatrix}$$

(c) $\begin{bmatrix} 1 & 0 & 0 \\ 1 & 1 & -1 \\ -1 & 0 & 1 \end{bmatrix}\begin{bmatrix} 1 & 0 & 0 \\ -2 & 1 & 2 \\ -2 & 0 & 3 \end{bmatrix}\begin{bmatrix} 1 & 0 & 0 \\ 0 & 1 & 1 \\ 1 & 0 & 1 \end{bmatrix}$

$$= \begin{bmatrix} 1 & 0 & 0 \\ 0 & 1 & 0 \\ 0 & 0 & 3 \end{bmatrix}$$

6. (a) $\begin{bmatrix} \frac{1}{\sqrt{2}} & \frac{1}{\sqrt{2}} \\ -\frac{1}{\sqrt{2}} & \frac{1}{\sqrt{2}} \end{bmatrix}\begin{bmatrix} 1 & 2 \\ 2 & 1 \end{bmatrix}\begin{bmatrix} \frac{1}{\sqrt{2}} & -\frac{1}{\sqrt{2}} \\ \frac{1}{\sqrt{2}} & \frac{1}{\sqrt{2}} \end{bmatrix}$

$$= \begin{bmatrix} 3 & 0 \\ 0 & -1 \end{bmatrix}$$

(b) $\begin{bmatrix} \frac{1}{\sqrt{5}} & \frac{2}{\sqrt{5}} \\ -\frac{2}{\sqrt{5}} & \frac{1}{\sqrt{5}} \end{bmatrix}\begin{bmatrix} 11 & 2 \\ 2 & 14 \end{bmatrix}\begin{bmatrix} \frac{1}{\sqrt{5}} & -\frac{2}{\sqrt{5}} \\ \frac{2}{\sqrt{5}} & \frac{1}{\sqrt{5}} \end{bmatrix}$

$$= \begin{bmatrix} 15 & 0 \\ 0 & 10 \end{bmatrix}$$

7. (a) $\begin{bmatrix} \frac{1}{\sqrt{2}} & \frac{1}{\sqrt{2}} \\ -\frac{1}{\sqrt{2}} & \frac{1}{\sqrt{2}} \end{bmatrix}\begin{bmatrix} 1 & 5 \\ 5 & 1 \end{bmatrix}\begin{bmatrix} \frac{1}{\sqrt{2}} & -\frac{1}{\sqrt{2}} \\ \frac{1}{\sqrt{2}} & \frac{1}{\sqrt{2}} \end{bmatrix}$

$$= \begin{bmatrix} 6 & 0 \\ 0 & -4 \end{bmatrix}$$

(c) $\begin{bmatrix} \frac{1}{\sqrt{10}} & \frac{3}{\sqrt{10}} \\ -\frac{3}{\sqrt{10}} & \frac{1}{\sqrt{10}} \end{bmatrix}\begin{bmatrix} 1 & 3 \\ 3 & 9 \end{bmatrix}\begin{bmatrix} \frac{1}{\sqrt{10}} & -\frac{3}{\sqrt{10}} \\ \frac{3}{\sqrt{10}} & \frac{1}{\sqrt{10}} \end{bmatrix}$

$$= \begin{bmatrix} 10 & 0 \\ 0 & 0 \end{bmatrix}$$

8. (a) $\begin{bmatrix} 0 & 0 & 1 \\ \dfrac{1}{\sqrt{2}} & \dfrac{1}{\sqrt{2}} & 0 \\ \dfrac{1}{\sqrt{2}} & -\dfrac{1}{\sqrt{2}} & 0 \end{bmatrix} \begin{bmatrix} 0 & 2 & 0 \\ 2 & 0 & 0 \\ 0 & 0 & 1 \end{bmatrix} \begin{bmatrix} 0 & \dfrac{1}{\sqrt{2}} & \dfrac{1}{\sqrt{2}} \\ 0 & \dfrac{1}{\sqrt{2}} & -\dfrac{1}{\sqrt{2}} \\ 1 & 0 & 0 \end{bmatrix}$

$= \begin{bmatrix} 1 & 0 & 0 \\ 0 & 2 & 0 \\ 0 & 0 & -2 \end{bmatrix}$

(b) $\begin{bmatrix} 0 & \dfrac{1}{\sqrt{2}} & \dfrac{1}{\sqrt{2}} \\ \dfrac{1}{\sqrt{3}} & \dfrac{-1}{\sqrt{3}} & \dfrac{1}{\sqrt{3}} \\ \dfrac{2}{\sqrt{6}} & \dfrac{1}{\sqrt{6}} & \dfrac{-1}{\sqrt{6}} \end{bmatrix} \begin{bmatrix} 9 & -3 & 3 \\ -3 & 6 & -6 \\ 3 & -6 & 6 \end{bmatrix}$

$\times \begin{bmatrix} 0 & \dfrac{1}{\sqrt{3}} & \dfrac{2}{\sqrt{6}} \\ \dfrac{1}{\sqrt{2}} & \dfrac{-1}{\sqrt{3}} & \dfrac{1}{\sqrt{6}} \\ \dfrac{1}{\sqrt{2}} & \dfrac{1}{\sqrt{3}} & \dfrac{-1}{\sqrt{6}} \end{bmatrix} = \begin{bmatrix} 0 & 0 & 0 \\ 0 & 15 & 0 \\ 0 & 0 & 6 \end{bmatrix}$

9. (a) $\begin{bmatrix} 872576 & 807040 \\ 807040 & 872576 \end{bmatrix}$

(d) $\begin{bmatrix} 32768.5 & -32767.5 \\ -32767.5 & 32768.5 \end{bmatrix}$

10. (a) $\begin{bmatrix} 64 & 0 & 0 \\ 0 & 64 & 0 \\ 0 & 0 & 1 \end{bmatrix}$

(b) $\begin{bmatrix} 258309 & -250533 & 250533 \\ -250533 & 254421 & -254421 \\ 250533 & -254421 & 254421 \end{bmatrix}$

11. (a) If A and B are similar, then $B = C^{-1}AC$, so that $|B| = |C^{-1}AC| = |C^{-1}||A||C| = |A|$.

(c) We know from Exercise 16, Section 2.3, that for $n \times n$ matrices E and F, $\mathrm{tr}(EF) = \mathrm{tr}(FE)$. Thus if $B = C^{-1}AC$, then $\mathrm{tr}(B) = \mathrm{tr}(C^{-1}(AC)) = \mathrm{tr}((AC)C^{-1}) = \mathrm{tr}(A(CC^{-1})) = \mathrm{tr}(A)$.

(f) If $B = C^{-1}AC$ and A is nonsingular, then B is also nonsingular and $B^{-1} = (C^{-1}AC)^{-1} = C^{-1}A^{-1}C$, so B^{-1} is similar to A^{-1}.

13. (a) Let $A = C^{-1}BC$ and let D be the diagonal matrix $D = E^{-1}AE$. Then $D = E^{-1}(C^{-1}BC)E = (CE)^{-1}B(CE)$, so B is diagonalizable to the same diagonal matrix D.

(b) If $A = C^{-1}BC$ then $A + kI = C^{-1}BC + kC^{-1}IC = C^{-1}BC + C^{-1}kIC = C^{-1}(B + kI)C$, so $B + kI$ and $A + kI$ are similar for any scalar k.

15. Not unique

17. If $B = C^{-1}AC = C^tAC$, then $B^t = (C^tAC)^t = C^tA^t(C^t)^t = C^tA^tC = C^tAC = B$. Thus B is symmetric.

Exercise Set 5.4

1. (a) $[x \quad y] \begin{bmatrix} 1 & 2 \\ 2 & 2 \end{bmatrix} \begin{bmatrix} x \\ y \end{bmatrix}$

(c) $[x \quad y] \begin{bmatrix} 7 & -3 \\ -3 & -1 \end{bmatrix} \begin{bmatrix} x \\ y \end{bmatrix}$

(e) $[x \quad y] \begin{bmatrix} -3 & -\frac{7}{2} \\ -\frac{7}{2} & 4 \end{bmatrix} \begin{bmatrix} x \\ y \end{bmatrix}$

2. (a) $\dfrac{x'^2}{4} + \dfrac{y'^2}{6} = 1$. Graph is an ellipse with the lines $y = 2x$ and $x = -2y$ as axes.

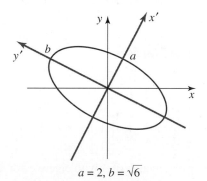

$a = 2,\ b = \sqrt{6}$

Sketch for Exercise 2(a)

(b) $\dfrac{x'^2}{3} + \dfrac{y'^2}{6} = 1$. Graph is an ellipse with the lines $y = x$ and $y = -x$ as axes.

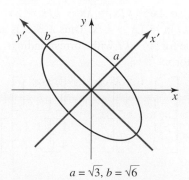

$a = \sqrt{3},\ b = \sqrt{6}$

Sketch for Exercise 2(b)

3. (a) $a_n = \dfrac{1}{3}(2^n + 2^{n-1}) = 2^{n-1}$. $a_{10} = 2^9 = 512$.

  (c) $a_n = (-1)^{n-1}$. $a_{12} = (-1)^{11} = -1$.

5. Mode 1: $\begin{bmatrix} x_1 \\ x_2 \end{bmatrix} = \cos(\alpha_1 t + \gamma_1)\begin{bmatrix} 1 \\ -1 \end{bmatrix}$,

where $\alpha_1 = \left(\dfrac{3T}{ma}\right)^{1/2}$, and

Mode 2: $\begin{bmatrix} x_1 \\ x_2 \end{bmatrix} = \cos(\alpha_2 t + \gamma_2)\begin{bmatrix} 1 \\ 1 \end{bmatrix}$,

where $\alpha_2 = \left(\dfrac{T}{ma}\right)^{1/2}$.

6. $\begin{bmatrix} x_1 \\ x_2 \end{bmatrix} = b_1\cos(\alpha_1 t + \gamma_1)\begin{bmatrix} 1 \\ 2 \end{bmatrix}$

$\qquad + b_2\cos(\alpha_2 t + \gamma_2)\begin{bmatrix} 2 \\ -1 \end{bmatrix}$,

where $\alpha_1 = \left(\dfrac{1}{M}\right)^{1/2}$ and $\alpha_2 = \left(\dfrac{6}{M}\right)^{1/2}$.

Chapter 5 Review Exercises

1. $(5 - \lambda)(1 - \lambda)(-2 - \lambda)$.

$\lambda = 5, r\begin{bmatrix} 1 \\ 1 \\ 1 \end{bmatrix}$. $\lambda = 1, s\begin{bmatrix} 0 \\ 1 \\ 1 \end{bmatrix}$. $\lambda = -2, t\begin{bmatrix} 1 \\ 0 \\ -1 \end{bmatrix}$.

2. Let λ be an eigenvalue of A with eigenvector $\mathbf{x}$. A is invertible so $\lambda \neq 0$. $A\mathbf{x} = \lambda x$, so $\mathbf{x} = A^{-1}A\mathbf{x} =$

$A^{-1}\lambda\mathbf{x} = \lambda A^{-1}\mathbf{x}$. Thus $\dfrac{1}{\lambda}\mathbf{x} = A^{-1}\mathbf{x}$, so the eigenvalues

for A^{-1} are the inverses of the eigenvalues for A and the corresponding eigenvectors are the same.

3. $A\mathbf{x} = \lambda\mathbf{x}$, so $A\mathbf{x} - kI\mathbf{x} = \lambda\mathbf{x} - kI\mathbf{x} = \lambda\mathbf{x} - k\mathbf{x}$, so $(A - kI)\mathbf{x} = (\lambda - k)\mathbf{x}$. Thus $\lambda - k$ is an eigenvalue for $A - kI$ with corresponding eigenvector $\mathbf{x}$.

4. $C^{-1}AC = \begin{bmatrix} 1 & -1 \\ -1 & 2 \end{bmatrix}\begin{bmatrix} 4 & -2 \\ 1 & 1 \end{bmatrix}\begin{bmatrix} 2 & 1 \\ 1 & 1 \end{bmatrix} = \begin{bmatrix} 3 & 0 \\ 0 & 2 \end{bmatrix}$

5. $C^{-1}AC = \begin{bmatrix} -1 & 1 \\ 2 & -1 \end{bmatrix}\begin{bmatrix} 1 & 1 \\ -2 & 4 \end{bmatrix}\begin{bmatrix} 1 & 1 \\ 2 & 1 \end{bmatrix} = \begin{bmatrix} 3 & 0 \\ 0 & 2 \end{bmatrix}$

6. $C^{-1}AC = \begin{bmatrix} \dfrac{1}{\sqrt{2}} & 0 & \dfrac{-1}{\sqrt{2}} \\ \dfrac{1}{\sqrt{3}} & \dfrac{1}{\sqrt{3}} & \dfrac{1}{\sqrt{3}} \\ \dfrac{1}{\sqrt{6}} & \dfrac{-2}{\sqrt{6}} & \dfrac{1}{\sqrt{6}} \end{bmatrix}\begin{bmatrix} 7 & -2 & 1 \\ -2 & 10 & -2 \\ 1 & -2 & 7 \end{bmatrix}$

$\times \begin{bmatrix} \dfrac{1}{\sqrt{2}} & \dfrac{1}{\sqrt{3}} & \dfrac{1}{\sqrt{6}} \\ 0 & \dfrac{1}{\sqrt{3}} & \dfrac{-2}{\sqrt{6}} \\ \dfrac{-1}{\sqrt{2}} & \dfrac{1}{\sqrt{3}} & \dfrac{1}{\sqrt{6}} \end{bmatrix} = \begin{bmatrix} 6 & 0 & 0 \\ 0 & 6 & 0 \\ 0 & 0 & 12 \end{bmatrix}$

7. $\begin{vmatrix} a - \lambda & b \\ b & c - \lambda \end{vmatrix} = (a - \lambda)(c - \lambda) - b^2 =$

$\lambda^2 - (a + c)\lambda + ac - b^2$. The characteristic equation $\lambda^2 - (a + c)\lambda + ac - b^2 = 0$ has roots

$\lambda = \dfrac{1}{2}(a + c \pm \sqrt{D})$, where $D = (a - c)^2 + 4b^2$

(from the quadratic formula). D is nonnegative for all values of a, b, and c, so the roots are real.

8. If A is symmetric then, A can be diagonalized, $D = C^{-1}AC$, where the diagonal elements of D are the eigenvalues of A. If A has only one eigenvalue, λ, then $D = \lambda I$, so $\lambda I = C^{-1}AC$. Thus $A = CC^{-1}ACC^{-1} = C\lambda IC^{-1} = \lambda CIC^{-1} = \lambda CC^{-1} = \lambda I$.

9. $\dfrac{x'^2}{2} - \dfrac{y'^2}{6} = 1$. The graph is a hyperbola with axes $y = -2x$ and $y = (\tfrac{1}{2})x$.

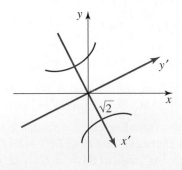

10. $a_n = \dfrac{1}{6}(13(-1)^{n-1} + 5^n)$. $a_{12} = 40{,}690{,}102$

CHAPTER 6

Exercise Set 6.1

1. $\begin{bmatrix} 8 \\ 1 \end{bmatrix}, \begin{bmatrix} 0 \\ 4 \end{bmatrix}, \begin{bmatrix} -8 \\ 7 \end{bmatrix}$

3. $\begin{bmatrix} 1 \\ 4 \\ 1 \end{bmatrix}, \begin{bmatrix} 8 \\ 7 \\ 8 \end{bmatrix}, \begin{bmatrix} 9 \\ 1 \\ 9 \end{bmatrix}$

4. $\begin{bmatrix} -1 & 0 \\ 0 & 1 \end{bmatrix}, \begin{bmatrix} -3 \\ 2 \end{bmatrix}$

5. (a) $\begin{bmatrix} 0 & -1 \\ 1 & 0 \end{bmatrix}, \begin{bmatrix} -1 \\ 2 \end{bmatrix}$

(c) $\begin{bmatrix} \dfrac{1}{\sqrt{2}} & \dfrac{-1}{\sqrt{2}} \\ \dfrac{1}{\sqrt{2}} & \dfrac{1}{\sqrt{2}} \end{bmatrix} \begin{bmatrix} \dfrac{1}{\sqrt{2}} \\ \dfrac{3}{\sqrt{2}} \end{bmatrix}$

(d) $\begin{bmatrix} -1 & 0 \\ 0 & -1 \end{bmatrix}, \begin{bmatrix} -2 \\ -1 \end{bmatrix}$

(f) $\begin{bmatrix} \dfrac{\sqrt{3}}{2} & \dfrac{-1}{2} \\ \dfrac{1}{2} & \dfrac{\sqrt{3}}{2} \end{bmatrix} \begin{bmatrix} \sqrt{3} - \dfrac{1}{2} \\ 1 + \dfrac{\sqrt{3}}{2} \end{bmatrix}$

6. $x^2 + y^2 = 9$

8. Vertices of the image of the unit square $((1, 0), (1, 1), (0, 1), (0, 0))$ are

(a) $(0, 1), (-1, 1), (-1, 0), (0, 0)$

(c) $(3, 1), (3, 5), (0, 4), (0, 0)$

9. Vertices of the image of the unit square $((1, 0), (1, 1), (0, 1), (0, 0))$ are

(a) $(-2, 0), (-5, 4), (-3, 4), (0, 0)$

(c) $(0, 2), (-2, 2), (-2, 0), (0, 0)$

10. (a) $\begin{bmatrix} -1 & -2 \\ 16 & 2 \end{bmatrix}, \begin{bmatrix} -9 \\ 84 \end{bmatrix}$
(b) $\begin{bmatrix} 6 & 10 & 2 \\ -3 & -3 & 3 \end{bmatrix} \begin{bmatrix} 16 \\ 6 \end{bmatrix}$

(c) $\begin{bmatrix} 6 & -2 \\ 3 & -3 \\ 0 & 4 \end{bmatrix} \begin{bmatrix} -22 \\ -15 \\ 8 \end{bmatrix}$

12. (a) $\begin{bmatrix} 0 & -0.5 \\ 0.5 & 0 \end{bmatrix} \begin{bmatrix} -0.5 \\ 1 \end{bmatrix}$ **(c)** $\begin{bmatrix} 0 & -1 \\ -1 & 0 \end{bmatrix} \begin{bmatrix} -1 \\ -2 \end{bmatrix}$

13. $\begin{bmatrix} 0 & -2 \\ 2 & 0 \end{bmatrix}$

20. Let $A = \begin{bmatrix} \cos\theta & -\sin\theta \\ \sin\theta & \cos\theta \end{bmatrix}$, the rotation matrix. Then

$$AA^t = \begin{bmatrix} \cos\theta & -\sin\theta \\ \sin\theta & \cos\theta \end{bmatrix} \begin{bmatrix} \cos\theta & \sin\theta \\ -\sin\theta & \cos\theta \end{bmatrix}$$

$$= \begin{bmatrix} \cos^2\theta + \sin^2\theta & \cos\theta\sin\theta - \sin\theta\cos\theta \\ \sin\theta\cos\theta - \cos\theta\sin\theta & \sin^2\theta + \cos^2\theta \end{bmatrix}$$

$$= \begin{bmatrix} 1 & 0 \\ 0 & 1 \end{bmatrix}$$

Similarly, $A^tA = I$. Thus $A^{-1} = A^t$. So A is orthogonal.

22. Triangle with vertices $(2, -3), (4, -1),$ and $(5, 1)$.

24. (a) $\begin{bmatrix} 1 \\ 0 \end{bmatrix} \mapsto \begin{bmatrix} 6 \\ 4 \end{bmatrix}, \begin{bmatrix} 1 \\ 1 \end{bmatrix} \mapsto \begin{bmatrix} 6 \\ 6 \end{bmatrix}, \begin{bmatrix} 0 \\ 1 \end{bmatrix} \mapsto \begin{bmatrix} 4 \\ 6 \end{bmatrix},$ and

$\begin{bmatrix} 0 \\ 0 \end{bmatrix} \mapsto \begin{bmatrix} 4 \\ 4 \end{bmatrix}.$

$x^2 + y^2 = 1 \mapsto (x - 4)^2 + (y - 4)^2 = 4$

Exercise Set 6.2

1. $T((x_1, y_1) + (x_2, y_2)) = T(x_1 + x_2, y_1 + y_2)$
$= (2(x_1 + x_2), x_1 + x_2 - y_1 - y_2)$
$= (2x_1 + 2x_2, x_1 - y_1 + x_2 - y_2)$
$= (2x_1, x_1 - y_1) + (2x_2, x_2 - y_2)$
$= T(x_1, y_1) + T(x_2, y_2)$

and $T(c(x, y)) = T(cx, cy)$
$= (2cx, cx - cy)$
$= c(2x, x - y) = cT(x, y),$

thus T is linear. $T(1, 2) = (2, -1)$ and $T(-1, 4) = (-2, -5)$.

3. $T((x_1, y_1, z_1) + (x_2, y_2, z_2))$
$= T(x_1 + x_2, y_1 + y_2, z_1 + z_2)$
$= (0, y_1 + y_2, 0)$
$= (0, y_1, 0) + (0, y_2, 0)$
$= T(x_1, y_1, z_1) + T(x_2, y_2, z_2)$

and $T(c(x, y, z)) = T(cx, cy, cz)$
$= (0, cy, 0)$
$= c(0, y, 0)$
$= cT(x, y, z).$

Thus T is linear. The image of (x, y, z) under T is the projection of (x, y, z) on the y-axis.

4. (a) $T(c(x, y, z)) = T(cx, cy, cz) = (3cx, (cy)^2) =$
$(3cx, c^2y^2) = c(3x, cy^2) \neq c(3x, y^2)$. Thus scalar
multiplication not preserved. T is not linear.

6. T is linear.

8. (a) T is linear.

  (b) T is not linear.

10. T is not linear.

11. T is linear.

12. (a) $\begin{bmatrix} 2 & 0 \\ 1 & -1 \end{bmatrix}$

13. (a) $\begin{bmatrix} 2 & -5 \\ 0 & 3 \end{bmatrix}$

16. $\begin{bmatrix} 1 & 0 \\ 0 & 0 \end{bmatrix}$

19. $\begin{bmatrix} a & 0 \\ 0 & b \end{bmatrix}$

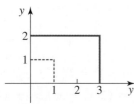

Sketch for Exercise 19

20. Line $y = 3x$

22. $\begin{bmatrix} 1 & c \\ 0 & 1 \end{bmatrix}$. If $c = 2$, then $\begin{bmatrix} 1 \\ 0 \end{bmatrix} \mapsto \begin{bmatrix} 1 \\ 0 \end{bmatrix}$, $\begin{bmatrix} 1 \\ 1 \end{bmatrix} \mapsto \begin{bmatrix} 3 \\ 1 \end{bmatrix}$,

  and $\begin{bmatrix} 0 \\ 1 \end{bmatrix} \mapsto \begin{bmatrix} 2 \\ 1 \end{bmatrix}$.

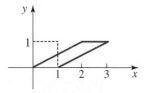

Sketch for Exercise 22

24. Line $16y = 3x$

25. (a) $\begin{bmatrix} 3 & 6 \\ 0 & 3 \end{bmatrix}$, $\begin{bmatrix} 21 \\ 6 \end{bmatrix}$ **(b)** $\begin{bmatrix} 0 & 2 \\ 3 & 0 \end{bmatrix}$, $\begin{bmatrix} 4 \\ 9 \end{bmatrix}$

27. Pairs that commute are D and R, D and F, D and S, and
D and H.

28. $\begin{bmatrix} 0 & 1 & 0 \\ -1 & 0 & 0 \\ 0 & 0 & 1 \end{bmatrix}$

30. $\begin{bmatrix} 0 & -1 & 6 \\ 1 & 0 & -4 \\ 0 & 0 & 1 \end{bmatrix}$. Image of the unit square is the square

  with vertices $(6, -3), (5, -3), (5, -4), (6, -4)$.

31. $T^{-1} = \begin{bmatrix} 1 & 0 & -h \\ 0 & 1 & -k \\ 0 & 0 & 1 \end{bmatrix}$

33. $SRT = \begin{bmatrix} 0 & 3 & -9 \\ -5 & 0 & -20 \\ 0 & 0 & 1 \end{bmatrix}$. Image of the given triangle

  is the triangle with vertices $(9, -25), (-9, -35)$,
and $(9, -40)$.

34. $T((ax + b) + (cx + d)) = T((a + c)x +$
$(b + d)) = (a + c + b + d)x = (a + b)x +$
$(c + d)x = T(ax + b) + T(cx + d)$

  $T(k(ax + b)) = T(kax + kb) = (ka + kb)x =$
$k(a + b)x = kT(ax + b)$. T is linear.

36. $T((ax + b) + (cx + d)) = T((a + c)x +$
$(b + d)) = (a + c)x^2 = ax^2 + cx^2 = T(ax + b) +$
$T(cx + d)$.

  $T(k(ax + b)) = T(kax + kb) = kax^2 = k(ax^2) =$
$kT(ax + b)$. T is linear.

38. $T((ax + b) + (cx + d)) = T((a + c)x +$
$(b + d)) = (a + c)x + 2 = ax + 2 + cx =$
$T(ax + b) + cx \neq T(ax + b) + T(cx + d)$. Addition
is not preserved. Not linear.

  Let us check scalar multiplication: $T(k(ax + b)) =$
$T(kax + kb) = kax + 2 \neq kT(ax + b)$. Not preserved.

41. $T((ax^2 + bx + c) + (px^2 + qx + r)) =$
$T((a + p)x^2 + (b + q)x + c + r) = (c + r)x^2 +$
$(a + p) = (cx^2 + a) + (rx^2 + p) =$
$T(ax^2 + bx + c) + T(px^2 + qx + r)$.

  $T(k(ax^2 + bx + c)) = T(kax^2 + kbx + kc) =$
$kcx^2 + ka = k(cx^2 + a) = kT(ax^2 + bx + c)$.

  T is linear. $T(3x^2 - x + 2) = 2x^2 + 3$.
$T(3x^2 + bx + 2) = 2x^2 + 3$ for any value of b.

43. $T((ax + b) + (cx + d)) = T((a + c)x +$
$(b + d)) = x + (a + c)$.

  But $T(ax + b) + T(cx + d) = (x + a) + (x + c) =$
$2x + (a + c)$. T is not linear.

  Further, $T(k(ax + b)) = T(kax + kb) = x + ka$. But
$kT(ax + b) = k(x + a) = kx + ka$.

  Neither addition or multiplication is preserved.

44. $J(f) = \int_0^1 f(x)dx.$ $J(f + g) = \int_0^1 (f(x) + g(x))dx =$

$\int_0^1 f(x)dx + \int_0^1 g(x)dx = J(f) + J(g)$ and

$J(cf) = \int_0^1 cf(x)dx = c\int_0^1 f(x)dx = cJ(f)$, so J is a

linear mapping of P_n to **R**.

45. (a) $T(c\mathbf{v}) = cT(\mathbf{v})$. Let $c = -1$. $T(-\mathbf{v}) =$
$T((-1)\mathbf{v}) = (-1)T(\mathbf{v}) = -T(\mathbf{v})$.

49. (a) True. $T((x_1, y_1) + (x_2, y_2)) = T(x_1 + x_2,$
$y_1 + y_2) = (x_1 + x_2, y_1 + y_2) = (x_1, y_1) +$
(x_2, y_2). $T(c(x_1, y_1)) = T((cx_1, cy_1)) =$
$(cx_1, cy_1) = c(x_1, y_1)$.

Thus linear. T is called the identity transformation.

(c) False. Any 2×3 matrix defines a linear transformation from **R**^{2} to **R**^{3}.

Exercise Set 6.3

1. (a) Kernel is zero vector. Range is **R**^{2}. Dim ker$(T) = 0$,
dim range$(T) = 2$, and dim domain$(T) = 2$,
so dim ker(T) + dim range(T) = dim domain(T).

(c) Kernel is $\{(-2r, r)\}$ and range $\{(r, 2r)\}$.
Dim ker$(T) = 1$, dim range$(T) = 1$, and
dim domain$(T) = 2$, so dim ker(T) +
dim range(T) = dim domain(T).

(e) Kernel is $\{(r, -2r, r)\}$. Range is **R**^{2}.
Dim ker$(T) = 1$, dim range$(T) = 2$, and
dim domain$(T) = 3$, so dim ker(T) +
dim range(T) = dim domain(T).

(g) Kernel is $\{(r, 0, 0)\}$. Range is $\{(a, 2a, b)\}$.
Dim ker$(T) = 1$, dim range$(T) = 2$, and
dim domain$(T) = 3$, so dim ker(T) +
dim range(T) = dim domain(T).

2. (a) The kernel is the set $\{(0, r, s)\}$ and the range is the set
$\{(a, 0, 0)\}$. Dim ker$(T) = 2$, dim range$(T) = 1$, and
dim domain$(T) = 3$, so dim ker(T) +
dim range(T) = dim domain(T).

(b) The kernel is the set $\{(r, -r, 0)\}$ and the range is **R**^{2}.
Dim ker$(T) = 1$, dim range$(T) = 2$, and
dim domain$(T) = 3$, so dim ker(T) +
dim range(T) = dim domain(T).

(e) The kernel is the zero vector and the range is the
set $\{(3a, a - b, b)\}$. Dim ker$(T) = 0$,
dim range$(T) = 2$, and dim domain$(T) = 2$, so
dim ker(T) + dim range(T) = dim domain(T).

3. (a) $T(\mathbf{u} + \mathbf{w}) = 5(\mathbf{u} + \mathbf{w}) = 5\mathbf{u} + 5\mathbf{w} = T(\mathbf{u}) + T(\mathbf{w})$
and $T(c\mathbf{u}) = 5(c\mathbf{u}) = 5c\mathbf{u} = c(5\mathbf{u}) = cT(\mathbf{u})$, so
T is linear. ker(T) is the zero vector and range(T) is
U (since for any vector **u** in U, $\mathbf{u} = T(.2\mathbf{u})$).

(b) $T(c\mathbf{u}) = 2(c\mathbf{u}) + 3\mathbf{v} \neq c(2\mathbf{u} + 3\mathbf{v}) = cT(\mathbf{u})$, so T
is not linear.

4. (a) Let $\mathbf{u}_1, \mathbf{u}_2, \ldots, \mathbf{u}_n$ be a basis for U.

The vectors $T(\mathbf{u}_1), T(\mathbf{u}_2), \ldots, T(\mathbf{u}_n)$ span
range(T): If **v** is a vector in range(T), then $\mathbf{v} = T(\mathbf{u})$
for some vector $\mathbf{u} = a_1\mathbf{u}_1 + a_2\mathbf{u}_2 + \cdots + a_n\mathbf{u}_n$ in U.
We get $\mathbf{v} = T(\mathbf{u}) = T(a_1\mathbf{u}_1 + a_2\mathbf{u}_2 + \cdots + a_n\mathbf{u}_n) =$
$a_1T(\mathbf{u}_1) + a_2 T(\mathbf{u}_2) + \cdots + a_n T(\mathbf{u}_n)$. The vectors
$T(\mathbf{u}_1), T(\mathbf{u}_2), \ldots, T(\mathbf{u}_n)$ are linearly independent.
Consider the identity $a_1T(\mathbf{u}_1) + a_2T(\mathbf{u}_2) + \cdots +$
$a_nT(\mathbf{u}_n) = \mathbf{0}$. The linearity of T gives
$T(a_1\mathbf{u}_1 + a_2\mathbf{u}_2 + \cdots + a_n\mathbf{u}_n) = \mathbf{0}$. Since
ker$(T) = \mathbf{0}$ this implies that $a_1\mathbf{u}_1 + a_2\mathbf{u}_2 + \cdots +$
$a_n\mathbf{u}_n = \mathbf{0}$. Since $\mathbf{u}_1, \mathbf{u}_2, \ldots, \mathbf{u}_n$ are linearly independent this means that $a_1 = a_2 = \cdots = a_n = 0$.
Thus $T(\mathbf{u}_1), T(\mathbf{u}_2), \ldots, T(\mathbf{u}_n)$ are a basis for
range (T). So dim range(T) = dim domain(T),
and since dim ker$(T) = 0$, the equality is proved.

6. dim ker(T) + dim range(T) = dim domain(T).
dim ker$(T) \geq 0$, so dim range$(T) \leq$ dim domain(T).

8. $T(x, y, z) = (x, y, 0) = (1, 2, 0)$ if $x = 1, y = 2, z = r$.
Thus the set of vectors mapped by T into $(1, 2, 0)$ is the
set $\{(1, 2, r)\}$.

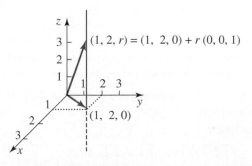

Figure for Exercise 8

10. $T(x, y) = (2x, 3x) = (4, 6)$ if $2x = 4$ and $3x = 6$, i.e., if $x = 2$. Thus the set of vectors mapped by T into $(4, 6)$ is the set $\{(2, r)\}$. This set is not a subspace of $\mathbf{R}^2$. It does not contain the zero vector.

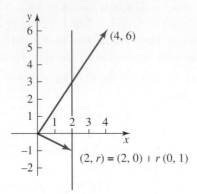

Figure for Exercise 10

12. $T(a_2x^2 + a_1x + a_0) = 0$ if $a_2 = a_1 = a_0 = 0$, so $\ker(T)$ is the zero polynomial and range(T) is P_2. A basis for P_2 is the set $\{1, x, x^2\}$.

14. $g(a_2x^2 + a_1x + a_0) = 0$ if $a_2 = a_1 = a_0 = 0$ so $\ker(g)$ is the zero polynomial and range$(g) = \{a_3x^3 + a_1x + a_0\}$. A basis for range$(g)$ is the set $\{1, x, x^3\}$.

15. $D(x^3 - 3x^2 + 2x + 1) = 3x^2 - 6x + 2$. $D(a_nx^n + \cdots + a_1x + a_0) = 0$ if $a_n = \cdots = a_1 = 0$, i.e., $\ker(D) = $ the set of constant polynomials. Range$(D) = P_{n-1}$, because every polynomial of degree less than or equal to $n - 1$ is the derivative of a polynomial of degree one larger than its own degree, and no polynomial of degree n is the derivative of a polynomial of degree n or less.

17. $D(a_nx^n + \cdots + a_1x + a_0) = na_nx^{n-1} + \cdots + 3a_3x^2 + 2a_2x + a_1 = 3x^2 - 4x + 7$ if $a_n = a_{n-1} = \cdots = a_4 = 0$, $a_3 = 1$, $a_2 = -2$, $a_1 = 7$, and $a_0 = r$, where r is any real number. Thus the set of polynomials mapped into $3x^2 - 4x + 7$ is the set $\{x^3 - 2x^2 + 7x + r\}$.

21. Let A be a 2×2 matrix. Then $\det(cA) = |cA|$ (See Theorem 3.4) $= c^2|A| = c^2\det(A) \neq c\det(A)$. Thus det is not linear.

22. (a) $T\left(\begin{bmatrix} 2 & 0 \\ 1 & 3 \end{bmatrix}\right) = \begin{bmatrix} 1 & 2 \\ 3 & 4 \end{bmatrix}\begin{bmatrix} 2 & 0 \\ 1 & 3 \end{bmatrix} = \begin{bmatrix} 4 & 6 \\ 10 & 12 \end{bmatrix}$.

24. (a) $T(A + C) = (A + C)^t = A^t + C^t = T(A) + T(C)$ and $T(cA) = (cA)^t = cA^t = cT(A)$, so T is linear. $A^t = O$ if and only if $A = O$, so $\ker(T) = O$ and range$(T) = U$.

(b) $T(A + C) = |A + C|$ and $T(A) + T(C) = |A| + |C| \neq |A + C|$, so T is not linear.

(f) $T(A + C) = a_{11} + c_{11} = T(A) + T(C)$ and $T(cA) = ca_{11} = cT(A)$, so T is linear.

$\text{Ker}(T) = \left\{\begin{bmatrix} 0 & a \\ b & c \end{bmatrix}\right\}$ and range$(T) = \mathbf{R}$.

(i) $T(A + C) = A + C + (A + C)^t = A + C + A^t + C^t = A + A^t + C + C^t = T(A) + T(C)$ and $T(cA) = cA + (cA)^t = cA + cA^t = c(A + A^t) = cT(A)$, so T is linear.

$\begin{bmatrix} a & b \\ c & d \end{bmatrix} + \begin{bmatrix} a & c \\ b & d \end{bmatrix} = \begin{bmatrix} 2a & b + c \\ b + c & 2d \end{bmatrix}$,

so $\ker(T) = \left\{\begin{bmatrix} 0 & r \\ -r & 0 \end{bmatrix}\right\}$ and range(T) is the set of all symmetric matrices.

25. This set is not a subspace because it does not contain the zero vector.

Exercise Set 6.4

1. (a) The dimension of the range of the transformation is the rank of the matrix $= 2$. The domain is $\mathbf{R}^3$, which has dimension 3, so the dimension of the kernel is 1. This transformation is not one-to-one.

(c) The dimension of the range of the transformation is the rank of the matrix $= 3$. The domain is $\mathbf{R}^4$, which has dimension 4, so the dimension of the kernel is 1. This transformation is not one-to-one.

(f) The dimension of the range of the transformation is the rank of the matrix $= 3$. The domain is $\mathbf{R}^3$, which has dimension 3, so the dimension of the kernel is 0. This transformation is one-to-one.

2. (a) $|A| = 12 \neq 0$, so the transformation is nonsingular and therefore one-to-one.

(c) $|C| = 0$, so the transformation is not one-to-one.

(f) $|F| = 212$, so the transformation is one-to-one.

3. (a) The set of fixed points is the set of all points for which $(x, y) = (x, 3y)$. This is the set $\{(r, 0)\}$.

(c) This transformation has no fixed points since there are no solutions to the equation $y = y + 1$.

(f) The set of fixed points is the set of all points for which $(x, y) = (x + y, x - y)$. This set contains only the zero vector.

4. (a) $T^{-1}(x, y) = (-2x + 3y, -5x + 7y)$. $T^{-1}(2, 3) = (5, 11)$.

(c) T^{-1} does not exist.

5. (a) $T^{-1}(x, y, z) =$
$(x + y + z, 3x + 5y + 4z, 3x + 6y + 5z)$.
$T^{-1}(1, -1, 2) = (2, 6, 7)$.

(c) T^{-1} does not exist.

6. T is one-to-one if and only if $\ker(T)$ is the zero vector if and only if $\dim \ker(T) = 0$ if and only if $\dim \text{range}(T) = \dim \text{domain}(T)$.

8. Converse: Let $T: U \to V$ be a linear transformation. If T preserves linear independence then it is one-to-one.

Proof: Suppose $T(\mathbf{u}) = T(\mathbf{w})$. Then
$\mathbf{0} = T(\mathbf{u}) - T(\mathbf{w}) = T(\mathbf{u} - \mathbf{w})$.

Let $\mathbf{u} - \mathbf{w} = a_1\mathbf{u}_1 + a_2\mathbf{u}_2 + \cdots + a_n\mathbf{u}_n$, where $\{\mathbf{u}_1, \mathbf{u}_2, \ldots, \mathbf{u}_n\}$ is a basis for U. Thus
$\mathbf{0} = T(\mathbf{u} - \mathbf{w}) = T(a_1\mathbf{u}_1 + a_2\mathbf{u}_2 + \cdots + a_n\mathbf{u}_n) = a_1T(\mathbf{u}_1) + a_2T(\mathbf{u}_2) + \cdots + a_nT(\mathbf{u}_n)$. But $\{T(\mathbf{u}_1), T(\mathbf{u}_2), \ldots, T(\mathbf{u}_n)\}$ is a linearly independent set in V so $a_1 = a_2 = \cdots = a_n = 0$. Therefore $\mathbf{u} - \mathbf{w} = \mathbf{0}$, $\mathbf{u} = \mathbf{w}$, and T is one-to-one.

9. (a) $T(\mathbf{u}_1) = \mathbf{v}_1$ and $T(\mathbf{u}_2) = \mathbf{v}_2$. $T(\mathbf{u}_1) + T(\mathbf{u}_2) = \mathbf{v}_1 + \mathbf{v}_2$. $T(\mathbf{u}_1 + \mathbf{u}_2) = \mathbf{v}_1 + \mathbf{v}_2$ since T is linear.

Thus $T^{-1}(\mathbf{v}_1 + \mathbf{v}_2) = \mathbf{u}_1 + \mathbf{u}_2$.

(b) $T(\mathbf{u}) = \mathbf{v}$. $cT(\mathbf{u}) = c\mathbf{v}$. $T(c\mathbf{u}) = c\mathbf{v}$, since T is linear. $T^{-1}(c\mathbf{v}) = c\mathbf{u}$.

Exercise Set 6.5

1. $(r, r, 1) = r(1, 1, 0) + (0, 0, 1)$

3. $(r + 1, 2r, r) = r(1, 2, 1) + (1, 0, 0)$

5. $(-4r + 3s - 2, -5r + 5s - 6, r, s) =$
$r(-4, -5, 1, 0) + s(3, 5, 0, 1) + (-2, -6, 0, 0)$

7. (a) Solutions exist   **(d)** No solutions

8. Any particular solution of the system will do in place of $\mathbf{x}_1$; e.g., $(2, 3, 1)$.

10. $x_1 + 2x_2 + x_3 = 3$
$x_2 + 2x_3 = 7$
$x_1 + x_2 - x_3 = -4$

13. (a) Basis $= \{e^{-2x}, e^x\}$. Ker $= \{ae^{-2x} + be^x\}$.

(c) Basis $= \{e^{-4x}, e^{2x}\}$. Ker $= \{ae^{-4x} + be^{2x}\}$.

14. (a) $y = re^{-3x} + se^{-2x} + \dfrac{4}{3}$

(c) $y = re^{3x} + se^{4x} + 2$

Exercise Set 6.6

1. $\begin{bmatrix} 2 \\ -3 \end{bmatrix}$ **2.** $\begin{bmatrix} 2 \\ 1 \end{bmatrix}$ **5.** $\begin{bmatrix} 4 \\ 0 \\ -2 \end{bmatrix}$

6. $\begin{bmatrix} 2 \\ -1 \\ 3 \end{bmatrix}$ **10.** $\begin{bmatrix} 2 \\ -4 \end{bmatrix}$ **12.** $\begin{bmatrix} 3 \\ 2 \\ -4 \end{bmatrix}$

13. $\begin{bmatrix} 0 \\ 11 \\ -2 \end{bmatrix}$ **15.** $\begin{bmatrix} 2 \\ 5 \\ \sqrt{2} \\ -\dfrac{3}{\sqrt{2}} \end{bmatrix}$

16. $P = \begin{bmatrix} 2 & 1 \\ 3 & 2 \end{bmatrix}$, $\mathbf{u}_{B'} = \begin{bmatrix} 4 \\ 7 \end{bmatrix}$, $\mathbf{v}_{B'} = \begin{bmatrix} 5 \\ 7 \end{bmatrix}$, $\mathbf{w}_{B'} = \begin{bmatrix} 8 \\ 14 \end{bmatrix}$

18. $P = \begin{bmatrix} 1 & 2 \\ 1 & -3 \end{bmatrix}$, $\mathbf{u}_{B'} = \begin{bmatrix} 2 \\ -3 \end{bmatrix}$, $\mathbf{v}_{B'} = \begin{bmatrix} -3 \\ 7 \end{bmatrix}$, $\mathbf{w}_{B'} = \begin{bmatrix} 5 \\ 0 \end{bmatrix}$

20. $P = \begin{bmatrix} 2 & -3 \\ -3 & 4 \end{bmatrix}$, $\mathbf{u}_{B'} = \begin{bmatrix} -1 \\ 1 \end{bmatrix}$, $\mathbf{v}_{B'} = \begin{bmatrix} 6 \\ -9 \end{bmatrix}$,
$\mathbf{w}_{B'} = \begin{bmatrix} 2 \\ -4 \end{bmatrix}$

21. Transition matrix $\begin{bmatrix} 2 & -3 \\ -3 & 5 \end{bmatrix}$, $\mathbf{u}_{B'} = \begin{bmatrix} -19 \\ 31 \end{bmatrix}$

23. Transition matrix $\begin{bmatrix} -4 & 6 \\ 3 & -3 \end{bmatrix}$, $\mathbf{u}_{B'} = \begin{bmatrix} 24 \\ -15 \end{bmatrix}$

25. Transition matrix $\begin{bmatrix} \frac{1}{3} & 0 & 0 \\ 0 & 1 & 0 \\ 0 & \frac{1}{4} & \frac{1}{4} \end{bmatrix}$, coordinate vectors $\begin{bmatrix} 1 \\ 4 \\ 3 \end{bmatrix}$,
$\begin{bmatrix} 2 \\ 0 \\ 1 \end{bmatrix}, \begin{bmatrix} 0 \\ 8 \\ 5 \end{bmatrix}, \begin{bmatrix} 1 \\ 4 \\ 2 \end{bmatrix}$

26. Transition matrix $\dfrac{1}{3}\begin{bmatrix} 3 & 0 \\ -2 & 1 \end{bmatrix}$, coordinate vectors
$\begin{bmatrix} 3 \\ -1 \end{bmatrix}, \begin{bmatrix} 6 \\ -4 \end{bmatrix}, \begin{bmatrix} 6 \\ -1 \end{bmatrix}, \begin{bmatrix} 12 \\ -9 \end{bmatrix}$

27. $T\left(\begin{bmatrix} a & 0 \\ 0 & b \end{bmatrix}\right) = (a, b)$. T is linear, one-to-one, and onto. Thus T is an isomorphism.

Exercise Set 6.7

1. $(1, -3)$ **2.** 18

4. $11x - 6$ **6.** $\begin{bmatrix} 2 & 4 \\ 3 & -1 \end{bmatrix}$, $24\mathbf{v}_1 + \mathbf{v}_2$

8. $\begin{bmatrix} 1 & 3 & 1 \\ 1 & -2 & 2 \\ 1 & 0 & -1 \end{bmatrix}$, $4\mathbf{v}_1 - 11\mathbf{v}_2 + 8\mathbf{v}_3$

9. (a) $\begin{bmatrix} 1 & 0 & 0 \\ 0 & 0 & 1 \end{bmatrix}$, $(1, 3)$

(c) $\begin{bmatrix} 1 & 1 & 0 \\ 2 & -1 & 0 \end{bmatrix}$, $(3, 0)$

10. (a) $\begin{bmatrix} 1 & 0 & 0 \\ 0 & 2 & 0 \\ 0 & 0 & 3 \end{bmatrix}$, $(-1, 10, 6)$

(c) $\begin{bmatrix} 1 & 0 & 0 \\ 0 & 0 & 0 \\ 0 & 0 & 0 \end{bmatrix}$, $(-1, 0, 0)$

11. $\begin{bmatrix} -2 & -2 & 1 \\ 1 & 3 & 2 \end{bmatrix}$, $(-7, 3)$

13. $\begin{bmatrix} 2 & 0 \\ 1 & 1 \end{bmatrix}$, $(-2, 2)$

14. $\begin{bmatrix} 0 & 0 & 0 \\ 4 & 0 & 0 \\ 0 & -1 & 0 \end{bmatrix}$, $6x - 2$

16. (a) $\begin{bmatrix} 0 & 1 & 1 \\ 0 & 1 & -1 \\ 0 & 0 & 0 \end{bmatrix}$ **(b)** $\begin{bmatrix} 0 & 1 \\ 1 & 0 \\ 0 & 1 \end{bmatrix}$

17. $\begin{bmatrix} 1 & 0 & 0 \\ 0 & 0 & 1 \end{bmatrix}$, $3x - 1$

19. $\begin{bmatrix} 1 & 1 \\ 0 & -1 \end{bmatrix}, \dfrac{1}{2}\begin{bmatrix} 1 & 1 \\ 3 & -1 \end{bmatrix}$

20. $\begin{bmatrix} 2 & 0 \\ 1 & 1 \end{bmatrix}, \begin{bmatrix} 2 & 2 \\ 0 & 1 \end{bmatrix}$

22. (a) Yes if $\dim(W) \geq \dim(V) - \dim(U)$, no if $\dim(W) < \dim(V) - \dim(U)$.

(b) Let T be the linear transformation given by $T(1, 3 -1) = (0, 0)$, $T(1, 0, 0) = (1, 0)$ and $T(0, 1, 0) = (0, 1)$.

23. Let T be the linear transformation given by $T(2, -1) = (0, 0)$ and $T(1, 0) = (1, 0)$.

25. Let $B = \{\mathbf{u}_1, \mathbf{u}_2, \ldots, \mathbf{u}_n\}$ be a basis for U. $T(\mathbf{u}_i) = \mathbf{u}_i$, so if A is the matrix of T with respect to B, its ith column will have a 1 in the ith position and zeros elsewhere. Thus $A = I_n$.

28. No

29. (a) $B' = \left\{\left(\dfrac{1}{\sqrt{2}}, \dfrac{1}{\sqrt{2}}\right)\right\}, \left\{\left(\dfrac{-1}{\sqrt{2}}, \dfrac{1}{\sqrt{2}}\right)\right\}$. The matrix

representation of T relative to B' is $A' = \begin{bmatrix} 6 & 0 \\ 0 & 2 \end{bmatrix}$.

B' defines an $x'y'$ coordinate system, rotated $45°$ counterclockwise from the xy system. T is a scaling in the $x'y'$ system with factor 6 in the x' direction and factor 2 in the y' direction.

30. (a) $B' = \{(1, 1), (2, 3)\}$. The matrix representation of T

relative to B' is $A' = \begin{bmatrix} 2 & 0 \\ 0 & -1 \end{bmatrix}$. B' defines an $x'y'$

coordinate system, which is not rectangular. The transformation T is a scaling in the $x'y'$ system with factor 2 in the x' direction and factor 1 in the y' direction followed by a reflection about the x' axis.

(c) $B' = \{(2, 1), (1, 1)\}$. The matrix representation of T

relative to B' is $A' = \begin{bmatrix} 1 & 0 \\ 0 & -1 \end{bmatrix}$. B' defines an $x'y'$

coordinate system, which is not rectangular. The transformation T is a reflection about the x' axis.

Chapter 6 Review Exercises

1. $\begin{bmatrix} 0 \\ 1 \end{bmatrix}, \begin{bmatrix} 1 \\ 18 \end{bmatrix}$

2. $\begin{bmatrix} 1 & 2 \\ 3 & -1 \end{bmatrix}$

3. $\begin{bmatrix} \dfrac{3\sqrt{3}}{2} & \dfrac{-3}{2} \\ \dfrac{3}{2} & \dfrac{3\sqrt{3}}{2} \end{bmatrix}$

4 $\begin{bmatrix} 0 & -1 \\ -1 & 0 \end{bmatrix}$

5. $\begin{bmatrix} \dfrac{1}{2} & -\dfrac{1}{2} \\ -\dfrac{1}{2} & \dfrac{1}{2} \end{bmatrix}$

6. $y = -2x + 2$

7. $y = 5x + 3$

8. $\begin{bmatrix} -2 & 0 \\ 3 & 1 \end{bmatrix}$

9. $T((x_1, y_1) + (x_2, y_2)) = T(x_1 + x_2, y_1 + y_2)$
$= (2(x_1 + x_2), x_1 + x_2 + 3(y_1 + y_2))$
$= (2x_1 + 2x_2, x_1 + 3y_1 + x_2 + 3y_2)$
$= (2x_1, x_1 + 3y_1) + (2x_2, x_2 + 3y_2)$
$= T(x_1, y_1) + T(x_2, y_2)$

and $T(c(x, y)) = T(cx, cy)$
$= (2cx, cx + 3cy)$
$= c(2x, x + 3y) = cT(x, y)$.

Thus T is linear. $T(1, 2) = (2(1), 1 + 6) = (3, 7)$.

10. (a) T is linear   **(b)** T is not linear

11. $T((ax^2 + bx + c) + (px^2 + qx + r))$
$$= T((a + p)x^2 + (b + q)x + c + r)$$
$$= 2(a + p)x + (b + q)$$
$$= (2ax + b) + (2px + q)$$
$$= T(ax^2 + bx + c) + T(px^2 + qx + r).$$
$$T(k(ax^2 + bx + c)) = T(kax^2 + kbx + kc)$$
$$= 2kax + kb$$
$$= k(2ax + b)$$
$$= kT(ax^2 + bx + c).$$

12. Kernel is $\left\{\left(r, -\dfrac{3r}{2}\right)\right\}$ and range is $\{(2r, r)\}$.
Dim $\ker(T) = 1$, dim range$(T) = 1$ and
dim domain$(T) = 2$, so dim $\ker(T) + $ dim range$(T) = $
dim domain(T).

13. Kernel is $\{(-2r, r, r)\}$ with basis $\{(-2, 1, 1)\}$, and a
basis for the range is the set $\{(1, 0, 2), (1, 1, 3)\}$.

14. (a) Not one-to-one   **(b)** One-to-one

15. The kernel is the set $\{(0, r, r)\}$ with basis $\{(0, 1, 1)\}$
and the range is the set $\{(a, 2a, b)\}$ with basis
$\{(1, 2, 0), (0, 0, 1)\}$.

16. Ker(g) is the set of all polynomials $a_2x^2 + a_1x + a_0$
with $a_2 - a_1 = 0$, $a_1 = 0$, and $a_0 = 0$; i.e.,
$a_2 = a_1 = a_0 = 0$, so $\ker(g)$ is the zero polynomial.
Range(g) is the set of all polynomials $ax^3 + bx + c$.
The set $\{x^3, x, 1\}$ is a basis for range(g).

17. Only polynomial mapped into $12x - 4$ is the polynomial
$12x + 20$.

18. $\begin{bmatrix} 2 \\ 3 \end{bmatrix}$ **19.** $\begin{bmatrix} 3 \\ -2 \\ 4 \end{bmatrix}$ **20.** $\begin{bmatrix} 5 \\ -12 \\ -9 \end{bmatrix}$

21. $P = \begin{bmatrix} 1 & 5 \\ 3 & 2 \end{bmatrix}$, and $\mathbf{u}_{B'} = P\mathbf{u}_B = \begin{bmatrix} 8 \\ 11 \end{bmatrix}$, $\mathbf{v}_{B'} = P\mathbf{v}_B = \begin{bmatrix} -5 \\ 11 \end{bmatrix}$, and $\mathbf{w}_{B'} = P\mathbf{w}_B = \begin{bmatrix} 9 \\ 14 \end{bmatrix}$

22. $\begin{bmatrix} 4 & -3 \\ 3 & 2 \end{bmatrix}$, $\mathbf{u}_{B'} = \dfrac{1}{17}\begin{bmatrix} 23 \\ 42 \end{bmatrix}$

23. $(-1, 32)$

24. $\begin{bmatrix} 1 & 3 \\ 5 & -1 \\ -2 & 2 \end{bmatrix}$, $(-7, 13, -10)$

25. $\begin{bmatrix} 2 & 0 & 0 \\ 0 & -3 & 0 \end{bmatrix}$, $(2, -6)$

26. $\begin{bmatrix} 1 & -1 & 0 \\ 0 & 0 & 2 \\ 0 & 0 & 0 \end{bmatrix}$, $3x^2 + 6x$

27. $\begin{bmatrix} 3 & 0 \\ 1 & -1 \end{bmatrix}$, $\begin{bmatrix} -11 & -20 \\ 7 & 13 \end{bmatrix}$

CHAPTER 7

Exercise Set 7.1

1. $\mathbf{u} = (x_1, x_2)$, $\mathbf{v} = (y_1, y_2)$, $\mathbf{w} = (z_1, z_2)$, and c is a scalar.
$$\langle \mathbf{u}, \mathbf{v} \rangle = 4x_1y_1 + 9x_2y_2 = 4y_1x_1 + 9y_2x_2 = \langle \mathbf{v}, \mathbf{u} \rangle,$$
$$\langle \mathbf{u} + \mathbf{v}, \mathbf{w} \rangle = 4(x_1 + y_1)z_1 + 9(x_2 + y_2)z_2$$
$$= 4x_1z_1 + 4y_1z_1 + 9x_2z_2 + 9y_2z_2$$
$$= 4x_1z_1 + 9x_2z_2 + 4y_1z_1 + 9y_2z_2$$
$$= \langle \mathbf{u}, \mathbf{w} \rangle + \langle \mathbf{v}, \mathbf{w} \rangle,$$
$$\langle c\mathbf{u}, \mathbf{v} \rangle = 4cx_1y_1 + 9cx_2y_2$$
$$= c(4x_1y_1 + 9x_2y_2) = c\langle \mathbf{u}, \mathbf{v} \rangle,$$
and $\langle \mathbf{u}, \mathbf{u} \rangle = 4x_1x_1 + 9x_2x_2 = 4x_1^2 + 9x_2^2 \geq 0$
and equality holds if and only if $4x_1^2 = 0$ and $9x_2^2 = 0$,
i.e., if and only if $x_1 = 0$ and $x_2 = 0$.

3. $\langle \mathbf{u}, \mathbf{u} \rangle = 2x_1x_1 - x_2x_2 = 2x_1^2 - x_2^2 < 0$ for
the vector $\mathbf{u} = (1, 2)$, so condition 4 of the
definition is not satisfied.

4. $\mathbf{u} = \begin{bmatrix} a & b \\ c & d \end{bmatrix}$, $\mathbf{v} = \begin{bmatrix} e & f \\ g & h \end{bmatrix}$, and $\mathbf{w} = \begin{bmatrix} j & k \\ l & m \end{bmatrix}$.
$\langle \mathbf{u} + \mathbf{v}, \mathbf{w} \rangle$
$$= (a + e)j + (b + f)k + (c + g)l + (d + h)m$$
$$= aj + ej + bk + fk + cl + gl + dm + hm$$
$$= aj + bk + cl + dm + ej + fk + gl + hm$$
$$= \langle \mathbf{u}, \mathbf{w} \rangle + \langle \mathbf{v}, \mathbf{w} \rangle, \text{ so axiom 2 is satisfied.}$$

5. (a) -16

8. (a) $-\dfrac{1}{2}$ **(c)** $\dfrac{1}{12}$

9. (a) $\dfrac{2}{\sqrt{3}}$ **(c)** $\dfrac{7}{\sqrt{5}}$

10. $\langle f, g \rangle = \displaystyle\int_0^1 (x^2)(4x - 3)dx = \int_0^1 (4x^3 - 3x^2)dx = $
$[x^4 - x^3]_0^1 = 0$, so the functions are orthogonal.

12. $\dfrac{\sqrt{15}}{4}$

13. Any function $ax + b$ with $8a + 15b = 0$ will do.

14. $\sqrt{\dfrac{76}{3}}$

15. g is closer

16. (a) 9

17. (a) $\sqrt{30}$ **(c)** $\sqrt{66}$

18. (a) $\left\langle \begin{bmatrix} 1 & 2 \\ -1 & 1 \end{bmatrix}, \begin{bmatrix} 2 & 4 \\ 3 & -7 \end{bmatrix} \right\rangle = 2 + 8 - 3 - 7 = 0$, so the matrices are orthogonal.

19. Any matrix $\begin{bmatrix} a & b \\ c & d \end{bmatrix}$ with $a + 2b + 3c + 4d = 0$ will do.

20. (a) $3\sqrt{2}$

21. (a) $\langle \mathbf{u}, \mathbf{v} \rangle = 16 + 2i$, $\|\mathbf{u}\| = \|\mathbf{v}\| = 3\sqrt{2}$, $d(\mathbf{u}, \mathbf{v}) = 2$.
Not orthogonal.

     (d) $\langle \mathbf{u}, \mathbf{v} \rangle = 0$, $\|\mathbf{u}\| = \sqrt{26}$, $\|\mathbf{v}\| = \sqrt{2}$, $d(\mathbf{u}, \mathbf{v}) = 2\sqrt{7}$.
Orthogonal.

22. (a) $\langle \mathbf{u}, \mathbf{v} \rangle = 2 - 2i$, $\|\mathbf{u}\| = \|\mathbf{v}\| = \sqrt{19}$, $d(\mathbf{u}, \mathbf{v}) = \sqrt{34}$.
Not orthogonal.

     (c) $\langle \mathbf{u}, \mathbf{v} \rangle = 10 - 10i$, $\|\mathbf{u}\| = 2\sqrt{3}$, $\|\mathbf{v}\| = \sqrt{30}$,
$d(\mathbf{u}, \mathbf{v}) = \sqrt{22}$. Not orthogonal.

23. $\langle \mathbf{u}, \mathbf{v} \rangle = x_1 \overline{y}_1 + \cdots + x_n \overline{y}_n = \overline{y}_1 x_1 + \cdots$
       $+ \ \overline{y}_n x_n = \overline{y_1 \overline{x}_1} + \cdots + \overline{y_n \overline{x}_n} = \overline{\langle \mathbf{v}, \mathbf{u} \rangle} = \langle \mathbf{v}, \mathbf{u} \rangle$

24. $\langle \mathbf{u}, k\mathbf{v} \rangle = x_1 \overline{k} \, \overline{y}_1 + \cdots + x_n \overline{k} \, \overline{y}_n = \overline{k} x_1 \overline{y}_1 + \cdots$
       $+ \ \overline{k} x_n \overline{y}_n = \overline{k}(x_1 \overline{y}_1 + \cdots + x_n \overline{y}_n) = \overline{k} \langle \mathbf{u}, \mathbf{v} \rangle$

25. (a) Let $\mathbf{u}$ be any nonzero vector in the inner product
space. Then using axiom 1, the fact that
$0\mathbf{u} = \mathbf{0}$, axiom 3, and the fact that zero times
any real number is zero we have
$\langle \mathbf{v}, \mathbf{0} \rangle = \langle \mathbf{0}, \mathbf{v} \rangle = \langle 0\mathbf{u}, \mathbf{v} \rangle = 0\langle \mathbf{u}, \mathbf{v} \rangle = 0$.

26. (a) Axiom 1: $\langle \mathbf{u}, \mathbf{v} \rangle = \mathbf{u}A\mathbf{v}^t = (\mathbf{u}A) \cdot \mathbf{v} = \mathbf{v} \cdot (\mathbf{u}A) =$
$\mathbf{v}(\mathbf{u}A)^t = \mathbf{v}A\mathbf{u}^t = \langle \mathbf{v}, \mathbf{u} \rangle$

     Axiom 4: $\langle \mathbf{u}, \mathbf{u} \rangle = \mathbf{u}A\mathbf{u}^t > 0$ if $\mathbf{u} = 0$ (given), and
$\mathbf{u}A\mathbf{u}^t = 0$ if $\mathbf{u} = 0$.

     (c) Using A, $\langle \mathbf{u}, \mathbf{v} \rangle = 0$, $\|\mathbf{u}\| = \sqrt{2}$, $\|\mathbf{v}\| = \sqrt{3}$, and
$$d(\mathbf{u}, \mathbf{v}) = \sqrt{5}$$

     Using B, $\langle \mathbf{u}, \mathbf{v} \rangle = 2$, $\|\mathbf{u}\| = 1$, $\|\mathbf{v}\| = \sqrt{3}$, and
$$d(\mathbf{u}, \mathbf{v}) = 0$$

     Using C, $\langle \mathbf{u}, \mathbf{v} \rangle = 3$, $\|\mathbf{u}\| = \sqrt{2}$, $\|\mathbf{v}\| = \sqrt{5}$, and
$$d(\mathbf{u}, \mathbf{v}) = 1$$

     (d) $\begin{bmatrix} 1 & 0 \\ 0 & 4 \end{bmatrix}$

28. (a) True: $\langle \mathbf{u}, \mathbf{v} \rangle = 0$. Thus $c\langle \mathbf{u}, \mathbf{v} \rangle = 0$, $\langle \mathbf{u}, c\mathbf{v} \rangle = 0$.

     (c) False: $\langle f, g \rangle = \displaystyle\int_0^1 f(x)g(x)dx$.

$$\langle 3x, 3x \rangle = \int_0^1 9x^2 dx = [3x^3]_0^1 = 3.$$
$$\|3x\| = \sqrt{3}.$$

Exercise Set 7.2

   1. $x_1^2 + 4x_2^2 = 1$

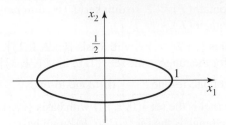

Figure for Exercise 1

2. (a) $\|(1, 0)\| = 2$, $\|(0, 1)\| = 3$, $\|(1, 1)\| \, \sqrt{13}$,
$\|(2, 3)\| = \sqrt{97}$

     (c) $d((1, 0), (0, 1)) = \sqrt{13}$

3. (a) $\|(1, 0)\| = 1$, $\|(0, 1)\| = 4$, $\|(1, 1)\| = \sqrt{17}$

     (c) $d((5, 0), (0, 4)) = \sqrt{281}$. In Euclidean space the
distance is $\sqrt{41}$.

     (d) $x_1^2 + 16x_2^2 = 1$

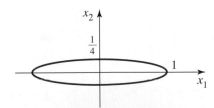

Figure for Exercise 3

4. $\langle (x_1, x_2), (y_1, y_2) \rangle = \dfrac{1}{4} x_1 y_1 + \dfrac{1}{25} x_2 y_2$

6. $d(R, Q)^2 = \|(4, 0, 0, -5)\|^2 =$
$|-16 - 0 - 0 + 25| = 9$, so $d(R, Q) = 3$.

8. $\langle (2, 0, 0, 1), (1, 0, 0, 2) \rangle = -2 - 0 - 0 + 2 = 0$, so the vectors are orthogonal.

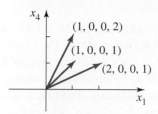

Figure for Exercise 8

9. The equations of the circles with radii $a = 1, 2, 3$ and center at $(0, 0)$ are $|-x_1^2 + x_4^2| = a^2$.

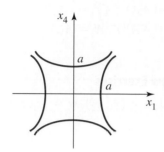

Figure for Exercise 9

10. 12 years, 0.8 speed of light

12. More than 8 centuries

Exercise Set 7.3

1. $g(x) = \dfrac{1}{3}$

2. $\left\{ 1, 2\sqrt{3}\left(x - \dfrac{1}{2} \right) \right\}$ is an orthonormal basis for P_1 $[0, 1]$, $g(x) = 4e - 10 + (18 - 6e)x$

4. $\left\{ \pi^{-\frac{1}{2}}, 2\sqrt{3}, \pi^{-\frac{3}{2}}, \left(x - \dfrac{\pi}{2} \right) \right\}$ is an orthonormal basis for $P_1[0, \pi]$, $g(x) = -24\pi^{-3} x + 12\pi^{-2}$

5. (a) $\left\{ \dfrac{1}{\sqrt{2}}, \dfrac{\sqrt{3}}{\sqrt{2}} x, \dfrac{3\sqrt{5}}{2\sqrt{2}} \left(x^2 - \dfrac{1}{3} \right) \right\}$ is an orthonormal basis for $P_2 [-1, 1]$,

$g(x) = -\dfrac{3}{4}e + \dfrac{33}{4}e^{-1} + 3e^{-1}x + \dfrac{15}{4}(e - 7e^{-1})x^2$

6. $\left\{ 1, 2\sqrt{3}\left(x - \dfrac{1}{2} \right), 6\sqrt{5}\left(x^2 - x + \dfrac{1}{6} \right) \right\}$ is an ortho-

normal basis for $P_2 [0, 1]$, $g(x) = -\dfrac{4}{7}x^2 + \dfrac{48}{35}x + \dfrac{6}{35}$

8. $\left\{ \pi^{-\frac{1}{2}}, 2\sqrt{3}\, \pi^{-\frac{3}{2}} \left(x - \dfrac{\pi}{2} \right), 6\sqrt{5}\, \pi^{-\frac{5}{2}} \left(x^2 - \pi x + \dfrac{\pi^2}{6} \right) \right\}$

is an orthonormal basis for $P_2 [0, \pi]$,

$g(x) = 180\pi^{-5}\left(-4 + \dfrac{\pi^2}{3} \right) x^2$

$- 180\pi^{-4}\left(-4 + \dfrac{\pi^2}{3} \right) x$

$- 120\pi^{-3} + 12\pi^{-1}$

9. The vectors that form an orthonormal basis over $[-\pi, \pi]$ also form an orthonormal basis over $[0, 2\pi]$,

$g(x) = \pi - 2\left(\sin x + \dfrac{1}{2}\sin 2x + \dfrac{1}{3}\sin 3x + \dfrac{1}{4}\sin 4x \right)$

11. $g(x) = \dfrac{4}{3}\pi^2$

$+ 4\left(\cos x + \dfrac{1}{4}\cos 2x + \dfrac{1}{9}\cos 3x + \dfrac{1}{16}\cos 4x \right)$

$- 4\pi\left(\sin x + \dfrac{1}{2}\sin 2x + \dfrac{1}{3}\sin 3x + \dfrac{1}{4}\sin 4x \right)$

13. $(1, 0, 0, 0, 0, 1, 1), (0, 1, 0, 0, 1, 0, 1),$
$(0, 0, 1, 0, 1, 1, 0), (0, 0, 0, 1, 1, 1, 1),$
$(0, 1, 1, 1, 1, 0, 0), (1, 0, 1, 1, 0, 1, 0),$
$(1, 1, 0, 1, 0, 0, 1), (1, 1, 1, 0, 0, 0, 0),$
$(1, 1, 0, 0, 1, 1, 0), (1, 0, 1, 0, 1, 0, 1),$
$(1, 0, 0, 1, 1, 0, 0), (1, 1, 1, 1, 1, 1, 1),$
$(0, 0, 1, 1, 0, 0, 1), (0, 1, 0, 1, 0, 1, 0),$
$(0, 1, 1, 0, 0, 1, 1), (0, 0, 0, 0, 0, 0, 0)$

14. (a) Center $= (0, 0, 1, 0, 1, 1, 0)$
$(1, 0, 1, 0, 1, 1, 0), (0, 1, 1, 0, 1, 1, 0),$
$(0, 0, 0, 0, 1, 1, 0), (0, 0, 1, 1, 1, 1, 0),$
$(0, 0, 1, 0, 0, 1, 0), (0, 0, 1, 0, 1, 0, 0),$
$(0, 0, 1, 0, 1, 1, 1)$

(b) Center $= (1, 1, 0, 1, 0, 0, 1)$
$(0, 1, 0, 1, 0, 0, 1), (1, 0, 0, 1, 0, 0, 1),$
$(1, 1, 1, 1, 0, 0, 1), (1, 1, 0, 0, 0, 0, 1),$
$(1, 1, 0, 1, 1, 0, 1), (1, 1, 0, 1, 0, 1, 1),$
$(1, 1, 0, 1, 0, 0, 0)$

15. (a) $(0, 0, 0, 0, 0, 0, 0)$ **(c)** $(0, 0, 1, 0, 1, 1, 0)$

16. (a) Since each vector in V_{23} has 23 components and there are 2 possible values (0 and 1) for each component, there are 2^{23} vectors.

(c) There are 23 ways to change exactly one component of the center vector, $\dfrac{(23 \times 22)}{2} = 253$ ways to change exactly two components, and $\dfrac{(23 \times 22 \times 21)}{6} = 1771$ ways to change exactly three components. $1 + 23 + 253 + 1771 = 2048$ vectors.

Exercise Set 7.4

1. $\dfrac{1}{27}\begin{bmatrix} 11 & -1 & -16 \\ 5 & 2 & 5 \end{bmatrix}$

3. Does not exist

5. $\dfrac{1}{14}\begin{bmatrix} 1 & 5 & 3 \\ 4 & -1 & -2 \end{bmatrix}$

6. There is no pseudoinverse. This is not the coefficient matrix of an overdetermined system of equations.

8. $\text{pinv}\begin{bmatrix} 1 & 2 \\ -3 & 5 \end{bmatrix} = \left(\begin{bmatrix} 1 & -3 \\ 2 & 5 \end{bmatrix}\begin{bmatrix} 1 & 2 \\ -3 & 5 \end{bmatrix}\right)^{-1}\begin{bmatrix} 1 & -3 \\ 2 & 5 \end{bmatrix}$

$= \dfrac{1}{11}\begin{bmatrix} 5 & -2 \\ 3 & 1 \end{bmatrix}$

9. $y = -\dfrac{10}{3} + \dfrac{7}{2}x$

11. $y = -3 + 4x$

13. $y = 10 - \dfrac{7x}{2}$

15. $y = -\dfrac{3}{2} + 2x$

17. $y = \dfrac{23}{2} - \dfrac{5x}{2}$

19. $y = -3 + \dfrac{12x}{5}$

21. $y = 12 - 9x + 2x^2$

23. $y = -\dfrac{29}{4} + \dfrac{223}{20}x - \dfrac{9}{4}x^2$

25. (a) $L = 3.8 + 1.06F.$ 19.7 inches

    (b) $L = 4.2 + 1.565F.$ 27.675 inches

26. $G = 23.545 - 0.1351S.$ 16.1145 miles per gallon

28. $U = \dfrac{127955000 + 10545D}{262500}$, 688 units

30. $\% = \dfrac{-1632.75 + 97.65Y}{4900}$, 0.125

32. Public: $y = 2850.4286 + 128.4286x$,
Private: $y = 12175 + 634x$.

For 2005 $(x = 36)$, Public enrollment = 51.5903 million, Private = 6.6326 million

For 2010 $(x = 41)$, Public enrollment = 56.9953 million, Private = 7.3871 million.

33. (a) $\dfrac{1}{3}\begin{bmatrix} 2 & 1 & 1 \\ 1 & 2 & -1 \\ 1 & -1 & 2 \end{bmatrix}\left(\dfrac{4}{3}, \dfrac{5}{3}, -\dfrac{1}{3}\right)$

    (c) $\dfrac{1}{6}\begin{bmatrix} 5 & 2 & -1 \\ 2 & 2 & 2 \\ -1 & 2 & 5 \end{bmatrix}(1, 1, 1)$

34. (a) $\dfrac{1}{8}\begin{bmatrix} 8 & 0 & 0 \\ 0 & 4 & -4 \\ 0 & -4 & 4 \end{bmatrix}(-2, -1, 1)$

36. $A\mathbf{x} = \mathbf{y}$, so $A^tA\mathbf{x} = A^t\mathbf{y}$, and if $(A^tA)^{-1}$ exists then $\mathbf{x} = (A^tA)^{-1}A^tA\mathbf{x} = (A^tA)^{-1}A^t\mathbf{y}$.

    $\mathbf{x} = (A^tA)^{-1}A^t\mathbf{y}$ solves the system $A^tA\mathbf{x} = A^t\mathbf{y}$ but does not solve the original system unless A^t (and therefore A) is an invertible matrix. The first step of multiplying by A^t is not reversible unless A^t is invertible.

39. (a) $\text{pinv}(cA) = ((cA)^t(cA))^{-1}(cA)^t = (c^2A^tA)^{-1}cA^t$

$= \dfrac{1}{c^2}(A^tA)^{-1}cA^t = \dfrac{1}{c}(A^tA)^{-1}A^t$

41. (a) $P^t = (A(A^tA)^{-1}A^t)^t = (A^t)^t((A^tA)^{-1})^tA^t$

$= A((A^tA)^t)^{-1}A^t = A(A^tA)^{-1}A^t = P$

Chapter 7 Review Exercises

1. $\mathbf{u} = (x_1, x_2)$, $\mathbf{v} = (y_1, y_2)$, $\mathbf{w} = (z_1, z_2)$, and c is a scalar.

$\langle \mathbf{u}, \mathbf{v} \rangle = 2x_1y_1 + 3x_2y_2 = 2y_1x_1 + 3y_2x_2 = \langle \mathbf{v}, \mathbf{u} \rangle$,

$\langle \mathbf{u} + \mathbf{v}, \mathbf{w} \rangle = 2(x_1 + y_1)z_1 + 3(x_2 + y_2)z_2$

$= 2x_1z_1 + 2y_1z_1 + 3x_2z_2 + 3y_2z_2$

$= 2x_1z_1 + 3x_2z_2 + 2y_1z_1 + 3y_2z_2$

$= \langle \mathbf{u}, \mathbf{w} \rangle + \langle \mathbf{v}, \mathbf{w} \rangle$,

$\langle c\mathbf{u}, \mathbf{v} \rangle = 2cx_1y_1 + 3cx_2y_2 = c(2x_1y_1 + 3x_2y_2) = c\langle \mathbf{u}, \mathbf{v} \rangle$, and $\langle \mathbf{u}, \mathbf{u} \rangle = 2x_1x_1 + 3x_2x_2 = 2x_1^2 + 3x_2^2 \geq 0$ and equality holds if and only if $2x_1^2 = 0$ and $3x_2^2 = 0$, i.e., if and only if $x_1 = x_2 = 0$. Thus the given function is an inner product on $\mathbf{R}^2$.

2. $4, 1, \dfrac{\sqrt{97}}{\sqrt{3}}, 2\dfrac{\sqrt{19}}{\sqrt{3}}$

3. $\sqrt{30}, 2\sqrt{14}$

4. $2x_1^2 + 3x_2^2 = 1$

5. $g(x) = 2x - \dfrac{2}{3}$

6. $g(x) = 2\pi - 1 + \displaystyle\sum_{k=1}^{4} -\dfrac{4}{k}\sin kx$

$= 2\pi - 1 - 4\left(\sin x + \dfrac{1}{2}\sin 2x + \dfrac{1}{3}\sin 3x + \dfrac{1}{4}\sin 4x\right)$

7. $(1, 1, 1, 0, 0, 1, 1)$, $(0, 0, 1, 0, 0, 1, 1)$,
$(0, 1, 0, 0, 0, 1, 1)$, $(0, 1, 1, 1, 0, 1, 1)$,
$(0, 1, 1, 0, 1, 1, 1)$, $(0, 1, 1, 0, 0, 0, 1)$,
$(0, 1, 1, 0, 0, 1, 0)$

8. $\dfrac{1}{65}\begin{bmatrix} -17 & 14 & 10 \\ 39 & -13 & 0 \end{bmatrix}$

9. $y = 12.5 - 8.8x + 2x^2$

10. $B = \text{pinv} A = (A^t A)^{-1} A^t$. If A is $n \times m$, then B is $m \times n$.

(a) $ABA = A(A^t A)^{-1} A^t A = A I_m = A$

(b) $BAB = (A^t A)^{-1} A^t AB = I_m B = B$

(c) AB is the projection matrix for the subspace spanned by the columns of A. This matrix was shown to be symmetric in Exercise 41(a) of Section 7.4.

(d) $BA = (A^t A)^{-1} A^t A = I_m$ and is therefore symmetric.

CHAPTER 8

Exercise Set 8.1

1. (a) Yes (c) Yes

2. (a) Yes (b) No

3. $x_1 = 1, x_2 = 2, x_3 = 3$

5. No solution

6. $x_1 = 2 + 2r - s - t, x_2 = 1 + 3r + s - 3t,$
$x_3 = r, x_4 = s, x_5 = t$

7. Eight operations are saved using Gaussian elimination.

9. Yes

Exercise Set 8.2

1. $x_1 = 1, x_2 = 4, x_3 = -1$

3. $x_1 = -2, x_2 = 1, x_3 = 1$

4. $x_1 = \frac{1}{2}, x_2 = \frac{1}{2}, x_3 = 2$

6. $x_1 = 1, x_2 = -1, x_3 = 7$

7. (a) $L = \begin{bmatrix} 1 & 0 & 0 \\ 1 & 1 & 0 \\ -1 & -1 & 1 \end{bmatrix}$

(c) $L = \begin{bmatrix} 1 & 0 & 0 \\ -\frac{1}{2} & 1 & 0 \\ -\frac{1}{5} & 1 & 1 \end{bmatrix}$

8. (a) R2 − 2R1, R3 + 3R1, R3 − 5R2

(c) R3 − $\dfrac{1}{7}$R1, R3 + $\dfrac{3}{4}$R2

9. $L = \begin{bmatrix} 1 & 0 & 0 \\ -2 & 1 & 0 \\ 1 & -1 & 1 \end{bmatrix}$,

$U = \begin{bmatrix} 1 & 2 & -1 \\ 0 & 3 & 1 \\ 0 & 0 & -2 \end{bmatrix}, \mathbf{x} = \begin{bmatrix} -1 \\ 2 \\ 1 \end{bmatrix}$

11. $L = \begin{bmatrix} 1 & 0 & 0 \\ -1 & 1 & 0 \\ 3 & 8 & 1 \end{bmatrix}$,

$U = \begin{bmatrix} 3 & -1 & 1 \\ 0 & 1 & 2 \\ 0 & 0 & -22 \end{bmatrix}, \mathbf{x} = \begin{bmatrix} 3 \\ 0 \\ 1 \end{bmatrix}$

14. $L = \begin{bmatrix} 1 & 0 & 0 \\ 2 & 1 & 0 \\ -4 & 3 & 1 \end{bmatrix}$,

$U = \begin{bmatrix} -2 & 0 & 3 \\ 0 & 3 & -1 \\ 0 & 0 & 4 \end{bmatrix}, \mathbf{x} = \begin{bmatrix} 0 \\ 2 \\ 1 \end{bmatrix}$

16. $L = \begin{bmatrix} 1 & 0 & 0 \\ 2 & 1 & 0 \\ -5 & 4 & 1 \end{bmatrix}$,

$U = \begin{bmatrix} 2 & -3 & 1 \\ 0 & 1 & 4 \\ 0 & 0 & -2 \end{bmatrix}, \mathbf{x} = \begin{bmatrix} -43 \\ -24 \\ 9 \end{bmatrix}$

18. $L = \begin{bmatrix} 1 & 0 & 0 \\ 2 & 1 & 0 \\ -1 & 1 & 1 \end{bmatrix}$,

$U = \begin{bmatrix} 1 & 2 & -1 \\ 0 & 1 & 3 \\ 0 & 0 & 0 \end{bmatrix}, \mathbf{x} = \begin{bmatrix} 4 + 7r \\ -1 - 3r \\ r \end{bmatrix}$

19. $L = \begin{bmatrix} 1 & 0 & 0 \\ -1 & 1 & 0 \\ 2 & -3 & 1 \end{bmatrix}, U = \begin{bmatrix} 4 & 1 & -2 \\ 0 & 3 & 1 \\ 0 & 0 & -2 \end{bmatrix}$, no solution

20. $L = \begin{bmatrix} 1 & 0 & 0 & 0 \\ 1 & 1 & 0 & 0 \\ -1 & -1 & 1 & 0 \\ 0 & 2 & -\frac{1}{2} & 1 \end{bmatrix}$,

$U = \begin{bmatrix} 1 & 1 & -1 & 2 \\ 0 & 2 & 3 & 0 \\ 0 & 0 & -2 & 8 \\ 0 & 0 & 0 & 2 \end{bmatrix}, \mathbf{x} = \begin{bmatrix} 1 \\ 1 \\ -1 \\ 2 \end{bmatrix}$

24. $\begin{bmatrix} 6 & -2 \\ 12 & 8 \end{bmatrix} = \begin{bmatrix} 1 & 0 \\ 2 & 1 \end{bmatrix}\begin{bmatrix} 6 & -2 \\ 0 & 12 \end{bmatrix}$

$= \begin{bmatrix} 2 & 0 \\ 4 & 2 \end{bmatrix}\begin{bmatrix} 3 & -1 \\ 0 & 6 \end{bmatrix}$

26. 13 arithmetic operations are required to obtain U and no arithmetic operations are required for L. To solve $L\mathbf{y} = \mathbf{b}$ requires 6 operations and to solve $U\mathbf{x} = \mathbf{y}$ requires 9 operations.

Exercise Set 8.3

1. (a) 6   **(c)** 10

2. (a) 21   **(c)** 143   **(e)** $148\frac{1}{3}$

3. (a) 0.26×10^1, one significant digit less accurate

    (c) 0.217007×10^5, five significant digits less accurate

    (e) 0.36×10^2, two significant digits less accurate

4. (a) 0.3×10^1. The solution of the system of equations can have one fewer significant digits of accuracy than the elements of the matrix.

5. 4606

6. $k < -99, 0.96 < k < 1, 1 < k < 1.0417, k > 96.9583$

8. (a) $I^{-1} = I$, so both have 1-norm of 1 and $C(I) = 1$.

    (b) $1 = \|AA^{-1}\| \leq \|A\|\|A^{-1}\| = c(A)$.

10. If A is a diagonal matrix with diagonal elements a_{ii}, then A^{-1} is a diagonal matrix with diagonal elements $\dfrac{1}{a_{ii}}$.

$$c(A) = \|A\|\|A^{-1}\| = (\max\{a_{ii}\})\left(\frac{1}{\min}\{a_{ii}\}\right).$$

11. $x_1 = 6, x_2 = -2$, and $x_3 = 1$

12. $x_1 = -5.46, x_2 = 4.93$, and $x_3 = -1.2$
(exact solution: $x_1 = -\frac{82}{15}, x_2 = \frac{74}{15}, x_3 = -\frac{6}{5}$)

15. $x_1 = 1.000, x_2 = -1.000$, and $x_3 = -29.999$
(exact solution: $x_1 = 1, x_2 = -1$, and $x_3 = -30$)

17. $x_1 = -2.500, x_2 = 0.500$, and $x_3 = -4.500$

Exercise Set 8.4

1. $x = 2.40, y = 0.40$, and $z = 2.00$

3. $x = 4.51, y = 5.24$, and $z = 2.68$
(exact solution: $x = \frac{185}{41}, y = \frac{215}{41}, z = \frac{110}{41}$)

5. $x = 7.00, y = -0.43$, and $z = 2.29$
(exact solution: $x = 7, y = -\frac{3}{7}, z = \frac{16}{7}$)

6. $x = 2.29, y = 1.98, z = 5.94$, and $w = 2.33$
(exact solution: $x = \frac{7310}{3199}, y = \frac{6330}{3199}, z = \frac{19010}{3199}, w = \frac{7440}{3199}$)

Exercise Set 8.5

1. dominant eigenvalue $= 8$,

dominant eigenvector $= r\begin{bmatrix} 1 \\ 0 \\ -1 \end{bmatrix}$

3. dominant eigenvalue -6,

dominant eigenvector $= r\begin{bmatrix} 1 \\ 0 \\ 1 \end{bmatrix}$

5. dominant eigenvalue $= 6$,

dominant eigenvector $= r\begin{bmatrix} 0 \\ 0 \\ 1 \\ 1 \end{bmatrix}$

7. dominant eigenvalue $= 84.311$

dominant eigenvector $= r\begin{bmatrix} 1 \\ 0.011 \\ -0.011 \\ 0.040 \end{bmatrix}$

9. eigenvalues are 10 and 1(dim 2) with corresponding

eigenvectors $\begin{bmatrix} 1 \\ 1 \\ 0.5 \end{bmatrix}$ and $\begin{bmatrix} -1 \\ 0.8 \\ 0.4 \end{bmatrix}$

11. eigenvalues are 12, 2(dim 2) with corresponding

eigenvectors $\begin{bmatrix} 1 \\ 0 \\ 1 \end{bmatrix}$ and $\begin{bmatrix} 0 \\ 1 \\ 0 \end{bmatrix}$

12. eigenvalues are 10, 6, and -2(dim 2) with

corresponding eigenvectors $\begin{bmatrix} 1 \\ 0 \\ 0 \\ 1 \end{bmatrix}, \begin{bmatrix} 0 \\ 1 \\ -1 \\ 0 \end{bmatrix}$, and $\begin{bmatrix} -\frac{1}{3} \\ 1 \\ 1 \\ \frac{1}{3} \end{bmatrix}$

14. (a) $0.191, 0.309, 0.309, 0.191$

    (b) $\frac{1}{8}, \frac{1}{8}, \frac{1}{4}, \frac{1}{4}, \frac{1}{8}, \frac{1}{8}$

Chapter 8 Review Exercises

1. $x_1 = 2$, $x_2 = -1$, $x_3 = 1$

2. $L = \begin{bmatrix} 1 & 0 & 0 \\ \frac{3}{2} & 1 & 0 \\ -\frac{2}{7} & 4 & 1 \end{bmatrix}$

3. $x_1 = 1$, $x_2 = 3$, $x_3 = 4$

4. Number of multiplications $= \dfrac{2n^3 + 3n^2 + n}{6}$, number of

additions $= \dfrac{2n^3 - 3n^2 + n}{6}$, total number of

operations $= \dfrac{2n^3 + n}{3}$.

If the elements on the diagonal of L are all 1, the number

of multiplications $= \dfrac{n^3 - n}{3}$ and the total number of

operations $= \dfrac{4n^3 - 3n^2 - n}{6}$.

5. 0.3135216×10^4, four significant digits less accurate

6. (a) 5, one significant digit less accurate

(b) 12.5, two significant digits less accurate

7. $c(A) = \|A\|\|A^{-1}\| = \|A^{-1}\|\|A\|$
$= \|A^{-1}\|\|(A^{-1})^{-1}\| = c(A^{-1})$

8. Not a linear mapping

9. (a) $x_1 = \dfrac{1}{50}$, $x_2 = 0$, and $x_3 = 100$

(b) $x_1 = -12$, $x_2 = -1$, and $x_3 = 5000$

10. $x = 4.80$, $y = 1.20$, and $z = 3.00$

11. Dominant eigenvalue $= 11.716$, dominant

eigenvector $= r \begin{bmatrix} 0.652 \\ 0.997 \\ 1 \end{bmatrix}$

12. Eigenvalues are 6, 4, 0 with corresponding eigenvectors

$\begin{bmatrix} 1 \\ 2 \\ -1 \end{bmatrix}, \begin{bmatrix} 1 \\ 0 \\ 1 \end{bmatrix}, \begin{bmatrix} -1 \\ 1 \\ 1 \end{bmatrix}$

CHAPTER 9

Exercise Set 9.1

1. Maximum is 24 at $(6, 12)$

3. Maximum is 20 at every point on the line segment joining $(5, 0)$ and $(3, 4)$

5. Maximum is 6 at $(1, 2)$

6. Minimum is -350 at $(100, 50)$

8. Minimum is -4 at $(4, 0)$

9. Maximum profit is \$6400 when the company manufactures 400 of $C1$ and 300 of $C2$

11. Maximum profit is \$2400 when the company ships x refrigerators from town X and $120 - 2x$ refrigerators from town Y, where $0 \le x \le 30$

13. Maximum income is \$5040 when the tailor makes 24 suits and 32 dresses

15. Maximum is 54 cars moved from B to C, with 146 cars moved from B to D, 66 cars from A to C, and 34 cars from A to D

17. Maximum is 150 tons when 50 tons of X and 100 tons of Y are manufactured

18. Minimum is 56 cents when 7 oz. of item M and no item N is served

20. (a) Maximum profit is \$800 when the shipper carries no packages for Pringle and 2000 packages for Williams

Exercise Set 9.2

1. Maximum is 24 when $x = 6$ and $y = 12$

2. Maximum is 4 when $x = 4$ and $y = 0$

4. Maximum is 1600 when $x = 160$ and $y = 0$

6. Maximum is 22500 when $x = 0$, $y = 100$, and $z = 50$

7. No maximum

10. Maximum is \$1740 when 30 of X, no Y, and 120 of Z are produced

12. Maximum profit is \$7200 when 600 washing machines are transported from A to P and none are transported from B or C to P

13. Maximum profit is \$2080 if no Aspens are manufactured and the number of Alpines manufactured plus one-half the number of Cubs manufactured is 260

15. Minimum is -5 at $x = 3$, $y = 1$

16. Minimum is $-\dfrac{15}{4}$ at $x = 0$, $y = 0$, $z = \dfrac{15}{4}$

Exercise Set 9.3

1. Optimal solution is $f = 24$ at $x = 6$, $y = 12$

3. Optimal solution is $f = 6$ at $x = 0$, $y = 3$, $z = 0$ (x and z are nonbasic variables)

6. Optimal solution is $f = 50$ at $x = 0$, $y = 15$, $z = 5$, $w = 0$ (x and w are nonbasic variables)

Chapter 9 Review Exercises

1. Maximum value is 13 at $(2, 3)$

2. Maximum is 48 at every point on the line segment joining $(8, 0)$ and $(4, 6)$

3. Minimum is 0 at $(0, 0)$

4. Maximum profit is \$27000 when 30 acres of strawberries and 10 acres of tomatoes are planted

5. Maximum volume is 1456 when 16 lockers of type X and 20 of type Y are purchased

6. Maximum is 40 when $x = 20$, $y = 0$, and $z = 0$

7. Maximum daily profit is \$240 when the number of X tables plus one-half the number of Y tables finished is 30

8. Minimum is -2 for all x, y, z where $z = 0$ and $y = 1 + \dfrac{x}{2}$, with $0 \le x \le \dfrac{4}{3}$

APPENDIX A

Exercise Set Appendix A

1. (a) $(8, -7, 2)$

   (c) $(-8, 4, 0)$

   (e) $(3, 10, 23)$

2. (a) $(10, 22, -6)$

   (c) $(-10, -17, 6)$

   (e) $(-80, 28, -54)$

3. (a) $10\mathbf{i} - 9\mathbf{j} + 7\mathbf{k}$

   (c) $21\mathbf{i} - 17\mathbf{j} + 9\mathbf{k}$

   (e) $-86\mathbf{i} - 93\mathbf{j} + 25\mathbf{k}$

4. (a) $(0, -14, -7)$

   (c) -28

   (d) $(4, -36, -26)$

   (f) -196

5. (a) $\sqrt{329}$

   (c) $3\sqrt{69}$

6. (a) 130

   (c) 30

8. $(\mathbf{i} \times \mathbf{j}) \cdot \mathbf{k} =$
$(0 \times 0 - 0 \times 1, 0 \times 0 - 1 \times 0, 1 \times 1 - 0 \times 0) \cdot$
$(0, 0, 1) = (0, 0, 1) \cdot (0, 0, 1) = 0 + 0 + 1 = 1$

12. $c(\mathbf{u} \times \mathbf{v}) = c(u_2v_3 - u_3v_2, u_3v_1 - u_1v_3, u_1v_2 - u_2v_1) =$
$(cu_2v_3 - cu_3v_2, cu_3v_1 - cu_1v_3, cu_1v_2 - cu_2v_1) = c\mathbf{u} \cdot \mathbf{v}$.

14. Hint: Use the result of Exercise 11.

15. $\mathbf{u} \cdot (\mathbf{v} \times \mathbf{w}) = \begin{vmatrix} u_1 & u_2 & u_3 \\ v_1 & v_2 & v_3 \\ w_1 & w_2 & w_3 \end{vmatrix} = 0$ if and only if the row

vectors $\mathbf{u}$, $\mathbf{v}$, and $\mathbf{w}$ are linearly dependent, i.e., if and only if one vector lies in the plane of the other two.

APPENDIX B

Exercise Set Appendix B

1. (a) Point-normal form:
    $(x - 1) + (y + 2) + (z - 4) = 0$
    General form: $x + y + z - 3 = 0$

   (c) Point-normal form: $x + 2y + 3z = 0$
    General form: $x + 2y + 3z = 0$

2. (a) $6x + y + 2z - 10 = 0$

   (c) $3x - 2y + 1 = 0$

3. $(3, -2, 4)$ is normal to the plane $3x - 2y + 4z - 3 = 0$ and $(-6, 4, -8)$ is normal to the plane $-6x + 4y - 8z + 7 = 0$. $(-6, 4, -8) = -2(3, -2, 4)$, so the normals are parallel and therefore the planes are parallel.

5. Point-normal form:
    $2(x - 1) - 3(y - 2) + (z + 3) = 0$
    General form: $2x - 3y + z + 7 = 0$

6. (a) Parametric equations: $x = 1 - t$, $y = 2 + 2t$, $z = 3 + 4t$, $-\infty < t < \infty$

Symmetric equations: $-x + 1 = \dfrac{y - 2}{2} = \dfrac{z - 3}{4}$

(c) Parametric equations: $x = -2t$, $y = -3t$, $z = 5t$, $-\infty < t < \infty$

Symmetric equations: $-\dfrac{x}{2} = -\dfrac{y}{3} = \dfrac{z}{5}$

7. $x = 1 + 2t$, $y = 2 + 3t$, $z = -4 + t$, $-\infty < t < \infty$

9. $x = 4 + 2t$, $y = -1 - t$, $z = 3 + 4t$, $-\infty < t < \infty$

11. The points P_1, P_2, and P_3 are collinear. There are many planes through the line containing these three points.

12. $-x + 2y - 4z + 26 = 0$

13. The points on the line are of the form $(1 + t, 14 - t, 2 - t)$. These points satisfy the equation of the plane, so the line lies in the plane.

16. The directions of the two lines are $(-4, 4, 8)$ and $(3, -1, 2)$. $(-4, 4, 8) \cdot (3, -1, 2) = 0$, so the lines are orthogonal. The point of intersection is $x = 1$, $y = 2$, $z = 3$.

17. The directions of the two lines are $(4, 5, 3)$ and $(1, -3, -2)$. $(4, 5, 3) \neq c(1, -3, -2)$, so the lines are not parallel. They do not intersect, so they are skew.

18. (a) A basis for the space must consist of two linearly independent vectors that are orthogonal to $(2, 3, -4)$. Two such vectors are $(0, 4, 3)$ and $(2, 0, 1)$.

(b) The point $(0, 0, 0)$ is not on the plane so the set of all points on the plane is not a subspace of $\mathbf{R}^3$.

APPENDIX D

MATLAB Exercises

D.2 Solving Systems of Linear Equations (Sections 1.1–1.3)

1. (a) $x = 2$, $y = -1$, $z = 1$

(b) $x = 1 - 2r$, $y = r$, $z = -2$

2. (a) $x = 2$, $y = 5$

(b) No solution

3. (a) $x = 4$, $y = -3$, $z = -1$

(c) $x = 2$, $y = 3$, $x = -1$

9. (a) 8 mult, 9 add , 0 swap

(b) 11 mult, 12 add, 0 swap

Creation of zeros and leading 1s are not counted as operations—they are substitutions. Gauss gains one multiplication and one addition in the back substitution. No operations take place on the $(1, 3)$ location when a zero is created in the $(1, 2)$ location.

11. (a) $y = 2x^2 + x + 5$

D.3 Matrix Operations (Sections 2.1, 2.2)

1. (a) $\begin{bmatrix} 4 & 5 \\ 3 & 1 \end{bmatrix}$

(c) $\begin{bmatrix} 1 & 0 \\ 2 & -1 \end{bmatrix}$, that is, B

(d) $\begin{bmatrix} 3 & 5 \\ 5 & 8 \end{bmatrix}$

2. (a) $\begin{bmatrix} -9 & 35 \\ 1 & 10 \end{bmatrix}$

4. 29

5. (a) »$X = A(3, :); Y = A(:, 4); X*A'*Y$» ans $= 821$

D.4 Computational Considerations (Section 2.2)

3. (a) $A(BC)$ 3,864 $(AB)C$ 7,395

4. (a) $(AB)C$ 4,320; $A(BC)$ 6,345

(c) $((AB)B)B$ 12,150; $A(B^3)$ 186,300

D.5 Inverse of a Matrix (Section 2.4)

2. (b) $(B^{194})^{-1} = I^{-1} = I$

3. (a) $A^{-1} = \begin{bmatrix} 1 & 0 \\ -2 & 1 \end{bmatrix}$

(b) $B^{-1} = \begin{bmatrix} 0.15116279069767 & 0.10465116279070 \\ 0.12790697674419 & -0.01162790697674 \end{bmatrix} = \begin{bmatrix} -\dfrac{13}{86} & \dfrac{9}{86} \\ \dfrac{11}{86} & -\dfrac{1}{86} \end{bmatrix}$

4. (a) $B = \begin{bmatrix} 2 & 0 \\ 0 & 1 \end{bmatrix}$ **(c)** $B = \begin{bmatrix} 6 & 14 & -7 \\ -5 & -11 & 6 \\ -3 & -6 & 5 \end{bmatrix}$

D.6 Solving Systems of Equations Using Matrix Inverse (Section 2.4)

1. $x = 10$, $y = -2$

3. $x = 9$, $y = -14$

D.7 Cryptography (Section 2.4)

1. (a)

1 27 6 15 18 27 20 8 9 19 27 5 24 5 18 3 9 19 5
A – F O R – T H I S – E X E R C I S E

D.8 Leontief I/O Model (Section 2.5)

1. $X = \begin{bmatrix} 165 \\ 480 \\ 250 \end{bmatrix}$

3. $D = \begin{bmatrix} 4 \\ 0.9 \\ 0.65 \end{bmatrix}$

D.9 Markov Chains (Sections 2.6, 5.2)

1. (a) $X_{2008} = \begin{bmatrix} 243.59 \\ 53.41 \end{bmatrix} \begin{matrix} \text{City} \\ \text{Suburb} \end{matrix}$

$X_{2009} = \begin{bmatrix} 242.2223 \\ 54.7777 \end{bmatrix}$,

$X_{2010} = \begin{bmatrix} 240.8956 \\ 56.1044 \end{bmatrix}$,

$X_{2011} = \begin{bmatrix} 239.6088 \\ 57.3912 \end{bmatrix}$,

$X_{2012} = \begin{bmatrix} 238.3605 \\ 58.6395 \end{bmatrix}$.

(b) $\begin{bmatrix} 0.99 & 0.02 \\ 0.01 & 0.98 \end{bmatrix}^5 = \begin{bmatrix} 0.9529 & 0.0942 \\ 0.0471 & 0.9058 \end{bmatrix}$

$p_{11}^{(5)} = 0.9529$

6. (a) $X_{2008} = \begin{bmatrix} 81.1300 \\ 162.4600 \\ 53.4100 \end{bmatrix} \begin{matrix} \text{City} \\ \text{Suburb} \\ \text{Nonmetro} \end{matrix}$

$X_{2009} = \begin{bmatrix} 80.3105 \\ 161.9117 \\ 54.7777 \end{bmatrix}$,

$X_{2010} = \begin{bmatrix} 79.5389 \\ 161.3567 \\ 56.1044 \end{bmatrix}$,

$X_{2011} = \begin{bmatrix} 78.8125 \\ 160.7963 \\ 57.3912 \end{bmatrix}$,

$X_{2012} = \begin{bmatrix} 78.1288 \\ 160.2317 \\ 58.6395 \end{bmatrix}$.

(b) $\begin{bmatrix} 69.3000 \\ 128.7000 \\ 99.0000 \end{bmatrix} \begin{matrix} \text{City} \\ \text{Suburb} \\ \text{Nonmetro} \end{matrix}$

(c) City: 0.2333, suburb: 0.4333, nonmetro: 0.3333.

D.10 Digraphs (Section 2.7)

1. Distance = 3. Two paths $2 \to 4 \to 5 \to 1$ and $2 \to 4 \to 3 \to 1$.

2. (a)

$$D = \begin{bmatrix} 0 & 1 & 2 & 3 & 1 \\ 3 & 0 & 1 & 2 & 4 \\ 2 & 3 & 0 & 1 & 3 \\ 1 & 2 & 3 & 0 & 2 \\ 4 & 1 & 2 & 3 & 0 \end{bmatrix} \begin{matrix} 7 \\ 10 \\ 9 \\ 8 \\ 10 \end{matrix}$$

Most to least influential: M_1, M_4, M_3, with M_2 and M_5 equally uninfluential. M_3 becomes most influential person by influencing M_1.

D.11 Determinants (Sections 3.1–3.3)

1. (a) $|A| = 7$, $M(a_{22}) = -5$, $M(a_{31}) = 5$.

2. All determinants are zero.

(a) Row 3 is 3 times row 1.

(c) Columns 2 and 3 are equal.

3. (a)

$$\begin{vmatrix} 1 & -1 & 0 & 2 \\ -1 & 1 & 0 & 0 \\ 2 & -2 & 0 & 1 \\ 3 & 1 & 5 & -1 \end{vmatrix} \begin{matrix} = \\ R2 + R1 \\ R3 + (-2)R1 \\ R4 + (-3)R1 \end{matrix}$$

$$\begin{vmatrix} 1 & -1 & 0 & 2 \\ 0 & 0 & 0 & 2 \\ 0 & 0 & 0 & -3 \\ 0 & 4 & 5 & -7 \end{vmatrix} \underset{R2 \leftrightarrow R4}{=} - \begin{vmatrix} 1 & -1 & 0 & 2 \\ 0 & 4 & 5 & -7 \\ 0 & 0 & 0 & -3 \\ 0 & 0 & 0 & 2 \end{vmatrix}$$

D.12 Cramer's Rule (Section 3.3)

1. (a) $-\dfrac{1}{3} \begin{bmatrix} -7 & 9 & 1 \\ 8 & -9 & -2 \\ -4 & 3 & 1 \end{bmatrix}$

(c) $-\dfrac{1}{4} \begin{bmatrix} -6 & 2 & -2 \\ -3 & 1 & 1 \\ -8 & 4 & 0 \end{bmatrix}$

2. (a) $1, 2, -1$ **(c)** $0.5, 0.25, 0.25$

D.13 Dot Product, Norm, Angle, Distance (Section 4.2)

1. $\mathbf{u} \cdot \mathbf{v} = 2$, $\|\mathbf{u}\| = 6.3246$, angle between $\mathbf{u}$ and
$\mathbf{v} = 87.2031°$, $d(X, Y) = 5.7446$.

3. $\|\mathbf{u}\| = 3.7417$, $\|\mathbf{v}\| = 11.0454$,
$\|\mathbf{u} + \mathbf{v}\| = 11.5758 < 3.7417 + 11.0454$.

D.14 Linear Combinations, Dependence, Basis, Rank (Sections 4.5–4.8)

1. (a) $(-3, 3, 7) = 2(1, -1, 2) - (2, 1, 0) + 3(-1, 2, 1)$

(b) Not a combination

2. (a) $2(-1, 3, 2) - 3(1, -1, -3) - (-5, 9, 13) = 0$

3. (a) Linearly independent, $\det(A) = 2268 \neq 0$.
$\operatorname{rank}(A) = 4$

5. (a) $\{(1, 0, 1), (0, 1, 1)\}$

D.15 Projection, Gram-Schmidt Orthogonalization (Section 4.9)

1. $(0.1905, -0.2381)$

3. (a) $\{(0.5345, 0.8018, 0.2673), (0.3841, -0.5121, 0.7682)\}$

D.16 Eigenvalues and Eigenvectors (Sections 5.1, 5.2)

1. $\lambda_1 = 1, \mathbf{v}_1 = \begin{bmatrix} -0.6017 \\ 0.7453 \\ -0.2872 \end{bmatrix}$, $\lambda_2 = 1, \mathbf{v}_2 = \begin{bmatrix} 0.4399 \\ 0.0091 \\ -0.8980 \end{bmatrix}$,

$\lambda_3 = 10, \mathbf{v}_3 = \begin{bmatrix} 0.6667 \\ 0.6667 \\ 0.3333 \end{bmatrix}$.

4. The eigenvectors of $\lambda = 1$ are vectors of the form $r\begin{bmatrix} 2 \\ 1 \end{bmatrix}$.

If there is no change in total population $2r + r = 245 + 52 = 297$, so $r = \frac{297}{3}$. Thus the long-term prediction is that population in metropolitan areas will be $2r = 198$ million and population in nonmetropolitan areas will be $r = 99$ million.

D.17 Transformations Defined by Matrices (Sections 6.1, 6.2)

1. » map ([cos (2*π/3) − sin (2*π/3);
sin (2*π/3)cos (2*π/3)])

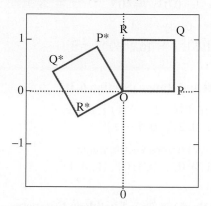

$O* = (0, 0)$, $P* = (-0.5000, 0.8600)$,
$Q* = (-1.3660, -1.5000)$, $R* = (-0.8660, -0.5000)$

3. $A = \begin{bmatrix} 3 & 0 \\ 0 & 3 \end{bmatrix}$ $O* = (0, 0)$, $P* = (3, 0)$,
$Q* = (3, 3)$, $R* = (0, 3)$

5. (a) $O* = (0, 0)$, $P* = (3, 1)$, $Q* = (3, 5)$,
$R* = (0, 4)$

6. (a) $A = \left[\cos\left(\frac{\pi}{2}\right) - \sin\left(\frac{\pi}{2}\right); \sin\left(\frac{\pi}{2}\right) \cos\left(\frac{\pi}{2}\right) \right]$,
$B = \begin{bmatrix} 2 & 0 \\ 0 & 2 \end{bmatrix}$, $O\# = (0, 0)$, $P\# = (0, 2)$,
$Q\# = (-2, 2)$, $R\# = (-2, 0)$

7. (a) $A = \begin{bmatrix} 3 & 0 \\ 0 & 3 \end{bmatrix}$, $B = \begin{bmatrix} 1 & 2 \\ 0 & 1 \end{bmatrix}$
$O\# = (0, 0)$, $P\# = (3, 0)$,
$Q\# = (9, 3)$, $R\# = (6, 3)$

9. (a) $(2, 2)$

D.18 Fractals (Section 6.2)

4. (a) Let $\begin{bmatrix} 0.86 & 0.03 \\ -0.03 & 0.86 \end{bmatrix} = \begin{bmatrix} k & 0 \\ 0 & k \end{bmatrix} \begin{bmatrix} \cos\alpha & -\sin\alpha \\ \sin\alpha & \cos\alpha \end{bmatrix}$.
$k\cos\alpha = 0.86$, $k\sin\alpha = -0.03$.
$k^2 = 0.86^2 + 0.03^2$. $k = 0.8605230967$.
$\alpha = -1.998°$, to three decimal places.
Dilation factor $= 0.8605230967$,
angle of rotation $= -1.998°$.

5. Tip of fern will be point where $T_1(x, y) = (x, y)$.

$$\begin{bmatrix} 0.86 & 0.03 \\ -0.03 & 0.86 \end{bmatrix} \begin{bmatrix} x \\ y \end{bmatrix} + \begin{bmatrix} 0 \\ 1.5 \end{bmatrix} = \begin{bmatrix} x \\ y \end{bmatrix}.$$

$-0.14x + 0.03y = 0$

$0.03x + 0.14y = 1.5$

$x = 2.1951, y = 10.2439$

D.19 Kernel and Range (Section 6.3)

1. (a) Kernel $\{(-2, -3, 1)^t\}$, range
$\{(1, 0, 2)^t, (0, 1, -1)^t\}$

(c) Kernel is zero vector, range
$\{(1, 0, 0)^t, (0, 1, 0)^t, (0, 0, 1)^t\}$

D.20 Inner Product, Non-Euclidean Geometry (Sections 7.1, 7.2)

1. (a) $\|(0, 1)\| = 2$

(b) $90°$. The vectors $(1, 1)$, and $(-4, 1)$ are at right angles.

(c) $\text{dist}((1, 0), (0, 1)) = \sqrt{5}$

(d) $\text{dist}((x, y), (0, 0)) = 1, 2, 3$;
$<(x, y), (x, y)> = 1, 4, 9$;
$[x \quad y]A[x \quad y]^t = 1, 4, 9; x^2 + 4y^2 = 1, 4, 9.$

4. (a), (b) All vectors of form $a(1, 1)$ or $b(1, -1)$ i.e., lie on the cone through the origin.

(c) Any two points of the form (x, y) and
$(x + 1, y + 1)$ or (x, y) and $(x - 1), (y + 1)$.

(d) $-x^2 + y^2 = 1, 4, 9$

D.21 Space–Time Travel (Section 7.2)

1. Earth time 120 yrs; spaceship time 79.3725 yrs.

D.22 Pseudoinverse and Least Squares Curves (Section 7.4)

2. (a) $\begin{bmatrix} 0.3333 & 0 & 0.6667 \\ 0 & 0.1667 & -0.1667 \end{bmatrix}$

5. (a) $x = 1.7333, y = 0.6$

6. $y = 4x - 3$

8. $y = 0.0026x + 1.7323$. When $x = 670, y = 3.5$.

D.23 *LU* Decomposition (Section 8.2)

1. (a) $L = \begin{bmatrix} 1 & 0 \\ 0.3333 & 1 \end{bmatrix}, U = \begin{bmatrix} 3 & 5 \\ 0 & 0.3333 \end{bmatrix}$

2. (a) $X = \begin{bmatrix} 1.0000 & 1.0000 \\ 1.0000 & -1.0000 \\ 2.0000 & 1.0000 \end{bmatrix}$

D.24 Condition Number of a Matrix (Section 8.3)

1. (a) 21

(c) 65

4. $y = 25 - 26\frac{1}{3}x + 9x^2 - \frac{2}{3}x^3$

$$A = \begin{bmatrix} 1 & 1 & 1 & 1 \\ 1 & 2 & 4 & 16 \\ 1 & 3 & 9 & 27 \\ 1 & 4 & 16 & 64 \end{bmatrix}, \text{ a Vandermonde matrix.}$$

$c(A) = 2000$, system is ill-conditioned.

D.25 Jacobi and Gauss-Seidel Iterative Methods (Section 8.4)

1. (a) Solution $x = 1, y = 0.5, z = 0$. With tolerance of 0.0001, Jacobi takes 11 iterations to converge; Gauss-Seidel takes 4 iterations.

D.26 The Simplex Method in Linear Programming (Section 9.2)

1. $f = 24$ at $x = 6, y = 12$

3. $f = 16.5$, at $x = 1.25, y = 2.25$

D.27 Cross Product (Appendix A)

1. (a) $(36, -27, -26)$

2. (a) 108

Index